Milestones in Drug Therapy
MDT

Cognitive Enhancing Drugs

Edited by J.J. Buccafusco

Springer Basel AG

Editors
Prof. Dr. Jerry J. Buccafusco
Medical College of Georgia
Department of Pharmacology and Toxicology
Alzheimer's Research Center
1120 – 15th Street
Augusta, GA 30912-2300
USA

Library of Congress Cataloging-in-Publication Data
Cognitive enhancing drugs / edited by J.J. Buccafusco.
p. ; cm. -- (Milestones in drug therapy)
Includes bibliographical references and index.
ISBN 978-3-0348-9603-0 ISBN 978-3-0348-7867-8 (eBook)
DOI 10.1007/978-3-0348-7867-8
1. Nootropic agents--Mechanism of action. I. Buccafusco Jerry J. II. Series.
[DNLM: 1. Nootropic Agents--therapeutic use. 2. Cognition--drug effects. 3. Memory--drug effects. 4. Memory Disorders--drug therapy. QV 76.5 C676 2003]
RM334.C64 2003
615'.78--dc22

2003063677

Bibliographic information published by Die Deutsche Bibliothek
Die Deutsche Bibliothek lists this publication in the Deutsche Nationalbibliografie; detailed bibliographic data is available in the internet at http://dnb.ddb.de

ISBN 978-3-0348-9603-0

Originally published by Birkhäuser Verlag in 2004
Softcover reprint of the hardcover 1st edition 2004

Printed on acid-free paper produced from chlorine-free pulp. TFC ∞
Cover illustration: Newly developed compounds act on multiple drug targets in several higher brain regions to result in synergistic actions on cognition and memory. (The cover artwork was produced by Sharon K. Barnes, Medical Illustration and Design, Medical College of Georgia, USA)

ISBN 978-3-0348-9603-0
9 8 7 6 5 4 3 2 1

www.birkhäuser-science.com

Dedication

This book is dedicated to my wife Regina, our children Chris and Marty; to my parents Dominick and Rose, and to my Aunt Angie – with love and appreciation for their constant encouragement and support, without which this book would not be possible.

Acknowledgement and appreciation also goes to my office manager Vanessa Cherry for her assistance in the organization of the manuscript and to two excellent primate behavior technical assistants, Nancy Kille and Ritu Duhan.

Contents

List of contributors

Jerry J. Buccafusco, Alzheimer's Research Center, Medical College of Georgia, 1120 – 15th Street, Augusta, Georgia 30912-2300, and Veterans Administration Medical Center, Augusta, Georgia 30904, USA; e-mail: jbuccafu@mail.mcg.edu

Gerard B. Fox, Neuroscience Research, Global Pharmaceutical Research and Development, R4N5 AP9A/LL, Abbott Laboratories, 100 Abbott Park Road, Abbott Park, IL 60064-6115, USA; e-mail: gerard.b.fox@abbott.com

Ezio Giacobini, University Hospitals of Geneva, Department of Geriatrics, University of Geneva Medical School, CH-1226 Thonex/Geneva, Switzerland; e-mail: Ezio.Giacobini@hcuge.ch

Arthur A. Hancock, Neuroscience Research, Global Pharmaceutical Research and Development, R4MN AP9A/3, Abbott Laboratories, 100 Abbott Park Road, Abbott Park, IL 60064-6125, e-mail: Art.A.Hancock@Abbott.com

Stephen C. Heinrichs, Boston College, Department of Psychology, 140 Commonwealth Avenue, Chestnut Hill, MA 02467, USA; e-mail: heinrics@bc.edu

David Kennedy, Human Cognitive Neuroscience Unit, Division of Psychology, Northumbria University, Newcastle upon Tyne, NE1 8ST, UK; e-mail: david.kennedy@unn.ac.uk

Edward D. Levin, Duke University Medical Center, Department of Psychiatry and Behavioral Sciences, Neurobehavioral Research Laboratory, Box 3412, Durham, NC 27710, USA; e-mail: e-mail: edlevin@duke.edu

Joe L. Martinez, Jr., Cajal Neuroscience Research Center, Associate Vice Provost for Research, Department of Biology, University of Texas at San Antonio, 6900 N Loop 1604 W., San Antonio, TX 78249, USA; e-mail: JMartinez@utsa.edu

Melanie P. McFadyen-Leussis, Boston College, Department of Psychology, 140 Commonwealth Avenue, Chestnut Hill, MA 02467, USA; e-mail: mcfadyme@bc.edu

William S. Messer, The University of Toledo, Departments of Pharmacology and Medicinal & Biological Chemistry, 2801 W. Bancroft Street, Toledo, OH 43606, USA; e-mail: wmesser@utnet.utoledo.edu

Andrew Scholey, Human Cognitive Neuroscience Unit, Division of Psychology, Northumbria University, Newcastle upon Tyne, NE1 8ST, UK, e-mail: a.scholey@unn.ac.uk

Alvin V. Terry Jr., Program in Clinical and Experimental Therapeutics, University of Georgia College of Pharmacy (Augusta Campus), CJ-1020, Medical College of Georgia, Augusta, GA 30912-2450, USA; e-mail:

aterry@mail.mcg.edu

Kenira Thompson, Department of Biology, University of Texas at San Antonio, 6900 N Loop 1604 W., San Antonio, TX 78249, USA; e-mail: kthompson@utsa.edu

Keith A. Wesnes, Cognitive Drug Research Ltd, Holder of the Queen's Award for Enterprise 2002, CDR House, 24 Portman Road, Reading, RG30 1EA, UK, e-mail: Keithw@cdr.org.uk

Thomas Weiser, Department CNS Research, Boehringer Ingelheim Pharma GmbH & Co KG, Birkendorfer Strasse, D-88400 Biberach, Germany; e-mail: thomas.weiser@bc.boehringer-ingelheim.com

Andrea Wevers, Department of Anatomy – Neuroanatomy, University of Cologne, Joseph-Stelzmann Str. 9, D-50931 Cologne, Germany; e-mail: wevers.anatomie@uni-koeln.de

Diana S. Woodruff-Pak, Research and Technology Development, Korman Research Pavilion, Room 100, Albert Einstein Healthcare Network, 5501 Old York Road, Philadelphia, PA 19141, and Temple University, Philadelphia, PA 19122 USA; e-mail: woodrufd@einstein.edu

Moussa B.H. Youdim, Centers of Excellence for Neurodegenerative Diseases, Technion-Faculty of Medicine, Haifa, Israel, e-mail: Youdim@tx.technion.ac.il

Andrea Zangara, Cognitive Drug Research Ltd, CDR House, 24 Portman Road, Reading RG30 1EA, UK, e-mail: Andreaz@cdr.org.uk

Preface

Improving cognitive power – the dream of retaining vast amounts of information and having instant access to that information – the photographic memory, each has been part of the realm of imagination and fiction. This is with good reason. Currently our knowledge pertaining to the neural construct of intelligence and to the make-up of the basic engram is in its infancy. We know that memory involves protein synthesis within neural cells, and possibly the restructuring of the three-dimensional interactions among dendritic processes. We have uncovered the electrophysiological basis of conditioned cellular behavior, e.g., long-term potentiation, and we have determined some of the cellular signaling events involved. Yet this knowledge, prodigious as it is, has not moved us significantly closer to the dream of greater cognitive power. Yet as anyone subject to the advertising media knows, there are drugs and natural product preparations that purport to do just that. Whereas some of these products are often considered harmless nutritionally-based brain boosters, the possibility of achieving significant improvement in cognitive power or of increasing memory and enhancing learning by standard pharmaceutical means in otherwise cognitively normal individuals is sometimes viewed as ethically suspect. Thus, most serious scientific inquiry has been directed at the reversal of cognitive impairment associated with disease syndromes, particularly Alzheimer's disease. There is no doubt that drugs effective in humans with dementia or in animal impairment models are often effective in cognitively normal subjects; and it is likely – if not inevitable – that pharmaceutical products, like their natural product counterparts, will be used to enhance cognitive power in multiple settings.

In developing this book, contributors were sought who have years of experience working with drugs and natural substances that have been suggested or observed to improve aspects of cognition and memory in experimental animals and in human beings. The first two chapters underscore the cholinergic hypothesis of memory and describe the advent and ultimate use of long-acting inhibitors of acetylcholinesterase for the treatment of the symptoms associated with Alzheimer's disease. The subsequent chapters specialize in areas less advanced clinically, but with promise for future applications. One of the key themes that reverberates through the chapters and culminates in the final chapter pertains to the possibility of targeting multiple brain substrates to develop additive or synergistic positive mnemonic actions. Some work has already been accomplished to support this approach. New hybrid molecules are now being tested that were specifically designed with this hypothesis in mind.

This book also is relevant for those interested in how animal models are used in the development of cognition-enhancing agents. Since many of the compounds discussed have binary modes of action that often include both positive mnemonic effects and neuroprotective actions, the methodology described in these chapters includes cell and molecular techniques, tissue culture procedures, various studies in normal and transgenic mouse strains, rat models, and studies in non-human primates.

Perhaps most importantly, this book will present to the reader some of the most well-studied molecular targets for cognition enhancement. Potential targets have been partitioned based on primary neurotransmitter systems, although significant overlap among systems exists. There will be some areas not covered in this book, but the approaches to the development of novel drug entities will be similar to those described here. Ultimately the goal is to prepare the reader for a virtually wide open field with the hope that new discoveries will proceed from our readership in the coming years.

Jerry J. Buccafusco Augusta, September 2003

Cognitive Enhancing Drugs
Edited by J.J. Buccafusco

The cholinergic hypothesis – past and present

Jerry J. Buccafusco

Alzheimer's Research Center, Medical College of Georgia, 1120 – 15th Street, Augusta, Georgia, USA

The past

It is difficult to pinpoint the birth of cognitive pharmacology, i.e., a study of the effect of drugs on the behavioral and neurochemical processes we attribute to higher order thinking, learning, and memory. Moreover, it is not possible here to discuss and critique the various dominant theories of cognition. Rather it would be more relevant and in-keeping with the focus of this book to consider some of the past work, primarily pharmacological, that led to current concepts that will be described in detail in some of the following chapters. One of these concepts, the "cholinergic hypothesis" of learning and memory, has undergone the most vigorous scientific scrutiny relative to other theories. The culmination of the cholinergic hypothesis is represented by the current application of the theory to the treatment of Alzheimer's disease (AD) in the form of centrally-acting inhibitors of acetylcholinesterase (AChE), the enzyme that degrades synaptically-released acetylcholine. Despite this important application of the hypothesis, the origin of the cholinergic theory likely derives mainly from the effects produced by anticholinergic drugs, chiefly natural products like atropine and scopolamine which are active ingredients of various medicinal and cosmetic preparations of antiquity. Perhaps one of the oldest published studies pertaining to the physiological and behavioral effects of the administration of atropinic drugs in man was the account of Gauss in 1906 pertaining to the use of hyoscine in the induction of "twilight sleep" during labor [1]. Early rigorous experimentation with centrally-acting antimuscarinic agents largely confirmed the clinical observations [2–4]. Over subsequent decades scopolamine was in routine use for induction of sedation and the production of amnesia during labor and delivery and during surgical procedures. Scopolamine and other antimuscarinic agents are particularly effective amnestic drugs in inhibitory avoidance tasks performed by rodents. Scopolamine is particularly effective when administered pre-training i.e., during the acquisition phase of the task (e.g., [5]). The scopolamine-induced amnestic action in rodents is one of most widely utilized models during early stages of new drug development for therapeutic indications that require cognitive enhancement [6]. The antimuscarinic agent as used in various memory-related behavioral paradigms such as inhibitory avoidance, spatial reference and working memory, fear conditioning, and sensory gating is useful to a degree in the prediction of positive mnemonic efficacy in human beings; i.e., drug candidates that effectively reversed scopolamine-induced task impairment were shown to enhance cognition during clinical trials. In fact, the role of cholinergic neurons in the processes subserving learning and memory had been appreciated years prior to the first studies linking the loss of basal forebrain cholinergic neurons to the symptoms of AD. In fact, the *post mortem* neurochemical findings associated with AD pathology seemed to finally validate the cholinergic hypothesis of AD, as the loss of nigro-striatal dopaminergic neurons validated the dopaminergic hypothesis of Parkinson's disease.

The present

Despite this body of supporting evidence there have been recent challenges to the cholinergic hypothesis of AD. For example, Davis and colleagues [7] recently reported that the activities of AChE and choline acetyltransferase (ChAT) were not reduced in *postmortem* neocortical tissues obtained from individuals recently diagnosed with mild AD. These findings cast doubt as to the role of cholinergic neuron loss in mediating the early symptoms associated with AD. Likewise, DeKosky and co-workers [8] failed to detect a reduction in ChAT activity in cortical regions studied in patients diagnosed with mild cognitive impairment (a pre-AD state?) Finally, little cholinergic cell loss was detected in brain samples derived from individuals suffering near the time of death from mild cognitive impairment or early AD [9]. This impressive body of work was performed by using very rare tissue samples, and the results have had profound implications for the study of AD. However, they do not necessarily spell the death knell for the cholinergic hypothesis of AD. In fact, as early as 1999 Bartus and Emerich [10] pointed out that since neither ChAT nor AChE are rate limiting in the biosynthesis of acetylcholine, they are unlikely to accurately reflect active cholinergic function. And in fact, a variety of systems important for the dynamic function of cholinergic neurons could be compromised well before the degradation or loss of the more static cholinergic markers. Moreover, it should be considered that the collection of postmortem human tissues available for analysis often involves delays from a few hours to days as opposed to animals studies in which the *postmortem* intervals often involves a matter of minutes. The unavoidable degradation of tissue viability contributing to variability in the data poses a significant challenge. It may require the development of *in vivo* imaging methods capable of assessing the dynamic function of central cholinergic neurons in living patients suffering from early stages of AD to resolve these issues.

As age currently represents the most potent of the known risk factors for AD, it seems relevant to ask whether the function of central cholinergic neurons is impaired in the aged. Challenges to the cholinergic hypothesis of AD appear to ignore the body of evidence in support of the relationships between aging, cholinergic impairment and cognitive decline. A study of the effect of advanced age on brain cholinergic function began in earnest in the early 1980s when chemical enzymatic methods were developed with the specificity and sensitivity to measure the dynamic aspects of transmitter function. Methods for the rapid stabilization of brain levels of acetylcholine and choline by near freezing or focused microwave irradiation were also introduced for routine use. For example, Gibson and co-workers [11] examined the whole brain synthesis of acetylcholine in aged mice from 3–30 months of age. They reported that the biosynthesis of acetylcholine (measured by injection of a radio-labeled precursor) declined by up to 75% in the 30-month-old animals. Mild hypoxia further decreased acetylcholine synthesis by 90%. Moreover, aged cholinergic neurons were more impaired in their ability to release acetylcholine following potassium stimulation than they were in their ability to synthesize the transmitter [12]. Subsequent *in vivo* microdialysis methods largely confirmed these early findings (e.g., 13]. The concept that aged brain cholinergic neurons function relatively normally until stressed has been supported through experiments that used various methods to increase acetylcholine output [14–17], or which damage cholinergic neurons [18, 19]. It seems reasonable to conclude, therefore, that any sustained insult to forebrain cholinergic or hippocampal cholinergic neurons could interfere with the ability of these cells to provide sufficient transmitter release for normal function. This possibility was tested directly in a longitudinal series of experiments in which chemical lesions of basal forebrain cholinergic neurons were induced in young rats with the aim of producing only limited loss of the cholinergic cells [19]. The rats had been previously well trained in the performance of a sustained attention task. Whereas initially there was a similar degree of task performance in both experimental groups, a significant dissociation between lesioned and control rats in terms of task efficiency did not occur until the animals reached 31 months of age when the lesioned group exhibited significant task impairment. The results of these studies in aged rodents are relevant to the topic of this review considering the observation that most of the age-related changes pertained specifically to dynamic aspects of brain cholinergic neurons. In many of the studies cited above and in many other reports, indirect measures of standard cholinergic markers (as might be determined from autopsied tissues) often do not show such dramatic age-related differences.

As mentioned above, the cholinergic hypothesis has engendered the potential use of cholinergic agonists such as the AChE inhibitors for the treatment of the symptoms of AD. It should be pointed out, however, that this drug class may also improve cognitive performance in younger, non-impaired individuals. Non-human primate data (e.g., [20, 21]) as well as the results obtained in young human subjects [22] certainly support this possibility. AChE inhibitors also may exhibit efficacy in non-AD syndromes in which cognitive impairment is an accompanying symptom, including schizophrenia [23, 24],

Parkinson's disease [25, 26], Lewy-body dementia [27], traumatic brain injury [28], and others [29, 30]. And of course AChE inhibitors are effective in improving memory-related task performance in aged non-human primates, even in animals that are not that impaired. Indeed, it has been our experience that there is little difference in effectiveness in terms of task improvement to AChE inhibitors between young and aged subjects. In fact, as Bartus [21] pointed out, and as we have experienced [31, 32], young animals often provide a less variable and a more dose-dependent model for revealing the cognitive-enhancing actions of new therapeutic agents. In such studies, however, task variables such as retention intervals are optimized for all subjects to allow for drug-induced improvements to become manifest [31]. Drug-induced memory enhancement in young subjects is not relegated to cholinergic drugs, but is apparent for drugs of different pharmacological classes [32]. These observations speak to what is likely generally accepted, but not often discussed: that pharmacological enhancement of cognition can occur in normal individuals, particularly if the individual is stressed to his mnemonic limits. The burgeoning sales in over-the-counter "memory aids" has yet to benefit the prescription drug market. However, it is likely that many physicians who treat AD, as well as many Alzheimer's research centers, have received inquiries (as our Center has) regarding the acquisition of prescription AD therapeutic agents as memory aids for use in young individuals hoping to gain an advantage during scholastic examinations.

Notwithstanding the obvious ethical issues implied in the use of cognition-enhancing agents in non-impaired individuals, the cholinergic hypothesis has provided a framework for the development of drugs that have the potential to impact the most complex of brain functions. Despite the lack of disease specificity for the AChE inhibitors, the cholinergic hypothesis is not diminished. At greater issue is the possibility that central cholinergic neurons play an important role in at least some aspect of learning and memory that may be targeted pharmaceutically. It is also likely that other non-cholinergic compounds that have exhibited effectiveness in improving performance in memory-related tasks exert their effects indirectly via cholinergic pathways. For example certain biogenic amine-mimetics have been demonstrated to increase the release of endogenous acetylcholine. Even drugs acting at nicotinic receptors may owe their positive mnemonic effects to their ability to cause brain acetylcholine release [33]. It may be of interest to consider the fact that many, if not most, of the drugs demonstrated to offer improvement in the performance of memory-related tasks, irrespective of their pharmacological class, were initially examined in a model that used scopolamine to induce task impairment. The less often utilized experimental paradigm would be to ask whether pretreatment with scopolamine or other cholinergic antagonist (at non-amnestic doses) would block the positive effects of a non-cholinergic cognitive-enhancing agent in, for instance, an aging model. We used this approach to study the mechanism of the nicotine-induced increase in delayed matching-to-sample accuracy in macaques [33]. In this study low non-amnestic doses of scopolamine were shown to block the increased accuracy produced by nicotine. The data were interpreted to suggest that nicotine increased acetylcholine release (as was known from *in vitro* studies) which in turn acted on muscarinic receptors involved in the mnemonic pathway. It would be of interest to know whether this situation would apply to the various non-cholinergic compounds known to improve memory-related task performance, particularly since a wide variety of neurotransmitters and neuropeptides have been shown to directly or indirectly modulate the release of acetylcholine within the CNS. Many of these interactions were shown to have implications for learning and memory (for review, [34–44]).

In 1974, R.D. Myers asked how a transformation arising from a cytoplasmaic protein molecule could almost instantaneously retrieve an old memory by way of activating the release of a substance such as acetylcholine (rephrased). He termed this conundrum "...a puzzle to end all puzzlements" [45]. During those years there was no mechanism that could account for the millisecond requirement for memory retrieval. More recently, Nancy Woolf and her colleagues [46–48] have put forth a cogent and intriguing argument for the role of cortical cholinoceptive neural cells in human memory. Her work and that of others cited in her papers indicate that cortical acetylcholine release represents a primary mechanism for the cortical response adaptation associated with behavioral conditioning. Cortical cholinoceptive cells also are enriched in microtubule-associated protein-2 (MAP-2), a protein whose expression is highly correlated with the consolidation of contextual memory as measured behaviorally in animals. She also describes anatomical modules within the cytoarchitectonically-defined regions of the cerebral cortex, hippocampus, and amygdala that likely contain the mnemonic engrams. Memory storage and retrieval are visualized as dendritic modifications (altered dendrite length and/or arborization) within cholinoceptive cells induced by the destabilization or degradation of MAP-2 protein. MAP-2 degradation is subsequent to increased neuronal activity (LTP?), neurotropin (e.g., NGF) release, and enhanced acetylcholine release. Subsequent activation of cholinoceptive cells and their linked G-proteins results in calcium influx and activation of calcium-sensitive proteases that easily degrade MAP-2 protein thus allowing for dendritic modifications. Dr. Woolf points out that in AD, early in the syndrome, degenerative pathology is relegated to distal dendrites of hippocampal pyramidal cells – the potential sites for the storage of recent memories, and in fact, most cells possessing neurofibrillary tangles are cholinoceptive. She goes on to propose the interesting possibility that the selective vulnerability of cholinergic neurons in AD is related to the repeated restructuring of cholinoceptive cells due to the continuous encoding of long-term representations. It is possible that enhanced acetylcholine release associated with contextual memory results in the delayed degradation of MAP-2 in cholinoceptive neurons which must continually increase axonal growth to suitable dendritic segments for memory encoding. Apparently this increasing requirement for plasticity in basal forebrain cholinergic neurons as they are continually challenged by the need to seek out unmodified cholinoceptive dendritic segments leads to vulnerability.

From a pharmacological perspective Dr. Woolf concludes that a more effective treatment modality for AD could be developed by simultaneously inhibiting neuronal growth within AD-damaged areas like the hippocampus while applying cholinergic agonists to selectively enhance memory encoding and retrieval. Alternatively cholinergic function within the neocortex might be requisitioned to replace the damaged hippocampal function. This latter possibility might be achieved by simultaneous inhibition of muscarinic receptor function within the hippocampus and muscarinic receptor stimulation within the neocortex. The potential for such a pharmacologic scenario could reside in new drugs designed to be selective for muscarinic receptor subtypes. Whereas there is not that much difference in the overall expression of M1 receptors between cortex and hippocampus, within the hippocampus, M1 cholinoceptive cell bodies are relegated almost exclusively to the CA1 region and the dentate gyrus; and M3 receptor expression may be slightly greater in the cortex [49, 50]. In AD it might be possible to inhibit neural growth in already-damaged hippocampal regions with the judicious use of a selective M1 receptor antagonist, while simultaneously activating M3 receptors which are relegated to more unaffected cortical regions. Another approach might include the combination of an M1 antagonist with an α_2-adrenergic (or D1 dopaminergic) agonist. Attention and memory may be enhanced by adrenergic agonists, particularly via their actions within the prefrontal cortex ([46] and see Chapter by Edward D. Levin). This latter possibility could be more easily tested because the receptor-selective compounds already exist.

Finally, nicotinic receptor agonists, including nicotine itself, have the potential for inducing the release of a variety of neurotransmitters and neuropeptides. Cortical nicotinic acetylcholine, norepinephrine, and dopamine release along with low M1 muscarinic receptor blockade within damaged hippocampal regions might provide another useful approach. In fact, Buccafusco and Jackson [51] first reported that nicotine can produce a protracted improvement in task accuracy in monkeys trained in the performance of a delayed matching-to-sample task. The protracted feature of nicotine's beneficial mnemonic action was unexpected, particularly in view of the short plasma half-life of these drugs in rhesus monkeys. Levin and colleagues [52] reported similar findings for nicotine in rats. In fact, they demonstrated that this protracted beneficial effect of nicotine on radial arm maze performance was not dependent upon the presence of the drug at the time of behavioral training. In monkeys, the improvement in task efficiency that was measured on the day after nicotine administration generally occurred for trials associated with long delay intervals as they had for the previous day's session. This pattern (the retention interval receiving the greatest improvement on the day of testing was the same retention interval receiving the greatest improvement upon testing 24 hours later) was generally maintained for other compounds evaluated under similar conditions [53]. Nicotinic drugs are by no means the only class that appears to exhibit protracted improvement in task performance. For example, we have noted similar responses with the α_2-adrenergic agonist clonidine [54], and with the muscarinic (M1-preferring) receptor agonist WAY-132983 [55]. It is tempting at this point to consider the possibility that these drugs set into motion pharmacodynamic processes linked to the factors discussed by Dr. Woolf (*vide supra*). All of these compounds have the potential to affect either cholinergic or adrenergic processes within the frontal cortex and it would be of interest to determine whether their protracted positive mnemonic actions are related to distal dendritic modifications within the relevant cortical regions.

The future

It seems, therefore, that the cholinergic hypothesis of memory will continue to drive drug discovery and basic experimentation for some time to come. In the pages to follow we hope to provide an overview of the potential neural systems that have, and continue to be, exploited for cognition enhancement. Although there is much to rejoice in the wide variety of potential drug targets, the challenge will continue to be to develop drugs more selective for memory, and which are less apt to affect the other myriad brain functions which often translate to side-effects or toxicity. It is without doubt that the perfect cholinergic drug for cognition does not yet exist; and it is somewhat disappointing to read (in trade journals and reviewers' comments) about the 'limitations' of the current crop of clinically useful AChE inhibitors (as if these compounds represented the potential epitome of the class), and by way of association to downplay the future of cholinergic pharmacology in general. However, as you will read in subsequent chapters, cholinergic drugs have more to offer than mere cognition enhancement. Both muscarinic and nicotinic drugs have the ability to alter amyloid processing to the betterment of the brain. Nicotinic drugs are potent neuroprotective agents; and nicotine itself has been shown to prevent amyloid deposition in the brains of transgenic mice developed to over-express Aβ1-42 [56]. As such, cholinergic drugs may exhibit both cognitive-enhancing and disease modifying effects in AD individuals. Even the lowly cholinesterase inhibitors have shown their efficacy in slowing the progress of AD and significantly delaying nursing home placement.

Acknowledgements
The author would like to acknowledge his long-term colleagues Dr. William J. Jackson (retired) and Dr. Alvin V. Terry, Jr. for their valued contributions to many of the cognition-related studies we performed over the last 15 years. Much of the work cited for the author's studies was supported by the Office of Research and Development, Medical Research Service, Department of Veterans Affairs, and by the Alzheimer's Association.

References

1 Gauss CJ (1906) Geburten in künstlichem Dämmerschlaf. *Arch Gynaek* 78: 579–631
2 Safer DG, Allen RP (1971) The central effects of scopolamine in man. *Biol Psychiat* 3: 347–355
3 Drachman DA, Leavitt J (1974) Human memory and the cholinergic system. A relationship to ageing? *Arch Neurol* 30: 113–121
4 Crow TJ, Ross DG (1976) The specificity of the action of hyoscine on human learning. *Brit J Clin Pharmacol* 2: 367P–368P
5 Elrod K, Buccafusco JJ (1988) An evaluation of the mechanism of scopolamine-induced impairment in two passive avoidance protocols. *Pharmacol Biochem Behav* 29: 15–21
6 Porsolt RD, McArthur RA, Lenègre A (1993) Psychotropic screening procedures. In: H van Harran (ed.): *Methods in behavioral pharmacology*. Elsevier, New York, chap. 2
7 Davis KL, Mohs RC, Marin D, Purohit DP, Perl DP, Lantz M, Austin G, and Haroutunian V (1999) Cholinergic markers in elderly patients with early signs of Alzheimer disease. *J Am Med Assoc* 281: 1401–1406
8 DeKosky ST, Ikonomovic MD, Styren SD, Beckett L, Wisniewski S, Bennett DA, Cochran EJ, Kordower JH, Mufson EJ (2002) Up-regulation of choline acetyltransferase activity in hippocampus and frontal cortex of elderly subjects with mild cognitive impairment. *Ann Neurol* 51: 145–155
9 Gilmor ML, Erickson JD, Varoqui H, Hersh LB, Bennett DA, Cochran EJ, Mufson EJ, Levey AI (1999) Preservation of nucleus basalis neurons containing choline acetyltransferase and the vesicular acetylcholine transporter in the elderly with mild cognitive impairment and early Alzheimer's disease. *Comp Neurol* 411: 693–704
10 Bartus RT, Emerich DF (1999) Cholinergic markers in Alzheimer's disease. *JAMA* 282: 2208–2209
11 Gibson GE, Peterson C, and Sansone J (1981) Neurotransmitter and carbohydrate metabolism during aging and mild hypoxia. *Neurobiol Aging* 2: 165–172
12 Gibson GE, Peterson C (1981) Aging decreases oxidative metabolism and the release and synthesis of acetylcholine. *J Neurochem* 37: 978–984
13 Wu CF, Bertorelli R, Sacconi M, Pepeu G, and Consolo S (1988) Decrease of brain acetylcholine release in aging freely-moving rats detected by microdialysis. *Neurobiol Aging* 9: 357–361
14 Meyer EM, Crews FT, Otero DH, and Larsen K (1986) Aging decreases the sensitivity of rat cortical synaptosomes to calcium inotophore-induced acetylcholine release. *J Neurochem* 47: 1244–1246
15 Gilad GM, Rabey JM, Tizabi Y, Gilad VH (1987) Age-dependent loss and compensatory changes of septohippocampal cholinergic neurons in two rat strains differing in longevity and response to stress. *Brain Res* 436: 311–322
16 Giovannelli L, Giovannini MG, Pedata F, Pepeu G (1988) Purinergic modulation of cortical acetylcholine release is decreased in aging rats. *Exp Gerontol* 23: 175–181
17 Moore H, Stuckman W, Sarter M, Bruno JP (1996) Potassium, but not atropine-stimulated cortical acetylcholine efflux is reduced in aged rats. *Neurobiol Aging* 17: 565–571
18 Zawia N, Arendash GW, Wecker L (1992) Basal forebrain cholinergic neurons in aged rat brain are more susceptible to ibotenate-induced degeneration than neurons in young adult brain. *Brain Res* 589: 333–337
19 Burk JA, Herzog CD, Porter MC, Sarter M (2002) Interactions between aging and cortical cholinergic deafferentation on attention. *Neurobiol Aging* 23: 467–477
20 Terry AV Jr, Jackson WJ, Buccafusco JJ (1993) Effects of concomitant cholinergic and adrenergic stimulation on learning and memory performance by young and aged monkeys. *Cerebral Cortex* 3: 304–312
21 Bartus RT (2000) On neurodegenerative diseases, models, and treatment strategies: lesions learned and lessons forgotten a generation following the cholinergic hypothesis. *Exp Neurol* 163: 495–529
22 Davis K, Mohs R, Tinklenberg J, Pfefferbaum A, Hollister L, Kopell B (1978) Physostigmine: improvement of long-term memory processes in normal humans. *Science* 201: 946
23 MacEwan GW, Ehmann TS, Khanbhai I, Wrixon C (2001) Donepezil in schizophrenia – is it helpful? An experimental design case study. *Acta Psychiatrica Scand* 104: 469–472
24 Stryjer R, Strous RD, Bar F, Werber E, Shaked G, Buhiri Y, Kotler M, Weizman A, Rabey JM (2003) Beneficial effect of donepezil augmentation for the management of comorbid schizophrenia and dementia. *Clin Neuropharmacol* 26: 12–17

25 Werber EA, Rabey JM (2001) The beneficial effect of cholinesterase inhibitors on patients suffering from Parkinson's disease and dementia. *J Neural Transm* 108: 1319–1325
26 Aarsland D, Laake K, Larsen JP, Janvin C (2002) Donepezil for cognitive impairment in Parkinson's disease: a randomised controlled study. *J Neurol Neurosurgery Psychiat* 72: 708–712
27 Liberini P, Valerio A, Memo M, Spano P (1996) Lewy-body dementia and responsiveness to cholinesterase inhibitors: A paradigm for heterogeneity of Alzheimer's disease. *TIPS* 17: 155–160
28 Whelan FJ, Walker MS, Schultz SK (2000) Donepezil in the treatment of cognitive dysfunction associated with traumatic brain injury. *Annals Clin Psychiat* 12: 131–135
29 Litvan I, Phipps M, Pharr VL, Hallett M, Grafman J, Salazar A (2001) Randomized placebo-controlled trial of donepezil in patients with progressive supranuclear palsy. *Neurology* 57: 467–473
30 Barton D, Loprinzi C (2002) Novel approaches to preventing chemotherapy-induced cognitive dysfunction in breast cancer: the art of the possible. *Clin Breast Cancer* 3(Suppl 3): S121–S127
31 Paule MG, Bushnell PJ, Maurissen JPJ, Wenger GR, Buccafusco JJ, Chelonis JJ, Elliott R (1998) Symposium overview: The use of delayed matching-to-sample procedures in studies of short-term memory in animals and humans. *Neurotoxicol Teratol* 20: 493–502
32 Buccafusco JJ, Terry AV Jr (2000) Multiple CNS targets for eliciting beneficial effects on memory and cognition. *J Pharmacol Exp Ther* 295: 438–446
33 Terry AV Jr, Buccafusco JJ, Jackson WJ (1993) Scopolamine reversal of nicotine enhanced delayed matching-to-sample performance by monkeys. *Pharmacol Biochem Behav* 45: 925–929
34 Raiteri M, Marchi M, Maura G, Bonanno G (1989) Presynaptic regulation of acetylcholine release in the CNS. *Cell Biology Internat Rpts* 13: 1109–1118
35 Sarter M (1994) Neuronal mechanisms of the attentional dysfunctions in senile dementia and schizophrenia: Two sides of the same coin? *Psychopharmacology* 114: 539–550
36 Sastry BV (1995) Neuropharmacology of nicotine: effects on the autoregulation of acetylcholine release by substance P and methionine enkephalin in rodent cerebral slices and toxicological implications. *Clin Exp Pharmacol Physiol* 22: 288–290
37 Richter-Levin G, Segal M (1996) Serotonin, aging and cognitive functions of the hippocampus. *Reviews in the Neurosciences* 7: 103–113
38 Saito H, Matsumoto M, Togashi H, Yoshioka M (1996) Functional interaction between serotonin and other neuronal systems: focus on *in vivo* microdialysis studies. *Japan J Pharmacol* 70: 203–205
39 Govantes C, Marin J (1996) Effect of angiotensin converting enzyme inhibitors on quality of life in hypertensive patients. Pharmacodynamic basis. *Fund Clin Pharmacol* 10: 400–405
40 Maurice T, Lockhart BP (1997) Neuroprotective and anti-amnesic potentials of sigma (sigma) receptor ligands *Prog Neuro-Psychopharmacol Biol Psychiat* 21: 69–102
41 Feuerstein TJ, Seeger W (1997) Modulation of acetylcholine release in human cortical slices: possible implications for Alzheimer's disease. *Pharmacol Ther* 74: 333–347
42 Ogren SO, Schott PA, Kehr J, Yoshitake T, Misane I, Mannstrom P, Sandin J (1998) Modulation of acetylcholine and serotonin transmission by galanin. Relationship to spatial and aversive learning. *Annals NY Acad Sci* 863: 342–363
43 Hiramatsu M, Kameyama T (1998) Roles of kappa-opioid receptor agonists in learning and memory impairment in animal models. *Methods Findings Exp Clin Pharmacol* 20: 595–599
44 Bacciottini L, Passani MB, Mannaioni PF, Blandina P (2001) Interactions between histaminergic and cholinergic systems in learning and memory. *Behav Brain Res* 124: 183–194
45 Myers RD (1974) Learning and memory. In: Myers RD (ed.) *Handbook of drug and chemical stimulation of the brain*. Van Nostrand Reinhold, New York, p 596–647
46 Woolf NJ (1998) A structural basis for memory storage in mammals. *Prog Neurobiol* 55: 59–77
47 Woolf NJ, Zinnerman MD, Johnson GV (1999) Hippocampal microtubule-associated protein-2 alterations with contextual memory. *Brain Res* 821: 241–249
48 Woolf NJ, Milov AM, Schweitzer ES (2001) Roghani A. Elevation of nerve growth factor and antisense knockdown of TrkA receptor during contextual memory consolidation. *J Neurosci* 21: 1047–1055
49 Wei J, Walton EA, Milici A, Buccafusco JJ (1994) m1–m5 Muscarinic receptor distribution in rat CNS by RT-PCR and HPLC. *J Neurochem* 63: 815–821
50 Tohyama M, Takatsuji K (eds) (1998) *Atlas of neuroactive substances and their receptors in the rat*. Oxford University Press, New York
51 Jackson WJ, Buccafusco JJ (1991) Clonidine enhances delayed matching-to-sample performance by young and aged monkeys. *Pharmacol Biochem Behav* 39: 79–84

52 Levin ED (1992) Nicotinic systems and cognitive function. *Psychopharmacology* 108: 417–431
53 Buccafusco JJ, Terry AV Jr (2002) Nicotine and cognition in young and aged non-human primates. In: ED Levin (ed.): *Nicotinic receptors in the nervous system.* CRC Press, New York, 179–197
54 Jackson WJ, Buccafusco JJ (1991) Clonidine enhances delayed matching-to-sample performance by young and aged monkeys. *Pharmacol Biochem Behav* 39: 79–84
55 Bartolomeo AC, Morris H, Buccafusco JJ, Kille N, Rosenzweig-Lipson S, Husbands MG, Sabb AL, Abou-Gharbia M, Moyer JA, Boast CA (2000) The preclinical pharmacological profile of WAY-132983, a potent M1 Preferring Agonist. *J Pharmacol Exp Ther* 292: 584–596
56 Nordberg A, Hellström-Lindahl E, Lee M, Johnson M, Mousavi M, Hall R, Perry E, Bednar I (2002) Chronic nicotine treatment reduces b-amyloidosis in the brain of a mouse model of Alzheimer's disease (APPsw). *J Neurochem* 81: 655–658

Cognitive Enhancing Drugs
Edited by J.J. Buccafusco

Drugs that target cholinesterases

Ezio Giacobini

University Hospitals of Geneva, Department of Geriatrics, University of Geneva Medical School, CH-1226 Thonex, Geneva, Switzerland

Development of cholinesterase inhibitors in Alzheimer's therapy

Currently, cholinesterase inhibitors (ChEI) represent the treatment of choice for Alzheimer's disease (AD). Following the introduction in the 1980s of a first generation of drugs such as physostigmine and tacrine, a second generation of more suitable compounds was developed in the 1990s. These drugs are clinically more efficacious and produce less severe side-effects at effective doses. Contrary to the discovery of other neurotransmitter-based CNS drugs such as neuroleptics, tricyclic anti-depressants and anxiolytics, the clinical application of ChEI in the treatment of cognitive deficits in AD was neither accidental nor serendipitous. Its rationale was solidly founded on data derived from experimental physiology and behavioral pharmacology of the cholinergic system in animals and humans. Clinical results on the effect of these drugs on cognition (memory, attention and concentration) and more recently on behavioral symptoms in AD (apathy, hallucinations and motor agitation) confirmed predictions of potential clinical efficacy based on laboratory data.

Historically, the first ChEI to be used against AD was physostigmine administered under various modes [1]. Under those conditions, the effect of physostigmine was found to be too short lasting and the drug too toxic, therefore it was followed by oral tacrine [2]. Subsequently, metrifonate [3] and galantamine [4] were also tested orally.

Changes in cholinesterase activity related to Alzheimer disease and the role of butyrylcholinesterase in brain

Table 1 shows the changes in cholinesterase activity in the cortex of AD patients. Acetylcholinesterase (AChE) activity decreases progressively in certain brain regions from mild to severe stages of the disease to reach 10 to 15% of normal values while butyrylcholinesterase (BuChE) activity is unchanged or even increased by 20% [5–7].

In spite of the general reduction in brain AChE activity, the enzyme appears to be increased within and around neuritic plaques. In the plaques, AChE is

Table 1. Variation in cholinergic enzyme activity determined in autopsied human cortex from individuals with late stage Alzheimer's disease relative to normal controls

Enzyme	Localization in brain	Activity (% Control)	Molecular form
AChE	Neuronal	10–15	50–70% decrease, mainly G4
BuChE*	Glia-neuritic plaques	120	20% decrease in G4; 30–60% increase inG1
ChAT**	Neuronal	10–15	

*BuChE = butyrylcholinesterase; **ChAT = Choline acetyltransferase

closely associated with β-amyloid. As examples of regional difference in changes, BuChE/AChE ratio increases from 0.6 to 0.9 in the frontal cortex but from 0.6 to 11 in the enthorinal cortex [8]. This change may reflect a combination of reactive gliosis following severe neuronal damage (glial cells having preponderantly BuChE) and of an accumulation of BuChE in neuritic plaques, which contain both enzymes [9]. As the disease progresses and concentration of synaptic AChE (in particular the membrane-anchored G4 form) decreases [10], ChEI probably increase acetylcholine (ACh) concentrations to levels which may be inhibitory for AChE activity. This increase in substrate concentration may trigger glial BuChE to hydrolyze ACh and could thus represent a compensatory mechanism to the loss of neuronal AChE activity. Given the close spatial relationship between glial cell protoplasm and the synaptic gap, it is likely that extracellularly diffusing ACh may come into contact with glial BuChE and be effectively hydrolyzed, as demonstrated in our experiments in the rat with intracerebral microdialysis (Fig. 1) [11] and by the administration of rivastigmine (an AChE-BuChE-mixed inhibitor) in AD patients [12]. In these patients, CSF BuChE inhibition significantly correlates with cognitive benefit measured with a comprehensive computerized neuropsychological test (CNTB) (Fig. 2) [12].

Molecular forms of acetylcholinesterase in the human brain

Human brain AChE exists in multiple molecular forms as defined by their different sedimentation coefficient. Based on their shapes, collagen-tailed asymmetric forms and globular forms can be separated. Studies on whole brain fractions suggest that 60–90% of the tetrameric (G4) form is extracellular and membrane-located while the monomeric (G1) form is 90% intracellular and cytoplasmic [10, 13]. The different effect of certain inhibitors may be primarily related to the localization of the enzyme and the penetration of the inhibitor rather than to pharmacological or tissue selectivity. Selective loss of the membrane-bound G4 form has been reported in AD, suggesting a pre-synaptic

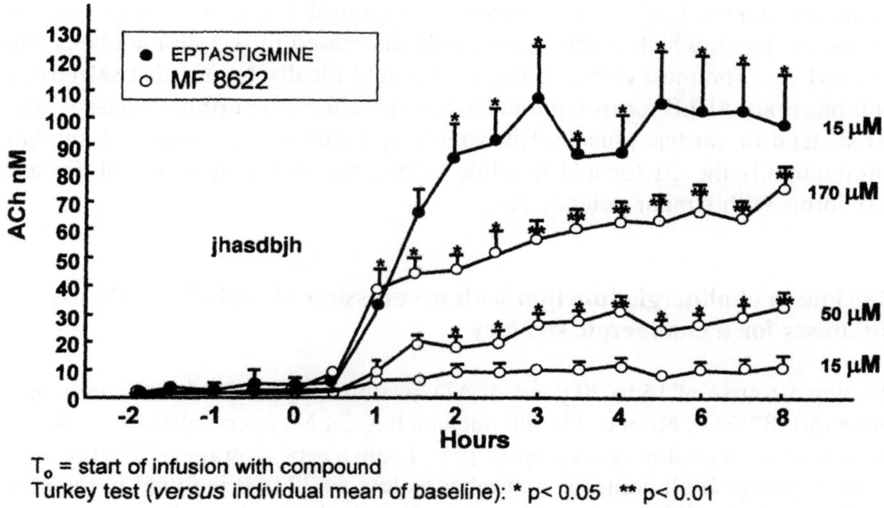

Figure 1. Extracellular levels of ACh measured in the rat cortex with microdialysis without adding a second ChEI. Following intracortical administration of eptastigmine, a non-selective ChE inhibitor, or MF 8622, a selective BuChE inhibitor at two different concentrations, a significant and prolonged increase in ACh levels is seen for both drugs lasting several hours.

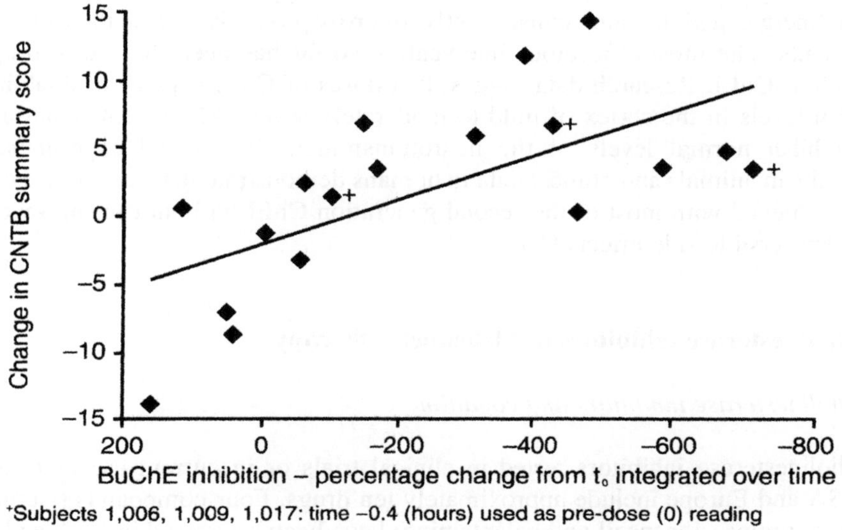

Figure 2. BuChE inhibition in CSF correlates significantly with cognitive improvement in AD patients administered rivastigmine. Cognitive changes have been measured with the Computerized Neuropsychological Test Battery (CNTB). Modif. from [12].

localization (Tab. 1). In severe Alzheimer patients, the membrane-bound G4 form is decreased in the frontal (–71%), the parietal cortex (–45%) and in the caudate putamen (–47%) in comparison to control levels. The exact function of the G1 form, which is not significantly decreased in AD, has not been elucidated yet. The most effective inhibitor would ideally be one that selectively inhibits brain AChE forms without having any effect on peripheral tissues such as skeletal or cardiac muscle. Rivastigmine, a carbamate compound, inhibits preferentially the G1 form [13], while tacrine and metrifonate, inhibit G4 and G1 forms with similar potency [14].

Decline in cholinergic function with progression of Alzheimer disease: premises for a cholinergic strategy

During a course of 15 to 20 years of AD, a continuous loss of cholinergic neurons (50–87%) is observed in the nucleus basalis Meynert (nbM) as well as a loss of cortical cholinergic synapses [15]. From a total average of 350,000 neurons in young adult controls, a number as low as 72,000 is found in the nbM of AD patients. This profound loss in subcortical nuclei results in a progressive cortical cholinergic denervation [8]. It is still controversial whether or not early decline in cognition in AD is associated with a decrease in cortical choline-acetyltransferase (ChAT) activity or with other changes in cholinergic function, such as selective choline uptake, ACh vesicular storage and release or ACh synthesis (Tab. 1). A cholinomimetic strategy should therefore increase cholinergic activity and consequently improve cognitive capacities of AD patients. The most efficacious intervention so far has been the use of drugs such as ChEI. Research data suggest that doses of ChEI capable of doubling ACh levels in the cortex of mild to moderately severe AD patients could re-establish normal levels of the neurotransmitter. Pre-clinical experimental results in animals and clinical data in humans demonstrate that such a goal can be achieved with most of the second generation ChEI without causing severe or irreversible side-effects [16].

Cholinesterase inhibitors in Alzheimer's therapy

Cholinesterase inhibitors and cognition

Cholinesterase inhibitors tested in clinical trials or in current use in Japan, USA and Europe include approximately ten drugs. Four compounds (tacrine, rivastigmine, donepezil and galantamine) have been registered in USA and in Europe. Tacrine, a ChEI of the first generation, has been withdrawn from the market. Two new ChEI are in clinical Phase II; Huperzine A (a chinese natural product) and phenserine (a carbamate). Galantamine (galantamine hydrobromide), the latest introduced drug, is a reversible inhibitor of AChE and also an allosteric modulator of nicotinic acetylcholine receptors [17]. Because galantamine binds to a site on nicotine acetylcholine receptors which is different than the acetylcholine binding site, it has been suggested that this provides galantamine with an additional mechanism of action which may activate non-cholinergic pathways impaired in AD [18]. Further clinical development of two other compounds, eptastigmine and metrifonate, were suspended because of side-effects (bone marrow suppression for eptastagmine and muscular weakness for metrifonate). To replace tacrine, the second generation ChEI (donepezil, rivastigmine, galantamine) had to fulfill specific requirements such as lower toxicity (hepatic) and easier administration besides demonstrable clinical efficacy [15].

There are differences between the tested compounds with regard to efficacy, percentage of treatable patients and responders, drop outs, severity and incidence of side-effects. Table 2 compares the effect of six ChEI using "intention to treat" criteria (ITT) [19–25, 27]. Pharmacologically, these drugs represent either reversible (tacrine, eptastigmine, donepezil and galantamine) or pseudo-irreversible or irreversible (rivastigmine, metrifonate) ChEI. The duration of these Phase III clinical trials varied from 24 to 30 weeks and over 10,000 patients were included in 26 different countries. The six most extensively clinically tested ChEI (tacrine, eptastigmine, donepezil, rivastigmine, metrifonate and galantamine) all produced statistically significant improvement in multiple clinical trials using similar standardized and internationally validated measures of both cognitive and non-cognitive functions. The most frequently used instrument for the evaluation of cognition, the ADAS-Cog, measures memory, orientation, language and praxis with a total score of 70 points. The mean annual change in ADAS-Cog scores in untreated AD patients was estimated to be approximately nine points per year in longitudinal studies. Obviously, there are large variations among patients, as the level of change seems to be dependent on the stage of the illness. The magnitude of cognitive effects measured with the ADAS-Cog scale for all six drugs – either expressed as the difference between drug- and placebo-treated patients or as the difference between drug-treated patients and baseline – is similar under present treatment conditions. This similarity after 26–30 weeks of treatment suggests a "ceiling effect" of approximately five ADAS-Cog points on average for approximately one third of patients in mild to moderate stages of the disease. It should be pointed out that this gain becomes more substantial, both clinically and economically, when evaluated after one year (8–9 points or more). Differences in effect between the drugs may be related partly to the rate of deterioration of the placebo group which can vary from trial to trial. The results obtained with some irreversible compounds suggest that the maximal clinical effect has not been reached yet. On the other hand, cholinergic toxicity related to maximal tolerated doses indicates a limit in safe achievable levels of ChE inhibition. Analysis of results also imply that both very mild and more severe cases need to be studied. Furthermore, the data showed wide variations of effect among patients; in some patients the gain was twice as large as in average. Cholinergic side-effects were transient, reversible and similar for all drugs. The percentage of improved patients varied from 25% to 50% with an average of 34%. This indicates that

Table 2. The effects of cholinesterase inhibitors on the ADAS-COG Test (ITT)

Drug	Dose mg/day	Duration of study weeks	Treatment difference from placebo*/baseline**	Improved patients (%)	Drop out %	Side effects %
Tacrine	120–160	30	4.0–5.3/0.8–2.8	30–50	55–73	40–58
Eptastigmine	45	25	4.7/1.83	12	35	
Donepezil	5–10	24	2.8–4.6/0.7–1	40–58	5–13	6–13
Rivastigmine	6–12	24	1.9–4.9/0.7–1.2	25–37	15–36	15–28
Metrifonate	25–75–80 60–80	12–26 26	2.8–3.1–3.2/0.75–0.5 3.9/2.24	35–40 15	2–21–8 7	2–12
Galantamine	24 32	20–24 24	3.1–3.9/1.73 3.8–3.9/1.6	10–23 34	10–13 32	 13–16

ADAS-cog = AD Assessment Scale-cognitive subscale; ITT = Intention to treat; * study end point *versus* Placebo; ** study end point *versus* baseline. Modified from [20].

more that one third of treated patients showed a significant clinical response to ChEI. This effect can be maintained for five drugs (tacrine, donepezil, rivastigmine, galantamine and metrifonate) for at least one year representing a high impact value for patients and caregivers. A smaller percentage (about 10–15%) of patients did not improve on the ADAS-Cog with any of the used drugs, while a second group of patients (5% or more) showed a response significantly higher than 5 points. The similarity in clinical efficacy of the tested drugs is underlined by a practically identical effect on global scales such as the Clinicians Interview-Based Impression of Change-plus (CIBIC-plus).

An even more important result of the six or 12 month clinical data is the observation that patients treated with the active compound changed little cognitively and behaviorally from the beginning of the study to the end (Fig. 3). This suggests a stabilization effect of disease-related deterioration, which is clinically more significant than expected symptomatic improvement. Placebo-controlled studies also indicate that this effect can be prolonged for at least one year (Fig. 3). Four drugs (donepezil, tacrine, metrifonate and rivastigmine) have shown a long-term effect extending from 1 to 4.5 years [28]. Differences between responders (i.e., stable or improved patients) *versus* non-responders may reflect the level of cholinergic damage present in the brain, genetic factors (such as the presence of APOE-ε4 alleles), gender or too low levels of ChE inhibition in the brain [20, 29, 30]. The question whether rivastigmine has a protective effect on subjects suffering from mild cognitive impairment is presently evaluated in on-going trials. Direct evidence of such an effect would modify the present definition of ChEI as drugs with symptomatic effects only.

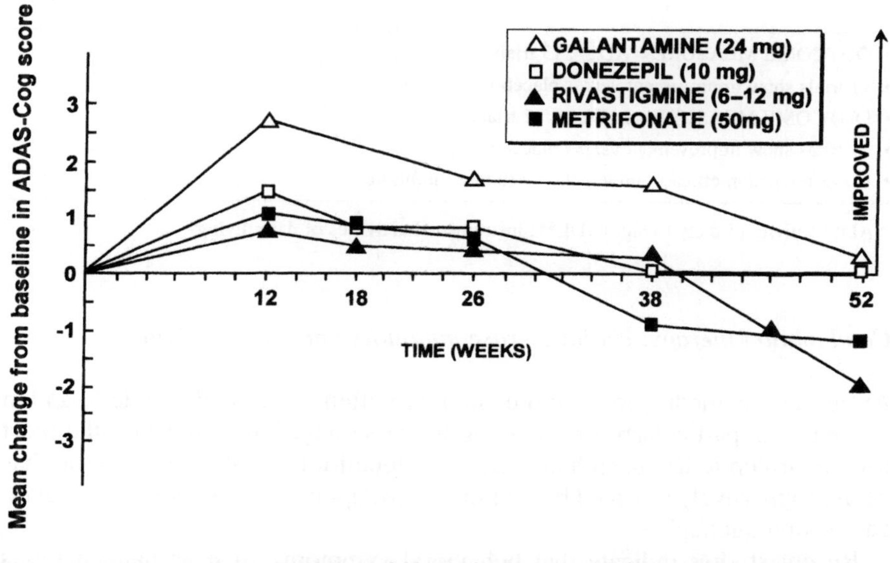

Figure 3 Stabilization effect (ADAS-Cog) of 12 month treatment with four cholinesterase inhibitors. The patients change little cognitively (ADAS-Cog) from baseline during this period.

Effects of cholinesterase inhibitors on behavior

Neuropsychiatric symptoms and functional deficits contribute greatly to the disability associated with AD. Neuropsychiatric symptoms occur in almost 90% of AD patients and are the most important factor of stress in caregivers. Biochemical and pharmacological considerations suggest the involvement of cholinergic deficiency in the mechanism of psychotic symptoms in AD. Ameliorating cholinergic function with ChEI should therefore decrease these problems.

Several studies have confirmed the beneficial effects of ChEI on neuropsychiatric symptoms in AD patients as measured by changes in the Neuropsychiatric Inventory (NPI) Scores [23, 27, 31, 32]. Significant improvement was seen for four drugs (rivastigmine, donepezil, galantamine and metrifonate) in delusions, hallucinations, apathy, motor agitation, depression and anxiety. Based on meta-analyses of 29 randomized controlled trials (16 with neuropsychiatric outcomes and 18 with functional outcomes) it is seen that ChEI exert a beneficial effect in treating neuropsychiatric symptoms and reducing functional impairment in patients with mild to moderate AD living in the community (Tab. 3) [33]. Thus, these studies support the hypothesis of a cholinergic link between cognitive and behavioral deficits in AD. They also demonstrated that the attenuation of such symptoms reduces caregiver burden, delays nursing home placement and in consequence could decrease the costs of patients care.

Table 3. The effect of cholinesterase inhibitors on functional impairment of Alzheimer's patients: A summary of 25 trials [33]

- OUTCOME MEASURE = ADL*, 13 trials
- 11 trials show improvement *versus* placebo
- OUTCOME MEASURE = IADL**, 12 trials
- 11 trials show improvement *versus* placebo
- No difference in efficacy among cholinesterase inhibitors

* ADL: activity of daily living; IADL**: instrumental activities of daily living

Combination therapy: cholinesterase inhibitors and anti-psychotics

Antipsychotic medication is more and more often prescribed in elderly and in AD patients particularly in a nursing home setting. Up to recently the most used neuroleptic has been haloperidol. Haloperidol a typical neuroleptic, has been progressively replaced by atypical neuroleptics such as risperidone, olanzapine and quetiapine.

Recent studies indicate that behavioral symptoms such as hallucinations and agitation can be treated with a combination of antipsychotics and ChEI and in some cases with ChEI only [35] by combining rivastigmine with either antipsychotics, anxiolytics or antidepressants were able to reduce or terminate the use of these medications after 52 weeks of rivastigmine treatment in a nursing home study. In a second US study [36], over 50% of institutionalized patients, receiving antipsychotics, antidepressants or anxiolytics in combination with rivastigmine, reduced the dose or ceased the treatment. In a third study of patients with severe AD, discontinuation of atypical antipsychotics (risperidone, olanzapine and quetiapine) was possible after eight months treatment with rivastigmine [35]. In a randomized, 20 week open label trial with AD and vascular dementia (VD) patients with behavioral disturbances being treated with rivastigmine, a combination of rivastigmine with risperidone or risperidone alone, showed that the improvement in behavior was greatest when rivastigmine was co-administered with risperidone than when the drugs were given individually, suggesting a synergistic effect of the two drugs [37].

These results suggest some benefits of a concomitant use of atypical neuroleptics together with ChEI; however, a recent short communication reports of extrapyramidal side-effects seen with a combination of risperidone and donepezil [38].

Differences in cholinergic effects among cholinesterase inhibitors

As mentioned above, cholinomimetic therapy is based on the principle that brain ChE inhibition increases synaptic ACh levels which may lead to cognitive improvement. Ideally, the correlation between cognitive effects and level of AChE or BuChE inhibition would be best observed in the brain or in the cerebrospinal fluid (CSF) [12] (Fig. 2). However, as CSF monitoring is difficult to achieve, peripheral inhibition of AChE in red blood cells (RBC) or plasma BuChE has been studied as an indirect measure of the drug effect. Inhibition varies between 30% and 80% depending on dose and pharmacokinetic characteristics of the compound. For some drugs (donepezil and metrifonate) the mean level of peripheral ChE inhibition is 65–70% and could be safely brought to as high as 90%. For other drugs, such as tacrine, the practical limit of inhibition can be as low as 30% and may be increased only at the expenses of severe side-effects. There is little correlation between central AChE inhibition and side-effects; the severity of side-effects is mainly due to peripheral inhibition [20, 39].

A direct clinical implication of this relationship is that drugs producing high levels of central AChE inhibition (or BuChE inhibition) at a low dose with a short 1/2 life (see below) will produce only mild peripheral cholinergic side-effects. A high increase of brain ACh may be achieved within a full range of therapeutic potency. As an example, rivastigmine at doses of 6 mg (corresponding to 62% AChE inhibition in CSF) produces a significantly greater improvement in cognitive function than at doses between 1 to 4 mg [40].

Pharmacological properties of cholinesterase inhibitors

Pharmacological properties and differences of ChEI affect both safety and clinical efficacy of these agents. These properties are related to the way the inhibitor interact with the enzyme, to the characteristic of the enzyme and its localization in brain. Cholinesterases hydrolyze choline-esters into the respective acid (such as acetic acid for ACh) and choline. These enzymes are widely distributed in different tissues of vertebrates and invertebrates. An important feature distinguishing BuChE present in serum and glia from AChE present in neurons and erythrocytes is its kinetics toward concentrations of ACh. BuChE is less substrate-specific for ACh than AChE. Butyrylcholinesterase catalyzes the hydrolysis of both ACh and BuCh as well as of their analogues; however, its affinity for ACh is less than that for BuCh. A wide range of inhibition (measured in rat brain and plasma or in human plasma and erythrocytes) of the two types of ChEI is seen for various ChEI with a different substrate specificity (Tab. 4).

Particularly interesting from the clinical point of view is the relative rate of inhibition shown by several ChEI for BuChE and AChE. We observe (Tab. 5) that most inhibitors presently utilized for AD therapy with the exception of donepezil, are not selective for AChE. However, they all show a similar degree

Table 4. Selectivity of cholinesterase inhibitors in humans

Compound	AChE[a] IC50[c]	BuChE[b] IC50	BuChE/AChE[d]	Clinical dose (mg/day)
BW 284 C51	18.8	48 000	2 553	
Huperzine	47	30 000	638	0.15–0.8
Donepezil	22	4 150	186	5–10
Phenserine	22	1 560	70	
Metrifonate	800	18 000	22.5	25–80
Galantamine	800	7 300	9	30
Rivastigmine	48 000	54 000	1.1	6–12
Physostigmine	28	16	0.6	36
Tacrine	190	47	0.25	80–160
Eptastigmine	20	5	0.25	45–60
Cymserine	758	50	0.7	
Iso-ompa	34 000	980	0.03	
Bisnorcymserine	110	1.0	0.009	
Hetopropazine	260 000	300	0.001	
Phenylethylcymserine	30 000	6	0.0002	
Bambuterol	30 000	3	0.0001	
MF-8622	100 000	9	0.00009	

[a]Human erythrocytes; [b]Human plasma, [c]IC50 = concentration of drug required to inhibit enzyme activity by 50%; [d]The higher the ratio the greater the selectivity for AChE.

Table 5. Pharmacological properties of cholinesterase inhibitors

Compound	Type of inhibition	Specificity of substrate	Chemical structure
Physostigmine	Pseudo-irreversible	ACh-BuCh	Carbamate
Tacrine	Reversible	ACh-BuCh	Acridine
Metrifonate	Irreversible	ACh-BuCh	Organophosphate
Donepezil	Reversible	ACh	Piperidine
Rivastigmine	Pseudo-irreversible	ACh-BuCh	Carbamate
Galantamine	Reversible	ACh-BuCh	Phenantrene
Huperzine A	Reversible	ACh	Pyridine

of clinical efficacy. Considering the drastic decrease in AChE activity taking place in the brain of advanced cases of AD patients (reaching 5% AChE levels at autopsy in some regions), and the large pool of BuChE available in glia neurons and neuritic plaques, it may not be an advantage for a ChEI to be selective for AChE. On the contrary, a good balance between AChE and BuChE inhibition may result in higher efficacy and allow a longer therapeutic use throughout the course of the disease. A very few selective inhibitors of BuChE are known and none has been tested clinically in placebo controlled trials. As it can be seen in Table 5, ChEI used in therapy belong to different classes with distinct differences in their chemical structure. ChEI can also be classified on the basis of the nature of the bonds formed between the agent and the enzyme in the complex (reversible, non-covalent) or the conjugate (carbamoyl or phosphoryl) (Tab. 5). Tacrine, donepezil and huperzine-A are high affinity, non-covalent inhibitors binding at the active center of the enzyme and occupying the choline binding subsite. Carbamoylating agents such as physostigmine and rivastigmine react with the active center serine to form a carbamoyl ester. On the other hand, phosphorylating agents such as metrifonate (DDVP is the active inhibitor) react covalently with the enzyme to form an inactive phosphoryl enzyme. This classification provides a clearer definition of the mechanism and the duration of action of the inhibitor.

The present knowledge of the molecular configuration of the two enzymes active sites allow to design compounds possessing well-balanced AChE-BuChE specificity, high CNS penetration and low peripheral and central cholinergic toxicity. Some of the second generation ChEI have demonstrated interesting beta-amyloid inhibitory characteristics.

Pharmacokinetic differences

A summary of pharmacokinetic properties of five ChEI is presented in Table 6 [17, 20, 25, 40–42]. Several important differences are apparent with regard to metabolism as well as other characteristics. While tacrine, galantamine and

Table 6. Comparison of pharmacokinetic properties of cholinesterase inhibitors after oral dosage in humans*

Drug	Plasma conc. (ug/L) C-max	Time to peak (h) T-max	Elimin. 1/2 life (h) T 1/2	Metabolism
Tacrine	–	1–2	2–4	hepatic (P450)
Donepezil	30–60	3–4	73	hepatic (P450)
Rivastigmine	114	1.7–1	5	non-hepatic
Metrifonate	500	0.5	2	non-hepatic
Galantamine	543	0.5	4.4–5–7	hepatic (P450)

*Modified from [20].

donepezil are metabolized through the hepatic route (P-450), rivastigmine and metrifonate are not. This difference is clinically important since elderly patients show decreased hepatic metabolism and therefore drugs not hepatically metabolized should be preferred. Another important characteristic is the difference in drug elimination with a half life (T1/2) between 2 to 73 hours. Such differences are important as a short T1/2 reduces the time of exposure of the peripheral pool of ChE to the inhibitor decreasing side-effects. Galantamine and metrifonate have maximal bioavailability (100% and 90% respectively) and lowest plasma protein binding (10% and 20% respectively). With 96%, plasma protein binding is highest for donezepil. Since elderly patients are generally treated with several drugs simultaneously, this factor is of particular interest in relation to drug interactions.

Pharmacokinetic properties may also cause important differences in efficacy and severity of side-effects. To maximize therapeutic central effects and to minimize peripheral (bronchial, muscular, gastro-intestinal and cardiac) side-effects, elimination half life should be short (around 1–2 hours). The effective dose should be low but able to produce a substantial CNS enzyme inhibition (60–80%) and a steady (small diurnal/nocturnal variations) and long-lasting (several days) level of inhibition. Irreversible ChEI (such as metrifonate) satisfy such criteria more closely than reversible ones.

Long-term effects of cholinesterase inhibitors

Prolonged clinical efficacy of ChEI is deduced from the two observations. Firstly, if drug treatment is interrupted, the cognitive effect may continue for three to four weeks even in the absence of ChE inhibition. Secondly, as mentioned above, treatment benefits can be maintained in a number of patients for at least one year (Fig. 3). This suggests that ChEI effect may not be related exclusively to an elevation in brain acetylcholine levels but also to a modification of the amyloid-linked pathology in AD.

Clinical data support a stabilizing effect of cholinesterase inhibitors

The benefit of ChEI treatment has been previously considered to be exclusively symptomatic and cognitive. It has now been demonstrated that improvement involves cognitive as well as behavioral symptoms [43, 44]. In many patients, cognitive improvement is significant up to 12 months (Fig. 3). Several clinical studies have demonstrated that the drug effect in many patients can be seen for as long as two years (Fig. 3, Table 7) [45–56]. This long-term effect translates into improved activity of daily living of the patient and reduced emotional impact for the caregiver as well as in reduction in care costs. Two year open-label data from a donepezil trial reveal a decline in ADAS-Cog from the base line 50% lower than the predicted value [56]. Untreated patients progress more rapidly than treated ones and the treatment effect seems to be related to the dose [56]. Average annual rate of decline for patients with a higher dose of rivastigmine is almost 50% lower (4.5 ADAS-Cog points/year) than that of patients treated with a lower dose (8.2 ADAS-Cog points/year) [56]. Increasing the dose of rivastigmine reduces the rate of cognitive decline over a three year period suggesting a reduction in the rate of progression of cognitive deterioration [50, 55]. Clinical data also indicate that rapidly progressing patients show the strongest drug effect, therefore, both disease stage and dose of the ChEI seem to play a role in altering the course of the disease [56]. Stabilization of cognitive deterioration suggests either a protective and struc-

Table 7. Long-term efficacy of cholinesterase inhibitors in Alzheimer's disease patients

References	Drug	Number of participants	Maximum treatment duration/year	Cognitive test paradigms	Benefit difference
[45, 54, 55]	Donepezil	1600	2	ADAS-COG	positive
[46]		133	4.9	ADAS-COG, CDR	Positive
[47]		431	1	ADFAC-CDR	Positive
[48]		286	1	GBS-MMSE	Positive
[49]	Tacrine	25	1	MMSE-EEG	Positive
[50]	Metrifonate	432	1	ADAS COG-CIBIC plus, MMSE	Positive
[51, 56]	Rivastigmine	2149	2	ADAS-COG, MMSE, CIBIC+, GDS,	Positive
[52]	Galantamine	44	3	ADAS-COG	Positive
[53]		636	1	ADAS-COG	Positive

Studies [47, 48, 50] are prospective, placebo controlled, double blind.
Total number of study participants = 4258. "Positive" indicates statistically significant clinical improvement from baseline for that specific measure.

tural effect or a long-term improvement of cholinergic synaptic function. The gradual return to the predicted deterioration-line after wash-out of the drug also suggests additional non-cholinergic effects [51].

Long-term stabilizing effects of cholinesterase inhibitors

Recent data from 12–24 month open trials and one randomized placebo-controlled trials suggest that optimization and maintenance of clinical effects for two years or more is a feasible goal in many patients (Tab. 7). Figure 3 reports the effect on the mean change in ADAS-Cog score of a 12 month treatment with four ChEI in clinical use (donepezil [44], galantamine [52], rivastigmine [50], metrifonate [49]). At twelve months, the data show no statistical difference from base line for all four ChEI (Fig. 3). The results of several clinical studies (placebo-controlled and open label) for periods longer than one year (up to three years) are reported in Table 7 [45–56]. These data indicate that benefit differences can be maintained in a number of patients for up to 12–24 months for five different inhibitors (donepezil, tacrine, metrifonate, rivastigmine and galantamine). In terms of global improvement in the ADAS-Cog score, this may sum to a total 15–20 point gain which represents a 18–24 month difference in disease history from placebo-treated patients. How to interpret this improvement? Is it the result of slowing down increase of disability or is it an expression of delaying progression of the disease?

Pre-clinical data supporting a non symptomatic effect of cholinesterase inhibitors

The amyloid precursor protein (APP) pathway which generates (A-beta)-amyloid is regulated by the sequential action of three enzymes (alpha, beta and gamma secretases). Alpha secretase cleaves APP within the A-beta sequence and releases soluble N-terminal non-aggregating fragments (sAPP). Numerous studies have shown that stimulation of sAPP release is associated with reduced formation of amyloidogenic peptides. Muscarinic-agonist induce sAPP secretion through activation by carbachol of m1 and m^3 (but not m^2 and m4) receptor subtypes increases sAPP release in human embryonic cell lines [59]. Activation of the pathway that cleaves APP decreases the release of beta-A fragments and may slow down amyloid formation in the brain. On the basis of results obtained from rat superfused cortical slices demonstrating an increased release of sAPP in response to muscarinic stimulation, we proposed an effect of ChEI on sAPP secretion acting through the same pathway [57].

Racchi et al. [58] using neuroblastoma cells and Pakaski et al. [61] using primary cultures of rat basal forebrain neurons have shown that short-term treatment with reversible and irreversible ChEI such as ambenomium and metrifonate or its metabolite DDVP increases sAPP release into the conditioned media and elevates levels of protein kinase C (PKC). These studies have demonstrated that this effect on APP is consistent with AChE inhibition and with indirect muscarinic-mediated cholinergic stimulation. In addition, short-term or long-term stimulation do not result in changes in APP mRNA expression either in cortical slices or neuroblastoma cells [57–59] and in down-regulation of the response to cholinergic stimulation of muscarinic receptors [60]. These results suggest that ChEI promote the non-amyloidogenic route of APP processing through a stimulation of alpha-secretase activity mediated through PKC [61]. This demonstrated feature of ChEI and of muscarinic agonists and their ability to enhance the release of non-amyloidogenic soluble derivatives of APP *in vitro* and *in vivo* suggests a slowing down in the formation of amyloidogenic compounds in brain of AD patients [20, 57, 59, 62].

Effects of cholinesterase inhibitors on Aβ metabolism independent of cholinesterase inhibition

Recent data demonstrate that the effect of ChEI on A-beta metabolism does not necessarily relate to ChE inhibition [63]. Phenserine, a carbamate-type of ChEI, reduces beta-APP levels *in vivo* and also decreases secretion of soluble beta-APP and beta-A into the conditioned media of human neuroblastoma cells. However, phenserine action is neither mediated through classical receptor signalling pathway, involving extracellular signal-regulated kinase or phosphatidylinositol 3-kinase activation, nor is associated with ChE inhibition. Phenserine reduces A-beta levels by regulating beta-APP translation by a putative interleukin-1 or TGF-beta responsive element located within the 5'-UTR [63]. This effect is translated in the cortex of transgenic mice (APPswe + PSI) with a 53% reduction of A-beta 1-42 peptide which is evident following a three week treatment with phenserine [64]. These results suggest that phenserine regulates APP protein expression at the post-transcriptional level as it suppresses APP protein expression without altering message levels.

Figure 4 summarizes the relationship between AChE and beta-amyloid in the brain of AD patients [62]. Accordingly, AChE which is present in a glycosylated form associated with the amyloid core of neuritic plaques [64, 65] is stimulated in its synthesis by beta-amyloid. A-beta stimulates beta-amyloid accumulation in or near to the plaque [65–68] (Fig. 4). An indirect support for this hypothesis is the finding of Rees et al. [69] that double transgenic mice over-expressing human amyloid precursor protein and human AChE develop earlier and larger plaque burden than single transgenic animals. On the other hand, AChE is increased in the brains of transgenic mice expressing the C-terminal fragment of the A-beta precursor protein [70]. These results suggest that a direct interaction between A-beta and AChE can accelerate and promote amyloid deposition in brain. Therefore, drugs designed to target this particular protein-protein interaction could retard progression of AD.

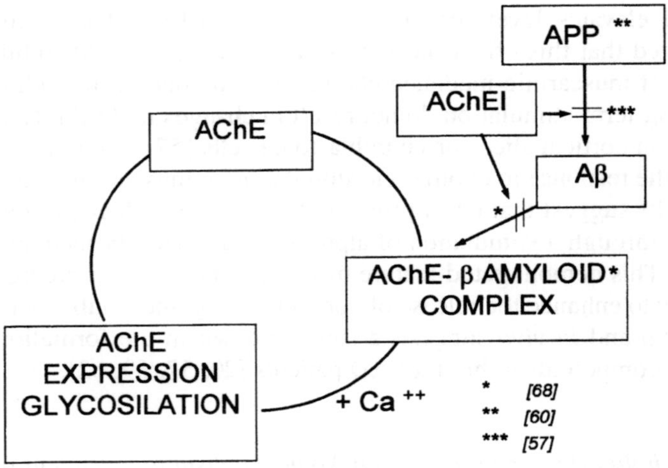

Figure 4. Proposed beta-amyloid cycle. AChE co-localizes with beta-amyloid and accelerates beta-amyloid formation and deposition in AD brain. *Beta*-amyloid increases AChE and in brain. Inhibition of AChE activity inhibiting amyloid precursor protein (APP) release, reduces beta-amyloid deposition. Inhibition of AChE activity by decreasing APP release, or by AChE-beta-amyloid interaction reduces beta-amyloid deposition (see text). This mechanism could contribute to the patient long-term cognitive stabilization seen during AChEI treatment.

New indications for cholinesterase inhibitor therapy

Treatment of mild cognitive impairment (MCI)

A natural challenge to the findings of a prolonged effect of ChEI is to investigate whether or not early treatment may alter the course of the disease by delaying its clinical onset. This approach could be of interest in subjects "at risk" such as asymptomatic members of familial AD pedigrees or in persons with mild cognitive impairment [71, 72]. As the definition of MCI is going to differentiate into various sub-groups corresponding to various types of diagnosis, such as VD-MCI (MCI in vascular dementia) or PDD-MCI (MCI in Parkinson Disease Dementia), the treatment will also vary accordingly. Presently, only results of small studies are available suggesting a possible delayed onset of AD and VD with ChEI such as donepezil [73]. Two major multicenter studies on MCI and ChEI are in progress in US and Europe with donepezil and rivastigmine.

Dementia with Lewy bodies

Dementia with Lewy bodies (DLB) is thought to be the second most common cause of dementia after AD [74]. The disorder is characterized by progressive fluctuating of cognitive impairment, visual hallucinations and motor features of parkinsonism. The disease presents itself from dementia with subsequent parkinsonian syndrome to parkinsonian syndrome with dementia. Studies have shown that neocortical cholinergic activity is more severely depleted in DLB than in AD, and that this deficit also affects the caudate nucleus, the thalamus and the brain stem [74]. It is likely this pattern is related to the occurrence of hallucinations characteristic for the disease. As typical neuroleptic treatment is contra-indicated in DLB, cholinergic therapy with rivastigmine has been investigated in 120 patients in a multicenter study with statistically and clinically significant effects on behavior [75]. At least 10 other smaller studies during the period 1998–2002 are supporting this beneficial effect of ChEI on DLB patients. Clinical data show that tolerability of ChEI in DLB appears similar to AD with some gastrointestinal effects and muscle cramps. Tremor may be present at higher dose. Particularly responsive to ChEI treatment are hallucinations and delusions.

Parkinson's dementia

Parkinson disease (PD) patients may suffer from cognitive impairment and behavioral problems such as apathy, personality changes and visual hallucinations. There is currently no recommended treatment for PD dementia (PDD). The observation that PDD patients have extensive cholinergic deficits led to the hypothesis that ChEI may provide benefits for patients with this condition. The treatment of PDD with ChEI (tacrine, donepezil, rivastigmine) has shown variable results from general improvement of cognition and psychotic symptoms to no change or even worsening of motor responses. Although some authors have concluded that ChEI have a beneficial effect on the cognitive state of PDD patients and do not aggravate motor function, larger studies are needed in order to demonstrate a clear benefit [76].

Vascular dementia

Vascular dementia (VD) accounts for approximately 20–30% of dementia cases and there is a large degree of clinical and pathological overlap between VD and AD. The presence of a cholinergic deficit in VD, similar to that observed in AD, is suggested by reductions in cholinergic markers. Enhancing the availability of endogenous acetylcholine by inhibition of cholinesterase increases cerebral blood flow. Clinical trials with all three inhibitors (donepezil, rivastigmine and galantamine) have shown that this effect is possible (Tab. 8).

One first attempt to treat vascular dementia with a ChEI (rivastigmine) was published with encouraging results [77] and, more recently, positive findings have been reported in the treatment of vascular and mixed dementia with galantamine [78], donepezil [79] and rivastigmine for 12 months [80].

Table 8. Vascular dementia: Treatment with cholinesterase inhibitors

Drug	Clinical effect	Reference
Rivastigmine	Improvement in cognition and behavior	[77, 80]
Donepezil	Cognitive effect	[80]
Galantamine	Cognitive and functional effect	[79]

However, clinical evaluation of these cases remains difficult in the absence of consensus on valid criteria for the diagnosis of vascular dementia.

Down's syndrome

Cholinesterase inhibitor treatment might also be indicated in Down's syndrome [81]. Genetic (chromosome 21), neuropathological and neurochemical similarities between Down's syndrome and AD as well as the presence of cognitive impairment, have motivated the use of cholinergic therapy in this disorder. Four published trials on a small number of patients (3–30) for a period of 8–40 weeks demonstrated a decrease in confusion and improvement of cognition [82–85] (Tab. 9).

Table 9. Down's syndrome: Effect of donepezil

Reference	Number of patients; duration; type of trial	Clinical effect	Side effects
[82]	3 patients; 8–24w; Open	Decreased confusion	Agitation incontinence
[83]	9 patients; 40w; open	Cognitive improvement	
[84]	30 patients; 24w; double blind placebo	Cognitive improvement	G.I.
[85]	6 patients; 24w; double blind placebo	Cognitive improvement	G.I.

Traumatic brain injury

Traumatic brain injury is the most common cause of death in subjects under the age of 40. Loss of hippocampal cells and reduction of acetylcholine levels and of muscarinic receptors can be attenuated in experimental animals by using ChEI such as rivastigmine [86]. The drug improves blood perfusion in ischemic areas and cholinergic transmission in cortex and hippocampus by increasing cholinergic activity in cerebral vessels [86]. This is the same mechanism invoked for treatment of VD. Several small trials with acute physostigmine treatment [87, 88] or chronic treatment (three week to two years) with donepezil have shown improvement in attention, verbal memory, general cognition and behavior [89–91] (Tab. 10).

Table 10. Traumatic brain injury: Treatment with cholinesterase inhibitors

Reference	Drug	Number of patients and duration of treatment	Clinical effect
[87]	Physostigmine	1 patient, acute	Improved verbal memory
[88]	Physostigmine	16 patients/6 weeks	Improved attention
[89]	Donepezil	2 patients/3 weeks	Improved memory
[90]	Donepezil	53 patients/2 years	Improved cognition
[91]	Donepezil	4 patients/16 weeks	Improved memory and behavior

Korsakoff's disease

Wernicke-Korsakoff disease is characterized by an amnestic state resulting from selective lesions in the limbic system with impairment of episodic memory. Basal forebrain nuclei are also affected. The aetiology is thiamine deficiency in alcoholic patients. In two of the three studies performed with donepezil, an improvement in memory has been reported [92–94]. However, the number of treated patients is too small to show evidence of clinical effect.

Delirium

Delirium is a common complication of dementia with fluctuation of attention and consciousness which may produce considerable morbidity. It is not always reversible and there is no specific treatment. The "scopolamine type" of dementia is an acute cognitive disorder which could be classified as a scopolamine-induced delirium in old age triggered by a muscarinic antagonistic effect. Most often, delirium is a consequence of a general medical condition and the direct cause of this syndrome is unknown. Cases of post-narcotic delirium, somnolence or coma being interpreted as central cholinergic syndromes have been reversed by using physostigmine [95]. More recently delirium of various origin (dementia, opioids, lithium etc.) has been reversed with ChEI such as donepezil or rivastigmine [96–99] (Tab. 11). Based on these results, it would be of great interest to test the effect of ChEI administered prior to general anaesthesia to prevent the occurrence of delirium.

Table 11. Delirium: Treatment with cholinesterase inihibitors

Reference	Drug	Etiology	Clinical effect
[95]	Physostigmine	Midazolam, fentanyl, other	5 patients, reversal
[99]	Donepezil	Dementia	1 patient, reversal
[96]	Donepezil	Opioids	6 patients, decreased sedation
[97]	Rivastigmine	Lithium	Reversal
[98]	Various ChEI	Various origins	Reversal

Migraine

The antinoceptive activity of donepezil which is prevented by muscarinic antagonists has been demonstrated in experimental animals. The analgesic effect of donepezil was investigated in patients suffering from migraine. The drug was effective in reducing the number of attacks and severity of pain [100].

Can cholinesterase inhibitors be switched or combined?

According to the report of the British National Institute for Clinical Excellence (NICE) about 50% of the patients treated with ChEI (donepezil, rivastigmine and galantamine) show evidence of improvement. This is probably an underestimation depending on the low sensitivity of the outcome measures particularly when they are applied to short-term (three months or shorter) observations. The decision to discontinue the treatment is based on early intolerable side-effects or lack of an initial therapeutic effect, loss of therapeutic effect during long-term treatment or occurrence of tolerability problems. As a result, patients are often tried on one drug only and if unsuccessful, this is a definitive trial. This decision is taken in spite of the fact that clinical experience shows that when a particular drug is ineffective or poorly tolerated, another agent belonging to the same pharmacological class may still result effective and safe. An open retrospective study of patients in the United Kingdom showed that 55% of patients who could not tolerate donepezil due to side-effects or did not show a sustained therapeutic effect, benefited from treatment with rivastigmine [101]. These results suggest that both therapeutic effect and side-effects of donepezil and rivastigmine may be independent of one another.

Several switching studies have confirmed that over 50% of patients may benefit from treatment with a second ChEI [102, 103]. A *post hoc* analysis of a five-month trial showed that a previous failure to respond to a ChEI did not predict response to a second one (galantamine) [104]. The same result was obtained switching from memantine, an NMDA antagonist, to donepezil supporting the idea that the mechanism of action of the two drugs may be different [105].

Switching from one ChEI to another is worthwhile when intolerable side-effects are present or the first drug has failed to prove effective following a period of time of several months testing.

Switching should not be considered when the patient responds and have no tolerability problems. Before switching, it is important to adjust the dose to an appropriate range.

Clinical experience has shown that switching to attain these potential benefits can take place without the need for an extensive washout period, beginning with the recommended starting dose of the second agent. If the patient is experiencing tolerability problems from the first drug, a washout period may be necessary. This procedure provides continuous cholinergic stimulation and ensures that AD patients are offered the maximum benefit of the drugs currently available. Little is known about benefits of switching between drugs of different pharmacological classes or combination. A recent study suggests that combination of a ChEI (donepezil) with memantine (an NMDA antagonist) is well tolerated and may show improved efficacy. These results need to be verified by more extensive double-blind studies.

General conclusions

This review illustrates that the number of therapeutic indications for ChEI has been growing since the first use of these drugs as cognitive enhancing treatments in AD almost two decades ago. Our knowledge of ChEI has also been growing in parallel with the increased clinical use of these drugs in AD patients. First, the possibility of using either selective BuChE inhibitors or dual-action AChE-BuChE inhibitors for cognitive improvement in AD or for newer indications represents a most recent development based on new knowledge of the role of BuChE in brain function. A second possibility of switching from one ChEI to another in order to increase the number of responders is now being investigated. A third possibility is to combine ChEI with non-cholinergic drugs such as memantine with the hope of achieving potentiation of mnemonic effects (also see Chapter by Jerry L. Buccafusco and Moussa B.H. Youdim). Why is therapy with ChEI efficacious in various types of CNS disorders having different etiologies?

Cholinergic systems are more numerous and more widely distributed in the brain than any other neurotransmitter system. As an example, the nucleus basalis of Meynert represents the most substantial regulatory system of the cerebral cortex, and the ascending cholinergic system alone is capable of maintaining the neocortex operative. In particular, cholinergic projections control selective attention and conscious awareness. The multifaceted effect of ChEI in various syndromes and diseases is therefore not surprising considering the major involvement of cholinergic transmission in CNS function at various brain localizations and neurotransmitters levels. Due to the central role of the cholinergic system and the variety of cholinergic functions that are impaired in a number of brain disorders, it is plausible to consider that in the future, ChEI might be revealed to be more efficacious in CNS disorders other than in AD.

References

1 Davis KL, Mohs RC (1979) Enhancement of memory by physostigmine. *New Engl J Med* 301: 946–956

2 Summer WK, Viesselman JO, Marsh GM, Candelora K (1981) Use of THA in treatment of Alzheimer-like dementia: Pilot study in twelve patients. *Biol Psychiatry* 16: 145–153

3 Becker E, Giacobini E (1988) Mechanisms of cholinesterase inhibition in senile dementia of the Alzheimer type. *Drug Development Research* 12: 163–195

4 Rainer M, Mark TH, Haushofer A (1989) Galantamine hydrobromide in the treatment of senile dementia of Alzheimer's type. In: T Kewitz, L Thomsen, A Bickel (eds): *Pharmacological interventions on central mechanisms in senile dementia.* Zuckschwerd, Munchen, 53–71

5 Atack JR, Perry EK, Bonham JR, Candy JM, Perry RH (1987) Molecular forms of butyrylcholinesterase in the human neocortex during development and degeneration of the cortical cholinergic system. *J Neurochem* 48: 1687–1692

6 Giacobini E, DeSarno P, Clark B, McIlhany M (1989) The cholinergic receptor system of the human brain – neurochemical and pharmacological aspects in aging and Alzheimer. In: A Nordberg, K Fuxe, B Holmstedt (eds): *Progress in Brain Research.* Elsevier, Amsterdam, 335–343

7 Perry EK, Perry RH, Blessed G, Tomlinson BE (1978) Changes in brain cholinesterases in senile dementia of Alzheimer type. *Neuropathol Appl Neurobiol* 4: 273–277

8 Geula C, Mesulam MM (1994) Cholinergic systems and related neuropathological predilection patterns in Alzheimer disease. In: RD Terry, R Katzman, KL Bick (eds): *Alzheimer Disease.* Raven Press, New York, 263–291

9 Wright CI, Geula C, Mesulam MM (1993) Neuroglial cholinesterases in the normal brain and in Alzheimer's disease: Relationship to plaques, tangles and patterns of selective vulnerability. *Annals of Neurology* 34: 373–384

10 Ogane N, Giacobini E, Struble R (1992) Differential inhibition of acetylcholinesterase molecular forms in normal and Alzheimer disease brain. *Brain Research* 589: 307–331

11 Cuadra G, Summers K, Giacobini E (1994) Cholinesterase inhibitor effects on neurotransmitters in rat cortex *in vivo*. *J Pharmacol Exp Ther* 270, 277–284

12 Giacobini E, Spiegel R, Enz A, Veroff AE, Cutler RN (2002) Inhibition of acetyl-and butyrylcholinesterase in the cerebrospinal fluid of patients with Alzheimer's disease by rivastigmine: Correlation with cognitive benefit. *J Neural Transm* 109: 1053–1065

13 Enz A, Florsheim P (1996) Cholinesterase Inhibitors: An overview of their mechanism of action. In: E Giacobini, R Becker (eds): *Alzheimer's Disease: From Molecular Biology to Therapy.* Birkhäuser, Boston, 211–215

14 Ogane N, Giacobini E, Struble R (1992) Differential inhibition of acetylcholinesterase molecular forms in normal and Alzheimer's disease brain. *Brain Research* 589: 307–331

15 Giacobini E (1994) Cholinomimetic therapy of Alzheimer disease: Does it slow down deterioration? In: G Racagni, N Brunello, SZ Langer (eds): *Recent Advances in the Treatment of Neurodegenerative Disorders and Cognitive Dysfunctions.* Karger, New York, 51–57

16 Giacobini E, Cuadra G (1994) Second and third generation cholinesterase inhibitors: From preclinical studies to clinical efficacy. In: E Giacobini, R Becker (eds): *Alzheimer Disease: Therapeutic Strategies.* Boston, Birkhäuser, 155–171

17 Bores GM, Huger FP, Petko W, Mutlib AE, Camacho F, Rush DK, Selk DE, Wolf V, Kosley RW, Davis L et al. (1996) Pharmacological evaluation of novel Alzheimer's disease therapeutics: Acetylcholinesterase inhibitors related to galantamine. *J Pharmacol Exp Ther* 277: 728–738

18 Francis PT, Palmer AM, Snape M, Wilcock GK (1999) The cholinergic hypothesis of Alzheimer's disease: A review of progress. *J Neurol Neurosurg Psychiatry* 66: 137–147

19 Becker RE, Colliver JA, Markwell SJ, Moriearty PL, Unni LK, Vicari S (1996) Double-blind, placebo-controlled study of metrifonate, an acetylcholinesterase inhibitor for Alzheimer disease. *Alzheimer Disease and Associated Disorders* 1: 24–131

20 Giacobini E (1998) Cholinergic foundations of Alzheimer's disease therapy. *Journal de Physiologie (Paris)* 92: 283–287
21 Imbimbo BP, Lucchelli PE (1994) A pharmacodynamic strategy to optimize the clinical response to eptastigmine. In: R Becker, E Giacobini (eds): *Alzheimer Disease Therapeutic Strategies*. Birkhäuser, Boston, 223–230
22 Knapp MJ, Knopman DS, Solomon PR, Pendlebury WW, Davis CS, Gracon SI (1994) A 30 week randomized controlled trial of high-dose tacrine in patients with Alzheimer's disease. *JAMA* 271: 985–991
23 Morris JC, Cyrus PA, Orazem J, Mas J, Bieber F, Ruzicka BB, Gulanski B (1998) Metrifonate benefits cognitive, behavioral, and global function in patients with Alzheimer's disease. *Neurology* 50: 1222–1230
24 Raskind MA, Peskind ER, Wessel T (2000) Galantamine in AD: A sixth month randomized, placebo-controlled trial with a six-month extension. *Neurology* 54: 2261–2268
25 Rogers SL, Farlow MR, Doody RS, Mohs R, Friedhoff LT (1998) A 24 week, double-blind placebo-controlled trial of donepezil in patients with AD. *Neurology* 50: 136–145
26 Rosler M, Anand R, Cicin-Sain A, Gauthier S, Agid Y, Dal-Bianco P, Stahelin HB, Hartman R, Gharabawi M (1999) Efficacy and safety of rivastigmine in patients with Alzheimer's disease: International randomized controlled trial. *British Medical J* 318: 633–658
27 Tariot PN, Solomon PR, Morris JC, Kershaw P, Lilienfel FCP, Ding C, The Galantamine USA-10 Study Group (2000) A 5-month, randomized, placebo-controlled trial of galantamine in AD. *Neurology* 54: 2269–2276
28 Giacobini E (2000) Cholinesterase inhibitors stabilize Alzheimer's disease. *Annals of the New York Academy of Sciences* 920: 321–327
29 Farlow MR, Cyrus PA, Nadel A, Lahiri DK, Brashear A, Gulanski B (1999) Metrifonate treatment of AD: Influence of APOE genotype. *Neurology* 53: 2010–2016
30 MacGowan SH, Wilcock G, Scott M (1998) Effect of gender and apolipoprotein E genotype on response to anticholinesterase therapy in Alzheimer's disease. *Int J Geriat Psych* 13: 625–630
31 Cummings JL, Masterman DK (1998) Assessment of treatment-associated changes in behavior and cholinergic therapy of neuropsychiatric symptoms in Alzheimer's disease. *J Clin Psychiatry* 59 (Suppl. 13): 23–30
32 Levy ML, Cummings JL, Kahn-Rose R (1999) Neuropsychiatric symptoms and cholinergic therapy for Alzheimer's disease. *Gerontology* 45: 15–22
33 Trinh NH, Hoblyn J, Mohanty S, Yaffe K (2003) Efficacy of cholinesterase inhibitors in the treatment of neuropsychiatric symptoms and functional impairment in Alzheimer disease: a meta-analysis. *JAMA* 8;289(2): 210–216
34 Messina J, Koumaras B, Hartman R (2000) Treatment with rivastigmine reduces the need for psychotropic medications in patients with Alzheimers'disease residing in long-term facilities. *3rd Int Meet Coll Psychiatric Neurological Pharmacists*. Washington DC, Poster presentation
35 Etemad B (2001) Behavioural and cognitive benefits of rivastigmine in nursing home patients – a 26-week follow-up. *Int Psychogeriatr* 13: 241(abstr. p246)
36 Shua-Haim J, Smith Amin J (2002) Slow dose escalation of rivastigmine (Exelon®) treatment of agitation in patients with Alzheimer's disease: An eight month prospective study. Poster presentation at *Seventh International Geneva/Springfield Symposium on Advances in Alzheimer therapy*
37 Weiser M, Davidson M, Hartmann R (2002) A pilot randomized, open-label trial assessing safety and pharmacokinetic parameters of co-administration of rivastigmine with risperdone in dementia patients with behavioural disturbances. *Int J Geriatr Psychiatry* 17: 343–346
38 Liu HC, Lin SK, Sung SM (2002) Extra-pyramidal side-effect due to drug combination of risperdone and donepezil. *Psychiatry Clin Neurosci* 5(4): 479
39 Becker E, Giacobini E (1988) Mechanisms of cholinesterase inhibition in senile dementia of the Alzheimer type. *Drug Development Research* 12: 163–195
40 Anand R, Hartman RD, Hayes PE (1996) An overview of the development of SDZ ENA 713, a brain selective cholinesterase inhibitor disease. In: R Becker, E Giacobini (eds): *Alzheimer Disease: From molecular biology to therapy*. Birkhäuser, Boston, 239–243
41 Lilienfels S, Parys W (2000) Galantamine: Additional Benefits to patients with Alzheimer's disease. *Dementia and Geriatric Cognitive Disorders* 11 (Suppl. 1): 19–27
42 Pettigrew LC, Bieber F, Lettieri J, Wermeling DP, Schmitt FA, Tikhtman AJ, Ashford JW, Smith CD, Wekstein DR, Markesbery WR et al. (1998) A study of the pharmacokinetics, pharmacodynamics and safety of metrifonate in Alzheimer's disease patients. *J Clin Pharmacol* 38: 236–245

43 Cummings JL (1997) Changes in neuropsychiatric symptoms as outcome measures in clinical trials with cholinergic therapies for Alzheimer disease. *Alz Dis Ass Dis* 11: S1–S9
44 Raskind MA, Cyrus PA, Ruzicka BB (1999) The effects of Metrifonate on the cognitive, behavioral and functional performances of Alzheimer's disease patients. *J Clin Psychiatry* 60: 318–325
45 Matthews HP (2000) Donepezil in Alzheimer's disease: eighteen month results from Southampton memory clinic. *Int J Ger Psychiatry* 15: 713–720
46 Rogers SL (2000) Long-term efficacy and safety of donezepil in the treatment of Alzheimer's disease: Final analysis of a US multicentre open-label study. *Eur Neuropsychopharmacol* 10: 195–203
47 Winblad B, Engedal K, Soininen H (1999) Donepezil enhances global function, cognition and activities of daily living compared with placebo one year. *12th Congr. ECNP*, London Abst. 30
48 Mohs R, Doody R, Morris J (1999) Donepezil preserves functional status in Alzheimer's disease patients. *Eur Neuropsych* Suppl 5 Abst. S328
49 Jelic V, Amberla K, Almkvist O (1998) Long-term tacrine treatment slows the increase of theta power in the EEG of mild Alzheimer patients compared to untreated controls. *Fifth Int Geneva/Springfield Symposium on Advances in Alzheimer Therapy, Geneva*; Abst. P.147
50 Bernhardt T, Woelk H (2000) Metrifonate demonstrates sustained improvement in cognition and global functioning in a 12-month, double-blind placebo-controlled trial. *Eur Neur Soc Meeting* (Jerusalem), p 36
51 Anand R, Hartman R, Messina J (1998) Long-term treatment with rivastigmine continue to provide benefits for up to one year. *Fifth Int Geneva/Springfield Symposium on Advances in Alzheimer Therapy,* Geneva Abst. p.18
52 Rainer M, Mucke HAM (1999) Long-term cognitive benefit from galantamine in Alzheimer's disease *Int J Geriatr Psychiatry* 1: 197–201
53 Raskind MA, Peskind ER, Wessel T (2000) Galantamine in AD: A sixth month randomized, placebo-controlled trial with a six-month extension. *Neurology* 54: 2261–2268
54 Doody RS, Pratt RD, Perdomo CA (1999) Clinical benefits of donepezil: Results from a long-term Phase III extension trial. *Neurology* 52 Suppl 2:A 174
55 Rogers SL, Friedhoff L (1998) Long-term efficacy and safety of donepezil in the treatment of Alzheimer's disease. *Europ Neuropsychopharmacol* 8(1): 67–75
56 Farlow M (2000) New approaches in assessing delay of progression of AD. *Symp. Pivotal Research. World Alzheimer Congress*, Washington D.C. Abst p.10–11
57 Mori F, Lai CC, Fusi F, Giacobini E (1995) Cholinesterase inhibitors increase secretion of APPs in rat brain cortex. *Neurol Rep* 6: 633–636
58 Racchi M, Schmidt B, Koenig G (1999) Treatment with metrifonate promotes soluble amyloid precursor protein release from SH-SY5Y neuroblastoma cells. *Alz Dis Assoc Dis* 13: 679–688
59 Giacobini E (1996) Cholinesterase inhibitors do more than inhibit cholinesterase. In: R. Becker, E. Giacobini (eds): *Alzheimer Disease: From molecular biology to therapy*. Birkhäuser, Boston, 187–204
60 Nitsch RM (1992) Release of Alzheimer amyloid precursor derivatives stimulated by activation of muscarinic acetylcholine receptors. *Science* 258: 304–307
61 Pakaski M, Rakonczay Z, Kasa P (2001) Reversible or irreversible cholinesterase inhibitors cause changes in neuronal amyloid percursor processing and protein kinase C level *in vitro*. *Neurochem Int* 38: 219–226
62 Giacobini E (2000) Cholinesterase inhibitors: from the calabar bean to Alzheimer therapy. In: E Giacobini (eds): *Cholinesterase and Cholinesterase Inhibitors. From molecular biology to therapy*. Martin Dunitz, London, 181–122
63 Shaw KT, Utsuki T, Rogers J (2001) Phenserine regulates translation of beta-amyloid precursor protein mRNA by a putative interleukin-1 responsive element, a target for drug development. *Proc Natl Acad Sci* USA 19;98(13): 7605–7610
64 Lahiri DK, Farlow MR, Ge YW (2002) Phenserine: A new generation of cholinesterase inhibitors with amyloid-modifying properties. *7th Int. Geneva Springfield Symposium,* Abst p. 80
65 Saez-Valero J, Sberna G, McLean CA, Small DH (1999) Molecular isoform distribution and glycosylation of acetylcholinesterase are altered in brain and cerebrospinal fluid of patients with Alzheimer's disease. *J Neurochem* 72: 1600–1608
66 Small DH, Sberna G, Li QX (1998) The beta-amyloid protein influence acetylcholinesterase expression, assembly and glycosylation. *Sixth Int. Conf. On Alzheimer's disease and Related Disorders*. Amsterdam Abst. 880,S.209

67 Alvarez A (1997) Acetylcholinesterase promotes the aggregation of amyloid B-peptide fragments by forming a complex with the growing fibrils. *J Mol Biol* 272: 348–361
68 Inestrosa N, Alvarez A, Reyes A, De Ferrari G. (2000) Acetylcholinesterase-amyloid peptide interaction and WnT signalling involvement in A-*Beta* neurotoxicity. *Acta Neurol Scand* Suppl. 176: 53–59
69 Rees T, Hammond P, Younkin S, Soreq H, Brimijoin S (2003) Acetylcholinesterase facilitates amyloid deposition in a mouse model of Alzheimer's disease. *Neurobiology of Aging* 24: 777–787
70 Sberna G, Saez-Valero J, Li QX (1998) Acetylcholinesterase is increased in the brains of transgenic mice expressing C-terminal fragment of the beta-amyloid protein precursor of Alzheimers' disease. *J Neurochem* 71: 723–731
71 Fox NC, Warrington EK, Seiffer AL, Agnew SK, Rossor MN (1998) Presymptomatic cognitive deficits in individuals at risk of familial Alzheimer's disease. A longitudinal prospective study. *Brain* 121: 1631–1639
72 Sunderland T (1998) Cholinergic Therapy. *The American Journal of Geriatric Psychiatry* 6 (Suppl. 1): 57–63
73 Stirling Meyer J, Li Y, Xu G, Thornby J, Chowdhury M, Quach M (2002) Feasibility of treating mild cognitive impairment with cholinesterase inhibitors. *Int J Geriatr Psychiatry* 17: 586–588
74 Perry EK, McKeith IG, Perry RH (1998) Dementia with Lewy bodies: A common cause of dementia with therapeutic potential. *Inter J of Geriatric Pharmacology* 1: 120–125
75 McKeith I, Del Ser T, Spano P, Emre M, Wesnes K, Anand R, Cicin-Sain A, Ferrara R, Spiegel R (2000) Efficacy of rivastigmine in dementia with Lewy bodies: A randomised, double-blind, placebo-controlled international study. *Lancet* 356: 2031–2036
76 Werber EA, Rabey JM (2001) The beneficial effect of cholinesterase inhibitors on patients suffering from Parkinson's disease and dementia. *J Neural Transm* 108(11): 1319–1325
77 Kumar V, Anand R, Messina J, Hartman R, Veach J (2000) An efficacy and safety analysis of Exelon in Alzheimer's disease patients with concurrent vascular risk factors. *Eur J Neurology* 7: 159–169
78 Erkinjuntti T, Kurz A, Gauthier S, Bullock R, Lilienfeld S, Damaraju CV (2002) Efficacy of galantamine in probable vascular dementia and Alzheimer's disease combined with cerebrovascular disease: A randomised trial. *Lancet* 359: 1283–1290
79 Pratt RD, Perdomo CA, the donepezil 307VaD Study Group (2002) Donepezil improves cognition in patients with vascular dementia: Results from study 307, a 24-week, randomized, double-blind, placebo-controlled trial. *7th International Geneva/Springfield Symposium on Advances in Alzheimer Therapy*. Geneva, Switzerland
80 Moretti R, Torre P, Antonello RM, Cazzato G (2001) Rivastigmine in subcortical vascular dementia: A comparison trial on efficacy and tolerability for 12 months follow-up. *Eur J Neurol* 8: 361–362
81 Kishnani PS, Sullivan JA, Walter BK, Spiridigliozzi GA, Doraiswamy PM, Krishnan KRR (1999) Cholinergic therapy for Down's syndrome. *Lancet* 353: 1064–1065
82 Hemingway-Eltomey JM, Lerner AJ (1999) Adverse effects of donepezil in treating Alzheimer's disease associated with Down's syndrome. *Am J Psychiatry* 156(9): 1470
83 Kishnani PS, Spiridigliozzi GA, Heller JH, Sullivan JA, Doraiswamy PM, Krishnan KRR (2001) Donepezil for Down's Syndrome. *Am J Psychiatry* 158: 143
84 Prasher VP, Huxley A, Haque MS, The Down syndrome ageing study group (2002) A 24-week, double-blind, placebo-controlled trial of donepezil in patients with Down syndrome and Alzheimer's disease – Pilot study. *Int J Geriat Psychiatry* 17: 270–278
85 Lott IT, Osann K, Doran E, Nelson L (2002) Down syndrome and Alzheimer disease: Response to donepezil. *Arch Neurol* 59(7): 1133–1136
86 Chen Y, Shohami E, Bass R, Weinstock M (1998) Cerebro-protective effects of ENA713, a novel acetylcholinesterase inhibitor, in closed head injury in the rat. *Brain Res* 784: 18–24
87 Goldberg E, Gerstman LJ, Mattis S, Hugues JEO, Sirio CA, Bilder RM Jr (1982) Selective effects of cholinergic treatment on verbal memory in Posttraumatic amnesia. *J Clin Neuropsychol* 4: 219–234
88 Levin HS, Peters BH, Kalisky Z, High WH Jr, von Laufen A, Eisenberg HM (1986) Effects of oral physostigmine and lecithin on memory and attention in closed head injured patients. *Cent Nerv Syst Trauma* 3: 333–342
89 Taverni JP, Seliger G, Lichtman SW (1998) Donepezil medicated memory improvement in tramatic brain injury post acute rehabilitation. *Brain Inj* 12(1): 77–80

90 Whelan FJ, Walker MS, Schultz SK (2000) Donepezil in the treatment of cognitive dysfunction associated with traumatic brain injury. *Ann Clin Psychiatry* 12(3): 131–135
91 Masanic CA, Bayley MT, van Reekum R, Simard M (2001) Open-label study of Donepezil in traumatic brain injury. *Arch Phys Med Rehabil* 82: 896–901
92 Iga JI, Araki M, Ishimoto Y, Ohmori T (2001) A case of Korsakoff's syndrome improved by high doses of donepezil. *Alcohol Alcohol* 36(6): 553–555
93 Angunawela II, Barker A (2001) Anticholinesterase drugs for alcoholic Korsakoff syndrome. *Int J Geriat Psychiatry* 16(3): 338–339
94 Sahin HA, Gurvit IH, Bilgic B, Hanagasi HA, Emre M (2002) Therapeutic effects of an acetylcholinesterase inhibitor (donepezil) on memory in Wernicke-Korsakoff's disease. *Clin Neuropharmacol* 25(1): 16–20
95 Jost U, Schmid A, Ruppert M (1982) Physostigmine reversal of central anticholinergic syndrome induced by midazolam/fentanyl, benzoctamine/buprenorphine and etomidate/carticaine or by atropine/promethazine/pethidine for premedication? *Anaesthesist* 31(1): 21–24
96 Slatkin NE, Rhiner M, Maluso Bolton T (2001) Donepezil in the treatment of opioid-induced sedation: report of six cases. *J of Pain Symptom Manage* 21(5): 425–438
97 Fischer P (2001) Successful treatment on non-anticholinergic delirium with a cholinesterase inhibitor. *J Clin Psychopharmacol* 21(1): 118
98 Gustafson Y, Lundstrom M, Bucht G, Edlund A (2002) Delirium in old age can be prevented and treated. *Tidsskr Nor Laegeforen* 122(8): 810–814
99 Wengel SP, Roccaforte WH, Burke WJ (1998) Donepezil improves symptoms of delirium in dementia: Implications for future research. *J Geriatr Psychiatry Neurol* 11(3): 159–161
100 Nicolodi M, Galeotti N, Ghelardini C, Bartolini A, Sicuteri F (2002) Central cholinergic challenging of migraine by testing second-generation anticholinesterase drugs. *Headache* 42(7): 596–602
101 Bullock R, Connolly C (2002) Switching cholinesterase inhibitor therapy in Alzheimer's disease-donepezil to rivastigmine, is it worth it? *Int J Geriat Psychiatry* 17: 288–289
102 Emre M (2002) Switching cholinesterase inhibitors in patients with Alzheimer's disease. *Int J Clin Pract* 127: 64–72
103 Auriacombe S, Pere JJ, Loria-Kanza Y, Vellas B (2002) Efficacy and safety of rivagstigmine in patients with Alzheimer's disease who failed to benefit from treatment with donepezil. *Current Medical Research and Opinion* 18: 129–138
104 Morris JC (2001) Therapeutic continuity in Alzheimer's disease: switching patients to galantamine. Panel discussion: Recommendations for prescribers. *Clinical therapeutics* 23: A31–A39
105 Mielke R, Winkel M, Goebel C, Berger F (2002) Switching from memantine to donepezil: Effects on cognition and quality of life. *18th international conf of Alzheimer's disease International (ADI) October 23–26, Barcelona, Spain*

Cognitive Enhancing Drugs
Edited by J.J. Buccafusco

Drugs that target muscarinic cholinergic receptors

William S. Messer Jr.

Departments of Pharmacology and Medicinal & Biological Chemistry, College of Pharmacy, The University of Toledo, 2801 W. Bancroft St., Toledo, OH 43606, USA

Cholinergic pathways

Six groups of cholinergic neurons have been identified in the central nervous system (Ch1–Ch6) [1]. Cholinergic neurons innervating the hippocampus and cerebral cortex arise from basal forebrain nuclei, including the medial septum (MS, Ch1 group), the vertical and horizontal limbs of the diagonal band (Ch2 and Ch3) and the nucleus basalis of Meynert (NBM, Ch4) [2]. Cholinergic neurons arising from the MS project mainly to the hippocampus, a brain region that plays an important role in learning and memory function. The diagonal band projections innervate the anterior cingulate cortex and the olfactory bulb, while the NBM projects to the amygdala and the cerebral cortex. The cholinergic projection from the NBM is widespread, covering the entire cortical mantle [3]. Basal forebrain cholinergic pathways have been implicated in attention, learning, memory and cognitive function. Moreover, the degeneration of basal forebrain cholinergic neurons is a consistent neurochemical hallmark of Alzheimer's disease.

Other important cholinergic projections originate in the laterodorsal tegmental nucleus (LDTN, Ch5) and the pedunculopontine tegmental nucleus (PPTN, Ch6) and project to various brainstem and midbrain regions, including the ventral tegmental area, a brain region that regulates forebrain dopamine pathways. Although the basal forebrain system has received much attention due to its involvement in Alzheimer's disease, the brainstem cholinergic system has been found to play important roles in sensory-motor gating and cognitive function [2].

Muscarinic receptors

Muscarinic receptors play important roles in the regulation of physiological responses to acetylcholine. Five subtypes of muscarinic receptor have been identified and each subtype has a unique tissue distribution and physiological function [4]. Although there are relatively few selective ligands for individual muscarinic receptor subtypes, the development of mice lacking specific muscarinic receptors has helped clarify the roles of each subtype [5]. M_1 receptors are found in the cerebral cortex and hippocampus and have been proposed to modulate cognitive function and memory processes. M_2 receptors are found in the heart, where they mediate the negative inotropic effects of acetylcholine. In the central nervous system, M_2 receptors serve as autoreceptors regulating acetylcholine release and mediate the analgesic effects of muscarinic agonists [6]. M_3 receptors stimulate contractions of smooth muscle and promote secretion from exocrine glands. M_4 receptors are found in the central nervous system and regulate the release of neurotransmitters, including dopamine [6]. M_5 receptors are present in very low abundance in the brain. Although the M_5 receptor protein has been difficult to identify in brain, M_5 receptor mRNA is found in the ventral tegmental area and substantia nigra [7, 8].

Muscarinic receptors belong to the large family of G protein-coupled receptors. M_1, M_3 and M_5 receptors couple to the stimulation of phospholipase C through the $G_{q/11}$ family of G proteins. Activation of M_1, M_3 and M_5 receptors results in the activation of protein kinase C through formation of 1,2-diacylglycerol and the release of Ca^{2+} from the endoplasmic reticulum through generation of inositol-1,4,5-trisphosphate. In contrast M_2 and M_4 receptors inhibit adenylyl cyclase activity through activation of the $G_{i/o}$ family of G proteins. Thus, M_1, M_3 and M_5 receptors are generally excitatory while M_2 and M_4 receptors are generally inhibitory, although the net effect of muscarinic receptor activation depends largely upon the integration of acetylcholine signaling within distinct brain pathways. For example, activation of M_4 receptors in the striatum results in increased dopamine release through the inhibition of striatal GABA release [9]. Because muscarinic receptors mediate their effects through activation of second messenger systems, muscarinic receptor stimulation results in a slower rate of onset and longer duration of action as compared with nicotinic receptor activation. As a result, muscarinic receptors are considered to play a modulatory rather than a direct stimulatory role in the central nervous system.

Neural networks and physiology

The central cholinergic system is a critical component of the cortical neural network. The basal forebrain cholinergic innervation of the cerebral cortex is diffuse and extends in a roughly topographical manner throughout the cerebral cortex [10]. Muscarinic receptors, particularly the M_1 receptor subtype, are found throughout the forebrain, including the neocortex [11]. Activation of cholinergic pathways results in the modulation of excitatory and inhibitory synapses, through regulation of on-going activity in glutamate and GABA systems. For example, activation of muscarinic receptors in the rat prefrontal cortex leads to enhancement of $GABA_A$ receptor currents through activation of a protein kinase C-dependent tyrosine kinase pathway [12]. The increased amplitude of $GABA_A$ receptor currents is likely mediated by either M_1 or M_4 receptors. The activation of muscarinic receptors has been proposed to shift the dynamics of cortical networks into a mode whereby afferent inputs exert a stronger influence than intracortical pathways on neural activity [13, 14].

Muscarinic receptor subtypes are found throughout the hippocampus, with M_1 receptors expressed predominantly in pyramidal neurons of the hippocampus and granule cells of the dentate gyrus [15]. The cholinergic projection from the MS helps drive the expression of the theta rhythm revealed through electrophysiological recordings of the rodent hippocampus. The theta rhythm is most pronounced during active exploratory movement and REM sleep. Lesions of the MS and muscarinic antagonists impair the expression of the theta rhythm. Induction of the theta rhythm with cholinergic agonists such as carbachol enhances synaptic plasticity [16], suggesting a potential role for muscarinic agonists in the storage of new information.

Memory function

The role of acetylcholine in memory function has been well documented. Classical muscarinic antagonists such as scopolamine and atropine impair memory function in experimental animals and in humans. Lesions of basal forebrain cholinergic pathways with specific neurotoxins such as AF64A and 192 IgG-saporin also impair memory function [17–22]. The role of individual muscarinic receptor subtypes in memory and cognitive function has been limited by the availability of agonists and antagonists selective for muscarinic receptor subtypes.

With the development of mice that lack specific muscarinic receptor subtypes, the role of individual muscarinic receptor subtypes in cognitive and memory function is currently being explored, and some recent studies provide some insight. Nathanson's group studied cognitive and memory function in mice with a null mutation of the gene coding the M_1 receptor [23]. Interestingly, the mice exhibited a phenotype characterized by both enhancements and deficits in memory function. Mice lacking the M_1 receptor exhibited impairments in non-matching-to-sample working memory tasks, yet displayed normal or enhanced performances in matching-to-sample tasks. For example, memory was impaired in mice lacking M_1 receptors on the eight-arm radial maze task as evidenced by a significant increase in errors following a 60-minute delay as compared to wild-type controls. In contrast, mice lacking M_1 receptors performed as well as wild-type mice in the Morris water maze, a paradigm that measures spatial memory function. In addition, hippocampal long-term potentiation (LTP) associated with theta burst stimulation was significantly lower in mice lacking M_1 receptors as compared with wild-type controls. Other forms of LTP were relatively unaffected in mice lacking M_1 receptors.

Neurological disorders

Muscarinic receptors have been implicated in the pathophysiology of Alzheimer's disease and other neurological disorders, including schizophrenia and drug abuse. The cholinergic hypothesis of geriatric memory dysfunction focused attention on the role of acetylcholine in memory and cognition [24]. Numerous studies indicated a consistent loss of basal forebrain cholinergic neurons in Alzheimer's patients that appeared to correlate with cognitive and memory deficits [24, 25]. Recent studies have raised questions regarding the role that cholinergic degeneration plays in the disease process, since the loss of cholinergic enzymes such as choline acetyltransferase does not precede changes in cognition and memory function in early stage Alzheimer's patients [26]. Despite such concerns, compounds that activate cholinergic systems represent a viable approach for the treatment of Alzheimer's disease [25]. As of this writing, acetylcholinesterase inhibitors are the only compounds approved for the treatment of the symptoms of Alzheimer's disease. The development of compounds that activate post-synaptic receptors directly could provide a distinct advantage over acetylcholinesterase inhibitors since activity is dependent on the integrity of cholinergic nerve terminals for the production of acetylcholine.

M_1 receptors have been a major target for drug development over the past decade based on their preferential localization in the forebrain, the potential utility of muscarinic agonists in the treatment of Alzheimer's disease and the adverse side-effects associated with activation of other receptor subtypes [24, 27–29]. Although acetylcholinesterase inhibitors have been utilized in treating the symptoms of Alzheimer's disease, efforts to develop selective muscarinic agonists have met with limited success.

Two main problems have been encountered with the muscarinic agonists tested in the clinic. First, many of the compounds lacked efficacy at M_1 receptors in the central nervous system [30]. A number of compounds were identified using muscle preparations from peripheral tissue such as rabbit vas deferens, where a population of M_1 receptors is located [31]. Activity in the rabbit vas deferens assay did not always translate into activity at M_1 receptors in brain [30]. As a result, some compounds did not exhibit efficacy in clinical studies. Second, many compounds exhibiting strong M_1 agonist activity lacked functional selectivity for M_1 receptors. These compounds produced significant side-effects in clinical studies, most notably nausea, salivation and diaphoresis due to activation of M_3 receptors. One compound, xanomeline, displayed relatively high efficacy and selectivity for M_1 receptors. Although xanomeline exhibited some utility in treating the symptoms of Alzheimer's disease in clinical trials, side-effects associated with metabolism of the parent compound limited its utility.

Despite the difficulties, selective muscarinic agonists could be useful in treating not only the symptoms, but also the underlying cause(s) of Alzheimer's disease. Muscarinic agonists promote the release of soluble forms of amyloid precursor protein through the activation of α-secretase [32–34]. Furthermore, muscarinic agonists activate the Akt/protein kinase B cell signaling pathway [35, 36], which leads to protection from apoptosis, limits glycogen synthase kinase-3β activity and prevents hyperphosphorylation of tau proteins [37]. Clinical studies with talsaclidine and cevimeline have demonstrated that muscarinic agonists can alter cerebrospinal fluid (CSF) levels of Aβ in Alzheimer's patients [38, 39]. Unfortunately, all of the muscarinic agonists tested thus far suffer from a lack of efficacy and/or selectivity, which greatly limits their therapeutic potential. The high degree of amino acid homology among the five receptor subtypes, particularly within the transmembrane domains, has likely contributed to difficulties in developing selective muscarinic agonists.

Beyond potential usefulness in the treatment of Alzheimer's disease, muscarinic agonists could be useful in treating other neurological disorders, including schizophrenia and chronic pain [40, 41]. The M_1/M_4 agonist xanomeline produced modest effects on cognitive function, and dramatically decreased the incidence of hallucinations in Alzheimer's patients. Unfortunately, xanomeline is rapidly metabolized *in vivo*, resulting in non-selective effects associated with a high level of side-effects mediated by M_3 receptors.

The potential utility of muscarinic agonists in treating cognitive deficits associated with schizophrenia has received increased interest. Cognitive deficits such as behavioral praxis, verbal memory and verbal fluency impairments are relatively common in schizophrenia patients. Muscarinic agonists might be particularly helpful in treating cognitive deficits associated with schizophrenia. As noted by Hyman and Fenton [42], the cognitive aspects of schizophrenia have been largely ignored by the pharmaceutical industry as evidenced by the low number of clinical trials addressing cognitive function in patients with schizophrenia. The development of muscarinic agonists as cognitive enhancers may provide important symptomatic relief for patients with schizophrenia.

Muscarinic agonists

Classical muscarinic agonists such as arecoline and pilocarpine improve memory and cognitive function in experimental animals. Unfortunately, most of the classical muscarinic agonists are non-selective and suffer from a short duration of action. Over the past twenty years, several selective muscarinic agonists have been identified and some have been tested in clinical trials for their ability to enhance memory and cognitive function in Alzheimer's patients.

Cevimeline, among the first selective M_1 agonists to be identified, is a quinuclidine derivative that enhances memory function in both rodents [43] and primates [44]. Cevimeline has been approved for the treatment of xerostomia associated with Sjögren's disease. Derivatives of cevimeline also display cog-

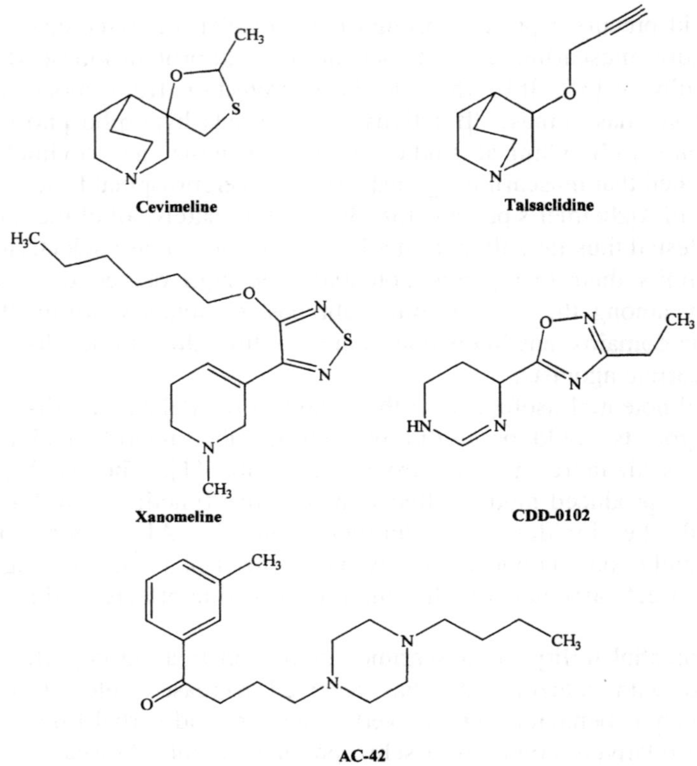

Figure 1.

nitive enhancing effects in rodents exhibiting cognitive impairments [45]. Another quinuclidine derivative, talsaclidine, also enhances memory function in the non-match-to-sample eight arm radial maze task in rodents [46] and in a delayed match-to-sample task in primates [46, 47].

Xanomeline shows functional selectivity for M_1 receptors [48], and enhances memory function on a paired-run alternation task in rats [22]. Xanomeline produced some cognitive benefits in Alzheimer's disease patients, particularly on the cognitive subscale of the Alzheimer's disease Assessment Scale (ADAS-Cog) [49, 50], although unwanted side-effects greatly limited its therapeutic utility. The novel amidine derivative, 5-(3-ethyl-1,2,4-oxadiazol-5-yl)-1,4,5,6-tetrahydropyrimidine (CDD-0102), also exhibits functional selectivity for M_1 receptors [21, 51] and enhances memory function on a paired-run alternation task in rats [48]. CDD-0102 displays a low side-effect profile in rodents, but its functional selectivity in humans remains untested.

The clinical utility of muscarinic agonists will depend on the development of compounds with improved selectivity for individual muscarinic receptor subtypes. Recent studies with bivalent derivatives of xanomeline have suggested a new approach for the design and development of truly selective mus-

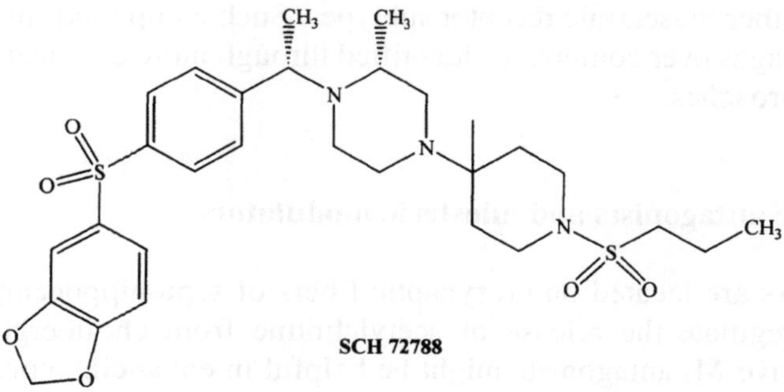

SCH 72788

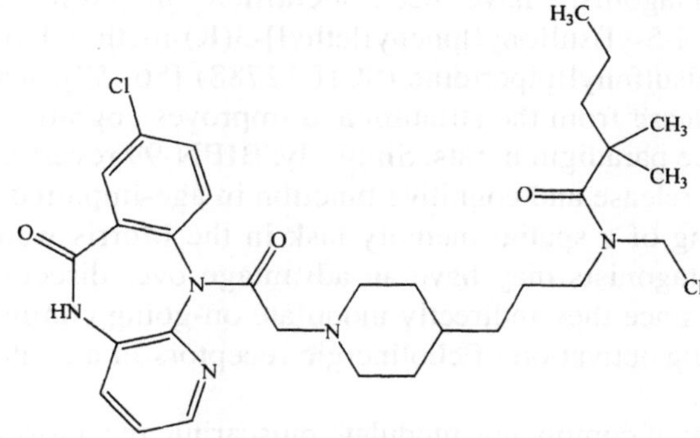

BIBN-99

Figure 2.

carinic agonists [52, 53] by taking advantage of the relatively low homology within the extracellular loops of muscarinic receptor subtypes. While small molecules interact mainly with highly conserved residues buried within the transmembrane domains, larger molecules can interact with potentially unique accessory binding sites on different receptor subtypes. Small structural modifications of the bivalent ligands yield compounds with agonist activity and selectivity for M_1 *versus* M_3 or M_5 receptors [52, 53]. These compounds serve as useful lead compounds for the next generation of selective muscarinic agonists.

Using a high-throughput functional screening approach to identify M_1 agonists, Spalding and colleagues [54] have characterized a unique compound (4-n-butyl-1-[4-(2-methylphenyl)-4-oxo-1-butyl]-piperidine, AC-42) that appears to bind to an ectopic site on M_1 receptors. The screening approach utilized assesses agonist activity by measuring cell growth, rather than screening for interaction with receptors by displacement of radiolabeled antagonists in traditional receptor binding assays. In this manner, compounds can be identified that do not compete for the acetylcholine binding site, and thus may exhibit receptor subtype selectivity. The compound identified through the novel screening approach activates M_1 receptors yet does not display significant activity at other muscarinic receptor subtypes. Such compounds may have distinct advantages over compounds identified through more classical drug development approaches.

Muscarinic antagonists and allosteric modulators

M_2 receptors are located on presynaptic fibers of septohippocampal neurons [15], and regulate the release of acetylcholine from cholinergic terminals [55]. Selective M_2 antagonists might be helpful in enhancing cognitive function by elevating acetylcholine release. Over the past several years, a few selective M_2 antagonists have been identified, including 4-[4-[1(S)-[4-[(1,3-benzodioxol-5-yl)sulfonyl]phenyl]ethyl]-3(R)-methyl-1-piperazinyl]-4-methyl-1-(propylsulfonyl)piperidine (SCH 72788) [56, 57], which enhances acetylcholine release from the striatum and improves cognitive function in a passive avoidance paradigm in rats. Similarly, BIBN-99 reverses impairments of acetylcholine release and cognitive function in age-impaired rats [58], and enhances learning of a spatial memory task in the Morris water maze [59]. Selective M_2 antagonists may have an advantage over directly acting muscarinic agonists since they indirectly modulate on-going cholinergic activity, thereby permitting activation of cholinergic receptors in a context-dependent manner [60, 61].

A wide variety of compounds modulate muscarinic receptors through interaction with allosteric binding sites, which are separate from the binding site for acetylcholine and other muscarinic agonists and antagonists. Binding to an allosteric site can enhance (positive co-operativity) or reduce (negative co-operativity) the affinity of other molecules for the receptor. For example, gallamine exhibits strong negative co-operativity with muscarinic agonists and antagonists at M_2 receptors [62]. Over the past few years, a few strychnine and brucine derivatives have been identified with positive co-operativity at muscarinic receptor subtypes [63]. Although additional work is needed to optimize affinity, selectivity and CNS penetration, such compounds could be useful in enhancing on-going cholinergic activity in a selective manner, and serve as cognitive enhancers [64].

Summary

In summary, muscarinic receptors play important roles in the expression of memory and cognitive function. Selective muscarinic agonists may be useful in enhancing cognitive function in normal individuals and in patients with neurological disorders such as Alzheimer's disease and schizophrenia. In addition, M_2-selective antagonists could be beneficial as cognitive enhancers through the stimulation of acetylcholine release, while allosteric modulators might enhance on-going activity at muscarinic receptor subtypes. The development of selective muscarinic agonists with improved selectivity for M_1 and M_4 receptors remains an important goal for drug discovery efforts.

References

1 Mesulam MM, Mufson EJ, Wainer BH, Levey AI (1983) Central cholinergic pathways in the rat: an overview based on an alternative nomenclature (Ch1-Ch6). *Neuroscience* 10(4): 1185–1201
2 Kobayashi Y, Isa T (2002) Sensory-motor gating and cognitive control by the brainstem cholinergic system. *Neural Netw* 15(4–6): 731–741
3 Auld DS, Kornecook TJ, Bastianetto S, Quirion R (2002) Alzheimer's disease and the basal forebrain cholinergic system: relations to beta-amyloid peptides, cognition, and treatment strategies. *Prog Neurobiol* 68(3): 209–245
4 Caulfield MP, Birdsall NJ (1998) International Union of Pharmacology. XVII. Classification of muscarinic acetylcholine receptors. *Pharmacol Rev* 50(2): 279–290
5 Bymaster FP, McKinzie DL, Felder CC, Wess J (2003) Use of M1–M5 muscarinic receptor knockout mice as novel tools to delineate the physiological roles of the muscarinic cholinergic system. *Neurochem Res* 28(3–4): 437–442
6 Gomeza J, Zhang L, Kostenis E, Felder CC, Bymaster FP, Brodkin J, Shannon H, Xia B, Duttaroy A, Deng CX et al. (2001) Generation and pharmacological analysis of M2 and M4 muscarinic receptor knockout mice. *Life Sci* 68(22–23): 2457–2466
7 Weiner DM, Levey AI, Brann MR (1990) Expression of muscarinic acetylcholine and dopamine receptor mRNAs in rat basal ganglia. *Proc Natl Acad Sci USA* 87(18): 7050–7054
8 Vilaro MT, Palacios JM, Mengod G (1990) Localization of m5 muscarinic receptor mRNA in rat brain examined by *in situ* hybridization histochemistry. *Neurosci Lett* 114(2): 154–159
9 Zhang W, Yamada M, Gomeza J, Basile AS, Wess J (2002) Multiple muscarinic acetylcholine receptor subtypes modulate striatal dopamine release, as studied with M1–M5 muscarinic receptor knock-out mice. *J Neurosci* 22(15): 6347–6352
10 Lucas-Meunier E, Fossier P, Baux G, Amar M (2003) Cholinergic modulation of the cortical neuronal network. *Pflugers Arch* 446(1): 17–29
11 Levey AI (1996) Muscarinic acetylcholine receptor expression in memory circuits: implications for treatment of Alzheimer disease. *Proc Natl Acad Sci USA* 93(24): 13541–13546
12 Ma XH, Zhong P, Gu Z, Feng J, Yan Z (2003) Muscarinic potentiation of GABA(A) receptor currents is gated by insulin signaling in the prefrontal cortex. *J Neurosci* 23(4): 1159–1168
13 Kimura F (2000) Cholinergic modulation of cortical function: A hypothetical role in shifting the dynamics in cortical network. *Neurosci Res* 38(1): 19–26
14 Hasselmo ME (1995) Neuromodulation and cortical function: modeling the physiological basis of behavior. *Behav Brain Res* 67(1): 1–27
15 Levey AI, Edmunds SM, Koliatsos V, Wiley RG, Heilman CJ (1995) Expression of m1–m4 muscarinic acetylcholine receptor proteins in rat hippocampus and regulation by cholinergic innervation. *J Neurosci* 15(5 Pt 2): 4077–4092
16 Huerta PT, Lisman JE (1993) Heightened synaptic plasticity of hippocampal CA1 neurons during a cholinergically induced rhythmic state. *Nature* 364(6439): 723–725
17 Walsh TJ, Tilson HA, DeHaven DL, Mailman RB, Fisher A, Hanin I (1984) AF64A, a cholinergic neurotoxin, selectively depletes acetylcholine in hippocampus and cortex, and produces long-term passive avoidance and radial-arm maze deficits in the rat. *Brain Res* 321(1): 91–102
18 Messer WS Jr, Stibbe JR, Bohnett M (1991) Involvement of the septohippocampal cholinergic system in representational memory. *Brain Res* 564(1): 66–72
19 Walsh TJ, Herzog CD, Gandhi C, Stackman RW, Wiley RG (1996) Injection of IgG 192-saporin into the medial septum produces cholinergic hypofunction and dose-dependent working memory deficits. *Brain Res* 726(1–2): 69–79
20 Walsh TJ, Kelly RM, Dougherty KD, Stackman RW, Wiley RG, Kutscher CL (1995) Behavioral and neurobiological alterations induced by the immunotoxin 192-IgG-saporin: cholinergic and non-cholinergic effects following *i.c.v.* injection. *Brain Res* 702(1–2): 233–245
21 Messer Jr WS, Abuh YF, Ryan K, Shepherd MA, Schroeder M, Abunada S, Sehgal R, El-Assadi AA (1997) Tetrahydropyrimidine derivatives display functional selectivity for M_1 muscarinic receptors in brain. *Drug Dev Res* 40: 171–184

22 Messer WS Jr, Bachmann KA, Dockery C, El-Assadi AA, Hassoun E, Haupt N, Tang B, Li X (2002) Development of CDD-0102 as a selective M_1 agonist for the treatment of Alzheimer's disease. *Drug Dev Res* 57(4): 200–213
23 Anagnostaras SG, Murphy GG, Hamilton SE, Mitchell SL, Rahnama NP, Nathanson NM, Silva AJ (2003) Selective cognitive dysfunction in acetylcholine M1 muscarinic receptor mutant mice. *Nat Neurosci* 6(1): 51–58
24 Bartus RT, Dean RL 3rd, Beer B, Lippa AS (1982) The cholinergic hypothesis of geriatric memory dysfunction. *Science* 217(4558): 408–414
25 Bartus RT (2000) On neurodegenerative diseases, models, and treatment strategies: Lessons learned and lessons forgotten a generation following the cholinergic hypothesis. *Exp Neurol* 163(2): 495–529
26 DeKosky ST, Ikonomovic MD, Styren SD, Beckett L, Wisniewski S, Bennett DA, Cochran EJ, Kordower JH, Mufson EJ (2002) Up-regulation of choline acetyltransferase activity in hippocampus and frontal cortex of elderly subjects with mild cognitive impairment. *Ann Neurol* 51(2): 145–155
27 Moos WH, Davis RE, Schwarz RD, Gamzu ER (1988) Cognition activators. *Med Res Rev* 8(3): 353–391
28 Hollander E, Mohs RC, Davis KL (1986) Cholinergic approaches to the treatment of Alzheimer's disease. *Br Med Bull* 42(1): 97–100
29 Coyle JT, Price DL, DeLong MR (1983) Alzheimer's disease: A disorder of cortical cholinergic innervation. *Science* 219(4589): 1184–1190
30 Bymaster FP, Carter PA, Peters SC, Zhang W, Ward JS, Mitch CH, Calligaro DO, Whitesitt CA, DeLapp N, Shannon HE et al. (1998) Xanomeline compared to other muscarinic agents on stimulation of phosphoinositide hydrolysis *in vivo* and other cholinomimetic effects. *Brain Res* 795(1–2): 179–190
31 Eltze M (1994) Pathways involved in muscarinic M1 and M2 receptor stimulation in rabbit vas deferens. *Eur J Pharmacol* 263(1–2): 31–37
32 Nitsch RM, Slack BE, Wurtman RJ, Growdon JH (1992) Release of Alzheimer amyloid precursor derivatives stimulated by activation of muscarinic acetylcholine receptors. *Science* 258(5080): 304–307
33 Wolf BA, Wertkin AM, Jolly YC, Yasuda RP, Wolfe BB, Konrad RJ, Manning D, Ravi S, Williamson JR, Lee VM (1995) Muscarinic regulation of Alzheimer's disease amyloid precursor protein secretion and amyloid beta-protein production in human neuronal NT2N cells. *J Biol Chem* 270(9): 4916–4922
34 Buxbaum JD, Ruefli AA, Parker CA, Cypess AM, Greengard P (1994) Calcium regulates processing of the Alzheimer amyloid protein precursor in a protein kinase C-independent manner. *Proc Natl Acad Sci USA* 91(10): 4489–4493
35 Murga C, Laguinge L, Wetzker R, Cuadrado A, Gutkind JS (1908) Activation of Akt/protein kinase B by G protein-coupled receptors. A role for alpha and beta gamma subunits of heterotrimeric G proteins acting through phosphatidylinositol-3-OH kinasegamma. *J Biol Chem* 273(30): 0–5, 1998
36 Murga C, Fukuhara S, Gutkind JS (2069) A novel role for phosphatidylinositol 3-kinase beta in signaling from G protein-coupled receptors to Akt. *J Biol Chem* 275(16): 1–73, 2000
37 Sadot E, Gurwitz D, Barg J, Behar L, Ginzburg I, Fisher A (1996) Activation of m1 muscarinic acetylcholine receptor regulates tau phosphorylation in transfected PC12 cells. *J Neurochem* 66(2): 877–880
38 Hock C, Maddalena A, Heuser I, Naber D, Oertel W, von der Kammer H, Wienrich M, Deng M, Growdon JH, Nitsch RM *(2000)* Treatment with the selective muscarinic agonist talsaclidine decreases cerebrospinal fluid levels of Ab in patients with Alzheimer's disease. In: *Soc. Neurosci. abstr., New Orleans, LA,*, pp 679.13. Soc. Neurosci
39 Nitsch RM, Deng M, Tennis M, Schoenfeld D, Growdon JH (2000) The selective muscarinic M1 agonist AF102B decreases levels of total A[beta] in cerebrospinal fluid of patients with Alzheimer's disease. *Ann Neurol* 48: 913–918
40 Felder CC, Porter AC, Skillman TL, Zhang L, Bymaster FP, Nathanson NM, Hamilton SE, Gomeza J, Wess J, McKinzie DL (2001) Elucidating the role of muscarinic receptors in psychosis. *Life Sci* 68(22–23): 2605–2613
41 Bernardini N, Roza C, Sauer SK, Gomeza J, Wess J, Reeh PW (2002) Muscarinic M2 receptors on peripheral nerve endings: A molecular target of antinociception. *J Neurosci* 22(12): RC229

42 Iadanza M, Holtje M, Ronsisvalle G, Holtje HD (2002) Kappa-opioid receptor model in a phospholipid bilayer: molecular dynamics simulation. *J Med Chem* 45(22): 4838–4846
43 Fisher A, Brandeis R, Karton I, Pittel Z, Gurwitz D, Haring R, Sapir M, Levy A, Heldman E (1991) (+–)-cis-2-methyl-spiro(1,3-oxathiolane-5,3')quinuclidine, an M1 selective cholinergic agonist, attenuates cognitive dysfunctions in an animal model of Alzheimer's disease. *J Pharmacol Exp Ther* 257(1): 392–403
44 O'Neill J, Fitten LJ, Siembieda D, Halgren E, Kim E, Fisher A, Perryman K (1998) Effects of AF102B and tacrine on delayed match-to-sample in monkeys. *Prog Neuropsychopharmacol Biol Psychiatry* 22(4): 665–678
45 Fisher A, Brandeis R, Bar-Ner RH, Kliger-Spatz M, Natan N, Sonego H, Marcovitch I, Pittel Z (2002) AF150(S) and AF267B: M1 muscarinic agonists as innovative therapies for Alzheimer's disease. *J Mol Neurosci* 19(1–2): 145–153
46 Bartolomeo AC, Morris H, Buccafusco JJ, Kille N, Rosenzweig-Lipson S, Husbands MG, Sabb AL, Abou-Gharbia M, Moyer JA et al. (2000) The preclinical pharmacological profile of WAY-132983, a potent M1 preferring agonist. *J Pharmacol Exp Ther* 292(2): 584–596
47 Terry AV Jr, Buccafusco JJ, Borsini F, Leusch A (2002) Memory-related task performance by aged Rhesus monkeys administered the muscarinic M(1)-preferring agonist, talsaclidine. *Psychopharmacology (Berl)* 162(3): 292–300
48 Shannon HE, Bymaster FP, Calligaro DO, Greenwood B, Mitch CH, Sawyer BD, Ward JS, Wong DT, Olesen PH, Sheardown MJ et al. (1994) Xanomeline: a novel muscarinic receptor agonist with functional selectivity for M1 receptors. *J Pharmacol Exp Ther* 269(1): 271–281
49 Bodick NC, Offen WW, Levey AI, Cutler NR, Gauthier SG, Satlin A, Shannon HE, Tollefson GD, Rasmussen K, Bymaster FP et al. (1997) Effects of xanomeline, a selective muscarinic receptor agonist, on cognitive function and behavioral symptoms in Alzheimer disease. *Arch Neurol* 54(4): 465–473
50 Veroff AE, Bodick NC, Offen WW, Sramek JJ and Cutler NR (1998) Efficacy of xanomeline in Alzheimer disease: cognitive improvement measured using the Computerized Neuropsychological Test Battery (CNTB). *Alzheimer Dis Assoc Disord* 12(4): 304–312
51 Messer WS Jr, Abuh YF, Liu Y, Periyasamy S, Ngur DO, Edgar MA, El-Assadi AA, Sbeih S, Dunbar PG, Roknich S et al. (1997) Synthesis and biological characterization of 1,4,5,6- tetrahydropyrimidine and 2-amino-3,4,5,6-tetrahydropyridine derivatives as selective m1 agonists. *J Med Chem* 40(8): 1230–1246
52 Christopoulos A, Grant MK, Ayoubzadeh N, Kim ON, Sauerberg P, Jeppesen L, El-Fakahany EE (2001) Synthesis and pharmacological evaluation of dimeric muscarinic acetylcholine receptor agonists. *J Pharmacol Exp Ther* 298(3): 1260–1268
53 Rajeswaran WG, Cao Y, Huang XP, Wroblewski ME, Colclough T, Lee S, Liu F, Nagy PI, Ellis J, Levine BA et al. (2001) Design, Synthesis, and Biological Characterization of Bivalent 1-Methyl-1,2,5,6-tetrahydropyridyl-1,2,5-thiadiazole Derivatives as Selective Muscarinic Agonists. *J Med Chem* 44(26): 4563–4576
54 Spalding TA, Trotter C, Skjaerbaek N, Messier TL, Currier EA, Burstein ES, Li D, Hacksell U, Brann MR (2002) Discovery of an ectopic activation site on the M(1) muscarinic receptor. *Mol Pharmacol* 61(6): 1297–1302
55 Slutsky I, Wess J, Gomeza J, Dudel J, Parnas I, Parnas H (1954) Use of knockout mice reveals involvement of M2-muscarinic receptors in control of the kinetics of acetylcholine release. *J Neurophysiol* 89(4): –67, 2003
56 Lachowicz JE, Duffy RA, Ruperto V, Kozlowski J, Zhou G, Clader J, Billard W, Binch H 3rd, Crosby G, Cohen-Williams M et al. (2001) Facilitation of acetylcholine release and improvement in cognition by a selective M2 muscarinic antagonist, SCH 72788. *Life Sci* 68(22–23): 2585–2592
57 Vannucchi MG, Scali C, Kopf SR, Pepeu G, Casamenti F (1997) Selective muscarinic antagonists differentially affect *in vivo* acetylcholine release and memory performances of young and aged rats. *Neuroscience* 79(3): 837–846
58 Quirion R, Wilson A, Rowe W, Aubert I, Richard J, Doods H, Parent A, White N, Meaney MJ (1995) Facilitation of acetylcholine release and cognitive performance by an M(2)-muscarinic receptor antagonist in aged memory-impaired. *J Neurosci* 15(2): 1455–1462
59 Tombaugh GC, Rowe WB, Chow AR, Michael TH, Rose GM (2002) Theta-frequency synaptic potentiation in CA1 *in vitro* distinguishes cognitively impaired from unimpaired aged Fischer 344 rats. *J Neurosci* 22(22): 9932–9940
60 Sarter MF, Bruno JP (1994) Cognitive functions of cortical ACh: Lessons from studies on trans-

synaptic modulation of activated efflux. *Trends Neurosci* 17(6): 217–221
61 Sarter M, Bruno JP (1997) Cognitive functions of cortical acetylcholine: Toward a unifying hypothesis. *Brain Res Brain Res Rev* 23(1–2): 28–46
62 Stockton JM, Birdsall NJ, Burgen AS, Hulme EC (1983) Modification of the binding properties of muscarinic receptors by gallamine. *Mol Pharmacol* 23(3): 551–557
63 Gharagozloo P, Lazareno S, Popham A, Birdsall NJ (1999) Allosteric interactions of quaternary strychnine and brucine derivatives with muscarinic acetylcholine receptors. *J Med Chem* 42(3): 438–445
64 Birdsall NJ, Farries T, Gharagozloo P, Kobayashi S, Lazareno S, Sugimoto M (1999) Subtype-selective positive cooperative interactions between brucine analogs and acetylcholine at muscarinic receptors: functional studies. *Mol Pharmacol* 55(4): 778–786

Cognitive Enhancing Drugs
Edited by J.J. Buccafusco

Rationale and prospects for drugs that target nicotinic acetylcholine receptors

Andrea Wevers[1] and Diana S. Woodruff-Pak[2]

[1] *Department II of Anatomy – Neuroanatomy, University of Cologne, 50931 Cologne, Germany*
[2] *Department of Psychology, Temple University, Philadelphia, PA 19122 and Department of Neurology, Albert Einstein Healthcare Network, Philadelphia, PA 19141 USA*

Introduction

Nicotinic acetylcholine receptors (nAChRs) are implicated in a variety of disorders of the human central nervous system including addiction to nicotine, Alzheimer's disease, anxiety, autism, depression, epilepsy, Parkinson's disease, schizophrenia, and Tourette's syndrome [1, 2]. Mechanisms of nAChR impairment in this disparate group of syndromes are poorly understood. Additionally, in healthy organisms nAChRs play a significant role in a number of cognitive processes including learning and memory [3, 4]. Because nAChRs are involved in normal cognitive processes as well as a complex range of central nervous system disorders, it is important to define the means by which these receptors exert their action in the brain and interact with disease-related neuropathology. It is also imperative to explore the prospects of therapeutic manipulations of nAChRs in human central nervous systems disorders.

The characteristic structure of nAChRs is a ring of five subunits arranged around a ligand-gated excitatory ion channel. Neuronal nAChR subunits are classified as α (α2 to α10) and β (β2 to β4) subunits. The two main neuronal categories that have been identified on the basis of function and pharmacology are: (1) heterologous pentamers, constructed from combinations of α- and β-subunits [5], and (2) homologous pentamers, constructed from one subunit type, α7, α8, and α9 [6]. Besides the α4β2 nAChRs that constitute the high affinity binding sites for nicotine [7], the α7 nAChRs being highly permeable for Ca^{2+} are expressed widely and abundantly in the mammalian brain [8, 9]. The α7 nAChRs produce multiple effects at the cellular level including presynaptic modulation of neurotransmitter release [10] and post-synaptic generation of depolarizing currents [11].

The various types of nAChRs have characteristic patterns of distribution in the brain, and they have several loci on neurons, including on terminals, soma, and dendrites [12]. Neuromodulation of communicative processes in the brain is one role for nAChRs [13]. The demonstrated involvement of nAChRs in cognitive processes, the dramatic cognitive impairment apparent in Alzheimer's disease (AD), and the mounting number of AD patients in the world are among the compelling rationales for investigating nAChRs in AD. The role of nAChRs in AD and the potential for drug therapies in this neurodegenerative disease is the focus of this chapter. Studies of nAChR subtypes in the brains of patients with AD indicate that nAChRs that participate in high-affinity nicotine binding (e.g., α4β2) are substantially reduced but those binding to α-bungarotoxin (e.g., α7) are better maintained (for a review, see [14]). However, neuropathology may interact with the nAChR subtypes that survive in the AD brain and render them functionally ineffective.

Relationship of β-amyloid (Aβ) and nicotinic acetylcholine receptors (nAChRs)

AD is neurochemically characterized by cholinergic dysfunction, in particular affecting the nAChRs, and neuropathologically by intracellular neurofibrillary tangles and the extracellular β-amyloid (Aβ) plaques (for a review, see [14]). However, finding the possible link between these markers has remained elusive. Although increasing numbers of reports suggest that Aβ can potently inhibit various cholinergic transmitter functions independently of apparent Aβ neurotoxicity (for review cf. [15]), little attention has been paid to the cholinoceptive site. Nowadays there are numerous indications that chronic exposure to nicotine may protect against Aβ pathology (cf. [16]). In addition, it was shown recently that nicotine not only dose-dependently inhibited fibril formation from $A\beta_{1-40}$ and $A\beta_{1-42}$ but also disrupted preformed fibrillary $A\beta_{1-42}$ *in vitro* [17]. Although epidemiological studies of tobacco use and risk of developing AD are somewhat contradictory [18, 19], a significant lower plaque density has been observed in autopsy cortical tissue of smokers as compared to non-smokers [20, 21].

In vitro studies have also indicated protective effects of nicotine against Aβ toxicity ([22, 23], for review cf. [24, 25]). Initial evidence for an interaction of Aβ with nAChRs was provided by the finding that stimulation of the α7 and α4β2 nAChR subtype in cultured rat cortical neurons was able to inhibit Aβ cytotoxicity [22, 23]. Furthermore, Alzheimer patients carrying the Swedish amyloid precursor protein 670/671 mutation and early and excessive accumulation of Aβ plaques were shown to display nAChR deficits [26]. In the meantime, numerous studies have investigated the relationship of Aβ and nAChR expression and the mechanisms that may be involved in this interaction.

So far only few reports on human brain have addressed potential links between AD neuropathology and cholinergic dysfunction. Our own previous studies on autoptic human cortex combining *in situ* hybridization and immunohistochemistry revealed that in the frontal cortex neurons expressed α4 as well as α7 mRNA even in the vicinity of Aβ-immunoreactive plaques [27]. The density of α4 and α7 transcript expressing neurons was not altered in AD. This finding is confirmed by other studies [28, 29] and may point to a lack of the impact of Aβ plaques on nAChR expression. By contrast to the findings on the mRNA level there is a decrease of nAChR subunit proteins in AD. Several groups by using immunohistochemistry, Western blot, and binding studies reported on a massive reduction of the α4 subunit, whereas less consensus exists regarding the decrease of the α7 subunit protein (for review see [27, 30]).

To investigate the possible impact of Aβ on nAChR protein expression we approached co-localization studies by two different ways. First, we combined silver staining techniques to visualize plaques with immunohistochemistry to show nAChR expressing cells. Second, we applied a double-immunohistochemical protocol. With both methods we were able to demonstrate the presence of nAChR-immunoreactive neurons in close proximity to and sometimes even within plaques. Therefore it seems to be rather unlikely that plaques interfere with nAChR expression. This result is confirmed by an investigation demonstrating no correlation of the plaque load in AD patients and nicotine binding, whereas an inverse correlation between [^{3}H]epibatidine binding and $A\beta_{1-42}$ levels was found [21]. In addition, studies in transgenic mice [31–34] support the emerging view for a plaque-independent Aβ toxicity in the development of synaptic deficits in AD. The positive correlation between the decrease of synaptophysin immunoreactivity and [^{3}H]epibatidine binding [35] complements the picture.

The increasing evidence that soluble Aβ plays a more important role in the development of AD and the general acceptance that $A\beta_{1-42}$ is the pathogenic Aβ peptide [36] led to the development of various *in vitro* model systems to study the impact of Aβ on nAChR expression under standardized conditions. Using primary hippocampal cultures grown under defined serum-free conditions it was shown that incubation for three days with 1 μM $A\beta_{1-42}$ caused severe morphological alterations: Neuronal cell bodies appeared to be shrunken and a remarkable retraction of dendrites as well as a loss of dendritic density was observed. The reduction of labelled processes was most prominent in the α4-immunoreactive neurons, whereas α7-expressing neurons seemed to be less affected (cf. [30]). Comparable morphological changes occurred after incubation with 0.5 μM $A\beta_{1-42}$. This different behaviour may indicate a subunit-dependent impact of $A\beta_{1-42}$ on nAChR expression.

Interaction of Aβ with nAChRs

The underlying mechanism of nicotinic transmission and Aβ interaction remained unclear until recently when investigations of Wang and co-workers shed light on this issue [37, 38]. They found that $A\beta_{1-42}$ binds specifically and with picomolar affinity to the neuronal α7 nAChRs. So far numerous studies provide evidence for a functional block of acetylcholine (ACh)-evoked current responses by $A\beta_{1-42}$ via the α7 nAChR-subtype in various types of cells and in α7-transfected oocytes [39–41]. At complete variance with these data, 10 pM of non-fibrillar $A\beta_{1-42}$ have been reported to activate rat α7 nAChRs expressed in *Xenopus* oocytes, although only upon the very first exposure, whereas a blockade of the α7 nAChR only occurred after application of 100 nM $A\beta_{1-42}$ [42]. An activation of α7 nAChRs by $A\beta_{1-42}$ was described for *Xenopus* oocytes transfected with mutant rat and human α7 cDNA carrying a point mutation within the pore-forming region (rat: L250T [42], human: L248T [41]). The latter finding is in good agreement with data obtained from studies with mutant α7 nAChR receptors carrying the particular threonine-for-leucine substitution showing that several antagonists of chick and human wild-type α7 nAChRs behave like agonists (for review see [41]).

Consensus exists on the ability of Aβ1–42 to block α-bungarotoxin (α-Bgt) sensitive nAChR subtypes like the α7 nAChR [37, 39–43]. Its interaction with non-α-Bgt sensitive nAChR subtypes, however, is not clear yet. In the initial study it was demonstrated on hippocampal membranes from sporadic AD brains by immunoprecipitation and Western blot analysis that the nAChR α7-subunit co-immunoprecipitated with Aβ1–42, whereas the nAChR α1-, α3-, α4-, α5-, α8-, or β2-subunit failed to yield any detectable co-immunoprecipitate with anti-Aβ1–42 [37]. In functional studies some groups did not find any blockade of non-α-Bgt sensitive nAChR subtypes by Aβ1–42 [40, 41], whereas other researchers found a binding affinity for the α4β2 nAChR subtype in the range of 20–30 nM [38] or a partial blockade of non-α7 nAChRs at higher concentrations [39]. In one paper even a higher depression of α4β2 than of α7 nAChR currents upon Aβ1–42 incubation was reported [43]. In PC12 cells nanomolar concentrations of Aβ1–40 and Aβ25–35 significantly decreased $[^3H]$epibatidine and $[^{125}I]$α-Bgt binding sites as well as the mRNA and protein amounts of nAChR α3-, α7-, and β2-subunits [44]. A possible explanation of the inconsistencies in these findings may be the different Aβ1–42 concentrations used or the different cell types studied [40].

Another controversial issue is the nature of the interaction between $A\beta_{1-42}$ and α7 nAChRs, as $A\beta_{1-42}$ has been reported to compete with α-Bgt for binding to α7 nAChRs [37], whereas no competitive displacement of α-Bgt by $A\beta_{1-42}$ was observed neither in intact mouse hippocampal neurons nor in neurons derived from chick ciliary ganglia [40]. Also Guan et al. [44] failed to displace $[^3H]$epibatidine and $[^{125}I]$α-Bgt by $A\beta_{1-42}$, $A\beta_{1-40}$, and $A\beta_{25-35}$ in PC12 cells. Further functional characterization of the interaction of $A\beta_{1-42}$ with α7 nAChRs revealed that the blockade of α7 nAChR function does not require the presence of an agonist. On the contrary, co-application of $A\beta_{1-42}$ and ACh resulted in a failure to induce blockade [40].

The formation of an α7 nAChR-$A\beta_{1-42}$ complex can be efficiently suppressed by $A\beta_{12-28}$, implying that the binding epitope for α7 nAChRs resides in the amino acid 12–28 sequence region of $A\beta_{1-42}$ [37]. Further studies using chimeric α7/$5HT_3$ receptors suggested that $A\beta_{1-42}$ interacts with the extracellular N-terminal domain of the α7 nAChR [40]. However, little information exists on the physical state of active $A\beta_{1-42}$ that binds to nAChRs. In brain slices incubation with freshly prepared $A\beta_{1-42}$ led to a rapid onset of inhibition, which speaks in favour of a soluble, oligomeric rather than a fibrillar form of $A\beta_{1-42}$ [39].

nAChR mediated signal transduction in respect to Aβ

The question remains whether the observed $A\beta_{1-42}$-induced nAChR functional changes affect synaptic transmission. It has been shown that $A\beta_{1-42}$ affects I_{ACh} with an IC_{50} around 100 nM (~ 450 ng ml^{-1}) [39, 41, 42], although an IC_{50} of about 7.5 nM has been described for rat hippocampal neurons [40]. The concentration of Aβ peptides in brain tissue from AD patients (2–20 μM) is higher than that required for maximal α7 nAChR blockade [45], but most Aβ peptides are concentrated in the Aβ deposits, and their exchange with interstitial fluid and their proximity to receptors are difficult to estimate. It is known, however, that mice genetically engineered to express Aβ peptides can display synaptic toxicity that correlates with $A\beta_{1-42}$ level in the 10–100 nM range [31, 32], concentrations that are effective at blocking α7 nAChRs.

In a recent review it was summarized that Aβ peptide can be directly neurotoxic, induce oxidative stress, incite an inflammatory response, and alter calcium homeostasis. These events might be mediated by direct interaction of Aβ aggregates with cellular membranes, or by binding of Aβ to microglial and neuronal cellular receptors [46]. High levels of $A\beta_{1-42}$ in AD may promote interactions with cholinergic neurons that express α7 nAChRs. The high affinity binding of $A\beta_{1-42}$ to α7 nAChRs may result in the inhibition of ACh release and may alter Ca^{2+}-homeostasis for the affected cholinergic neuron. These significant and chronic physiological perturbations may lead to stress and even neurodegeneration.

An important role for α7 nAChRs in facilitating the entry and intraneuronal accumulation of $A\beta_{1-42}$ via endocytosis has been described [47]. The rate and extent of $A\beta_{1-42}$ internalization was directly related to the amount of α7 nAChR protein and effectively blocked by α-Bgt and by the endocytosis inhibitor phenylarsine oxid. Furthermore, the α7 nAChR was co-localized with $A\beta_{1-42}$ in prominent intracellular aggregates [47]. Based on these data the authors [47] suggest that internalization of $A\beta_{1-42}$ occurs predominantly in neurons expressing the α7 nAChR and may be facilitated by the high-affinity binding of $A\beta_{1-42}$ to α7 nAChRs on neuronal cell surfaces followed by endocytosis of the resulting complex. According to the authors their data provide a plausible explanation for the selective vulnerability of neurons expressing the nAChR α7 subtype in AD brain. The data are also interpreted to suggest that $A\beta_{1-42}$ is the dominant Aβ peptide in intracellular accumulation and amyloid plaques [47]. However, the nAChR subtype affected in AD most severely is the α4β2-subtype providing the high-affinity nicotine binding sites.

Further evidence for the view that high affinity α7 nAChR-$A\beta_{1-42}$ interaction may be a critical step leading to AD pathology arises from the findings of the nAChR-mediated protection against glutamate- and Aβ-induced neurotoxicity (for review see [24]). Stimulation of nAChRs with nicotine or epibatidine resulted in a prevention of Aβ-induced cytotoxicity [22, 23, 37]. Evidence for a nAChR-mediated protection against the Aβ-enhanced glutamate neurotoxicity was provided by the finding that α-Bgt suppressed this protection which therefore seems to be mediated via the α7 nAChR subtype. It is supposed that nicotine activates α7 nAChRs to stimulate the Src gene family, which in turn activates phosphatidylinositol 3-kinase (PI-3-K). PI-3-K phosphorylates Akt, which causes upregulation of Bcl-2 and Bcl-x preventing cells from neuronal death induced by Aβ and glutamate [24, 48]. Further insights into the underlying mechanisms of the nAChR-mediated neuroprotection were given recently by the findings of Shaw et al. [49] in PC12 cells. The authors provide evidence that nicotine stimulation of the α7 nAChRs transduce signals to PI-3-K and Akt via Janus kinase 2 (JAK2) in a cascade that results in neuroprotection. Exposure to Aβ leads to an activation of the apoptotic enzyme caspase-3 and cleavage of the DNA-repairing enzyme poly-(ADP-ribose) polymerase. This apoptotic cascade is inhibited by nicotine through JAK2 activation. Pre-treatment of cells with angiotensin II was able to block the nicotine-induced activation of JAK2 via the AT_2 receptor and completely prevented α7 nAChR-mediated neuroprotective effects [49]. These findings identify novel mechanisms of receptor interactions relevant to neuronal viability and suggest novel therapeutic strategies to optimise neuroprotection [49].

A possible link between $A\beta_{1-42}$ binding to α7 nAChRs and cognitive impairment in AD was suggested by a paper showing that $A\beta_{1-42}$ is able to promote MAP kinase activation by inducing Ca^{2+} influx through mutant L250T α7 nAChRs, thereby interfering with long-term potentiation processes [50]. Contrary to the wild-type receptor that is blocked by $A\beta_{1-42}$, the mutant receptor is activated. Therefore, the significance of the interaction between $A\beta_{1-42}$ and α7 nAChRs for the etiology or the pathogenesis of AD is unclear [41] and needs further elucidation.

Current therapy strategies targeting Aβ in the brain

Besides AChE-inhibitor approaches the most intensive therapeutic efforts thus far have been directed toward a decrease of Aβ in the brain. Procedures inhibiting Aβ formation or modulating Aβ production, assembly and/or removal might be useful as treatments for preventing AD or having a meaningful impact on disease progression. Immunization with Aβ of aged transgenic PDAPP-over-expressing mice led to a reduction in Aβ deposits, whereas immunization of young mice prevented formation of Aβ deposits [51]. In addition, in TgCRND8 mice (K670N/M671/L and V717F human βAPP_{695}) a reduction of the cerebral fibrillar Aβ and of cognitive dysfunction was documented, without, however, altering the total level of Aβ in the brain after Aβ immunization [52]. Passive immunization of PDAPP transgenic mice with m266, an antibody directed to the central part of the Aβ, can rapidly reverse memory deficits but without altering brain Aβ burden [53]. These promising results have led to the development of an $A\beta_{1-42}$-compound (AN-1792, Elan Pharmaceuticals), as a potential vaccine to treat AD. AN-1792 was undergoing Phase IIa trials in 360 AD patients, as it had to be stopped in January 2002 due to meningio-encephalitic presentation in about 5% of the study group participants [46, 54]. Very recently after the death of the first patient participating in this study an analysis of human neuropathology became available [55]. Some of the findings strongly resemble the changes seen after Aβ-immunotherapy in mouse models of AD. First, there were extensive areas with a low density of Aβ plaques without plaque-associated dystrophic neurites and GFAP-immunoreactive astrocytes. Second, remaining Aβ-immunoreactivity was associated with microglia in areas devoid of plaques. Third, there was persistence of cerebrovascular amyloid. Additional features that were not predicted by the mouse models of Aβ-immunotherapy comprise (1) a CD4+ lymphocytic meningoencephalitis, (2) persistence of neurofibrillary tangles and neuropil threads in areas devoid of plaques, and (3) extensive macrophage infiltration of cerebral white matter [55].

A possible alternative for plaque clearance by immunization with Aβ may be the stimulation of nAChRs. Chronic treatment using high doses of nicotine in the diet of APPsw transgenic mice carrying the Swedish mutation of the human amyloid precursor protein [Tg(Hu.App695.K670N-M671L)2576] led to a selective reduction in $A\beta_{1-42}$ positive plaques by more than 80%, but there were no changes in soluble $A\beta_{1-40}$ or $A\beta_{1-42}$ levels [56]. However, there was also significant weight loss in these animals. Their findings led the authors to conclude that chronic nicotine administration can be added to an increasing number of treatments that are able to reduce Aβ deposition in animal models and that nicotinic drugs directed towards select nAChR subtypes represent a feasible neuroprotective therapeutic approach in AD [56]. Whether the chronic nicotine treatment of APPsw mice has any effect on cognitive improvement, however, needs to be further elucidated.

Designing compounds that distinguish individual receptor subtypes is a highly desirable therapeutic strategy for redressing some of the degenerative effects associated with AD. To the extent that α7 nAChRs represent an early molecular casualty of the disease, they should be considered a high-priority target for drug design [40].

An early attempt to target α7 nAChRs used a compound designed from a naturally-occurring substance: A synthesized analog of the marine natural product anabasine [57] called GTS-21 (3-(2,4-dimethoxybenzylidene) anabaseine) that preferentially interacts with α7 nAChRs. When GTS-21 was introduced to cultured rat cortical neurons, it protected neurons against Aβ-induced death [22]. These results suggest that α7 nAChR activation can play an important role in neuroprotection against Aβ neurotoxicity. Activation of α7 nAChRs may be able to protect neurons from degeneration induced by Aβ and may have effects that counter the progression of AD. In a subsequent study,

Kihara et al. [23] reported that nicotine neuroprotection could be blocked by an α4β2 nAChR antagonist, suggesting a neuroprotective effect for α4β2 nAChRs as well as α7 nAChRs.

Current therapy strategies with nAChRs targeting learning and memory

A role for nAChRs has been demonstrated in a number of forms of cognition including attention [58, 59], sensorimotor gating [60–62], and learning and memory [63–66]. These studies have been carried out in several mammalian species, including humans. Because of the central role that memory loss plays in AD, the focus here is on nAChR involvement in learning and memory. Among the novel approaches to cognition enhancement is the application of agonists to nAChRs, in particular the α7 nAChR, and allosteric modulation of nAChRs.

Role for nAChRs in working memory

Repeatedly it has been shown that working memory is facilitated by both acute injections of nicotine [67, 68] and chronic infusion of nicotine [69–72]. Nicotine's facilitatory effect on working memory is blocked by the nAChR anatagonist, mecamylamine [67, 68, 72, 73]. Mecamylamine is not selective to one specific nAChR subtype. It appears to inhibit preferentially first α3, then α4, and finally, α7 subunits and may likewise block β2 and β4 nAChR subunits [74].

Initial behavioral pharmacology research on the nicotinic cholinergic system used this broad-spectrum approach with agonists (nicotine) and antagonists (mecamylamine). More recent research has used a receptor-targeted approach. In the case of working memory as assessed by the 8- and 16-arm radial maze in rats, the approach using focal brain infusions and antagonizing the broad-spectrum agonist, nicotine with receptor-targeted antagonists has demonstrated that both α7 and α4β2 nAChRs in the ventral hippocampus are critical [71]. The work of Levin and associates demonstrates the utility of infusing drugs focally in the brain to identify sites of drug action. Experiments infusing antagonists specific to α7 nAChRs (methyllycaconitine) or to α4β2 nAChRs (dihydro-beta-erythroidine) into the ventral hippocampus demonstrated that both types of nAChRs are critical for working memory function measured by the radial arm maze in rats [69–71].

Role of nAChRs in associative learning and memory

Brain structures and systems of demonstrated involvement in eyeblink classical conditioning in rabbits and humans are compromised during the progression of AD. Patients diagnosed with probable AD are severely impaired in eyeblink conditioning beyond the impairment observed in normal aging [75–77]. For a decade, it has been our working hypothesis that disruption of the septohippocampal cholinergic system and selective loss of hippocampal pyramidal cells impair acquisition of eyeblink classical conditioning in AD beyond the impairment observed in normal aging. Additional evidence that septohippocampal lesions disrupt conditioning in humans comes from patients with a common cerebral aneurysm: aneurysm of the anterior communicating artery. The anterior communicating artery vascularizes the basal forebrain, and survivors of an aneurysm of this artery often display some degree of anterograde amnesia. Anterior communicating artery aneurysm survivors often sustain basal forebrain lesions that include lesions of the medial septum. Six patients with such lesions were tested in the delay eyeblink conditioning procedure and showed significant impairment [78].

In addition to septohippocampal disruption, AD-related injury to the cerebellum may play a role in impairing delay eyeblink classical conditioning in AD. Purkinje cell loss in the AD cerebellum is significantly greater than in age-matched control brains [79, 80], and there is exceptional vulnerability of the ACh system and ACh-receptive cells in AD that may be associated with nAChRs. With the mounting evidence for a role for ACh in the cerebellum as well as the hippocampus, the possibility exists that impaired eyeblink conditioning in AD occurs at least in part because of impairment in ACh modulation and nAChR function in the essential cerebellar circuitry.

Experiments with muscarinic ACh receptor agonists [81] and antagonists [82] and nicotinic ACh receptor agonists [83] and antagonists [84] have demonstrated a role for ACh in the model system of eyeblink classical conditioning in the rabbit. Eyeblink conditioning reveals natural age-related deficits in several non-human mammals, and the similarities between age differences in eyeblink conditioning in these animal species and humans are striking. Moreover, delay eyeblink conditioning is impaired profoundly in patients with AD, making the procedure relevant for preclinical studies of cognition-enhancing drugs. In addition to parallels with human behavior and neurobiology, the model system of eyeblink classical conditioning possesses a considerable advantage over the behavioral models commonly used preclinically: The essential neural circuitry in the cerebellum has been identified along with modulatory circuits in hippocampus and cortex.

The standard format for the presentation of stimuli in eyeblink classical conditioning is called the delay procedure. The subject is presented with a neutral stimulus such as a tone or light, called the conditioned stimulus (CS), for a short duration usually less than one second. Before the CS expires, the unconditioned stimulus (US) is presented concurrently, and the briefly coinciding CS and US co-terminate 50 to 100 ms later. The US, either a shock to the infra-orbital region of the eye or a corneal airpuff, always elicits from the organism an eyeblink or nictitating membrane (NM) unconditioned response (UR). With the repeated pairing of the CS and the US, the subject learns to blink to the tone before the onset of the US. This learned response is called a conditioned response (CR).

The site essential for acquisition and retention of the classically conditioned eyeblink response in rabbits is the cerebellar interpositus nucleus ipsilateral to the eye receiving the US. In humans, this nucleus becomes two deep cerebellar nuclei (emboliform and globose). Cerebellar cortex ipsilateral to the US also contributes to the process of acquisition, such that an intact cerebellar cortex enables acquisition to occur at a faster rate. The hippocampus itself is normally involved during acquisition in the delay procedure, however, in a complex modulatory role. The role of the hippocampus during acquisition in delay eyeblink conditioning seems paradoxical in that conditioning proceeds normally in animals with bilateral removal of the hippocampus, but manipulation of hippocampal function (in an intact hippocampus) with drugs can facilitate or impair acquisition considerably. For example, the muscarinic cholinergic antagonist scopolamine impairs acquisition in the delay procedure only when the hippocampus is intact [85]. Likewise, the cognition-enhancing drug, nefiracetam ameliorates learning impairment in older rabbits in the delay procedure only when the hippocampus is intact [86]. This modulatory role for the hippocampus may be particularly significant in AD, since, in humans, AD appears to alter hippocampal neuronal function and cause a major disruption of the brain cholinergic system. Eyeblink conditioning impairment in AD may reflect medial-temporal lobe atrophy and central nervous system cholinergic dysfunction that occurs early in disease progression.

Identification of nAChRs as the cholinergic receptors impaired in AD led to a test of an antagonist to nAChRs in the animal model of eyeblink classical conditioning. Mecamylamine is a central nervous system nicotinic antagonist that binds to a site on the receptor other than the ACh recognition site. Papke et al. [74] demonstrated with electrophysiological recordings of nAChRs expressed in *Xenopus* oocytes that the residual inhibition produced by 10 μM mecamylamine was greatest for human β2-containing receptors and least for α7 nAChRs.

Using a 0.5 mg/kg dose of mecamylamine, a role for nAChRs in eyeblink conditioning in young rabbits was demonstrated [84]. The acquisition of conditioned eyeblink responses was severely disrupted so that young rabbits learned at a rate comparable to older rabbits. The deleterious effect of mecamylamine on eyeblink conditioning was not accompanied by a measurable change in brain nAChR concentration. These results in combination with studies using the muscarinic acetylcholine receptor (mAChR) antagonist, scopolamine, suggest that nAChRs as well as mAChRs are involved in the modulation of eyeblink classical conditioning.

Partial agonism of α7 nAChRs

Results with the nAChR antagonist, mecamylamine led to the prediction that nAChR agonists would facilitate eyeblink conditioning in older rabbits. The effect of GTS-21, a selective nicotinic agonist acting primarily at the α7 nAChR subtype [87, 88], on learning was tested using the 750-ms delay eyeblink conditioning procedure [89]. There were 15 daily subcutaneous injections administered 15 minutes before behavioral training. At dosage levels of 0.5 and 1.0 mg/kg, GTS-21 acted as a cognition-enhancing agent in older rabbits, resulting in eyeblink conditioning performance comparable to young rabbits. The cognition-enhancing effect of GTS-21 upon eyeblink conditioning was not accompanied by a measurable change in brain nAChRs.

The effect of GTS-21 on acquisition, retention, and relearning was also examined with a focus on the duration of the effect of GTS-21 as assessed by retention and relearning [90]. First there were 15 sessions of acquisition training with injections of 0.5 mg/kg GTS-21 or vehicle. Then drug administration ended and older rabbits were tested for retention and relearning six weeks and 13 weeks after the beginning of the experiment. Acquisition of CRs was significantly better in GTS-21-treated rabbits. During the first tone-alone retention session in week six of the experiment, rabbits initially treated with GTS-21 produced significantly more CRs than vehicle-treated rabbits. There were no group differences in retention at the 13-week retest. Differences in relearning were numerically greater for GTS-21 treated older rabbits, but these effects did not attain statistical significance. Results indicated that treatment with GTS-21 ameliorated learning and memory beyond the period when the drug was actually administered.

Given the results outlined above confirming that nAChRs are involved in the modulation of acquisition and retention in eyeblink classical conditioning, reversal of the nAChR antagonist mecamylamine with nicotine or GTS-21 was explored [91]. Young rabbits were injected with 0.5 mg/kg mecamylamine in combination with nicotine or GTS-21 and compared to vehicle-treated rabbits. Control groups were tested in the explicitly unpaired condition. Both GTS-21 and nicotine reversed the deleterious effect of mecamylamine on acquisition of CRs enabling the rabbits treated with mecemylamine and the agonists to perform at the level of vehicle-treated control rabbits (Fig. 1). Combinations of GTS-21 or nicotine and mecamylamine did not cause sensitization or habituation in the explicitly unpaired condition. That fact that mecamylamine inhibits preferentially first α3, then α4, and finally α7 subunits and may likewise block β2 and β4 nAChR subunits [74] suggests that the reversal by GTS-21 and nicotine of mecamylamine's antagonistic effect was via α7 nAChRs. There is some evidence that GTS-21 actually antagonizes α4β2 nAChRs in addition to serving as an agonist for α7 nAChRs [87]. If α7 nAChRs are of primary importance in eyeblink conditioning, then selective α7 antagonists such as methyllycaconitine should impair acquisition to a greater extent than selective α4β2 antagonists such as dihydro-beta-erythroidine. Demonstration that eyeblink conditioning relies more on α7 nAChRs than on α4β2 nAChRs would make this behavior a sensitive index of the drugs that are α7 agonists under development to treat AD.

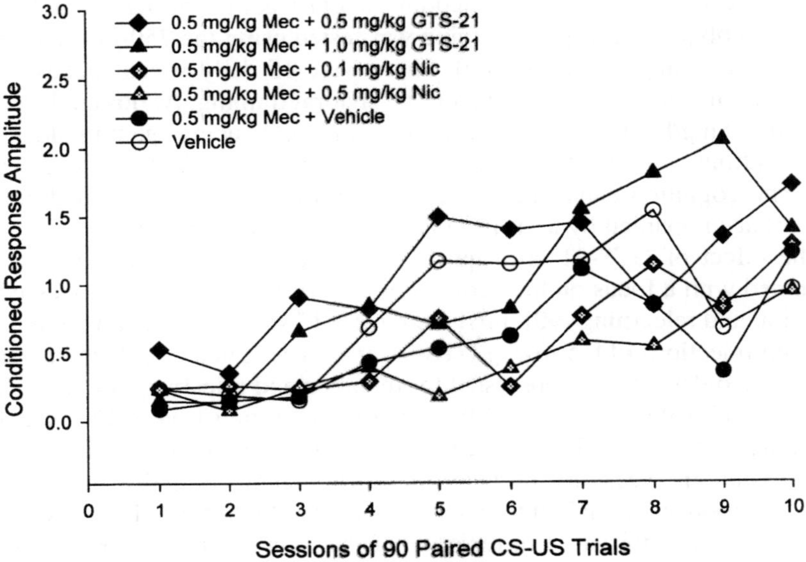

Figure 1. Mean amplitude of conditioned responses (CRs) in 58 three-month-old rabbits for the various drug treatment groups for the first 10 daily training sessions of 90 trials/session. Each group consisted of a minimum of eight rabbits. Rabbits received daily subcutaneous injections of sterile saline vehicle, 0.5 mg/kg mecamylamine (Mec) alone, or 0.5 mg/kg mecamylamine in combination with 0.5 or 1.0 mg/kg GTS-21 or 0.25 or 0.5 mg/kg nicotine (Nic) 15 min before training. The group treated with 0.5 mg/kg Mec had significantly fewer CRs than the vehicle-treated group ($p = 0.005$). Early in training (Sessions 2–3) rabbits treated with mecamylamine and GTS-21 actually had numerically higher CR amplitudes than all other groups, but by Session 4 vehicle-treated animals began to show superior performance to the mecamylamine-alone group. Rabbits treated with agonist (GTS-21 or nicotine) antagonist (mecamylamine) combinations performed similarly to vehicle-treated rabbits and better than rabbits treated with the antagonist alone [91].

Allosteric modulation of nAChRs

Allosteric modulators, such as galantamine (Reminyl®; Inteligen, Janssen) and physostigmine employ a dual action at the cholinergic synapse. First, they act as acetylcholinesterase (AChE) inhibitors. Second, they also act as agonists at presynaptic nAChRs, enhancing the release of ACh. Functionally unique features of allosterically potentiating ligands include the ability as assessed with patch-clamp recordings to induce single-channel activity indistinguishable from the single-channel activity induced by ACh. With allosteric potentiation, galantamine and physostigmine induced single-channel activity in excised patches from various cells [92–94] that could not be blocked by established nAChR antagonists like mecamylamine.

Given that an α7 nAChR partial agonist, GTS-21 ameliorated learning deficits in older rabbits, the aim was to determine if nicotinic agonism using a different mechanism of action would be effective in the eyeblink classical conditioning paradigm. With physostigmine, a wide range of doses (0.0005 to 2.0 mg/kg) were effective in ameliorating eyeblink conditioning deficits in older rabbits [95]. Control tests of rabbits in explicitly unpaired conditions demonstrated that non-associative factors could not account for the results. This allosteric modulator had dramatic effects on associative learning in older rabbits (see Fig. 2).

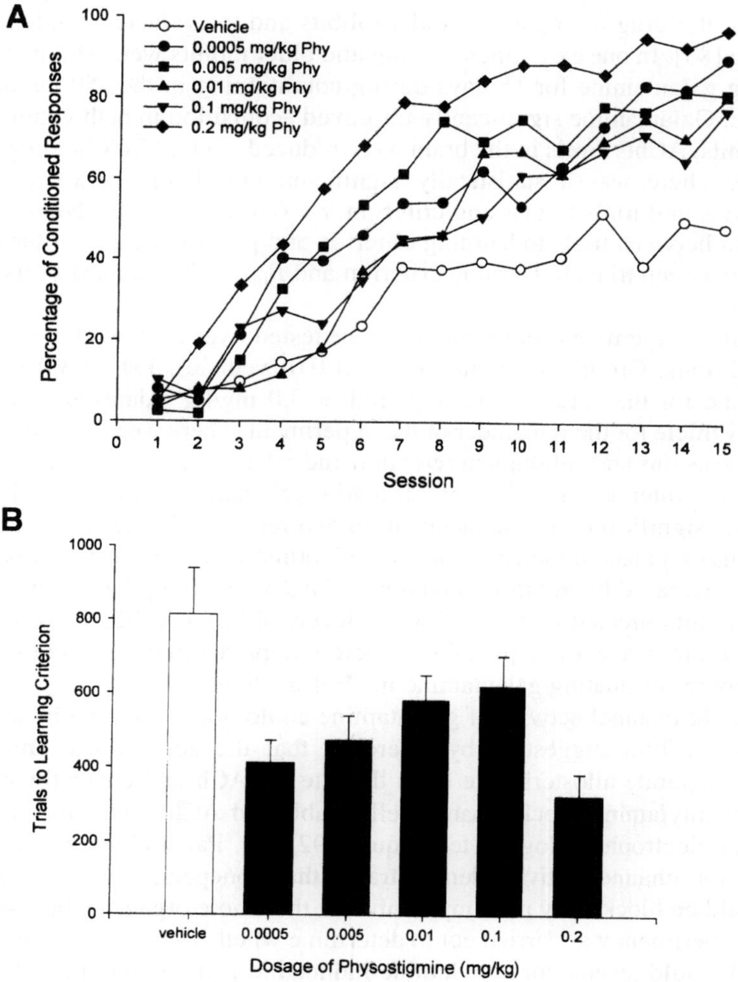

Figure 2. (A) Percentage of conditioned responses (CRs) of a total of 59 older rabbits (mean age = 28 mo., s.d. = 5.0 mo) treated with five doses of physostigmine (Phy) or vehicle over 15 daily training sessions in the 750 ms delay eyeblink classical conditioning procedure. There were a minimum of seven older rabbits in each of the Phy-treated groups and 12 rabbits in the vehicle-treated group. (B) The number of training trials to a learning criterion of eight CRs in nine consecutive trials in the 59 rabbits shown in the left panel. Physostigmine at a range of doses significantly improved learning in older rabbits [95].

Galantamine doses of 0.0, 1.0, 2.0, 3.0, and 4.0 mg/kg were tested in 10 daily sessions in old rabbits in the 750 ms delay eyeblink classical conditioning procedure [96]. A dose of 3.0 mg/kg galantamine was effective in improving conditioning in older rabbits, enabling them to achieve learning criterion rapidly and to produce a very high level of learning performance. Like physostigmine, galantamine had no effects on sensitization, habituation, or motor aspects of the task.

Additional experiments with galantamine were carried out to compare the efficacy of the drug in young and older rabbits and to evaluate retention and relearning [83]. In one experiment, young and older rabbits were administered 3.0 mg/kg galantamine for 15 days during conditioning in the 750 ms delay procedure. Galantamine significantly improved acquisition in both young and older rabbits. AChE levels in the brain were reduced, and nAChR binding was increased. There was a statistically significant correlation between brain AChE levels and trials to learning criterion, $r = 0.621$, $p = 0.007$. Neither the correlation between trials to learning criterion and plasma AChE, nor the correlations between trials to learning criterion and B_{max} or K_D attained statistical significance.

In another experiment, older rabbits were tested over a 15-week period in four conditions. Groups of rabbits received 0.0 (vehicle), 1.0, or 3.0 mg/kg galantamine for the entire 15-week period or 3.0 mg/kg galantamine for 15 days and vehicle for the remainder of the experiment. There were 15 daily conditioning sessions and subsequent retention and relearning assessments spaced at one-month intervals. The dose of 3.0 mg/kg galantamine ameliorated learning deficits significantly during acquisition and retention in the group receiving 3.0 mg/kg galantamine continuously. Nicotinic receptor binding was significantly increased in rabbits treated for 15 days with 3.0 mg/kg galantamine. All galantamine-treated rabbits had lower levels of brain AChE. The efficacy of galantamine in a learning paradigm severely impaired in AD was consistent with outcomes evaluating galantamine in clinical studies.

The single-channel activity of galantamine could not be blocked by antagonists to nAChRs suggesting by inference that the activity was induced through a separate allosteric site from the site for ACh and competitive ligands. Mecamylamine blocks many well-established AChE inhibitors tested with these electrophysiological techniques [92, 94]. Patch-clamp recordings of single ion channel activity demonstrated that donepezil, but not galantamine, could be blocked by mecamylamine. At the whole organism, behavioral level an experiment was carried out to determine whether galantamine, but not donepezil, could reverse mecamylamine-induced learning impairment. Young rabbits received delay eyeblink conditioning after one of six drug treatments: 0.5 mg/kg mecamylamine, 3.0 mg/kg galantamine, 3.0 mg/kg donepezil, 0.5 mg/kg mecamylamine plus 3.0 mg/kg galantamine, 0.5 mg/kg mecamylamine plus 3.0 mg/kg donepezil, or sterile saline vehicle [97]. Galantamine, but not donepezil, facilitated learning in young rabbits. However, both galantamine and donepezil reversed the deleterious effects of mecamylamine on learning. Significant differences in plasma (but not brain) AChE levels were detected among the drug treatment groups. Fifteen daily injections of mecamylamine, galantamine, or donepezil, alone or in combination did not produce statistically significant changes in nAChR binding. One possible interpretation of these results is that donepezil affected nAChRs by raising the synaptic level of ACh and hence, the probability of receptor activation, whereas galantamine bound to distinct allosteric sites not blocked by mecamylamine. It may be possible to facilitate learning in young rabbits with allosteric modulation (galantamine), but not with AChE inhibition alone (donepezil).

Summary and conclusions

The widespread involvement of nAChRs in disorders of the human nervous system makes these receptors likely targets for drug therapy. In this chapter we focused on the involvement of nAChRs in AD and the relationships between Aβ and nAChRs. Our demonstration in brain autopsy tissue from humans with AD of the presence of nAChR-immunoreactive neurons in close proximity to and sometimes even within plaques makes it appear rather unlikely that plaques interfere with nAChR expression. This perspective receives confirmation from a number of additional studies investigating relationships between plaque load and nicotine binding in humans and transgenic mice.

Whereas Aβ plaques may not impair nAChRs, there is increasing evidence that soluble Aβ interacts with nAChRs to cause morphological alterations and impairment of function. Consensus exists on the ability of $A\beta_{1-42}$ to block α-Bgt sensitive nAChR subtypes like α7 nAChR. The formation of an α7 nAChR-$A\beta_{1-42}$ complex can be efficiently suppressed by $A\beta_{12-28}$, implying that the binding epitope for α7 nAChR resides in the amino acid 12–28 sequence region of $A\beta_{1-42}$. The interaction of soluble Aβ with non-α-Bgt sensitive nAChR subtypes, however, is not yet clear.

It is likely that the observed $A\beta_{1-42}$-induced functional changes in nAChRs affect synaptic transmission. For example, the high affinity binding of $A\beta_{1-42}$ to α7 nAChRs may result in the inhibition of ACh release and may alter Ca^{2+}-homeostasis for the affected cholinergic neuron. Chronic physiological perturbations such as inhibition of ACh release and altered Ca^{2+}-homeostatic stress neuronal function and even lead to neurodegeneration. High affinity α7 nAChR-$A\beta_{1-42}$ interaction may be a critical step leading to AD pathology. In this regard, using a broad-spectrum agonist (nicotine) and an α7 receptor-targeted agonist (GTS-21) is neuroprotective against Aβ-enhanced glutamate neurotoxicity.

A possible alternative for plaque clearance by immunization with Aβ may be the stimulation of nAChRs. In particular, chronic stimulation of α7 nAChRs has significant therapeutic potential. In addition to its potential to affect AD neuropathology, chronic stimulation of α7 nAChRs has potential to reduce cognitive impairment in AD. Learning and memory are among the cognitive processes affected early in AD. Animal studies demonstrated that both α7 and α4β2 nAChRs in the ventral hippocampus are critical for normal working memory. Functional blockade of α7 nAChRs and loss of α4β2 nAChRs likely contribute significantly to the documented impairment of attention and working memory in AD patients. Associative learning as assessed by eyeblink classical conditioning is severely impaired in AD, and preclinical experiments with a partial α7 agonist and with allosteric modulators of nAChRs demonstrate dramatic therapeutic effects. Targeting α7 nAChRs to treat AD has a range of potential benefits, as does the targeting of selected individual receptor subtypes affected in diseases such as addiction to nicotine, anxiety, autism, depression, epilepsy, Parkinson's disease, schizophrenia, and Tourette's syndrome.

Acknowledgements
Research reported here was supported by local funding sources of the University of Cologne (Frau Maria Pesch-Stiftung, Imhoff-Stiftung) to AW and by grants from NIH (AGO9752), Alzheimer's Association (IIRG-99-1690), Daiichi Pharmaceutical Co., Ltd., Tokyo, Japan, Janssen Pharmaceutica, Titusville, NJ, and Taiho Pharmaceutical Co., Ltd., Tokyo, Japan to DSW-P.

References

1 Newhouse PA, Kelton M (2000) Clinical aspects of nicotinic agents: Therapeutic application in central nervous system disorders. In: F Clementi, D Fornasari, C Gotti (eds): *Neuronal nicotinic receptors: Experimental pharmacology*. Springer, Berlin, 779–812
2 Weiland S, Bertrand D, Leonard S (2000) Neuronal nicotinic acetylcholine receptors: From gene to disease. *Behav Brain Res* 113: 43–56
3 Gould TJ, Wehner JM (1999) Nicotine enhancement of contextual fear conditioning. *Behav Brain Res* 102: 31–39
4 Levin ED (1992) Nicotinic systems and cognitive function. *Psychopharmacol* 108: 417–431
5 Conroy WG, Vernallis AB, Berg DK (1992) The α5 gene product assembles with multiple acetylcholine receptor subunits to form distinctive receptor subtypes in brain. *Neuron* 9: 679–691
6 Lindstrom J, Schoepfer R, Conroy W, Whiting P, Das M, Saedi M, Anand R (1991) The nicotinic acetylcholine receptor gene family: Structure of nicotinic receptors from muscle and neurons and neuronal α-bungarotoxin-binding proteins. *Adv Exp Med Biol* 287: 255–278
7 Flores CM, Rogers SW, Pabreza LA, Wolfe BB, Kellar KJ (1992) A subtype of nicotinic cholinergic receptor in rat brain is composed of α4 and β2 subunits and is up-regulated by chronic nicotine treatment. *Mol Pharmacol* 41: 31–37
8 Couturier S, Bertrand D, Matter JM, Hernandez MC, Bertrand S, Millar N, Valera S, Barkas T, Ballivet M (1990) A neuronal nicotinic acetylcholine receptor subunit (α7) is developmentally regulated and forms a homo-oligomeric channel blocked by α-BTX. *Neuron* 5: 847–856
9 Schoepfer R, Conroy W, Whiting P, Gore M, Lindstrom J (1990) Brain α-bungarotoxin binding protein cDNAs and mAbs reveal subtypes of this branch of the ligand-gated ion channel gene superfamily. *Neuron* 5: 35–48
10 McGehee DS, Heath MJ, Gelber S, Devay P, Role LW (1995) Nicotine enhancement of fast excitatory synaptic transmission in CNS by presynaptic receptors. *Science* 269: 1692–1696
11 Frazier CJ, Buhler AV, Weiner JL, Dunwiddie TV (1998) Synaptic potentials mediated via α-bungarotoxin-sensitive nicotinic acetylcholine receptors in rat hippocampal interneurons. *J Neurosci* 18: 8228–8235
12 Court JA, Perry EK (1995) Distribution of nicotinic receptors in the CNS. In: TW Stone (ed.): *CNS neurotransmitters and neuromodulators*. CRC Press, London, 85–104
13 Lindstrom J (1997) Nicotinic acetylcholine receptors in health and disease. *Mol Neurobiol* 15: 193–222

14 Wevers A, Schröder H (1999) Nicotinic acetylcholine receptors in Alzheimer's disease. *J Alzheimer's Disease* 1: 207–219

15 Auld DS, Kar S, Quirion R (1998) β-Amyloid peptides as direct cholinergic neuromodulators: a missing link? *TINS* 21: 43–49

16 Zeng H, Zhang Y, Peng L, Shao H, Menon NK, Yang J, Salomon AR, Freidland RP, Zagorski MG (2001) Nicotine and amyloid formation. *Biol Psychiatry* 49: 248–257

17 Ono K, Hasegawa K, Yamada M, Naiki H (2002) Nicotine breaks down preformed Alzheimer's β-amyloid fibrils *in vitro*. *Biol Psychiatry* 52: 880–886

18 Lee PN (1994) Smoking and Alzheimer's disease: a review of epidemiological evidence. *Neuroepidemiology* 13: 131–144

19 Merchant C, Tang MX, Albert S, Manly J, Stern Y, Mayeux R (1999) The influence of smoking on the risk of Alzheimer's disease. *Neurology* 52: 1408–1412

20 Ulrich J, Johannson-Locher G, Seiler WO, Stähelin HB (1997) Does smoking protect from Alzheimer's disease? Alzheimer-type changes in 301 unselected brains from patients with known smoking history. *Acta Neuropathol (Berlin)* 94: 450–454

21 Perry E, Martin-Ruiz C, Lee M, Griffiths M, Johnson M, Piggott M, Haroutunian V, Buxbaum JD, Nasland J, Davis K et al. (2000) Nicotinic receptor subtypes in human brain ageing, Alzheimer and Lewy body disease. *Eur J Pharmacol* 393: 215–222

22 Kihara T, Shimohama S, Sawada H, Kimura J, Kume T, Kochiyama H, Maeda T, Akaike A (1997) Nicotinic receptor stimulation protects neurons against beta-amyloid toxicity. *Ann Neurol* 42: 159–163

23 Kihara T, Shimohama S, Urushitani M, Sawada H, Kimura J, Kume T, Maeda T, Akaike A (1998) Stimulation of alpha4beta2 nicotinic acetylcholine receptors inhibits beta-amyloid toxicity. *Brain Res* 792: 331–334

24 Shimohama S, Kihara T (2001) Nicotinic receptor-mediated protection against β-amyloid neurotoxicity. *Biol Psychiatry* 49: 233–239

25 Zamani MR, Allen S (2001) Nicotine and its interaction with β-amyloid protein: A short review. *Biol Psychiatry* 49: 221–232

26 Marutle A, Warpman U, Bogdanovic N, Lannfelt L, Nordberg A (1999) Neuronal nicotinic receptor deficits in Alzheimer patients with the Swedish amyloid precursor protein 670/671 mutation. *J Neurochem* 72: 1161–1169

27 Wevers A, Monteggia L, Nowacki S, Bloch W, Schütz U, Lindstrom J, Pereira EFR, Eisenberg H, Giacobini E, de Vos RAI et al. (1999) Expression of nicotinic acetylcholine receptor subunits in the cerebral cortex in Alzheimer's disease – Histotopographical correlation with amyloid plaques and hyperphosphorylated tau-protein. *Eur J Neurosci* 11: 2551–2565

28 Terzano S, Court JA, Fornasari D, Griffiths M, Spurden DP, Lloyd S, Perry RH, Perry EK, Clementi F (1998) Expression of the alpha3 nicotinic receptor subunit mRNA in aging and Alzheimer's disease. *Mol Brain Res* 63: 72–78

29 Hellström-Lindahl E, Mousavi M, Zhang X, Ravid R, Nordberg A (1999) Regional distribution of nicotinic receptor subunit mRNAs in human brain: Comparison between Alzheimer and normal brain. *Mol Brain Res* 66: 94–103

30 Wevers A, Witter B, Moser N, Burghaus L, Banerjee C, Steinlein OK, Schütz U, de Vos RAI, Jansen Steur ENH, Lindstrom J et al. (2000) Classical Alzheimer features and cholinergic dysfunction: towards a unifying hypothesis? *Acta Neurol Scand Suppl* 176: 42–48

31 Hsia AY, Masliah E, McConlogue L, Yu GQ, Tatsuno G, Hu K, Kholodenko D, Malenka RC, Nicoll RA, Mucke L (1999) Plaque-independent disruption of neural circuits in Alzheimer's disease mouse models. *Proc Natl Acad Sci USA* 96: 3228–3233

32 Mucke L, Masliah E, Yu GQ, Mallory M, Rockenstein EM, Tatsuno G, Hu K, Kholodenko D, Johnson-Wood K, McConlogue L (2000) High-level neuronal expression of Aβ 1-42 in wild-type human amyloid protein precursor transgenic mice: Synaptotoxicity without plaque formation. *J Neurosci* 20: 4050–4058

33 Bednar I, Paterson D, Marutle A, Pham TM, Svedberg M, Hellström-Lindahl E, Mousavi M, Court J, Morris C, Perry E et al. (2002) Selective nicotinic receptor consequences in APP(SWE) transgenic mice. *Mol Cell Neurosci* 20: 354–365

34 Apelt J, Kumar A, Schliebs R (2002) Impairment of cholinergic neurotransmission in adult and aged transgenic Tg2576 mouse brain expressing the Swedish mutation of human β-amyloid precursor protein. *Brain Res* 953: 17–30

35 Reid RT, Sabbagh MN, Corey-Bloom J, Tiraboschi P, Thal LJ (2000) Nicotinic receptor losses in

dementia with Lewy bodies: Comparisons with Alzheimer's disease. *Neurobiol Aging 21*: 741–746
36 Younkin SG (1998) The role of Aβ42 in Alzheimer's disease. *J Physiol Paris* 92: 289–292
37 Wang HY, Lee DHS, D'Andrea MR, Peterson PA, Shank RP, Reitz AB (2000) β-amyloid$_{1-42}$ binds to α7 nicotinic acetylcholine receptor with high affinity. Implications for Alzheimer's disease pathology. *J Biol Chem* 275: 5626–5632
38 Wang HY, Lee DH, Davis CB, Shark RR (2000) Amyloid peptide Abeta (1-42) binds selectively and with picomolar affinity to alpha 7 nicotinic acetylcholine receptor. *J Neurochem* 75: 1155–1161
39 Pettit DL, Shao Z, Yakel JL (2001) beta-Amyloid (1-42) peptide directly modulates nicotinic receptors in the rat hippocampal slice. *J Neurosci* 21: RC120 (1–5)
40 Liu Q, Kawai H, Berg DK (2001) beta -Amyloid peptide blocks the response of alpha 7-containing nicotinic receptors on hippocampal neurons. *Proc Natl Acad Sci USA* 98: 4734–4739
41 Grassi F, Palma E, Tonini R, Amici M, Ballivet M, Eusebi F (2003) Amyloid β$_{1-42}$ peptide alters the gating of human and mouse α-bungarotoxin-sensitive nicotinic receptors. *J Physiol* 547: 147–157
42 Dineley KT, Bell KA, Bui D, Sweatt JD (2002) β-Amyloid peptide activates α7 nicotinic acetylcholine receptors expressed in *Xenopus* oocytes. *J Biol Chem* 277: 25056–25061
43 Tozaki H, Matsumoto A, Kanno T, Nagai K, Nagata T, Yamamoto S, Nishizaki T (2002) The inhibitory and facilitatory actions of amyloid-beta peptides on nicotinic ACh receptors and AMPA receptors. *Biochem Biophys Res Commun* 294: 42–45
44 Guan Z-Z, Miao H, Tian J-Y, Unger C, Nordberg A, Zhang X (2001) Suppressed expression of nicotinic acetylcholine receptors by nanomolar β-amyloid peptides in PC12 cells. *J Neural Transm* 108: 1417–1433
45 Gravina S, Ho Libin, Eckman CB, Long KE, Otvos L, Younkin LH, Suzuki N, Younkin SG (1995) Amyloid β protein (Ab) in Alzheimer's disease brain. Biochemical and immunocytochemical analysis with antibodies specific for forms ending at Aβ40 or Aβ42(43). *J Biol Chem* 270: 7013–7016
46 Golde TE (2003) Alzheimer disease therapy: Can amyloid cascade be halted? *J Clin Invest* 111: 11–18
47 Nagele RG, D'Andrea MR, Anderson WJ, Wang HY (2002) Intracellular accumulation of β-amyloid$_{1-42}$ in neurons is facilitated by the α7 nicotinic acetylcholine receptor in Alzheimer's disease. *Neuroscience* 110: 199–211
48 Kihara T, Shimohama S, Sawada H, Honda K, Nakamizo T, Shibasaki H, Kume T, Akaike A (2001) α7 nicotinic receptor trancduces signals to phosphatidylinositol 3-kinase to block A beta-amyloid-induced neurotoxicity. *J Biol Chem* 276: 13541–13546
49 Shaw S, Bencherif M, Marrero B (2002) Janus kinase 2, an early target of α7 nicotinic acetylcholine receptor-mediated neuroprotection against Aβ-(1-42) amyloid. *J Biol Chem* 277: 44920–44924
50 Dineley KT, Westerman M, Bui D, Bell K, Ashe KH, Sweatt JD (2001) *Beta*-amyloid activates the mitogen-activated protein kinase cascade via hippocampal alpha7 nicotinic acetylcholine receptors: *In vitro* and *in vivo* mechanisms related to Alzheimer's disease. *J Neurosci* 21: 4125–4133
51 Schenk D, Barbour R, Dunn W, Gordon G, Grajeda H, Guido T, Hu K, Huang J, Johnson-Wood K, Khan K et al. (1999) Immunization with amyloid-β attenuates Alzheimer-disease-like pathology in the PDAPP mouse. *Nature* 400: 173–177
52 Janus C, Pearson J, McLaurin J, Mathews PM, Jiang Y, Schmidt SD, Chishti MA, Horne P, Heslin D, French J et al. (2000) Aβ peptide immunization reduces behavioural impairment and plaques in a model of Alzheimer's disease. *Nature* 408: 979–982
53 Dodart J-C, Bales KR, Gannon KS, Green SJ, DeMattos RB, Mathis C, DeLong CA, Wu S, Wu X, Holtzman DM et al. (2002) Immunization reverses memory deficits without reducing brain Aβ burden in Alzheimer's disease model. *Nature Neurosci* 5: 452–457
54 Birmingham K, Frantz S (2002) Set back to Alzheimer vaccine studies. *Nature Med* 8: 199–200
55 Nicoll JAR, Wilkinson D, Holmes C, Stert P, Markham H, Weller RO (2003) Neuropathology of human Alzheimer disease after immunization with amyloid-β peptide: A case report. *Nature Med* 9: 448–452
56 Nordberg A, Hellström-Lindahl E, Lee M, Johnson M, Mousavi M, Hall R, Perry E, Bednar I, Court J (2002) Chronic nicotine treatment reduces β-amyloidosis in the brain of a mouse model of Alzheimer's disease (APPsw). *J Neurochem* 81: 655–658

57 Kem WR, Abbott BC, Coates RM (1971) Isolation and structure of a hoplonemertine toxin. *Toxicol* 9: 15–22
58 Grottick AJ, Higgins GA (2000) Effect of subtype selective nicotinic compounds on attention as assessed by the five-choice serial reaction time task. *Behav Brain Res* 117: 197–208
59 Jones G, Sahakian B, Levy P, Warburton D, Gray J (1992) Effects of acute subcutaneous nicotine on attention, information processing, and short-term memory. *Psychopharmacol (Berl)* 108: 485–494
60 Acri JB, Morse DE, Popke EJ, Grunberg NE (1994) Nicotine increases sensory gating measured as inhibition of the acoustic startle reflex in rats. *Psychopharmacol* 114: 369–374
61 Adler LE, Olincy A, Waldo M, Harris JG, Griffith J, Stevens K, Flach K, Nagamoto H, Bickford P, Leonard S et al. (1998) Schizophrenia, sensory gating, and nicotinic receptors. *Schizophrenia Bull 24*: 189–202
62 Schreiber R, Dalmus M, De Vry J (2002) Effects of alpha4/beta2- and alpha 7-nicotine acetylcholine receptor agonists on prepulse inhibition of the acoustic startle response in rats and mice. *Psychopharmacology (Berl)* 159: 248–257
63 Blondel A, Sanger DJ, Moser PC (2000) Characterisation of the effects of nicotine in the five-choice serial reaction time task in rats: Antagonist studies. *Psychopharmacol* 149: 293–305
64 Rezvani AH, Bushnell PJ, Levin ED (2002) Effects of nicotine and mecamylamine on choice accuracy in an operant visual signal detection task in female rats. *Psychopharmacol (Berl)* 164: 369–375
65 Schildein S, Huston JP, Schwarting RK (2002) Open field habituation learning is improved by nicotine and attenuated by mecamylamine administered post-trial into the nucleus accumbens. *Neurobiol Learn Mem* 77: 277–290
66 Ueno K, Togashi H, Matsumoto M, Ohashi S, Saito H, Yoshioka M (2002) Alpha4beta2 nicotinic acetylcholine receptor activation ameliorates impairment of spontaneous alternation behavior in stroke-prone spontaneously hypertensive rats, an animal model of attention deficit hyperactivity disorder. *J Pharmacol Exp Ther* 302: 95–100
67 Brown RW, Beale KS, Jay Frye GD (2002) Mecamylamine blocks enhancement of reference memory but not working memory produced by post-training injection of nicotine in rats tested on the radial arm maze. *Behav Brain Res* 134: 259–265
68 Brown RW, Gonzalez CL, Whishaw IQ, Kolb B (2001) Nicotine improvement of Morris water task performance after fimbria-fornix lesion is blocked by mecamylamine. *Behav Brain Res* 119: 185–202
69 Bancroft A, Levin ED (2000) Ventral hippocampal alpha4beta2 nicotinic receptors and chronic nicotine effects on memory. *Neuropharmacol* 39: 2770–2778
70 Bettany JH, Levin ED (2001) Ventral hippocampal alpha 7 nicotinic receptor blockade and chronic nicotine effects on memory performance in the radial-arm maze. *Pharmacol Biochem Behav* 70: 467–474
71 Levin ED, Bradley A, Addy N, Sigurani N (2002) Hippocampal alpha 7 and alpha4 beta 2 nicotinic receptors and working memory. *Neurosci* 109: 757–765
72 Levin ED, Rose JE (1990) Anticholinergic sensitivity following chronic nicotine administration as measured by radial-arm maze performance in rats. *Behav Pharmacol* 1: 511–520
73 Levin ED, Briggs SJ, Christopher NC, Rose JE (1993) Chronic nicotinic stimulation and blockade effects on working memory. *Behav Pharmacol* 4: 179–182
74 Papke RL, Sanberg, PR, Shytle, RD (2001) Analysis of mecamylamine stereoisomers on human nicotinic receptor subtypes. *J Pharmacol Exp Ther* 297: 646–656
75 Solomon PR, Levine E, Bein T, Pendlebury WW (1991) Disruption of classical conditioning in patients with Alzheimer's disease. *Neurobiol Aging* 12: 283–287
76 Woodruff-Pak DS, Finkbiner RG, Sasse DK (1990) Eyeblink conditioning discriminates Alzheimer's patients from non-demented aged. *Neuroreport* 1: 45–49
77 Woodruff-Pak DS, Papka M, Romano S, Li Y-T (1996) Eyeblink classical conditioning in Alzheimer's disease and cerebrovascular dementia. *Neurobiol Aging* 17: 505–512
78 Myers CE, DeLuca J, Schultheis MT, Schnirman GM, Ermita BR, Diamond B, Warren SG, Gluck MA (2001) Impaired delay eyeblink classical conditioning in individuals with anterograde amnesia resulting from anterior communicating artery aneurysm rupture. *Behav Neurosci* 115: 560–570
79 Fukutani Y, Cairns NJ, Rossor MN, Lantos PL (1996) Purkinje cell loss and astrocytosis in the cerebellum in familial and sporadic Alzheimer's disease. *Neurosci Lett* 214: 33–36
80 Fukutani Y, Cairns NJ, Rossor MN, Lantos PL (1997) Cerebellar pathology in sporadic and famil-

ial Alzheimer's disease APP 717 (Val–Ile) mutation cases: A morphometric investigation. *J Neurol Sci* 149: 177–184
81 Weiss C, Preston AR, Oh MM, Schwarz RD, Welty D, Disterhoft JD (2000) The M1 muscarinic agonist CI-1017 facilitates trace eyeblink conditioning in aging rabbits and increases the excitability of CA1 pyramidal neurons. *J Neurosci* 20: 783–790
82 Moore JW, Goodell NA, Solomon PR (1976) Central cholinergic blockade by scopolamine and habituation, classical conditioning, and latent inhibition of the rabbit's nictitating membrane response. *Physiolog Psychol* 4: 395–399
83 Woodruff-Pak DS, Vogel RW III, Wenk GL (2001) Galantamine: Effect on nicotinic receptor binding, acetylcholinesterase inhibition, and learning. *PNAS* 98: 2089–2094
84 Woodruff-Pak DS, Li Y-T, Kazmi A, Kem WR (1994) Nicotinic cholinergic system involvement in eyeblink classical conditioning in rabbits. *Behav Neurosci* 108: 486–493
85 Solomon PR, Solomon SD, van der Schaaf E, Perry HE (1983) Altered activity in the hippocampus is more detrimental to classical conditioning than removing the structure. *Science* 220: 329–331
86 Woodruff-Pak DS, Li Y-T, Hinchliffe RM, Port RL (1997) Hippocampus in delay eyeblink classical conditioning: Essential for nefiracetam amelioration of learning in older rabbits. *Brain Res* 747: 207–218
87 De Fiebre CM, Meyer EM, Henry JC, Muraskin SI, Kem WR, Papke RL (1995) Characterization of a series of anabaseine-derived compounds reveals that the 3-(4)-dimethylaminocinnamylidine derivative is a selective agonist at neuronal nicotinic $\alpha7/^{125}$-α-bungarotoxin receptor subtypes. *Mol Pharmacol* 47: 164–171
88 Kem WR, Mahnir VM, Papke RL, Lingle CJ (1997) Anabaseine is a potent agonist on muscle and neuronal α-bungarotoxin-sensitive nicotinic receptors. *Pharmacol Exp Ther* 283: 979–992
89 Woodruff-Pak DS, Li Y-T, Kem WR (1994) A nicotinic agonist (GTS-21), eyeblink classical conditioning, and nicotinic receptor binding in rabbit brain. *Brain Res* 645: 309–317
90 Woodruff-Pak DS, Green JT, Coleman-Valencia C, Pak JT (2000) A nicotinic cholinergic drug (GTS-21) and eyeblink classical conditioning: Acquisition, retention, and relearning in older rabbits. *Expt Aging Res* 26: 323–336
91 Woodruff-Pak, DS (2003) Mecamylamine reversal by nicotine and by a partial alpha 7 nicotinic acetylcholine receptor agonist (GTS-21) in rabbits tested with delay eyeblink classical conditioning. *Behav Brain Res* 143: 159–167
92 Pereira EFR, Alkondon M, Reinhardt S, Maelicke A, Peng X, Lindstrom J, Whiting P, Albuquerque EX (1994) Physostigmine and galantamine: Probes for a novel binding site on the $\alpha4\beta2$ subtype of neuronal nicotinic acetylcholine receptors stably expressed in fibroblast cells. *J Pharmacol Exp Ther* 270: 768–778
93 Pereira EFR, Reinhardt-Maelicke S, Schrattenholz A, Maelicke A, Albuquerque EX (1993) Identification and functional characterization of a new agonist site on nicotinic acetylcholine receptors of cultured hippocampal neurons. *J Pharmacol Exp Ther* 265: 1474–1491
94 Storch A, Schrattenholz A, Cooper JC, Abdel Ghani M, Gutbrod O, Weber K-H, Reinhardt S, Lobron C, Hermsen B, Soskic V et al. (1995) Physostigmine, galantamine and codeine act as noncompetitive nicotinic agonists on clonal rat pheochromocytoma cells. *European J Pharmacol* 290: 207–219
95 Woodruff-Pak DS, Ewers M, Vogel RW, Shiotani T, Watabe S, Tanaka M, Wenk GL (2003) Nefiracetam and physostigmine: Separate and combined effects on learning in older rabbits. *Neurobiol Aging*; *in press*
96 Woodruff-Pak DS, Santos I (2000) Nicotinic modulation in an animal model of a form of associative learning impaired in Alzheimer's disease. *Behav Brain Res* 113: 11–19
97 Woodruff-Pak DS, Vogel RW III, Wenk GL (2003) Mecamylamine interactions with galantamine and donepezil: Effects on learning, acetylcholinesterase, and nicotinic acetylcholine receptors. *Neurosci* 117: 439–447

Cognitive Enhancing Drugs
Edited by J.J. Buccafusco

Drugs that target catecholaminergic receptors

Edward D. Levin

Department of Psychiatry and Behavioral Sciences, Duke University Medical Center, Durham, NC 27710, USA

Introduction

Several small catecholaminergic (CA) nuclei in the midbrain contain neurons project to broad areas of the diencephalon and telencephalon. These neurons using the neurotransmitters, dopamine (DA) and norepinepherine (NE) play important roles in a wide variety of neurobehavioral functions including cognition. DA neurons from the substantia nigra (SN) project primarily to the striatum, while DA neurons from the nearby ventral tegmental area (VTA) project more broadly to the nucleus accumbens, thalamus, hippocampus and frontal cortex. NE neurons from the locus coeruleus (LC) project to broad areas of the cortex, limbic system and thalamus. These projections serve as the basis of CA involvement in cognitive function particularly in aspects of the relationship of arousal and attention to salient sensory cues [1]. CA neurons and their post-synaptic receptors are targets for therapeutic treatments for a variety of cognitive disorders [2].

Disruptions of CA systems are known to cause cognitive impairment. Parkinson's disease, in which there is a substantial dying off of midbrain DA neurons, is most well-known as a disorder of motor function, however, there are often pronounced cognitive impairments in Parkinson's disease. Schizophrenia is also known to involve disruptions of DA innervation. This classic psychotic disorder has substantial components of cognitive dysfunction as well. Unfortunately, the DA blockers often used for schizophrenia can cause further cognitive impairment. Attention deficit hyperactivity disorder (ADHD) has been found to involve alterations of DA and NE, particularly in the frontal cortex. Cognitive impairment is a hallmark of this syndrome.

CA agonists have been shown to improve cognitive function. Parkinson's disease medications and DA agonists can improve cognitive function as well as motor function. Methylphenidate and amphetamine indirect CA agonists significantly improve attentional function in people with ADHD; NE agonists have been found to improve cognitive function in a variety of disorders including Alzheimer's disease, schizophrenia and Parkinson's disease. Better basic understanding of the receptor subtypes and their locations critical for cognitive improvement will be valuable for further development of therapeutics.

Preclinical research

Animal models have contributed greatly to our knowledge of the role of CA systems in cognition. Both rodent and primate studies have used lesion and genetic methods as well as neurobehavioral pharmacology to determine the role of DA and NE neural systems in memory and attentional function. Many tasks, including active and passive avoidance procedures, spatial memory paradigms such as the radial-arm maze in rodents and delayed matching to sample in monkeys and tests of selective and sustained attention, have been used to illustrate the involvement of DA and NE systems in cognition [3–6]. Local infusion studies using receptor subtype selective agonists and antagonists have given insight into the roles CA systems play in the neural basis of cognition. These experimental studies help form the knowledge base for development of new drug therapies for cognitive impairment.

Transmitter selective lesions and more recently genetic knockout methods have been key in demonstrating the importance of CA systems in cognitive function. Lesion-induced disruptions of CA systems have been shown to cause cognitive impairment. DA lesions in the frontal cortex using 6-hydroxydopamine (6-OHDA) have been shown to cause memory and attentional impairment [7, 8]. 1-Methyl-4-phenyltetrahydropyridine (MPTP) lesions in rats, which caused reduced DA in both the striatum or frontal cortex, causes specific working, but not reference memory impairments [9].

Specific involvement of CA receptor subtypes can be determined with the use of genetic knockout studies. Involvement of D_2 and D_3 DA receptors in working memory was shown in a study of mice with genetic knockout of D_2 or D_3 receptors. D_2 or D_3 KO mice had significant working memory impairments in spatial alternation [10]. Methamphetamine treatment reversed the working memory impairment in the D_2 knockout mice but not the D_3 knockout mice. Future studies with conditional knockout and knockdown mice can lend further insight into the CA mechanisms in cognitive function.

Neurobehavioral pharmacology experimental animal studies have contributed greatly to our understanding of the roles DA and NE play in cognitive function and can aid the development of CA-based therapeutics for cognitive dysfunction. A wide variety of rodent studies have shown that DA and NE antagonists generally impair learning and memory while some DA and NE agonists can improve learning and memory [11]. DA systems are particularly important for reward associations with learning [12]. NE systems play a prominent role in emotionally relevent memory [13]. Both CA transmitters have been shown to be involved in memory function. Local infusions of D_1 but not D_2 antagonists in the frontal cortex of monkeys significantly impair memory [14]. Lesions of DA projections to the hippocampus impaired spatial memory function [15]. In rats we have found that local infusion of D_2 but not D_1 antagonists impaired working memory [16].

DA drug effects may be altered by frontal cortical brain damage. Lesions of frontal DA projections significantly increase DA in the nucleus accumbens

[17]. Lesions of the medial frontal cortex significantly impaired choice accuracy on the 5-choice attentional task [18]. This impairment was significantly reversed by systemic administration of the D_2 antagonist sulpiride at a dose that significantly impaired choice accuracy in sham operated rats.

DA systems in the ventral hippocampus are involved in working memory function. D_2 drugs had very clear and consistent effects on choice accuracy [16]. The D_2 agonist quinpirole caused a significant dose-related improvement in radial-arm maze choice accuracy over the 1.1–10 µg/side dose range (Fig. 1). In a complementary fashion, the D_2 antagonist raclopride caused a significant dose-related impairment in radial-arm maze choice accuracy over the 0.19–1.67 µg/side dose range (Fig. 1). In contrast, infusions of D_1 drugs (the D_1 agonist dihydrexidine or the D_1 antagonist SCH 23390) into the ventral hippocampus at the dose ranges examined did not have any discernible effects on choice accuracy in the radial-arm maze. The pharmacological study showing the positive relationship between D_2 hippocampal activation and better memory performance is supported by receptor binding experiments. Higher levels of D_2 binding in the hippocampus are correlated with better memory performance on the radial-arm maze [19]. Interestingly, a similar

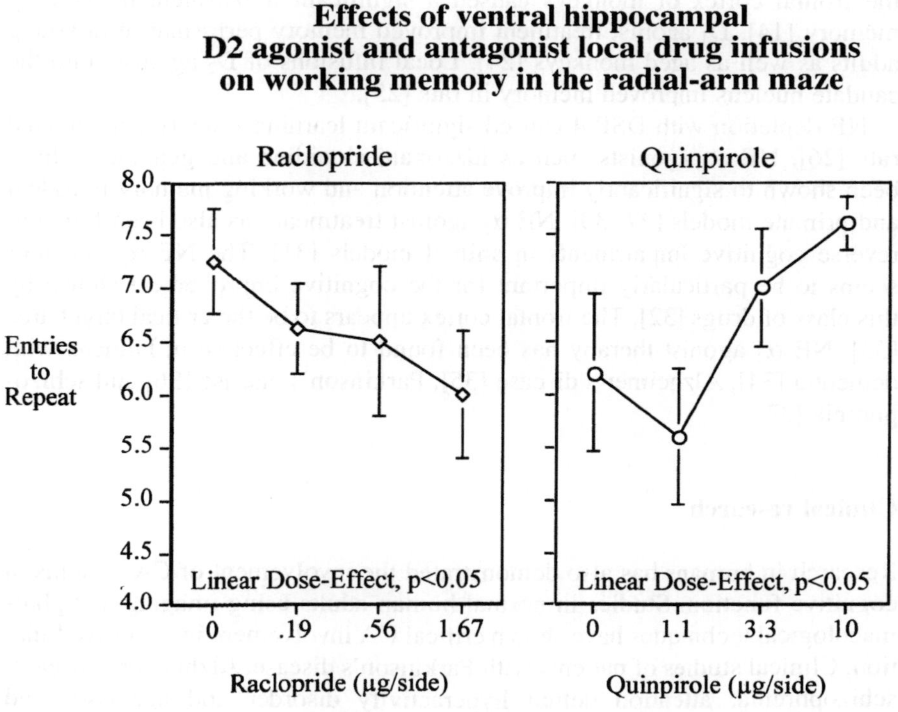

Figure 1. Working memory choice accuracy in the 8-arm radial maze (entries to repeat, mean ± sem) after local infusions of the D_2 antagonist raclopride or the D_2 agonist quinpirole [16].

correlation was also seen for D_2 receptors in the striatum another DA projection area which has clssically been thought to be only involved in motor function.

As more receptor subtype selective ligands are developed, further animal studies will be important in determining the neural bases of CA involvement in cognitive function. In particular, local infusion methods can be used to determine the circuit basis for CA interactions with other transmitter systems in the programming of cognitive function. These same animal models represent key elements in determining the safety and efficacy of novel CA ligands before they can be used in human clinical populations. Haloperidol-induced memory impairment is clearly seen in experimental rat models. Haloperidol significantly impaired working memory performance on the radial-arm maze. This effect is seen with either acute injection or chronic infusion [20–22]. The atypical antipsychotic clozapine has also been shown in rat models to cause memory impairment [20]. Very exciting is the finding that direct DA agonists could be useful in attenuating the degeneration in Parkinson's disease. The DA agonist pramipexole has shown efficacy in reducing neurotoxicity after exposure of experimental animals to MPTP or 6-OHDA [23].

Working memory was impaired by DA depletion of the hippocampal formation via injections of 6-OHDA [15]. Local infusions of D_1 antagonists into the frontal cortex of monkeys caused a significant impairment in working memory [14]. D_1 agonist treatment improved memory performance in young adults as well as aged monkeys [24]. Local infusions of D_2 agonists into the caudate nucleus improved memory in rats [25].

NE depletion with DSP-4 caused significant learning impairments in aged rats [26]. NE α_2 agonists such as idazoxan, clonidine and guanfacine have been shown to significantly improve attention and working memory in rodent and primate models [27–30]. NE α_2 agonist treatment has also been found to reverse cognitive impairments in animal models [31]. The NE α_{2a} receptor seems to be particularly important for the cognitive improvement shown by this class of drugs [32]. The frontal cortex appears to be the critical target area [33]. NE α_2 agonist therapy has been found to be effective in humans with dementia [34], Alzheimer's disease [35], Parkinson's disease [36] and schizophrenia [37].

Clinical research

Research in humans has also demonstrated the involvement of CA systems in cognitive function. Studies in normal human adults using imaging and pharmacological techniques have shown critical CA involvement in cognitive function. Clinical studies of patients with Parkinson's disease, Alzheimer's disease, schizophrenia, attention deficit hyperactivity disorder and age-associated memory impairment have shown CA involvement in cognitive dysfunction and the usefulness of CA-based therapeutics.

Parkinson's disease

Parkinson's disease is classically thought of as a motor disease. However, there are also often prominent cognitive impairments [38–41]. Parkinson's disease patients show deficits in visuospatial processing, motor planning, executive function, selective attention, mental flexibility and at later stages memory function [42]. Often these symptoms do not respond well to traditional DA therapy [43]. Direct DA agonist therapy such as pergolide, a non-specific DA receptor agonist, has been found to have mixed effects on cognitive performance in normal adults [44]. It improved working memory performance in already high performing subjects, but lowered scores in subjects with low baseline performance. In addition, the same dose of pergolide significantly impaired verbal fluency. Possibly DA receptor subtype-selective drugs would have a more pervasive cognitive enhancing effect.

Alzheimer's disease

Cholinergic systems are not the only systems which are impaired in Alzheimer's disease. DA systems also show decline. DA treatments, which have long been used in Parkinson's disease, are being found also to have cognitive-improving effects [45]. Both ACh and DA systems are disrupted by β-amyloid [46]. NE systems have also been shown to be impaired in patients with Alzheimer's disease. The decline in NE in the mid-temporal cortex has been shown to significantly associate with cognitive impairment [47]. Decreased NA cell number in the locus coerelus was also shown to be related to behavioral aggressiveness in demented patients.

Attention deficit hyperactivity disorder

ADHD has been shown to involve impairments in frontal CA stimulation. The use of indirect CA agonists such as methylphenidate and amphetamine are the standard treatment for ADHD and significantly improve symptoms of inattention [42]. Long known in adolescents and more recently adults with ADHD [48, 49]

Schizophrenia

Cognitive impairments are seen in schizophrenics both as a result of the disease process and its treatment. Attentional deficits have long been known to accompany schizophrenia [50]. Problems in disregarding irrelevant stimuli seems to be of particular significance [42]. This cognitive impairment could even contribute to the hallmark psychotic symptoms. Unfortunately classic antipsychotic drugs, and to a lesser degree atypical antipsychotics which are effective in reducing hallucinations, can also impair cognitive function. Classic treatments for schizophrenia have pronounced DA antagonist effects. Drugs like haloperidol have been shown to cause significant attentional and memory impairment in schizophrenia patients [51].

The use of CA agonists, either direct receptor agonists or indirect agonists such as methylphenidate or amphetamine, which have been shown to improve cognitive function in other syndromes of cognitive impairment, may have adverse effects in patients with schizophrenia because they can aggravate psychotic symptoms. This may be due to increased DA activity in the limbic system. Avoiding this action, NE α_2 agonists may be useful for improving cognitive function in people with schizophrenia [52]. In addition, further development of DA agonists, which have selective actions in the frontal cortex without increasing limbic DA activity, is warranted.

Age-associated memory impairment

In the non-demented elderly there can be a modest decline in memory function, which has been termed age associated memory impairment or mild cognitive impairment [53, 54]. DA function has been shown in imaging studies to be associated with declines in cognitive function including learning and memory [55]. It may be possible to provide symptomatic relief of this modest cognitive impairment with direct or indirect CA agonists, although safety and efficacy trials need to be conducted first.

Conclusions

Direct and indirect agonists of the CA systems DA and NE have been shown to improve cognitive function including attention and memory. They are useful therapeutic treatments for syndromes of cognitive impairment such as ADHD, Alzheimer's disease, Parkinson's disease and schizophrenia. Further development of medications acting on CA systems will depend on receptor subtype selectivity, so that the cognitive enhancement can be achieved with minimal side-effects of motor disruption, appetite dysfunction and abuse liability. CA systems play vital roles in the neural substrate of cognitive function. Disruptions of these systems are known to cause cognitive impairment. CA agonists can effectively reverse this dysfunction.

References

1 Berridge CW, Waterhouse BD (2003) The locus coeruleus-noradrenergic system: Modulation of behavioral state and state-dependent cognitive processes. *Brain Res* Rev 42: 33–84

2 Buccafusco JJ, Terry AV Jr (2000) Multiple central nervous system targets for eliciting beneficial effects on memory and cognition. *J Pharmacol Exp Ther* 295: 438–446
3 Baratti CM, Introini IB, Huygens P, Gusovsky F (1983) Possible cholinergic-dopaminergic link in memory facilitation induced by oxotremorine in mice. *Psychopharmacology* 80: 161–165
4 Helper DJ, Wenk GL, Cribbs BL, Olton DS, Coyle JT (1985) Memory impairments following basal forebrain lesions. *Brain Res* 346: 8–14
5 Levin ED, Galen D, Ellison GD (1987) Chronic haloperidol effects on radial-arm maze performance and oral movements in rats. *Pharmacol Biochem Behav* 26: 1–6
6 McGurk S, Levin ED, Butcher LL (1988) Cholinergic-dopaminergic interactions in radial-arm maze performance. *Behav Neural Biol* 49: 234–239
7 Bubser M, Schmidt WJ (1990) 6-hydroxydopamine lesion of the rat prefrontal cortex increases locomotor activity, impairs acquisition of delayed alternation tasks, but does not affect uninterrupted tasks in the radial maze. *Behav Brain Res* 37: 157–168
8 Crofts HS, Dalley JW, Collins P, Van Denderen JC, Everitt BJ, Robbins TW, Roberts AC (2001) Differential effects of 6-ohda lesions of the frontal cortex and caudate nucleus on the ability to acquire an attentional set. *Cerebral Cortex* 11: 1015–1026
9 Miyoshi E, Wietzikoski S, Camplessei M, Silveira R, Takahashi RN, Da Cunha C (2002) Impaired learning in a spatial working memory version and in a cued version of the water maze in rats with MPTP-induced mesencephalic dopaminergic lesions. *Brain Res Bull* 58: 41–47
10 Glickstein SB, Hof PR, Schmauss C (2002) Mice lacking dopamine d2 and d3 receptors have spatial working memory deficits. *J Neurosci* 22: 5619–5629
11 Myhrer T (2003) Neurotransmitter systems involved in learning and memory in the rat: A meta-analysis based on studies of four behavioral tasks. *Brain Res Rev* 41: 268–287
12 Wilson C, Nomikos G, Collu M, Fibiger H (1995) Dopaminergic correlates of motivated behavior: Importance of drive. *J Neurosci* 15: 5169–5178
13 Southwick SM, Davis M, Horner B, Cahill L, Morgan CA, 3rd, Gold PE, Bremner JD, Charney DC (2002) Relationship of enhanced norepinephrine activity during memory consolidation to enhanced long-term memory in humans. *Am J Psychiatry* 159: 1420–1422
14 Sawaguchi T, Goldman-Rakic PS (1994) The role of D1-dopamine receptor in working memory: Local injections of dopamine antagonists into the prefrontal cortex of rhesus monkeys performing an oculomotor delayed-response task. *J Neurophysiol* 71: 515–528
15 Gasbarri A, Sulli A, Innocenzi R, Pacitti C, Brioni JD (1996) Spatial memory impairment induced by lesion of the mesohippocampal dopaminergic system in the rat. *Neuroscience* 74: 1037–1044
16 Wilkerson A, Levin ED (1999) Ventral hippocampal dopamine D1 and D2 systems and spatial working memory in rats. *Neuroscience* 89: 743–749
17 Fernandez-Espejo E (2003) Prefrontocortical dopamine loss in rats delays long-term extinction of contextual conditioned fear, and reduces social interaction without affecting short-term social interaction memory. *Neuropsychopharmacology* 28: 490–498
18 Passetti F, Levita L, Robbins TW (2003) Sulpiride alleviates the attentional impairments of rats with medial prefrontal cortex lesions. *Behav Brain Res* 138: 59–69
19 Levin ED, Torry D, Christopher NC, Yu X, Einstein G, Schwartz-Bloom R (1997) Is binding to nicotinic acetylcholine and dopamine receptors related to working memory in rats? *Brain Res Bull* 43: 295–304
20 Addy N, Levin ED (2002) Nicotine interactions with haloperidol, clozapine and risperidone and working memory function in rats. *Neuropsychopharmacology* 27: 534–541
21 Levin ED (1997) Chronic haloperidol administration does not block acute nicotine-induced improvements in radial-arm maze performance in the rat. *Pharmacol Biochem Behav* 58: 899–902
22 McGurk SR, Levin ED, Butcher LL (1989) Nicotinic-dopaminergic relationships and radial-arm maze performance in rats. *Behav Neural Biol* 52: 78–86
23 Schapira AH (2002) Dopamine agonists and neuroprotection in Parkinson's disease. *Eur J Neurology* 9: 7–14
24 Arnsten AF, Cai JX, Murphy BL, Goldman-Rakic PS (1994) Dopamine d1 receptor mechanisms in the cognitive performance of young adult and aged monkeys. *Psychopharmacology (Berl)* 116: 143–151
25 White NM, Viaud M (1991) Localized intracaudate dopamine D2 receptor activation during the post-training period improves memory for visual or olfactory conditioned emotional responses in rats. *Behav Neural Biol* 55: 255–269
26 Sirvio J, Riekkinen P Jr, Valjakka A, Jolkkonen J, Riekkinen PJ (1991) The effects of noradrener-

gic neurotoxin, dsp-4, on the performance of young and aged rats in spatial navigation task. *Brain Res* 563: 297–302
27 Arnsten AF, Contant TA (1992) Alpha-2 adrenergic agonists decrease distractibility in aged monkeys performing the delayed response task. *Psychopharmacology* 108: 159–169
28 Franowicz JS, Arnsten AF (1999) Treatment with the noradrenergic alpha-2 agonist clonidine, but not diazepam, improves spatial working memory in normal young rhesus monkeys. *Neuropsychopharmacology* 21: 611–621
29 Rama P, Linnankoski I, Tanila H, Pertovaara A, Carlson S (1996) Medetomidine, atipamezole, and guanfacine in delayed response performance of aged monkeys. *Pharmacol Biochem Behav* 55: 415–422
30 Sara SJ, Devauges V (1989) Idazoxan, an alpha-2 antagonist, facilitates memory retrieval in the rat. *Behav Neural Biol* 51: 401–411
31 Birnbaum SG, Podell DM, Arnsten AF (2000) Noradrenergic alpha-2 receptor agonists reverse working memory deficits induced by the anxiogenic drug, FG7142, in rats. *Pharmacol Biochem Behav* 67: 397–403
32 Franowicz JS, Kessler LE, Borja CM, Kobilka BK, Limbird LE, Arnsten AF (2002) Mutation of the alpha2a-adrenoceptor impairs working memory performance and annuls cognitive enhancement by guanfacine. *J Neuroscience* 22: 8771–8777
33 Li BM, Mao ZM, Wang M, Mei ZT (1999) Alpha-2 adrenergic modulation of prefrontal cortical neuronal activity related to spatial working memory in monkeys. *Neuropsychopharmacology* 21: 601–610
34 Coull JT, Sahakian BJ, Hodges JR (1996) The alpha(2) antagonist idazoxan remediates certain attentional and executive dysfunction in patients with dementia of frontal type. *Psychopharmacology* 123: 239–249
35 Franowicz JS, Arnsten AF (1998) The alpha-2a noradrenergic agonist, guanfacine, improves delayed response performance in young adult rhesus monkeys. *Psychopharmacology* 136: 8–14
36 Riekkinen M, Riekkinen P Jr (1999) Alpha2-adrenergic agonist clonidine for improving spatial working memory in Parkinson's disease. *J Clin Psychopharmacol* 19: 444–449
37 Friedman JI, Adler DN, Temporini HD, Kemether E, Harvey PD, White L, Parrella M, Davis KL (2001) Guanfacine treatment of cognitive impairment in schizophrenia. *Neuropsychopharmacology* 25: 402–409
38 Ebmeier KP, Calder SA, Crawford JR, Stewart L, Cochrane RHB, Besson JAO (1991) Dementia in idiopathic Parkinson's disease: Prevalence and relationship with symptoms and signs of Parkinsonism. *Psychol Med* 21: 69–76
39 Faglioni P, Botti C, Scarpa M, Ferrari V, Saetti MC (1997) Learning and forgetting processes in Parkinson's disease: A model-based approach to disentangling storage, retention and retrieval contributions. *Neuropsychologia* 35: 767–779
40 Palazzini E, Soliveri P, Filippini G, Fetoni V, Zappacosta B, Scigliano G, Monza D, Caraceni T, Girotti F (1995) Progression of motor and cognitive impairment in Parkinson's disease. *J Neurol* 242: 535–540
41 Sawamoto N, Honda M, Hanakawa T, Fukuyama H, Shibasaki H (2002) Cognitive slowing in Parkinson's disease: A behavioral evaluation independent of motor slowing. *J Neurosci* 22: 5198–5203
42 Nieoullon A (2002) Dopamine and the regulation of cognition and attention. *Prog Neurobiol* 67: 53–83
43 Pillon B, Dubois B, Bonnet A-M, Esteguy M, Guimaraes J, Vigouret J-M, Lhermite F, Agid Y (1989) Cognitive slowing in Parkinson's disease fails to respond to levodopa treatment: The 15-objects tests. *Neurology* 39: 762–768
44 Kimberg DY, D'Esposito M (2003) Cognitive effects of the dopamine receptor agonist pergolide. *Neuropsychologia* 41: 1020–1027
45 Nisoli E, Missale C, Memo M, Spano P (1990) Dopaminergic system in the aged brain: An overview. In: Aging Drain and Dementia: New trends in diagnosis and therapy. New York: Alan R. Liss, 189–202
46 Itoh A, Nitta A, Nadai M, Nishimura K, Hirose M, Hasegawa T, Nabeshima T (1996) Dysfunction of cholinergic and dopaminergic neuronal systems in beta-amyloid protein-infused rats. *J Neurochem* 66: 1113–1117
47 Matthews KL, Chen CP, Esiri MM, Keene J, Minger SL, Francis PT (2002) Noradrenergic changes, aggressive behavior, and cognition in patients with dementia. *Biolog Psychiatry* 51: 407–416

48 Wender PH, Wood DR, Reimherr FW (1985) Pharmacological treatment of attention deficit disorder residual type (ADD, RT, “minimal brain dysfunction”, “hyperactivity”) in adults. *Psychopharmacol Bull* 21: 215–222
49 Wender P (1995) Attention-deficit hyperactivity disorder in adults. New York: Oxford University Press
50 Cornblatt BA, Keilp JG (1994) Impaired attention, genetics, and the pathophysiology of schizophrenia. *Schiz Bull* 20: 31–46
51 Levin E, Wilson W, Rose J, McEvoy J (1996) Nicotine-haloperidol interactions and cognitive performance in schizophrenics. *Neuropsychopharmacology* 15: 429–436
52 Friedman JI, Temporini H, Davis KL (1999) Pharmacologic strategies for augmenting cognitive performance in schizophrenia. *Biolog Psychiatry* 45: 1–16
53 Crook T, Bartus T, Ferris S, Whitehouse P (1986) Age associated memory impairment: Proposed diagnostic criteria and measures of clinical change-report of a national institute of mental health work group. *Develop Neurobiol* 2: 261–276
54 Crook T, Feher E, Larrabee G (1992) Assessment of memory complaints in age-associated memory impairment: The MAC-Q. *Intern Psychogeriatrics* 4: 165–175
55 Kaasinen V, Rinne JO (2002) Functional imaging studies of dopamine system and cognition in normal aging and Parkinson’s disease. *Neurosci Biobehav Rev* 26: 785–793

Cognitive Enhancing Drugs
Edited by J.J. Buccafusco

Drugs that target serotonergic receptors

Alvin V. Terry Jr.

Program in Clinical and Experimental Therapeutics, University of Georgia College of Pharmacy (Augusta Campus) and Small Animal Behavior Core, Medical College of Georgia, Augusta, GA 30912-2450, USA

Introduction

Of the neurotransmitter substances characterized to date, serotonin (5-hydroxytryptamine-5-HT) is perhaps one of the most extensively studied. Initial reports of its actions on smooth muscle appeared more than 50 years ago and it is now known to have widespread peripheral and central physiologic actions. In the periphery, the amine is released from platelets to activate clotting cascades and blood vessel constriction; in the gastrointestinal tract it inhibits gastric acid production and stimulates the contractions of smooth muscle [1]. In the central nervous system, 5HT is known to regulate or influence such diverse processes as activity patterns and circadian rhythm, food intake, sleep-wake cycles, aggressive behaviors, locomotion, thermoregulation, nociception, and sexual activity [2].

In humans, 5HT is well-known to play a key role in mood and emotion, and a number of psychiatric conditions including anxiety, depression, schizophrenia, and anorexia nervosa. While undoubtedly involved in attention, learning, and memory [3] these roles of 5HT are probably the least understood when compared to those of other major neurotransmitters, particularly other monoamines [4]. The complex interactions between 5HT neurons and other neuronal phenotypes, 5HT receptor heterogeneity, and the conflicting results of behavioral experiments have made functional assessments of 5HT in cognitive processes particularly difficult. There have been numerous conflicting or confusing findings regarding the effects of serotonin. For example, stimulation of the dorsal raphe and central injection of 5HT was found to impair learning and memory [5, 6], whereas elevated synaptic 5HT with re-uptake inhibitors was found to enhance memory [5, 7]. In contrast, Jaffard et al. [8] reported that 5HT re-uptake inhibitors caused memory impairments. Neuronal 5HT depletion from electrolytic and neurotoxic lesions have impaired, [9, 10] facilitated, [11, 12] or had little effect [13, 14] on memory processes. An array of ligands synthesized to function as agonists or antagonists at 5HT subtypes have been evaluated in recent years in an effort to resolve these controversies. However, due to the lack of adequate subtype specificity, in many cases the compounds have only served as a further source of confusion. Nevertheless, there is now a credible body of evidence suggesting that several of the 5HT receptor subtypes could indeed serve as therapeutics targets for memory enhancement. The purpose of this chapter is to provide a brief overview of these therapeutic targets within the 5HT system and the pharmacologic approaches (with a few representative drug examples) designed to enhance memory function. For a more comprehensive overview of the role of 5HT and its receptors in memory function see reviews [3, 15].

5HT receptor subtypes as potential drug targets

The mammalian 5HT receptor superfamily currently consists of 14 cloned and sequenced receptors divided into seven classes designated as 5HT1–7 [16]. The following text provides a brief overview of the receptor subtypes that have been targeted specifically for memory enhancement based on their anatomical distribution and in some cases based on alterations known to occur with aging and Alzheimer's disease (AD).

Table 1. Serotonin receptors and drug development. Representative compounds with memory enhancing properties

Subtype	Regional distribution (Memory-related areas)	Effective ligands	Action
5HT1A	septum	WAY 1000635	agonist
	hippocampus	MDL 73005	agonist
	entorhinal cortex		
	cingulate cortex		
5HT1 B	hippocampus	GR 127935	antagonist
	subiculum		
	striatum		
5HT2A	frontal cortex	MDL 100907	antagonist
	entorhinal cortex	EMD 281014	antagonist
	pyriform cortex		
	claustrum		
	nucleus accummbens		
	hippocampus		
5HT3	hippocampus	ondansetron	antagonist
	amygdala	granisetron	antagonist
	striatrum	tropisetron	antagonist
	cortex	RS 56812	antagonist
5HT4	frontal cortex	RS 17017	agonist
	hippocampus	SL 65.0155	partial agonist
5HT6	nucleus accumbens	Ro 04-6790	antagonist
	cortex	SB-271046-A	antagonist
	hippocampus	SB-357134-A	antagonist
5HT7	hippocampus	SB-258741?	antagonist
	thalamus		

$5HT_1$ receptors

With regard to memory-related brain area distribution, $5HT_{1A}$ receptors are concentrated in the limbic system, especially the hippocampus, lateral septum, cingulate, and entorhinal cortex. They are known to interact with a number of other neurotransmitter systems including noradrenergic, dopaminergic, GABAergic, and cholinergic systems. Their interactions with cholinergic systems may have particular relevance in dementing illnesses such as AD. For example, $5HT_{1A}$ receptors are expressed by cholinergic neurons in the medial septum and in the diagonal band of Broca [17]. They occur as somatodendritic receptors at high concentrations in the hippocampus [18], and are reduced in the *post mortem* brains of patients with AD-like dementia [19, 20]. Agonists of these receptors promote the growth and branching of cholinergic neurites in primary cultures of septal neurons [21], a finding that could have relevance to drug development strategies for AD. Further, $5HT_{1A}$ agonists MDL 73005, WAY 1000635, and WAY 1000135 prevent impairments induced by cholinergic antagonists [22]. However, the $5HT_{1A}$ agonists (8-OH-DPAT) and partial agonists (buspirone, tandospirone) disrupt passive avoidance, water maze spatial learning, and water avoidance learning tasks in rodents [23–26]. Therefore, at present it is unclear whether agonists or antagonists at this subtype offer the best approach to treating conditions such as AD.

$5HT_{1B}$ receptors appear to be located in several important memory-related areas including the subiculum, hippocampus, and striatum. They are thought to reside on axon terminals [27] in the hippocampus where they may function to inhibit the release of ACh [28] and possibly the release of other neurotransmitters [29, 30]. The $5HT_{1B}$ agonist CP-93,129 was found to disrupt radial arm maze performance in rodents [30], whereas administration of the $5HT_{1B}$ antagonist, GR127935, at higher doses was found to increase the consolidation of learning [31]. Furthermore, $5HT_{1B}$ knock-out mice performed better in a spatial memory task compared to controls [32]. Collectively, the data summarized here support a therapeutic strategy of using functionally selective antagonists at $5HT_{1B}$ receptors in treating memory disorders.

$5HT_2$ receptors

The functional role of $5HT_2$ receptors in the prefrontal cortex (PFC) and their contribution to working memory function and attention has received considerable attention recently. Preclinical as well as clinical evidence indicates that alterations in PFC function may at least partially underlie the cognitive and affective symptoms observed in depression and schizophrenia [33]. While it has been generally accepted that $5HT_2$ antagonists have a potential to enhance memory function, the $5HT_2$ subtype is now known to be more heterogeneous than originally thought, thus requiring further experimentation before firm conclusions can be made about the activity of specific ligands. The receptor is now subclassified as $5HT_{2A}$, $_{2B}$, or $_{2C}$ and the compounds previously thought to be selective for the $5HT_2$ subtype are now known to have various affinities for these (further classified) subtypes. The antagonists, ketanserin and ritanserin, are now designated as $5HT_{2A/2C}$ antagonists. Ketanserin improves memory of previously learned inhibitory responses and experimentally induced amnesia in rodents [34, 35], and enhances visual recognition memory in primates [36].

The $5HT_{2A}$ receptor subtype (specifically) has received considerable attention recently as a therapeutic target for the cognitive dysfunction associated with schizophrenia. The subtype is widely expressed in the prefrontal cortex and other memory-related brain regions including the entorhinal and pyriform cortex, claustrum, nucleus accumbens, and hippocampus. It has been shown in non-human primate studies to serve an important role in working memory function [37] and, further, antagonist activity at this subtype has been suggested to at least partially underlie the therapeutic advantages of atypical neuroleptics over typical agents. Support of this latter premise lies in the accumulating evidence that several atypical antipsychotics (which are known to antagonize $5HT_{2A}$ receptors) appear not only to attenuate positive symptoms of schizophrenia and to ameliorate negative symptoms, but also to enhance cognitive function in this patient population [38]. Accordingly, efforts to synthesize more selective $5HT_{2A}$ receptor antagonists are underway. MDL 100907, the first such agent introduced, is a $5HT_{2A}$ receptor antagonist with slightly greater than 100-fold selectivity for $5HT_{2A}$ receptors as compared to $5HT_{2C}$ receptors in radioligand binding studies [39]. The compound advanced from preclinical to clinical trials for schizophrenia [40], but has since been discontinued, possibly due to an inadequate separation between $5HT_{2A}$ and $5HT_{2C}$ receptors. In comparison to MDL 100907, the novel serotonin $5HT_{2A}$ ligand, EMD 281014, demonstrates a much higher selectivity for $5HT_{2A}$ *versus* $5HT_{2C}$ receptors with $5HT_{2A}$ antagonistic properties (Bartoszyk GD, personal communication, 2003). The effects of this more selective compound on memory function are unknown at present, but are currently the focus of behavioral studies.

$5HT_3$ receptors

$5HT_3$ receptors, the most extensively characterized of the serotonin receptor subtypes, are unique in that they are ligand-gated ion channels as opposed to second messenger-coupled receptors. They occur in relatively high concentrations in the amygdala and hippocampus (and to a lesser extent the cortex) of rodents and humans, and also appear to mediate inhibition of ACh release [41–44]. Considerable data now clearly support the assertion that antagonism of $5HT_3$ receptors offers a rational approach to the therapy of some cognitive disorders. For example, whereas the $5HT_3$ receptor agonist mCPBG impairs retention of an associative learning task in rats [45], ondansetron, a $5HT_3$

receptor antagonist, improves performance in several rodent and primate memory tasks while lacking cholinergic side-effects [42, 46, 47]. Other selective $5HT_3$ receptor antagonists, granisetron, tropisetron, and itasetron (DAU 6215), also improve memory in rodents [48–50].

Several years ago, we evaluated the (R) and (S) isomers of a potent serotonin ($5HT_3$) receptor ligand, RS-56812, to assess their effects on working memory performance in macaques trained to perform a delayed matching-to-sample (DMTS) task. A representative figure summarizing the results of these studies is presented in Figure 1A. While both isomers enhanced certain aspects of task performance, the (R) isomer produced more systematic and reproducible improvements [51] as indicated by the maintenance of improvements after repeated exposures to optimal doses. These results further support the potential therapeutic role for $5HT_3$ receptor antagonists in disorders involving cognitive decline.

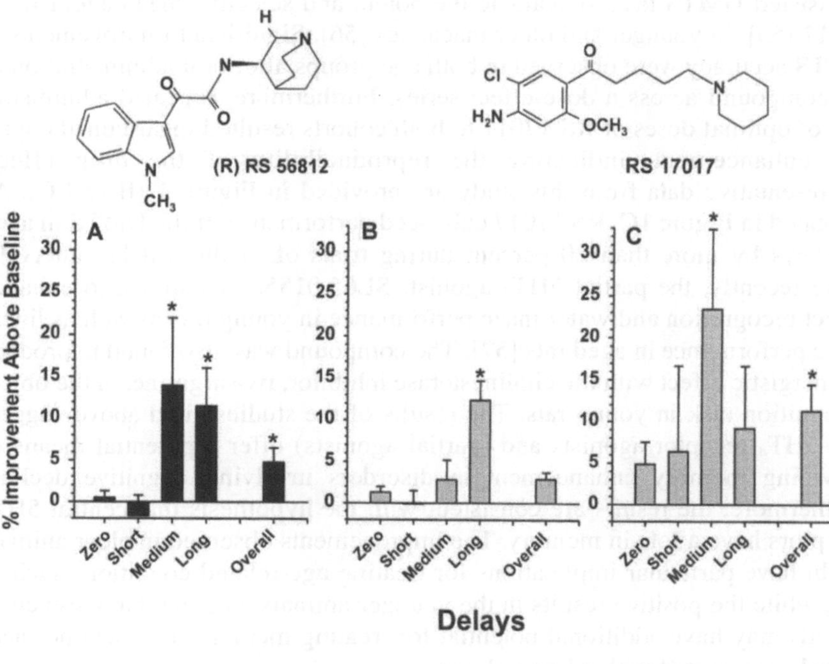

Figure 1. Improvements in accuracy above baseline levels of performance in a delayed matching to sample task (DMTS) task (employing 96 trials/session) by monkeys following administration of specific serotonin receptor ligands. (A) Performance associated with the (R) isomer of the $5HT_3$ antagonist RS 56812 in young adult macaques. (B) DMTS performance associated with the $5HT_4$ agonist RS 17017 in young adult macaques. (C) DMTS performance associated with the $5HT_4$ agonist RS 17017 in aged macaques. The bars represent the mean of at least 2–4 replicates of the individualized optimal doses in percentage points above baseline ± S.E.M. * = significantly different from placebo control value (Two-way repeated measures ANOVA, $p < 0.05$).

$5HT_4$ receptors

$5HT_4$ receptors are most dense in limbic system structures including the frontal cortex, hippocampus, and amygdala (areas which also contain a high density of cholinergic neurons or terminals) [52]. These receptors are markedly reduced in AD, and therefore have been identified as potential therapeutic targets for the disease. This hypothesis is supported by experiments in which activation of $5HT_4$ receptors resulted in enhanced release of acetylcholine in rat frontal cortex [53], and in studies in which the non-selective $5HT_4$ agonists BIMU-1, BIMU-8 and RS 66331 reversed experimentally induced amnesia in rodents. In addition, the selective $5HT_4$ agonist RS 67333 improved atropine-impaired performance in spatial learning in rodents; and its cognitive enhancing effects were reversed by the selective $5HT_4$ antagonist, RS 67532. [54]

Previously, experiments were conducted in our laboratory using a computer-assisted DMTS task to evaluate the potent and selective $5HT_4$ agonist, RS 17017 [55], in younger and older macaques [56]. Significant improvements in DMTS accuracy were observed in both age groups after oral administration of the compound across a dose-effect series. Furthermore, repeated administration of optimal doses of RS 17017 to both cohorts resulted in sustained cognitive enhancement, indicating the reproducibility of the drug effect. Representative data from this study are provided in Figure 1 (B and C). As indicated in Figure 1C, RS 17017 enhanced performance of the DMTS in aged monkeys by more than 20 percent during trials of medium delay intervals. More recently, the partial $5HT_4$ agonist, SL65.0155, was shown to enhance object recognition and water maze performance in young rats as well as linear maze performance in aged rats [57]. The compound was also found to produce a synergistic effect with the cholinesterase inhibitor, rivastigmine, in the object recognition task in young rats. The results of the studies cited above suggest that $5HT_4$ receptor agonists and (partial agonists) offer a potential means of providing memory enhancement in disorders involving cognitive decline. Furthermore, the results are consistent with the hypothesis that central $5HT_4$ receptors have a role in memory. The improvements observed in older animals might have particular implications for treating age-related conditions such as AD, while the positive results in the younger animals indicate that these compounds may have additional potential for treating memory disorders not necessarily associated with advanced age.

$5HT_6$ receptors

The recently characterized $5HT_6$ receptor, subtype appears to be almost exclusively expressed in the brain. In regard to limbic and memory-related areas, the subtype appears to be expressed in relatively high levels in the nucleus accumbens, cerebral cortex, and subfields of the hippocampus [58]. While the function of this subtype is currently unknown, the distribution of this receptor, combined with its affinity for certain antipsychotic (e.g., clozapine, olanzapine), as well as antidepressant drugs (e.g., amoxipine, and amitriptyline), supports the hypothesis that $5HT_6$ ligands may have a therapeutic role in conditions such as schizophrenia and depression. Recent data also suggest that $5HT_6$ ligands may have memory enhancing capabilities. For example, the 5HT6 receptor selective antagonists AO and RO 04-6790, SB-271046-A, and SB-357134-A have all been associated with positive effects in rodent water maze tasks [59].

$5HT_7$ receptors

The $5HT_7$ subtype is the most recently cloned serotonin receptor, although functional responses now attributed to this receptor have been published for several years. The receptor has been identified as important in circadian rhythms and sleep and some recent electrophysiologic studies appear to suggest that the receptor could serve as a target for anticonvulsant drugs [16]. As in the case of the $5HT_6$ subtype, $5HT_7$ receptors also bind several antidepressants (e.g., mianserin, maprotiline) and antipsychotics (clozapine, risperidone) with high affinity, indicating that this receptor may represent a therapeutic target for schizophrenia and other psychiatric disorders. In fact, the selective $5HT_7$ antagonist SB-258741 is currently being evaluated in preclinical studies as a potential treatment for schizophrenia [60]. With regard to its CNS distribution in memory-related areas, the receptor is found in relatively high concentrations in the hippocampus and thalamus, with generally lower levels found in the cortex and amygdala. To date several reports and reviews [15] suggested (based on receptor distribution and preliminary pharmacologic analyses) that this receptor might represent another serotonergic target for memory enhancement. At present, this hypothesis awaits further experimentation.

Conclusion

While many aspects regarding the functional role of serotonin (and its complex array of receptor subtypes) are yet to be learned, few would argue against the importance of this neurotransmitter system in memory function. Through the combined use of modern molecular biology, transgenic animal models, and other more traditional research methods such as medicinal chemistry and classical pharmacology, a clearer picture of the role of serotonin and its receptors in mnemonic processes is likely to emerge in the near future. Considerable data now support the argument that selective ligands at 5HT receptor subtypes could indeed serve as therapeutic agents designed to enhance cognitive function in both age-related CNS diseases and in diseases encountered by younger patients.

References

1 Fozard JR (ed.) (1989) The peripheral actions of 5-hydroxytryptamine. Oxford University Press, New York
2 Lucki I (1998) The spectrum of behaviors influenced by serotonin. *Biol Psychiatry* 44: 151–162
3 Buhot MC, Martin S, Segu L (2000) Role of serotonin in memory impairment. *Ann Med* 32: 210–221
4 Ellis KA, Nathan PJ (2001) The pharmacology of human working memory. *Int J Neuropsychopharmacol* 4: 299–313
5 Altman HJ, Normile HJ (1988) What is the nature of the role of the serotonergic nervous system in learning and memory? Prospects for development of an effective treatment strategy for senile dementia. *Neurobiol Aging* 9: 627–638
6 Briley M (1990) Biochemical strategies in the search for cognition enhancers. *Pharmacopsychiatry* 23(suppl): 75–80
7 Flood JF, Cherkin A (1987) Fluoxetine enhances memory in mice. *Psychopharmacology* 93: 36–43
8 Jaffard R, Mocaer E, Poignant JC, Micheau J, Marighetto A, Meunier M, Beracochea D (1991) Effects of tianeptine on spontaneous alternation, simple and concurrent spatial discrimination learning and on alcohol-induced alternation deficits in mice. *Behav Pharmacol* 2: 37–46
9 Lipska BK, Jaskiw GE, Arya A, Weinberger DR (1992) Serotonin depletion causes long-term reduction of exploration in the rat. *Pharmacol Biochem Behav* 43: 1247–1252
10 Wogar MA, Bradshaw C, Szabadi E (1992) Impaired acquisition of temporal differentiation performance following lesions of the ascending 5-hydroxytryptaminergic pathways. *Psychopharmacology* 111: 373–378
11 Altman HJ, Normile HJ, Galloway MP, Ramirez A, Azmitia EC (1990) Enhanced spatial discrimination learning in rats following 5,7-DHT-induced serotonergic deafferentation of the hippocampus. *Brain Res* 518: 61–66
12 Graham S, Ho M-Y, Bradshaw CM, Szabadi E (1994) Facilitated acquisition of a temporal discrimination following destruction of the ascending 5-hydroxytryptaminergic pathways. *Psychopharmacology* 116: 373–378
13 Asin KE, Wirtshafter D, Fibiger HC (1985) Electrolytic, but not 5,7-dihydroxytryptamine, lesions of the nucleus medianus raphe impair acquisition of a radial L-maze task. *Behav Neural Biol* 44: 415–424
14 Ricaurte GA, Markowska AL, Wenk GL, Hatzidimitriou G, Wlos J, Olton DS (1993) 3,4-Methylenedioxymethamphetamine, serotonin and memory. *J Pharmacol Exp Ther* 266: 1097
15 Meneses A (1999) 5HT system and cognition. *Neurosci Biobehav Rev* 23: 1111–1125
16 Barnes NM, Sharp T (1999) A review of central 5HT receptors and their function. *Neuropharmacology* 38: 1083–1152
17 Kia HK, Brisorgueil MJ, Daval G, Langlois X, Hamon M, Verge D (1996) Serotonin1A receptors are expressed by a subpopulation of cholinergic neurons in the rat medial septum and diagonal band of broca – a double immunocytochemical study. *Neuroscience* 75: 143–154
18 Marcinkiewicz M, Verge D, Gozlan H, Pichat L, Hamon M (1984) Autoradiographic evidence for the heterogeneity of $5HT_1$ sites in the rat brain. *Brain Res* 291: 159
19 Middlemiss DN, Palmer AM, Edel N, Bowen DM (1986) Binding of the novel serotonin agonist 8-hydroxy-2-(di-n-propylamino)tetralin in normal and Alzheimer brain. *J Neurochem* 46: 993–996
20 Cross AJ, Slater P, Perry E, Perry RH (1988) An autoradiographic analysis of serotonin receptors in human temporal cortex: Changes in Alzheimer-type dementia. *Neurochem Int* 13: 89–96
21 Riad M, Emerit MB, Hamon M (1994) Neurotrophic effects of ipsapirone and other 5HT1A receptor agonists on septal cholinergic neurons in culture. *Dev Brain Res* 82: 245–258
22 Bertrand F, Lehmann O, Galani R, Lazarus C, Jeltsch H, Cassel JC (2001) Effects of MDL 73005 on water-maze performances and locomotor activity in scopolamine-treated rats. *Pharmacol Biochem Behav* 68: 647–660
23 Bass EW, Means LW, McMillen BA (1992) Buspirone impairs performance of a three-choice working memroy water escape task in rats. *Brain Res Bull* 28: 455
24 Carli M, Samanin R (1992) 8-Hydroxy-2-(di-n-propylamino) tetralin impairs spatial learning in a water maze: role of post-synaptic $5HT_{1A}$ receptors. *Br J Pharmacol* 105: 720
25 Mendelson SD, Quartermain D, Francisco T, Shemer A (1993) $5HT_{1A}$ receptor agonists induce

anterograde amnesia in mice through a postsynaptic mechanism. *Eur J Pharmacol* 236: 177–182
26 Buhot MC, Naili S (1995) Changes in exploratory activity following stimulation of hippocampal 5HT1A and 5HT1B receptors in the rat. *Hippocampus* 5: 198–208
27 Boschert U, Amara DA, Segu L, Hen R (1994) The mouse 5-hydroxytryptamine1B receptor is localized predominantly on axon terminals. *Neuroscience* 58: 167–182
28 Maura G, Raiteri M (1986) Cholinergic terminals in rat hippocampus possess $5HT_{1B}$ receptors mediating inhibition of acetylcholine release. *Eur J Pharmacol* 129: 333–337
29 Ait Amara D, Segu L, Naili S, Buhot MC (1995) Serotonin 1B receptor regulation after dorsal subiculum deafferentation. *Brain Res Bull* 38: 17
30 Buhot MC, Patra SK, Naili S (1995) Spatial memory deficits following stimulation of hippocampal 5HT1B receptors in the rat. *Eur J Pharmacol* 285: 221–228
31 Meneses A, Terron JA, Hong E (1997) Effects of the 5HT receptor antagonists GR127935 (5HT1B/1D) and MDL100907 (5HT2A) in the consolidation of learning. *Behav Brain Res* 89: 217–223
32 Malleret G, Hen R, Guillou JL, Segu L, Buhot MC (1999) 5HT1B receptor knock-out mice exhibit increased exploratory activity and enhanced spatial memory performance in the Morris water maze. *J Neurosci* 19: 6157–6168
33 Winstanley CA, Chudasama Y, Dalley JW, Theobald DE, Glennon JC, Robbins TW (2003) Intra-prefrontal 8-OH-DPAT and M100907 improve visuospatial attention and decrease impulsivity on the five-choice serial reaction time task in rats. *Psychopharmacology* (Berl) 167: 304–314
34 Altman HJ, Normile HJ (1987) Different temporal effects of serotonergic antagonists on passive avoidance retention. *Pharmacol Biochem Behav* 28: 353–359
35 Strek KF, Spencer KR, DeNoble VJ (1989) Manipulation of serotonergic protects against a hypoxia-induced deficit of a passive avoidance response in rats. *Pharmacol Biochem Behav* 33: 241–244
36 DeNoble VJ, Schrack LM, Reigel AL, DeNoble KF (1991) Visual recognition memory in squirrel monkeys: Effects of serotonin antagonists on baseline and hypoxia-induced performance deficits. *Pharmacol Biochem Behav* 39: 991–996
37 Williams GV, Rao SG, Goldman-Rakic PS (2002) The physiological role of 5HT2A receptors in working memory. *J Neurosci* 22: 2843–2854
38 Meltzer HY (1999) The role of serotonin in antipsychotic drug action. *Neuropsychopharmacology* 21 (2 Suppl): 106S–115S
39 Kehne JH, Baron BM, Carr AA, Chaney SF, Elands J, Feldman DJ, Frank RA, van Giersbergen PL, McCloskey TC, Johnson MP et al. (1996) Preclinical characterization of the potential of the putative atypical antipsychotic MDL 100,907 as a potent 5HT2A antagonist with a favorable CNS safety profile. *J Pharmacol Exp Ther* 277: 968–981
40 Offord SJ, Wong DF, Nyberg S (1999) The role of positron emission tomography in the drug development of M100907, a putative antipsychotic with a novel mechanism of action. *J Clin Pharmacol* Suppl: 17S-24S
41 Kilpatrick GJ, Jones BJ, Tyers MB (1987) Identification and distribution of $5HT_3$ receptors in rat brain using radioligand binding. *Nature* 330: 746–768
42 Barnes JM, Barnes NM, Costall B, Naylor RJ, Tyers MB (1989) $5HT_3$ receptors mediate inhibition of acetylcholine release in cortical tissue. *Nature* 338[6218]: 762–763
43 Waeber C, Hoyer D, Palacios JM (1989) 5-Hydroxytryptamine$_3$ receptors in the human brain: Autoradiographic visualization using [^{3}H]CS 205.030. *Neuroscience* 31: 393–400
44 Maura G, Andrioli GC, Cavazzani P, Raiteri M (1992) 5-Hydroxytryptamine$_3$ receptors sited on cholinergic axon terminals of human cerebral cortex mediate inhibition of acetylcholine release. *J Neurochem* 58[6]: 2334–2337
45 Hong E, Meneses A (1996) Systemic injection of p-chloroamphetamine eliminates the effect of the 5HT3 compounds on learning. *Pharmacol Biochem Behav* 53: 765–769
46 Carey GJ, Costall B, Domeney AM, Gerrard PA, Jones DN, Naylor RJ, Tyers MB (1992) Ondansetron and arecoline prevent scopolamine-induced cognitive deficits in the marmoset. *Pharmacol Biochem Behav* 42[1]: 75–83
47 Fontana DJ, Daniels SE, Henderson C, Eglen RM, Wong EHF (1995) Ondansetron improves cognitive performance in the Morris water maze spatial navigation task. *Psychopharmacology* 120: 409–417
48 Chugh Y, Saha N, Sankaranarayanan A, Datta H (1991) Enhancement of memory retrieval and attenuation of scopolamine-induced amnesia following administration of $5HT_3$ antagonist ICS 205-930. *Pharmacol Toxicol* 69: 105–106

49 Chugh Y, Saha N, Sankaranarayanan A, Sharma P (1991) Memory enhancing effects of granisetron (BRL 43694) in a passive avoidance task. *Eur J Pharmacol* 203: 121–123
50 Pitsikas N, Brambilla A, Borsin F (1994) Effect of DAU 6215, a novel $5HT_3$ receptor antagonist, on scopolamine-induced amnesia in the rat in a spatial learning task. *Pharmacol Biochem Behav* 47: 95–99
51 Terry Jr, AV, Buccafusco JJ, Prendergast MA, Jackson WJ, Fontana DL, Wong EHF, Whiting RL, Eglen RM (1996) The $5HT_3$ receptor antagonist, RS-56812, enhances delayed matching performance in monkeys. *NeuroReport* 8: 49–54
52 Paxinos G (1985) The rat nervous system. Volume 1. *Forebrain and midbrain*. Academic Press, San Diego
53 Eglen RM, Wong EHF, Dumuis A, Bockaert J (1995) Central $5HT_4$ receptors. Trends *Pharmacol Sci* 16: 391–398
54 Eglen RM (1997) 5-Hydroxytryptamine (5HT4) receptors and central nervous system function: An update. In: Jucker E (ed.): *Progress in Drug Reseach* 49: 9–21
55 Clark RD, Jahangir A, Langston JA, Weinhardt KK, Miller AB, Leung E, Eglen RM (1994) Ketones related to the benzoate $5HT_4$ receptor antagonist RS-23597 are high affinity partial agonists. *Bio Org Med Chem Lett* 4: 2477–2480
56 Terry AV, Buccafusco JJ, Jackson WJ, Prendergast MA, Fontana DJ, Wong EH, Bonhaus DW, Weller P, Eglen RM (1998) Enhanced Delayed Matching Performance in Younger and Older Macaques administered the 5HT4 Receptor Agonist, RS-17017, *Psychopharmacology* 135: 407–415
57 Moser PC, Bergis OE, Jegham S, Lochead A, Duconseille E, Terranova JP, Caille D, Berque-Bestel I, Lezoualc'h F, Fischmeister R et al. (2002) SL65.0155, a novel 5-hydroxytryptamine(4) receptor partial agonist with potent cognition-enhancing properties. *J Pharmacol Exp Ther* 302: 731–741
58 Dawson LA, Nguyen HQ, Li P (2001) The 5HT(6) receptor antagonist SB-271046 selectively enhances excitatory neurotransmission in the rat frontal cortex and hippocampus. *Neuropsychopharmacology* 25: 662–668
59 Woolley ML, Bentley JC, Sleight AJ, Marsden CA, Fone KC (2001) A role for 5HT6 receptors in retention of spatial learning in the Morris water maze. *Neuropharmacology* 41: 210–219
60 Pouzet B, Didriksen M, Arnt J (2002) Effects of the 5HT(7) receptor antagonist SB-258741 in animal models for schizophrenia. *Pharmacol Biochem Behav* 71: 655–665

Cognitive Enhancing Drugs
Edited by J.J. Buccafusco

Drugs that target ionotropic excitatory amino acid receptors

Thomas Weiser

Department CNS Research, Boehringer Ingelheim Pharma GmbH&Co KG, Birkendorfer Strasse, D-88400 Biberach, Germany

Introduction

The idea of glutamate as a neurotransmitter in the mammalian CNS emerged around the middle of the 20th century [1–3], and over the following years it became evident that multiple receptor subtypes were responsible for the physiological properties of glutamatergic neurotransmission [4]. Thanks to the combined efforts of physiology, molecular biology and pharmacology we gained a complex picture of the various types and flavours of ionotropic and metabotropic glutamate receptors. Today we possess the biological and chemical tools to characterize glutamate receptor function from the molecular level to the situation *in vivo*. Based on this knowledge, the clinical use of advanced glutamate receptor modulators for the treatment of e.g., cognition deficits comes into view. In this chapter I will focus on the two main subtypes of ionotropic glutamate receptors, namely those activated by the specific agonists AMPA and NMDA, and the pharmacological intervention which should improve cognition in man. For details about receptor subtype classification and structure the reader is referred to recent reviews (e.g., [5]).

Physiological functions of glutamate receptors

In the prototypic glutamatergic synapse AMPA- and NMDA-receptor gated ion channels act in concert: Upon the release from the presynaptic terminal glutamate reaches millimolar concentrations in the synaptic cleft and binds to both receptor types. NMDA receptor channels, however, are blocked by extracellular Mg^{2+} ions, and this voltage-dependent blockade must first be relieved by the post-synaptic depolarization caused by AMPA receptor channel activation. Thus, the post-synaptic potentials (as well as the underlying membrane currents) are biphasic, and the two components can be discriminated biophysically and pharmacologically [6]. Upon high-frequent presynaptic stimulation the influx of Ca^{2+} via NMDA receptor channels can trigger further downstream processes, which can result in plasticity processes like long-term potentiation (LTP), which is regarded as prerequesite for memory formation [7, 8]. In recent years evidence has accumulated to further indicated that activation of AMPA receptor channels is important for synaptic bouton maintenance [9]; moreover the concentrations of neurotrophic factors in brain tissue are affected [10]. Based on these findings it is not surprising that drugs which target ionotropic glutamate receptors can have effects on cognition, memory formation and higher brain functions. The following sections will highlight some of the most relevant findings.

AMPA receptor modulators

AMPA receptor agonists and antagonists

Due to the physiological role of AMPA receptors it is not surprising that their ligands can affect cognition. Flood and co-workers [11] investigated the effects of agonists (quisqualate and kainate), as well as antagonists (GAMS and DNQX) of AMPA receptor channels on memory processing in a T-maze model in mice. Not unexpectedly, performance was improved by the agonists and worsened by the antagonists. The use of AMPA receptor agonists as medications in the treatment of disorders of cognition, however, is prevented by the neurotoxicity of such compounds. Domoic acid, for example, is an orally-active AMPA receptor agonist which can accumulate in sea mussels under certain environmental conditions. Consumption of these poisonous mussels can lead to intoxications resulting in seizures [12]. The administration of AMPA itself can be lethal, and a rodent assay based on this AMPA cytotoxicity is used as a simple *in vivo* test for the effects of potential neuroprotective drugs [13]. These non-physiological agonists can cause excitotoxicity, thereby mimicking elevated levels of glutamate which are known to occur in conditions of global or focal ischemia. Indeed, it has been postulated that excitotoxicity represents an underlying cause for the slow neurodegeneration that occurs in Alzheimer's disease [14].

Positive allosteric AMPA receptor modulators (PAARMs)

The prototype AMPA receptor channel modulator aniracetam was reported to improve performance in rodent models of learning and memory [15]. This compound had been used clinically for the treatment of cognitive dysfunction [16]. The mechanism of action of aniracetam was identified in the early 1990s, and it was shown that the drug reduced the desensitisation of AMPA receptor mediated membrane currents, it slowed the decay of excitatory synaptic signals, and it augmented LTP in hippocampal slices [17–22]. During subsequent years, a variety of new structures from different chemical classes having AMPA receptor modulating properties were identified (for review: [23]). The most completely characterized of these are the "ampakines" (Cortex Pharmaceuticals), and IDRA 21 (Fidia-Georgetown Institute of the Neurosciences). The latter compound will be described in more detail.

IDRA 21 is structurally related to cyclothiazide, and it produced similar effects on AMPA receptors in thin slices of rat brain. Membrane currents in response to glutamate application were augmented, and the desensitisation of glutamate responses in excised neuronal membrane patches was removed [24]. Our own experiments using HEK 293 cells which recombinantly expressed human GluR1/2 receptor channels support these findings (Fig. 1). In electro-

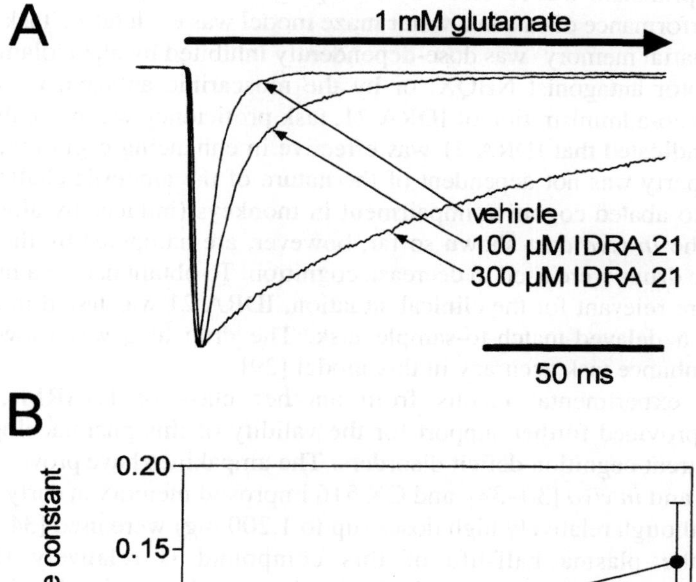

Figure 1. Effect of IDRA 21 on glutamate-induced ion currents in cells expressing recombinant human GluR1/2 receptors. Cells were investigated in the whole-cell, voltage-clamp configuration of the patch-clamp technique. (A) Original recordings from a cell to which 1 mM glutamate was applied together with vehicle, 100, or 300 µM IDRA 21. Traces were normalized to the peak current of the vehicle response. IDRA 21 dose-dependently inhibited inactivation. Holding potential was −80 mV. (B) Concentration-dependent increase of the time constant of inactivation. Data are shown as the means of 4–13 cells per IDRA 21 concentration. Recording conditions were as in (A).

physiological whole-cell, voltage-clamp experiments IDRA 21 reduced desensitisation of membrane currents in response to the application of 1 mM glutamate (Fig. 1A), and this effect was concentration-dependent (Fig. 1B). Thus, IDRA 21 is also active on human AMPA receptors.

The duration of synaptic responses in hippocampal cultures was increased by IDRA 21, and the development of LTP was promoted [25, 26]. The overall potency *in vitro*, however, was relatively low, and the effective concentrations in the various models were between 200 and 1,000 µM. Nevertheless, IDRA 21 was effective *in vivo* at relatively low doses. In a passive avoidance test in rats, behavioural impairment induced by the administration of the $GABA_A$ modulator alprazolam was reversed by 10 mg/kg IDRA 21 [27]. In the same study, the performance of rats in a water maze model was evalutated. Task performance (spatial memory) was dose-dependently inhibited by alprazolam, the AMPA receptor antagonist NBQX, or by the muscarinic antagonist scopolamine. With co-administration of IDRA 21, task proficiency was normalized. These data indicated that IDRA 21 was effective in enhancing cognition, and that this property was not dependent of the nature of the amnestic challenge. IDRA 21 also abated cognitive impairment in monkeys (induced by alprazolam [28]). The *in vivo* data shown so far, however, are hampered by the fact that amnestic drugs were used to decrease cognition. To obtain data in a model which is more relevant for the clinical situation, IDRA 21 was tested in aged monkeys in a delayed-match-to-sample task. The drug also was shown to effectively enhance task accuracy in this model [29].

The first experimental results from another class of PAARMs, the ampakines, provided further support for the validity of this pharmacological approach to treat cognition deficit disorders. The ampakines have proven efficacy *in vitro* and *in vivo* [30–33], and CX 516 improved memory in early clinical trials, although relatively high doses (up to 1,200 mg) were used [34–36]. Moreover, the plasma half-life of this compound is relatively short. Nevertheless, CX 516 is reported to be in development for use in Alzheimer's disease, schizophrenia, mild cognitive impairment, and attention deficit hyperactivity disorder [37]. Newer compounds with improved properties have been described (e.g., LY 404187, [38]; or S 18986, [39]), but no clinical data are thus far available. Taken together there exists compelling evidence that PAARMs could be a useful approach to treat disorders involving cognition deficits in man.

Open questions

One still unresolved issue is the large discrepancy between the concentrations of PAARMs which are active *in vitro*, and their plasma/brain concentrations *in vivo*. Concentrations of IDRA 21 to measurably affect AMPA receptor channel kinetics or synaptic transmission are in the range of 200 to 1,000 µM. Effective doses in animal experiments (~ 10 mg/kg) induce peak brain levels in the range of 10 to 30 μM (K. Klinder, personal communication). Similarily, Lynch and co-workers reported plasma levels of up to 16 μM after the administration of 1,200 mg CX 516 to healthy volunteers [35], and it can be assumed that brain levels are within the same range since in rats, the plasma/brain concentration ratio is ~1 (K. Klinder, personal communication). Such low concentrations do not show any observable effect *in vitro*! Since there is no doubt that PAARMs do possess cognition-enhancing properties, it is likely that slight modulation of glutamatergic synapses can account for their robust effects observed *in vivo*. This situation renders the conclusion that the *in vitro* identification and characterization of PAARMs may only be achieved at high, physiologically irrelevant, concentrations.

Another issue concerns the possible neurotoxicity of AMPA receptor modulators. One could postulate that facilitation of glutamate receptor function could worsen the clinical outcome in excitotoxicity-related neurodegenerative disorders. Indeed it has been reported that both the ampakine 1-BCP, and IDRA 21 potentiate the toxicity to glutamate in a cell culture model [40, 41]. Moreover, IDRA 21 enhanced hippocampal damage *in vivo* in a model of global ischemia [40].

Conversely, PAARMs were also shown to be neuroprotective *in vivo,* and are discussed as medication to promote recovery after traumatic brain injury or stroke [42, 43]. Thus, future will have to tell whether or not, or under which conditions, PAARMs can be neurotoxic.

NMDA receptor agonists

Due to the prominent role of NMDA receptors in mediating CNS plasticity, NMDA receptor agonists have the potential to serve as cognition-enhancing drugs, and indeed this has been demonstrated in rodent experiments [11]. Any beneficial effects to be gained in the clinical situation, however, would likely be masked by the inherent toxicity of NMDA receptor agonists, which are potent excitotoxic compounds. Alternatively, the use of allosteric NMDA receptor channel modulators might reduce the risk of unwanted side-effects. This may be the case as Schwartz and co-workers [44] reported that administration of d-cycloserine (a partial agonist at the modulatory glycine site of the NMDA receptor) to Alzheimer's patients resulted in the augmentation of implicit memory.

NMDA receptor antagonists

NMDA receptor channel antagonists like phencyclidine or ketamine produce acquisition deficits in animal models [45] and they can mimic certain symptoms of schizophrenia in animals [46], as well as in man [47, 48]. Thus, it appears contra-intuitive that receptor blockers could also exert positive effects on cognition. This very action was demonstrated, however, for the non-competitive NMDA receptor blocking drug memantine. This compound binds to the same binding site as phencyclidine or dizocilpine, but its affinity is lower, and the kinetics of binding and unbinding are much faster [49]. Surprisingly Frankiewicz and Parsons [50] reported that memantine increased LTP in a rodent hippocampus slice preparation at concentrations which were clinically relevant.

In Europe, memantine is approved for the treatment of Alzheimer's disease, and two mechanisms of action can be hypothetized. The drug can be postulated to reduce the progression of slow excitotoxicity-related neurodegeneration [51]. The second effect directly affects neurotransmission: In resting neurons, NMDA receptor channels are blocked by extracellular Mg^{2+} ions, which by a voltage-dependent process, occlude the channel pore. Upon depolarization (by the activation of AMPA receptors) this block is released, and downstream processes are initiated, which can culminate in LTP and memory formation. However, under conditions of chronic neurodegeneration and compromised energy metabolism, extracellular levels of glutamate should be slightly increased, and the neuronal membrane potential depolarized. This could lead to partial relief of the Mg^{2+} block of the NMDA receptor. As a consequence, the physiological processes underlying LTP would be functioning abnormally. Memantine might possibly replace the physiological function of Mg^{2+}, thus re-enabling LTP.

Conclusions

Ionotropic glutamate receptors are cornerstones of brain function, and biophysical and pharmacological investigations have left no doubt as to their importance for cognition. Future studies will determine whether it will be possible to develop potent glutamate receptor modulating drugs to treat cognitive disorders.

Acknowledgement
The skillful technical assistance of R. Ewen and S. Kurtze is highly acknowledged. I would also wish to thank Dr. Klaus Klinder (Drug Discovery Support, BI Pharma GmbH & Co KG) for his pharmacokinetic investigations.

References

1 Hayashi T (1952) A physiological study of the epileptic seizures following cortical stimulation in animals and its application to human clinics. *Jpn J Physiol* 3: 46–64
2 Hayashi T (1954) Effects of sodium glutamate on the nervous system. *Keio J Med* 3: 183–192
3 Curtis DR, Phillis JW, Watkins, JC (1959) Chemical excitation of spinal neurons. *Nature* 183: 611–612
4 Davies J, Watkins JC (1979) Selective antagonism of amino acid-induced and synaptic events in the cat spinal cord. *J Physiol* 297: 621–635
5 Parsons CG, Danysz W, Lodge D (2002) Introduction to glutamate receptors, their function and

physiology. In: Lodge D, Danysz W, Parsons CG (eds): Ionotropic glutamate receptors as therapeutic targets. Graham Publishing, Mountain Home, 1–30
6 Forsythe ID, Barnes-Davies M, Brew HM (1995) The calyx of Held: A model for transmission at mammalian glutamatergic synapses. In: Wheal HV, Thomson AM (eds): Excitatory amino acids and synaptic transmission. Academic Press, London, 133–144
7 Kauer JA, Malenka RC, Nicoll RA (1988) A persistent postsynaptic modification mediates long-term potentiation in the hippocampus. *Neuron.* 1: 911–917
8 Bliss TV, Collingridge GL (1993) A synaptic model of memory: Long-term potentiation in the hippocampus. *Nature* 361: 31–39
9 McKinney RA, Capogna M, Durr R, Gahwiler BH, Thompson SM (1999) Miniature synaptic events maintain dendritic spines via AMPA receptor activation. *Nat Neurosci* 2: 44–49
10 Xiong H, Futamura T, Jourdi H, Zhou H, Takei N, Diverse-Pierluissi M, Plevy S, Nawa H (2002) Neurotrophins induce BDNF expression through the glutamate receptor pathway in neocortical neurons. *Neuropharmacology* 42: 903–912
11 Flood JF, Baker ML, Davis JL (1990) Modulation of memory processing by glutamic acid receptor agonists and antagonists. *Brain Res* 521: 197–202
12 Tryphonas L, Iverson F (1990) Neuropathology of excitatory neurotoxins: The domoic acid model. *Toxicol Pathol* 18: 165–169
13 Wienrich M, Brenner M, Loscher W, Palluk R, Pieper M, Potschka H, Weiser T (2001) *In vivo* pharmacology of BIIR 561 CL, a novel combined antagonist of AMPA receptors and voltage-dependent Na(+) channels. *Br J Pharmacol* 133: 789–796
14 Greenamyre JT, Maragos WF, Albin RL, Penney JB, Young AB (1988) Glutamate transmission and toxicity in Alzheimer's disease. *Prog Neuropsychopharmacol Biol Psychiatry* 12: 421–430
15 Cumin R, Bandle EF, Gamzu E, Haefely WE (1982) Effects of the novel compound aniracetam (Ro 13-5057) upon impaired learning and memory in rodents. *Psychopharmacology* 78: 104–111
16 Lee CR, Benfield P (1994) Aniracetam. An overview of its pharmacodynamic and pharmacokinetic properties, and a review of its therapeutic potential in senile cognitive disorders. *Drugs Aging* 4: 257–273
17 Ito I, Tanabe S, Kohda A, Sugiyama H (1990) Allosteric potentiation of quisqualate receptors by a nootropic drug aniracetam. *J Physiol* 424: 533–543
18 Tang CM, Shi QY, Katchman A, Lynch G (1991) Modulation of the time course of fast EPSCs and glutamate channel kinetics by aniracetam. *Science* 254: 288–290
19 Isaacson JS, Nicoll RA (1991) Aniracetam reduces glutamate receptor desensitization and slows the decay of fast excitatory synaptic currents in the hippocampus. *Proc Natl Acad Sci USA* 88: 10936–10940
20 Ozawa S, Iino M, Abe M (1991) Excitatory synapse in the rat hippocampus in tissue culture and effects of aniracetam. *Neurosci Res* 12: 72–82
21 Xiao P, Staubli U, Kessler M, Lynch G (1991) Selective effects of aniracetam across receptor types and forms of synaptic facilitation in hippocampus. *Hippocampus.* 1: 373–380
22 Nicoletti F, Casabona G, Genazzani AA, Copani A, Aleppo G, Canonico PL, Scapagnini U (1992) Excitatory amino acids and neuronal plasticity: modulation of AMPA receptors as a novel substrate for the action of no-otropic drugs. *Funct Neurol* 7: 413–422
23 Grove SJA, Rogers GA, Zhang MQ (2000) Positive modulators of the AMPA receptor. *Exp Opin Ther Patents* 10: 1539–1548
24 Bertolino M, Baraldi M, Parenti C, Braghiroli D, DiBella M, Vicini S, Costa E (1993) Modulation of AMPA/kainate receptors by analogues of diazoxide and cyclothiazide in thin slices of rat hippocampus. *Receptors Channels* 1: 267–278
25 Yamada KA, Hill MW, Hu Y, Covey DF (1998) The diazoxide derivative 7-chloro-3-methyl-3,4-dihydro-2H-1,2,4-benzothiadiazine-S,S-dioxide augments AMPA- and GABA-mediated synaptic responses in cultured hippocampal neurons. *Neurobiol Dis* 5: 196–205
26 Arai A, Guidotti A, Costa E, Lynch G (1996) Effect of the AMPA receptor modulator IDRA 21 on LTP in hippocampal slices. *Neuroreport* 7: 2211–2215
27 Zivkovic I, Thompson DM, Bertolino M, Uzunov D, DiBella M, Costa E, Guidotti A (1995) 7-Chloro-3-methyl-3-4-dihydro-2H-1,2,4 benzothiadiazine S,S-dioxide (IDRA 21): a benzothiadiazine derivative that enhances cognition by attenuating DL-alpha-amino-2,3-dihydro-5-methyl-3-oxo-4-isoxazolepropanoic acid (AMPA) receptor desensitization. *J Pharmacol Exp Ther* 272: 300–309
28 Thompson DM, Guidotti A, DiBella M, Costa E (1995) 7-Chloro-3-methyl-3,4-dihydro-2H-1,2,4-benzothiadiazine S,S-dioxide (IDRA 21), a congener of aniracetam, potently abates phar-

macologically induced cognitive impairments in patas monkeys. *Proc Natl Acad Sci USA* 92: 7667–7671
29 Buccafusco JJ, Weiser T, Winter K, Klinder K, Terry AV Jr (2003) The effects of IDRA 21, a positive modulator of the AMPA receptor, on delayed matching performance by young and aged rhesus monkeys. *Neuropharmacology; in press*
30 Arai A, Kessler M, Rogers G, Lynch G (1996) Effects of a memory-enhancing drug on DL-alpha-amino-3-hydroxy-5-methyl-4-isoxazolepropionic acid receptor currents and synaptic transmission in hippocampus. *J Pharmacol Exp Ther* 278: 627–638
31 Staubli U, Rogers G, Lynch G (1994) Facilitation of glutamate receptors enhances memory. *Proc Natl Acad Sci USA* 91: 777–781
32 Hampson RE, Rogers G, Lynch G, Deadwyler SA (1998) Facilitative effects of the ampakine CX516 on short-term memory in rats: Enhancement of delayed-nonmatch-to-sample performance. *J Neurosci* 18: 2740–2747
33 Hampson RE, Rogers G, Lynch G, Deadwyler SA (1998) Facilitative effects of the ampakine CX516 on short-term memory in rats: correlations with hippocampal neuronal activity. *J Neurosci* 18: 2748–2763
34 Lynch G, Granger R, Ambros-Ingerson J, Davis CM, Kessler M, Schehr R (1997) Evidence that a positive modulator of AMPA-type glutamate receptors improves delayed recall in aged humans. *Exp Neurol* 145: 89–92
35 Lynch G, Kessler M, Rogers G, Ambros-Ingerson J, Granger R, Schehr RS (1996) Psychological effects of a drug that facilitates brain AMPA receptors. *Int Clin Psychopharmacol* 11: 13–19
36 Ingvar M, Ambros-Ingerson J, Davis M, Granger R, Kessler M, Rogers GA, Schehr RS, Lynch G (1997) Enhancement by an ampakine of memory encoding in humans. *Exp Neurol* 146: 553–559
37 Danysz W (2002) CX-516 Cortex pharmaceuticals. Curr Opin Investig Drugs 3: 1081–1088
38 Quirk JC, Nisenbaum ES (2002) LY404187: A novel positive allosteric modulator of AMPA receptors. *CNS Drug Rev* 8: 255–282
39 Lebrun C, Pilliere E, Lestage P (2000) Effects of S 18986-1, a novel cognitive enhancer, on memory performances in an object recognition task in rats. *Eur J Pharmacol* 401: 205–212
40 Yamada KA (1998) AMPA receptor activation potentiated by the AMPA modulator 1-BCP is toxic to cultured rat hippocampal neurons. *Neurosci Lett* 249: 119–122
41 Yamada KA, Covey DF, Hsu CY, Hu R, Hu Y, He YY (1998) The diazoxide derivative IDRA 21 enhances ischemic hippocampal neuron injury. *Ann Neurol* 43: 664–669
42 Bahr BA, Bendiske J, Brown QB, Munirathinam S, Caba E, Rudin M, Urwyler S, Sauter A, Rogers G (2002) Survival signaling and selective neuroprotection through glutamatergic transmission. *Exp Neurol* 174: 37–47
43 Dicou E, Rangon CM, Guimiot F, Spedding M, Gressens P (2003) Positive allosteric modulators of AMPA receptors are neuroprotective against lesions induced by an NMDA agonist in neonatal mouse brain. *Brain Res* 970: 221–225
44 Schwartz BL, Hashtroudi S, Herting RL, Schwartz P, Deutsch SI (1996) d-Cycloserine enhances implicit memory in Alzheimer patients. *Neurology* 46: 420–424
45 Danysz W, Archer T (1994) Glutamate learning, and dementia-selection of evidence. *Amino Acids* 7: 147–163
46 Sams-Dodd F (1999) Phencyclidine in the social interaction test: An animal model of schizophrenia with face and predictive validity. *Rev Neurosci* 10: 59–90
47 Murray JB (2002) Phencyclidine (PCP): A dangerous drug, but useful in schizophrenia research. *J Psychol* 136: 319–327
48 Krystal JH, Karper LP, Seibyl JP, Freeman GK, Delaney R, Bremner JD, Heninger GR, Bowers MB Jr, Charney DS (1994) Subanesthetic effects of the non-competitive NMDA antagonist, ketamine, in humans. Psychotomimetic, perceptual, cognitive, and neuroendocrine responses. *Arch Gen Psychiatry* 51: 199–214
49 Parsons CG, Gruner R, Rozental J, Millar J, Lodge D (1993) Patch clamp studies on the kinetics and selectivity of N-methyl-D-aspartate receptor antagonism by memantine (1-amino-3,5-dimethyladamantan). *Neuropharmacology* 32: 1337–1350
50 Frankiewicz T, Parsons CG (1999) Memantine restores long-term potentiation impaired by tonic N-methyl-D-aspartate (NMDA) receptor activation following reduction of Mg^{2+} in hippocampal slices. *Neuropharmacology* 38: 1253–1259
51 Jain KK (2000) Evaluation of memantine for neuroprotection in dementia. *Exp Opin Invest Drugs* 9: 1397–1406

Cognitive Enhancing Drugs
Edited by J.J. Buccafusco

Cognitive enhancing effects of drugs that target histamine receptors

Arthur A. Hancock and Gerard B. Fox

Neuroscience Research, Abbott Laboratories, 100 Abbott Park Road, Abbott Park, Illinois 60064-6125, USA

Introduction

While histamine effects have been studied for decades in many classical experiments of basic pharmacology, and histamine is often used as a tool to elucidate fundamental pharmacological principles of drug-receptor interactions, it is only within recent years that its importance as a neurotransmitter has been recognized. This is despite findings of the substance within the CNS in the early 1900s, with a differential distribution and enzymes necessary for synthesis (for review, see [1]). As will be described below, new findings suggest important therapeutic roles for histamine and therapeutic potential for drugs that interdict the histaminergic pathways in various neurological diseases, particularly those having a cognitive deficit that these drugs may be anticipated to ameliorate (see Fig. 1 for chemical structures of key compounds discussed).

Histamine as a neurotransmitter

Histamine, a key neurotransmitter in the CNS, is synthesized in hypothalamic neurons (for review see [1, 2]). These neurons are large, localized to the tuberomammillary nucleus and have been shown to have a regular rate of spontaneous firing [2, 3]. Nerve terminals from these cell bodies project mostly ipsilaterally throughout the CNS, especially to the hippocampus and cortex, areas important for memory and cognition, but also to other regions where histamine may play a key role in various physiological and homeostatic functions such as the hypothalamus, basal forebrain and amygdala [2–4]. The neuroanatomical architecture and electrophysiology of the histaminergic system, therefore, are much like those of other biogenic amine neurotransmitters such as dopamine, serotonin, or norepinephrine [2, 4].

Histamine is synthesized from L-histidine, and is degraded by histamine N-methyl transferase [1, 5]. However, the metabolic fate of histamine differs from dopamine, serotonin, or norepinephrine. These amines are normally

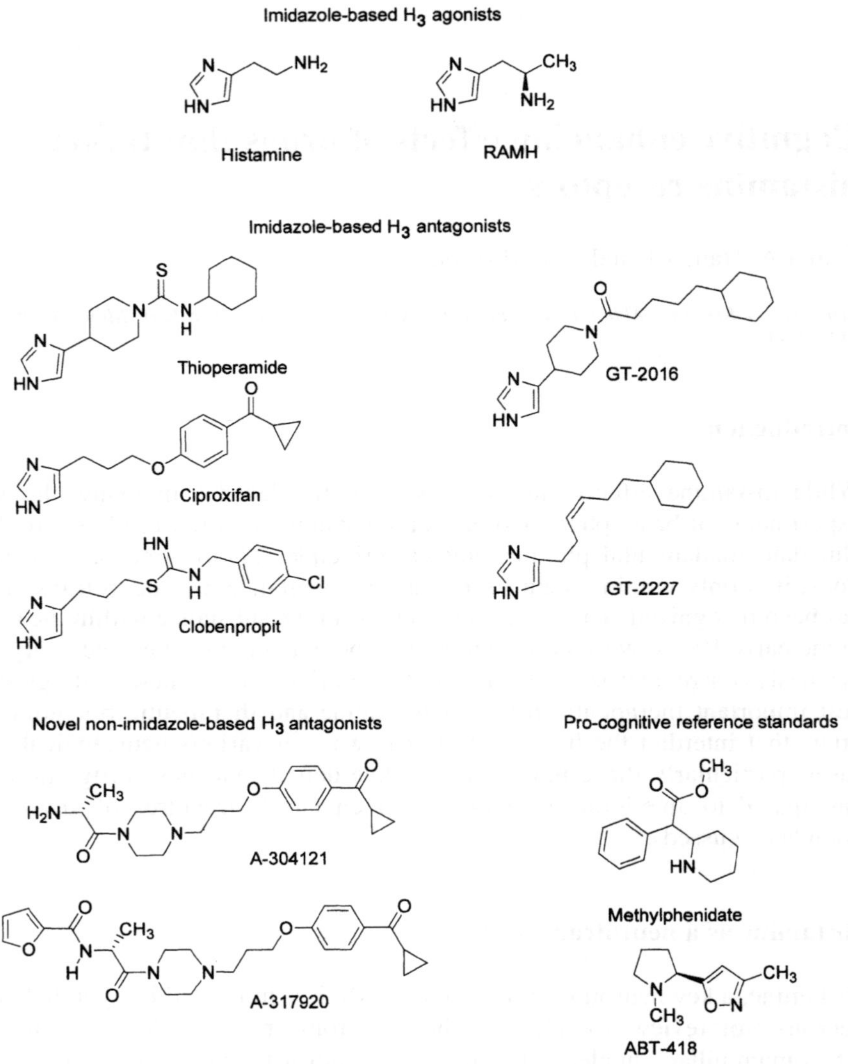

Figure 1. Chemical structures of H_3 receptor ligands and pro-cognitive reference standards discussed in the text.

removed from the synapse by very high affinity re-uptake pumps, sequestered in the presynaptic neuron where they may be repackaged for reuse and/or metabolized by inactivating enzymes such as monoamine oxidase. For histamine, there is no active uptake pump, and histamine is thought to diffuse from the synapse where it is inactivated by extrasynaptic histamine N-methyl transferase [1, 5].

Histaminergic receptor subtypes

Neuronally-released histamine can potentially interact with four subtypes of histaminergic receptors, H_1, H_2, H_3 and H_4, although there is little evidence for H_4 receptors within the CNS [6]. All four of these receptor types have been successfully cloned [7] and shown to belong to family 1 of the superfamily of G-protein coupled receptors (GPCRs), although they differ substantially in their protein sequences, their signal transduction pathways, their pharmacological characteristics and their anatomic distribution and function [7, 8]. H_1 receptors are known for their functional role in atopy and can be blocked by antiallergic "antihistamines". Their signal transduction pathway appears to be primarily the stimulation of calcium mobilization via a G_q protein [9]. H_2 receptors are important for acid secretion within the stomach and are blocked by antiulcer H_2 antagonists. These receptors preferentially couple to G_s to enhance cyclic AMP production [9], whereas H_3 and H_4 receptors appear to couple to Gi, the G-protein linked to inhibition of cyclic AMP production [7].

Within the CNS, H_1 receptors are chiefly neuronal and have a high density within the human neocortex and various limbic structures, although this is not the case in all species [1]. Administration of selective agonists and antagonists (and the well-known clinical somnolence of classical antiallergic antihistamines) have shown an important role for histaminergic stimulation of H_1 receptors in vigilance and alertness, most likely at sites within the ventrolateral posterior hypothalamus [1]. The autoradiographic analysis of H_1 receptor density in the rat brain also shows high densities in thalamic nuclei and the cerebellum [1].

CNS H_2 receptors are distributed in discrete regions, with highest densities in caudate, putamen, nucleus accumbens and cortical areas, with low levels in the hippocampus and globus pallidus [10]. Like H_1 receptors, H_2 receptors are primarily associated with neurons within the CNS [1]. Their functional role within the CNS is unclear. Early electrophysiological studies indicated that stimulation of H_2 receptors might inhibit firing rate in a number of brain regions, although the interpretation of that data has been questioned by the lack of specificity of some of the antagonists employed [1]. In comparison to functions known to be associated with H_1 receptor stimulation, H_2 receptor activation plays a lesser role, both in the number of functional behaviors linked to H_2 receptors, but also in the relative importance of H_2 compared to H_1 receptors in those functions where both receptors may be involved [3]. Knockout studies have similarly failed to elucidate an obvious role for H_2 receptors, since the behavioral phenotype of H_2 knockout animals (decreased spontaneous locomotor activity) is similar to that of H_1 receptor, H_3 receptor and histidine-decarboxylase knockout animals [11].

H_3 receptors have also a diverse CNS distribution pattern among species, as observed for the H_1 receptor [1]. Highest densities are observed in the cerebral cortex, with lesser amounts in the hippocampus, amygdala, anterior olfactory nuclei, nucleus accumbens, striatum and other regions [1]. A recent report has shown similar distribution patterns of H_3 receptors in the rat brain based on either autoradiographic data or the analysis of mRNA levels [12]. H_3 receptors have also been detected by autoradiographic, *in situ* hybridization and functional activation assays in human brain [13].

H_1 and H_2 receptors appear to be localized primarily post-synaptically, where they respond to histamine as described above to enhance vigilance (H_1) or to inhibit neuronal firing (H_2). In contrast, H_3 receptors are located presynaptically as autoreceptors or heteroreceptors to modulate neurotransmitter release [1, 2, 5]. In their role as autoreceptors, H_3 receptor stimulation has been shown to inhibit the synthesis of histamine [2, 5]. In their role as heteroreceptors, H_3 receptor stimulation has also been shown to modulate the release of a variety of neurotransmitters such as acetylcholine, norepinephrine, dopamine, serotonin and gamma-amino butyric acid (GABA), [1, 2, 5].

H_3 receptors display a number of splice variants that differ between species [14]. The differences among the splice variants are located at regions of the receptor that are thought to be important for different signal transduction pathways [14] such that drug actions at different splice isoforms could have different pharmacological results. In either rat or human, the long form of the receptor is predominant in the hippocampus and cortex, areas important for cognition and learning, while the short form has a higher expression level in the hypothalamus [15–17]. In addition, there are minor differences between the amino acid sequences of H_3 receptors from various species [14, 18–20] that are responsible for substantial pharmacological differences between species. Such differences at the pharmacological level have not been noted for either H_1 or H_2 receptors.

Another complexity of histaminergic H_3 receptors is the finding that the H_3 receptor is constitutively active [21–23], meaning that even in the absence of exogenous histamine, the receptor is activated. In the case of an inhibitory receptor such as the H_3 receptor, this inherent activity causes activation of downstream signal transduction pathways leading to tonic inhibition of neurotransmitter release. Therefore, it may be found that a desirable property of any H_3 antagonist used clinically would be to act also as an inverse agonist to not only block the effects of exogenous histamine but to also shift the receptor from its constitutively active (inhibitory) state to a neutral state lacking the tonic inhibition of neurotransmitter release.

Histaminergic neurons in the tuberomammillary region of the hypothalamus receive afferent input from a number of prefrontal and infralimbic regions of the cortex, from all septal regions and from other areas of the hypothalamus, particularly the preoptic and anterior areas [4]. Such afferent input results in the release of a number of neurotransmitters that have the potential to alter histaminergic neuronal function, including orexin, glutamate, serotonin, acetylcholine, GABA, galanin, and histaminergic neurons also receive inputs mediated by NMDA (N-methyl-D-aspartic acid) and AMPA (γ-amino-3-hydroxy-5-methyl-4-isoxazolepropionic acid) receptors [4, 24].

Histaminergic drugs that influence cognitive processes

The central histaminergic system has been shown to play an important role preclinically in rodent cognitive processing. For example, early work demonstrated that histamine administered intracerebroventricularly (*icv*) enhanced consolidation of passive avoidance recall [25]. In contrast, interruption of histaminergic neurotransmission following *icv*, intravenous (*iv*) or oral (*po*) administration of the classic H_1 receptor antagonists promethazine, diphenhydramine and pyrilamine negatively impacted the acquisition and retention of active avoidance responding in rats, effects that were attenuated by activation of the histaminergic system following *icv* administration of histamine itself or, intriguingly, by acetylcholine [26, 27]. Similarly, *icv* injection of histamine has been shown to attenuate hippocampal lesion-induced deficits in passive and active avoidance responding in adult rats [28, 29] and improved performance of aged rats in an active avoidance task [30]. Intracerebroventricular administration of histidine or histamine was also shown to improve social recognition in adult rats [31]. These early experiments demonstrated that histaminergic neurotransmission is involved in higher learning and memory processing in the rat. Many subsequent studies by research groups working in this field have focused on indirectly facilitating the histaminergic system by blocking H_3 auto- and/or heteroreceptors. Consequently, much of the remainder of this section will focus on recent advances in elucidating the role of H_3 receptors in cognition and the identification and testing of new, highly selective and potent H_3 receptor blockers.

H_3 receptor antagonists and arousal/attention

Initial evidence that histamine neurotransmission is involved in arousal, vigilance and attention was derived from clinical observations that 'antihistamine' H_1 receptor antagonists prescribed for treatment of allergic disorders sometimes induced sleepiness and cognitive deficits [32, 33]. The severity of this impairment was found to be correlated with H_1 receptor occupancy in the human brain, as revealed by positron emission tomography (PET) [33]. Preclinically, the arousal effects of hypocretin/orexin were not observed in H_1 receptor knockout mice [34], suggesting the histaminergic system is mediating the arousal effects of orexin. Hypothalamic histaminergic neurons are also known to fire in a rhythmic, synchronous pacemaker-like fashion, a property intrinsic to individual neurons [35] and a circadian rhythm in the release of histamine from the anterior and posterior hypothalamus has been shown in rats [36, 37]. Further, histaminergic neuronal firing has been shown to be correlated with behavioral state in rats, cats and monkeys, with maximal firing corresponding to periods of wakefulness, and reduced activity corresponding to periods of sleep [38].

This association of histaminergic neuronal firing with behavior state has led to significant interest in developing drugs that might modulate attention, vigilance and arousal in humans, potentially without the liabilities of current, clinically-used medications such as stimulants for attention deficit hyperactivity disorder (ADHD). Indeed, the discovery of the selective H_3 receptor agonist (*R*)-α-methylhistamine and the relatively selective antagonists thioperamide, ciproxifan, GT-2016 and GT-2227 has allowed a better understanding of the role of histamine in attention and arousal. For example, during natural arousal in response to handling, H_3 autoreceptors have been shown to modulate histamine release in the rat frontal cortex, with microdialysis probe-infused (*R*)-α-methylhistamine suppressing endogenous histamine release and thioperamide potentiating release [39]. In separate studies in the conscious rat and cat, activation of H_3 receptors with (*R*)-α-methylhistamine increased slow wave EEG activity (indicative of a sleep-like pattern) while opposite effects with thioperamide were interpreted to enhance wakefulness [40, 41]. Similarly, the potent H_3 receptor antagonist ciproxifan significantly decreased slow wave cortical activity following oral administration in the conscious cat [42], which has also been demonstrated following *ip* administration in the conscious rat [43]. In these latter two studies, ciproxifan also improved response accuracy in an adult rat 5-choice serial reaction time task [42] or enhanced performance in a rat pup 5-trial inhibitory avoidance task (attenuated by (*R*)-α-methylhistamine), both of which model aspects of attention and impulsivity (Fig. 2).

Both GT-2016 and GT-2227 have also been shown to improve acquisition of a 10-trial inhibitory avoidance task [44]. A potential confound of some of these early data with antagonists is binding at additional receptors such as $\alpha 2_c$ or at recently discovered H_4 receptors. Highly selective antagonists for H_3 receptors such as A-304121 and A-317920 were recently described by our laboratories; these have been found to be at least as efficacious as methylphenidate (Ritalin®) and ABT-418, both clinically effective for treating ADHD, in the rat pup 5-trial inhibitory avoidance task, offering improved therapeutic ratios over thioperamide and ciproxifan [43, 45, 46] (Fig. 3).

Interestingly, these pro-cognitive effects were observed at doses below the threshold for affecting slow wave activity, indicating that alteration of EEG patterns is not a prerequisite for efficacy in cognition models. In the recently described H_3 receptor knockout mouse [47] thioperamide had no effect on wakefulness, non-REM or REM sleep whereas thioperamide increased wakefulness and decreased non-REM sleep in wild type controls. Taken together, these data confirm the important role of the H_3 receptors as a mediator of wakefulness and support the idea that blockade of H_3 receptor function with highly selective compounds will promote wakefulness and improve attention *in vivo*.

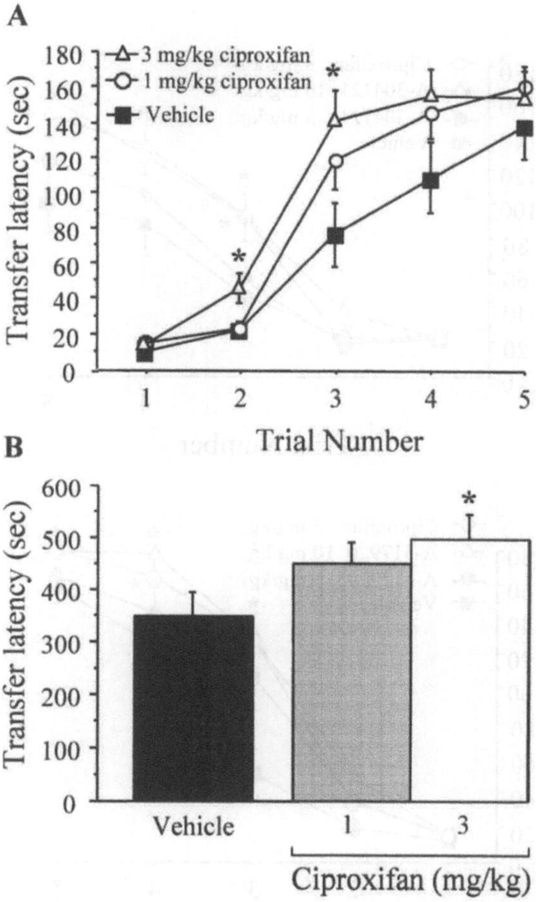

Figure 2. The H_3 receptor antagonist ciproxifan enhances performance of SHR pups in a 5-trial, repeated acquisition, inhibitory avoidance task, reaching significance (* $p < 0.05$ when compared with vehicle-treated controls) as early as the second trial (A). When data for trials 2–5 are summed and meaned, a significant improvement in performance over vehicle-treated controls was evident at 3 mg/kg (B). Rat pups were dosed with antagonist 30 min prior to the first trial. Data are represented by mean ± S.E.M for clarity: statistical calculations used non-parametric analyses. Reproduced with permission from [45].

H_3 receptor antagonists and short-term/working memory

Clear evidence for the involvement of histaminergic neurotransmission in short-term and working memory was demonstrated in a number of early studies utilizing several different animal behavior models. Prast and colleagues [31] showed that *icv* injection of histamine facilitated short-term social memory in the adult rat. Further, histidine, a histamine precursor, significantly increased histamine levels in frontal cortex, hypothalamus and other brain

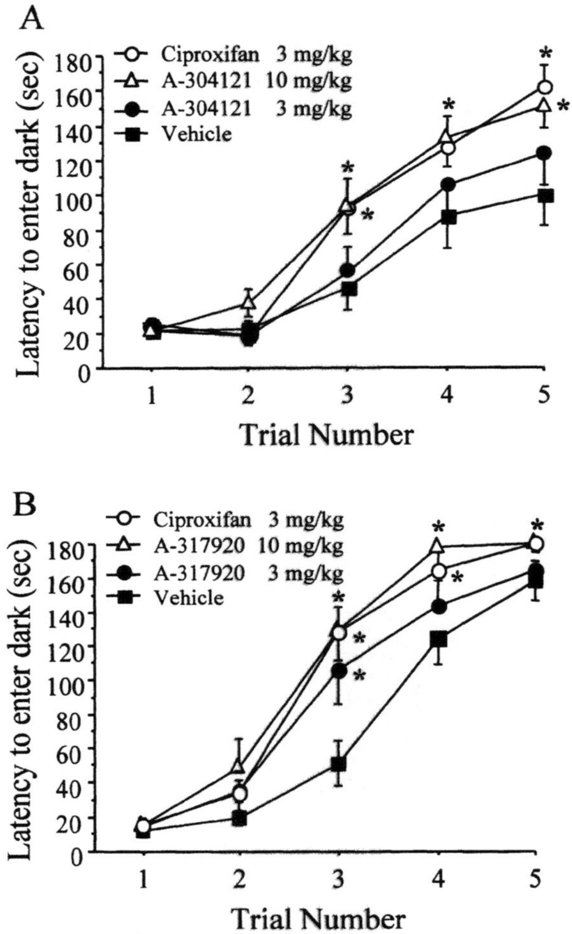

Figure 3. Enhancement of acquisition of a 5-trial inhibitory avoidance response in SHR pups with A-304121 (A) and A-317920 (B), with ciproxifan as a positive control in both instances (* $p < 0.05$ with respect to vehicle-treated controls). Rat pups were dosed with antagonist 30 min prior to the first trial. Data are represented by mean ± S.E.M for clarity: statistical calculations used non-parametric analyses. Reproduced with permission from [43].

regions and enhanced social memory. Taken together, these data indicate that histamine release from histaminergic neurons is involved in this particular task. Also in these studies, social memory impairment was produced by inhibiting synthesis of histamine with *icv* administration of the histidine decarboxylase inhibitor α-fluoromethylhistidine (α-FMH), which also significantly decreased brain histamine levels. A similar impairment was found following *icv* administration of the H_3 receptor agonist immepip, whereas the H_3 receptor antagonist, thioperamide, improved social memory, indicating that modulation of histamine release through blockade of the H_3 autoreceptor is an effective strategy for facilitating short-term memory in the rat. In the radial arm maze, Chen and colleagues [48] utilized a paradigm that addressed the effects of modulating histamine levels on spatial working and reference memory in adult rats. Significant performance deficits were observed following *icv* administration of α-FMH, with a strong correlation between total errors and decreased histamine levels in the cortex and hippocampus, regions of the brain known to be involved in performing this task. Histamine and thioperamide both attenuated the deficits induced by α-FMH in these studies. H_1 receptor antagonists have also been shown to impair radial arm maze performance [49]. In another model of short-term memory, Blandina and co-workers [50] reversed scopolamine-induced deficits in an object recognition task with thioperamide, but showed impairment in a similar task following *ip* administration of the H_3 receptor agonists imetit or (*R*)-α-methylhistamine [51]. Similarly, Orsetti et al. [52, 53] demonstrated enhanced short-term spatial recall in a hippocampus-dependent, two-trial, place recognition Y-maze task in rats following post acquisition intraperitoneal (*ip*) or intra-nucleus basalis injection of thioperamide. Even more recently, we have demonstrated significant enhancement of short-term memory in an adult rat social recognition model with methylphenidate and ciproxifan, as well as with the highly selective H_3 receptor antagonists A-304121 and A-317920 (Fig. 4) [43], a task that is also dependent on an intact hippocampus. As mentioned earlier, A-304121 and A-317920 were also effective in a five-trial inhibitory avoidance task that models aspects of attention and impulsivity in SHR pups, supporting a potential therapeutic role for specific H_3 receptor antagonism in neurological disorders such as attention deficit hyperactivity disorder. Interestingly, A-304121 and A-317920 did not exhibit any of the stimulant-like effects of methylphenidate on EEG or on locomotor activity in habituated or non-habituated mice [43].

H_3 receptor antagonists and spatial reference memory

Histaminergic neurotransmission has been shown to play an important role in delayed recall up to 24 h following acquisition in an inhibitory avoidance model in adult rats and in H_3 receptor knockout mice [47, 51] or 24 h following acquisition in a modified version of an elevated plus maze test in mice [54, 55]. Both of these tasks tap into a form of episodic memory that is dependent on intact frontal cortex and hippocampus. The effects of histamine and H_3 receptor ligands are less clear with respect to long-term, spatial reference memory, however. For example, there is a paucity of information on the effects of H_3 receptor ligands in the water maze, despite this model being perhaps the most commonly employed test of spatial learning and memory. An early report on the effects the agonist (*R*)-α-methylhistamine described, surprisingly, an attenuation of a scopolamine-induced deficit in a hidden platform version of the task [56]. The authors suggested a possible post-synaptic H_3 receptor

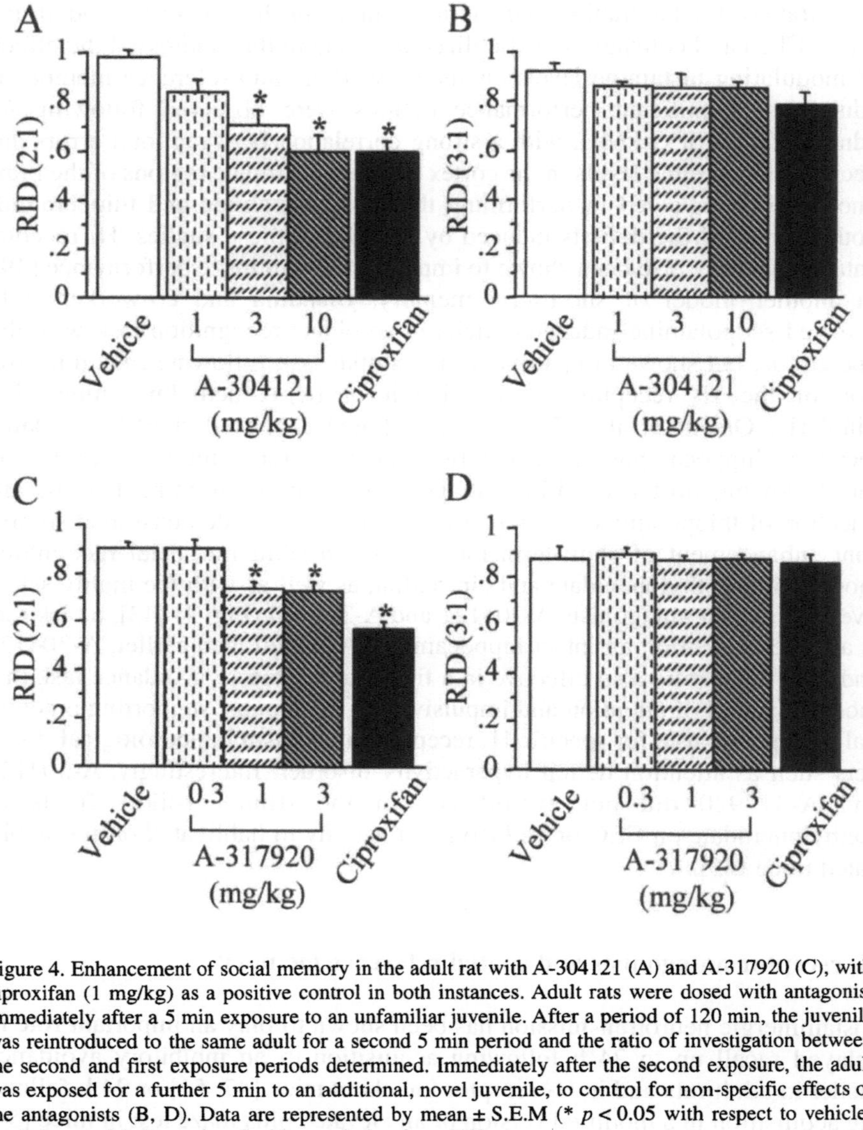

Figure 4. Enhancement of social memory in the adult rat with A-304121 (A) and A-317920 (C), with ciproxifan (1 mg/kg) as a positive control in both instances. Adult rats were dosed with antagonist immediately after a 5 min exposure to an unfamiliar juvenile. After a period of 120 min, the juvenile was reintroduced to the same adult for a second 5 min period and the ratio of investigation between the second and first exposure periods determined. Immediately after the second exposure, the adult was exposed for a further 5 min to an additional, novel juvenile, to control for non-specific effects of the antagonists (B, D). Data are represented by mean ± S.E.M (* $p < 0.05$ with respect to vehicle-treated controls). Reproduced with permission from [43].

mechanism, but no evidence has ever been shown for this. Since blockade of H_3 receptors causes increased release of acetylcholine [57], models that specifically address the interactions of the histaminergic and cholinergic systems are of particular interest. We have developed a different version of the water maze task, a two platform, visual discrimination test in which an adult rat is required to distinguish, using extramaze spatial cues, a stable escape platform from a inescapable, floating platform. Components of spatial reference and working memory can be assessed using this task, which is dependent on an intact cholinergic input to the hippocampus [58, 59]. In our hands, both thioperamide and ciproxifan reversed a scopolamine-induced deficit when administered *ip* once daily over the five testing days in this model (Fig. 5). This represents the first time, we believe, that procognitive effects of H_3 receptor blockade have been shown in the water maze. Further, we have also utilized the Barnes circular maze to evaluate the role of H_3 receptors in spatial reference memory. In this task, mice are trained to navigate a large, brightly lit circular plastic disk for a dark tunnel hidden below one of forty holes located around the perimeter of the maze. Unlike the water maze in which escape from water is the primary motivation to find the escape platform, in the Barnes maze the animal's motivation stems from an inherent survival instinct to avoid open, brightly lit spaces [60]. In the Barnes maze, ciproxifan (3 mg/kg *ip*) significantly attenuated a scopolamine-induced performance deficit in the acquisition of this task when administered to mice once daily for the four-day duration of the test. In addition, ciproxifan-treated mice that did not receive scopolamine exhibited a tendency toward improved acquisition of the Barnes maze task compared with vehicle controls. When these animals were reevaluated drug-free eight days after the last training session, a significant improvement in recall of the tunnel location was observed compared with animals that had previously received vehicle alone (Fig. 5).

H_3 receptor antagonists and sensory processing

Deficits in sensorimotor gating are frequently observed in patients presenting with Huntington's disease [61], Tourette's syndrome, obsessive-compulsive disorder [62], and schizophrenia (for review, see [63]). It is possible that deficits in attention and cognition also observed in these disorders may be related to impairments in sensorimotor processing. Recent evidence also suggests a role for the histaminergic system in psychosis. For example, overdose of first generation H_1R antagonists produced toxic psychoses with hallucinations resembling those observed in schizophrenia [64]. Central histaminergic systems have also recently been implicated in the pathophysiology of schizophrenia [65], although the receptor subtype involved is not clear. Given that blockade of H_3Rs consistently improves cognition, particularly with attention components, it seems plausible that H_3Rs may play a role in sensorimotor processing. One way of assessing deficits in sensorimotor gating is to measure prepulse inhibition (PPI) of a startle response, a phenomenon observed across species in which a startle response to an acoustic tone is attenuated when preceded by a weaker stimulus tone [66]. Deficits in PPI are believed to reflect the deficient sensorimotor gating underlying the sensory flooding and disrupted cognition in schizophrenia [63]. We have recently evaluated the H_3R antagonists ciproxifan and thioperamide in two mouse strains exhibiting a naturally

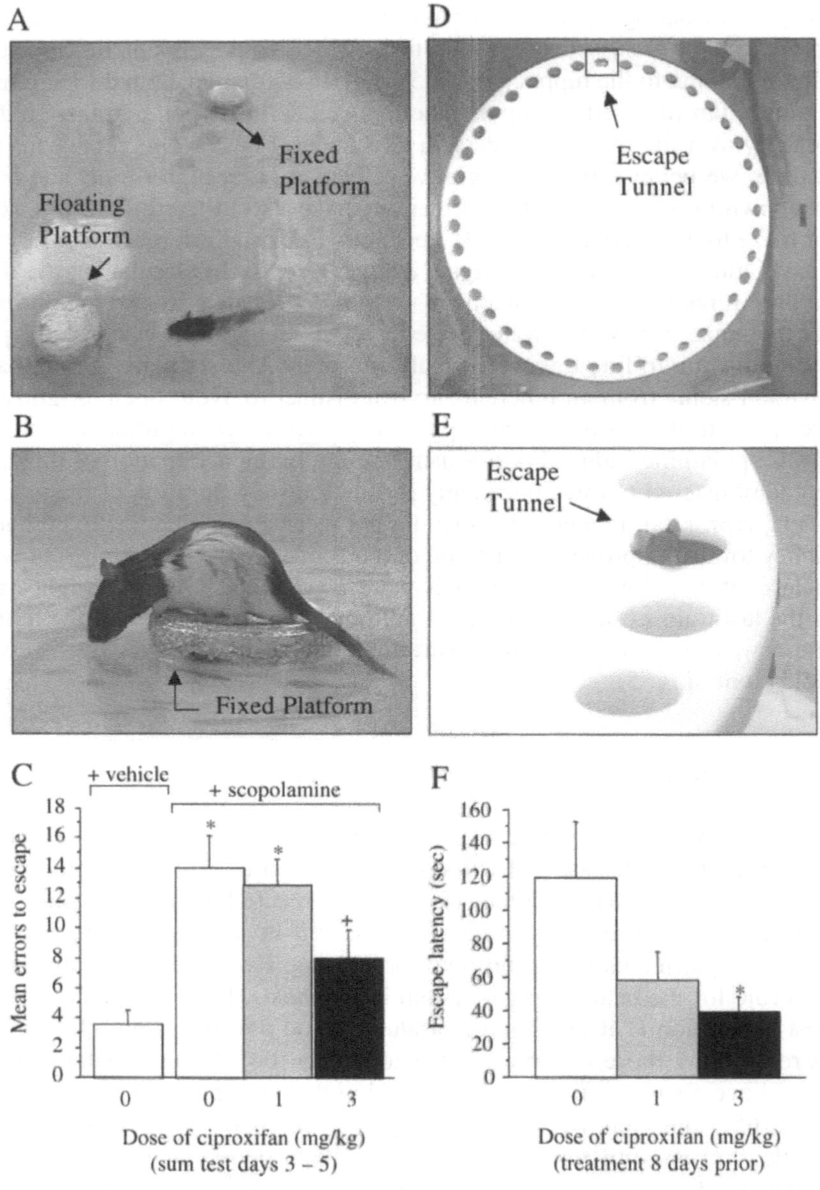

Figure 5. The rat water maze with two visible platforms, one stable and one fixed, is shown (A). Escape from the water is possible only for the fixed platform (B). Ciproxifan administered prior to acquisition sessions over five days attenuated scopolamine-induced increase in mean error number, as shown for data collapsed over days 3–5 (C). The mouse Barnes maze is also shown (D) with bright, overhead illumination, motivation for finding the darkened escape tunnel hidden beneath one of forty holes around the perimeter (E). Ciproxifan administered during acquisition sessions over four days enhanced recall for the tunnel location when the same mice were re-evaluated for recall in a drug-free state eight days after the last acquisition session (F). (* $p < 0.05$ with respect to vehicle-treated controls). Adapted from Society for Neuroscience (2003), abstract 938.3.

occurring deficit in PPI. Interestingly, both compounds enhanced PPI in DBA/2 mice, while thioperamide also enhanced PPI in C57Bl/6 mice to a degree comparable with the atypical antipsychotic risperidone (unpublished observations). Taken together, these preliminary data suggest that drugs targeting blockade of H_3Rs have additional potential in neurological disorders with deficits in sensorimotor processing.

H_3 receptor antagonists and contextual fear conditioning

Very recent evidence from the laboratory of Patrizio Blandina suggests that the histaminergic system and H_3Rs may play an important role in fear-associated memory. For example, direct injection of the H_3R antagonists ciproxifan, clobenpropit and thioperamide into the basolateral amygdala inhibited the release of acetylcholine and adversely affected memory for contextual fear conditioning [67], whereas direct injection of the H_3R agonists (*R*)-α-methylhistamine or immepip produced the opposite effects [68]. These data are the first to suggest that modulation of cholinergic tone in the amygdala through an interaction with H_3Rs can affect the consolidation of memory associated with a fear response and indicate a potential role for H_3R ligands in treating disorders such as social phobias or generalized anxiety disorders.

Potential clinical utility of histaminergic-related cognition enhancers

To date, no drugs with specific effects on histaminergic systems and cognition enhancing effects are clinically available. Historically, drugs acting via histaminergic pathways have, if anything, had negative effects on CNS function in the context of soporific effects of classical H_1 receptor antagonists, although there has been the suggestion of a positive effect of the H_2 receptor antagonist, famotidine, in schizophrenia [65]. In addition, there is not a clear relationship between changes in histamine levels (either in the CNS or periphery) and cognitive disorders [65]. For example, in Alzheimer's disease (AD), various studies have reported either increased or diminished histamine content in the brains of patients with AD (for review, see [65]). On the other hand, schizophrenic patients have been reported to be hyporesponsive to histamine, to have elevated H_2 receptor density in areas implicated in their disease, to have a reduced number of H_1 receptors and to have elevated levels of the histamine metabolite, *tele*-methylhistamine (an indicator of CNS histaminergic activation), in their cerebrospinal fluid [65]. Finally, in an animal model of neurodegeneration [69], injections of β-amyloid reduced histamine levels in cortex and/or hypothalamus (but not hippocampus), and animals exhibited cognitive deficits, with reversal of both parameters by the procognitive compound S12024 [69]. These data are consistent with findings of disruption of histaminergic neurotransmission as a result of neurodegenerative pathology in Alzheimer's disease [70].

Lesion studies and/or chemical disruption of histamine synthesis have shown discrepant effects on cognitive behavior in a variety of animal models [71]. Linkage to H_3 receptors was suggested by the effects of several specific H_3 agonists that disrupted cognitive behavior in several animal models of learning and memory, while pro-cognitive effects have been observed in other animal models [71]. In addition, several studies have suggested a potential role for the histaminergic system to modulate a variety of neurotransmitters that may be linked to attention and vigilance and may be important components of the cognitive deficits found in patients with ADHD [71]. Moreover, data cited above shows a potential for H_3 antagonists to ameliorate cognitive and attentional deficits in animal models that may be suggestive for a role of these compounds in ADHD [44, 45, 71].

Looking toward the future, a number of compounds have been recently described that illustrate the potential for agents that interact with the histaminergic system of the brain. These come from a broad variety of antagonists and inverse agonists of the H_3 receptor that seem to offer the most promise as new agents to treat cognitive disorders. These compounds are emerging from a number of laboratories and chemical classes. While imidazole-based compounds were the first potent and selective ligands at H_3 receptors [42, 72–78], newer agents are becoming known from diverse chemical series that lack the imidazole moiety [79–89] known to be a liability for drug-drug interactions [90]. Thus, there exists the potential that compounds similar to those described above will have useful pharmacological and therapeutic activity in the treatment of the serious cognitive deficiencies in ADHD, AD and schizophrenia.

References

1 Schwartz JC, Arrang JM, Garbarg M, Pollard H, Ruat M (1991) Histaminergic transmission in the mammalian brain. *Physiol Rev* 71: 1–51

2 Blandina P, Bacciottini L, Giovannini M, Mannaioni P (1998) H3 receptor modulation of the release of neurotransmitters *in vivo*. In: R Leurs, H Timmerman (eds): *The histamine H3 receptor*, Elsevier: Amsterdam, 27–40

3 Onodera K, Yamatodani A, Watanabe T, Wada H (1994) Neuropharmacology of the histaminergic neuron system in the brain and its relationship with behavioral disorders. *Prog Neurobiol* 42: 685–702

4 Brown RE, Stevens DR, Haas HL (2001) The physiology of brain histamine. *Prog Neurobiol* 63: 637–672

5 Barnes W, Boyd D, Hough L (2001) Dynamics of histamine H3 receptor antagonists on brain histamine metabolism: do all histamine H3 receptor antagonists act at a single site? *Eur J Pharmacol* 431: 215–221

6 Nakamura T, Itadani H, Hidaka Y, Ohta M, Tanaka K (2000) Molecular cloning and characterization of a new human histamine receptor, HH4R. *Biochem Biophys Res Comm* 279: 615–620

7 Hough L (2001) Genomics meets histamine receptors: New subtypes, new receptors. *Mol Pharmacol* 59: 415–419

8 Lovenberg TW, Roland BL, Wilson SJ, Jiang X, Pyati J, Huvar A, Jackson MR, Erlander MG (1999) Cloning and functional expression of the human histamine H3 receptor. *Mol Pharmacol* 55: 1101–1107

9 Esbenshade TE, Kang CH, Krueger KM, Miller TR, Witte DG, Roch JM, Masters JN, Hancock

AA (2003) Differential activation of dual signaling responses by human H1 and H2 histamine receptors. *J Recept Signal Transduct Res* 23: 17–31

10 Honrubia MA, Vilaro MT, Palacios JM, Mengod G (2000) Distribution of the histamine H2 receptor in monkey brain and its mRNA localization in monkey and human brain. *Synapse* 38: 343–354

11 Watanabe T (2001) Studies on histamine with L-histidine decarboxylase, a histamine-forming enzyme, as a probe: from purification to gene knockout. *Nippon Yakurigaku Zasshi* 118: 159–169

12 Pillot C, Heron A, Cochois V, Tardivel-Lacombe J, Ligneau X, Schwartz JC, Arrang JM (2002) A detailed mapping of the histamine H3 receptor and its gene transcripts in rat brain. *Neurosci* 114: 173–193

13 Anichtchik OV, Peitsaro N, Rinne JO, Kalimo H, Panula P (2001) Distribution and modulation of histamine H3 receptors in basal ganglia and frontal cortex of healthy controls and patients with Parkinson's disease. *Neurobiol Dis* 8: 707–716

14 Hancock AA, Esbenshade TE, Krueger KM, Yao BB (2003) Genetic and pharmacological aspects of histamine H3 receptor heterogeneity. *Life Sci* 73: 3043–3072

15 Wellendorph P, Goodman MW, Burstein ES, Nash NR, Brann MR, Weiner DM (2002) Molecular cloning and pharmacology of functionally distinct isoforms of the human histamine H3 receptor. *Neuropharmacol* 42: 929–940

16 Morisset S, Sasse A, Gbahou F, Heron A, Ligneau X, Tardivel-Lacombe J, Schwartz JC, Arrang JM (2001) The rat H3 receptor: Gene organization and multiple isoforms. *Biochem Biophys Res Comm* 280: 75–80

17 Drutel G, Peitsaro N, Karlstedt K, Wieland K, Smit M, Timmerman H, Panula P, Leurs R (2001) Identification of rat H3 receptor isoforms with different brain expression and signaling properties. *Mol Pharmacol* 59: 1–8

18 Lovenberg TW, Pyati J, Chang H, Wilson SJ, Erlander MG (2000) Cloning of rat histamine H3 receptor reveals distinct species pharmacological profiles. *J Pharmacol Exp Ther* 293: 771–778

19 Ligneau X, Morisset S, Tardivel-Lacombe J, Gbahou F, Ganellin CR, Stark H, Schunack W, Schwartz J-C, Arrang JM (2000) Distinct pharmacology of rat and human histamine H3 receptors; role of two amino acids in the third transmembrane domain. *Br J Pharmacol* 131: 1247–1250

20 Yao B, Hutchins CW, Carr TL, Cassar S, Miller TR, Kang CH, Black LA, Faghih R, Bennani YL, Esbenshade TE, Hancock AA (2003) Direct ligand interactions with transmembrane domain 3 of the histamine H3 receptor protein. *Neuropharmacol* 44: 773–786

21 Morisset S, Rouleau A, Ligneau X, Gbahou F, Tardivel-Lacombe J, Stark H, Schunack W, Ganellin CR, Schwartz JC, Arrang JM (2000) High constitutive activity of native H3 receptors regulates histamine neurons in brain. *Nature* 408: 860–864

22 Rouleau A, Ligneau X, Tardivel-Lacombe J, Morriset S, Gbahou F, Schwartz J-C, Arrang J-M (2002) Histamine H3-receptor mediated [^{35}S]GTPgS binding: Evidence for constitutive activity of the recombinant and native rat and human H3 receptors. *Br J Pharmacol* 135: 383–392

23 Wieland K, Bongers G, Yamamoto Y, Hashimoto T, Yamatodani A, Menge WMBP, Timmerman H, Lovenberg TW, Leurs R (2001) Constitutive activity of histamine H3 receptors stably expressed in SK-N-MC cells: Display of agonism and inverse agonism by H3 antagonists. *J Pharmacol Exp Ther* 299: 908–914

24 Schlicker E, Kathmann M (1998) Modulation of *in vitro* neurotransmission in the CNS and in the retina via H3 heteroreceptors. In: R Leurs, H Timmerman (eds): *The histamine H3 receptor.* Elsevier, Amsterdam, 13–26

25 de Almeida MA, Izquierdo I (1986) Memory facilitation by histamine. *Arch Int Pharmacodyn Ther* 283: 193–198

26 Kamei C, Chung YH, Tasaka K (1990) Influence of certain H1-blockers on the step-through active avoidance response in rats. *Psychopharmacol* 102: 312–318

27 Kamei C, Tasaka K (1991) Participation of histamine in the step-through active avoidance response and its inhibition by H1-blockers. *Jpn J Pharmacol* 57: 473–482

28 Chen Z, Sugimoto Y, Kamei C (1997) Effects of histamine and its related compounds on impairment of passive avoidance response following hippocampal lesions in rats. *J Brain Sci* 23: 225–240

29 Kamei C, Chen Z, Nakamura S, Sugimoto Y (1997) Effects of intracerebroventricular injection of histamine on memory deficits induced by hippocampal lesions in rats. *Methods Find Exp Clin Pharmacol* 19: 253–259

30 Kamei C, Tasaka K (1993) Effect of histamine on memory retrieval in old rats. *Biol Pharm Bull* 16: 128–132

31 Prast H, Argyriou A, Philippu A (1996) Histaminergic neurons facilitate social memory in rats. *Brain Res* 734: 316–318
32 Kay GG (2000) The effects of antihistamines on cognition and performance. *J Allergy Clin Immunol* 105: S622–S627
33 Tashiro M, Mochizuki H, Iwabuchi K, Sakurada Y, Itoh M, Watanabe T, Yanai K (2002) Roles of histamine in regulation of arousal and cognition: Functional neuroimaging of histamine H1 receptors in human brain. *Life Sci* 72: 409–414
34 Huang ZL, Qu WM, Li WD, Mochizuki T, Eguchi N, Watanabe T, Urade Y, Hayaishi O (2001) Arousal effect of orexin A depends on activation of the histaminergic system. *Proc Natl Acad Sci* 98: 9965–9970
35 Stevens DR, Eriksson KS, Brown RE, Haas HL (2001) The mechanism of spontaneous firing in histamine neurons. *Behav Brain Res* 124: 105–112
36 Mochizuki T, Yamatodani A, Okakura K, Horii A, Inagaki N, Wada H (1992) Circadian rhythm of histamine release from the hypothalamus of freely moving rats. *Physiol Behav* 51: 391–394
37 Prast H, Dietl H, Philippu A (1992) Pulsatile release of histamine in the hypothalamus of conscious rats. *J Auton Nerv Syst* 39: 105–110
38 Tuomisto L, Lozeva V, Valjakka A, Lecklin A (2001) Modifying effects of histamine on circadian rhythms and neuronal excitability. *Behav Brain Res* 124: 129–135
39 Westerink BHC, Cremers T, De Vries JB, Liefers H, Tran N, De Boer P (2002) Evidence for activation of histamine H-3 autoreceptors during handling stress in the prefrontal cortex of the rat. *Synapse* 43: 238–243
40 Lin JS, Sakai K, Vanni-Mercier G, Arrang JM, Garbarg M, Schwartz JC, Jouvet M (1990) Involvement of histaminergic neurons in arousal mechanisms demonstrated with H3-receptor ligands in the cat. *Brain Res* 523: 325–330
41 Monti JM, Jantos H, Boussard M, Altier H, Orellana C, Olivera S (1991) Effects of selective activation or blockade of the histamine H3 receptor on sleep and wakefulness. *Eur J Pharmacol* 205: 283–287
42 Ligneau X, Lin J, Vanni-Mercier G, Jouvet M, Muir JL, Ganellin CR, Stark H, Elz S, Schunack W, Schwartz J (1998) Neurochemical and behavioral effects of ciproxifan, a potent histamine H3-receptor antagonist. *J Pharmacol Exp Ther* 287: 658–666
43 Fox GB, Pan JB, Radek RJ, Lewis AM, Bitner RS, Esbenshade TA, Faghih R, Bennani YL, Williams M, Yao BB et al. (2003) Two novel and selective non-imidazole H3 receptor antagonists A-304121 and A-317920: II. *In vivo* behavioral and neurophysiological characterization. *J Pharmacol Exp Ther* 305: 887–896
44 Tedford C (1998) Clinical application of histamine H3 receptor antagonists in learning and memory disorders. In: R Leurs, H Timmerman (eds): *The histamine H3 receptor*. Elsevier, Amsterdam, 269–286
45 Fox GB, Pan JB, Esbenshade TA, Bennani YL, Black LA, Faghih R, Hancock AA, Decker MW (2002) Effects of histamine H3 receptor ligands GT-2331 and ciproxifan in a repeated acquisition avoidance response in the spontaneously hypertensive rat pup. *Behav Brain Res* 131: 151–161
46 Komater VA, Browman KE, Curzon P, Hancock AA, Decker MW, Fox GB (2003) H3 receptor blockade by thioperamide enhances cognition in rats without inducing locomotor sensitization. *Psychopharmacol* 167: 363–372
47 Toyota H, Dugovic C, Koehl M, Laposky AD, Weber C, Ngo K, Wu Y, Lee DH, Yanai K, Sakurai E et al. (2002) Behavioral characterization of mice lacking histamine H3 receptors. *Mol Pharmacol* 62: 389–397
48 Chen Z, Sugimoto Y, Kamei C (1999) Effects of intracerebroventricular injection of alpha-fluoromethylhistidine on radial maze performance in rats. *Pharmacol Biochem Behav* 64: 513–518
49 Chen Z, Chen JQ, Kamei C (2001) Effect of H1-antagonists on spatial memory deficit evaluated by 8-arm radial maze in rats. *Acta Pharmacol Sin* 22: 609–613
50 Giovannini MG, Bartolini L, Bacciottini L, Greco L, Blandina P (1999) Effects of histamine H3 receptor agonists and antagonists on cognitive performance and scopolamine-induced amnesia. *Behav Brain Res* 104: 147–155
51 Blandina P, Giorgetti M, Bartolini L, Cecchi M, Timmerman H, Leurs R, Pepeu G, Giovannini MG (1996) Inhibition of cortical acetylcholine release and cognitive performance by histamine H3 receptor activation in rats. *Brit J Pharmacol* 119: 1656–1664
52 Orsetti M, Ghi P, Di Carlo G (2001) Histamine H3-receptor antagonism improves memory reten-

tion and reverses the cognitive deficit induced by scopolamine in a two-trial place recognition task.PG - 235-42. *Behav Brain Res* 124:
53 Orsetti M, Ferretti C, Gamalero R, Ghi P (2002) Histamine H3-receptor blockade in the rat nucleus basalis magnocellularis improves place recognition memory. PG - 133-7. *Psychopharmacology (Berl)* 159: 133–137
54 Miyazaki S, Imaizumi M, Onodera K (1995) Effects of thioperamide, a histamine H3-receptor antagonists, on a scopolamine-induced learning deficit using an elevated plus-maze test in mice. *Life Sci* 57: 2137–2144
55 Onodera K, Miyazaki S, Imaizumi M, Stark H, Schunack W (1998) Improvement by FUB 181, a novel histamine H3-receptor antagonist, of learning and memory in the elevated plus-maze test in mice. *Naunyn Schmied Arch Pharmacol* 357: 508–513
56 Smith CP, Hunter AJ, Bennett GW (1994) Effects of (R)-alpha-methylhistamine and scopolamine on spatial learning in the rat assessed using a water maze. *Psychopharmacol* 114: 651–656
57 Hill SJ, Ganellin CR, Timmerman H, Schwartz JC, Shankley NP, Young JM, Schunack W, Levi R, Haas HL (1997) International Union of Pharmacology. XIII. Classification of histamine receptors. *Pharmacol Rev* 49: 253–278
58 Decker MW, Curzon P, Brioni JD (1995) Influence of separate and combined septal and amygdala lesions on memory, acoustic startle, anxiety, and locomotor activity in rats. *Neurobiol Learn Mem* 64: 156–168
59 Decker MW, Curzon P, Brioni JD, Arneric SP (1994) Effects of ABT-418, a novel cholinergic channel ligand, on place learning in septal-lesioned rats. *Eur J Pharmacol* 261: 217–222
60 Fox GB, Fan L, LeVasseur RA, Faden AI (1998) Effect of traumatic brain injury on mouse spatial and nonspatial learning in the Barnes circular maze. *J Neurotrauma* 15: 1037–1046
61 Swerdlow NR, Paulsen J, Bacciottini L, Butters N, Geyer MA, Swenson MR (1995) Impaired prepulse inhibition of acoustic and tactile startle in patients with Huntington's disease. *J Neurol Neurosurg Psych* 58: 192–200
62 Castellanos FX, Fine EJ, Kaysen D, Marsh WL, Rapoport JL, Hallett M (1996) Sensorimotor gating in boys with Tourette's syndrome and ADHD: Preliminary results. *Biol Psych* 39: 33–41
63 Geyer MA, Krebs-Thomson K, Braff DL, Swerdlow NR (2001) Pharmacological studies of prepulse inhibition models of sensorimotor gating deficits in schizophrenia: A decade in review. *Psychopharmacol* 156: 117–154
64 Sangalli BC (1997) Role of the central histaminergic neuronal system in the CNS toxicity of the first generation H1-antagonists. *Prog Neurobiol* 52: 145–157
65 Fernandez-Novoa L, Cacabelos R (2001) Histamine function in brain disorders. *Behav Brain Res* 124: 213–233
66 Fox GB, Curzon P, Decker MW (2001) The behavioral assessment of sensorimotor processes in the mouse: Acoustic startle, locomotor activity, rotarod and beam walking. In: J. Buccafusco (eds): *Methods of behavior analysis in neuroscience*. CRC Press, New York, 27–49
67 Passani MB, Cangioli I, Baldi E, Bucherelli C, Mannaioni PF, Blandina P (2001) Histamine H3 receptor-mediated impairment of contextual fear conditioning and *in vivo* inhibition of cholinergic transmission in the rat basolateral amygdala. *Eur J Neurosci* 14: 1522–1532
68 Cangioli I, Baldi E, Mannaioni PF, Bucherelli C, Blandina P, Passani MB (2002) Activation of histaminergic H3 receptors in the rat basolateral amygdala improves expression of fear memory and enhances acetylcholine release. *Eur J Neurosci* 16: 521–528
69 Miguel-Hidalgo JJ, Vecino B, Fernandez-Novoa L, Alvarez A, Cacabelos R (1998) Neuroprotective role of S12024 against neurodegeneration in the rat dentate gyrus. *Eur Neuropsychopharmacol* 8: 203–208
70 Higuchi M, Yanai K, Okamura N, Meguro K, Arai H, Itoh M, Iwata R, Ido T, Watanabe T, Sasaki H (2000) Histamine H1 receptors in patients with Alzheimer's disease assessed by positron emission tomography. *Neurosci* 99: 721–729
71 Passani MB, Bacciottini L, Mannaioni PF, Blandina P (2000) Central histaminergic system and cognition. *Neurosci Biobehav Rev* 24: 107–113
72 Menge WMPB, H Timmerman (1998) Substituted imidazoles, the key to histaminergic receptors. In: In: R Leurs, H Timmerman (eds): *The histamine H3 receptor*. Elsevier, Amsterdam, 145–148
73 Yates SL, Phillips JG, Gregory R, Pawlowski GP, Fadnis L, Khan MA, Ali SM, Tedford CE (1999) Identification and pharmacological characterization of a series of new 1H-4-substituted-imidazoyl histamine H3 receptor ligands. *J Pharmacol Exp Ther* 289: 1151–1159
74 Tedford CE, Yates SL, Pawlowski GP, Nalwalk JW, Hough LB, Khan MA, Phillips JG, Durant GJ,

Frederickson RC (1995) Pharmacological characterization of GT-2016, a non-thiourea-containing histamine H3 receptor antagonist: *in vitro* and *in vivo* studies. *J Pharmacol Exp Ther* 275: 598–604

75 Grassmann S, Sadek B, Ligneau X, Elz S, Ganellin CR, Arrang JM, Schwartz JC, Stark H, Schunack W (2002) Progress in the proxifan class: Heterocyclic congeners as novel potent and selective histamine H3-receptor antagonists. *Eur J Pharm Sci* 15: 367–378

76 Stark H, Sadek B, Krause M, Huls A, Ligneau X, Ganellin CR, Arrang JM, Schwartz JC, Schunack W (2000) Novel histamine H3-receptor antagonists with carbonyl-substituted 4-(3-(phenoxy)propyl)-1H-imidazole structures like ciproxifan and related compounds. *J Med Chem* 43: 3987–3994

77 Stark H, Ligneau X, Sadek B, Ganellin CR, Arrang JM, Schwartz JC, Schunack W (2000) Analogues and derivatives of ciproxifan, a novel prototype for generating potent histamine H3-receptor antagonists. *Bioorg Med Chem Lett* 10: 2379–2382

78 Kathmann M, Schlicker E, Marr I, Werthwein S, Stark H, Schunack W (1998) Ciproxifan and chemically related compounds are highly potent and selective histamine H3-receptor antagonists. *Naunyn Schmied Arch Pharmacol* 358: 623–627

79 Ganellin CR, Leurquin F, Piripitsi A, Arrang JM, Garbarg M, Ligneau X, Schunack W, Schwartz JC (1998) Synthesis of potent non-imidazole histamine H3-receptor antagonists. *Archiv der Pharmazie* 331: 395–404

80 Walczynski K, Guryn R, Zuiderveld OP, Timmerman H (1999) Non-imidazole histamine H3 ligands. Part I. Synthesis of 2-(1-piperazinyl)- and 2-(hexahydro-1H-1,4-diazepin-1-yl)benzothiazole derivatives as H3-antagonists with H1 blocking activities. *Farmaco* 54: 684–694

81 Walczynski K, Guryn R, Zuiderveld OP, Timmerman H (1999) Non-imidazole histamine H3 ligands, Part 2: New 2-substituted benzothiazoles as histamine H3 antagonists. *Archiv der Pharmazie* 332: 389–398

82 Linney I, Buck I, Harper E, Kalindjian S, Pether M, Shankley N, Watt G, Wright P (2000) Design, synthesis, and structure-activity relationships of novel non-imidazole histamine H3 receptor antagonists. *J Med Chem* 43: 2362–2370

83 Meier G, Apelt J, Reichert U, Grassmann S, Ligneau X, Elz S, Lerquin F, Ganellin C, Schwartz J, Schunack W et al. (2001) Influence of imidazole replacement in different structural classes of histamine H3-receptor antagonists. *Eur J Pharm Sci* 13: 249–259

84 Faghih R, Dwight W, Gentles R, Phelan K, Esbenshade T, Ireland L, Miller T, Kang C, Fox GB, Gopalkrishnan SM (2002) Structure-activity relationships of non-imidazole H3 receptor ligands: Part 1. *Bioorg Med Chem Lett* 12: 2031–2034

85 Faghih R, Dwight W, Black L, Liu H, Gentles R, Phelan K, Esbenshade T, Ireland L, Miller T, Kang C et al. (2002) Structure-activity relationships of non-imidazole H3 receptor ligands. Part 2: Preference for D-amino acids motifs. *Bioorg Med Chem Lett* 12: 2035–2037

86 Vasudevan A, Conner SE, Gentles RG, Faghih R, Liu H, Dwight W, Ireland L, Kang CH, Esbenshade TA, Bennani YL et al. (2002) Synthesis and evaluation of potent pyrrolidine H3 antagonists. *Bioorg Med Chem Lett* 12: 3055–3058

87 Faghih R, Dwight W, Vasudevan A, Dinges J, Conner SE, Esbenshade TA, Bennani YL, Hancock AA (2002) Aminoalkoxybiphenylnitriles as histamine-3 receptor ligands. *Bioorg Med Chem Lett* 12: 3077–3079

88 Faghih R, Dwight W, Pan JB, Fox GB, Krueger KM, Esbenshade TA, McVey JM, Marsh K, Bennani YL, Hancock AA (2003) Synthesis and SAR of aminoalkoxy-biaryl-4-carboxamides: novel and selective histamine H3 receptor antagonists. *Bioorg Med Chem Lett* 13: 1325–1328

89 Liedtke S, Flau K, Kathmann M, Schlicker E, Stark H, Meier G, Schunack W (2003) Replacement of imidazole by a piperidine moiety differentially affects the potency of histamine H3-receptor antagonists. *Naunyn Schmied Arch Pharmacol* 367: 43–50

90 Yang R, Hey JA, Aslanian R, Rizzo CA (2002) Co-ordination of histamine H3 receptor antagonists with human adrenal cytochrome P450 enzymes. *Pharmacol* 66: 128–135

Cognitive Enhancing Drugs
Edited by J.J. Buccafusco

Peptide and steroid hormone receptors as drug targets for enhancement of learning and memory performance

Joe L. Martinez Jr.[1], Kenira Thompson[1], Melanie P. McFadyen-Leussis[2] and Stephen C. Heinrichs[2]

[1] *University of Texas at San Antonio, Department of Biology, 6900 N. Loop 1604 West, San Antonio, TX 78249, USA*
[2] *Boston College, Department of Psychology, 140 Commonwealth Avenue, Chestnut Hill, MA 02467, USA*

Introduction

"Then Dr Strauss said Charlie even if this fales your making a grate contribyushun to sience. This experimint has been successful on lots of animals but its never bin tried on a humen being. You will be the first. After the operashun Im gonna try to be smart. Im gonna try awful hard." (sic) p. 8, Flowers For Angernon [1]

While therapeutic enhancement of cognitive function was once a topic for science fiction [1], the neuroscience and clinical literature of the past few decades provides ample documentation of the basic brain mechanisms for, and potential clinical utility of, neuroendocrine steroids and hormones tested in tasks sensitive to learning and memory performance [2–6]. Endocrine hormones represent a highly evolved messenger molecule system in vertebrates, and may guide and even be part of the normal machinery of a complex cognitive process such as learning and memory [7, 8]. Because there are many hormones that interact with neurotransmitters in order to influence learning and memory, Martinez and McGaugh (1981) proposed that these hormones be termed "learning modulatory hormones" [9].

The list of hormones that are reported to influence memory is impressively long and includes catecholamines, pituitary secretagogues, opioids and gut-brain peptides. The present review is by no means exhaustive and instead focuses upon a selected set of steroid/hormone mechanisms which share three attributes: 1) recent introduction or prominent representation in published literature over the past five years, 2) evidence from electrophysiological, genetic or other non-pharmacological dependent measures to validate the claim that changes in learning and memory performance are functionally specific, and 3) an indication that the therapeutic utility of the target can be critically examined in human patients suffering from some form of cognitive impairment or dementia.

The reader experienced in pre-clinical psychopharmacology will appreciate the conservative nature of the selection criteria listed above. These criteria serve to ascribe behavioral changes to underlying learning and memory processes only if alternative hypotheses related to non-specific general arousing or affective properties of experimental treatments, for example, can be convincingly discounted [10]. For instance, the putative learning modulatory actions of vasopressin, one of the peptide hormones to be discussed below, can be challenged based upon orthogonal affective and motor properties of peptide administration which exert generalized changes in arousal in a non-specific manner, or in peripheral organs entirely separate from the brain [11–13]. One way to effectively meet this challenge is to advance increasingly stringent criteria for documentation of cognitive enhancement which can be assumed *a priori* to be unproven [6, 12]. For example, information about how hormones actually modulate long-lasting synaptic connectivity and efficacy would be a welcome adjunct to functional testing methods [14]. Accordingly, recent advances in centrally acting neuropeptide and steroid hormone mechanisms relevant for animal and human learning and memory functions are presented in this review. In addition, we address specific examples of convergent electrophysiological [15] and behavioral genetic [16] methodologies which are likely to yield future advances in this field.

Pharmacological tools for learning and memory enhancement

Peptide hormone receptor targets

Classification of signaling molecules is somewhat arbitrary in that a single peptide can act as a classical neurotransmitter, as a neuromodulator, or serve neural and endocrine hormone functions simultaneously [8]. The common property shared by the present peptide hormone grouping is a chemical structure consisting of amino acids joined by peptide bonds. A great many putative peptide and hormone modulators of learning and memory performance are known at the present time (Tab. 1), although all require further characterization if we are to fully understand their role in learning and memory.

Neurohypophysical hormones

Vasopressin (AVP), sometimes labeled anti-diuretic hormone, is secreted by the hypothalamus and stored in and released from the posterior pituitary gland in order to increase blood pressure and the rate at which the kidneys absorb water. Extra-hypothalamic AVP pathways mediate learning and memory predominantly via receptors distributed in limbic structures such as the hippocampus, amygdala and septum. AVP can act centrally through AVP type 1 receptors, which activate protein kinase C and increase cytosolic Ca^{2+}, AVP

Table 1. Potential peptide and hormone targets for learning and memory performance enhancement

Peptide	Receptor	Species	Main findings	Reference
Growth hormone releasing hormone	GHRHR	Rat	Chronic GHRH improves age-related spatial memory impairments, and this corresponds to an increase in insulin-like growth factor 1 release in old age.	[243]
Leptin	OB	Rat	Leptin facilitates the induction of hippocampal synaptic plasticity, by enhancing NMDA-receptor mediated Ca^{2+} influx.	[244]
		Mouse	Leptin receptor deficient mice exhibit impaired LTP and spatial memory.	[245]
Nociceptin/ Orphanin FQ	ORL1	Mouse	At low doses (10–100 fmol), nociceptin reduces scopolamine-induced deficits in passive avoidance and spontaneous alternations in a Y-maze.	[246]
Orexin A	Orexin A	Rat	Orexin A facilitates learning, consolidation and retrieval processes in passive avoidance.	[247]
		Mouse	Orexin A improves retention in T-maze footshock avoidance and step-down passive avoidance tests in CD-1 mice. In young and aged SAMP8 mice, which overproduce amyloid *Beta*, Orexin A similarly improves retention.	[248]
Substance P	NK-1	Rat	Subtance P (10 ng, but not 100 ng) infusion into the basolateral amygdala post-shock in a passive avoidance paradigm enhances learning.	[249]
		Rat	Both systemic administration of SP and injection into the ventral pallidum improve memory, and in the latter case also result in the release of acetylcholine in the frontal cortex.	[250]
Neurotrophin-3	TrkC	Rat	NT-3 reverses cholinergic deficits and performance of a spatial memory task in aged rats when administered over a period of four weeks by ICV injection.	[36]
Insulin-like growth factor-1	NMDAR2A and R2B	Rat	IGF-1 reverses age-related deficits in working memory in a repeated acquisition task and an object recognition task, and reference memory in a place discrimination task.	[251]

type 2 receptors, which stimulate adenylate cyclase, and oxytocin receptors. Central vasopressinergic neural systems are widely distributed in many parts of the brain where they and their breakdown products have neuromodulatory effects completely independent of the neurohypophysis [17, 18].

AVP and certain AVP analogues that have no anti-diuretic action may improve long-term memory in a variety of different tasks [19]. Intra-cerebral administration of AVP restores learning and memory function in rats with *diabetes insipidus* and anti-AVP antiserum impairs learning in the rat [20]. An AVP analogue that is devoid of hormonal activity, AVP (4–9), enhances radial maze performance in rats, producing a faster rate of acquisition of reference and working memory. Likewise, AVP (4–9) inhibits scopolamine-induced memory deficits in the radial maze in rats [21, 22]. This enhancement is likely mediated by AVP type 1 receptors, whose activity stimulates acetylcholine release in hippocampal slices, while AVP type 2 receptors do not appear to be involved [23]. A second AVP analogue, AVP (4–8), improves concept learning in a win-stay/loose-shift paradigm in rats with hippocampal damage but not in rats with prefrontal cortex lesions [24]. Another AVP derivative, NC-1900, enhances place learning in rats with cycloheximide-induced hippocampal lesions [25]. Taken together, these findings suggest that increased signaling via AVP type 1 receptors is capable of counteracting pharmacologically-induced and tissue damage-related learning impairments in rodents via a mechanism which is dissociable from the classical hormonal (pressor) actions of AVP [17]. AVP has also been reported to have clinical efficacy following human administration in alleviating post-traumatic amnesia [26].

Oxytocin (OXT) is a peptide hormone that is produced by the posterior lobe of the pituitary gland and induces contraction of the smooth muscle of the uterus and myoepithelial cells of the mammary gland. In contrast to AVP, OXT impairs memory retention in an inhibitory avoidance paradigm, and this also occurs via modulation of the cholinergic system [27]. Both of these peptides influence social recognition, but different effects have been noted across different rodent species, different genders within the same species, and different brain regions of interest [28, 29]. For example, although oxytocin (OXT) has complex effects on social memory in rats, mice with a null mutation of the OXT gene are completely socially amnestic without other cognitive deficits [30]. As OXT given centrally before, but not after, the initial encounter restores social recognition in these mutant mice, the neuropeptide appears critical for the acquisition rather than the consolidation phase of memory. These findings support the hypothesis that OXT is essential for social memory, although it may modulate different mnemonic processes and different neural systems depending on the organism.

Trophic factors

Nerve Growth Factor (NGF) is a multimeric protein, the beta subunit of which is required for the proper development and maintenance of the sensory neurons of the dorsal root ganglion and of the post-ganglionic sympathetic neurons. NGF has been implicated in several forms of learning and memory. In aged rats, chronic, four week intracerebroventricular infusion of exogenous NGF improved memory performance in a delayed non-matching-to-position task, and this restoration of memory performance persisted for at least four-weeks after NGF infusions ceased [31, 32]. In a water maze spatial memory task, exogenous NGF improved performance in aged rats [33] as well as in rats withdrawn from chronic ethanol exposure [34]. In the latter case, improved water maze performance was accompanied by a restoration of septohippocampal cholinergic projections [34], while in the former case, the number of synaptophysin immunoreactive pre-synaptic terminals increased in the frontal cortex [33]. The link between NGF and cholinergic activity is further supported by the observation that the effects of NGF on recent memory in the delayed non-matching-to-position task correlate with changes in the cholinergic system, including increased size of cholinergic neurons, and a change in the terminal fields of these same neurons [35]. Additionally, the characteristic cholinergic atrophy observed in aged rats can be reversed by chronic, four-week infusion of NGF, and a concurrent improvement in spatial memory performance is observed [36].

The generality of NGF-induced facilitation of learning and memory performance is supported by efficacy of NGF treatment in a variety of other animal species and testing contexts. In a classical fear conditioning paradigm, endogenous NGF is reported to increase one week after training (i.e., during the consolidation phase), but infusion of tyrosine kinase A (TrkA) antisense into the hippocampus one week post-training can block this effect, impairing contextual retention [37]. In a simple passive avoidance paradigm, combined central administration of NGF and epidermal growth factor resulted in improved learning in aged mice, but had no effect in normal adult mice [38]. Neither trophic factor administered alone had any effect on learning in this paradigm. In developing CD-1 mice, a single intracerebroventricular administration of NGF at post-natal day 15 resulted in adult-like spatial novelty discrimination in males but not females tested at post-natal day 18, although increased choline acetyltransferase activity was observed in both sexes as a result of NGF treatment [39]. By post-natal day 28, no behavioral or neurochemical effects of NGF administration are observed. Importantly, NGF administration in infrahuman primates is reported to attenuate age and lesion-induced forebrain cholinergic degeneration [40, 41]. Taken together, these results suggest that central administration of exogenous NGF can facilitate learning performance and ameliorate age and pharmacologically-induced learning impairment while remodeling cholinergic brain areas thought to subserve learning and memory functions.

Brain-derived neurotrophic factor (BDNF), a NGF-related neurotrophin with high affinity for the TrkB receptor, is known to have numerous roles in learning and memory, and contributes to the process of hippocampal long-term potentiation. Two excellent reviews provide a very comprehensive review of our understanding of the role of BDNF in learning and memory [42, 43]. BDNF contributes to the functional decline that occurs with aging [44, 45]. Mnemonic effects of BDNF are found in rodents, and in primates, where the peptide is up-regulated in the inferior temporal cortex during visual pair-association learning [46], and in day-old chicks, where anti-sense administration impairs memory consolidation in a one-trial inhibitory avoidance paradigm [47].

The research highlighted above provides evidence that NGF and BDNF, in addition to their effects on neuronal survival and differentiation, enhance learning and memory processes in a number of different species. Both trophic factors can influence the cholinergic system [48], and may represent potential clinical therapies.

Hypothalamo-pituitary-adrenal axis peptides

It is widely documented in both animal and clinical studies that certain neuropathological states, including human and animal models of dementia, are accompanied by an altered endocrine stress axis. The endocrine stress axis can be defined as the biological interface for neural and humoral communication between the central nervous system and peripheral glands or organs responsible for mobilizing the stress response. Largely correlational results link some measure of cognitive capacity in man or indices of information plasticity in animal studies with a circulating marker of stress-like activation such as plasma glucocorticoid levels. These findings can now be critically examined due to the availability of a variety of targeted mutant mice in which specific components of the biological stress axis are neutralized or made constitutively active (Tab. 2). These new research tools provide leverage in clarifying a long-standing issue of causality: are endocrine markers of stress merely diagnostic of dementia-like disorders or do they instead constitute the primary defect of homeostasis which provides a mechanism for increased vulnerability to cognitive decline?

Stress and behavioral plasticity are interrelated; levels of alertness correspond to success in performance of a learning task in what can be described as

Table 2. Learning task performance of CRF system and glucocorticoid receptor gene knockdown and mutant mice

Targeted gene	Knockdown	Knockout	Over-expression
CRF	Impairment [252]	No effect [253]	Impairment [254]
CRF-R1	No effect [255]	Impairment [256]	ND
GR	Impairment [257]	Impairment [258]	ND

CRF – Corticotropin-releasing factor; CRF-R1 – Type I CRF receptor; GR – Glucocorticoid receptor; ND – not yet determined the stress-cognition axis [49]. Stressors, neuroendocrine peptides activated by stressors, and circulating stress-related hormones all modulate learning and memory storage processes. For example, brief, mild activation of brain, autonomic or endocrine stress systems coincident with an emotional experience as well as long-term or intense activation of the hypothalamo-pituitary-adrenocortical (HPA) axis, both alter learning and memory capacity [50]. It is important to note that an inverted U-shaped function governs the relationship between HPA axis stimulation and cognitive function. In particular, elevating stress hormone dose levels, either acutely or chronically, or prolonging the duration of exposure to low or high stress hormone levels would be expected either to enhance or impair learning and memory functions in both animal models [51, 52] and man [53].

Many HPA axis neuropeptides which act as neurotransmitters within the central nervous system, including corticotropin-releasing factor (CRF), adrenocorticotropin (ACTH), β-endorphin, and α-melanocyte-stimulating hormone can modulate learning and memory through central, extrahypophysiotropic mechanisms [54]. For example, rodent pharmacological studies of learned avoidance behavior employing administration of CRF itself [55] the pituitary product of CRF secretagogue action, ACTH [56], and the adrenocortical product of ACTH secretagogue action, corticosterone (or cortisol in humans) [57], into either brain or periphery consistently reveal modification of learned performance when the hormone/steroid is administered prior to training, immediately following training or prior to retention testing [56]. Depending on the interaction between the intensity of training stimulus employed and the dose of steroid or neuropeptide administered, either enhancement or impairment of learned behavior is observed [56]. An extensive literature implicates deficiency in ACTH-related peptides in defects in the acquisition of and the more rapid extinction of learned behaviors [58]. Similar to the results described for AVP above, central administration of ACTH or ACTH analogues that have no adrenal cortex-stimulating activity restores learning ability in rats, whereas the results of studies of the effects of ACTH-related peptides on human memory are conflicting. A typical study reports that human administration of an ACTH analogue devoid of endocrine effects prior to completing a range of performance tests including a complicated serial reaction task, running memory span, verbal learning and non-verbal mental ability tests, increases sustained attention [59]. The important implication of these findings is that an optimal level of performance associated with a moderate, fine-tuned degree of activation is modulated by deviations in HPA activity.

Corticotropin-releasing factor (CRF) is widely recognized as part of a neuropeptide system whose activation is a necessary component of the biological response to stressor exposure [60]. Characteristic features of brain CRF system activation include behavioral vigilance, suspension of appetite and pituitary-adrenocortical stimulation [61]. Previous research suggests that the consequences of stress neuropeptide activation in an information processing context can be predicted according to an arousal/performance heuristic [62, 63].

One of the hallmarks of the arousal/performance interaction is that the direction of change produced by a particular treatment in a learning task depends on the exact context in which the treatment is administered [5]. For example, whereas acute stress neuropeptide activation facilitates active avoidance performance, chronic activation has an impairing effect [64]. The bimodal effects of CRF administration in an inhibitory avoidance task also suggest that mild arousal facilitates performance whereas over-arousal engenders memory impairment [62]. This conclusion is supported by the finding that many other hypophysiotropic neuropeptides of the pituitary gland, such as ACTH, are also capable of exerting bimodal impairing and enhancing effects on learning and memory performance [65]. The multiple endocrine, autonomic and behavioral actions of CRF and urocortin systems complicate the task of identifying the exact mechanism underlying stress-related memory modulation [66]. However, emerging evidence suggests that central actions of brain stress neuropeptide systems provide an autonomous mechanism for exerting memory modulatory consequences of stress-related activation [67, 68].

Several lines of evidence support the present identification of a physiological role for CRF systems in information processing functions of the central nervous system. First, steady-state levels of endogenous CRF family neuropeptide receptor agonists appear sufficient to modulate learning and memory functions since pharmacological dissociation of CRF and a related neuropeptide, urocortin, from their binding protein enhances performance in appetitive- and aversively- motivated memory tasks [69–72]. Second, central CRF administration exerts electrophysiological and neurochemical activation of hippocampal circuits relevant for learning and memory processes in several species [73–75]. Finally, brain and cortical CRF levels are significantly reduced in patients with both mild and severe dementia [76]. Moreover, cerebrospinal fluid levels of CRF correlate with degrees of cognitive impairment in dementia sufferers [77]. Thus, CRF decrements may serve as a potential neurochemical marker of early dementia and possibly early Alzheimer's disease [76].

Homeostatic peptides

The following homeostatic neuropeptides/hormones are considered in some depth because plausible overlap between their brain substrates of action and brain learning and memory mechanisms is articulated. Note that some of the following memory modulatory neuropeptides/hormones, angiotensin, cholecystokinin, galanin and neuropeptide Y, are known primarily for their prominent role in homeostatic regulatory systems, controlling eating and drinking behaviors. This least-common functional denominator may be relevant for generating hypotheses to explain effects of these peptides in learning and memory contexts as discussed below.

The angiotensins are a group of blood pressure-regulating peptides that act as powerful vasopressors and stimulators of aldosterone secretion by the adrenal cortex. At least four angiotensin receptor subtypes subserve a variety of different physiological functions [78]. Angiotensin IV (Ang IV) binds specifically to AT4 receptors in the brain, which are not activated by Ang I–III, although the AT4 receptor can be activated by certain fragments including Ang II (2–7) and Ang I (3–10). Full-length Ang II binds to AT1 and AT2 receptor subtypes.

Ang IV administration improves recall in a passive avoidance task in a dose-dependent manner [79]. Similarly, Ang II (2–7) and Ang I (3–10) also improve performance in both active and passive avoidance tasks [80–82]. Intracerebroventricular injection of an Ang IV agonist increases the rate of acquisition in the Morris water maze, while an Ang IV receptor antagonist impairs the rate of acquisition in the same task [83]. Ang IV, and LVV-hemorphin-7, an AT4 receptor ligand, both potentiate depolarization-induced acetylcholine release from rat hippocampal slices in a concentration-dependent manner [84]. This effect is blocked by an AT4 receptor antagonist, but not by AT1 or AT2 receptor antagonists. The distribution of Ang IV binding sites in the human brain is similar to that found in other species, with a high density of AT4 receptors found in numerous regions including the hippocampus, entorhinal, prefrontal and cingulate cortices, and some thalamic nuclei. This distribution supports multiple roles for AT4 binding sites in the central nervous system, including facilitation of memory retention and retrieval [85].

Ang II inhibits potassium-induced release of acetylcholine from fresh slices of human temporal cortex obtained at surgery, but this effect is blocked by an Ang II receptor antagonist [86]. Ang II inhibition of potassium-evoked release of acetylcholine occurs in a concentration-dependent manner in slices of rat entorhinal cortex [87]. It is hypothesized that the potential cognitive enhancing effects of acetylcholinesterase inhibitors are modulated through the removal of the Ang II-mediated inhibition of cholinergic function. In general, it appears that Ang II facilitates learning and memory in conditioned avoidance and inhibitory avoidance tasks, whatever role it may have in other learning and memory tasks [80].

Cholecystokinin (CCK) is a polypeptide hormone of the duodenum released in response to increased amounts of free fatty acids in the intestinal lumen in order to promote contraction of the gallbladder and secretion of pancreatic enzymes. CCK-8 (an octapeptide derivative of CCK) is the most abundant form in the brain and it activates both CCK-A and CCK-B receptors, although only the latter are widely distributed in the CNS. Systemic CCK-8 injections help prevent memory deficits induced by age [88], electroconvulsive shock [89], NMDA receptor antagonists [90], scopolamine [91], or protein kinase inhibitors [92] in the inhibitory avoidance test. There is evidence to suggest a functional balance between CCK-A and CCK-B receptors in the mediation of memory tasks. CCK-A-specific antagonists are detrimental to memory processes [93], and rats lacking CCK-A receptors exhibit learning and memory impairments in a radial maze [94]. In contrast, the role of CCK-B receptors in learning and memory processes has not been unequivocally shown. For instance, two selective CCK-B agonists (BC264 and BC197) produce opposite effects on working memory in a Y-maze paradigm [95], an effect that appears to be modulated by differential dopamine activation [96]. Others have also reported either enhanced [97–99] or impaired [100] learning with other CCK-B receptor agonists. More evidence is needed to clearly elucidate the complete role of CCK in learning and memory, especially for effects mediated by CCK-B receptors.

CCK-8 administration induces a dose- and time-dependent increase in NGF levels in the hypothalamus, pituitary and hippocampus. Pre-treatment with a selective CCK-A receptor antagonist blocks the increase in NGF in the hypothalamus and pituitary but not in the hippocampus, while pre-treatment with a selective CCK-B receptor antagonist blocks the increase in NGF in the hippocampus only [101]. In unlesioned mice, CCK-8 injections increase choline acetyltransferase activity in the forebrain, while in fimbria-fornix lesioned mice, which exhibit reduced levels of choline acetyltransferase in the septo-hippocampal circuit, CCK-8 injections counteract the choline acetyltransferase deficits. This effect on cholinergic cells is mediated through the synthesis and release of NGF [102]. Similarly, continuous intracerebroventricular infusion of CCK-8 prevents the degeneration of cortical cholinergic neurons following basal forebrain lesions in rats [103]. Taken together, these results suggest that the cognitive enhancement resulting from CCK-8 exposure may be partially mediated by effects on the cholinergic system.

The neuropeptide galanin enjoys widespread expression in the nervous and endocrine systems and is reported to alter performance in learning and memory tasks. There is now a substantial body of work to indicate that galanin plays an important biological role as a regulator of neurotransmitter and hormone release in the adult organism. Studies demonstrate that galanin acts as a developmental and trophic factor to subsets of neurons in the nervous and neuroendocrine systems [111]. Galanin has been shown to impair learning and memory in a variety of paradigms [104–106]. However, a dual nature of galanin is evident as infusions into the ventral hippocampus produce bi-phasic, dose-dependent effects on spatial learning in rats [107]. The learning and memory effects of galanin are predominantly mediated through changes in cholinergic systems, as galanin administration decreases acetylcholine release in the ventral hippocampus and cerebral cortex [106, 107], but increases release in the dorsal hippocampus and striatum [108]. Thus, galanin's functional role appears to depend on the brain region in question. Further, galanin co-localizes with choline acetyltransferase in a subset of cholinergic neurons in the rodent basal forebrain.

Although the role of galanin in Alzheimer's disease is controversial, galanin expression is up-regulated in forebrain nuclei in Alzheimer's tissue and innervation of the remaining basal forebrain cholinergic neurons is likewise augmented [116]. Thus, it has been postulated that this galanin hyper-innervation results in depression of the septo-hippocampal circuits involved in learning and memory [117]. However, evidence suggests that galanin may compensate for the loss of cholinergic neurons in this region through an excitatory action on the remaining cells, thus augmenting the release of acetylcholine, and ultimately leading to a delay in the progression of Alzheimer's disease [118].

Neuropeptide Y (NPY) is one the most abundant neuropeptides in the mammalian brain where it exerts control over endocrine hypothalamic and pituitary functions, hypothalamic control of food intake and circadian rhythm, and limbic emotional integration [109]. Within the mammalian brain, Y2 is the predominant NPY receptor and can be found in the hippocampus, hypothalamus, thalamus, amygdala, and brainstem as well as in other regions, although species differences do exist in Y2 receptor localization and quantity [110–112]. Activation of the Y2 receptor, a G-protein coupled receptor, leads to inhibition of adenylate cyclase. NPY modulates memory retention in both inhibitory and active avoidance tasks, where regional injections of NPY into the rostral portion of the hippocampus and septum enhance memory retention while injections into the caudal hippocampus and amygdala impair retention [113]. NPY injections alleviate MK-801 induced learning impairments in a step-down inhibitory avoidance task [114].

While there is great interest in developing NPY receptor antagonists for the treatment of obesity-related disorders [115], application in dementing disorders can also be considered. Significant decreases in NPY immunoreactivity occur in conjunction with Alzheimer's disease, and it is thought these may be related to the cholinergic deficits that are observed in the same individuals [116]. Activation of the Y2 receptor may stimulate the release of AVP and OXT [117], thereby providing one potential indirect mechanism for cognitive improvement. A review of NPY receptors with a focus on their potential as therapeutic drug targets for cognitive dysfunction is available [118].

Opioid peptides

Opioid peptides in the neocortex, hippocampus, and amygdala play an important role in learning and memory. Derived from the proteolytic cleavage of the precursor proteins preproenkephalin, preprodynorphin, proopiomelanocortin (POMC), they produce enkephalins, dynorphins (A and B), and β-endorphins, respectively [119–121]. Binding of opioid peptides to μ, δ, and κ opioid receptors results in varied behavioral and physiological responses.

Enkephalins are neuronally produced and released. They are also released in conjunction with adrenaline in response to stress. β-endorphin and ACTH are also released from the anterior pituitary in response to stress. Numerous studies have outlined the hormonal/neuromodulatory actions of enkephalins and β-endorphins on learning and memory. Most frequently, β-endorphins and enkephalins are associated with impairments in acquisition and retention of learning tasks, although some studies report memory enhancement [122].

Peripheral administration of enkephalins results in impairments in inhibitory avoidance retention in rodents [123, 124]. Similar impairments result from systemic injection of β-endorphin. Thus, it seems that enkephalins and endorphins share a modulatory role in learning and memory-related events.

The dynorphins also affect memory formation. For example, injection of dynorphin A enhances rodent retention of avoidance responses, and impairs aversive and appetitive learning in chicks [125, 126]. Hippocampal injections with dynorphin impair spatial learning/water maze performance, and working memory in the radial arm maze [127]. Spatial memory deficits observed in aged rats are partly attributed to elevated levels of dynorphin A [128]. The existing evidence suggests that opioid peptides can enhance or inhibit memory formation. This modulation is dependent on the strength of training, and peripheral *versus* central administration. Systemic administration results in enkephalin, endorphin, or dynorphin action on opioid receptors located outside of the blood-brain barrier in order to alter learning and memory mechanisms. Central administration, on the contrary, results in direct effect of the opioid peptides on central nervous system activity. These results indicate that opioid peptides might influence the strength of a memory via parallel central and peripheral mechanisms.

Steroid hormone receptor targets

The steroid hormones are synthesized from cholesterol, predominantly in the gonads in the case of estrogen and progesterone, or in the adrenal cortex in the case of glucocorticoids. Although classical steroid hormone actions are mediated predominantly through nuclear receptors, many steroid hormone effects on learning and memory are exerted via membrane receptors that regulate ion channel function.

Estrogen

The estrogens are a family of steroid hormones that regulate and sustain female sexual development and reproductive function. Besides affecting the hypothalamus and other brain areas related to reproduction, ovarian steroids have widespread effects throughout the brain, on serotonin pathways, catecholaminergic neurons, the basal forebrain cholinergic system, and the hippocampal formation [130]. In rats, intrahippocampal infusions of estradiol potentiate acetylcholine- and glutamate-mediated memory retention in an avoidance learning task [129]. Estrogen is also important for performance during acquisition training, as ovariectomized females exhibit a slower rate of acquisition in a delayed matching-to-position task and this deficit can be overcome by chronic administration of estrogen [130]. In intact male and female mice, as well as ovariectomized females, chronic estrogen treatment improves radial arm maze working memory performance [131]. Additionally estrogen-mediated improvement in radial maze working memory is dependent on acetylcholine acting through M2 muscarinic receptors to increase N-methyl-D-aspartate (NMDA) receptor binding in the hippocampus [132]. Preliminary evidence suggests that mice deficient in either estrogen receptor subtype, ERα or ERβ, show impaired learning, which indicates that both estrogen receptor subtypes are important for normal cognitive functioning [133, 134].

Ovarian hormones regulate synapse turnover in the CA1 region of the hippocampus during the four- to five-day estrous cycle of the female rat [135]. Formation of new excitatory synapses is induced by estradiol, involves NMDA receptors and is mediated by acetylcholine [132]. Although NMDA receptor activation is required for synapse formation, inhibitory interneurons may play a pivotal role as they express nuclear ERα. It is also likely that estrogens may locally regulate events at the sites of synaptic contact in the excitatory pyramidal neurons where the synapses form [136]. Estrogen interacts with the rat cholinergic system in numerous ways, such as enhancing cortical cholinergic enervation and preserving synaptic density following excitotoxic lesions in the basal forebrain [137]. Estrogen replacement in aged female mice similarly increases hippocampal synaptophysin immunoreactivity, a marker for synaptic density, and this effect correlates with an increase in spatial reference memory [138].

Ovarian steroids have measurable effects on affective state as well as cognition, with implications for dementia [136]. In particular, replacement of estrogen in post-menopausal women protects against cognitive effects of aging assessed using several measures of cognitive function. In a large sample of non-demented post-menopausal women assessed longitudinally either with or without estrogen and/or progestin replacement therapy, hormone supplementation reduced the risk of contracting Alzheimer's disease and protected against a decline in verbal memory task performance [139]. One meta-analysis of 42 studies examined the effects of estrogen replacement therapy (ERT) on memory and cognition in non-demented post-menopausal women [140]. While some studies report no performance benefit of ERT, the preponderance of significant findings favor efficacy of ERT as evidenced, for example, by a consistent beneficial effect of ERT on verbal memory [140]. The observational studies suggest that there may be a long-lasting effect of continued ERT on cognitive functioning. ERT is associated with a decreased risk for dementia, but there is little evidence for a positive effect of estrogen therapy on cognition in women with Alzheimer's disease [140].

Neurosteroids

Neurosteroids, synthesized in the central and peripheral nervous systems from cholesterol or steroidal precursors [141], can rapidly alter neuronal excitability by non-genomic mechanisms such as GABA-A, NMDA, and sigma 1 receptors [142]. Pregnenolone sulfate (PREGS) and dehydroepiandrosterone (DHEAS) act as antagonists at GABA-A receptors, and positively modulate NMDA receptor responses. DHEAS also acts through sigma receptors [143, 144].

Recent evidence suggests a role for PREGS and DHEAS in improving hippocampally-mediated memory tasks such as spatial recognition [145, 146], Y-maze [147, 148], visual discrimination go/no-go [149], and motivated lever-press learning [148]. PREGS and DHEAS are both effective at increasing learning and retention when administered pre- and post-training, but not when administered just prior to retention testing in a passive avoidance paradigm [150]. In addition, PREGS blocks memory impairments induced by scopolamine [149] and D-2-amino-5-phosphonovalerate [148] while DHEAS blocks dizocilpine-induced learning deficits [144]. In aged rats, PREGS levels correlate with performance in the Morris water maze and in a Y-maze task, such that low levels were indicative of poor task performance [151]. In addition, the deficits of aged rats were transiently ameliorated by either intraperitoneal or bilateral intra-hippocampal injections of PREGS, which was shown to stimulate acetylcholine release in the hippocampus [145, 151, 152]. Central PREGS administration up-regulates neurogenesis in the dentate gyrus of adult and aged rats, which occurs via blockade of the GABA-A receptor [152]. These findings suggest that PREGS and DHEAS play a role in preserving or enhancing cognitive abilities, and that this may occur, at least in part, through modulation of the cholinergic system [153].

There are several mechanisms that could contribute to the promnestic actions of PREGS. One possibility is that PREGS enhances central cholinergic function, a major system involved in attention and memory processing. This concept is supported by the observation that administration of PREGS in the nucleus basalis magnocellularis, the main source of cortical cholinergic innervation, improves memory performance of young rats [146]. Additionally, central administration of PREGS increases extracellular acetylcholine concentrations in the hippocampus [145, 151].

PREGS modulates several ligand-gated ion channels, with NMDA and GABA-A receptors being the most potently affected (Fig. 1). PREGS enhances NMDA-activated currents and inhibits GABA-mediated currents in cultured rat hippocampal neurons [154]. These *in vitro* results are consistent with neuronal excitatory and convulsant effects of PREGS *in vivo* [155]. In addition, PREGS could influence NMDA and GABA-A receptor functions by a non-specific action such as altering membrane fluidity [156].

Glucocorticoids

The glucocorticoids are a group of adrenocortical steroid hormones whose metabolic effects include stimulation of gluconeogenesis, increased catabolism of proteins, mobilization of free fatty acids and potent inhibition of the inflammatory response. In addition, the effect on learning capacity of chronic activation of the hypothalamo-pituitary-adrenal axis has been characterized

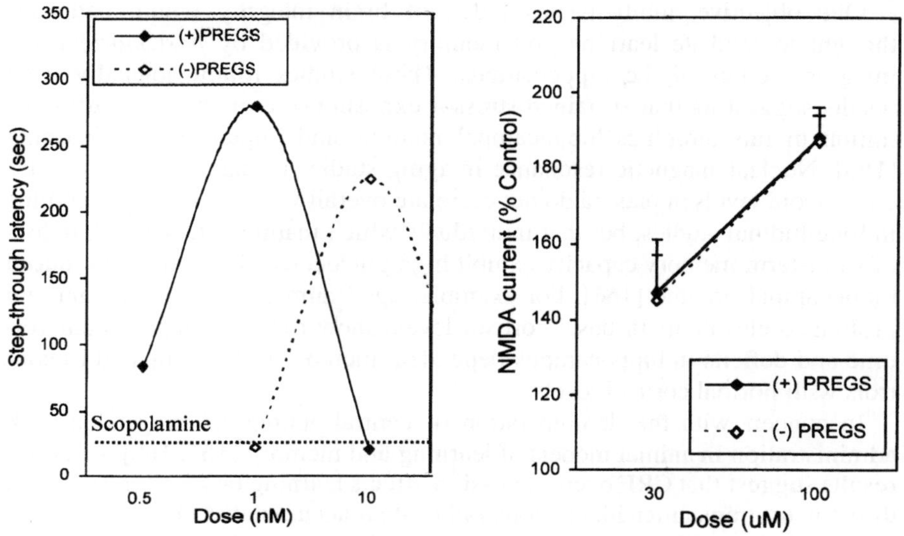

Figure 1. The effects of pre-training injection of two pregnenolone sulfate (PREGS) enantiomers in combination with scopolamine on retention performance after 24 hours in a passive avoidance paradigm is depicted in the left panel. Performance was differentially affected in a inverted U-shape dose-dependent manner for each of the enantiomers, with a peak in step-through latency (best performance) for (+) PREGS occurring at a dose of 1 nmol and for (–) PREGS occurring at a dose of 10 nmol. The step-through latency after scopolamine administration, but in the absence of either PREGS enantiomer is denoted by the dotted line. The same PREGS enantiomers were tested for their ability to potentiate NMDA currents in hippocampal neurons, summarized in the right panel. Both enantiomers modulated the NMDA current to the same extent regardless of dose. These results indicate that the cognitive effects of PREGS are not mediated through an effect on NMDA currents. (Modified from [259])

using long-term peripheral administration of glucocorticoids in mice, rats, monkeys and man [157]. Administration of these stress hormones alters acquisition of a previously unlearned task in a dose-related, inverted U-shaped fashion [158]. Brains from chronic corticosterone-treated animals reveal morphological changes, usually cell loss, in hippocampal areas of the brain believed to subserve learning and memory [159]. Note that while a significant literature appears to exclude corticosteroids from the set of stress hormones which act to modulate retrieval processes [160, 161], other evidence suggests that corticosterone administration prior to retention testing can indeed alter retrieval of spatial memory [162]. One prediction from this correlational link between the level of arousal and performance of learned behaviors is that intrinsic over-activation and/or long-term stimulation of neurobiological and endocrine substrates of the stress response would have the effect of producing learning and memory deficits. In contrast, short-term exposure to physiological levels of exogenous glucocorticoids could be expected to enhance performance in a learning and memory context and this hypothesis is supported by animal and human clinical studies [163, 164].

One objective, multi-species index of brain integrity within pathways thought to mediate learning and memory is provided by *post mortem* and imaging studies of the hippocampus. These studies follow logically from results suggesting that sustained stressor exposure or corticosterone administration in rats atrophies hippocampal neurons and impairs spatial learning [165]. Nuclear magnetic resonance imaging studies in man suggest that glucocorticoid levels in plasma do not correlate overall with hippocampal volume in longitudinal studies, but that individuals which manifest dementia-like loss of short-term memory capacity exhibit high glucocorticoid levels and reduced hippocampal volume [166]. For example, aged humans with significant and prolonged elevations in basal cortisol levels show reduced hippocampal volume and deficits in hippocampus-dependent memory tasks compared to controls with normal cortisol levels.

In concert with the demonstration of central actions of exogenous CRF administration in animal models of learning and memory [167, 168] the above results suggest that CRF over-expression affects learning task performance via direct neurotransmitter-like actions rather than accumulated changes in gross neuropathology. This conclusion is supported by results indicating that CRF, ACTH and corticosterone continue to be significant modulators of learning and memory processes when either the organism or the treatment itself is rendered incapable of HPA activation [64].

Neurophysiological approach to developing novel targets

The previous sections outlined important hormones and their respective targets. In this section, we will discuss neurophysiological approaches used to determine such targets, and we will address the role that hormones play in the modulation of learning and memory-related processes. We know that the brain can communicate with, and be influenced by, other systems, particularly endocrine systems. Hormone-triggered responses can serve to strengthen memories for particular events by providing input to the memory trace through hormonal pathways (Fig. 2) [169].

Hormones that modulate learning and memory, for instance through hormone-neurotransmitter interactions, also play an important role in the modulation of long-term potentiation (LTP). It is believed that modifications at synaptic connections within the nervous system may underlie memory [170]. LTP refers to a relatively long-lasting change in synaptic strength that remains as an ideal candidate model for the cellular mechanisms that underlie learning and memory [15]. LTP is expressed in multiple forms by neurons in the hippocampus, a brain structure that has been widely implicated in long-term memory processes [170]. Because LTP is typically recognized as a model for memory storage, vast research has focused on the role that either endogenous or exogenous compounds exert on its induction and maintenance.

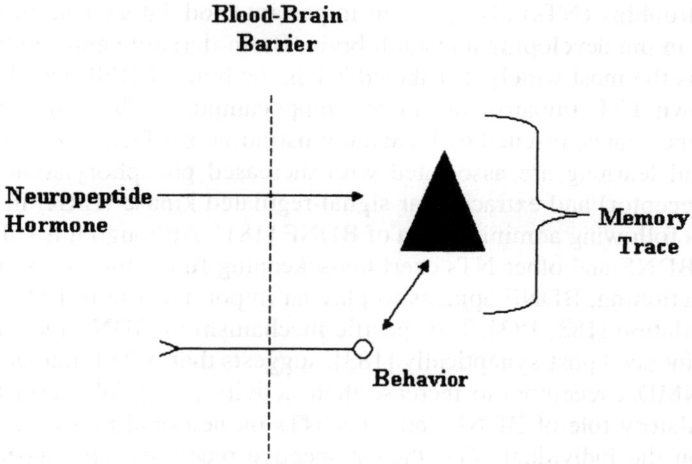

Figure 2. Diagram depicts a model of how neuropeptides are thought to modulate the memory trace. (Modified from [5])

As noted in a previous section, the neurons of the posterior pituitary produce oxytocin (OXT) and arginine vasopressin (AVP). Co-localization of these hormones with other neuropeptides is typical, and they are released selectively during stress [171]. Because OXT and AVP can act as both hormones and neurotransmitters, it is not surprising that they mediate fast neuronal responses by altering the permeability of both pre- and post-synaptic ion channels [172, 173]. AVP and AVP derivatives such as AVP(4–9) facilitate LTP *in vitro* [174] and *in vivo* [175]. There is evidence to suggest that AVP potentiates LTP via intracellular mechanisms [176]. Further work is in progress to clearly identify the effect that AVP has on cell membranes to facilitate this potentiation.

Oxytocin, typically known as an essential hormone during mammalian labor and lactation, is also involved in the development of social recognition. There is evidence to suggest that pregnancy, birth, and lactation improve spatial memory in the female rat [177]. This type of memory becomes essential when the mother rat wanders to find food and water, yet must rely on spatial memory to remember the location of her offspring. OXT results in hippocampal plasticity and LTP, and this may account for the observed memory enhancements during motherhood, which occur as a result of activation of the mitogen-activated protein (MAP) kinase cascade and CREB phosphorylation [178]. It follows that OXT-induced memory improvement is regulated by the action of OXT as a neuromodulator, and not by the induction of receptor expression. Although current research is still on-going to confirm the role of OXT in memory, this information has provided insight into the development of therapies to selectively transduce OXT into the brain, as OXT or OXT agonists may be useful in the therapeutic treatment of memory deficits.

Neurotrophins (NTs) also play an important modulatory role in neuronal plasticity in the developing and adult brain. Brain-derived neurotrophic factor (BDNF) is the most widely distributed NT in the brain. BDNF-knockout mice have shown LTP impairments in the hippocampus [179, 180], and these impairments can be rescued by local administration of BDNF [181]. Both LTP and spatial learning are associated with increased phosphorylation of TrkB (BDNF receptor) and extracellular signal-regulated kinase (ERK) in the dentate gyrus following administration of BDNF [181]. Although it is still unclear whether BDNF and other NTs exert housekeeping functions to maintain neuronal functioning, BDNF appears to play an important role in LTP induction and modulation [182, 183]. The specific mechanism of BDNF-mediated LTP, which is induced post-synaptically [183], suggests that BDNF interacts directly with NMDA receptors to increase their activity [184]. Whatever the case, the modulatory role of BDNF and other NTs on neuronal plasticity seems to depend on the individual NTs, their respective receptors, and appears to be mediated by local increases in Ca^{2+}.

CRF modulates learning, food intake, arousal, and fear responses [185–187]. More recently, it has been implicated as an enhancer of synaptic efficacy in the rodent hippocampus *in vivo* [188, 189]. In contrast, sustained CRF administration actually blocks hippocampal LTP in a dose-dependent manner [190]. CRF produces a protein synthesis-dependent LTP in the dentate gyrus [188], and improves retention in various memory tasks, making the peptide a potentially important player in the mechanisms underlying stress and cognition. CRF's actions are mediated by activation of Ca^{2+}/calmodulin-dependent kinase II, and this may represent an essential mechanism by which CRF contributes to memory storage [191].

The angiotensin system, widely implicated in neural plasticity and memory, facilitates LTP, spatial, and associative memory. In particular, agonists of the AT4 receptor augment CA1 LTP *in vitro* [174, 192], and *in vivo* [193]. Ang Is correlated with increases in hippocampal MMP-9 levels. MMP-9 is a gelatinase that serves as a plasminogen activator. Previous literature indicates that the tissue plasminogen activator (tPA) is a serine protease that plays an important role in tissue remodeling and LTP. The existing literature indicates that tPA serves as an immediate-early gene and is induced in the hippocampus during seizures, kindling, and LTP [194]. tPA knockout mice show a decrease in late-phase LTP [195], and show deficits in two-way avoidance tasks. Additionally, over-expression of tPA results in enhanced CA1 LTP and learning [196]. To date, the role of tPA on hippocampal function is not clear. One possibility is that tPA converts plasminogen, which is the enzyme's main substrate and is known to be found in the hippocampus, to the protease plasmin, which in turn can cleave many other extracellular substrates (for example, laminin) to result in alterations of hippocampal structure and function [197]. Other studies have found binding of tPA to the low-density lipoprotein receptor-related protein in hippocampal neurons enhances the activity of cAMP-dependent protein kinase, a key molecule in LTP [198]. Overall, the angiotensin-tPA interaction appears to play an important role in the successful acquisition of a new memory and needs to be further examined [199].

Alterations in estrogen levels influence many behaviors, even those not necessarily linked to sexual behavior. Among these are memory-related alterations, which are not clearly understood. Research indicates that the administration of estradiol to ovarietomized rats results in synaptic modifications in the hypothalamus [200]. Additional work indicates that hippocampal LTP is facilitated by increased levels of circulating estrogen, as evidenced by the finding that cyclical changes in endogenous estrogen levels can augment LTP [201, 202]. These studies support the hypothesis that estrogen is a regulator of learning-related mechanisms, although some contradictory findings exist. For instance, estrogen administration results in improved performance in avoidance tasks but not in the Morris water maze [203]. However, a different study indicates that estrogen can improve performance in the water maze task [204]. LTP enhancement by estrogen is mediated by both mitogen-activated protein kinase-dependent and independent components [205]. Further, evidence suggests that estrogen increases NMDA receptor activity, and this is likely a further mechanism through which it enhances LTP [206]. Further studies are needed to elucidate the exact mechanism through which estrogen modulates hippocampal LTP and learning and memory.

Neurogenomic techniques in the development of novel targets

It is widely thought that long-term alterations of cell function may mediate learning and memory in the brain. These long-term changes must involve gene expression and resultant protein production. Thus, for every sustained memory there is likely a chain of events leading from the initiation of activity at a synaptic receptor, to the activity of second messenger systems, to intermediate early gene induction, and to secondary gene induction in every cell that participates in the memory network. The same is likely true for LTP [207].

A number of research groups are endeavoring to trace the chain of cellular events that underlie induction and maintenance of LTP [208]. In these studies single genes, controlling what are hoped to be specific events within cells, can be eliminated and the resultant effect can be studied simultaneously in whole animals minus one gene, so-called knockouts. In this method the gene of interest, usually a well-characterized gene, is cloned and in most cases altered so that important regulatory regions of the gene are non-functional.

One reason to target genes is that these genetic procedures have the potential to overcome the current limitations of pharmacology. Numerous studies have evaluated hormone involvement in learning and memory by genetic manipulations. For instance, OXT knockout mice show complete social amnesia, but exhibit no other cognitive deficits [28]. Research using CRF peptide, CRF post-synaptic receptor and CRF binding-protein transgenic and knockout mouse models allows for critical analysis of hypotheses relating HPA axis tone and brain CRF system activation in animal models to a variety of clinical psychopathologies. While the putative role of brain CRF in affective disorders has been the primary focus of research, learning and memory capacities of CRF over-expressing and knockout mice and CRF1 receptor knockdown and knockout mice can now be assessed in these animals, which exhibit targeted defects in homeostasis. In Table 2, results from antisense oligonucleotide studies are tabulated alongside gene targeting studies since gene knockdown via translational arrest could be expected to produce functional, albeit transient, consequences similar to a null mutation.

Galanin knockout and over-expressing transgenic mice have recently been generated to facilitate understanding of galanin activity in basal forebrain function. Galanin knockout mice have fewer cholinergic basal forebrain neurons and show behavioral memory deficits [209]. On the other hand, mice over-expressing galanin exhibit hyper-enervation of the basal forebrain, but still have functional memory deficits [210]. These data highlight the need to explore the putative mechanisms by which galanin signaling might be beneficial or deleterious to cholinergic cell survival and activity within the basal forebrain [119]. Mice carrying a targeted loss-of-function mutation in the galanin gene exhibit age-dependent deficits in stimulated acetylcholine release, water maze performance, and induction of long-term potentiation in the CA1 region of the hippocampus, as well as a decrease in the number of cholinergic neurons in the medial septum and vertical limb diagonal band [112]. This suggests galanin plays a trophic role in the development and function of septo-hippocampal cholinergic neurons. The following sections will outline new techniques, other than knockout technology, currently used to identify genes that play an important role in the modulation of memory and cognition.

As mentioned previously, the hippocampus is widely recognized as an important structure involved in learning and memory. The hippocampus displays two forms of LTP: an NMDA receptor-dependent form [211], and an NMDA receptor-independent form [211–216]. Of particular interest to this section is the mossy fiber projection from granule cells in the dentate gyrus to the *stratum lucidum* layer of area CA3 of the hippocampus, which displays the NMDA-independent form of LTP, and is dependent on the activation of opioid receptors. This activation is thought to be facilitated by opioid peptides contained in, and released by, the mossy fibers and has been confirmed both *in vivo* and *in vitro* [212–218].

A number of mechanisms are implicated in LTP induction and maintenance [219]. Studies indicate that an influx in Ca^{2+} activates a series of second messenger cascades, including the calcium/calmodulin-dependent adenylyl cyclase pathway [220], results in an increase in cAMP, and activates the cAMP-dependent protein kinase [219]. Additional work has shown that protein synthesis inhibition blocks the induction of LTP *in vivo* in the mossy fiber-CA3 pathway of the hippocampus, which has been previously implicated in spatial memory [221]. Despite the mounting evidence supporting NMDA-receptor independent LTP in the mossy fiber-CA3 pathway, the specific mechanisms by which opioid peptides facilitate LTP in this region have not yet been elucidated. Alterations in gene expression are important for LTP induction and maintenance in multiple hippocampal pathways. Because relatively little is known regarding the actual mechanisms involved in LTP in the mossy fiber-CA3 pathway, one laboratory examined the regulatory processes underlying LTP induction in this pathway by focusing on alterations in gene expression after LTP induction using Affymetrix microarray technology [222].

Microarrays allow one to monitor the expression patterns of numerous genes simultaneously. DNA oligonucleotides are synthesized directly onto the array by means of photolithography. A given gene is represented by 15–20 different 25mer oligonucleotides that serve as unique, sequence-specific detectors. An additional control element on these arrays is the use of mis-match control oligonucleotides that are identical to their perfect match patterns except for a single base difference in a central position. The presence of the mismatched oligonucleotide allows cross-hybridization and local background to be estimated and subtracted from the perfect match signal. Hybridized probes are detected by incorporated fluorescent nucleotide analogs. Thus the DNA Microarray is very suitable to study global gene expression profiles. Thompson et al. (2003) used Affymetrix oligonucleotide arrays that contained probe sets corresponding to 1,200 genes (Rat Neurobiology Array, RN-U34) to identify changes in hippocampal gene expression associated with LTP induction in the mossy fiber-CA3 pathway. They found that genes involved in synaptic plasticity, neurotransmission, transcription factors, cell survival, trafficking, and ion channels are altered in the hippocampus following LTP induction in the mossy fiber-CA3 pathway (Tab. 3).

Opioid-related genes altered in the LTP group are known to be involved in enhancing neurotransmission. For example, proenkephalin was found to be up-regulated in the LTP group and this corresponds with previous literature showing that enkephalin peptides released from hippocampal mossy fibers lower the threshold for induction of LTP at mossy fiber synapses [223]. Neuropeptide Y (NPY), which was also up-regulated in our LTP animals, has been previously linked to inhibition of glutamate release and LTP in the dentate gyrus [224]. tPA, which interacts with the angiotensin system and is known to play an important role in synaptic remodeling, was similarly up-regulated following LTP induction in the MF-CA3 pathway.

Growth factor changes were also noted. BDNF, for example, was up-regulated in the group in which LTP was blocked. Previously linked to LTP induction in the hippocampus, BDNF is thought to trigger long-lasting synaptic strengthening through MEK/ERK [219]. Other growth factors such as VGF are also up-regulated in both the LTP and non-LTP groups, suggesting a possible role in synaptic modification. One possibility is that the up-regulated growth factors interact with down-regulated IGF-1 and endothelin receptor to regulate cell survival during synaptic alterations.

Microarray technology is still in its infancy and, therefore, it is essential to verify findings using other mechanisms, such as Real-time PCR (polymerase

Table 3 - List of significantly altered (up or down-regulated) genes in the rodent hippocampus following LTP induction in the MF-CA3 pathway (see Thompson et al. [222] for details)

Accession number	P value	Direction	Description	Function
AJ006710_at	0.007	Down	*Rattus norvegicus* mRNA for IP3-kinase	Signal transduction
L39018_at	0.048	Down	*Rattus norvegicus* sodium channel protein 6 (SCP6) mRNA	Ion channel
S65355_g_at	0.02	Down	non-selective endothelin receptor	Neuronal survival
L36884_at	0.02	Down	*Rattus norvegicus* protein tyrosine phosphatase (OST-PTP) mRNA	Phosphorylation
AJ001641_at	0.01	Down	*Rattus norvegicus* mRNA for Brain-1 (Brn-1) protein	Transcription factor
AJ007632_s_at	0.008	Down	*Rattus norvegicus* mRNA for ELK channel 3, partial	Transcription factor
M10244_at	0.049	Down	Rat tyrosine hydroxylase mRNA	Signal transduction
L13040_s_at	0.048	Up	*Rattus norvegicus* calcitonin receptor C1b mRNA	Signal transduction
X07729exon#5	0.044	Up	*R. norvegicus* gene encoding neuron-specific enolase	Neuronal marker
K02248cds_s_at	0.032	Up	Rat somatostatin-14 gene, complete cds	Synaptic transmission
S49491_s_at	0.02	Up	Proenkephalin	Synaptic transmission
X06655_at	0.018	Up	Rat mRNA for p38	Signaling
M15880_at	0.017	Up	Rat Neuropeptide Y	Inhibit glutamate release and LTP
M74223_at	0.015	Up	Rat VGF mRNA	Synaptic transmission

chain reaction). The findings to date provide new information regarding gene alterations that occur in the hippocampus following stimulation of the mossy fiber-CA3 pathway, thus enhancing knowledge of the potential mechanisms underlying opioid-dependent hippocampally-dependent learning.

Conclusions and future directions

An increasing number of structurally heterogeneous compounds, which may act via very different neuronal mechanisms, have been proposed to facilitate attention and acquisition, storage and retrieval of information, and/or to attenuate the impairments of such cognitive functions associated with age or dementia [225]. It should be noted that while the present review describes many elaborate neurobiological mechanisms for enhancing performance in animals and humans tested in learning and memory contexts, very simple approaches already exist that are capable of achieving the same feat. For example, glucose administration regulates many neural and behavioral processes in rodents, including learning and memory [226]. Glucose ingestion can improve memory in highly functioning populations [227] and glucose restriction can have an impairing effect on cognitive performance [228]. Similarly, glucose enhances performance on specific measures in an elderly population, particularly on those tasks where mild age-related deficits appear (e.g., verbal declarative memory) [226, 229]. Thus, one could rightly question whether direct neuropharmacological actions of peptide or steroid hormones which regulate energy balance, such as neuropeptide Y and adrenal steroids, exert their long-term effects on cognitive function by indirect stimulation of glucose availability [230]. The often cited link in the present review between facilitated learning performance on the one hand and enhanced cholinergic neurotransmission on the other, is another double-edged sword which does suggest favorable correlates of functional efficacy for a particular peptide/steroid hormone in a behavior assay while rendering the actual mechanism of action more ambiguous and imprecise from one hormone to the next. This problem of specificity has been called one of the most difficult in learning and memory research [10] and can be pursued in the future by defining the number and identity of unique mechanisms for optimizing learning and memory capability [231].

The search for drugs that enhance cognition requires the development of behavioral tests for animals [232]. These tests must be able to identify potentially therapeutic drugs and reject ineffective drugs. Therefore, a coherent conceptual and experimental framework is needed to organize future research in this area. Unfortunately, previous pre-clinical research strategies appear to have focused on the demonstration of drug effects in a wide variety of tests of uncertain validity, rather than on determination of the specific psychological and neurobiological processes affected by putative cognition enhancers. For example, some sort of noxious stimulus, such as electric shock delivered unexpectedly to rodent paw pads, is typically used to motivate learning in the widely-used avoidance conditioning context in spite of the fact that shock exposure produces an unconditioned affective arousal state which confounds interpretation of learning performance in the task [2]. A further disincentive for employing alarming and traumatic stimuli in the conditioning environment is provided by behavioral and cognitive neuroscience studies demonstrating that the affective salience of stimuli can bias encoding and retrieval of learned information in an automatic manner [233, 234]. Thus, future efforts require explicit identification of the goals of the research, the cognitive process, the neural systems and cellular gene products involved in the process, the selectivity and sensitivity of tasks that measure the process and the validity of the behavioral tasks as a model to predict the effects of the drug in humans. For example, a hierarchical, multi-task approach employed to phenotype learning and memory capabilities in mutant mice appears to be suitable for wider adoption [235, 236].

The combined use of genetic tools, such as genetically engineered mice and microarray technology, together with more classical pharmacological methods can greatly enhance our ability to detect novel therapeutic targets. For instance, peptides such as proenkephalin and neuropeptide Y, which are known to play a role in learning and memory, are up-regulated after LTP induction [222]. Moreover, genes identified in other studies of learning and memory-related genetic regulation could provide unforeseen, previously unidentified drug targets for cognitive enhancement. Pharmacological and behavioral characterization of these targets will confirm or refute their ability to modulate learning and memory processes, and ultimately their therapeutic potential in clinical populations.

The conceptual foundations of research aimed at the determination of potential neuronal, neuropharmacological, genetic, and behavioral/cognitive mechanisms mediating drug-induced cognition enhancement require definition [237]. For example, peptides and steroid hormones of the HPA axis such as CRF and glucocorticoids are presumed to be the neurochemical mediators of enhanced long-term memory for stressful or emotionally arousing experiences [238]. The pharmacological [239], neurobiological [240] and clinical [241] evidence necessary to support this claim convincingly is only now being assembled. In particular, the behavioral phenotype of learning and memory impairments described in the present review for CRF and glucocorticoid receptor mutant mice supports the view that pharmacological normalization of HPA axis tone would have therapeutic benefits for dementing disorders [242]. Nonetheless, the systematic development of a psychopharmacology of cognition enhancement may require concerted application of behavioral, genetic, cognitive, pharmacological and electrophysiological techniques.

References

1 Keyes D (1959) *Flowers for Algernon*. New York: Bantam Books

2 Koob GF (1987) Neuropeptides and memory. In: LL Iversen, SD Iversen (eds): *Handbook of Psychopharmacology*, Plenum, New York, 531–573

3 White NM, Salinas JA (1998) Pharmacology approaches to the study of learning and memory. In: JL Martinez Jr, R Kesner (eds): *Neurobiology of learning and memory*, Academic Press, San Diego, 143–176

4 Lupien SJ, de Leon M, de Santi S, Convit A, Tarshish C, Nair NPV, Thakur M, McEwen BS, Hauger RL, Meaney MJ (1998) Cortisol levels during human aging predict hippocampal atrophy and memory deficits. *Nature Neurosci* 1(1): 69–73

5 Martinez JL Jr, Schulteis G, Weinberger SB (1991) How to increase and decrease the strength of memory traces: The effects of drugs and hormones. In: JLj Martinez, RP Kesner (eds): *Learning and memory: A biological view*, Academic Press: New York, 149–198

6 Squire LR, Davis HP (1981) The pharmacology of memory: A neurobiological perspective. *Ann Rev Pharmacol Toxicol* 21(323): 323–356

7 Martinez JLj (1986) Memory: Drugs and hormones. In: JLj Martinez, RP Kesner (eds): *Learning and memory: A biological view*, Academic Press: New York
8 Krieger DT (1983) Brain peptides: What, where, and why? *Science* 222 (4627): 975–985
9 Martinez JL Jr, Jensen RA, McGaugh JL (1981) Attenuation of experimentally-induced amnesia. *Prog Neurobiol* 16(2): 155–186
10 Martinez JL Jr, Jensen RA, McGaugh JL (1983) *Facilitation of memory consolidation*. In: JA Deutsch (ed.): *The Physiological Basis of Memory*. Academic Press, New York, 49–70
11 Ettenberg A, Van der Kooy D, Le Moal M, Koob GF, Bloom FE (1983) Can aversive properties of (peripherally-injected) vasopressin account for its putative role in memory. *Behav Brain Res* 7: 331–350
12 Koob GF (1987) Learning and Memory: Principles and findings with a focus on vasopressin. Durham, North Carolina
13 Sahgal A (1984) A critique of the vasopressin-memory hypothesis. *Psychopharmacol (Berl)* 83(3): 215–228
14 Squire LR (1987) *Memory and Brain*. Oxford: Oxford University Press
15 Martinez JL Jr, Derrick BE (1996) Long-term potentiation and learning. *Annu Rev Psychol* 47: 173–203
16 Silva AJ (2003) Molecular and cellular cognitive studies of the role of synaptic plasticity in memory. *J Neurobiol* 54(1): 224–237
17 Meisenberg G, Simmons WH (1983) Centrally mediated effects of neurohypophyseal hormones. *Neurosci Biobehav Rev* 7(2): 263–280
18 Dantzer R, Bluthe RM (1992) Vasopressin involvement in antipyresis, social communication, and social recognition: A synthesis. *Crit Rev Neurobiol* 6(4): 243–255
19 Dantzer R (1998) Vasopressin, gonadal steroids and social recognition. *Prog Brain Res* 119: 409–414
20 De Wied D (1984) The importance of vasopressin in memory. *Trends in Neuroscience* 7(62): 62–64
21 Mishima K, Tsukikawa H, Inada K, Fujii M, Iwasaki K, Matsumoto Y, Abe K, Egawa T, Fujiwara M (2001) Ameliorative effect of vasopressin-(4–9) through vasopressin V(1A) receptor on scopolamine-induced impairments of rat spatial memory in the eight-arm radial maze. *Eur J Pharmacol* 427(1): 43–52
22 Taga C, Sugimoto Y, Nishiga M, Fujii Y, Kamei C (2001) Effects of vasopressin on histamine H(1) receptor antagonist-induced spatial memory deficits in rats. *Eur J Pharmacol* 423(2–3): 167–170
23 Tanabe S, Shishido Y, Nakayama Y, Furushiro M, Hashimoto S, Terasaki T, Tsujimoto G, Yokokura T (1999) Effects of arginine-vasopressin fragment 4–9 on rodent cholinergic systems. *Pharmacol Biochem Behav* 63(4): 549–553
24 Dietrich A, Taylor JT, Passmore CE (2001) AVP (4–8) improves concept learning in PFC-damaged but not hippocampal-damaged rats. *Brain Res* 919(1): 41–47
25 Hori E, Uwano T, Tamura R, Miyake N, Nishijo H, Ono T (2002) Effects of a novel arginine-vasopressin derivative, NC-1900, on the spatial memory impairment of rats with transient forebrain ischemia. *Brain Res Cogn Brain Res* 13(1): 1–15
26 Jolles J (1983) Vasopressin-like peptides and the treatment of memory disorders in man. *Prog Brain Res* 60: 169–182
27 Boccia MM, Baratti CM (2000) Involvement of central cholinergic mechanisms in the effects of oxytocin and an oxytocin receptor antagonist on retention performance in mice. *Neurobiol Learn Mem* 74(3): 217–228
28 Ferguson JN, Young LJ, Insel TR (2002) The neuroendocrine basis of social recognition. *Front Neuroendocrinol* 23(2): 200–224
29 Young LJ (2002) The neurobiology of social recognition, approach, and avoidance. *Biol Psychiatry* 51(1): 18–26
30 Ferguson JN, Young LJ, Insel TR (2002) The neuroendocrine basis of social recognition. *Front Neuroendocrinol* 23(2): 200–224
31 Frick KM, Price DL, Koliatsos VE, Markowska AL (1997) The effects of nerve growth factor on spatial recent memory in aged rats persist after discontinuation of treatment. *J Neurosci* 17(7): 2543–2550
32 Markowska AL, Price D, Koliatsos VE (1996) Selective effects of nerve growth factor on spatial recent memory as assessed by a delayed nonmatching-to-position task in the water maze. *J Neurosci* 16(10): 3541–3548

33 Chen KS, Masliah E, Mallory M, Gage FH (1995) Synaptic loss in cognitively impaired aged rats is ameliorated by chronic human nerve growth factor infusion. *Neurosci* 68(1): 19–27
34 Lukoyanov NV, Pereira PA, Paula-Barbosa MM, Cadete-Leite A (2003) Nerve growth factor improves spatial learning and restores hippocampal cholinergic fibers in rats withdrawn from chronic treatment with ethanol. *Exp Brain Res* 148(1): 88–94
35 Gustilo MC, Markowska AL, Breckler SJ, Fleischman CA, Price DL, Koliatsos VE (1999) Evidence that nerve growth factor influences recent memory through structural changes in septohippocampal cholinergic neurons. *J Comp Neurol* 405(4): 491–507
36 Fischer W, Sirevaag A, Wiegand SJ, Lindsay RM, Bjorklund A (1994) Reversal of spatial memory impairments in aged rats by nerve growth factor and neurotrophins 3 and 4/5 but not by brain-derived neurotrophic factor. *Proc Natl Acad Sci USA* 91(18): 8607–8611
37 Woolf NJ, Milov AM, Schweitzer ES, Roghani A (2001) Elevation of nerve growth factor and antisense knockdown of TrkA receptor during contextual memory consolidation. *J Neurosci* 21(3): 1047–1055
38 Fiore M, Triaca V, Amendola T, Tirassa P, Aloe L (2002) Brain NGF and EGF administration improves passive avoidance response and stimulates brain precursor cells in aged male mice. *Physiol Behav* 77(2–3): 437–443
39 Calamandrei G, Valanzano A, Ricceri L (2002) NGF induces appearance of adult-like response to spatial novelty in 18-day male mice. *Behav Brain Res* 136(1): 289–298
40 Koliatsos VE, Clatterbuck RE, Nauta HJ, Knusel B, Burton LE, Hefti FF, Mobley WC, Price DL (1991) Human nerve growth factor prevents degeneration of basal forebrain cholinergic neurons in primates. *Ann Neurol* 30(6): 831–840
41 Smith DE, Roberts J, Gage FH, Tuszynski MH (1999) Age-associated neuronal atrophy occurs in the primate brain and is reversible by growth factor gene therapy. *Proc Natl Acad Sci USA* 96(19): 10893–10898
42 Tyler WJ, Alonso M, Bramham CR, Pozzo-Miller LD (2002) From acquisition to consolidation: On the role of brain-derived neurotrophic factor signaling in hippocampal-dependent learning. *Learn Mem* 9(5): 224–237
43 Yamada K, Mizuno M, Nabeshima T (2002) Role for brain-derived neurotrophic factor in learning and memory. *Life Sci* 70(7): 735–744
44 Croll SD, Ip NY, Lindsay RM, Wiegand SJ (1998) Expression of BDNF and trkB as a function of age and cognitive performance. *Brain Res* 812(1–2): 200–208
45 Schaaf MJ, Workel JO, Lesscher HM, Vreugdenhil E, Oitzl MS, de Kloet ER (2001) Correlation between hippocampal BDNF mRNA expression and memory performance in senescent rats. *Brain Res* 915(2): 227–233
46 Tokuyama W, Okuno H, Hashimoto T, Xin Li Y, Miyashita Y (2000) BDNF up-regulation during declarative memory formation in monkey inferior temporal cortex. *Nature Neurosci* 3(11): 1134–1142
47 Johnston AN, Rose SP (2001) Memory consolidation in day-old chicks requires BDNF but not NGF or NT-3; an antisense study. *Brain Res Mol Brain Res* 88(1–2): 26–36
48 Takei N, Kuramoto H, Endo Y, Hatanaka H (1997) NGF and BDNF increase the immunoreactivity of vesicular acetylcholine transporter in cultured neurons from the embryonic rat septum. *Neurosci Lett* 226(3): 207–209
49 Heinrichs SC (1999) Stress-axis, coping and dementia: gene-manipulation studies. *Trends in Pharmacol Sci* 20: 311–315
50 Luine VN, Spencer RL, McEwen BS (1993) Effects of chronic corticosterone ingestion on spatial memory performance and hippocampal serotonergic function. *Brain Res* 616: 65–70
51 Gold PE, van Buskirk R (1976) Enhancement and impairment of memory processes with post-trial injections of adrenocorticotrophic hormone. *Behav Biol* 16: 387–400
52 Ohl F, Fuchs E (1999) Differential effects of chronic stress on memory processes in the tree shrew. *Brain Res Cogn Brain Res* 7(3): 379–387
53 Kirschbaum C, Wolf OT, May M, Wippich W, Hellhammer DH (1996) Stress- and treatment-induced elevations of cortisol levels associated with impaired declarative memory in healthy adults. *Life Sci* 58(17): 1475–1483
54 De Wied D, Jolles J (1982) Neuropeptides derived from pro-opiocortin: Behavioral, physiological and neurochemical effects. *Physiol Rev* 62: 976–1059
55 Kumar KB, Karanth KS (1996) Alpha-helical CRF blocks differential influence of corticotropin releasing factor (CRF) on appetitive and aversive memory retrieval in rats. *J Neural Transm* 103:

1117–1126
56 McGaugh JL (1989) Involvement of hormonal and neuromodulatory systems in the regulation of memory storage. *Ann Rev Neurosci* 12: 255–287
57 Dachir S, Kadar T, Robinzon B, Levy A (1993) Cognitive deficits induced in young rats by long-term corticosterone administration. *Behav Neural Biol* 60: 103–109
58 Martinez JL Jr, Jensen RA, Messing RB, Rigter H, McGaugh JL (1981) Endogenous Peptides and Learning and Memory Processes. In: JL McGaugh, JC Fentress, JP Hegmann (eds): *Behavioral Biology: An International Series*, Academic Press, New York
59 Gaillard AW, Varey CA (1979) Some effects of an ACTH 4-9 analog (Org 2766) on human performance. *Physiol Behav* 23(1): 79–84
60 Heinrichs SC, De Souza EB (2001) Corticotropin-releasing factor in brain: Executive gating of neuroendocrine and functional outflow. In: BS McEwen (ed.): *Handbook of Physiology*. Oxford University Press, New York, 125–137
61 Akwa Y, Purdy RH, Koob GF, Britton KT (1999) The amygdala mediates the anxiolytic-like effect of the neurosteroid allopregnanolone in rat. *Behav Brain Res* 106(1–2): 119–125
62 Sahgal A, Wright C, Edwardson JA, Keith AB (1983) Corticotrophin releasing factor is more potent than some corticotrophin-related peptides in affecting passive avoidance behaviour in rats. *Neurosci Lett* 36: 81–86
63 Deffenbacher KA (1983) The influence of arousal on reliability of testimony, Chapter 13. In: SMA Lloyd-Bostock, BR Clifford (eds): *Evaluating Witness Evidence*. John Wiley and Sons Ltd., New York, 235–249
64 Honour LC, White MH (1988) Pre- and postnatally administered ACTH, Organon 2766 and CRF facilitate or inhibit active avoidance task performance in young adult mice. *Peptides* 9(4): 745–750
65 De Wied D, Bohus B (1979) Modulation of memory processes by neuropeptides of hypothalamic-neurohypophyseal origin. *Raven Press, New York* (139): 139–149
66 Contarino A, Heinrichs SC, Gold LH (1999) Understanding CRF neurobiology: contribution from mutant mice. *Neuropeptides* 33: 1–12
67 Steckler T, Holsboer F (1999) Corticotropin-releasing hormone receptor subtypes and emotion. *Biol Psychiatry* 46(11): 1480–1508
68 Deak T, Nguyen KT, Ehrlich AL, Watkins LR, Spencer RL, Maier SF, Licinio J, Wong M-L, Chrousos GP, Webster E et al. (1999) The impact of the nonpeptide corticotropin-releasing hormone antagonist antalarmin on behavioral and endocrine responses to stress. *Endocrinology* 140(1): 79–86
69 Eckart K, Radulovic J, Radulovic M, Jahn O, Blank T, Stiedl O, Spiess J (1999) Actions of CRF and its analogs. *Current Medicinal Chemistry* 6(11): 1035–1053
70 Zorrilla EP, Schulteis G, Ling N, Koob GF, De Souza EB (2001) Performance enhancing effects of CRF-BP ligand inhibitors. *Neuroreport* 12: 1231–1234
71 Behan DP, De Souza EB, Lowry PJ, Potter E, Sawchenko P, Vale WW (1995) Corticotropin releasing factor (CRF) binding protein: A novel regulator of CRF and related peptides. *Front Neuroendocrinol* 16: 362–382
72 Heinrichs SC, Lapsansky J, Lovenberg TW, De Souza EB, Chalmers DT (1997) Corticotropin-releasing factor CRF1, but not CRF2, receptors mediate anxiogenic-like behavior. *Regulatory Peptides* 71: 15–21
73 Bonaz B, Rivest S (1998) Effect of a chronic stress on CRF neuronal activity and expression of its type 1 receptor in the rat brain. *American J Physiology* 275(5 PART 2): R1438-R1449
74 Wang HL, Wayner MJ, Chai CY, Lee EHY (1998) Corticotropin-releasing factor produces a long-lasting enhancement of synaptic efficacy in the hippocampus. *Eur J Neurosci* 10: 3428–3437
75 Fuchs E, Flügge G, Ohl F, Lucassen P, Vollmann-Honsdorf GK, Michaelis TM (2001) Psychosocial stress, glucocorticoids, and the structural alterations in the tree shrew hippocampus. *Physiol Behav* 73: 285–291
76 Buggy J, Hewitt CD, Kemeny G, Davis JM, Stock H, Hand GA (1999) Hypothalamic and amygdalar corticotropin-releasing factor (CRF) after fatiguing exercise. *Society for Neuroscience Abstracts* 25(1–2): 63
77 De Souza E, Whitehouse PJ, Kuhar MJ, Price DL, Vale W (1986) Reciprocal changes in corticotropin-releasing factor (CRF)-like Alzheimer's disease. *Nature* 319(593): 593–595
78 Wright JW, Harding JW (1995) Brain angiotensin receptor subtypes AT1, AT2, and AT4 and their functions. *Regul Pept* 59(3): 269–295

79 Wright JW, Harding JW (1997) Important role for angiotensin III and IV in the brain renin-angiotensin system. *Brain Res Rev* 25(1): 96–124
80 Gard PR (2002) The role of angiotensin II in cognition and behaviour. *Eur J Pharmacol* 438(1–2): 1–14
81 Karwowska-Polecka W, Kulakowska A, Wisniewski K, Braszko JJ (1997) Losartan influences behavioural effects of angiotensin II(3–7) in rats. *Pharmacol Res* 36(4): 275–283
82 Tchekalarova J, Kambourova T, Georgiev V (2002) Interaction of angiotensin II and III with adenosine A(1) receptor-related drugs in passive avoidance conditioning in rats. *Behav Brain Res* 129(1–2): 61–64
83 Wright JW, Stubley L, Pederson ES, Kramar EA, Hanesworth JM, Harding JW (1999) Contributions of the brain angiotensin IV-AT4 receptor subtype system to spatial learning. *J Neurosci* 19(10): 3952–3961
84 Lee J, Chai SY, Mendelsohn FA, Morris MJ, Allen AM (2001) Potentiation of cholinergic transmission in the rat hippocampus by angiotensin IV and LVV-hemorphin-7. *Neuropharmacol* 40(4): 618–623
85 Chai SY, Bastias MA, Clune EF, Matsacos DJ, Mustafa T, Lee JH, McDowall SG, Paxinos G, Mendelsohn FA, Albiston AL (2000) Distribution of angiotensin IV binding sites (AT4 receptor) in the human forebrain, midbrain and pons as visualised by *in vitro* receptor autoradiography. *J Chem Neuroanat* 20(3–4): 339–348
86 Barnes JM, Barnes NM, Costall B, Horovitz ZP, Ironside JW, Naylor RJ, Williams TJ (1990) Angiotensin II inhibits acetylcholine release from human temporal cortex: Implications for cognition. *Brain Res* 507(2): 341–343
87 Barnes JM, Barnes NM, Costall B, Horovitz ZP, Ironside JW, Naylor RJ, Williams TJ (1990) Angiotensin II inhibits cortical cholinergic function: Implications for cognition. *J Cardiovasc Pharmacol* 16(2): 234–238
88 Voits M, Hasenohrl RU, Huston JP, Fink H (2001) Repeated treatment with cholecystokinin octapeptide improves maze performance in aged Fischer 344 rats. *Peptides* 22(8): 1325–1330
89 Katsuura G, Itoh S (1986) Passive avoidance deficit following intracerebroventricular administration of cholecystokinin tetrapeptide amide in rats. *Peptides* 7(5): 809–814
90 Itoh S, Takashima A, Igano K, Inouye K (1989) Memory effect of caerulein and its analogs in active and passive avoidance responses in the rat. *Peptides* 10(4): 843–848
91 Itoh S, Takashima A, Katsuura G (1988) Effect of cholecystokinin tetrapeptide amide on the metabolism of 5-hydroxytryptamine in the rat brain. *Neuropharmacol* 27(4): 427–431
92 Takashima M, Katoh Y, Takeuchi Y, Takahashi K (1991) Effect of subchronic administration of methylcobalamin on the acetylcholine and choline content in the brain and locomotor activity in rats. *Jpn J Psychiatry Neurol* 45(1): 173–175
93 Itoh S, Takashima A, Maeda Y (1992) Protective effect of cerulein on memory impairment induced by protein synthesis inhibitors in rats. *Peptides* 13(5): 1007–1012
94 Nomoto S, Miyake M, Ohta M, Funakoshi A, Miyasaka K (1999) Impaired learning and memory in OLETF rats without cholecystokinin (CCK)-A receptor. *Physiol Behav* 66(5): 869–872
95 Lena I, Simon H, Roques BP, Dauge V (1999) Opposing effects of two CCK(B) agonists on the retrieval phase of a two-trial memory task after systemic injection in the rat. *Neuropharmacol* 38(4): 543–553
96 Lena I, Dh tel H, Garbay C, Dauge V (2001) Involvement of D2 dopamine receptors in the opposing effects of two CCK-B agonists in a spatial recognition memory task: Role of the anterior nucleus accumbens. *Psychopharmacol* 153(2): 170–179
97 Sebret A, Lena I, Crete D, Matsui T, Roques BP, Dauge V (1999) Rat hippocampal neurons are critically involved in physiological improvement of memory processes induced by cholecystokinin-B receptor stimulation. *J Neurosci* 19(16): 7230–7237
98 Taghzouti K, Lena I, Dellu F, Roques BP, Dauge V, Simon H (1999) Cognitive enhancing effects in young and old rats of pBC264, a selective CCK(B) receptor agonist. *Psychopharmacol* 143(2): 141–149
99 Weller A, Tsitolovskya L, Gispan IH, Rabinovitz S (2001) Examining the role of cholecystokinin in appetitive learning in the infant rat. *Peptides* 22(8): 1317–1323
100 Derrien M, Dauge V, Blommaert A, Roques BP (1994) The selective CCK-B agonist, BC 264, impairs socially reinforced memory in the three-panel runway test in rats. *Behav Brain Res* 65(2): 139–146
101 Tirassa P, Stenfors C, Lundeberg T, Aloe L (1998) Cholecystokinin-8 regulation of NGF con-

centrations in adult mouse brain through a mechanism involving CCK(A) and CCK(B) receptors. *Br J Pharmacol* 123(6): 1230–1236
102 Tirassa P, Aloe L, Stenfors C, Turrini P, Lundeberg T (1999) Cholecystokinin-8 protects central cholinergic neurons against fimbria-fornix lesion through the up-regulation of nerve growth factor synthesis. *Proc Natl Acad Sci USA* 96(11): 6473–6477
103 Sugaya K, Takahashi M, Kubota K (1992) Cholecystokinin protects cholinergic neurons against basal forebrain lesion. *Jpn J Pharmacol* 59(1): 125–128
104 Robinson JK, Crawley JN (1994) Analysis of anatomical sites at which galanin impairs delayed nonmatching to sample in rats. *Behav Neurosci* 108(5): 941–950
105 Crawley JN (1993) Functional interactions of galanin and acetylcholine: Relevance to memory and Alzheimer's disease. *Behav Brain Res* 57(2): 133–141
106 Sundstrom E, Archer T, Melander T, Hokfelt T (1988) Galanin impairs acquisition but not retrieval of spatial memory in rats studied in the Morris swim maze. *Neurosci Lett* 88(3): 331–335
107 Ogren SO, Kehr J, Schott PA (1996) Effects of ventral hippocampal galanin on spatial learning and on *in vivo* acetylcholine release in the rat. *Neuroscience* 75(4): 1127–1140
108 Ogren SO, Schott PA, Kehr J, Misane I, Razani H (1999) Galanin and learning. *Brain Res* 848(1–2): 174–182
109 Thorsell A, Heilig M (2002) Diverse functions of neuropeptide Y revealed using genetically modified animals. *Neuropeptides* 36(2–3): 182–193
110 Caberlotto L, Fuxe K, Rimland JM, Sedvall G, Hurd YL (1998) Regional distribution of neuropeptide Y Y2 receptor messenger RNA in the human *post mortem* brain. *Neuroscience* 86(1): 167–178
111 Dumont Y, Jacques D, Bouchard P, Quirion R (1998) Species differences in the expression and distribution of the neuropeptide Y Y1, Y2, Y4, and Y5 receptors in rodents, guinea pig, and primates brains. *J Comp Neurol* 402(3): 372–384
112 Gustafson EL, Smith KE, Durkin MM, Walker MW, Gerald C, Weinshank R, Branchek TA (1997) Distribution of the neuropeptide Y Y2 receptor mRNA in rat central nervous system. *Brain Res Mol Brain Res* 46(1–2): 223–235
113 Morley JE, Flood JF (1990) Neuropeptide Y and memory processing. *Ann N Y Acad Sci* 611: 226–231
114 Bouchard P, Maurice T, St-Pierre S, Privat A, Quirion R (1997) Neuropeptide Y and the calcitonin gene-related peptide attenuate learning impairments induced by MK-801 via a sigma receptor-related mechanism. *Eur J Neurosci* 9(10): 2142–2151
115 Parker E, Van Heek M, Stamford A (2002) Neuropeptide Y receptors as targets for anti-obesity drug development: Perspective and current status. *Eur J Pharmacol* 440(2–3): 173–187
116 Redrobe JP, Dumont Y, St-Pierre JA, Quirion R (1999) Multiple receptors for neuropeptide Y in the hippocampus: Putative roles in seizures and cognition. *Brain Res* 848(1–2): 153–166
117 Sheikh SP, Feldthus N, Orkild H, Goke R, McGregor GP, Turner D, Moller M, Stuenkel EL (1998) Neuropeptide Y2 receptors on nerve endings from the rat neurohypophysis regulate vasopressin and oxytocin release. *Neuroscience* 82(1): 107–115
118 Silva AP, Cavadas C, Grouzmann E (2002) Neuropeptide Y and its receptors as potential therapeutic drug targets. *Clin Chim Acta* 326(1–2): 3–25
119 Pasternak GW (1988) Studies of multiple morphine and enkephalin receptors: Evidence for mu1 receptors. *Adv Exp Med Biol* 236: 81–93
120 Pasternak GW (1988) Multiple morphine and enkephalin receptors and the relief of pain. *JAMA* 259(9): 1362–1367
121 Akil H, Watson SJ, Young E, Lewis ME, Khachaturian H, Walker JM (1984) Endogenous opioids: Biology and function. *Annu Rev Neurosci* 7: 223–255
122 Martinez JL Jr, Olson K, Hilston C (1984) Opposite effects of Met-enkephalin and Leu-enkephalin on a discriminated shock-escape task. *Behav Neurosci* 98(3): 487–495
123 Izquierdo I, Dias RD, Souza DO, Carrasco MA, Elisabetsky E, Perry ML (1980) The role of opioid peptides in memory and learning. *Behav Brain Res* 1(6): 451–468
124 Izquierdo I, Dias RD (1981) Retrograde amnesia caused by Met-, Leu- and des-Try-Met-enkephalin in the rat and its reversal by naloxone. *Neurosci Lett* 22(2): 189–193
125 Colombo PJ, Thompson KR, Martinez JL Jr, Bennett EL, Rosenzweig MR (1993) Dynorphin(1-13) impairs memory formation for aversive and appetitive learning in chicks. *Peptides* 14(6): 1165–1170

126 Colombo PJ, Martinez JL Jr, Bennett EL, Rosenzweig MR (1992) Kappa opioid receptor activity modulates memory for peck-avoidance training in the 2-day-old chick. *Psychopharmacology (Berl)* 108(1–2): 235–240
127 Sandin J, Nylander I, Georgieva J, Schott PA, Ogren SO, Terenius L (1998) Hippocampal dynorphin B injections impair spatial learning in rats: a kappa-opioid receptor-mediated effect. *Neuroscience* 85(2): 375–382
128 Jiang HK, Owyang VV, Hong JS, Gallagher M (1989) Elevated dynorphin in the hippocampal formation of aged rats: Relation to cognitive impairment on a spatial learning task. *Proc Natl Acad Sci USA* 86(8): 2948–2951
129 Farr SA, Banks WA, Morley JE (2000) Estradiol potentiates acetylcholine and glutamate-mediated post-trial memory processing in the hippocampus. *Brain Res* 864(2): 263–269
130 Gibbs RB (1999) Estrogen replacement enhances acquisition of a spatial memory task and reduces deficits associated with hippocampal muscarinic receptor inhibition. *Horm Behav* 36(3): 222–233
131 Heikkinen T, Puolivali J, Liu L, Rissanen A, Tanila H (2002) Effects of ovariectomy and estrogen treatment on learning and hippocampal neurotransmitters in mice. *Horm Behav* 41(1): 22–32
132 Daniel JM, Dohanich GP (2001) Acetylcholine mediates the estrogen-induced increase in NMDA receptor binding in CA1 of the hippocampus and the associated improvement in working memory. *J Neurosci* 21(17): 6949–6956
133 Fugger HN, Foster TC, Gustafsson J, Rissman EF (2000) Novel effects of estradiol and estrogen receptor alpha and beta on cognitive function. *Brain Res* 883(2): 258–264
134 Rissman EF, Heck AL, Leonard JE, Shupnik MA, Gustafsson JA (2002) Disruption of estrogen receptor beta gene impairs spatial learning in female mice. *Proc Natl Acad Sci USA* 99(6): 3996–4001
135 Woolley CS, Gould E, Frankfurt M, McEwen BS (1990) Naturally occurring fluctuation in dendritic spine density on adult hippocampal pyramidal neurons. *J Neurosci* 10(12): 4035–4039
136 McEwen B (2002) Estrogen actions throughout the brain. *Recent Progress in Hormone Research* 57: 357–384
137 Horvath KM, Hartig W, Van der Veen R, Keijser JN, Mulder J, Ziegert M, Van der Zee EA, Harkany T, Luiten PG (2002) 17beta-estradiol enhances cortical cholinergic innervation and preserves synaptic density following excitotoxic lesions to the rat nucleus basalis magnocellularis. *Neuroscience* 110(3): 489–504
138 Frick KM, Fernandez SM, Bulinski SC (2002) Estrogen replacement improves spatial reference memory and increases hippocampal synaptophysin in aged female mice. *Neuroscience* 115(2): 547–558
139 Resnick SM, Maki PM (2001) Effects of hormone replacement therapy on cognitive and brain aging. *Ann N Y Acad Sci* 949: 203–214
140 Zec RF, Trivedi MA (2002) The effects of estrogen replacement therapy on neuropsychological functioning in post-menopausal women with and without dementia: A critical and theoretical review. *Neuropsychol Rev* 12(2): 65–109
141 Baulieu EE, Robel P, Schumacher M (2001) Neurosteroids: Beginning of the story. *Int Rev Neurobiol* 46: 1–32
142 Paul SM, Purdy RH (1992) Neuroactive steroids. *FASEB J* 6: 2311–2311
143 Bergeron R, de Montigny C, Debonnel G (1996) Potentiation of neuronal NMDA response induced by dehydroepiandrosterone and its suppression by progesterone: Effects mediated via sigma receptors. *J Neurosci* 16(3): 1193–1202
144 Maurice T, Junien JL, Privat A (1997) Dehydroepiandrosterone sulfate attenuates dizocilpine-induced learning impairment in mice via sigma 1-receptors. *Behav Brain Res* 83(1–2): 159–164
145 Darnaudery M, Koehl M, Piazza PV, Le Moal M, Mayo W (2000) Pregnenolone sulfate increases hippocampal acetylcholine release and spatial recognition. *Brain Res* 852(1): 173–179
146 Pallares M, Darnaudery M, Day J, Le Moal M, Mayo W (1998) The neurosteroid pregnenolone sulfate infused into the nucleus basalis increases both acetylcholine release in the frontal cortex or amygdala and spatial memory. *Neuroscience* 87(3): 551–558
147 Akwa Y, Ladurelle N, Covey DF, Baulieu EE (2001) The synthetic enantiomer of pregnenolone sulfate is very active on memory in rats and mice, even more so than its physiological neurosteroid counterpart: distinct mechanisms? *Proc Natl Acad Sci USA* 98(24): 14033–14037
148 Mathis C, Vogel E, Cagniard B, Criscuolo F, Ungerer A (1996) The neurosteroid pregnenolone sulfate blocks deficits induced by a competitive NMDA antagonist in active avoidance and lever-

press learning tasks in mice. *Neuropharmacology* 35(8): 1057–1064
149 Meziane H, Mathis C, Paul SM, Ungerer A (1996) The neurosteroid pregnenolone sulfate reduces learning deficits induced by scopolamine and has promnestic effects in mice performing an appetitive learning task. *Psychopharmacology* 126(4): 323–330
150 Reddy DS, Kulkarni SK (1998) The effects of neurosteroids on acquisition and retention of a modified passive-avoidance learning task in mice. *Brain Res* 791(1–2): 108–116
151 Vallee M, Mayo W, Darnaudery M, Corpechot C, Young J, Koehl M, Le Moal M, Baulieu EE, Robel P, Simon H (1997) Neurosteroids: Deficient cognitive performance in aged rats depends on low pregnenolone sulfate levels in the hippocampus. *Proc Natl Acad Sci USA* 94(26): 14865–14870
152 Mayo W, Le Moal M, Abrous DN (2001) Pregnenolone sulfate and aging of cognitive functions: behavioral, neurochemical, and morphological investigations. *Horm Behav* 40(2): 215–217
153 Vallee M, Mayo W, Koob GF, Le Moal M (2001) Neurosteroids in learning and memory processes. *Int Rev Neurobiol* 46: 273–320
154 Bowlby MR (1993) Pregnenolone sulfate potentiation of N-methyl-D-aspartate receptor channels in hippocampal neurons. *Mol Pharmacol* 43: 813–819
155 Majewska MD, Bluet, Pajot M, T, Robel P, Baulieu EE (1989) Pregnenolone sulfate antagonizes barbituate-induced hypnosis. *Pharmacol Biochem Behav* 33(701): 701–703
156 Nilsson KR, Zorumski CF, Covey DF (1998) Neurosteroid analogues. 6. The synthesis and GABAA receptor pharmacology of enantiomers of dehydroepiandrosterone sulfate, pregnenolone sulfate, and (3alpha,5beta)-3-hydroxypregnan-20-one sulfate. *J Med Chem* 41(14): 2604–2613
157 Martignoni E, Costa A, Sinforiani E, Liuzzi A, Chiodini P, Mauri M, Bono G, Nappi G (1992) The brain as a target for adrenocortical steroids: Cognitive implications. *Psychoneuoroendocrinol* 17: 343–354
158 de Kloet ER, Grootendorst J, Karssen AM, Oitzl MS (2002) Gene × environment interaction and cognitive performance: animal studies on the role of corticosterone. *Neurobiol Learn Mem* 78(3): 570–577
159 Sapolsky RM (1993) Potential behavioral modification of glucocorticoid damage to the hippocampus. *Behav Brain Res* 57: 175–182
160 Oitzl MS, de Kloet ER (1992) Selective corticosteroid antagonists modulate specific aspects of spatial orientation learning. *Behav Neurosci* 106(1): 62–71
161 Sandi C, Rose SPR (1994) Corticosterone enhances long-term retention in one-day-old chicks trained in a weak passive avoidance learning paradigm. *Brain Res* 647: 106–112
162 de Quervain DJ, Roozendaal B, McGaugh JL (1998) Stress and glucocorticoids impair retrieval of long-term spatial memory. *Nature* 394: 787–790
163 McGaugh JL, Roozendaal B (2002) Role of adrenal stress hormones in forming lasting memories in the brain. *Curr Opin Neurobiol* 12(2): 205–210
164 Buchanan TW, Lovallo WR (2001) Enhanced memory for emotional material following stress-level cortisol treatment in humans. *Psychoneuroendocrinology* 26(3): 307–317
165 Porter NM, Landfield PW (1998) Stress hormones and brain aging: Adding injury to insult. *Nature Neuroscience* 1(1): 3–4
166 McEwen BS (1997) Possible mechanisms for atrophy of the human hippocampus. *Molecular Psychiatry* 2: 255–262
167 Heinrichs SC, Joppa M (2001) Dissociation of arousal-like from anxiogenic-like actions of brain corticotropin-releasing factor receptor ligands in rats. *Behav Brain Res* 122: 43–50
168 Heinrichs SC, Vale EA, Lapsansky J, Behan DP, McClure LV, Ling N, De Souza EB, Schulteis G (1997) Enhancement of performance in multiple learning tasks by corticotropin-releasing factor-binding protein ligand inhibitors. *Peptides* 18(5): 711–716
169 Martinez JL Jr, Janak PH, Weinberger SB, Schulteis G, Derrick BE (1990) Enkephalin influences on behavioral and neural plasticity: Mechanisms of action. *NIDA Res Monogr* 97: 48–78
170 Bliss TV, Collingridge GL (1993) A synaptic model of memory: Long-term potentiation in the hippocampus. *Nature* 361(6407): 31–39
171 Gibbs DM (1984) Dissociation of oxytocin, vasopressin and corticotropin secretion during different types of stress. *Life Sci* 35(5): 487–491
172 Buijs RM (1983) Vasopressin and oxytocin—their role in neurotransmission. *Pharmacol Ther* 22(1): 127–141
173 Buijs RM, De Vries GJ, Van Leeuwen FW, Swaab DF (1983) Vasopressin and oxytocin:

Distribution and putative functions in the brain. *Prog Brain Res* 60: 115–122
174 Chepkova AN, Kapai NA, Skrebitskii VG (2001) Arginine vasopressin fragment AVP(4–9) facilitates induction of long-term potentiation in the hippocampus. *Bull Exp Biol Med* 131(2): 136–138
175 Dubrovsky B, Tatarinov A, Gijsbers K, Harris J, Tsiodras A (2003) Effects of arginine-vasopressin (AVP) on long-term potentiation in intact anesthetized rats. *Brain Res Bull* 59(6): 467–472
176 Chen C, Diaz Brinton RD, Shors TJ, Thompson RF (1993) Vasopressin induction of long-lasting potentiation of synaptic transmission in the dentate gyrus. *Hippocampus* 3(2): 193–203
177 Kinsley CH, Madonia L, Gifford GW, Tureski K, Griffin GR, Lowry C, Williams J, Collins J, McLearie H, Lambert KG (1999) Motherhood improves learning and memory. *Nature* 402(6758): 137–138
178 Tomizawa K, Iga N, Lu YF, Moriwaki A, Matsushita M, Li ST, Miyamoto O, Itano T, Matsui H (2003) Oxytocin improves long-lasting spatial memory during motherhood through MAP kinase cascade. *Nat Neurosci* 6(4): 384–390
179 Korte M, Carroll P, Wolf E, Brem G, Thoenen H, Bonhoeffer T (1995) Hippocampal long-term potentiation is impaired in mice lacking brain-derived neurotrophic factor. *Proc Natl Acad Sci USA* 92(19): 8856–8860
180 Patterson SL, Abel T, Deuel TA, Martin KC, Rose JC, Kandel ER (1996) Recombinant BDNF rescues deficits in basal synaptic transmission and hippocampal LTP in BDNF knockout mice. *Neuron* 16(6): 1137–1145
181 Gooney M, Shaw K, Kelly A, O'Mara SM, Lynch MA (2002) Long-term potentiation and spatial learning are associated with increased phosphorylation of TrkB and extracellular signal-regulated kinase (ERK) in the dentate gyrus: Evidence for a role for brain-derived neurotrophic factor. *Behav Neurosci* 116(3): 455–463
182 Messaoudi E, Ying SW, Kanhema T, Croll SD, Bramham CR (2002) Brain-derived neurotrophic factor triggers transcription-dependent, late phase long-term potentiation *in vivo*. *J Neurosci* 22(17): 7453–7461
183 Kovalchuk Y, Hanse E, Kafitz KW, Konnerth A (2002) Postsynaptic Induction of BDNF-Mediated Long-Term Potentiation. *Science* 295(5560): 1729–1734
184 Mizuno M, Yamada K, He J, Nakajima A, Nabeshima T (2003) Involvement of BDNF receptor TrkB in spatial memory formation. *Learn Mem* 10(2): 108–115
185 Buwalda B, de Boer SF, Van Kalkeren AA, Koolhaas JM (1997) Physiological and behavioral effects of chronic intracerebroventricular infusion of corticotropin-releasing factor in the rat. *Psychoneuroendocrinology* 22(5): 297–309
186 Heinrichs SC, Menzaghi F, Schulteis G, Koob GF, Stinus L (1995) Suppression of corticotropin-releasing factor in the amygdala attenuates aversive consequences of morphine withdrawal. *Behav Pharmacol* 6(1): 74–80
187 Heinrichs SC, Menzaghi F, Merlo Pich E, Britton KT, Koob GF (1995) The role of CRF in behavioral aspects of stress. *Ann N Y Acad Sci* 771: 92–104
188 Wang HL, Tsai LY, Lee EH (2000) Corticotropin-releasing factor produces a protein synthesis—dependent long-lasting potentiation in dentate gyrus neurons. *J Neurophysiol* 83(1): 343–349
189 Wang HL, Wayner MJ, Chai CY, Lee EH (1998) Corticotrophin-releasing factor produces a long-lasting enhancement of synaptic efficacy in the hippocampus. *Eur J Neurosci* 10(11): 3428–3437
190 Rebaudo R, Melani R, Balestrino M, Izvarina N (2001) Electrophysiological effects of sustained delivery of CRF and its receptor agonists in hippocampal slices. *Brain Res* 922(1): 112–117
191 Blank T, Nijholt I, Eckart K, Spiess J (2002) Priming of long-term potentiation in mouse hippocampus by corticotropin-releasing factor and acute stress: Implications for hippocampus-dependent learning. *J Neurosci* 22(9): 3788–3794
192 Kramar EA, Armstrong DL, Ikeda S, Wayner MJ, Harding JW, Wright JW (2001) The effects of angiotensin IV analogs on long-term potentiation within the CA1 region of the hippocampus *in vitro*. *Brain Res* 897(1–2): 114–121
193 Wayner MJ, Armstrong DL, Phelix CF, Wright JW, Harding JW (2001) Angiotensin IV enhances LTP in rat dentate gyrus *in vivo*. *Peptides* 22(9): 1403–1414
194 Qian Z, Gilbert ME, Colicos MA, Kandel ER, Kuhl D (1993) Tissue-plasminogen activator is induced as an immediate-early gene during seizure, kindling and long-term potentiation. *Nature* 361(6411): 453–457
195 Carmeliet P, Schoonjans L, Kieckens L, Ream B, Degen J, Bronson R, De Vos R, van den Oord

JJ, Collen D, Mulligan RC (1994) Physiological consequences of loss of plasminogen activator gene function in mice. *Nature* 368(6470): 419–424
196 Madani R, Hulo S, Toni N, Madani H, Steimer T, Muller D, Vassalli JD (1999) Enhanced hippocampal long-term potentiation and learning by increased neuronal expression of tissue-type plasminogen activator in transgenic mice. *Embo J* 18(11): 3007–3012
197 Chen ZL, Strickland S (1997) Neuronal death in the hippocampus is promoted by plasmin-catalyzed degradation of laminin. *Cell* 91(7): 917–925
198 Zhuo M, Holtzman DM, Li Y, Osaka H, DeMaro J, Jacquin M, Bu G (2000) Role of tissue plasminogen activator receptor LRP in hippocampal long-term potentiation. *J Neurosci* 20(2): 542–549
199 Wright JW, Kramar EA, Meighan SE, Harding JW (2002) Extracellular matrix molecules, long-term potentiation, memory consolidation and the brain angiotensin system. *Peptides* 23(1): 221–246
200 Matsumoto A, Arai Y (1977) Precocious puberty and synaptogenesis in the hypothalamic arcuate nucleus in pregnant mare serum gonadotropin (PMSG) treated immature female rats. *Brain Res* 129(2): 375–378
201 Cordoba Montoya DA, Carrer HF (1997) Estrogen facilitates induction of long-term potentiation in the hippocampus of awake rats. *Brain Res* 778(2): 430–438
202 Good M, Day M, Muir JL (1999) Cyclical changes in endogenous levels of oestrogen modulate the induction of LTD and LTP in the hippocampal CA1 region. *Eur J Neurosci* 11(12): 4476–4480
203 Singh M, Meyer EM, Millard WJ, Simpkins JW (1994) Ovarian steroid deprivation results in a reversible learning impairment and compromised cholinergic function in female Sprague-Dawley rats. *Brain Res* 644(2): 305–312
204 O'Neal MF, Means LW, Poole MC, Hamm RJ (1996) Estrogen affects performance of ovariectomized rats in a two-choice water-escape working memory task. *Psychoneuroendocrinology* 21(1): 51–65
205 Kim JS, Kim HY, Kim JH, Shin HK, Lee SH, Lee YS, Son H (2002) Enhancement of rat hippocampal long-term potentiation by 17 beta-estradiol involves mitogen-activated protein kinase-dependent and -independent components. *Neurosci Lett* 332(1): 65–69
206 Gureviciene I, Puolivali J, Pussinen R, Wang J, Tanila H, Ylinen A (2003) Estrogen treatment alleviates NMDA-antagonist induced hippocampal LTP blockade and cognitive deficits in ovariectomized mice. *Neurobiol Learn Mem* 79(1): 72–80
207 Lisman J (1989) A mechanism for the Hebb and the anti-Hebb processes underlying learning and memory. *Proc Natl Acad Sci USA* 86(23): 9574–9578
208 Grant SG, O'Dell TJ, Karl KA, Stein PL, Soriano P, Kandel ER (1992) Impaired long-term potentiation, spatial learning, and hippocampal development in fyn mutant mice. *Science* 258(5090): 1903–1910
209 O'*Meara* G, Coumis U, Ma SY, Kehr J, Mahoney S, Bacon A, Allen SJ, Holmes F, Kahl U, Wang FH et al. (2000) Galanin regulates the postnatal survival of a subset of basal forebrain cholinergic neurons. *Proc Natl Acad Sci USA* 97(21): 11569–11574
210 Crawley JN, Mufson EJ, Hohmann JG, Teklemichael D, Steiner RA, Holmberg K, Xu ZQ, Blakeman KH, Xu XJ, Wiesenfeld-Hallin Z et al. (2002) Galanin over-expressing transgenic mice. *Neuropeptides* 36(2–3): 145–156
211 Harris EW, Cotman CW (1986) Long-term potentiation of guinea pig mossy fiber responses is not blocked by N-methyl D-aspartate antagonists. *Neurosci Lett* 70(1): 132–137
212 Derrick BE, Martinez JL Jr (1994) Frequency-dependent associative long-term potentiation at the hippocampal mossy fiber-CA3 synapse. *Proc Natl Acad Sci USA* 91(22): 10290–10294
213 Derrick BE, Martinez JL Jr (1994) Opioid receptor activation is one factor underlying the frequency dependence of mossy fiber LTP induction. *J Neurosci* 14(7): 4359–4367
214 Derrick BE, Martinez JL Jr (1996) Associative, bidirectional modifications at the hippocampal mossy fibre-CA3 synapse. *Nature* 381(6581): 429–434
215 Derrick BE, Rodriguez SB, Lieberman DN, Martinez JL Jr (1992) Mu opioid receptors are associated with the induction of hippocampal mossy fiber long-term potentiation. *J Pharmacol Exp Ther* 263(2): 725–733
216 Derrick BE, Weinberger SB, Martinez JL Jr (1991) Opioid receptors are involved in an NMDA receptor-independent mechanism of LTP induction at hippocampal mossy fiber-CA3 synapses. *Brain Res Bull* 27(2): 219–223

217 Do VH, Martinez CO, Martinez JL Jr, Derrick BE (2002) Long-term potentiation in direct perforant path projections to the hippocampal CA3 region *in vivo. J Neurophysiol* 87(2): 669–678
218 Jin W, Chavkin C (1999) Mu opioids enhance mossy fiber synaptic transmission indirectly by reducing GABAB receptor activation. *Brain Res* 821(2): 286–293
219 Ying SW, Futter M, Rosenblum K, Webber MJ, Hunt SP, Bliss TV, Bramham CR (2002) Brain-derived neurotrophic factor induces long-term potentiation in intact adult hippocampus: Requirement for ERK activation coupled to CREB and upregulation of Arc synthesis. *J Neurosci* 22(5): 1532–1540
220 Weisskopf MG, Zalutsky RA, Nicoll RA (1993) The opioid peptide dynorphin mediates heterosynaptic depression of hippocampal mossy fibre synapses and modulates long-term potentiation. *Nature* 362(6419): 423–427
221 Barea-Rodriguez EJ, Rivera DT, Jaffe DB, Martinez JL Jr (2000) Protein synthesis inhibition blocks the induction of mossy fiber long-term potentiation *in vivo. J Neurosci* 20(22): 8528–8532
222 Thompson KJ, Orfila JE, Archanta P, Martinez JJ (2003) *submitted to Cell and Molecular Biology*
223 Roberts LA, Large CH, O'Shaughnessy CT, Morris BJ (1997) Long-term potentiation in perforant path/granule cell synapses is associated with a post-synaptic induction of proenkephalin gene expression. *Neurosci Lett* 227(3): 205–208
224 Whittaker E, Vereker E, Lynch MA (1999) Neuropeptide Y inhibits glutamate release and long-term potentiation in rat dentate gyrus. *Brain Res* 827(1–2): 229–233
225 Sarter M (1991) Taking stock of cognition enhancers. *Trends Pharmacol Sci* 12(12): 456–461
226 Korol DL (2002) Enhancing cognitive function across the life-span. *Ann N Y Acad Sci* 959: 167–179
227 Benton D (1990) The impact of increasing blood glucose on psychological functioning. *Biol Psychol* 30(1): 13–19
228 Pollitt E, Lewis NL, Garza C, Shulman RJ (1982) Fasting and cognitive function. *J Psychiatr Res* 17(2): 169–174
229 Gold PE, Stone WS (1988) Neuroendocrine effects on memory in aged rodents and humans. *Neurobiol Aging* 9(5–6): 709–717
230 Wenk GL (1989) An hypothesis on the role of glucose in the mechanism of action of cognitive enhancers. *Psychopharmacology (Berl)* 99(4): 431–438
231 Lombardi WJ, Weingartner H (1995) Pharmacoloigcal treatment of impaired memory function. In: AD Baddeley, BA Wilson, FN Watts (eds): *Handbook of Memory Disorders*. Wiley, New York, 577–601
232 Olton DS, Wenk L (1990) The development of behavioral tests to assess the effects of cognitive enhancers. *Pharmacopsychiatry* 23 Suppl 2: 65–69
233 Cardinal RN, Parkinson JA, Hall J, Everitt BJ (2002) Emotion and motivation: the role of the amygdala, ventral striatum, and prefrontal cortex. *Neurosci Biobehav Rev* 26(3): 321–352
234 Ochsner KN (2000) Are affective events richly recollected or simply familiar? The experience and process of recognizing feelings past. *J Exp Psychol Gen* 129(2): 242–261
235 Gold LH (1999) Hierarchical strategy for phenotypic analysis in mice. *Psychopharmacology* 147: 2–4
236 Crawley JN, Paylor R (1997) A proposed test battery and constellations of specific behavior paradigms to investigate the behavioral phenotypes of transgenic and knockout mice. *Hormones and Behavior* 31: 197–211
237 Sarter M, Bruno JP, Givens B, Moore H, McGaughy J, McMahon K (1996) Neuronal mechanisms mediating drug-induced cognition enhancement: cognitive activity as a necessary intervening variable. *Brain Res Cogn Brain Res* 3(3–4): 329–343
238 Roozendaal B, Brunson KL, Holloway BL, McGaugh JL, Baram TZ (2002) Involvement of stress-released corticotropin-releasing hormone in the basolateral amygdala in regulating memory consolidation. *Proc Natl Acad Sci USA* 99(21): 13908–13913
239 Behan DP, Heinrichs SC, Troncoso JC, Liu X-J, Kawas CH, Ling N, De Souza EB (1995) Displacement of corticotropin releasing factor from its binding protein as a possible treatment for Alzheimer's disease. *Nature* 378: 284–287
240 Radulovic J, Fischer A, Katerkamp U, Spiess J (2000) Role of regional neurotransmitter receptors in corticotropin-releasing factor (CRF)-mediated modulation of fear conditioning. *Neuropharmacology.* 39(4): 707–710
241 Bernardi F, Lanzone A, Cento RM, Spada RS, Pezzani I, Genazzani AD, Luisi S, Luisi M,

Petraglia F, Genazzani AR (2000) Allopregnanolone and dehydroepiandrosterone response to corticotropin-releasing factor in patients suffering from Alzheimer's disease and vascular dementia. *European J Endocrinology* 142(5): 466–471

242 Behan DP, Grigoriadis DE, Lovenberg T, Chalmers D, Heinrichs S, Liaw C, De Souza EB (1996) Neurobiology of corticotropin releasing factor (CRF) receptors and CRF-binding protein: Implications for the treatment of CNS disorders. *Molecular Psychiatry* 1: 265–277

243 Thornton PL, Ingram RL, Sonntag WE (2000) Chronic [D-Ala2]-growth hormone-releasing hormone administration attenuates age-related deficits in spatial memory. *J Gerontol A Biol Sci Med Sci* 55(2): B106–112

244 Shanley LJ, Irving AJ, Harvey J (2001) Leptin enhances NMDA receptor function and modulates hippocampal synaptic plasticity. *J Neurosci* 21(24): RC186

245 Li XL, Aou S, Oomura Y, Hori N, Fukunaga K, Hori T (2002) Impairment of long-term potentiation and spatial memory in leptin receptor-deficient rodents. *Neuroscience* 113(3): 607–615

246 Hiramatsu M, Inoue K (2000) Improvement by low doses of nociceptin on scopolamine-induced impairment of learning and/or memory. *European J Pharmacology* 395(2): 149–156

247 Telegdy G, Adamik A (2002) The action of orexin A on passive avoidance learning. Involvement of transmitters. *Regulatory Peptides* 104(1–3): 105–110

248 Jaeger LB, Farr SA, Banks WA, Morley JE (2002) Effects of orexin-A on memory processing. *Peptides* 23(9): 1683–1688

249 Lenard L, Kertes E (2002) Influence of passive avoidance learning by substance P in the basolateral amygdala. *Acta Biol Hung* 53(1–2): 95–104

250 Hasenohrl RU, Souza-Silva MA, Nikolaus S, Tomaz C, Brandao ML, Schwarting RK, Huston JP (2000) Substance P and its role in neural mechanisms governing learning, anxiety and functional recovery. *Neuropeptides* 34(5): 272–280

251 Markowska AL, Mooney M, Sonntag WE (1998) Insulin-like growth factor-1 ameliorates age-related behavioral deficits. *Neuroscience* 87(3): 559–569

252 Wu HC, Chen KY, Lee WY, Lee EHY (1997) Antisense oligonucleotides to corticotropin-releasing factor impair memory retention and increase exploration in rats. *Neuroscience* 78(1): 147–153

253 Weninger SC, Dunn AJ, Muglia LJ, Dikkes P, Miczek KA, Swiergiel AH, Berridge CW, Majzoub JA (1999) Stress-induced behaviors require the corticotropin-releasing hormone (CRH) receptor, but not CRH. *Proc Natl Acad Sci USA* 96(14): 8283–8288

254 Heinrichs SC, Stenzel-Poore MP, Gold LH, Battenberg E, Bloom FE, Koob GF, Vale WW, Merlo Pich E (1996) Learning deficits in transgenic mice with central over-expression of corticotropin-releasing factor. *Neuroscience* 74(2): 303–311

255 Liebsch G, Landgraf R, Engelmann M, Loerscher P, Holsboer F (1999) Differential behavioural effects of chronic infusion of CRH 1 and CRH 2 receptor antisense oligonucleotides into the rat brain. *J Psychiatric Research* 33(2): 153–163

256 Contarino A, Dellu F, Koob GF, Smith GW, Lee KF, Vale W, Gold LH (1999) Reduced anxiety-like and cognitive performance in mice lacking the corticotropin-releasing factor receptor 1. *Brain Res* 835(1): 1–9

257 Pepin M-C, Pothier F, Barden N (1992) Impaired type II glucocorticoid-receptor function in mice bearing antisense RNA transgene. *Nature* 355: 725–728

258 Oitzl MS, Reichardt HM, Joels M, de Kloet ER (2001) Point mutation in the mouse glucocorticoid receptor preventing DNA binding impairs spatial memory. *Proc Natl Acad Sci USA* 98(22): 12790–12795

259 Vallee M, Shen W, Heinrichs SC, Zorumski CF, Covey DF, Koob GF, Purdy RH (2001) Steroid structure and pharmacological properties determine the anti-amnesic effects of pregnenolone sulphate in the passive avoidance task in rats. *Eur J Neurosci* 14(12): 2003–2010

Cognitive Enhancing Drugs
Edited by J.J. Buccafusco

Natural products as cognition enhancing agents

Keith A. Wesnes[1,2], Andrea Zangara[1,2], Andrew Scholey[1] and David Kennedy[1]

[1] *Human Cognitive Neuroscience Unit, Northumbria University, Newcastle-upon-Tyne NE1 8ST, UK*
[2] *Cognitive Drug Research Ltd, CDR House, 24 Portman Road, Reading RG30 1EA, UK*

Introduction

Natural products have a widespread public appeal that appears only to be growing. The fastest growth is in the western world although in lesser economically developed countries the demand remains as strong as ever. This appeal is aided by the almost universal, though completely irrational assumption, that if a product is natural it must be safe. This chapter will consider those naturally-occurring substances that are believed to beneficially affect cognitive function.

The definition of cognitive enhancement adopted for this chapter is an improvement in core aspects of cognitive function, which is crucial to the conduct of the activities of daily living. For normal volunteers, this represents identifying an improvement on a valid measure of cognitive function known to be important for everyday behaviour. The same is true for a patient population known to have a cognitive impairment, the point simply being that for enhancement to have taken place, cognitive function needs to be measurably better than it was before treatment. The improvements must be in domains of cognitive function which are widely recognised to be important to everyday behaviour, such as attention and memory. Further, the improvements must be assessed by objective tests that are appropriate, valid and sensitive instruments for assessing the areas of function of interest. Finally, the study design must reflect the current gold standard in psychopharmacology, i.e., be at the minimum randomised, double-blind and placebo controlled,

Not surprisingly there is considerable interest in natural products that may enhance human cognitive function. A thorough review of the field would fill this and probably several more volumes, even if restricted to human studies. The 'natural product' field is widely characterised as being an area of 'soft science', where trials are rarely conducted with the rigour required for publication in leading peer reviewed human cognitive psychopharmacology journals. Often claims are based on anecdotal evidence, uncontrolled trials or work with animals. While such evidence is helpful in helping to select a substance to study, this evidence cannot form part of the 'core' evidence to support a particular claim for its cognition-enhancing potential.

This chapter will consider a number of substances that have been the subject of repeated scientific scrutiny and which have shown cognition enhancing properties. The substances described will either have a clear mechanism of action for which the chemical structure of the active ingredient is known, or will be available in standardised preparations in which the amounts and ratios of the various ingredients can be guaranteed to be present within acceptable limits. Other promising substances with little experimental evidence to date are covered in Table 1. Trials of compounds for which such standardisation is not possible, at least for the known active ingredients, will not be covered, as such trials cannot be replicated using comparable formulations.

Ginseng

'Ginseng' is generally taken to refer to the dried root of several species in the plant genus *Panax* (Araliaceae family). The most widely used family member is *Panax ginseng*, which is indigenous to the Far East (most notably China and Korea). It was first cultivated around 11 BC, and has a medical history (as a wild herb) stretching back more than 5000 years [1]. Other members of the genus include *Panax quinquefolius* (American), *Panax notoginseng* and *Panax japonicus*. Given that ginseng has been estimated to have the second highest financial turnover of any herb (after *Ginkgo biloba*) in the US marketplace [2] and enjoys ubiquitous and undocumented use throughout a number of societies and traditional medicinal systems, it is potentially the most widely taken herbal product in the world. While ginseng is taken both as a general tonic and prophylactic agent it is also widely taken in western markets to ameliorate 'memory loss' and 'absentmindedness' [3].

The major active constituents of the *Panax genus* are thought to be triterpenoid glycosides or saponins, also known as ginsenosides, of which over 30 individual examples, many of which exist only in minute amounts, have been identified [4]. The ginsenoside content of ginseng extracts can vary depending on the species, the age and part of the plant, the preservation method, the season of harvest, and the extraction method [5–7] also notes that no herb is more subject to adulteration and misrepresentation. Currently, the only standardised extract that has attracted widespread research attention is G115 (Pharmaton SA) which is standardised to an invariable 4% of ginsenosides.

A number of *in vitro* and *in vivo* properties potentially relevant to the modulation of cognitive performance have been attributed to single and multiple ginsenosides and whole extracts of ginseng. These include; effects on vasoconstriction. [8, 9], a beneficial influence on blood flow through modulation of platelet aggregation [10, 11], roles in both cardio-protection [12–14] and neuroprotection following a number of insults [15–18], shifting of the hormonal balance of the hypothalamic-pituitary-adrenal system [19–21], modulation of a number of neurotransmitter systems [22–25], and modulation of blood glucose levels in both diabetic [26–28], and non-diabetic humans [27, 29].

Table 1. Summary of promising cognition-enhancing substances not as completely investigated as those discussed in the text

Name	Anecdotal use	Alleged actions	Mechanism of action	– Drug interactions – Side effects – Precautions	Usual dosage or content	The science
Bacopa monniera (Water yssop, Brahmi)	In Ayurveda, Brahmi is described as a brain tonic that promotes mental functioning and has general rejuvenative effects.	Facilitation of memory retention and alleviation of the symptoms of anxiety and convulsive disorders. Powerful antioxidant actions are also attributed to *Bacopa*.	A number of biologically active compounds are present in the plant including alkaloids, saponins and sterols, however those responsible for the memory enhancing effects are triterpenoid saponins called "bacosides".	May interact with CNS active drugs. Some aminoglycoside antibiotics such as clindamycin may enhance the neuromuscular relaxing action of Brahmi. No adverse effects are reported from clinical trials.	2–6 g per day.	*Bacopa* has demonstrated antioxidant properties [188] and the protection of mental function in epilepsy when drug phenytion is taken [189]. *Bacopa* administration improved learning [190] and had antidepressant activity [191] in rats. Recently bacopa improved learning and memory in healthy humans with maximal effects evident after 12 weeks [192].
Fava beans (*Vicia faba*)	A vegetable that is widely cultivated in China and the Mediterranean region.	The L-dopa in Fava and *Mucuna pruriens* beans facilitates memory, energy, sense of well-being, sex drive, and helps to control the symptoms of Parkinson's Disease.	Fava and *Mucuna pruriens* beans (and the entire fava plant) have high concentrations of L-dopa, an amino acid that is enzymatically converted to the neurotransmitter dopamine.	Taken in combination with monoamine oxidase inhibitors (MAOIs), blood pressure can increase dangerously. Nausea, dyskinesia and allergic reactions have been reported. People with Favism can develop a serious condition called hemolytic anemia [193].	L-dopa in *Mucuna pruriens* is from 3 to 7%, in Fava beans is below 0.5%.	*Mucuna* seed powder (HP-200) was given to 60 Parkinson's patients for 12 weeks, whilst a control group of 26 patients were given synthetic treatments. The Unified Parkinson's Disease Rating Scale showed that *Mucuna* was effective with minor adverse consequences (Hp-200). Similar results were obtained with fava beans [194, 195].
Mucuna pruriens	In the Ayurvedic system, *Mucuna* powder is used for the treatment of Parkinson's disease					

(Continued on next page)

Table 1. (Continued)

Name	Anecdotal use	Alleged actions	Mechanism of action	– Drug interactions – Side effects – Precautions	Usual dosage or content	The science
Ashwaganda (*Withania somnifera*)	Ashwagandha is a bush cultivated in India and North America. It has been used for thousands of years by Ayurvedic practitioners.	Ashwagandha is supposed to have anti-inflammatory, antitumor, anti-stress, antioxidant, mind-boosting, and rejuvenating properties.	Ashwagandha contains flavonoids and active ingredients of the withanolide class. Ashwagandha has GABA-like activity (anxiolityc activity) and can increase acetylcholine level in brain (cognition enhancement).	Should not be used in combination with sedative drugs. Potential source of hypoglycemic, diuretic and hypocholesterolemic agents.	100–200 mg standardised extract daily.	A research indicates that ashwagandha stimulates the growth of axons and dendrites and levels of antioxidantes [196]. Several studies in rodents showed Ashwagandha had memory boosting, antianxiety, anti-depression and testosterone-like effects [197–199].
Celastrus paniculatus (Jyotishmati)	Ayurvedic medicine in India considers *Celastrus paniculatus* (CP) as a brain tonic and aphrodisiac.	Cognitive enhancing effect and powerful antioxidant activities are attributed to extracts of CP	Some data suggested that active ingredients of CP (Bacosides A and B) cause an overall decrease in the turnover of norepinephrine, dopamine and serotonin [200] but the mechanism underlying the cognitive effect of CP is not yet very clear.	Reversible arrest of spermatogenesis and reversible liver necrosis (lesions were absent after 45 days post treatment) were found after chronic administration of CP in rats [201].	10 drops of the oil extracted from seeds.	Seeds of CP administered chronically in rats selectively reversed the impairment in spatial memory produced by scopolamine [202]. Strong antioxidant effects in rats brain was also demonstrated [203].

(Continued on next page)

Table 1. (Continued)

Name	Anecdotal use	Alleged actions	Mechanism of action	– Drug interactions – Side effects – Precautions	Usual dosage or content	The science
Origanum majorana	Traditionally used as an anti-microbial remedy. The Greeks used it as a remedy for narcotic poisoning, convulsions and dropsy.	Natural and synthetic Acetylcholinesterase inhibitors (AChE) increase the cholinergic tone in brain and are useful in Alzheimer's disease and for cognition enhancement.	Ursolic acid, one of the the active components of origanum majorana (OM) and other herbs from the Labiatae family, is a potent natural AChE inhibitor [204].	No evidence of drug interactions with OM has been found in the medical literature. OM should not be used during pregnancy as it stimulates menstruation and may cause miscarriage.	Dose for cognition enhancement is not established.	Ursolic acid demonstrated antioxidant action, reducing Abeta neurotoxicity in rat's brain [205].
Alpha lipoic acid (ALPA, thioctic acid)	This water and fat-soluble antioxidant has been suggested for liver health and liver intoxication, for diabetes-related neuropathy, as a possible HIV inhibitor and as an antioxidant.	Improves memory by reversing the damage that had been induced by oxidative stress to neurons and preventing them from lack of oxygen and poor blood supply.	Raises glutathione, vitamin C and E levels and decreases concentrations of reactive oxidants. It also lowers glucose levels and preserves the mitochondria [206].	Drugs that influence blood-sugar level may require a change in dosage if taken with ALPA.	200–600 mg a day.	Double blind placebo controlled trials support the antioxidant effect in various kind of poisoning [207], aging [208] and diabetes [209] damages. In animal studies, ALPA taken alone [210, 211] or with acetyl-L-carnitine [212] was able to reverse memory impairment and oxidative stress in aged mice.

This plethora of potential mechanisms is unmatched by solid evidence of efficacy, including that pertaining to cognitive function following chronic regimens of ginseng extracts administered to humans. However, to a great extent, the equivocal nature of this area can be attributed to the methodological limitations of the extant literature [3, 30–32].

Whilst human chronic dosage research has almost exclusively been conducted by carrying out testing only at the end of the dosing period, usually of 8 to 12 weeks, a series of recent studies suggests that single doses of a standardised *Panax ginseng* extract (G115) can modulate cognitive performance [33–35].

These studies shared the same randomised, double-blind, placebo-controlled, balanced-cross-over design, with 20 participants receiving three different single doses of the relevant extract and an identical placebo on separate occasions seven days apart. In the initial study [33], which utilised a computerised cognitive assessment system (the CDR system), all three doses of ginseng (200 mg, 400 mg, 600 mg) were associated with improvements on a factor analysis derived 'Secondary Memory' measure (comprised of % accuracy scores from four secondary memory tasks). However, these improvements were most pronounced for the middle (400 mg) dose. In contrast to these improvements, both of the less mnemonically active doses (200 mg and 600 mg) were associated at the later testing sessions with slowed performance on a 'Power of Attention' factor (comprising reaction times on three attention tasks). This improved secondary memory performance following 400 mg of G115 was replicated in the second, methodologically similar, study again using the CDR system which compared the effects of single doses of *Ginkgo biloba*, ginseng G115, and their combination against a placebo in a single cohort [34]. A further experiment also assessed the effects of the same doses of G115 on Serial subtraction mental arithmetic tasks. On the most demanding (Serial 7 s) task the 400 mg dose proved beneficial, with increased accuracy, but once again the 200 mg dose led to decrements, this time in terms of the number of responses made. A trial which has just been completed studied the effects of G115 in combination with 22 minerals and vitamins upon fatigue and cognitive decline induced by shift-work [36]. It was found that compared to nurses who took placebo over 12 weeks, nurses who took the active treatment showed smaller shift-work induced fatigue, mood changes and cognitive deficits. The clearest cognitive effect was seen on the CDR Quality of Memory factor.

The CNS effects of *Panax ginseng* have also been assessed in a double-blind, placebo controlled, balanced cross-over, topographic Electroencephalograph (EEG) experiment comparing the effects of both 360 mg of *Ginkgo biloba*, and 200 mg of *Panax ginseng* G115 in 15 healthy volunteers [37]. The results suggested that there were similarities in the topographic EEG effects elicited by both extracts (in comparison to placebo) with reduction in the power of 'eyes closed' frontal theta and beta wavebands. However, these effects were more marked for ginseng, and were accompanied by reductions in frontal alpha waveband activity and decreased latency of the P300 component of the auditory evoked potential.

Ginkgo biloba

The *Ginkgo biloba* tree is one of the oldest surviving tree species on earth [38] and has probably existed in its current form for up to 200 million years, leading to its description by Darwin as a "living fossil". The Ginkgo tree has a lifespan of up to a thousand years [39]. Extracts and infusions made from Ginkgo leaves have been used in traditional Chinese medicine for at least 5,000 years. Currently it is sold either as an 'over the counter' food supplement, or as a prescription medicine throughout the western world, and ranks as the best-selling herbal medication in western marketplaces [40, 41]. The popularity of Ginkgo can be attributed in some measure to the development in the mid 1960s of manufacturing processes capable of producing high quality standardised extracts. These extracts, derived by a complex drying process, are concentrated in a ratio of approximately 1 part extract to 50 part dried leaves, and are generally standardised to a content of 24%–25% flavonoids and 6% terpenoids (e.g., GK501, Egb761, LI 1370).

The most important active substances in ginkgo are thought to be the flavonoids – ginkgo-flavone glycosides of kaempferol, quercitin and isohamnetin, and the terpenoids – bilobalide and ginkgolides A, B,C and J [42]. Whole extracts and the active components have been attributed with a number of physiological effects potentially relevant to the enhancement of cognition. These include: specific antagonism of platelet activating factor [43–45], scavenging and inhibition of free radicals [46–48], modulation of a number of neurotransmitter systems [49–52], beneficial effects on blood circulation [53–57], both *in vitro* and *in vivo* protection against hypoxic challenges [58–60] and *in vivo* neuro-protective properties [61–64].

Accumulating evidence suggests that Ginkgo may be effective in its prescribed role in the amelioration of cognitive decline. As an example, a recent Cochrane review [65] meta-analysed the 33 extant studies involving cohorts suffering from dementia or age-related cognitive impairment that met their inclusion criteria. The results across the studies suggest improvements in cognitive performance following Ginkgo at all dose levels (< and > 200 mg/day) and time points (12, 24 and 52 weeks). The authors, while allowing the possibility that the overwhelmingly positive literature may reflect a publication bias, conclude both that Ginkgo is not associated with any more adverse events than placebo, and that "Overall there is promising evidence of improvement in cognition and function associated with Ginkgo".

There is also evidence of cognitive enhancement in healthy 'cognitively intact' populations following chronic administration of Ginkgo. In older adults Mix and Crews [66] reported improved speed of performance on a timed Stroop task, and trends towards increased speed of performance on three further tasks following six weeks administration of 180 mg/day Egb 761 or placebo to 40 healthy participants. Similarly, Crews and Mix [66] assessed the effects of six weeks administration of 180 mg/day of Egb 761 or placebo to 249 healthy older individuals and demonstrated superior self-rated memory performance and objectively assessed long-term recognition and recall. However, in contrast to these studies, Solomon and co-workers [67] administered 120 mg/day of GK501 or placebo for six weeks to 219 healthy elderly participants and found no treatment-related differences on any cognitive measure.

In healthy 'non elderly' adults Stough and co-workers [68] reported Ginkgo-related improvements on Digit Span Backwards, and speed on working memory and delayed auditory verbal learning tasks following 30 days administration of Ginkgo or placebo to 50 participants. In contrast, Moulton and co-workers [69] found no interpretable significant differences on a range of cognitive tasks in 60 healthy young males administered of 120 mg LI1370 or placebo for five days in a double-blind, between subjects experiment.

A number of studies have also assessed the effects of acute dosage of Ginkgo in healthy adults utilising a double-blind, counterbalanced cross-over design. In the earliest of these studies Hindmarch [70] reported shortened reaction times on a Sternberg short-term memory scanning task following the highest (600 mg) of three single doses of Ginkgo taken by a small cohort (8) of healthy females. Warot and co-workers [71] failed to replicate this effect on the Sternberg task but did generate an improvement in free recall score for one 600 mg dose of standardised extract. In a more adequately powered study, Rigney and co-workers [72] examined the effects of one and two day's administration regimens of four doses (120 to 300 mg) of ginkgo in 31 participants. Performance was only significantly improved on reaction times for the Sternberg numeric working memory task. Utilising the methodology described in relation to ginseng above, Kennedy and co-workers [73] demonstrated dose-dependent increases in speed across three attention tasks in 20 young participants administered single doses of 120 mg, 240 mg and 360 mg of GK 501. There was also evidence of improved secondary memory performance following the lowest dose. In a partial replication of this study [34] a different cohort of 20 young participants ingested single doses of Ginkgo GK501, ginseng (G115) and their combination. Whilst the speed of performing attention tasks was unaffected the results suggested that 360 mg of Ginkgo improved secondary memory performance.

Ginkgo biloba/Panax ginseng combination

A number of studies have shown improved cognitive performance following administration of a 60:100 combination of *Ginkgo biloba* GK501 and *Panax ginseng* G115.

In the first of two double-blind, placebo-controlled, chronic dosage studies utilising the CDR system, Wesnes and co-workers [74] assessed the effects of 80 mg, 160 mg and 320 mg or placebo, administered in two daily doses to 64 participants who satisfied the criteria for neurasthenia, an age related condition with a possible cerebro-vascular aetiology. Results showed significant improvements for all three doses at one hour past the morning dose on at least two of the three (1, 30 and 90 days) assessments on a 'Quality of Memory' measure (comprising scores from six memory tasks). However, occasional impairments were seen on the same measure after the second lunchtime dose. The second study [75] utilised the same measures and involved a cohort of 256 healthy middle-aged participants who received 320 mg of the combination or placebo. Testing took place four times daily (1 hour pre-dose and 1, 3, and 6 hours after the first dose) at pre-commencement of treatment and at 4, 8, 12 and 14 weeks post-commencement of treatment. Results showed improvements on the same memory measure across the study at the 1 and 6 hour post-dose testing sessions. The overall improvement in the Quality of Memory measure was 7.5%, which sets a minimum standard for compounds aimed at improving memory in healthy middle-aged volunteers to achieve.

Improvements in memory performance have also been shown in groups of 20 healthy young adults following single doses of the combination in two studies utilising the CDR system in the study paradigm described above [73, 33]. In the first [76] the pattern of results following the combination was similar to that following ginseng alone, with improved Quality of Memory for the highest dose (960 mg), but decrements in the speed of performing attention tasks for the mnemonically-inactive lowest dose (320 mg). In the second study [34], comparing single doses of Ginkgo, ginseng and the combination to placebo, the improvement in memory performance was confirmed for the 960 mg dose.

Salvia officinalis/Lavandulaefolia and *Melissa officinalis*

Several recent strands of evidence suggest that members of the Labiatae family may enhance cognitive performance. The potential utility of these plants was first suggested by a retrospective examination of historic pharmacopoeias and herbal texts with subsequent *in vitro* analysis of the plants that have had a historical role in improving mental function [77, 78]. This historical and *in vitro* analysis suggested that the two most promising candidates were the Salvias 'officinalis' and 'lavandulaefolia' (Sage) and *Melissa officinalis* (Lemon Balm). In the case of the former this was predicated on demonstrations of acetylcholinesterase inhibition in brain tissue both *in vitro* [78, 79] and *in vivo* [80], and in the case of the latter both nicotinic [78, 81] and muscarinic [81] cholinergic receptor binding in human brain tissue. Both plants also exhibited substantial anti-oxidant properties [82–84].

A series of double-blind, placebo-controlled, cross-over studies using the CDR battery have assessed the nootropic potential of these plants. Evidence from two studies suggested that single doses of *Salvia lavandulaefolia* could improve word recall in 20 healthy young participants [85], and that single doses of *Salvia officinalis* could improve the secondary memory and attention task performance of 20 elderly participants (mean age 72.9 years) [86].

With regards *Melissa officinalis,* findings of dose-dependent decrements in timed memory task performance along with increased calmness for a low dose (300 mg) and reduced alertness for a high dose (900 mg) of a concentrated manufactured extract administered to 20 healthy adults were broadly in line with current usage as a mild sedative and sleep aid. However, retrospective *in vitro* analysis showed that the extract lacked the expected substantial cholinergic binding properties [87]. In a subsequent study [37], a number of acquisitions of dried *Melissa officinalis* leaf were first assessed for cholinergic binding, with the cognitive and mood effects of single doses of the leaf with the most promising profile being investigated in 20 healthy adults. Once again decrements were seen on the timed memory tasks, but in this instance as dose increased (600 mg, 1000 mg, 1600 mg), these decrements decreased, with the highest dose engendering improved memory performance and increased calmness at all post-dose time points (1, 3 and 6 hours).

Vinpocetine

Vinpocetine is derived from the alkaloid vincamine found in the Lesser Periwinkle plant, *Vinca minor.* Vinpocetine is believed to be a promising treatment for cerebral vascular insufficiencies and may have clinical utility in stroke consequences. Vinpocetine increases cerebral metabolism and raises ATP levels in nerve cells; is a highly potent and safe vasodilator, acting by direct relaxation of the vascular smooth muscle; enhances cerebral blood flow and also has a great capacity to lower the viscosity of the blood [87a]. Vinpocetine is safer than other vasodilators because it selectively increases cerebral blood flow without a "stealing" effect, i.e., without removing blood from underperfused zones of ischemic damage, possibly due to its ability to lower blood viscosity. It has anticonvulsant actions related to its ability to maintain brain cell electrical conductivity and to protect against damage caused by excessive intracellular release of calcium. Vinpocetine also partially blocks hypoxic damage to brain tissue, and is a good scavenger of hydroxyl radicals [87b, 87c].

Other properties are its metal-chelating capacity [88], the enhancement of retinal microcirculation [89] and of the circulation of the inner ear [90]. Vinpocetine shares this latter property with *Ginkgo biloba*, thus also being a possible treatment for tinnitus. In animal models of anoxia, vinpocetine reduced cerebral oedema and prolonged survival [91].

Vinpocetine has cognitive-enhancing activities in an animal model of memory retrieval [92]. In double-blind clinical trials conducted with patients suffering from mild-to-moderate vascular dementia, vinpocetine enhanced memory and learning as well as clinical global measures of cognitive performance [93–95]. However, in one open-label trial with 15 Alzheimer's patients with doses increasing from 30 mg to 60 mg a day for a year there was no sign of improvement [96]. Clear evidence of the real benefits of this substance is still not consistently documented; however, vinpocetine appears safe and potentially helpful for cognitive improvement.

Acetyl-l-carnitine

Acetyl-l-carnitine (ALC) is a transport molecule that occurs naturally in the brain, liver, and kidney which plays a role in mitochondrial energy production. Levels of Acetyl-l-carnitine diminish with age. The acetyl group that is part of acetyl-L-carnitine contributes to the production of the neurotransmitter acetylcholine. ALC may increase B-endorphin levels and may increase nerve growth factor production within the hippocampal region of the brain [97]. ALC improves age-related changes in intracellular membranes and in dopamine, NMDA and glucocorticoid receptors; sustains mitochondrial metabolism and cholinergic neurotransmission, improving mental function [98]. Trials on acetylcholine deficit correction in rodents showed benefits [99], and protection from beta-amyloid damaging effects was also demonstrated in the rat cortex [100]. Memory in aging rats improved when they received a combination of ALC and lipoic acid [101].

Meta-analysis [102] of double-blind randomised controlled clinical trials of ALC *versus* placebo in the treatment of mild cognitive impairment and mild Alzheimer's disease supports a significant advantage for ALC compared to placebo on both the clinical scales and the psychometric tests.

Phosphatidylserine

Phosphatidylserine (PS) is a naturally-occurring phospholipid which maintains neuronal structure and helps the neuronal membrane to maintain its charged state. It is believed to increase the receptor number and promote dendritic branching. It is also thought to activate protein-chinase-C within the neuronal membrane, a substance involved in the regulation of major neurotransmitters (e.g., acetylcholine, dopamine and norepinephrine) and in the activation of memory-specific genes. Generally speaking, PS should allow neurons to communicate within each other more efficiently, thus improving mental functioning.

In animal studies PS has been shown to enhance cognitive function [103]. Properly controlled clinical studies have also shown benefits of PS [104–106]. Reviews [107] have supported the case for PS: preventing and even reversing memory loss associated with aging and dementia; enhancing memory in normal people, and relieving symptoms of anxiety and stress [108]. It must be noted that these studies have evaluated the role of Bovine Cortex phosphatidylserine administration. Because of potential viruses present in cow brain extracts, PS is now obtained from soy. As the PS in soy is chemically different than that found in the cow brain, it is possible the effects of soy-based

PS could be different. However, a recent open trial with plant-source derived phosphatydilserine [109] showed encouraging results, although these of course need to be confirmed with double-blind, placebo controlled studies.

Huperzine A

Huperzine A (HupA) is an alkaloid extracted from a club moss (*Huperzia serrata*). The average content of HupA in plants is 0.011%. Huperzine A, in the form of a remedy called Qian Ceng Ta, has been used for centuries in Chinese folk medicine to treat fever, blood disorders, swelling and schizophrenia [110]. It acts as a potent, highly-specific and reversible inhibitor of the enzyme acetylcholinesterase (AChE) over butyrylcholinesterase, which crosses the blood-brain barrier [111]. HupA fits into the active site of AChE where acetylcholine is broken down, and binds onto this site. This binding closes off the enzyme's "cutting" machinery and keeps acetylcholine in circulation cycle, as do other AChE inhibitors (donepezil, galanthamine, tacrine) approved for Alzheimer therapy. HupA also has an antagonistic effect on N-methyl-D-aspartate (NMDA) receptors, and increases the flux of calcium ions into the neurons [112].

Huperzine A has shown both neuroactive and neuroprotective effects in animals and humans. It has been demonstrated to be a safe and effective substance for improving cognitive function and quality of life in patients suffering from varying degrees of dementia, and is used as an over-the-counter cognitive enhancer for healthy people complaining of memory problems. Its clinical evaluation for Alzheimer's disease is now in Phase IV.

One major theory [113] proposes that memory loss and other cognitive deficits in Alzheimer's patients result from degeneration of the nerve cells that release the chemical messenger acetylcholine (ACh). HupA increases the level of the neurotransmitter ACh (particularly in the cerebral cortex), but also of norepinephrine and dopamine, suggesting action on different systems [114]. It has little effect on nicotinic and muscarinic receptors, but huprine X (a hybrid between HupA and tacrine) exhibited more activity at muscarinic receptors [115], promising a superior therapeutic advantages in dementia therapy.

Glutamate activates NMDA receptors and increases the flux of calcium ions in the neurons. Calcium at toxic levels can disrupt normal cognitive processes and kill the cells [116]. HupA non-competitively inhibits the passage of calcium ions through NMDA ion channels, protecting primary neuronal cells against the toxic consequences of the excitatory aminoacid (EAA)-induced over stimulation, which is implicated in a variety of acute and chronic neurodegenerative disorders [117]. As a pre-treatment, HupA may act as a prophylactic drug against soman and other nerve gas poisoning used as chemical weapons, protecting cerebral ACh from soman inhibition, and interfering with EAA-induced toxicity [118]. HupA could be used as a preventive agent for AD having similar properties to Memantidine, a new drug that protects the brain against the excess of glutamate observed in AD. HupA can also decrease the damage induced by free radicals [119], oxidative stress and by amyloid-beta-peptide (Abeta) [120]. The deposition of Abeta creates the plaques observed in the brains of AD patients.

In vitro experiments demonstrated that HupA had more power in the inhibition of ACh than tacrine and galantamine, but less than donepezil, with a pattern of inhibition of the mixed competitive type. Inhibition of BuChE revealed a different pattern of actions: HupA inhibited BuChE at a higher concentration than needed for AChE compared with donepezil [121]. This selectivity for AChE as opposed to BuChE may suggest a safer profile of side-effects [122] but a stronger inhibition of BuChE seems to be more useful in the later stage of AD [123]. *In vivo* tests revealed that the relative inhibitory activity of oral HupA on AChE was about 24-fold and 180-fold more potent, on an equimolar basis, than donepezil and tacrine respectively [124]. Maximal AChE inhibition in rat cortex and whole brain was reached 30–60 minutes following oral administration, and was maintained for 360 minutes [125]. After repeated doses of HupA there were no signs of declined AChE inhibition compared to a single dose, demonstrating that no tolerance to this alkaloid occurred [126].

HupA has been tested in various animal models of experimental cognitive impairment, showing memory-enhancing activities with a superior safety/efficacy ratio when compared with other AChEIs like physostigmine, galantamine and tacrine [127].

Its evaluation is now in Phase IV of clinical trials (AD), and approximately 800 AD patients have participated in studies with HupA, including double-blind and placebo-controlled studies [119, 128–130] with daily doses ranging from 0.05 to 0.10 mg. All trials found significant improvements in cognitive and behavioural function compared both to placebo and baseline. A study conducted using the Cognitive Drug Research computerised tests battery in 10 healthy elderly volunteers suggested that ZT-1, a HupA derivative, had the ability to partially reverse the cognitive impairment caused by scopolamine [131]. The longest open-label study with HupA (involving 33 patients) indicated a 66.7% improvement ($p < 0.05$) on cognitive and behavioural measures after the 48 week-long trial [132]. A study conducted with a non-patient population (68 secondary school students) suggested a potential use of HupA for cognitive-enhancing purposes [133]. The dose of HupA administered in this study was 0.05 mg twice daily and taken orally. Human testing at daily dosages ranging from 100 to 600 mcg for periods varying from 2 to 48 weeks showed no statistically significant side-effects with HupA. Following a supratherapeutic oral dose of 0.99 mg peak serum concentrations were reached in 79 minutes and half-life was 288 minutes; there were no side-effects with doses between 0.18 and 0.54 mg [134].

It is important to remember that the three compounds currently registered for the symptomatic treatment of AD (tacrine, galantamine and donepezil) also prevent the breakdown of ACh, so it would be advisable to avoid the use of HupA in combination with these drugs.

Glucose

In recent years there has been increasing focus on the potential for nutritional interventions to improve cognitive function. It remains to be established whether such agents have differential effects with respect to macronutrient composition, as opposed to being a function simply of calorific intake and absorption (see [135, 136] for reviews). Nevertheless, a starting point for such studies, and one which has attracted a good deal of attention, is to examine the effects of a simple carbohydrate such as glucose on cognitive function. The general finding is that ingestion of a glucose load (generally 25 or 50 g) can improve aspects of cognitive function in the minutes and hours following ingestions of the drink. Most work in this area has focused upon memory. Indeed it has been postulated that the effect may differentially affect memory processes (especially long-term episodic memory function). While there is reasonable evidence to support this contention, it is probably true to say that studies which have attempted to assess the impact of a glucose load on non-mnemonic processes have also found positive effects. An alternative hypothesis has been put forward which suggests that glucose has more profound effects upon more effortful cognitive processes and it remains to be established whether the glucose enhancement of cognition effect is domain- or demand-specific.

The background to this work is partly grounded in the assumption that brain possesses negligible stores of glycogen (which, in other tissue can be broken down to provide glucose to meet on-going energy demands). It should be noted that recent findings have challenged this assumption. For example the widely-held belief that astrocytes contain negligible, rapidly-depleted, static glycogen stores may not be true. It has recently been suggested that the astrocyte glycogen store is in a dynamic state of turnover may buffer glucose levels in response to on-going neural energy demands [137].

Nevertheless, current models of energetic brain function are based on the assumption that neural energy requirements are met almost exclusively by the oxidative breakdown of blood-borne glucose [138]. This model suggests that glucose is transported across the blood-brain barrier at a rate dictated by capillary surface area and changes in local metabolism [139]. Additionally it is assumed that there is a fairly tightly-coupled equilibrium between blood and brain glucose levels. It follows that any fluctuations in the availability of blood-borne metabolic substrates (i.e., glucose and oxygen) may modulate brain metabolism and consequently cognitive function. This contention is supported by evidence of transient cognitive impairment during hypoglycaemia [140, 141] and lowered by supra-hypoglycaemic glucose levels [142, 143]. Similarly, hypoxia is associated with cognitive impairment [144, 145] where such deficits at altitude can be reversed by supplemental oxygen breathing [146]. A number of studies have demonstrated that oxygen administration can improve cognitive performance in healthy volunteers [147–151]. Such improvement occurred only when concurrently measured hyperoxia coincided with encoding of target material in memory tasks, or with performance of a reaction time task [149, 150]. It should be noted that at least one study has failed to replicate these effects [152]. Nevertheless the overwhelmingly positive results are consistent with the notion that there is a strong relationship between the availability of metabolic substrates and cognitive enhancement during periods of cognitive demand.

Enhanced cognitive performance on a number of memory tasks results from the administration of glucose, including in the elderly [153, 154] and in young adults [155–158]. Some authors have proposed that there is a fractionation of the memory-enhancing effects of glucose, with tests evaluating long-term declarative memory showing the greatest facilitation [159–162]. However not all studies examining declarative memory have found a glucose enhancement effect [35, 163, 164]. Additionally certain tests of working memory appear to be reliably enhanced by glucose [35, 162, 165, 166] as does kinaesthetic memory [167]. Furthermore, there is an association between glucose and performance on a number of non-mnemonic cognitive measures including choice reaction time tasks [168], the Stroop paradigm [156, 157] vigilance tasks [169], and rapid information processing [156].

With regards to non-memory tasks, glucose tends to facilitate only those tasks, or parts of tasks, that would appear to require a relatively high cognitive load [35, 166]. Indeed in a paradigm where robust enhancement of verbal memory been reliably, during word presentation, participants were required to simultaneously perform self-generated alternating hand movement sequences reported [159–162]. When this additional load was removed, so was the glucose enhancement effect [161].

Where no direct group effects have been found there is nevertheless a relationship between changing blood glucose level and improved performance. The issue is not straightforward; falling blood glucose levels predicted performance on tasks of memory and reaction time [159] and a dichotic listening task [170], while an increase in blood glucose was associated with better memory [155]. On the other hand Kennedy and Scholey [73] found that falling blood glucose levels during task performance were correlated with performance of a heavily-loaded working memory/executive task in a glucose group only, whereas baseline glucose levels predicted performance in both groups. The reason for the disparity between these studies is not clear but may be attributable to the level of cognitive processing required of participants either immediately following a glucose load [171] or while blood glucose levels are raised [35, 166]. Moreover it has been demonstrated that heavily-loaded cognitive processing itself results in a measurable drop in blood glucose levels [35, 171]. In the latter case the task was differentially-enhanced by a glucose drink.

These observations support the suggestion, not only that effortful cognitive processes may be "fuel limited" and therefore augmented by the simple provision of supplemental metabolic substrates. This is supported to some degree by the fact that the ability to efficiently utilise excess blood glucose predicts cognitive performance. This contention is supported by the consistent finding that poorer regulators of blood glucose perform worse on memory tasks even without a glucose load [153]. Indeed there is some evidence that glucose may merely reverse memorial deficits in poor glucoregulators [172].

In a similar vein, Gold [173] has suggested that age-related memory deficits reflect "sub-optimal neuroendocrine regulation of memory storage". Wenk [174] also argues that many cognition-enhancing drugs work via the glycogenic action of adrenaline. The putative role of adrenal activation in cognitive functioning is also supported by studies demonstrating an arousal-induced facilitation of memory, and its subsequent attenuation by the administration of adrenergic antagonists, both in animals [175] and humans [176, 177]. The mechanisms underlying glucose-related enhancement of cognition are not presently know. One possibility is that insulin released in response to a glucose load promotes glucose uptake in specific brain areas including the hippocampus (which is differentially rich in insulin receptors). Another suggestion is that glucose breakdown promotes the synthesis of acetylcholine, since acetyl CoA, formed as a by product of glycolysis, is rate-limiting in the cholinergic synthetic pathway. However, neither of these proposals adequately explains the pattern of cognitive enhancement seen following a glucose load and it is possible that both of these, in combination with glucose's role as a metabolic fuel, act in conjunction to aid cognitive processing.

A relatively high cognitive load has itself has been shown to elicit physiological arousal. For instance, Fibiger and co-workers [178] reported increased levels of salivary cortisol and urinary catecholamines following a mental arithmetic task, with adrenaline levels differentially affected by task difficulty. Similarly, there appear to be increases in cardiac output above somatic requirements as a consequence of cognitive demand. Cardiovascular responses have also been found to be sensitive to levels of difficulty on a number of cognitive tasks, including mental arithmetic, Raven's matrices and sentence comprehension [179, 180]. Similarly heart rate, general metabolic rate and cerebral glucose utilisation increase with greater cognitive load of a working memory task [181, 182]. In some ways this mirrors both previous findings with regards glucose regulation, and also the results of Positron Emission Tomography (PET) studies which have demonstrated negative correlations between absolute cortical metabolic rate and performance on Raven's Matrices [183] and a Word Retrieval (Verbal Fluency) task [184]. Such findings suggest that there may be physiological individual differences which contribute to cognitive performance and, specifically, that a relationship may exist between physiological and cognitive efficiency.

From the above, it seems plausible to suggest that glucose regulation and utilisation, and physiological arousal due to cognitive demand are linked. The aforementioned autonomic responses to cognitive demand may reflect motivational changes, however it also seems feasible that certain processes associated with physiological arousal may serve as mechanisms to increase the delivery of metabolic substrates to the brain. On the other hand, a different interpretation has recently been forwarded by Gibson. In this model, cognitive testing acts as a psychosocial stressor, mobilising cortisol and inhibiting the uptake and disposal of circulating glucose. Thus poorer regulators of glucose will not benefit from a glucose load [136]. While this model is attractive and consistent with much of the data in this area, it has not been tested directly. It is also difficult to reconcile with recent findings of retrograde enhancement of memory function, which occurs even when task demands appear minimal [167].

Discussion

This chapter has confirmed that natural products can enhance cognitive function even in healthy young volunteers. Improvements have been seen to important domains such as attention and memory with a variety of substances having a diverse range of potential mechanisms. Having established that such effects are possible, for each substance attention should in future be focussed upon the more important questions, i.e., the domains of function which can be improved, the extent of improvement possible, the time-course of such effects, and the optimal populations for improvement.

Detecting enhancements to cognitive function is not as easy as detecting impairments. In normal populations, there is a much smaller window of opportunity to enhance function than there is to impair it. As we have seen in the preceding pages, and elsewhere in this book, improvements in normals are rarely greater than 15%, whereas, of course, impairments can be as large as 100%. The signal is therefore weaker in such trials, and if enhancements are to be detected, it is necessary to do everything reasonably possible to ensure that the power of the trial is maximised.

There are a number of simple steps which can be taken to maximise power in human cognitive psychopharmacology. The first is to reduce the variability due to inter-individual differences. There are basically two types of trials which can be conducted in this field, single-dose acute studies or studies which involve multiple dosing. To minimise inter-volunteer differences, cross-over designs using the volunteer as his or her own control are an ideal way to enhance the sensitivity of the study. In single dose trials, the sessions can be separated by a week or more, to avoid any problems with carry-over effects, and such cross-over designs are the current gold standard for single dose trials in this field. A number of examples of this design was seen in this chapter, and a body of consistent results has emerged from this work. Cross-over trials are also possible in multiple dosing trials, though the length of the dosing period is an important practical issue. If the dosing periods are relatively short, say two to four weeks, then at least one cross-over arm is possible. A washout period needs to be inserted between the two dosing periods, and for example with two successive four-week dosing periods separated by a two-week washout, the study will last 10 weeks. With shorter dosing periods (two weeks for example), two doses and a placebo could be tested and with two-week washouts, the study would again last for 10 weeks. This is not an unusual time to keep a volunteer or patient in a trial and such cross-overs are clearly possible. However, with dosing periods longer than a month it is rarely practical to have a cross-over, and most trials of this nature employ parallel group designs.

Whether in a cross-over or parallel group design, an important technique to further reduce the variability due to individual differences is to make repeated assessments of cognitive function. At the absolute minimum, each volunteer or patient should be assessed before treatment as well as after, and ideally such testing should be repeated to reflect the opportunity of the compound to enhance function. In acute trials, repeated testing over the hours immediately following administration can identify clear time-based profiles of effects as has been seen in this chapter [73]. In longer trials, testing can be repeated at various stages, e.g., monthly. A further enhancement to this design was seen in a trial in which testing was not only repeated regularly throughout a 14-week period, but was tested at 07:30, 09:30, 11:30 and 14:30 on each study day [185]. Improvements were only detected at two of the testing times but these effects were seen consistently throughout the study. This raises the stakes in trials in this field, as simply employing a single post-dosing test session in a trial lasting weeks or months may miss effects by testing at an inappropriate time (e.g., [67])

Another technique for maximising the power in such work is to properly train the volunteers or patients on the cognitive testing procedures prior to the start of the trial. It has long been known that performance improves with repeated testing on cognitive tests [186]. These improvements can occur at all ages and over extended periods. A recent trial in an elderly population in which various pencil and paper cognitive tests were administered annually has identified practice effects to persist until at least the third year [187]. Trials must have pre-study training on the tests employed to overcome such effects, otherwise training effects may obscure any treatment effects, or make the interpretation of any changes identified extremely difficult. Most long-term training effects are due to procedural learning, and as this type of learning is fairly robust over the age-span, it is rarely the target for enhancement. If improvements in active are seen which are greater than those on placebo, the active improvements could be interpreted to be the results of improvements to procedural learning, as the study will have shown the presence of it by the placebo enhancement. [185]. Such problems need to be avoided in this field. Having parallel forms of tests (i.e., differing lists of words etc.) for each successive session is an absolute minimum requirement to prevent training effects. Over and above this, pre-study training on the tasks should be mandatory to minimise any practise. Finally, tests or test systems which have known training profiles or, better still, no training effects will be more useful in this area of research.

Reviews in cognitive psychopharmacology consistently comment upon the lack of standardisation of testing used in different laboratories making precise comparisons between trials difficult. Certainly there are advantages to having standardised procedures in trials, and fifteen of the studies covered in this chapter have used the same computerised test system, which has enabled some general findings to emerge. Another useful technique in this field is to subject test batteries to factor analysis to identify the interrelationships between the various measures. This has been done for the CDR system and a number of factor scores have been derived. Some of these have shown great sensitivity to the effects of natural substances, for example the Quality of Memory Score, which combines the scores of the ability to retain and retrieve information from two working memory (spatial and articulatory) and four episodic secondary (verbal and pictorial) memory tasks into a single measure. Such measures become more powerful as they combine scores from a number of assessments, but also make trials more powerful as they can avoid the statistical problems involved in assessing multiple endpoints.

Overall, it is clear that some natural products can enhance cognitive function. The Quality of Memory factor from the CDR system has consistently been shown to respond positively to particular extracts of Ginkgo and ginseng, whether dosed individually or together. Such effects have occurred in four acute volunteer trials [33, 34, 73] and in multiple-dose trials in volunteers with neurasthenia [74], middle-aged volunteers [75] and nurses working night-shifts [36]. The sites for these trials have been located in the UK, Holland, Sweden and France. This replication in seven studies is one of the most consistent findings in the field of natural products. While further work is required to identify the precise roles of the two extracts in these benefits and the nature of any synergy between them, the consistency of the effects is striking. Besides using standardised extracts, the common features for all trials are the use of a widely-used and extensively-validated computerized cognitive test system, pre-study training on the tests, and repeated testing – both over the study day and for multiple dosing trials, during the dosing period as well. All trials were randomised, double-blind and placebo controlled. Such methodologies for acute cross-over trials and parallel group designs are clearly effective and should set the standard for future work in this area.

References

1 Yun TK (2001) *Panax ginseng*—a non-organ-specific cancer preventive? *Lancet Oncology* 2(1): 49–55

2 Blumenthal M (2001) Herb sales down 15 percent in mainstream market. *HerbalGram* 51: 69

3 Kennedy DO, Scholey AB (2003) Ginseng: Potential in the enhancement of cognitive performance and mood. *Pharmacol Biochem Behav* 75: 687–700

4 Tachikawa E, Kudo K, Harada K, Kashimoto T, Miyate Y, Kakizaki A, Takahashi E (1999) Effects of ginseng saponins on reponses induced by various receptor stimuli. *European J Pharmacology* 369: 23–32

5 Liberti LE, Der Manderosian A (1978) Evaluation of commercial ginseng products. *J Pharmaceutical Science* 67 (10): 1487–1489

6 Phillipson JD, Anderson LA (1984) Ginseng—quality safety and efficacy? *Pharmaceut Journal* 232: 161–165

7 Russo E (2001) *Handbook of psychotropic herbs. A scientific analysis of herbal remedies for psychiatric conditions.* Howarth Herbal Press, New York
8 Lee MO, Kim CY, Clifford DH (1981) Effect of ether, ethanol and aqueous extracts of ginseng on cardiovascular function in dogs. *Canadian J Comparative Medicine* 45(2): 182–187
9 Lei XL, Chiou GC (1986) Cardiovascular pharmacology of *Panax notoginseng* (Burk) F.H. Chen and *Salvia miltiorrhiza. American J Chinese Medecine* 14(3–4): 145–152
10 Shi L, Fan PS, Wu L, Fang JX, Han ZX (1990) Effects of total saponins of *Panax notoginseng* on increasing PGI2 in carotid artery and decreasing TXA2 in blood platelets. *Chung-Kuo Yao Li Hsueh Pao – Acta Pharmacologica Sinica* 11(1): 29–32
11 Jung KY, Kim DS, Oh SR, Lee IS, Lee JJ, Park JD, Kim SI, Lee HK (1998) Platelet activating factor antagonist activity of ginsenosides. *Biological and Pharmaceutical Bulletin* 21(1): 79–80
12 Chen X (1996) Cardiovascular protection by ginsenosides and their nitric oxide releasing action. *Clinical and Experimental Pharmacology and Physiology* 23(8): 728–732
13 Gillis CN (1997) *Panax ginseng* pharmacology: A nitric oxide link. *Biochemical Pharmacology* 4(1): 1–8
14 Zhan Y, Xu XH, Jiang YP (1994) Effects of ginsenosides on myocardial ischemia/reperfiusion damage in open-heart surgery patients. *Med J China* 74: 626–628
15 Rudakewich M, Ba F, Benishin CG (2001) Neurotrophic and neuroprotective actions of ginsenosides Rb(1) and Rg(1). *Planta Medica* 67(6): 533–537
16 Choi SR, Saji H, Iida Y, Magata Y, Yokoyama A (1996) Ginseng pre-treatment protects against transient global cerebral ischemia in the rat: Measuement of local cerebral glucose utilization by [^{14}C]deoxyglucose autoradiography. *Biological and Pharmaceutical Bulletin* 19(4): 644–646
17 Chen X, Salwinski S, Lee TJ (1997) Extracts of *Ginkgo biloba* and ginsenosides exert cerebral vasorelaxation via a nitric oxide pathway. *Clinical and Experimental Pharmacology and Physiology* 24(12): 958–959
18 Liao B, Newmark H, Zhou R (2002) Neuroprotective effects of ginseng total saponin and ginsenosides Rb1 and Rg1 on spinal cord neurons *in vitro*. *Experimental Neurology* 173(2): 224–234
19 Sonnenborn U, Proppert Y (1990) Ginseng (*Panax* Ginseng C A Meyer). *Z Phytother* 11: 35–49
20 Filaretov AA, Bogdanova TS, Podvigina TT, Bodganov AI (1988) Role of pituitary-adrenocortical system in body adaptation abilities. *Experimental and Clinical Endocrinology* 92(2): 129–136
21 Kim DH, Jung JS, Suh HW, Huh SO, Min SK, Son BK, Park JH, Kim ND, Kim YH, Song DK (1998) Inhibition of stress-induced plasma corticosterone levels by ginsenosides in mice: involvement of nitric oxide. *Neuroreport* 9(10): 2261–2264
22 Lewis R, Wake G, Court G, Court JA, Pickering AT, Kim YC Perry, EK (1999) Non-ginsenoside nicotinic activity in Ginseng species. *Phytotherapy Research* 13: 59–64
23 Sala F, Mulet J, Choi S, Jung SY, Nah SY, Rhim H, Valor LM, Criado M, Sala S (2002) Effects of ginsenoside Rg2 on human neuronal nicotinic acetylcholine receptors. *J Pharmacology and Experimental Therapeutics* 301(3): 1052–1059
24 Benishin CG (1992) Actions of ginsenoside Rb1 on choline uptake in central cholinergic nerve-endings. *Neurochemistry International* 21(1): 1–5
25 Kim HS, Hong YT, Oh KW, Seong YH, Rheu HM, Cho DH, Oh S, Park WK, Jang CG (1998) Inhibition by ginsenosides Rb1 and Rg1 of methamphetamine-induced hyperactivity, conditioned place preference and post-synaptic dopamine receptor supersensitivity in mice. *General Pharmacology* 30(5): 783–789
26 Sotaniemi EA, Haapakoski E, Rautio A (1995) Ginseng therapy in non-insulin-dependent diabetic patients. *Diabetes Care* 18(10): 1373–1375
27 Vuksan V, Sievenpiper JL, Koo VYY, Francis T, Beljan-Zdravkovic U, Xu Z, Vidgen E (2000a) American ginseng (*Panax quinquefolius* L) reduces postprandial glycemia in nondiabetic subjects and subjects with type 2 diabetes mellitus. *Archives of Internal Medici*ne 160(7): 1009–1013
28 Vuksan V, Stavro MP, Sievenpiper JL, Beljan-Zdravkovic U, Leiter LA, Josse RG, Xu Z (2000b) Similar postprandial glycemic reductions with escalation of dose and administration time of American ginseng in type 2 diabetes. *Diabetes Care* 23(9): 1221–1226
29 Vuksan V, Sievenpiper JL, Wong J, Xu Z, Beljan-Zdravkovic U, Arnason JT, Assinewe, Stavro MP, Jenkins AL, Leiter LA et al. (2001) (American ginseng (*Panax quinquefolius* L.) attenuates postprandial glycemia in a time-dependent but not dose-dependent manner in healthy individuals. *American J Clinical Nutrition* 73(4): 753–758
30 Bahrke MS, Morgan WP (1994) Evaluation of the ergogenic properties of ginseng. *Sports Medicine* 18(4): 229–248

31 Bahrke MS, Morgan WP (*2000)* Evaluation of the ergogenic properties of ginseng: An update. *Sports Medicine* 298(2): 113–133
32 Vogler BK, Pittler MH, Ernst E (1999) The efficacy of Ginseng. A systematic review of randomised clinical trials. *European J Clinical Pharmacology* 55(8): 567–575
33 Kennedy DO, Scholey AB, Wesnes KA (2001a) Differential, dose-dependent changes in cognitive performance and mood following acute administration of Ginseng to healthy young volunteers. *Nutritional Neuroscience* 4(4): 295–310
34 Kennedy DO, Scholey AB, Wesnes KA (2002a) Modulation of cognition and mood following administration of single doses of *Ginkgo biloba*, Ginseng and a Ginkgo/Ginseng combination to healthy young adults. *Physiology and Behavior* 75: 1–13
35 Scholey AB, Harper S, Kennedy DO (2001) Cognitive demand and blood glucose. *Physiol Behav 73: 585–592*
36 Wesnes K, Luthringer R, Ambrosetti L, Edgar C, Petrini O (2003) The effects of a combination of *Panax ginseng*, vitamins and minerals on mental performance, mood and physical fatigue in nurses working night shifts: A double-blind, placebo controlled trial. *Current Topics in Nutraceutical Research* 1: 169–174
37 Kennedy D, Scholey AB, Drewery L, Marsh VR, Moore B, Ashton H (2003) Electroencephalograph (EEG) effects of single doses of *Ginkgo biloba* and *Panax ginseng* in healthy young volunteers. *Pharmacol Biochem Behav* 75: 701–709
38 Major RT (1967) The ginkgo, the most ancient living tree. *Science* 15, 157(794): 1270–1273
39 Foster (1996) *Ginkgo*: *Ginkgo biloba* (botanical series 304) *American Botanical Society*, Austin, Texas
40 Ernst E. (2000) Herbal medicines: Where is the evidence? Growing evidence of effectiveness is counterbalanced by inadequate regulation. *British Medical Journal* 321 (7258): 395–396
41 Jacobs BP, Browner WS (2000) *Ginkgo biloba*: A living fossil. *American J Medicine* 108: 341–342
42 Kleijnen J, Knipschild P (1992) *Ginkgo biloba. Lancet* 12, 340(8833): 1474
43 Smith PF, Maclennan K, Darlington CL (1996) The neuroprotective properties of the *Ginkgo biloba* leaf: A review of the possible relationship to platelet-activating factor (PAF). *J Ethnopharmacology* 50(3): 131–139
44 Akiba S, Kawauchi T, Oka T, Hashizume T, Sato T (1998) Inhibitory effect of the leaf extract of *Ginkgo biloba* L. on oxidative stress-induced platelet aggregation. *Biochemistry and Molecular Biology International* 46(6): 1243–1248
45 Stromgaard K, Saito DR, Shindou H, Ishii S, Shimizu T, Nakanishi K (2002) Ginkgolide derivatives for photolabeling studies: Preparation and pharmacological evaluation. *J Medicinal Chemistry* 45(18): 4038–4046
46 Siddique MS, Eddeb F, Mantle D, Mendelow AD (2000) Extracts of *Ginkgo biloba* and *Panax ginseng* protect brain proteins from free radical induced oxidative damage *in vitro*. *Acta Neurochir* 76: 87–90
47 Hibatallah J, Carduner C, Poelman MC (1999) *In vivo* and *in vitro* assessment of the free-radical-scavenger activity of Ginkgo flavone glycosides at high concentrations. *J Pharmacy and Pharmacology* 51(12): 1435–1440
48 Droy-Lefaix MT (1997) Effect of the antioxidant action of *Ginkgo biloba* extract (Egb 761) on aging and oxidative stress. *Age* 20: 141–149
49 Kristoikova Z, Klaschka J (1997) *In vitro* effect of *Ginkgo biloba* extract (EGb 761) on the activity of presynaptic cholinergic nerve terminals in rat hippocampus. *Dementia and Geriatric Cognitive Disorders* 8(1): 43–48
50 Ramassamy C, Christen Y, Clostre F, Costentin J (1992) The *Ginkgo biloba* extract, EGb761, increases synaptosomal uptake of 5-hydroxytryptamine: *in vitro* and *ex vivo* studies. *J Pharmacy and Pharmacology* 44(11): 943–945
51 Taylor JE (1986) Neuromediator binding to receptors in the rat brain. The effect of chronic administration of *Ginkgo biloba* extract. *Presse Medicale* 15(31): 1491–1493
52 Huguet F, Tarrade T (1992) Alpha 2-adrenoceptor changes during cerebral ageing. The effect of *Ginkgo biloba* extract. *J Pharmacy and Pharmacology* 44(1): 24–27
53 Topp S, Knoefel WT, Schutte S, Brilloff S, Rogiers X, Gubdlach M (2001) *Ginkgo biloba* (Egb761) improves microcirculation after warm water ischemia of the rat liver. *Transplantation proceedings* 33: 979–981
54 Krieglstein J, Beck T, Seibert A (1986) Influence of an extract of *Ginkgo biloba* on cerebral blood

flow and metabolism. *Life Sciences* 39(24): 2327–2334

55 Chung HS, Harris A, Kristinsson JK, Ciulla TA, Kagemann C, Ritch R (1999) *Ginkgo biloba* extract increases ocular blood flow velocity. *J Ocular Pharmacology and Therapeutics* 15(3): 233–240

56 Jung F, Mrowietz C, Kiesewetter H, Wenzel E (1990) Effect of *Ginkgo biloba* on fluidity of blood and peripheral microcirculation in volunteers. *Arzneimittel-Forschung* 40(5): 589–593

57 Koltringer P, Langsteger W, Klima G, Reisecker F, Eber (1993) Hemorrheologic effects of *Ginkgo biloba* extract EGb 761. Dose-dependent effect of EGb 761 on microcirculation and viscoelasticity of blood. *Fortschritte der Medizin* 111(10): 170–172

58 Klein J, Chatterjee SS, Loffelholz K (1997) Phospholipid breakdown and choline release under hypoxic conditions: inhibition by bilobalide, a constituent of *Ginkgo biloba*. *Brain Res* 755(2): 347–350

59 Janssens D, Remacle J, Drieu K, Michiels C (1999) Protection of mitochondrial respiration activity by bilobalide. *Biochemical Pharmacology* 58(1): 109–119

60 Oberpichler H, Beck T, Abdel-Rahman MM, Bielenberg GW, Krieglstein J (1988) Effects of *Ginkgo biloba* constituents related to protection against brain damage caused by hypoxia. *Pharmacological Research Communications* 20(5): 349–368

61 Bruno C, Cuppini R, Sartini S, Cecchini T, Ambrogini P, Bombardelli E (1993) Regeneration of motor nerves in bilobalide-treated rats. *Planta Medica* 59(4): 302–307

62 Attella MJ, Hoffman SW, Stasio MJ, Stein (1989) *Ginkgo biloba* extract facilitates recovery from penetrating brain injury in adult male rats. *Experimental Neurology* 105(1): 62–71

63 Tadano T, Nakagawasai O, Tan-no K, Morikawa Y, Takahashi N, Kisara K (1998) Effects of *Ginkgo biloba* extract on impairment oflearning induced by cerebral ischemia in mice. *American J Chinese Medicine* 26(2): 127–132

64 Lee EJ, Chen HY, Wu TS, Chen TY (2002) Maynard KI. Acute administration of *Ginkgo biloba* extract (EGb 761) affords neuroprotection against permanent and transient focal cerebral ischemia in Sprague-Dawley rats. *J Neuroscience Research* 68(5): 636–645

65 Birks J, Grimley Evans J, Van Dongen M (2003) *Ginkgo biloba* for cognitive impairment and dementia (Cochrane Review). In: *The Cochrane Library*, Issue 1, Oxford, Update Software

66 Mix JA, Crews WD (2000) An examination of the efficacy of *Ginkgo biloba* extract EGb761 on the neuropsychologic functioning of cognitively intact older adults. *J Alternative and Complementary Medicine* 6(3): 219–229

67 Solomon PR, Adams F, Silver A, Zimmer J, DeVeaux R (2002) *Ginkgo* for memory enhancement: a randomized controlled trial. *JAMA* 21;288(7): 835–840

68 Stough C, Clarke J, Lloyd J, Nathan PJ (2001) Neuropsychological changes after 30-day *Ginkgo biloba* administration in healthy participants. International J *Neuropsychopharmacology* 4(2): 131–134

69 Moulton PL, Boyko NB, Fitzpatrick JL, Petros TV (2001) The effect of *Ginkgo biloba* on memory in healthy male volunteers. *Physiology and Behaviour* 73: 659–665

70 Hindmarch I (1986) Activity of *Ginkgo biloba* extract on short-term memory. *Presse Medicale* 15(31): 1592–1594

71 Warot D, Lacomblez L, Danjou P, Weiller E, Payan C, Puech AJ (1991) Comparative effects of *Ginkgo biloba* extracts on psychomotor performances and memory in healthy subjects. *Therapie* 46(1): 33–36

72 Igney U, Kimber S, Hindmarch I (1999) The effects of acute doses of standardised *Ginkgo biloba* extract on memory and psychomotor performance in volunteers. *Phytotherapy Research* 13(5): 408–415

73 Kennedy DO, Scholey AB, Wesnes KA (2000) The dose-dependent cognitive effects of acute administration of *Ginkgo biloba* to healthy young volunteers. *Psychopharmacology* 151: 416–423

74 Wesnes KA, Faleni RA, Hefting NR, Hoogsteen G, Houben JJG, Jenkins E, Jonkman JHG, Leonard J, Petrini O, van Lier JJ (1997) The cognitive, subjective, and physical effects of a *Ginkgo biloba/Panax ginseng* combination in healthy volunteers with neurasthenic complaints. *Psychopharmacology Bulletin* 33: 677–683

75 Wesnes KA, Ward T, McGinty A, Petrini O (2000) The memory enhancing effects of a *Ginkgo biloba/Panax ginseng* combination in healthy middle aged volunteers. *Psychopharmacology* 152: 353–361

76 Kennedy DO, Scholey AB, Wesnes KA. (2001b) Differential, dose-dependent changes in cognitive performance following acute administration of a *Ginkgo biloba/Panax ginseng* combination

to healthy young volunteers. *Nutritional Neuroscience* 4(5): 399–412
77 Perry EK, Pickering AT, Wang WW, Houghton PJ, Perry NSL (1999) Medicinal plants and Alzheimer's disease: From ethnobotany to phytotherapy. *J Pharmacy and Pharmacology* 51: 527–534
78 Perry N, Court G, Bidet N, Court J, Perry E (1996) European herbs with cholinergic activities: Potential in dementia therapy. *International J Geriatric Psychiatry* 11: 1063–1069
79 Perry NSL, Houghton P, Theobald A, Jenner P, Perry EK (2000) *In vitro* inhibition of human erythrocyte acetylcholinesterase by *Salvia lavandulaefolia* essential oil and constituent terpenes. *J Pharmacy and Pharmacology* 52: 895–902
80 Perry NSL, Houghton PJ, Jenner P, Keith A, Perry EK (2002) *Salvia lavandulaefolia* essential oil inhibits cholinesterase *in vivo*. *Phytomedicine* 9: 48–51
81 Wake G, Court J, Pickering A, Lewis R, Wilkins R Perry E (2000) CNS acetylcholine receptor activity in European medicinal plants traditionally used to improve failing memory. *J Ethnopharmacology* 69: 105–114
82 Mantle D, Eddeb F, Pickering A (2000) Comparison of relative antioxidant activities of British medicinal plant species *in vitro*. *J Ethnopharmacology* 72: 47–51
83 Hohmann J, Zupko I, Redei D, Csanyi M, Falkay G, Mathe I, Janicsak G (1999) Protective effects of the aerial parts of *Salvia officinalis*, *Melissa officinalis* and *Lavandula angustifolia* and their constituents against enzyme-dependent and enzyme-independent lipid peroxidation. *Planta Medica* 65: 576–578
84 Perry NSL, Houghton PJ, Sampson J, Theobald AE, Hart S, Lis-Balchin M, Hoult JRS, Evans P, Jenner P, Milligan S, Perry EK (2001) *In vitro* activities of *S. lavandulaefolia* (Spanish Sage) relevant to treatment of Alzheimer's disease. *J Pharmacy and Pharmacology* 53: 1347–1356
85 Tildesley NTJ, Kennedy DO, Perry EK, Ballard CG, Savalev S, Wesnes KA, Scholey AB (2003) *Salvia lavandulaefolia* (Spanish Sage) enhances memory in healthy young volunteers. *Pharmacol Biochem Behav* 75: 669–674
86 Tildesley NTJ, Kennedy DO, Ballard CG, Wesnes KA, Hylands P, Perry EK, Scholey AB Sage Improves Memory and Attention in Healthy Older Volunteers: A controlled double blind acute study; *submitted*
87 Kennedy DO, Scholey AB, Tildesley NTJ, Perry EK, Wesnes KA (2002b) Modulation of mood and cognitive performance following acute administration of single doses of *Melissa officinalis* (Lemon Balm). *Pharmacology, Biochemistry and Behavior* 72(4): 953–964
87a Kiss B, Karpati E (1996) Mechanism of action of vinpocetine. *Acta Pharm Hung* 66(5): 213–224
87b Orvisky E, Soltes L, Stancikera M (1997) High-molecular-weight hyaluronan – a valuable tool in testing the antioxidative activity of amphiphilic drugs stobadine and vinpocetine. *J Pharm Biomed Anal* 16(3): 419–424
87c Santos MS, Duarte AI, Moreira PI, Oliveira CR (2000) Synaptosomal response to oxidative stress: effect of vinpcetine. *Free Radic Res* 32(1): 57–66
88 Yasui M, Yano I, Ota K, Oshima A (1990) Contents of calcium, phosphorus and aluminum in central nervous system, liver and kidney of rabbits with experimental atherosclerosis – scavenger effects of vinpocetine on the deposition of elements. *No To Shinkei* 42(4): 325–331
89 Imre G (1986) Use of Cavinton drip infusion in ophthalmology. II. Effect of Cavinton drip infusion in eye diseases caused by vascular occlusion or stenosis. *Ther Hung* 34 (2): 84–86
90 Konopka W, Zalewski P, Olszewski J, Olszewska-Ziaber A, Pietkiewicz P (1997) Treatment results of acoustic trauma. *Otolaryngol Pol* 51 Suppl 25: 281–284
91 King GA (1987) Protective effects of vinpocetine and structurally related drugs on the lethal consequences of hypoxia in mice. *Arch Int Pharmacodyn Ther* 286(2): 299–307
92 DeNoble VJ (1987) Vinpocetine enhances retrieval of a step-through passive avoidance response in rats. *Int Clin Psychopharmacol* 2(4): 325–331
93 Bhatti JZ, Hindmarch I (1987) Vinpocetine effects on cognitive impairments produced by flunitrazepam. *Int Clin Psychopharmacol* Oct 2(4): 325–331
94 Balestreri R, Fontana L, Astengo F (1987) A double-blind placebo controlled evaluation of the safety and efficacy of vinpocetine in the treatment of patients with chronic vascular senile cerebral dysfunction. *J Am Geriatr Soc* 35(5): 425–430
95 Hindmarch I, Fuchs HH, Erzigkeit H (1991) Efficacy and tolerance of vinpocetine in ambulant patients suffering from mild to moderate organic psychosyndromes. *Int Clin Psychopharmacol* 6(1): 31–43
96 Thal LJ, Salmon DP, Lasker B, Bower D, Klauber MR (1989) The safety and lack of efficacy of

vinpocetine in Alzheimer's disease. *J the American Geriatrics Society* 37: 515–520
97 McKay Hart A, Wiberg M, Terenghi G (2002) Pharmacological enhancement of peripheral nerve regeneration in the rat by systemic acetyl-L-carnitine treatment. *Neurosci Lett* 16 334(3): 181–185
98 Castorina M, Ferraris L (1994) Acetyl-L-carnitine affects aged brain receptorial system in rodents. *Life Sci* 54(17): 1205–1214
99 Ando S, Tadenuma T, Tanaka Y, Fukui F, Kobayashi S, Ohashi Y, Kawabata T (2001) Enhancement of learning capacity and cholinergic synaptic function by carnitine in aging rats. *J Neurosci Res* 66(2): 266–271
100 Dhitavat S, Ortiz D, Shea TB, Rivera ER (2002) Acetyl-L-carnitine protects against amyloid-beta neurotoxicity: Roles of oxidative buffering and ATP levels. *Neurochem Res* 27(6): 501–505
101 Hagen TM, Liu J, Lykkesfeldt J, Wehr CM, Ingersoll RT, Vinarsky V, Bartholomew JC, Ames BN (2002) Feeding acetyl-L-carnitine and lipoic acid to old rats significantly improves metabolic function while decreasing oxidative stress. *Proc Natl Acad Sci USA* 99(4): 1870–1875
102 Montgomery SA, Thal LJ, Amrein R (2003) Meta-analysis of double blind randomized controlled clinical trials of acetyl-L-carnitine *versus* placebo in the treatment of mild cognitive impairment and mild Alzheimer's disease. *Int Clin Psychopharmacol* 18(2): 61–71
103 Corwin J, Dean RL, Bartus RT, Rotrosen J, Watkins DL (1985) Behavioral effects of phosphatidylserine in the aged Fischer 344 rat: Amelioration of passive avoidance deficits without changes in psychomotor task performance. *Neurobiol Aging* 6(1): 11–15
104 Crook TH, Tinklenberg J, Yesavage J, Petrie W, Nunzi MG, Massari DC (1991) Effects of phosphatidylserine in age-associated memory impairment. *Neurology* 41(5): 644–649
105 Delwaide PJ, Gyselynck-Mambourg AM, Hurlet A, Ylieff M (1986) Double-blind randomized controlled study of phosphatidylserine in senile demented patients. *Acta Neurol Scand* 73(2): 136–140
106 Engel RR, Satzger W, Gunther W, Kathmann N, Bove D, Gerke S, Munch U, Hippius H (1992) Double-blind cross-over study of phosphatidylserine *versus* placebo in patients with early dementia of the Alzheimer type. *Eur Neuropsychopharmacol* 2(2): 149–155
107 Pepeu G, Spignoli G (1989) No-otropic drugs and brain cholinergic mechanisms. Prog *Neuropsychopharmacol Biol Psychiatry* 13 Suppl: S77–88
108 Maggioni M, Picotti GB, Bondiolotti GP, Panerai A, Cenacchi T, Nobile P, Brambilla F (1990) Effects of phosphatidylserine therapy in geriatric patients with depressive disorders. *Acta Psychiatr Scand* 81(3): 265–270
109 Schreiber S, Kampf-Sherf O, Gorfine M, Kelly D, Oppenheim Y, Lerer B (2000) An open trial of plant-source derived phosphatydilserine for treatment of age-related cognitive decline. *Isr J Psychiatry Relat Sci* 37(4): 302–307
110 Liu JS, Zhu YL, Yu CM et al. (1986) The structure of huperzine A and B, two new alkaloids exhibiting marked anticholinesterase activity. *Can J Chem* 64: 837–839
111 Ashani Y, Peggins JO, Doctor BP (1992) Mechanism of inhibition of cholinesterase by huperzine A. *Biochem Biophys Res Comm* 184: 7719–7726
112 Gordon RK, Nigam SV, Weitz JA, Dave JR, Doctor BP, Ved HS (2001) The NMDA receptor ion channel: A site for binding of Huperzine A. *J Appl Toxicol* 21 (Suppl1): S47–51
113 Perry EK, Perry RH, Blessed G, Tomlison BE (1978a) Changes in brain cholinesterases in senile dementia of Alzheimer type. *Neuropathol Appl Neurobiol* 4(4): 273–277
114 Zhu XD, Giacobini E (1995) Second generation cholinesterase inhibitors: Effect of (L)-huperzine A on cortical biogenic amines. *J Neurosci Res* 41: 828–835
115 Roman S, Vivas NM, Badia A, Clos MV (2002) Interaction of a new potent anticholinesterasic compound (+/–)huprine X with muscarinic receptors in rat brain. *Neurosci Lett* 325(2): 103–106
116 Sattler R, Tymianski M (2000) Molecular mechanisms of calcium-dependent excitotoxicity. *J Mol Med* 78(1): 3–13
117 Choi, DW (1994) Calcium and excitotoxic neuronal injury. Ann N Y *Acad Sci* 747: 162–171
118 Grunwald J, Raveh L, Doctor BP, Ashani Y (1994) Huperzine A as a pre-treatment candidate drug against nerve agent toxicity. *Life Sci* 54: 991–997
119 Xu SS, Cai ZY, Qu ZW et al. (1999) Huperzine-A in capsules and tablets for treating patients with Alzheimer's disease. *Acta Pharmacol Sin* 20: 486–490
120 Xiao XQ, Wang R, Tang XC (2000) Huperzine A and tacrine attenuate beta-amyloid peptide-induced oxidative injury. *J Neurosci Res* 61: 564–569
121 Bai DL, Tang XC, He XC (2000) Huperzine A, a potential therapeutic agent for treatment of Alzheimer's disease. *Curr Med Chem* 7(3): 355–374

122 Scott LJ, Gao KL (2000) Galantamine a review of its use in Alzheimer's disease. *Drugs* 60(5): 1095–1122

123 Ballard CG (2002) Advances in the treatment of Alzheimer's disease: Benefits of dual cholinesterase inhibition. *Eur Neurol* 47: 64–70

124 heng DH, Tang XC (1998) Comparative studies of huperzine A, E2020, and tacrine on behavior and cholinesterase activities. *Pharmacol Biochem Behav* 60: 377–386

125 Tang XC, De Sarno P, Sugaya K, Giacobini E (1989) Effect of huperzine A, a new cholinesterase inhibitor, on the central cholinergic system of the rat. *J Neurosci Res* 24: 276–285

126 Laganiere S, Corey J, Tang XC, Wülfert E, Hanin I (1991) Acute and chronic studies with the anticholinesterase huperzine A: Effect on central nervous system cholinergic parameters. *Neuropharmacology* 30: 763–768

127 Tang XC, Han YF (1999) Pharmacological profile of huperzine A, a novel acetylcholinesterase inhibitor from Chinese herb. *CNS Drug Reviews* 5: 281–300

128 Zhang RW, Tang XC, Han YY et al. (1991) Drug evaluation of huperzine A in the treatment of senile memory disorders. *Acta Pharmacol Sin* 12: 250–252

129 Xu SS, Cai ZY, Qu ZW, Yang RM, Cai YL, Wang GQ, Su XQ, Zhong XS, Cheng RY, Xu WA, Li JX, Feng B (1995) Efficacy of tablet huperzine-A on memory, cognition and behavior in Alzheimer's disease. *Acta Pharmacol Sin* 16: 391–395

130 Zhang Z, Wang X, Chen Q, Shu L, Wang J, Shan G (2002) Clinical efficacy and safety of huperzine Alpha in treatment of mild to moderate Alzheimer disease, a placebo-controlled, double-blind, randomized trial. *Natl Med J China* 82(14): 941–944

131 Zangara A (2003) The psychopharmacology of Huperzine A: An alkaloid with cognitive enhancing and neuroprotective properties of interest in the treatment of Alzheimer's disease. *Pharmacol Biochem Behav* 75: 675–686

132 Jian YB, Huang S, Huang L (2002) The improvement of Huperzine A on cognitive functions and behaviour of Alzheimer's patients. *Clinical Med China* 18(9): 802–803 (from abstract, in Chinese)

133 Sun QQ, Xu SS, *Pan* JL, Guo HM, Cao WQ (1999) Efficacy of huperzine A capsules on memory and learning performance in 34 pairs of matched junior middle school students. *Acta Pharmacol Sin* 20: 601–603

134 Qian BC, Wang M, Zhou ZF, Chen K, Zhou RR, Chen GS (1995) Pharmacokinetics of tablet huperzine A in six volunteers. *Acta Pharmacol Sin* 16: 396–398

135 Dye L, Lluch A, Blundell J (2000) Macronutrients and performance. *Nutrition* 16: 1021–1034

136 Gibson EL, Green MW (2002) Nutritional influences on cognitive function: Mechanisms of susceptibility. *Nutr Res Rev*: 169–206

137 Fillenz M, Lowry JP, Boutelle MG, Fray AE (1999) The role of astrocytes and noradrenaline in neuronal glucose metabolism. *Acta Physiologica Scandinavia* 167: 275–284

138 Weiss V (1986) From memory span to the quantum mechanics of intelligence. *Personality Ind Diff* 7: 737–749

139 Roland PE (1993) *Brain Activation*. Wiley, New York

140 Gold PE, Macleod KM, Thompson IJ, Frier BM (1985) Hypoglaecemic-induced cognitive dysfunction in diabetes mellitus; effects of hypoglaecemic unawareness. *Physiol Behav* 58: 501–511

141 Holmes CS, Koepke KM, Thompson RG, Gyves PW, Weydert JA (1984) Verbal fluency and naming performance in type I diabetics at different blood glucose concentrations. *Diabetes Care 7*: 454–459

142 Taylor LA, Rachman SJ (1987) The effects of blood sugar level changes on cognitive function, affective state and somatic symptoms. *J Behav Med* 11: 279–291

143 De Feo P, Gallai V, Mazzotta G, Crispino G, Torlone E, Perriello G, Ventura M, Santusaneo F, Brunetti P, Bolli GB (1988) Modest decrements in plasma glucose concentrations cause early impairment in cognitive function and later activation of glucose counter-regulation in the absence of hypoglycaemic symptoms. *J Clin Investigation 82: 436–444*

144 Sarti C, Pantoni L, Bartolini L, Inzitari D (2002) Cognitive impairment and chronic cerebral hypoperfusion: What can be learned from experimental models. J the *Neurological Sciences* 203–204: 263–266

145 Kramer AF, Coyne JT, Strayer DL (1993) Cognitive function at high altitude. *Human Factors* 35: 329–344

146 Crowley JS, Wesenten NW, Kamimori G, Devine ME, Iwanyk E, Balkin T (1992) Effects of high terrestrial altitude and supplemental oxygen on human performance and mood. *Aviat Space Eviron Med* 63: 696–701

147 Moss MC, Scholey AB (1996) Oxygen administration enhances memory formation in healthy young adults. *Psychopharmacology* 124: 255–260
148 Moss MC, Scholey AB, Wesnes K (1998) Oxygen administration selectively enhances cognitive performance in healthy young adults: A placebo-controlled double blind crossover study. *Psychopharmacology* 138: 27–33
149 Scholey AB, Moss MC, Wesnes KA (1998) Oxygen and cognitive performance: the temporal relationship between hyperoxia and enhanced memory. *Psychopharmacology* 140: 123–126
150 Scholey AB, Moss MC, Neave N, Wesnes K (1999) Cognitive performance, hyperoxia and heart rate following oxygen administration in healthy young adults. *Physiol Behav* 67: 783–789
151 Winder R, Borrill J (1998) Fuels for memory: The role of oxygen and glucose in memory enhancement. *Psychopharmacology* 136: 349–356
152 Andersson J, Berggren P, Grönkvist M, Magnusson S, Svensson E (2002) Oxygen saturation and cognitive performance. *Psychopharmacology* 162: 119–128
153 Manning CA, Hall JL, Gold PE (1990) Glucose effects on memory and other neuropsychological tests in elderly humans. *Psychol Sci* 1: 307–311
154 Gonder-Frederick L, Hall JL, Vogt J, Cox DJ, Green J, Gold PE (1987) Memory enhancement in elderly humans: Effects of glucose ingestion. *Physiol Behav* 41: 503–504
155 Benton D, Owens D (1993) Is raised blood glucose associated with the relief of tension. J *Psychosom Res* 37: 1–13
156 Benton D, Owens DA Parker PY (1994) Blood glucose memory and attention. *Neuropsychologia* 32: 595–607
157 Craft S, Murphy C, Wernstrom J (1994) Glucose effects on complex memory and non-memory tasks: The influence of age, sex and glucoregulatory response. *Psychobiology* 22: 95–105
158 Benton D, Parker PY (1998) Breakfast, blood glucose and cognition. *Am J Clin Nutr* 67(suppl): 772–778
159 Foster JK, Lidder PG, Sünram S (1998) Glucose and memory: Fractionation of enhancement effects. *Psychopharmacology* 137: 259–270
160 Sünram-Lea SI, Foster JK, Durlach P, Perez C (2001) Examination of the relation of fast-duration, time of day and pre-consumption baseline plasma glucose levels on the glucose facilitation of cognitive performance effect. *Psychopharmacology* 157(1): 46–54
161 Sünram-Lea SI, Foster JK, Durlach P, Perez C (2002a) Investigation into the significance of task difficulty and divided allocation of resources on the glucose memory facilitation effect. *Psychopharmacology* 160: 387–397
162 Sünram-Lea SI, Foster JK, Durlach P, Perez C (2002b) The effect of retrograde and anterograde glucose administration on memory performance in healthy young adults. *Behav Brain Res* 134: 505–516
163 Azari NP (1991) Effects of glucose on memory processes in young adults. *Psychopharmacology* 105: 521–524
164 Ford CE, Scholey AB, Ayre G, Wesnes K (2002) The effect of glucose administration and the emotional content of words on heart rate and memory. *J Psychopharmacol* 16: 241–244
165 Martin PY, Benton D (1999) The influence of a glucose drink on a demanding working memory task. *Physiol Behav* 67: 69–74
166 Kennedy DO, Scholey AB (2000) Glucose administration, heart rate and cognitive performance: Effects of increasing mental effort. *Psychopharmacology* 149: 63–71
167 Scholey AB and Fowles K (2002) Retrograde enhancement of kinesthetic memory by low dose alcohol and by glucose. *Neurobiology of Learning and Memory* 78: 477–483
168 Owens DS, Benton D (1994) The impact of raising blood glucose on reaction times. *Neuropsychobiology* 30: 106–113
169 Benton D (1990) The impact of increasing blood glucose levels on psychological functioning. *Biol Psychol* 30: 13–19
170 Parker PY, Benton D (1995) Blood glucose levels selectively influences memory for words dichotically presented to the right ear. *Neuropsychologia* 33: 843–854
171 Donohoe RT, Benton D (1999) Declining blood glucose levels after a cognitively demanding task predict subsequent memory. *Nutrit Neuroscience* 2: 413–424
172 Awad N, Gagnon M, Desrochers A, Tsiakas M, Messier C (2002) Impact of peripheral glucoregulation on memory. *Behav Neurosci* 116: 691–702
173 Gold PE (1995) Role of glucose in regulating the brain and cognition. *Int J Clin Nutr* 61: 987–995

174 Wenk GL (1989) An hypothesis of the role of glucose in the mechanism of cognitive enhancers. *Psychopharmacology* 99: 431–438
175 Williams CL, McGaugh JL (1992) Reversible inactivation of the nucleus of the solitary tract impairs retention performance in an inhibitory avoidance task. *Behav Neurol Biol* 58: 20–40
176 Cahill L, Prins B, Weber M, McGaugh JL (1994) α-adrenergic activation and memory for emotional events. *Nature* 371: 702–704
177 Nielson KA, Radtke RC, Jensen RA (1996) Arousal induced modulation of memory storage processes in humans. *Neurobiology of Learning and Memory* 66(2): 133–142
178 Fibiger W, Evans O, Singer G (1986) Hormonal responses to graded mental workload. *Eur J Appl Physiol* 55: 339–343
179 Carroll D, Turner JR, Prasad R (1986) The effects of level of difficulty of mental arithmetic challenge on heart rate and oxygen consumption. *Int J Psychophysiol* 4: 167–173
180 Mulder G, Mulder LMJ (1981) Information processing and cardiovascular control. *Psychophysiology* 18: 392–402
181 Backs WR, Seljos KA (1994) Metabolic and cardiorespiratory measures of mental effort: the effects of level of difficulty in a working memory task. *Int J Psychophysiol* 16: 57–68
182 Jonides J, Schumacher EH, Smith EE, Lauber EK, Awh E, Satoshi M, Koeppe RA (1997) Verbal working memory load affects regional brain activation as measured by *PET J Cog Neurosci* 9: 462–475
183 Haier RJ, Siegel BV, MacLachlin A, Suderling E, Lottenberg S, Buchsbaum MS (1992) Regional glucose metabolic changes after learning a complex visuo-spatial/motor task: A positron Emission Tomographic study. *Brain Res* 570: 134–143
184 Parks RW, Loewenstein DA, Dodrill KL, Barker WW, Yoshii F, Chang JY, Emran A, Apicella A, Sherameta WA, Duara R (1988) Cerebral metabolic effects of a verbal fluency test: A PET scan study. *J Clin Exp Neuropsychol* 10: 565–575
185 Wesnes K, Pincock C (2002) Practice effects on cognitive tasks: a major problem? The *Lancet Neurology* 1: 473
186 McClelland GR (1987) The effects of practice on measures of performance. Human *Psychopharmacol* 210: 109–118
187 Wilson RS, Beckett LA, Barnes LL et al. (2002) Individual differences in rates of change in cognitive abilities of older persons. *Psychology and Aging* 17: 179–193
188 Tripathi YB, Chaurasia S, Tripathi E, Upadhyay A, Dubey GP (1996) *Bacopa monniera* Linn. as an antioxidant: Mechanism of action. *Indian J Exp Biol* 34(6): 523–526
189 Vohora D, Pal SN, Pillai KK (2000) Protection from phenytoin-induced cognitive deficit by *Bacopa monniera*, a reputed Indian nootropic plant. *J Ethnopharmacol* 71(3): 383–390
190 Das A, Shanker G, Nath C, Pal R, Singh S, Singh H (2002) A comparative study in rodents of standardized extracts of *Bacopa monniera* and *Ginkgo biloba*: anticholinesterase and cognitive enhancing activities. *Pharmacol Biochem Behav* 73(4): 893–900
191 Sairam K, Dorababu M, Goel RK, Bhattacharya SK (2002) Antidepressant activity of standardized extract of *Bacopa monniera* in experimental models of depression in rats. *Phytomedicine* 9(3): 207–211
192 Stough C, Lloyd J, Clarke J, Downey LA, Hutchison CW, Rodgers T, Nathan PJ (2001) The chronic effects of an extract of *Bacopa monniera* (Brahmi) on cognitive function in healthy human subjects. *Psychopharmacology* 156(4): 481–484
193 Mehta A, Mason PJ, Vulliamy TJ (2000) Glucose-6-phosphate dehydrogenase deficiency. *Baillieres Best Pract Res Clin Haematol* 13(1): 21–38
194 Rabey JM, Vered Y, Shabtai H, Graff E, Korczyn AD (1992) Improvement of parkinsonian features correlate with high plasma levodopa values after broad bean (*Vicia faba*) consumption. *J Neurol Neurosurg Psychiatry* 55(8): 725–727
195 Apaydin H, Ertan S, Ozekmekci S (2000) Broad bean (*Vicia faba*)—a natural source of L-dopa-prolongs "on" periods in patients with Parkinson's disease who have "on-off" fluctuations. *Mov Disord* 15(1): 164–166
196 Bhattacharya SK, Satyan KS, Ghosal S (1997) Antioxidant activity of glycowithanolides from withania somnifera. *Indian J Exp Biol* 35(3): 236–239
197 Bhattacharya SK, Bhattacharya A, Sairam K, Ghosal S (2000) Anxiolytic-antidepressant activity of *Withania somnifera* glycowithanolides: an experimental study. *Phytomedicine* 7(6): 463–469
198 Archana R and Namisivayam A (1999) Antistressor effects of *Withania somnifera*. *J Ethnopharmacol* 64 (1): 91–93

199 Abdel-Magied EM, Abdel-Rahman HA, Harraz FM (2001) The effect of aqueous extracts of *Cynomorium coccineum* and *Withania somnifera* on testicular development in immature Wistar rats. *J Ethnopharmacol* 75(1): 1–4
200 Nalina K, Karanth KS, Rao A, Aroor AR (1995) Effects of *Celastrus paniculatus* on passive avoidance performance and biogenic amine turnover in white rats. *J Ethnopharmacology* 47: 101–108
201 Bidwai PP, Wangoo D, Bhullar N (1990) Antispermatogenic action of *Celastrus paniculatus* seed extract in the rat with reversible changes in the liver. *J Ethnopharmacol* 28(3): 293–303
202 Gattu M, Boss K, Terry A, Buccafusco J (1997) Reversal of scopolamine-induced deficits in navigational memory performance by the seed oil of *Celastrus paniculatus*. *Pharmacology Biochemistry and Behavior* 57(4): 793–799
203 Kumar MH, Gupta YK (2002) Antioxidant property of *Celastrus paniculatus willd*: a possible mechanism in enhancing cognition. *Phytomedicine* 9: 302–311
204 Chung YK, Heo HJ, Kim EK, Kim HK, Huh TL, Lim YH, Kim SK, Shin DH (2001) Inhibitory effect of ursolic acid purified from *Origanum majorana* L. on the acetylcholinesterase. *Mol Cells* 11: 137–143
205 Heo HJ, Cho HY, Hong B, Kim HK, Heo TR, Kim EK, Kim SK, Kim CJ, Shin DH (2002) Ursolic acid of *Origanum majorana* L. reduces Abeta-induced oxidative injury. *Mol Cells* 13(1): 5–11
206 Liu J, Atamna H, Kuratsune H, Ames BN (2002a) Delaying brain mitochondrial decay and aging with mitochondrial antioxidants and metabolites. *Ann N Y Acad Sci* 959: 133–166
207 Deneke SM (2000) Thiol-based antioxidants. *Curr Top Cell Regul* 36: 151–180
208 Cantuti-Castelvetri I, Shukitt-Hale B, Joseph JA (2000) Neurobehavioral aspects of antioxidants in aging. *Int J Dev Neurosci* 18(4–5): 367–381
209 Halat KM, Dennehy CE (2003) Botanicals and dietary supplements in diabetic peripheral neuropathy. *J Am Board Fam Pract* 16(1): 47–57
210 Farr SA, Poon HF, Dogrukol-Ak D, Drake J, Banks WA, Eyerman E, Butterfield DA, Morley JE (2003) The antioxidants alpha-lipoic acid and N-acetylcysteine reverse memory impairment and brain oxidative stress in aged SAMP8 mice. *J Neurochem* 84 (5): 1173–1183
211 Stoll S, Hartmann H, Cohen SA, Muller WE (1993) The potent free radical scavenger alpha-lipoic acid improves memory in aged mice: Putative relationship to NMDA receptor deficits. *Pharmacol Biochem Behav* 46(4): 799–805
212 Liu J, Head E, Gharib AM, Yuan W, Ingersoll RT, Hagen TM, Cotman CW, Ames BN (2002b) Memory loss in old rats is associated with brain mitochondrial decay and RNA/DNA oxidation: Partial reversal by feeding acetyl-L-carnitine and/or R-alpha-lipoic acid. *Proc Natl Acad Sci USA* 99 (10): 7184–7185

Cognitive Enhancing Drugs
Edited by J.J. Buccafusco

Drugs with multiple CNS targets

Jerry J. Buccafusco[1] and Moussa B.H. Youdim[2]

[1] *Alzheimer's Research Center, Medical College of Georgia, 1120 – 15th Street, Augusta, Georgia 30912-2300, USA*
[2] *Finkelstein Professor of Life Sciences, Director of Eve Topf and National Parkinson Foundation (US), Centers of Excellence for Neurodegenerative Diseases, Technion-Faculty of Medicine, Haifa, Israel*

Diversity of neural systems important for cognition

A wide variety of clinical syndromes can manifest cognitive or memory dysfunction. These include head trauma, cerebrovascular accidents, convulsive disorders, nutritional deficits, drug-associated toxicity, etc. Although collectively these entities contribute significantly to the overall amalgamation of known memory disorders, by far the primary disease entity targeted by pharmaceutical research is Alzheimer's disease (AD). AD, which represents the most common form of dementia among individuals over 65 years of age, is now the third most expensive health care problem in the U.S. exceeded only by cancer and cardiovascular disease. It currently affects approximately 4 million Americans and imposes an annual economic burden estimated at between $80 and $100 billion. This devastating degenerative condition also inflicts an enormous emotional toll on patients, family members, and caregivers. As the geriatric population inexorably increases, the numbers of AD patients may increase to epidemic numbers (i.e., in excess of 9 million) by the middle of the twenty first century.

An additional concern of older adults is the perception that memory loss occurs as a natural result of aging. This apprehension has contributed at least in part to the enormous increase in sales of over the counter remedies and homeopathic products with claims of memory-enhancing properties. This demand from a large and ever increasing elderly population has also provided the basis for a rising interest in the development of pharmacological agents, not only for the treatment of AD, but also for the much more common, mild cognitive decline associated with normal, non-disease aging [1]. A measurable (albeit mild) decline in cognitive function can occur as a part of healthy aging in humans which begins at some point after the fifth decade of life. The changes observed in typical (non-pathologic) aging are manifested primarily as mild deficits in declarative memory which are thought to result as a consequence of a reduction in the speed of central processing necessary for encoding and retrieval of information [2]. Mild memory deficits that exceed those associated with normal aging, but that do not meet the (DSM IV) criteria for a diagnosis of dementia, have been referred to as "benign senescent forgetfulness" and "age associated memory impairment" (AAMI). Recent evidence suggests that patients with AAMI have an increased risk of developing dementia [1], a finding that has generated considerable concern and provided the impetus for rigorous investigation.

The study of AD, particularly of the neurochemistry of the post-mortem AD brain, has provided perhaps a greater level of indirect evidence for important components of cognitive and mnemonic pathways than has the study of the "normal" aging brain. Indeed, the well-known selective vulnerability of basal forebrain acetylcholine-containing neurons in AD has underscored the importance of this neurotransmitter system in memory and perhaps in other behavioral and cognitive functions affected by the disease (see first Chapter by J.J. Buccafusco). Among the host of degenerative processes occurring in AD, reproducible cholinergic deficits are consistently reported; they appear early in the disease process, and correlate well with the degree of dementia (for review, [3]). Moreover, abnormalities in cholinergic function are frequently reported in other degenerative conditions such as Parkinson's disease (PD), diffuse Lewy body dementia and Huntington's disease. As in AD, such cholinergic deficits often correlate with memory decline and dementia. Extensive AD-related neuropathology is also commonly found in areas normally rich in norepinephrine (locus ceruleus) and serotonin (dorsal raphe nucleus), particularly in the later stages of the disease [4]. Findings regarding the loss of gutamatergic and certain peptidergic pathways in the AD brain are also extant in the literature [4, 5]. These findings have prompted studies designed to reveal the potential for targeting these depleted neurotransmitter systems with respective receptor agonists or other synaptic signal-strengthening drugs.

The diversity of neurotransmitter substances involved in cognition is perhaps not too surprising, since it has been well-known for many years that memory is represented by several distinct processes, and different types of memory are relegated to different (but sometimes overlapping) brain regions. For example components of the hippocampal formation have been implicated in mediating or processing spatial, declarative, and episodic types of memory in humans, primates, and rodents (e.g., [6–10]). A reasonable argument has been made for the possibility that the hippocampus does not play as important a role in semantic memory [11, 12], with habit learning more dependent upon the striatum [13, 14]. Also, emotional or conditioning learning processes appear to reside within the amygdala [15]. Even within what has been termed working memory or episodic memory, there appears to exist separable and interacting components that may include acquisition (attention), consolidation, and retention (short and long-term); alternatively, encoding, retrieval, storage, and consolidation (see [7]). Certain amnestic agents, such as scopolamine, appear predominantly to affect the acquisition of new learning (e.g., [16]). Selectivity of action with regard to the components of memory has also been attributed to certain memory-enabling drugs, even within a pharmacological class [17]. Thus, it seems reasonable to conclude that there are several, if not numerous, potential targets for the pharmacological treatment of memory disorders, and that drugs that promote activity within different, but interacting components of cognitive function may be expected to act additively, if not synergistically, when administered together.

Potential multiple synergistic targets for memory enhancement

In recent years, much attention has been focused on the design of palliative agents (cholinergics, nootropics, etc.) that have the ability to offer subtle cognitive improvement. There has been much discussion as to the reason for the limitations in therapeutic efficacy noted for these classes of compounds. For cholinergic compounds demonstrated to improve the performance of cognitive tasks in animals, the potential effectiveness offered by them (cholinesterase inhibitors and direct cholinergic receptor agonists) in humans can be limited by the appearance of central and peripheral side-effects. The premise that high selectivity and high potency are the most desirable properties for a new therapeutic agent may not be the case for many drugs designed to treat brain disorders. For example, in PD activation of both D1 and D2 striatal dopaminergic neurons may be necessary for maximal drug efficacy in reducing motor symptoms. Also, in the treatment of major psychoses the most useful classes of agents are proving to be those 'atypical' drugs that often exhibit low potency and little selectivity. Similar pharmacological opportunities are available for AD drugs as well. As discussed in the preceding paragraphs, multiple neurotransmitter systems are affected to varying degrees in AD. Both noradrenergic neurons and cholinergic neurons have been shown to play a role in different components of learning and memory in rats. It may require combined therapy with adrenergic agonists such as clonidine and cholinergic agonists such as acetylcholinesterase (AChE) inhibitors to fully reverse the cognitive defects resulting from combined lesions of adrenergic and cholinergic neuronal pathways [18]. One other example of the concept of synergistic actions of different drug classes on memory-related task performance is a report in which the muscarinic M1-preferring receptor agonist, milameline, was shown to augment the ability of the AChE inhibitor tacrine to reverse a scopolamine-induced decrement in efficiency of maintaining a continuous performance task by Rhesus monkeys [19]. More recently it has been reported that the cognitive enhancement produced by cholinergic muscarinic agonists may involve septohippocampal GABAergic and hippocampal glutamatergic neurons [20]. The potential for combining drugs acting on the acetylcholine and glutamate systems is enhanced with the advent of the low affinity NMDA receptor antagonist memantine. This compound may prevent the excitatory amino acid neurotoxicity suggested to accompany AD without interfering with the actions of glutamate required for learning and memory [21]. Recent clinical trials have indicated that memtantine may improve cognition and result in the early improvement in behavior in AD [22].

As described throughout this book, many drugs and other natural substances derived from a wide variety of chemical and pharmacological classes have been shown to improve memory-related task performance in animals and humans. The clinical use of AChE inhibitors is likely to continue for some time into the future, and this pharmacological class continues to represent one of the most effective drugs tested in animal models [23]; and the use of AChE inhibitors may benefit quite dramatically from the addition of other pharmacological classes of cognitive-enhancing drugs. Our first indication that combinations of drugs might prove useful as a therapeutic approach to improving memory-related task performance was derived from one of our earlier studies in which we attempted to block the improvement in delayed matching-to-sample (DMTS) task efficiency to nicotine with the antagonist mecamylamine [24]. Generally, responsiveness to nicotine is relegated to one or two doses in a series of less than two log units. The highly individualized response to memory-enhancing drugs led us [25] to suggest that the effectiveness of a drug could mainly be determined by performing a dose-response series, but then selecting the individualized optimum dose or 'Best Dose'. Bartus used this approach to help identify non-responders. We have used the Best Dose as a means of comparison of drug effectiveness (see: [26]). However, the question remains as to the mechanism(s) contributing to the inverted-U-dose-response relationship. For nicotine-induced cognitive enhancement, higher doses may be associated with side-effects that could interfere with task motivation. In our initial study with mecamylamine (which was used to confirm that central nicotinic receptors mediated the positive mnemonic response to nicotine) we used the quaternary nicotinic antagonist hexamethonium to control for the potential peripheral actions (mainly ganglionic blockade) of mecamylamine on subjects performing the delayed matching-to-sample task. When the monkeys were pre-treated with hexamethonium, the nicotine-induced improvement in average task efficiency was further enhanced across all delays (although the effect was not statistically significant). Thus, it may be possible to expand the therapeutic window of certain agents like nicotine by preventing peripheral side-effects with the use of low levels of peripheral nicotinic receptor blockade. Along these lines, it is somewhat perplexing as to why low doses of selective peripherally-acting muscarinic antagonists such as methylatropine or glycopyrrolate have not been used in combination with cholinesterase inhibitors to help limit the latter drugs' side-effects.

Targeting brain AChE and α_2-adrenergic receptors

In addition to this heuristic approach to combination therapy, we also considered the possibility of pharmacologically exploiting at least two targets, central α_2-adrenergic receptors (with clonidine) and AChE with physostigmine [27]. Clonidine may target α_2-adrenergic receptors in the prefrontal cortex to evoke a moderate level of task improvement [28] (and see Chapter by Edward

D. Levin). From a different perspective there may be another rationale for considering that combined therapy may be superior to monotherapy. We have reported that clonidine is a potent inhibitor of the biosynthesis and the release of acetylcholine within specific brain regions (particularly in hypothalamic and hindbrain regions) in the rat and that the drug can inhibit the expression of cholinergic signs of toxicity to physostigmine and other cholinesterase inhibitors (see [29]). However, clonidine is only weakly effective in inhibiting cholinergic function within higher brain regions, presumably containing sites more relevant to the cognitive enhancing actions of AChE inhibitors [30]. We tested the possibility that combined treatment with clonidine and physostigmine could result in enhanced effects on DMTS accuracy by mature adult and aged macaques.

One of the most obvious effects of adding 0.5 µg/kg clonidine to the physostigmine regimen was that the animals were able to tolerate much higher doses of physostigmine. The individualized optimal ("Best") dose of physostigmine was determined for each animal as that dose which provided the greatest improvement in task accuracy averaged over the entire 96-trial session. The Best Doses determined for physostigmine alone ranged from 5–40 µg/kg (mean of 21.4 ± 4.5 µg/kg). Best Doses determined for physostigmine in the presence of 0.5 µg/kg clonidine ranged from 10–60 µg/kg (mean of (40.0 ± 6.9 µg/kg) – almost a two-fold increase. Despite the fact that doses used for physostigmine were maximal for each animal, when the two drugs were combined, a further improvement in performance was obtained. In fact, for the combination regimen, on the day after administration, performance accuracy continued to be elevated relative to baseline. Therefore, in this example the following factors appear to contribute to the enhancement of task performance in the combination: 1) a widening of the therapeutic window, most likely reflecting a reduction in AChE inhibitor-associated side-effects; 2) the targeting of separate neural substrates that each play a role in cognitive function; 3) the addition of clonidine extended the regimen's overall duration of action possibly through a unique pharmacodynamic action. In this particular study we used a fixed dose of clonidine previously determined to be optimal when used alone. It is possible that additional improvement might be obtained with further optimization of the regimen.

Targeting both brain AChE and nicotinic receptors

MHP-133 (3-(N,N-dimethylcarbamoyl)hydroxy-1-methyl-2-[[N-phenylamino-carbonyl)hydrazono]methyl]pyridinium chloride) is the lead compound of a novel series of analogs designed to target multiple brain substrates expected to have synergistic actions in the treatment of human cognitive disorders such as AD [31, 32]. MHP-133 was designed to target components of acetylcholine neurons that would act synergistically to enhance cholinergic function, including the stimulation of cholinergic receptors and the inhibition of AChE.

The strategy was to develop compounds with multiple targets relevant for enhancing cognition and memory, but avoiding the serious side-effects attributed to high potency cholinergic agonists. A preliminary assessment of the neurochemical properties exhibited by MHP-133 suggested that the drug may indeed fit this profile. For example, MHP-133 was shown in ligand binding studies to interact with subtypes of cholinergic (nicotinic and muscarinic, M1 and M2) receptors and to serve as a weak AChE inhibitor. Most importantly, however, MHP-133 was shown to enhance the performance of young and aged animals trained to perform memory-related tasks. Our study group of young and aged macaques was well-trained in the performance of an automated DMTS task. The monkeys were represented by individuals from three species, both genders, and two age groups, mature and aged. As indicated in Figure 1, MHP-133 significantly enhanced performance of all groups in the computer-assisted DMTS over a relatively wide range of doses. For the animals' individualized Best Dose (30–100 µg/kg, *i.m.*), performance of trials associated with long delay intervals increased on average by 32.2% over baseline performance for the entire group. Preliminary studies in a group of five aged rhesus monkeys demonstrated that MHP-133 was also effective at improving task performance after oral administration (data not shown).

MHP-133 has the potential to produce clinical improvement of the cognitive symptoms associated with AD and related disorders, and to inhibit the progress of the disease. The drug may offer multiple synergistic mechanisms

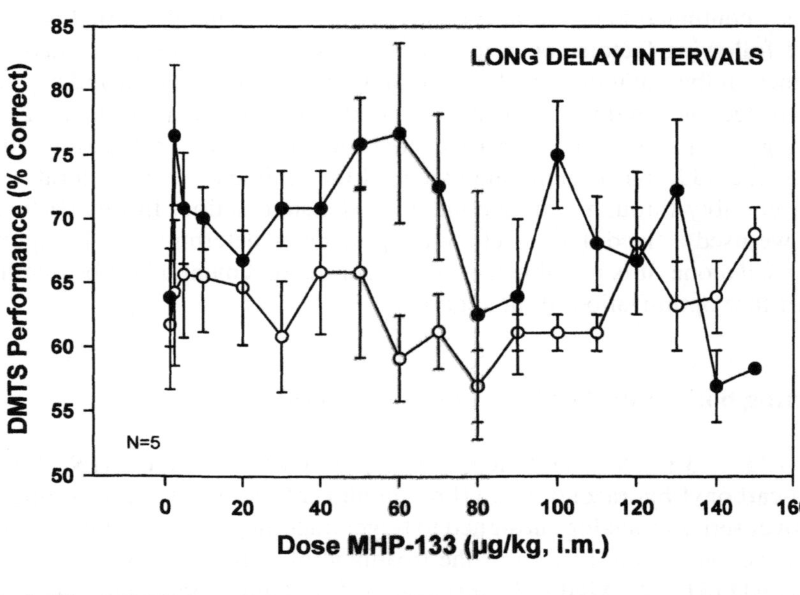

Figure 1. Effect of MHP-133 (filled circles) on DMTS performance efficiency by adult monkeys as a function of dose. Data are presented for trials associated with long delay intervals. Note the wide dose window for task improvement provided by the drug. Open circles = vehicle baseline values.

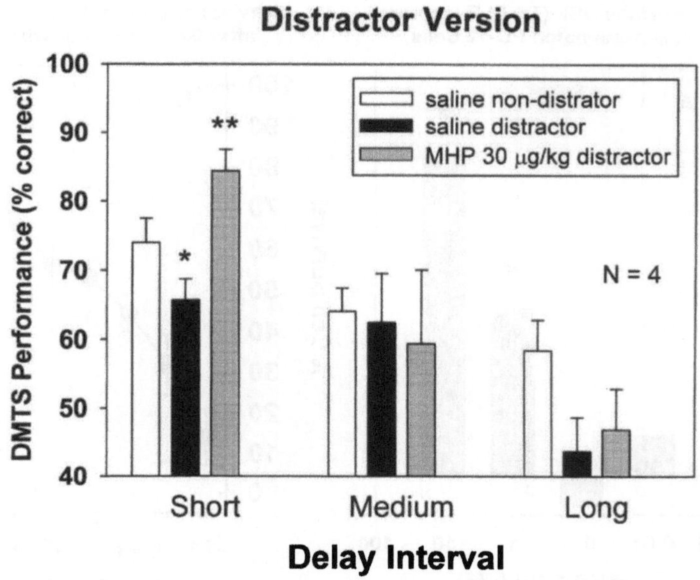

Figure 2. Mature (non-aged) Rhesus monkeys performed a modified DMTS task that included a randomly presented (20% of trials) distractor (flashing panel lights). Distractor trials were performed with significantly reduced efficiencies for short delay intervals. MHP-133 completely reversed the effect of the distractor, and further improved task performance above control (non-distractor levels).

of actions in this regard. This synergy of therapeutic actions may underlie both the marked effectiveness of the drug on memory, as well as the lowered potential for producing side-effects. As with nicotine ([33] and see Chapter by Andrea Wevers and Diana S. Woodruff-Pak), MHP-133 reversed distractor-induced (interference trials) performance decrements suggesting that part of its positive mnemonic action included improved levels of attention (Fig. 2). Also, like nicotine [34, 35], the drug enhanced nerve growth factor (NGF) – TrkA receptor expression in a neuronal cell line (Fig. 3A) indicating the potential for neuro-protective effects (Fig. 3B). MHP-133 and JWB-I-68-13 (a derivative of MHP-133) were also examined for their ability to alter the levels of amyloid precursor protein (APP) secreted by cultured astrocytes into the media during a 1 hr treatment with the cells. APP was measured by using a monoclonal antibody directed against the N-terminus of APP. Both MHP-133 and JWB-I-68-13 significantly increased the levels of secreted APP from cultured astrocytes over the concentration range of 10–100 μM. In general, the levels of secreted APP with MHP-133 or JWB-I-68-13 treatment were about 1.2–1.6-fold compared to the level of untreated control cells. This increase in APP secretion caused by MHP-133 and JWB-I-68-13 might be expected to decrease the cellular levels of amyloidogenic holoprotein. This possibility is in keeping with preliminary results showing that both compounds were associated with small

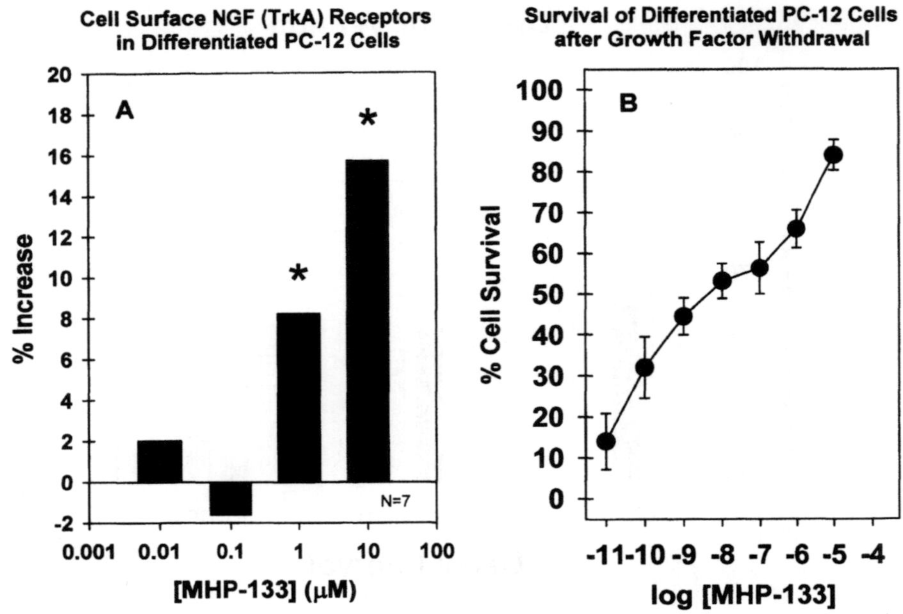

Figure 3. (A) cell-based ELISA was used to measure TrkA receptors on differentiated PC12 cells. Incubation of these cells for 24 hr with MHP-133 produced a significant concentration-dependent increase in TrkA receptor expression. * $p < 0.05$ compared to untreated control cells. (B) The ability of MHP-133 to protect differentiated PC12 cells from the cytotoxicity induced by (24 hr) NGF and serum (trophic factor) deprivation. Various concentrations of MHP-133 were incubated for 24 hr prior to trophic factor withdrawal. MHP-133 pretreatment produced a significant increase in cell viability compared with untreated control values (not shown), $p < 0.001$.

decreases in cell-associated APP holoprotein in cultured astrocytes and neurons (data not shown).

Despite the promise of MHP-133 and its analogs as potential therapeutic entities for AD and related disorders, these compounds are not yet optimized regarding the important pharmacological actions described above. Also, the concentrations required for neuroprotection and altered amyloid metabolism represent *in vivo* dose-ranges greater than that used for memory enhancement. Part of the challenge to developing bi- or multi-functional molecules is the ability to address the various targets with equivalent efficacies.

Targeting brain AChE and M2 muscarinic receptors

JWS-USC-75IX is a relatively potent AChE inhibitor (much more so than MHP-133), but it also exhibits high affinity antagonism for the muscarinic M2 muscarinic cholinergic receptor [36]. As AChE inhibitors have the potential of limiting their own actions through acetylcholine-induced feedback inhibition (mediated *via* activation of presynaptic M2 receptors), it was reasoned that M2 receptor antagonism could result both in the enhanced release of acetylcholine, and mitigation of the AChE inhibitor-induced feedback inhibition. JWS-USC-75IX was shown to improve the performance of rats in three different memory-related tasks, and in one of these, a delayed discrimination task, the drug was shown to exhibit repeatable improvements without the development of tolerance. The task was developed so that we could employ an operant paradigm (not unlike the primate DMTS) in rats. JWS-USC-75IX also exhibited a marked safety profile relative to drugs acting only to inhibit AChE [36].

At this point it is appropriate to point out that all efforts to combine multiple actions in one molecule have not met with success. An example is the compound RS66331 which (neurochemically) exhibits the properties of a $5HT_4$ agonist and a $5HT_3$ antagonist. Both properties have been associated with enhanced release of brain acetylcholine [37]. We studied this compound in aged rhesus monkeys and compared its effectiveness with that produced by individual administration of a $5HT_4$ agonist and a $5HT_3$ antagonist, both of which were demonstrated previously to enhance task performance in the same subjects. Rather than this combination of properties imbuing RS66331 with augmented memory-enhancing action, the effectiveness of the drug proved to be similar to that produced by the $5HT_3$ antagonist RS56812, but it was considerably reduced in effectiveness compared with the $5HT_4$ agonist RS17017 [23]. However, RS66331 was developed prior to our work with the individual compounds. There are many reasons for the failure of compounds to achieve expectations in memory paradigms, however, this may be one case wherein the information derived from the combined administration of various dose-regimens of RS56812 and RS17017 may have alerted us to the possibility that this is not a useful neural target combination, or to the possibility that different proportions of relative receptor activity were needed as compared to that inherent in RS66331.

Targeting brain monoamine oxidase (MAO) and cholinesterase

Relatively little attention has been paid to development of neuroprotective drugs for the treatment of AD. In some respects there is a better understanding of nigro-striatal dopaminergic neurodegeneration and neuroprotection mechanisms in PD because of the availability of relatively appropriate models. The mechanisms that may be involved in the process of neurodegeneration, particularly in AD, include oxidative stress, inflammatory processes, and accumulation of iron at the site of neurodegeneration. As such, antioxidants, MAO-B inhibitors, non-steroidal anti-inflammatory drugs and iron chelators have been suggested to exert neuroprotective actions, and they have been incompletely examined in AD. The MAO-B inhibitors selegiline and rasagiline [38] are anti-Parkinson's (anti-PD) drugs that have warranted more scrutiny as a consequence of their neuroprotective activity *in vitro* (neuronal cell cultures) and in animal studies. The established co-morbidities of (1) AD with extrapyramidal features; (2) PD with dementia; and (3) both AD and PD with depression stimulated the development of a series of novel bi-functional drugs possessing the MAO inhibitory activity and neuroprotective activity exhibited by anti-Parkinson drugs selegiline or rasagiline [39–41] along with the ability to inhibit AChE [42]. For these studies, the structural requirements for anti-PD activity as exhibited by rasagiline and selegiline, and the structural requirements for cholinesterase inhibition exhibited by rivastigmine, was initially considered.

Introduction of a carbamate moiety into the rasagiline molecule (Fig. 4) resulted in almost complete loss of brain MAO-B inhibitory activity for TV3326 and TV3279 as determined *in vitro*. However, chronic oral administration (12.5–76 mg/kg) in rats, mice and rabbits showed that TV3326, but not TV3279, is a CNS-selective inhibitor of MAO-A and MAO-B, with little inhibition of liver and small intestine MAO [43–45]. The possibility that TV3326 was serving as a prodrug in the brain to generate an active MAO inhibitory metabolite(s) is supported by the results of animal studies where, in contrast to the first order kinetic recovery of MAO activity from inhibition by rasagiline, TV3326 exhibited a biphasic action [45]. Also, an active metabolite of TV3294 has been identified. Both TV3326 and TV3279 inhibit AChE and butyrylcholinesterase, but with a slower time course and with slightly reduced effectiveness compared with rivastigmine. TV3326 is 100 times more potent against butyrylcholinesterase than it is against AChE. The cholineserase inhibitory activity of these drugs is consistent with their dose-related

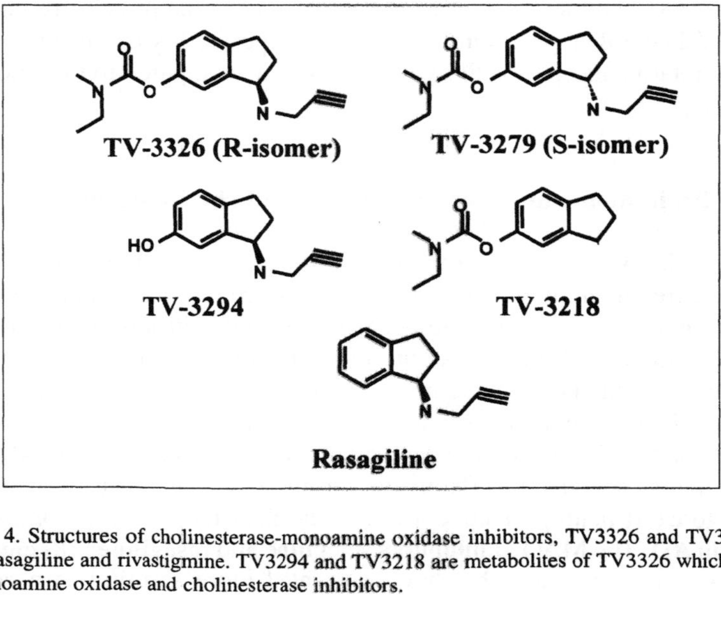

Figure 4. Structures of cholinesterase-monoamine oxidase inhibitors, TV3326 and TV3279 derived from rasagiline and rivastigmine. TV3294 and TV3218 are metabolites of TV3326 which respectively monoamine oxidase and cholinesterase inhibitors.

(12–26 mg/kg) antagonism of the spatial memory deficits induced by scopolamine in rats, indicating that they were able to increase brain acetylcholine levels sufficiently to compete with scopolamine for muscarinic receptors subserving memory [44].

Neuroprotective –antiapoptotic action of TV3326

The primary rationale for developing this class of bi-functional drugs was to combine the dopaminergic (anti-PD) and cholinergic (anti-AD) activities associated respectively with rasagiline and rivastigmine, with the neuroprotective effectiveness of rasagiline in a single molecule [46]. TV3326 exhibits both MAO inhibitory and anticholinesterase activities, although the drug's optical S-isomer TV3279 possesses only anticholinesterase activity. The profile of the neuroprotective activity of TV3326 as compared with rasagiline (Tab. 1) indicates that, to a large extent, TV3326 mimics the established neuroprotective

Table 1. Neuroprotective properties of rasagiline, TVP1022, TV3326 and TV3279

Increases SOD, catalase and BCL-2 activities by transcriptional and translational mechanisms in PC12 cells and rats (brain, heart and kidney)	[50, 51]
Prevents peroxynitrite-induced activation of caspase 3	[69]
Prevents peroxynitrite induced DNA laddering	[69, 58]
Prevents glutamate and NMDA induced neurotoxicity in hippocampal and cortical cell cultures [rasagiline only]	[67]
Prevent peroxynitrite and NM-(R)-sal induced fall in mitochondrial membrane potential and apoptosis	[51, 54, 69]
Protect against peroxynitrite (SIN-1) and 6-hydroxydopamine induced apoptosis	[53, 54]
Protect against cell death induced by ischemia and by glucose deprivation in PC12 cells	[67, 41]
Increased survival of dopaminergic neurons [rasagiline only]	[67]
Neuroprotective in closed head injury in mice	[41, 44, 47]
Prevents MPTP (mice) and 6-OHDA (PC12 cells and rats) neurotoxicity	[45]
Neuroprotection in models of motor and cognition disorders [rasagiline only]	[48]
APP preprocessing *in vitro* and *in vivo*	[65, 70]
Neuroprotection against meyelinated fiber damage and microgliosis induced in rat brain by *icv* STZ [TV3326]	[41]
Prevent serum and NGF induced apoptosis in PC-12 cells	[51]
Neuro-rescue after MPTP treatment in mice (rasagiline)	[45]
PKC activation and dependent amyloid precursor protein processing in PC-12 cells and *in vivo* (mice and rats)	[65, 70]

profile of rasagiline. Both rasagiline and TVP1022 prevent neuronal damage caused by closed head injury in mice [47], focal ischemia [48], and cytotoxicity in cultured PC12 cells induced by growth factor withdrawal [39] or glucose-oxygen deprivation [49]. They also prevent apoptosis induced by the neurotoxins N-methyl-R-salsolinol, 6-hydoxydopamine, the peroxynitrite donor, SIN-1, and by aggregated Aβ amylioid peptide (Aβ) in SHSY-5Y neuroblastoma cells. Rasagiline and TVP1022 (0.10 nM–1 mM) prevent the loss of intact nuclei normally observed in partially-differentiated PC12 cells after serum and NGF withdrawal. This cytoprotective activity is related to their anti-apoptotic action, since the drugs significantly diminished the percentage of cell nuclei with chromatin condensation (an index of apoptosis) over the same concentration range. The anti-apoptotic action of these compounds is dependent on the synthesis of new Bcl-2 and SOD proteins, and the response is prevented by transcriptional (actinomycin) and translational (cycloheximide) inhibitors as measured in partially neuronally-differentiated PC12 cells. A similar effect is obtained with the racemic form of TV3326, TV3219. *In vivo*, chronic treatment rasaguiline increase SOD and catalyse activities in the striatum hippocampus and cortex of rats [50]. Thus it can be inferred that TV3326 would also behave in a similar manner to rasagiline and TVP1022 [39, 51]. The mechanism of the anti-apoptotic effect and the identity of the proteins synthesized have not been fully determined, but they have been shown to be related to the ability of both drugs to prevent the decrease in Bcl-2 and Cu-Zn-SOD1 in response to growth factor withdrawal. In SHSY-5Y cells, rasagiline induces the expression of anti-apoptotic proteins as well as the mRNAs coding for Bcl-2 and Bcl-Xl, while simultaneously decreasing the pro-apoptotic Bax and Bad proteins [52]. In fact, significant evidence exists to suggest that the anti-apoptotic activity exhibited by rasagiline and its derivatives is associated with their ability (1) to prevent the collapse of the mitochondrial membrane potential by opening mitochondrial permeability transition pores that are part of the voltage-dependent anion channel, (Fig. 5); (2) to inhibit cytochrome c; and (3) to activate the caspase cascade that involves caspase 3 [52–55] (Fig. 6). The MAO-inhibitory action of rasagiline is not a prerequisite for its neuroprotective action since its optical isomer TVP1022 [40], which exhibits poor MAO inhibitory activity, is an equipotent neuroprotective agent [56, 57]. The neuroprotective activity of rasagiline, TVP1022 and selegiline resides in the propargylamine moiety, since propargylamine itself exerts similar neuroprotective/anti-apoptotic activities [40, 53–55, 58]. TV3326 and TV3279 retain the neuroprotective properties of rasagiline and TVP1022 (Tab. 1).

Antidepressant and anti-Parkinson's activities of TV3326 and TV3279

TV3326 (but not TV3279) as a brain selective inhibitor of MAO-A and B induces significant increases in striatal, hippocampal, brainstem and hypothalamus dopamine, serotonin and noradrenaline in rats and mice [45], a neuro-

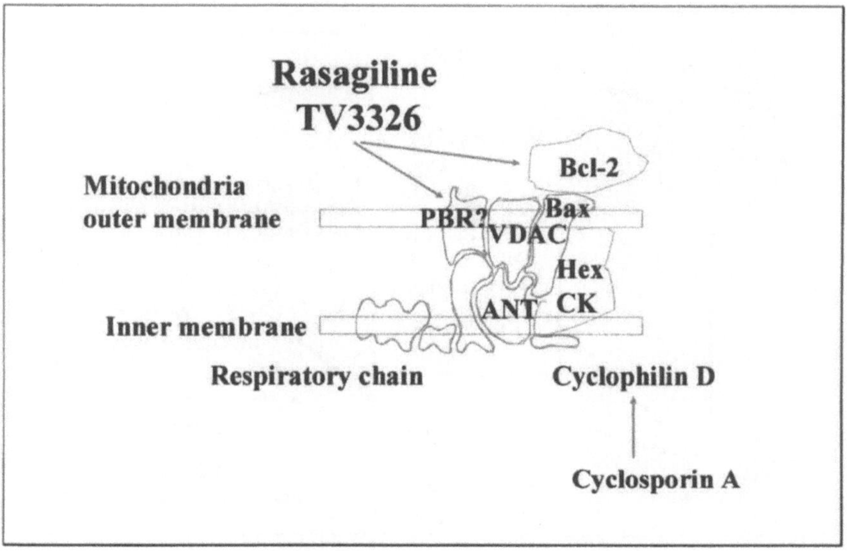

Figure 5. Possible site of action of rasagiline and its derivatives at the mitochondrial voltage dependent anion channel (VDAC), which is part of MPT. The exact protein constituents of MPT is not known but several of the proteins, such as anti and proapoptic proteins Bcl-2 and Bax respectively; porin; PBR, peripheral benzodiazepine receptor; ANT, adenosine nucleotide translocator; HEX, hexokinase and CK, createine kinase have been identified. In a number of respects mechanism of neuroprotective action of rasagiline and its interaction with MTP resembles that of cyclosporin A, a drug with neuroprotective activity [52].

chemical profile is suggestive of potential antidepressant activity (Tab. 2). Classical antidepressant drugs such as amitriptyline and moclobemide (selective reversible MAO-A inhibitor) reduce the duration of immobility behavior in the forced swim test in rats for potential antidepressant activity [59, 60]. Administration of TV-3326 (26 mg/kg/day for two weeks, or 52 mg/kg for one week), inhibited brain MAO-A and B by more than 65%, and the drug significantly reduced the immobility duration to the same extent obtained after chronic treatment with amitriptyline (10 mg/kg/day) or moclobemide (20 mg/kg/day) [61].

The MPTP (N-methyl-4-phenyl-1,2,3,6-tetrahydropyridine) neurotoxin model of PD is used routinely for the screening of potential anti-Parkinson's drugs. MPTP is inert but it is converted by MAO-B within microglia to the active neurotoxin MPP^+, which is then transported into striatal dopamine neurons where it induces neurotoxicity and depletion of dopamine. MAO inhibitors prevent MPTP neurotoxicity in monkeys and mice. As a non-selective MAO inhibitor TV3326 (but not TV3279) prevents the MPTP-induced degeneration of nigro-striatal dopaminergic neurons and the depletion of dopamine in mice [45]. In this regard the effects of TV3326 are similar to other MAO-B inhibitors.

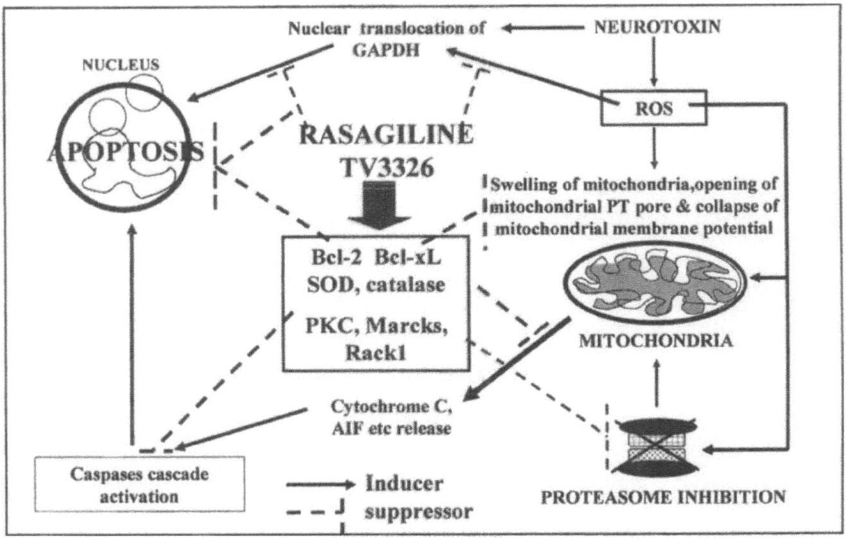

Figure 6. Mechanism of neuroprotective-antiapoptotic action of rasagiline and its anti-Alzheimer derivative cholinesterase-monoamine oxidase inhibitor, TV3326. Both drugs are N-propargyl-(1R)-aminoindan derivatives with TV3326 possessing a carbamate cholinesterase inhibitor moiety. It is the propargyl moiety in these drugs that confers the neuroprotective-antiapoptoic, Bcl-2 inducing activities and PKC activating properties. Rasagiline inhibits neurotoxin (SIN-1, NM-R-Sal) initiated apoptosis in SH-SY5Y and PC-12 cells by preventing the collapse of mitochondrial membrane potential, opening of the MPT, release of ubiquitin-proteasone dependent cytochrome C and caspase 3 activation resulting in its antiapoptotic activity. It also prevents the translocation of pro-apoptotic GAPDH in these cells. Its neuroprotective activity may also depend on its activation of SOD and catalase as has been observed *in vivo* in various tissues including brain and heart [52].

One major side-effect and limitation associated with the use of irreversible MAO-A inhibitors as antidepressant drugs is their ability to potentiate the sympathomimetic action of tyramine present in certain types of food. TV3326 exhibits limited tyramine potentiation and in this regard the drug appears to be equivalent (or superior) to other reversible MAO-A inhibitor antidepressant drugs [62]. This unique property of TV3326 is attributed to the fact that the drug appears to act selectively within the CNS, and as such it has limited effects on MAO-A and B activities in the periphery.

Cognitive enhancing property of TV3326

Recently we administered TV3326 to 7 year old Rhesus monkeys well trained to perform versions of a DMTS task [63]. An increasing dose regimen of TV3326 was administered orally according to a schedule that allowed the animals to perform the standard DMTS task and a self-titrating version of the DMTS task each week during the study. A distractor version of the task was

Table 2. Pharmacological properties of TV3326 and comparison with other cholinesterase and monoamine oxidase inhibitors

	TV3326	Riva-stigmine	Rasagiline	Clorgyline	Tranyl-cypromine
AChE inhibition	+	+	–	–	–
BuChE inhibition	+	+	–	–	–
MAO-A inhibition	+	–	–	–	–
MAO B inhibition	+	–	+	–	–
Increase brain acetylcholine	+	–	–	–	–
Increase brain dopamine	+	–	+	–	+
Increase brain norepinephrine	+	–	–	+	+
Increase brain serotonin	+	–	–	+	+
Tyramine potentiation	–	–	–	+	+
Antidepressant action	+	–	–	+	+
Hypothermic action	–	–	NC	NC	NC
Anti-Parkinson activity	+	–	+	–	+
Neuroprotection	+	–	+	–	–

NC, no change

administered during two of the doses of TV3326. Under the conditions of this experiment TV3326 failed to significantly affect accuracy on the standard DMTS task, however, the drug was very effective in improving the ability of subjects to titrate to longer duration delay intervals in the titrating version of the task. This version of the task is more sensitive to age-dependent cognitive impairment than is the standard DMTS task [64]. The maximal drug-induced extension of the self-titrated delay interval amounted to a 36.7% increase above baseline. TV3326 also significantly improved task accuracy during distractor (interference) sessions, a measure of attention deficit. The compound was effective enough to return group performance efficiency to standard DMTS vehicle levels of accuracy. Thus, TV3326 represents a new class of drug which is potentially suitable for the treatment of AD patients who require therapies that will delay the progression of the disease, and who suffer from impaired attention, impaired memory, extrapyramidal disorder and depression. The combination of the properties attributed to an MAO inhibitor and to a cholinesterase inhibitor may derive benefit from their combined cognitive enhancing properties, as well as from the ability of adrenergic/dopaminergic receptor activations to limit the side-effects of cholinesterase inhibition as discussed above for clonidine.

Amyloid precursor protein (APP) processing by TV3326

One of the debated current concepts regarding the neurotoxicity associated with AD is the processing of amyloid precursor protein (APP) by the three secretases, α, β and γ, and the formation of aggregated Aβ. The potential reduction of Aβ through the administration β and γ secretase inhibitors is one approach being addressed. However, certain cholinesterase inhibitors have been shown to induce the release of neuroprotective-neurotrophic soluble amyloid precursor protein alpha (sAPPα) by selectively enhancing the action of the zinc-metaloprotease, α-secretase. TV3326, TV3279, rasagiline and TVP1022 induce the release of sAPPα in PC-12 and SHSY-5Y cells through activation of α-secretase. The mechanism of sAPP release has been shown to be directly linked to the propargyl moiety group on these drugs, since propargylamine itself is as effective as TV3326, rasagiline and their S-isomer derivatives. Employing several signal transduction pathway inhibitors, it has been established that this process is mediated via the PKC-MAPK dependent pathway, as a consequence of activation of PKCα and PKCε and through ERK1/ERK2 phosphorylation [65] (Fig. 7). *In vivo*, chronic oral treatment with TV3326, TV3279 and rasagiline significantly reduced APP holoprotein in the hippocampus of rats and mice. The ability of these compounds to reduce Aβ in CHO cells and amyloid deposits in the CNS of transgenic mice that over-express this protein is being investigated.

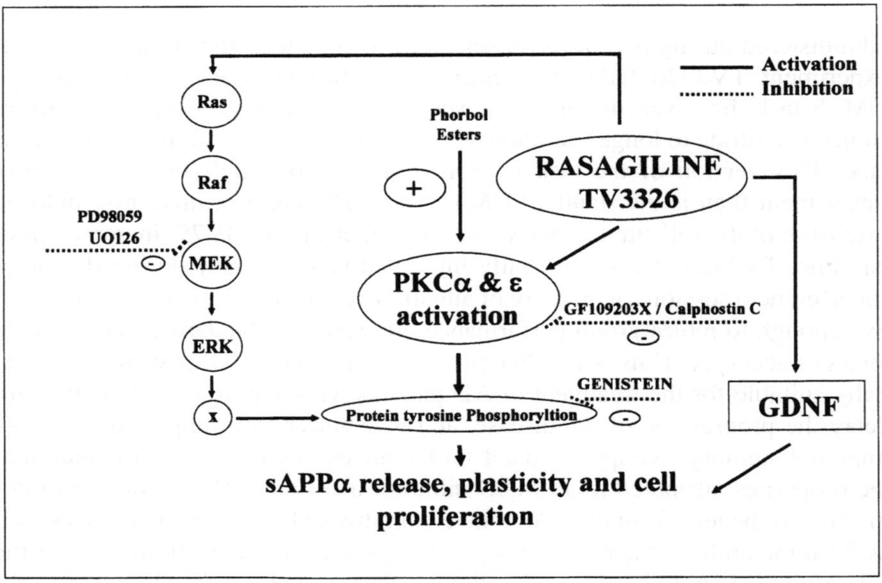

Figure 7. Signal transduction pathways mediating the activation of PKC dependent neuroprotection and plasticity by rasagiline and TV3326. Both drugs activate PKC and MAPK pathways in a time and concentration (0.1-10 μM) dependent manner in PC-12 and SH-SY5Y cells in culture, resulting in activation of α-secretase dependent release of sAPPα. Phorbol esters have similar action and inhibitors of PKC and MAPK pathways, as indicated, prevent rasagiline and TV3326 induced release of sAPPα [Yogev-Falach et al. 2002). *In vivo* both drugs activate mice and rat hippocampal PKCα and ε and promote their translocation from cytoplasm to the mitochondrial membrane [65].

The future

The potential for multi-functional drugs for the treatment of complex neurodegenerative diseases and perhaps even for the treatment of age-associated memory impairment has already been partially realized. The AChE inhibitor galantamine is in widespread use for the treatment of the symptoms of AD. Galantamine may offer additional potential for disease modification (neuroprotection) relative to its predecessors by virtue of its ability to allosterically activate nicotinic receptors [66], but has not been shown to possess the neuroprotective and APP processing properties of TV3326. Although, there continues to be debate as to the extent to which nicotinic receptor activation plays a role in the drug's profile of therapeutic benefit, the concept supports the continued development of bi-functional (AChE/nicotinic) drugs like MHP-133. There is no question that the development of multi-functional drugs presents additional problems for rationale drug design methodologies. Yet from our studies over the past decade in which we examined a wide variety of classes of pharmacological agents in basically the same non-human primate model it seems probable that drugs that target single functional components of cognition and memory will be limited in efficacy [26]. Moreover, it is likely that syndromes such as AD will require multiple drug therapy to address the varied pathological aspects of the disease. Even if the strategy of combining drugs with different therapeutic targets is workable, the development of multi-functional compounds will obviate the challenge of administering multiple single drug entities with potentially different degrees of bioavailability, pharmacokinetics, and metabolism. Also, the simplification of the therapeutic regimen for individuals with AD who have difficulty with compliance is important.

Acknowledgements
The author (JJB) would like to acknowledge his long-term colleague Dr. Alvin V. Terry, Jr. for his valued participation in many the studies described in this chapter. He also would like to thank primate technicians Nancy Kille and Ritu Duhan for their skillful contributions. Much of the work cited for the author's studies was supported by the Office of Research and Development, Medical Research Service, Department of Veterans Affairs, and by the Alzheimer's Association.
The author (MBHY) would like to thank Teva Pharmaceutical Co., Israel for their continued support. We are also grateful to the National Parkinson Foundation (Miami, USA), the Stein Foundation (Philadelphia, USA), the Friedman Foundation, and the Golding Parkinson's Disease Research Fund, Technion (Haifa, Israel).

References

1 Green RC (1995) Alzheimer's disease and other dementing disorders in adults. *Clinical Neurology* 3: 1–83
2 Morris JC, McKeel DW Jr, Storandt M, Rubin EH, Price JL, Grant EA, Ball MJ, Berg L (1991) Very mild Alzheimer's disease: Informant-based clinical, psychometric, and pathologic distinction from normal aging. *Neurology* 41: 469–478
3 Francis PT, Palmer AM, Snape M, Wilcock GK (1999) The cholinergic hypothesis of Alzheimer's disease: A review of progress. *J Neurol Neurosurg Psychiatry* 66: 137–147

4 Engelborghs S, De Deyn PP (1997) The neurochemistry of Alzheimer's disease. *Acta Neurologica Belgica* 97: 67–84
5 Myhrer T (1998) Adverse psychological impact, glutamatergic dysfunction, and risk factors for Alzheimer's disease. *Neurosci Biobehav Rev* 23: 131–139
6 Gilbert PE, Kesner, RP, DeCoteau WE (1998) Memory for spatial location: Role of the hippocampus in mediating spatial pattern separation. *J Neurosci* 18: 804–810
7 Riedel G, Micheau J, Lam AGM, Roloff EvL, Martin SJ, Bridge H, de Hoz L, Poeschel B, McCulloch J, Morris RGM (1999) Reversible neural inactivation reveals hippocampal participation in several memory processes. *Nature Neurosci* 2: 898–905
8 Zola SM, Squire LR, Teng E, Stefanacei L, Buffalo EA an Clark RE (2000) Impaired recognition memory in monkeys after damage limited to the hippocampal region. *J Neurosci* 20: 451–463
9 Vann SD, Brown MW, Erichsen JT, Aggleton JP (2000) Fos imaging reveals differential patterns of hippocampal and parahippocampal subfield activation in rats in response to different spatial memory tests. *J Neurosci* 20: 2711–2718
10 Sybirska E, Davachi L, Goldman-Rakic PS (2000) Prominance of direct entorhinal-CA1 pathway activation in sensorimotor and cognitive tasks revealed by 2-DG functional mapping in nonhuman primates. *J Neurosci* 20: 5827–5834
11 Eichenbaum H (1997) How does the brain organize memories. *Science* 277: 330–332
12 Vargha-Khadem F, Gadian DG, Watkins KE, Connelly A, Van Paesschen W, Mishkin M (1997) *Science* 277: 376–380
13 Knowlton BJ, Mangels JA, Squire LR (1996) A neostriatal habit learning system in humans. *Science* 273: 1399–1402
14 Teng E, Stefanacei L, Squire LR, Zola SM (2000) Contrasting effects on discrimination learning after hippocampal lesions and conjoint hippocampal-caudate lesions in monkeys. *J Neurosci* 20: 3853–3863
15 Bechara A, Tranei D, Damasio H, Adolphs R, Rockland C, Damasio AR (1995) Double dissociation of conditioning and declarative knowledge relative to the amygdala and hippocampus in humans. *Science* 269: 1115–1118
16 Elrod K, Buccafusco JJ (1988) An evaluation of the mechanism of scopolamine-induced impairment in two passive avoidance protocols. *J Pharmacol Biochem Behav* 29: 15–21
17 Buccafusco JJ, Prendergast M, Terry AV Jr, Jackson WJ (1996) Cognitive effects of nicotinic cholinergic agonists in non-human primates. *Drug Development Res* 38: 196–203
18 Haroutunian V, Kanof PD, Tsuboyama G, Davis KL (1990) Restoration of cholinomimetic activity by clonidine in cholinergic plus noradrenergic lesioned rats. *Brain Res* 507: 261–266
19 Callahan MJ (1999) Combining tacrine with milameline reverses a scopolamine-induced impairment of continuous performance in rhesus monkeys. *Psychopharmacology* 144: 234–238
20 Wu M, Shanabrough M, Leranth C, Alreja M (2000) Cholinergic excitation of septohippocampal GABA but not cholinergic neurons: Implications for learning and memory. *J Neurosci* 20: 3900–3908
21 Doraiswamy PM (2002) Non-cholinergic strategies for treating and preventing Alzheimer's disease. *CNS Drugs* 16: 811–824
22 Areosa SA, Sherriff F (2003) Memantine for dementia. *Cochrane Database of Systematic Reviews* 1:CD003154
23 Buccafusco JJ, Terry AV Jr (2000) Multiple CNS targets for eliciting beneficial effects on memory and cognition. *J Pharmacol Exp Ther* 295: 438–446
24 Elrod K, Buccafusco JJ, Jackson WJ (1988) Nicotine enhances delayed matching-to-sample performance by primates. *Life Sci* 43: 277–287
25 Bartus RT (2000) On neurodegenerative diseases, models, and treatment strategies: Lessons learned and lessons forgotten a generation following the cholinergic hypothesis. *Expl Neurol* 163: 495–529
26 Buccafusco JJ, Terry AV Jr (2000) Multiple CNS targets for eliciting beneficial effects on memory and cognition. *J Pharmacol Exp Ther* 295: 438–446
27 Terry AV Jr, Jackson WJ, Buccafusco JJ (1993) Effects of concomitant cholinergic and adrenergic stimulation on learning and memory performance by young and aged monkeys. *Cerebral Cortex* 3: 304–312
28 Jackson WJ, Buccafusco JJ (1991) Clonidine enhances delayed matching-to-sample performance by young and aged monkeys. *Pharmacol Biochem Behav* 39: 79–84
29 Buccafusco JJ (1992) Neuropharmacologic and behavioral actions of clonidine: Interactions with

central neurotransmitters. *International Rev Neurobiol* 33: 55–107
30 Buccafusco JJ (1984) Inhibition of regional brain acetylcholine biosynthesis by clonidine in spontaneously hypertensive rats. *Drug Devel Res* 4: 627–633
31 Powers JC, Buccafusco JJ, Starks K (1998) Pyridinium Compounds. Patent Number: 5,714,615; issue date: 02/03/98
32 Powers JC, Buccafusco JJ, Starks K (1998) Pyridinium Compounds. Patent Number: 5,726,314; issue date: 03/10/98
33 Prendergast MA, Jackson WJ, Terry AV, Decker MW, Arneric SA and Buccafusco JJ (1998) Central nicotinic receptor agonists ABT-418, ABT-089, and (–)-Nicotine reduce distractibility in young-adult monkeys. *Psychopharmacology* 136: 50–58
34 Terry AV Jr, Clark MS (1994) Nicotine stimulation of nerve growth factor receptor expression. *Life Sci* 55: PL91–98
35 Jonnala RR, Terry AV Jr, Buccafusco JJ (2002) Nicotine increases the expression of high affinity nerve growth factor receptors both *in vitro* and *in vivo*. *Life Sci* 70: 1543–1554, 2002
36 Terry AV Jr, Gattu M, Buccafusco JJ, Sowell JW, Kosh JW (1999b) Ranitidine analog, JWS-USC-75IX, enhances memory-related task performance in rats. *Drug Dev Res* 47: 97–106
37 Cassel JC, Jeltsch H (1995) Serotonergic modulation of cholinergic function in the central nervous system: Cognitive implications. *Neuroscience* 69: 1–41
38 Youdim, MBH (1989) Brain monoamine oxidase (MAO) B: A unique neurotoxin and neurotransmitter-producing enzyme. *Prog Neuropsychopharmacol Biol Psychiatry* 13: 363–371
39 Youdim MBH, Gross A, Finberg JP (2001) Rasagiline [N-propargyl-1R(+)-aminoindan], a selective and potent inhibitor of mitochondrial monoamine oxidase B. *Br J Pharmacol* 132: 500–506
40 Youdim MBH, Wadia A, Tatton W, Weinstock M (2001) The anti-Parkinson drug rasagiline and its derivatives exert neuroprotection unrelated to MAO inhibition in cell culture and *in vivo*. *Ann NY Acad Sci* 931: 450–459
41 Weinstock M, Kirschbaum-Slager N, Lazarovici P, Bejar C, Youdim MBH, Shoham S (2001) Neuroprotective effects of novel cholinesterase inhibitors derived from rasagiline as potential anti-Alzheimer drugs. *Ann NY Acad Sc.* 939: 148–162
42 Sterling J, Herzig Y, Goren T, Finkelstein N, Lerner D, Goldenberg W, Miskolczi I, Molnar S, Rantal F, Tamas T et al. (2002) Novel dual inhibitors of AChE and MAO derived from hydroxy aminoindan and phenethylamine as potential treatment for Alzheimer's disease. *J Med Chem* 45: 5260–5279
43 Weinstock M, Bejar C, Wang RH, Poltyrev T, Gross A, Finberg JPM, Youdim MBH (2000) TV3326, a novel neuroprotective drug with cholinesterase and monoamine oxidase inhibitory activities for the treatment of Alzheimer's disease. *J Neural Transm* 60: (suppl) 157–169
44 Weinstock M, Goren T, Youdim MBH (2000) Development of a novel neuroprotective drug (TV3326) for the treatment of Alzheimer's disease, with cholinesterase and monoamine oxidase inhibitory activities. *Drug Dev Res* 50: 216–222
45 Sagi Y, Weinstock M, Youdim MBH (2003) Attenuation of MPTP-induced dopaminergic neurotoxicity by TV3326, a cholinesterase-monoamine oxidase inhibitor. *J Neurochem*; *in press*
46 Youdim MB, Weinstock M (2002) Novel neuroprotective anti-Alzheimer drugs with anti-depressant activity derived from the anti-Parkinson drug, rasagiline. *Mechanisms Ageing Develop* 123: 1081–1086
47 Huang W, Chen Y, Shohami E, Weinstock M (1999) Neuroprotective effect of rasagiline, a selective monoamine oxidase-B inhibitor, against closed head injury in the mouse. *Eur J Pharmacol* 366: 127–135
48 Speiser Z, Levy R, Cohen S (1998) Effects of N-propargyl-1-(R)aminoindan (rasagiline) in models of motor and cognition disorders. *J Neural Transm* Suppl 52: 287–300
49 Abu-Raya S, Tabakman R, Blaugrund E, Trembovler V, Lazarovici P (2002) Neuroprotective and neurotoxic effects of monoamine oxidase-B inhibitors and derived metabolites under ischemia in PC12 cells. *Eur J Pharmacol* 434: 109–116
50 Carrillo MC, Minami C, Kitani K, Maruyama W, Ohashi K, Yamamoto T, Naoi M, Kanai S, Youdim MBH (2000) Enhancing effect of rasagiline on superoxide dismutase and catalase activities in the dopaminergic system in the rat. *Life Sciences* 67(5): 577–585
51 Youdim MBH, Weinstock M (2001) Molecular basis of neuroprotective activities of rasagiline and the anti-Alzheimer drug, TV3326, [(N-propargyl-(3r)aminoindan-5-yl)-ethyl methyl carbamate] *Cell Mol Biol* 21: 555–573
52 Youdim MBH, Amit T, Falach-Yogev M, Bar Am O, Maruyama W, Naoi M (2003) The essential-

ity of Bcl-2, PKc and proteasome-ubiquitin complex activation in the neuroprotective action of anti-Parkinson drug, rasdagiline. *Biochem Pharmacol* 66: 1635–1641
53 Maruyama W, Youdim MBH, Naoi M (2001) Anti-apoptotic properties of rasagiline, N-propargylamine-1(R)-aminoindan, and its optical (S)-isomer, TV1022. *Annals NY Acad Sci* 939: 320–329
54 Maruyama W, Boulton AA, Davis BA, Dostert P, Naoi M (2001) Enantio-specific induction of apoptosis by an endogenous neurotoxin, N- methyl(R)salsolinol, in dopaminergic SH-SY5Y cells: suppression of apoptosis by N-(2-heptyl)-N-methylpropargylamine. *J Neural Transm* 108: 11–24
55 Akao Y, Maruyama W, Shimizu S, Yi H, Nakagawa Y, Shamoto-Nagai M, Youdim MBH, Tsujimoto Y, Naoi M (2002) Mitochondrial permeability transition mediates apoptosis induced by N-methyl(R)salsolinol, an endogenous neurotoxin, and is inhibited by Bcl-2 and rasagiline, N-propargyl-1(R)-aminoindan. *J Neurochem* 82: 82: 913–923
56 Maruyama W, Akao Y, Carrillo M, Kitani K, Youdium M, Naoi M (2002) Neuroprotection by propargylamines in Parkinson's disease. Suppression of apoptosis and induction of prosurvival genes. *Neurotoxicol Teratol* 24: 675–682
57 Maruyama W, Takahashi T, Youdim MBH, Naoi M (2002) The anti-Parkinson drug, rasagiline, prevents apoptotic DNA damage induced by peroxynitrite in human dopaminergic neuroblastoma SH-SY5Y cells. *J Neural Transm* 109: 467–481
58 Maruyama W, SkaoY, Youdim MBH, Naoi M (2000) Neurotoxins induce apoptosis in dopamine neurons: protection by N-[ropargylamines-1®- and (S)-aminindan, rasagiline and TVP1022. *Adv Res Neurodegen* 8: 171–187
59 Porsolt RD, Bertin A, Blavet N, Deniel M, Jalfre M (1979) Immobility induced by forced swimming in rats: Effects of agents which modify central catecholamine and serotonin activity. *Eur J Pharmacol* 57: 201–210
60 Borsini F, Meli A (1988) Is the forced swimming test a suitable model for revealing antidepressant activity? *Psychopharmacology* 94: 147–160
61 Weinstock M, Poltyrev T, Bejar C, Youdim MB (2002) Effect of TV3326, a novel monoamine-oxidase cholinesterase inhibitor, in rat models of anxiety and depression. *Psychopharmacology* 160: 318–324
62 Weinstock M, Poltyrev T, Bejar C, Sagi Y, Youdim MBH (2002) TV3326, a novel cholinesterase and MAO inhibitor for Alzheimer's disease with co-morbidity of Parkinson's disease and depression. In: Y Mizuno, A Fisher, I Hanin eds: *Mapping the progress of Alzheimer's and Parkinson's disease*. Kluwer Academic/Plenum Publishers, New York, 199–204
63 Buccafusco JJ, Terry AV Jr, Goren T, Blaugrun E (2003) Potential cognitive actions of TV3326, a novel neuroprotective agent, as assessed in old Rhesus monkeys in their performance of versions of a delayed matching task. *Neuroscience* 119: 669–678
64 Buccafusco JJ, Terry AV Jr, Murdoch PB (2002) A computer assisted cognitive test battery for aged monkeys. *J Mol Neurosci* 19: 187–193
65 Yogev-Falach M, Amit T, Bar-Am O, Weinstock M, Youdim MB (2002) Involvement of MAP kinase in the regulation of amyloid precursor protein processing by novel cholinesterase inhibitors derived from rasagiline. *FASEB J* 16: 1674–1676
66 Corey-Bloom J (2003) Galantamine: A review of its use in Alzheimer's disease and vascular dementia. *Internat J Clin Practice* 57: 219–223
67 Finberg JP, Lamensdorf I, Weinstock M, Schwartz M, Youdim MB (1999) Pharmacology of rasagiline (N-propargyl-1R-aminoindan). *Adv Neurol* 80: 495–499
68 Abu-Raya S, Blaugrund E, Trembovler V, Shilderman-Bloch E, Shohami E. Lazarovici P (1999) Rasagiline, a monoamine oxidase-B inhibitor, protects NGF-differentiated PC12 cells against oxygen-glucose deprivation. *J Neurosci Res* 58: 456–463
69 Maruyama W, Weinstock M, Youdim MBH, Nagai M, Maoi M (2003) Antiapoptotic action of the anti-Alzheimer drug, TV3326 [(N-propagyl)-(3R)-aminoindan-5yl]ethyl methyl carbamate, a novelcholinesterase-monoamine oxidase inhibitor. *Neurosci Lett* 341: 331–334
70 Bar-Am O, Falch-Yogev M, Amit T, Youdim MBH (2003) The Importance of propargylamine moiety in the anti-Parkinson drug rasagiline and its derivatives in MAPK-dependent amyloid precursor protein processing. *FASEB J* 0.1096/fj. 03-007fje

Index

MIX
Papier aus verantwortungsvollen Quellen
Paper from responsible sources
FSC® C105338

If you have any concerns about our products,
you can contact us on
ProductSafety@springernature.com

In case Publisher is established outside the EU,
the EU authorized representative is:
Springer Nature Customer Service Center GmbH
Europaplatz 3, 69115 Heidelberg, Germany

Printed by Libri Plureos GmbH
in Hamburg, Germany

Lecture Notes in Information Systems and Organisation 81

Series Editors

Lecture Notes in Information Systems and Organization (LNISO) is a series of scientific books that explore the current state of information systems studies, in particular in the context of organizations, as well as behavioral and design aspects of information systems. The series offers an integrated publication platform for high-quality conferences, symposia and workshops proceedings. Books accepted for publication are selected based on a double-blind peer-review evaluation made by invited reviewers. The series is indexed in Scopus.

Badr-Eddine Boudriki Semlali ·
Ikram Ben Abdel Ouahab · Fabio Angeletti
Editors

Technological Innovations for Sustainable Development

Proceedings of the 6th International Conference DATA2025, Tangier, Morocco

Springer

Editors
Badr-Eddine Boudriki Semlali
Department of Computer Engineering
Abdelmalek Essaâdi University
Tangier, Morocco

Ikram Ben Abdel Ouahab
C3S Lab, FSTT
Abdelmalek Essaâdi University
Tangier, Morocco

Fabio Angeletti
LUISS Guido Carli University
Rome, Italy

ISSN 2195-4968 ISSN 2195-4976 (electronic)
Lecture Notes in Information Systems and Organisation
ISBN 978-3-032-06724-1 ISBN 978-3-032-06725-8 (eBook)
https://doi.org/10.1007/978-3-032-06725-8

This Springer imprint is published by the registered company Springer Nature Switzerland AG
The registered company address is: Gewerbestrasse 11, 6330 Cham, Switzerland

Preface

Digital technologies have become institutional drivers that shape, mediate, and transform contemporary organizations, governance systems, and societal structures. This volume reconceptualizes technology not as an isolated tool but as a dynamic force embedded in cultural, institutional, and economic logics and explores how technological affordances are put into action in varied contexts, enabling (or limiting) actors to reconfigure processes, roles, and values across domains. Its thematic architecture spans key sectors such as artificial intelligence and machine learning, public service innovation, sustainable manufacturing and agriculture, and educational technologies, highlighting the interplay between technological development, institutional adaptation, and socio-political consequences. From algorithmic management and platform governance to smart urbanism and data-driven education, the contributions critically examine the tensions and transformations arising from digitalization. In doing so, the volume positions digital transformation as a pivotal institutional shift that demands critical engagement and system thinking.

Badr-Eddine Boudriki Semlali
Ikram Ben Abdel Ouahab
Fabio Angeletti

Aim and Scope of the Book

This volume is a curated collection of peer-reviewed papers presented at the Sixth International Conference on Digital Age & Technological advances for Sustainable Development held in Tangier, Morocco on May 2025. The conference has been co-organized by the Faculty of Sciences and Techniques of Tangier with the Mediterranean Association of Sciences and Sustainable Development and partnered with the Commune of Tangier and Al-Mustaqbal University, Iraq.

The contemporary era is characterized by pervasive digitalization, a transformative process reshaping the ways in which individuals, organizations, and institutions operate and interact. While digitalization holds significant promise for improving efficiency, transparency, and responsiveness across sectors, it simultaneously introduces challenges related to governance, social equity, and ethical considerations. Digital technologies, from artificial intelligence (AI) to platform-based infrastructures, are not merely neutral instruments but they actively influence and redefine organizational practices, societal norms, and governance frameworks, often in complex and unintended ways. Understanding these transformations requires a nuanced exploration that goes beyond simplistic or unilateral perspectives, recognizing the interdependencies and mutual influences between technology, social systems, and institutional contexts.

Addressing the intricate impacts of digitalization necessitates a multifaceted analytical approach to capture complexity across several critical dimensions technological, organizational, institutional, and social. The *technological dimension* explores the characteristics, possibilities, and limitations inherent in digital technologies themselves, particularly emphasizing the concept of technological affordances—the action possibilities that technologies enable or constrain. The *organizational dimension* focuses on how digital technologies reshape organizational structures, workflows, cultures, and identities. The *institutional dimension* examines how broader governance structures, regulatory frameworks, and established societal norms co-evolve alongside technological change, influencing and being influenced by digital innovations in a reflexive fashion (Smith & Stirling, 2013). Finally, the *social dimension* addresses how digital transformation (DT) impacts social relations, identities, power dynamics, and inclusivity. The multidimensional approach is fundamental because digitalization inherently affects multiple interconnected domains simultaneously, making it impossible to fully understand its consequences through a single disciplinary lens and aligns with a sociotechnical view that considers technology not as an isolated artifact but as a constitutive element that co-evolves with institutional logics, economic imperatives, and cultural expectations, offering a comprehensive and integrated examination essential to adequately inform theory development, policymaking, and practical management strategies in an increasingly digitized world.

Through this lens, the volume positions itself at the intersection of information systems (IS), organizational theory, and digital innovation, emphasizing the complexity and hybridity of contemporary digital ecosystems. Beyond a mere description of current trends, it addresses the multifaceted consequences of digitization processes, from algorithmic management (Jarrahi et al., 2021) and platform governance (Halckenhaeusser et al., 2020) to datafication (Lycett, 2013) and surveillance capitalism (Zuboff, 2019) – mentioning just few aspects – and their systemic implications across diverse sectors.

Moreover, the book enriches the discourse by anchoring its exploration in several key theoretical conversations the role of digital platforms as infrastructure and intermediaries (Gawer & Cusumano, 2014), the interdependence of digital capabilities and innovation performance (Benitez et al., 2022), and the challenges of adaptation and orchestration in fluid digital ecosystems (Nambisan et al., 2019). By doing so, it embraces a pluralistic methodological approach that allows for both macro-structural and micro-interactional analyses, ultimately revealing how technological affordances (Volkoff & Strong, 2017) are mobilized within varied institutional and cultural contexts. The goal is to understand how individuals, organizations, and systems recognize, interpret, and act upon the possibilities that digital technologies offer, translating them into concrete practices and strategic actions. In this sense, affordances are not merely static features of technological artifacts, but dynamic, context-sensitive opportunities that must be mobilized through intentional engagement, situated knowledge, and embedded routines.

Themes

In the following, we will guide the reader through the most salient domains where DT is manifesting with the greatest intensity and impact using the thematic architecture depicted in Figure 1. This architecture emerged inductively from our review and interpretation of contributions submitted to the conference. Rather than predefining conceptual categories, the papers have been grouped according to evident patterns of shared concerns, analytical approaches, and empirical contexts. This inductive method ensures coherence and relevance, capturing the domains where DT is currently exerting the most pronounced impact. While each thematic cluster—Artificial Intelligence and Digital Transformation, Digital Governance and Urban Development, Sustainable Digital Manufacturing and Agriculture, Educational Technologies, and Integrated Digital Ecosystems—addresses distinct empirical areas, they collectively reflect broader underlying issues. Specifically, the clusters are connected through shared concerns about power dynamics, governance mechanisms, value creation processes, and technological mediation. This commonality provides a novel analytical framework to explore contemporary phenomena in IS research.

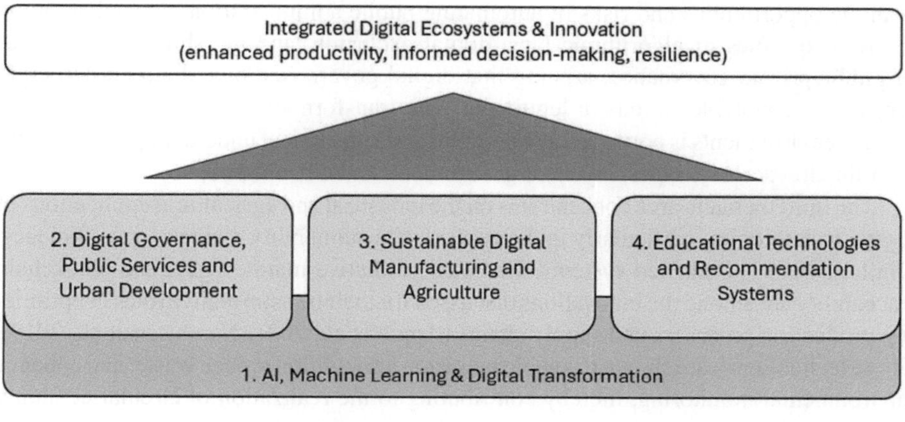

Figure 1. The thematic architecture identified.

AI, Machine Learning (ML), and DT have emerged as foundational drivers of innovation across public and private sectors. Their implementation enhances decision-making, automates operational processes, and supports personalized services across industries including healthcare, education, retail, and government. A recent study (Chui et al., 2023) highlights a sharp rise in generative AI adoption, reinforcing AI's strategic importance for competitiveness. AI-driven demand forecasting and fraud detection systems are now widely deployed, while ML algorithms support enhanced predictive analytics and improved business intelligence. These technologies empower firms to reduce waste, optimize supply chains, and anticipate customer behavior with greater accuracy (Brynjolfsson & McElheran, 2016). As organizations increasingly adopt cloud-native platforms, integrate sophisticated AI governance mechanisms, and embrace decentralized, agile workflows, AI and ML have become central to sociotechnical shifts reshaping organizational culture, workflows, and innovation capabilities. These shifts, however, introduce critical ethical considerations. As the influence of algorithmic decision-making grows, so do concerns about transparency, accountability, and potential biases embedded within AI systems. Frameworks around responsible AI—including guidelines for explainability, fairness, and data privacy—are gaining prominence, particularly in high-stakes domains such as medical diagnosis, financial services, and public governance. Addressing these ethical dimensions is essential to maintain trust, ensure equitable outcomes, and mitigate unintended societal impacts. Vial (Vial, 2021) notes that DT is inherently multi-dimensional, influencing not only firm performance but also stakeholder roles and institutional structures. In this light, AI and ML represent more than mere tools, they serve as transformative forces embedded in the broader architecture of digital society.

The second thematic cluster explores the intersection between digital technologies and institutional infrastructures, with a focus on public service innovation and smart urbanism. This section critically examines how digital platforms and data infrastructures reshape the roles and responsibilities of governments, enabling new forms of service delivery, citizen engagement, and algorithmic decision-making (Benitez et al., 2022; Gawer & Cusumano, 2014). Contributions analyze phenomena such as open data platforms, AI-based resource allocation, and digital twins in urban planning, emphasizing both the opportunities and risks of automating public administration. Special attention is given to issues of algorithmic bias, digital exclusion, and the shifting boundaries of public/private governance, arguing that digital governance must be both participatory and accountable to remain legitimate. The transformation of urban systems into 'smart' environments is portrayed as a complex socio-technical undertaking that requires coordination across sectors, layers of government, and citizen groups.

The third thematic area concentrates on the industrial and agricultural applications of digital technologies, particularly in the context of sustainability and resource efficiency. Digital twins, IoT-enabled systems, AI-based predictive maintenance, and blockchain traceability are among the innovations discussed for their transformative roles in optimizing production processes and supply chains (Appio et al., 2021; Nambisan et al., 2019). These technologies are shown to enable precision agriculture, reduce waste, and enhance environmental monitoring, thereby contributing to the realization of circular economy models. At the same time, the section critically interrogates the dependencies created by digital infrastructures and proprietary platforms, warning against the unchecked consolidation of technological power among a few actors. Sustainability is approached not only in environmental terms but also in relation to social and organizational resilience. The discussion foregrounds how digital capabilities can support smallholder farmers, local manufacturers, and SMEs by democratizing access to innovation and data. This approach aligns with global calls for sustainable development and responsible innovation, reinforcing the urgency of aligning technological advancement with long-term societal well-being.

The fourth group delves into the evolving landscape of educational technologies, highlighting how personalized learning environments, AI-driven tutors, and recommendation algorithms are redefining pedagogical experiences. Drawing from platform strategy and IS design theory (Appio et al., 2021; Spagnoletti et al., 2015), these chapters critically examine how adaptive systems tailor content delivery to individual learners while also raising concerns about surveillance, data ownership, and algorithmic transparency. Simultaneously, the growing integration of digital infrastructures into education exposes these systems to increasing cybersecurity threats, necessitating robust frameworks for data protection, system resilience, and ethical risk management. By connecting the themes of innovation and security, the contributions underscore the paradoxes of personalization and privacy in digital education. The section also considers the digital divide and equity in access to high-quality technological tools, underlining the need for inclusive strategies to mitigate structural inequalities in education.

Integrated digital ecosystems represent the next phase in the evolution of innovation environments, bringing together diverse stakeholders—public institutions, private enterprises, platform providers, and end-users—within shared digital infrastructures. These ecosystems are enabled by modular architectures, open APIs, and platform strategies that foster dynamic collaboration and rapid innovation. Rather than operating as isolated systems, organizations increasingly engage in co-creation of value through data-sharing partnerships, federated governance structures, and digitally mediated services. As these ecosystems grow, innovation becomes more distributed, agile, and network-based. The strength of digital ecosystems lies not just in technology, but in the institutions, norms, and governance frameworks that make integration and interoperability possible. As sectors such as smart cities, healthcare, agriculture, and education converge, ecosystem approaches support real-time coordination, inclusive participation, and resilience in the face of complexity. Innovation is no longer confined within organizational boundaries but extends into community engagement, cross-sector alliances, and globally connected value chains. This integrated approach promotes systemic transformation, enabling societies to address emerging challenges through collective intelligence, adaptability, and sustained technological evolution.

Content Overview

AI, Machine Learning, and Digital Transformation

The chapters in this thematic area collectively explore the profound influence of AI and ML on DT across industries. Several contributions emphasize the ethical and governance challenges accompanying AI implementation, such as the one of **Oussama El Azzouzy, Tarik Chanyour, and Said Jai Andaloussi**, in "Towards Robust and Transparent AI Addressing Adversarial Threats, Biases, and Governance Challenges" that presents a hybrid framework that enhances AI systems by addressing adversarial vulnerabilities and algorithmic biases for more equitable and secure decision-making. Practical applications featured include robust customer segmentation techniques provided by **Iman Squalli Houssaini and Daoud Miloud**, in "AI-Driven Customer Segmentation for Enhanced Loyalty Programs", where we can read about the use of clustering techniques like PCA and K-Means to optimize customer segmentation and improve engagement in the banking sector, as well as innovative machine-learning models that streamline early cancer diagnosis by **Ibtissam Samnane, Rachid Ed-daoudi, Samir Bouali, M'barek El Haloui, and Badia Ettaki**, in "Machine Learning-Based System for Colorectal Cancer Prediction", that shows a clinical ML model that predicts cancer type based on demographic and clinical factors. Further integrating these applications, other chapters demonstrate AI's diverse utility in commercial, financial, healthcare, and manufacturing domains. Still in the broad field of healthcare, **Kawtar Naim and Aziz Darouichi**, in "Enhanced Xception-Based Model for Binary Classification of Diabetic Retinopathy in Retinal Fundus Images", propose an advanced deep learning approach for detecting diabetic retinopathy using fundus images. **Sara Ahsain, Oumaima Louzar, M'hamed Ait Kbit, and Yasyn El Yusufi**, in "Enhancing Digital Marketing with AI A Machine Learning approach to identify fake reviews on Amazon", show how SVM and other models are used to identify fake Amazon reviews, achieving high accuracy and improving e-commerce trust. Advances in deep learning for handwritten text recognition are presented by **Burul Shambetova, Ruslan Isaev, Nazira Abdillaeva, and Mekia Shigute Gaso**, in "Comparative Analysis of Handwritten Text Recognition across Cyrillic-using Countries", where the authors examine the limitations and advances in HTR systems using deep learning across various Cyrillic-script regions. In finance, **Fatima Zahra Chaayra, Toufik Chaayra, Ismail Mhamdi, and Abdeljabar Rafiki**, within their work titled "Milstein-Based Maximum Likelihood Estimators for Discretely Observed Geometric Brownian Motion Stochastic Model with Application to Financial Risk Volatility", propose a refined stochastic model for estimating financial volatility and forecasting real interest rates, while **Nguyen Duy Phuong, Nguyen Minh Tuan, Le Dao Ngoc Thanh, and Bui Thanh Khoa**, in "The Digital Health with the Psychology of Fear of Missing Out in Electronic Commerce", analyze how FoMO affects consumer behavior, highlighting its dual impact on social commerce engagement and purchasing. Innovative methodological integrations also feature prominently. **Hajar Ezzahoud, Mohammed Ameksa,**

Asmaa Amzil, and Habibatou-Allah Amizmiz, in "Does Multi-Dataset Combination Impact Machine Learning Performance?", combine four datasets to improve emotion recognition accuracy in speech, demonstrating the power of integration for generalizable SER systems. **Fatima Zahra Didast, Rym Nassih, and Ilhame Ait Lbachir**, further illustrate methodological integration in "Improving Demand Forecasting Through an Ensemble Method", creating an adaptive forecasting system blending ML and traditional time-series models to boost retail prediction accuracy and market responsiveness. Meanwhile, **Hanae Abbassi, Saida El Mendili, and Youssef Gahi**, introduce quantum computing techniques in "Pioneering Quantum AI Applications for Fraud Detection in FinTechs", design a fraud detection system leveraging Quantum LSTM and quantum feature encoding, enhancing accuracy on complex transaction datasets. Lastly, **Basma Echaki, Mounir Boumhamdi, and Marouane Ikira**, apply spatial econometrics and AI analytics in "Exploring the Determinants of Business Performance", identifying productivity drivers in Moroccan manufacturing firms, showing regional and digital effects on performance.

Digital Governance, Public Services and Urban Development

This thematic cluster addresses how digital technologies reshape governance structures, enhance public service delivery, and influence urban and regional development. The chapters collectively provide insights into diverse technological and methodological innovations that contribute to more effective, inclusive, and strategic public administration. For instance, **Tihomir Dovramadjiev**, in his work "Digitalization of City and Coastal Environment Using Open Source Software", explores how digital tools and open-source GIS can enhance spatial analysis and planning in urban and coastal contexts, while **Marta Dmytryshyn, Tetyana Goran, and Roman Rusyn**, in "Digital Learning in Public Service", examine the prerequisites for effective digital learning implementation in public administration, emphasizing e-learning strategies to improve civil service capabilities. In parallel, **Ouaadi Ismail and Maoulainine Naama**, in "E-government educational IS, Did Moroccan e-gov system succeed?", apply an empirical approach to assess the effectiveness of e-government educational information systems in Morocco while **Imane Abouliatim and Karima Tounsi**, in "The Path to Adoption Insights into Morocco's Public E-Services", examine the key factors influencing the adoption of e-government services in Morocco, highlighting the role of trust, accessibility, and user readiness. Furthermore, the role of AI and digital tools in enhancing administrative processes and decision-making features prominently across chapters. **Ahmed Elkaabi, M'barek El Khoukhi, and Mina Elma Allam**, in "Towards an Improved Environmental Scanning Model Based on Artificial Intelligence Tools", propose an enhanced environmental scanning framework that integrates AI technologies to support strategic decision-making. Similarly, **Fatima Zahra El Karmaoui, El Ouali El Blaidi and Ahmed El Hammoumi**, in "The digitalization of territorial marketing in the service of development of southern regions, an empirical study", investigate how digital marketing contributes to territorial attractiveness and regional development in Morocco's southern regions. The adoption of digital procurement and effective management of financial risks through

AI-driven techniques also receives significant attention. **W.M.D.M. Dissanayaka and Prapon Sahapattana**, in "Determinants of Adoption Intention of Electronic Government Procurement A Conceptual Model for Sri Lanka", develop a conceptual framework to identify factors influencing public sector employees' willingness to adopt e-GP systems in Sri Lanka. In another critical area, **Safae Ndama, Oussama Ndama, and El Mokhtar En-Naimi**, in "A Comprehensive Evaluation of Data Balancing Techniques and Machine Learning Models for Credit Risk Assessment", evaluate various resampling techniques—SMOTE, ADASYN, RUS, and hybrids—across several ML models including Decision Tree, XGBoost, and Neural Networks.

Sustainable Digital Manufacturing and Agriculture

This thematic cluster addresses the role of digital technologies in fostering sustainability, efficiency, and resilience in manufacturing and agriculture. The chapters collectively illustrate the potential of digital innovation in optimizing resource use, enhancing operational transparency, and supporting sustainable development across sectors. Several contributions emphasize the practical application of digital technologies in industrial contexts, for example **Abdyreshit Sadyk, Yslam Orazov, Mekan Allyyev, Sahetnur Durdyyev, Mukam Ekayev, and Lachyn Hojamova**, in "Designing a Low-Cost CNC Milling Machine for Sustainable Digital Manufacturing", develop an affordable CNC milling machine using open-source components, targeting educational and small-scale manufacturing settings to promote sustainability in resource-constrained environments. Complementing this perspective, **Oualid Karim, Houda Lahyani, Ahmad Al Khatib, and Abdelfettah Sedqui**, in "Implementation of a Digital Twin in an Industrial Production Line Real-Time Visualization and Monitoring", propose a digital twin model incorporating real-time dashboards and simulation tools to enhance monitoring and decision-making in industrial production. Other chapters focus on the broader strategic implications of digitalization in supply chain and business process management. **Smina Nouhaila and Gharib Jihane**, in "Unified Data Management Framework for Supply Chain Optimization Challenges in Enterprise Architecture Models like TOGAF, RAMI 4.0 and IBM 4.0", introduce a tailored data management framework addressing real-time data integration and interoperability in supply chains, improving operational performance across value chains. In parallel, **Hind Bouifalioune, Majda Fikri, and Ghizlane Bencheikh**, in "Joint Optimization of Unrelated Parallel Machine Scheduling and Multi-Trips Vehicle Routing with Time Windows", propose a genetic algorithm-based solution to coordinate complex production and logistics scheduling, optimizing makespan and transportation costs in integrated systems. The integration of digital solutions with sustainability and resource efficiency objectives is also highlighted through innovative evaluation frameworks. **Oksana Makovoz and Sergii Lysenko**, in "Assessing Digital Transformation in Business Process Management Towards Sustainable Development", build an evaluation framework that combines BPM, digital transformation, and multi-criteria decision-making tools, enhancing strategic alignment with sustainability goals. Lastly, the chapters clearly demonstrate digital innovation in agricultural contexts, particularly emphasizing sustainable resource management in water-scarce environments. **Zaineb Bouswir, Salah Er-Raki, Saïd Khabba, Jamal Ezzahar, and Abdelghani Chehbouni**, in "Optimization of Winter Wheat Irrigation in a Semi-Arid Region Application of the SIMDualKc Model for Efficient Water Management", apply the SIMDualKc model to optimize irrigation scheduling for wheat, demonstrating water-saving benefits without compromising crop productivity. Further expanding on sustainable irrigation techniques, **Khaoula Ibba, Salah Er-Raki, and Rachid Hadria**, in "Evaluating the Benefits of Regulated and Sustained Deficit Irrigation for Yield, Water Use and Sustainability in the Moroccan Menara Olive Cultivar", assess various irrigation strategies and find that regulated deficit irrigation can enhance oil content and water productivity while maintaining sustainable yields.

Educational Technologies and Recommendation Systems

The last cluster explores innovative digital tools and methodologies shaping contemporary educational experiences, emphasizing personalized learning and recommendation systems.. The chapters collectively highlight how advanced AI and data-driven approaches are enhancing pedagogical effectiveness, improving educational accessibility, and addressing critical concerns. Several contributions examine personalized learning and recommendation systems aimed at enhancing educational outcomes. **Ali Mijlad and Yacine El Younoussi**, in "A Multi-level Arabic Text Diacritization System", develop a dual-layer encoder-decoder model using Luong attention to restore diacritics in Arabic texts, significantly reducing error rates and enhancing accessibility for language learners and visually impaired individuals. Building upon this theme of personalization, **Sandireddy Vyshnavi Prasad Chowdary, Atluri Sai Charan, Ashutosh Satapathy, Chinnam Sri Sai Prasanna, and Kokila Thottempudi**, in "An AI-Powered Chatbot Using A Transformer-based Language Model for Graduate Educational Schemes Recommendation", design a LLaMA 3.1-based chatbot to recommend government-funded educational programs to underprivileged students in India, achieving over 94% accuracy. Similarly, **Mohamed El Ghali, Issam Atouf, Kamal El Guemmat, Said Broumi, and Mohamed Talea**, in "Generative AI for Personalized E-Learning A Conversational and Adaptive Framework for Language Acquisition", introduce a transformer-based adaptive system that personalizes English language learning via real-time AI-generated feedback and exercises. In addition to personalization, the chapters emphasize the critical role of recommendation systems in guiding academic and career development. **Imran Khaled, Hanene Trichili, and Faiez Gargouri**, in "E-Learning Recommendation Systems Synthesis, Applications, and Future Trends", conduct a comprehensive review of e-learning recommendation systems, highlighting key methodologies, challenges like cold-start and sparsity, and emerging trends such as context-aware and emotion-aware recommendations. Complementing this, **Hasna Mahmoud, Mohamed Badouch, Mehdi Boutaounte, Omar Zioudi, and Es-Said Boulmane**, in "A Systematic Review of Recommender Systems for Student Academic and Career Guidance", analyze 21 recent studies on academic recommender systems, identifying a trend toward hybrid machine learning approaches and emphasizing future directions such as deep learning and labor market integration. **Jalal Lahiassi, Oussama Elwarraki, Souhaib**

Aammou, Youssef Jdidou, and Hind Ben Rahmoun, in "A KNN-Based Recommendation System for Adaptive Collaborative Learning", develop a KNN-based model for personalized collaborative learning activities, improving student participation and performance in a Master's course through targeted recommendations. Furthermore, **Fatima Zahra Abbadi, Mohamed Fourka, and Chahinaze Fikri Benbrahim**, in "Leveraging Ontological Models in Recommender Systems for Skills Assessment and Personalized Career Development", propose the use of ontologies in recommender systems to identify skill gaps and suggest personalized learning and career paths. The integration of educational technology (EdTech) and its impacts on institutional resilience and entrepreneurial motivation is also critically examined. **Adil Idrissi, Fatima-Zahrae Lakhlifi, Souad Habbani, and Mohammed Abdellaoui**, in "EdTech as a Lever of Resilience for Moroccan Educational Startups in the Context of Education 4.0", analyze the role of EdTech in enhancing the resilience of Moroccan educational startups, using SEM modeling to explore the relationship between digital adoption and crisis adaptability. Complementing this, **Hafsa Boutaqqa and Sanae El Maliki**, in "The Impact of ChatGPT on the Entrepreneurial Intention of Students at Sidi Mohamed Ben Abdellah University: A Gender-Based Approach", explore how ChatGPT influences entrepreneurial motivation among students in the Management Techniques program through a qualitative analysis. Lastly, **Jibin Jose P, Amutha Dhanaraj, and Pranav Prakash**, in "The Future of English Language Teaching: Investigating AI-Enhanced Personalisation in Digital Learning Contexts", examine how AI tools like chatbots and adaptive platforms transform English language education, highlighting benefits and ethical concerns. Cybersecurity emerges as a significant concern alongside educational innovation. **Mouhssine Yassine, Hicham Boutrachech, and Aniss Moumen**, in "Bridging Theory and Practice: The Educational Impact of Cybersecurity Competitions", analyze 88 studies to assess the value of cybersecurity competitions like CTF and Hack the Box in enhancing student skills and professional readiness. Addressing data security directly, **Najlaa Ghazi Thajeel, Hamid Jassam Mohammed, Ali Abdulwahhab Mohammed, Saif Safaa Hussein, and Ali H. Abdulwahhab**, in "A Secure File Sharing Using Computer Engineering Techniques", develop a secure file-sharing application with advanced encryption and error-checking, addressing challenges in data security and transmission efficiency.

Conclusions

Through this volume, the reader will explore contemporary sociotechnical transformations by positioning technology as a powerful institutional force that shapes, mediates, and evolves alongside organizational, societal, and governance structures. The contributions presented challenge traditional separations between technology and institutions, illuminating the intricate and multi-layered changes driven by digitalization and offering fresh critical views. The authors illustrate how cross-domain innovation and contextualized knowledge integration are essential for navigating modern complexities, highlighting the interwoven nature of digital infrastructures, organizational practices, and societal structures. These resources can inform the redesign of institutions, guide strategic decision-making, and foster responsible governance of digital transformations, offering scholars, policymakers, and practitioners both critical insights and actionable conceptual tools. Emphasizing the normative, ethical, and strategic dimensions of digitalization, the chapters collectively advocate for adaptive, inclusive, and reflexive policy-making that are essential for aligning technological progress with broader societal values—transparency, sustainability, equity, and resilience—and addressing urgent global challenges such as climate change, digital divides, and risks mitigation.

References

Appio, F.P., Frattini, F., Petruzzelli, A.M., Neirotti, P.: Digital transformation and innovation management: A synthesis of existing research and an agenda for future studies. J. Prod. Innov. Manage. **38**(1), 4–20 (2021)

Benitez, J., Arenas, A., Castillo, A., & Esteves, J.: Impact of digital leadership capability on innovation performance: the role of platform digitization capability. Inf. Manage. **59**(2), 103590 (2022)

Brynjolfsson, E., McElheran, K.: The rapid adoption of data-driven decision-making. Am. Econ. Rev. **106**(5), 133–139 (2016)

Chui, M., Yee, L., Hall, B., Singla, A.: The state of AI in 2023: Generative AI's breakout year (2023)

Gawer, A., & Cusumano, M. A. (2014). Industry platforms and ecosystem innovation. J. Prod. Innov. Manage. **31**(3), 417–433 (2023)

Halckenhaeusser, A., Forderer, J., Heinzl, A.: Platform governance mechanisms: An integrated literature review and research directions. In: Proceedings of the 28th European Conference on Information Systems (ECIS), pp. 15–17 (2020)

Jarrahi, M.H., Newlands, G., Lee, M.K., Wolf, C.T., Kinder, E., Sutherland, W.: Algorithmic management in a work context. Big Data Soc. **8**(2), 20539517211020332 (2021)

Lycett, M.: 'Datafication': making sense of (big) data in a complex world. Eur. J. Inf. Syst. **22**(4), 381–386 (2013). https://doi.org/10.1057/ejis.2013.10

Nambisan, S., Wright, M., Feldman, M.: The digital transformation of innovation and entrepreneurship: Progress, challenges and key themes. Res. Policy. **48**(8), 103773 (2019)

Smith, A., Stirling, A.: Moving outside or inside? Objectification and reflexivity in the governance of socio-technical systems. In: Governance for sustainable development, pp. 159–181. Routledge(2013)

Spagnoletti, P., Resca, A., Lee, G.: A design theory for digital platforms supporting online communities: a multiple case study. J. Inf. Technol. **30**(4), 364–380 (2015)

Vial, G.: Understanding digital transformation: A review and a research agenda. Manage. Digital Transf. 13–66 (2021)

Volkoff, O., Strong, D.M.: Affordance theory and how to use it in IS research. In: The Routledge Companion to Management Information Systems, pp. 232–245. Routledge (2017)

Zuboff, S.: The age of surveillance capitalism: The fight for a human future at the new frontier of power, edn. PublicAffairs, New York (2019)

Committees

Conference Chairs

Badr-Eddine Boudriki Semlali	FST Tangier, UAE University, Morocco
Ikram Ben abdel ouahab	FST Tangier, UAE University, Morocco

Conference Steering Committee

Fabio Angeletti	Sapienza University of Rome, Italy
Luis Bragança	University of Minho, Portugal
Mohamed Ben Ahmed	FST Tangier, UAE University, Morocco
Anouar Abdelhakim Boudhir	FST Tangier, UAE University, Morocco
Hasan Majdi	Al-Mustaqbal University, Iraq
Mazin Hadi Kzar	Al-Mustaqbal University, Iraq
Mohammed Bouhorma	FST Tangier, UAE University, Morocco
Lotfi Elaachak	FST Tangier, UAE University, Morocco
Chaker El Amrani	FST Tangier, UAE University, Morocco
Aicha Chorak	ENSI Tangier, Morocco

Local Organizing Committee

Aziz Mahboub	FST Tangier, UAE University, Morocco
Abdelhadi Fennan	FST Tangier, UAE University, Morocco
El Mokhtar En-naimi	FST Tangier, UAE University, Morocco
Yasyn Elyusufi	FST Tangier, UAE University, Morocco
Zouhair Abdelhamid	FST Tangier, UAE University, Morocco
Hicham Gibet Tani	FP Larache, UAE University, Morocco
Fatiha Elouaai	FST Tangier, UAE University, Morocco
Mohamed Elbrak	FST Tangier, UAE University, Morocco

Technical Program Committee

Abderrahim El Mhouti	UAE University, Morocco
Adnane El Hanjri	EMSI Tangier, Morocco
Adriana Estokova	Technical university of Kosice, Slovakia

Aicha Chorak	ENSI Tangier, Morocco
Ali Aghazadeh Ardebili	University of Salento, Italy
Ali Shakir Majdi	Al-Mustaqbal University, Iraq
Anass Belcaid	ENSATe, UAE University, Morocco
Assaad El Makhloufi	EST, USMBA, Morocco
Bassam Zafar	King Abdulaziz University, Jeddah, Saudi Arabia
Bouchra Ait Hssaine	Center for remote sensing applications, UMP6, Morocco
Chaimae Azroumahli	FST Settat, Morocco
Cristiana Croitoru	Technical University of Civil Engineering Bucharest, Romania
Dounia Chaatouf	FST Tangier, UAE University, Morocco
Dehbi Youness	University of Bonn, Germany
Elsadig Musa Ahmed Mohammed	Multimedia University, Malacca
El Meouche Rani	ESTP Paris, France
Fabio Angeletti	Sapienza University of Rome, Italy
Fatima Zahra Fagroud	Hassan II University of Casablanca
Fatma Zohra Bessai-Mechmache	CERIST, Algeria
Hicham Gibet Tani	FP Larache, UAE University, Morocco
Idriss Moumen	EST, Hassan II University of Casablanca
Jai Woo Lee	Korea University
Jesús Rosa-Bilbao	Universidad de Cádiz, Spain
Jihane Abdessadak	ENSI Tangier, Morocco
Jihane Gharib	Ibn Tofail University
Joel Rodrigues	INATEL, Brazil
Lamya Alabdulkarim	King Saud University, Saudi Arabia
Larysa Vdovenko	State Tax University
Loubaba Rida	ENSI Tangier, Morocco
Louise Hayes	UCL, UK
Mahbubul Bhuiyan	Southeast University, Bangladesh
Md Adnan Rahman	Uttara University, Bangladesh
Mohamed Zarid	Ibn-Tofail University, Morocco
Mohammed Aldulaimi	Al-Mustaqbal University, Iraq
Mohammed Chekour	ESEF Kenitra, Ibn Tofail University, Morocco
Mohammed Loukili	Naval Academy Brest, France
Nouhaila El Koufi	Hassan II University, Casablanca, Morocco
Oksana Makovoz	National Technical University, Ukraine
Osvaldo Santos	IPCB, Portugal
Paraskevi Tsoutsa	University of Thessaly, Greece
Pedro Ribeiro Mucharreira	University of Lisbon, Portugal
Raida Ktari	Higher institute of computer science and multimedia of Sfax, Tunis

Rochdi Rais	FSTH, UAE University, Morocco
Salwa Belaqziz	UIZ, Morocco
Sanjay Agal	Parul University, India
Savo Tomovic	University of Montenegro, Montenegro
Souhaib Aammou	ENS Tetouan, UAE University, Morocco
Soukaina Zahm	Université Chouaib Doukkali, Morocco
Tanveer Hasan	Daffodil International University, Bangladesh
Theuns Pelser	Toyota Wessels Institute for Manufacturing Studies, South Africa
Vicky Zilvan	Reseach Center for AI and Cybersecurity - National Research and Innovation Agency, Jakarta
Yasin Kaya	ankara yıldırım beyazıt university, Turkey
Yasyn Elyusufi	FST Tangier, UAE University, Morocco
Youssef Menchafou	ENSAK, Ibn tofail University Kenitra
Yslam Orazov	Oguz han Engineering and technology university of Turkmenistan, Turkey
Zouhair Abdelhamid	FST Tangier, UAE University, Morocco

Keynote Speakers

Remote Sensing, IoT and IA For Water Resources Management

Abdelghani Chehbouni

Director of the Center for Remote Sensing Applications, UM6P, Morocco

Dr. Abdelghani Chehbouni is a doctor in hydrology and remote sensing, he received his B.S. in fluids Mechanics from University Paul Sabatier, Toulouse, France in 1987 and his M.S. in Environment Science from the National Polytechnic Institute of Toulouse in 1989, and a PhD in Hydrology and Remote Sensing from University Paul Sabatier in Toulouse in 1992.

He has been working at the University of Arizona for 18 months and the Jet Propulsion Laboratory for 2 years. He is co-investigator of several Remote Sensing programs (VEGETATION, ERS2/ATSR2, EOS).

His main research interests include the application of remotely sensed data in land-surface-atmosphere models especially in arid semi-arid regions.

He participates in several international experiments (MONSOON'90, Hapex Sahel) and he was a co-leader of the Semi-Arid-Land-Surface-Atmosphere (SALSA) international program. He joined CESBIO in 2000 as the leader of hydrology group. He is currently the Director of the Center for Remote Sensing Applications, UM6P, Morocco.

Toward a Holistic Approach to Smart and Sustainable Mobility

Mohamed Essaaidi

President of MedAI and Ex-Dean of ENSIAS

Prof. Mohamed Essaaidi is a Professor and past Dean of ENSIAS school of Computer Science, Mohammed V University, Rabat, Morocco, and is CEO of the Moroccan School of Engineering Science (EMSI) Group, (2023–2024), Chief of Party of Interactive Digital & XR Center Morocco at UM6P (2020–2025), Past Director of International Cooperation at the Ministry of General Affairs & Governance, Morocco (2019), and past faculty member of the of Abdelmalek Essaadi University, Tetuan, Morocco (1993-2011).

He is an IEEE Senior Member, he was the IEEE Morocco Section Founding Chair (2004), IEEE MTT-S/AP-S Morocco Chapter Founding Chair (2008), IEEE Global Cities Alliance, Middle East & Africa Chairman (2020–2022), and he has been the IEEE Humanitarian Technologies Board Programs Committee Chair and IEEE Special Interest Group on Humanitarian Technologies Global Chair since January (2023–2024), IEEE Public Safety Technology, Education Chair (2021–22).

Integrating Green Education and AI at Al-Mustaqbal University A Framework for Sustainable Higher Education

Hassan Majdi

President of Al-Mustaqbal University, Babil, Iraq

Prof. Dr. Hasan Shakir Majdi, born in 1957 in Babylon, Iraq, is a distinguished chemical engineer and academic leader. He earned his Bachelor of Science in Chemical Engineering in 1978 and later obtained his Ph.D. in Chemical Thermodynamics from the University of London in 1991. Dr. Majdi has dedicated his career to academia and research, with a focus on renewable energy and heat transfer. He has held the position of President at Al-Mustaqbal University in Babylon, Iraq, where he has also been associated with the Department of Chemical Engineering and Petroleum Industries.

Throughout his academic tenure, Dr. Majdi has contributed significantly to scientific literature. His research encompasses topics such as adsorption techniques for dye removal, advancements in absorption cooling cycles, and the development of heterogeneous catalysts for hydro isomerization and hydrocracking processes. His work has garnered substantial citations, reflecting his influence in the field. In recognition of his leadership and contributions to higher education, Dr. Majdi was honored as the Best President of a Private University in Iraq for the year 2023.

Dr. Majdi's career exemplifies a commitment to advancing chemical engineering education and research, particularly in the realm of sustainable and renewable energy solutions.

Green and Sustainable Computing Using Remote Sensing in Saudi Arabia

Sultan Alsultan

The Environmental Center Of Remote Sensing (TECRS), Riyadh, Saudi Arabia

Dr. SULTAN ALSULTAN main professional interest has been on Remote Sensing for Desert Area Development as well as City and Environmental Land Management. Within that field of interest Dr Sultan obtained his PhD from Tokyo Institute of Technology, Japan just after graduating from MIT Massachusetts Institute of Technology, and previously from George Washington University, both in the USA. Dr Sultan has been senior researcher with NASA and the USGS both in the USA, and has been researcher in JAXA, RESTEC-Japan. At present, he is (CEO, TECRS, Saudi Arabia) a leading Environmental Consultant of Remote Sensing Institution and its parcener of Al-Qassim University. And as Saudi Arabia Coordinator for the Mission Idea Contest.

Contents

Digital Manufacturing for Sustainable Development

Green and Sustainable Computing for Sustainable Development

Artificial Intelligence for Sustainable Development

About the Editors

Badr-Eddine

is a Computer Science Professor at the Faculty of Sciences and Techniques of Tangier, Abdelmalek Essaâdi University, Morocco. He has expertise in system administration, network management, distributed systems, IoT, cloud computing, software development, machine learning, and satellite data processing. Badr-Eddine has engaged in various international activities in research and innovation, including fellowships and programs in Romania, Taiwan, Spain, Belgium, and Saudi Arabia. He has authored over 18 high-quality, Scopus-indexed papers, participated in more than 20 international conferences, and is a reviewer for numerous journals. As a member of ASAF and IEEE, he actively contributes to technical program committees at several international conferences.

Ikram Ben Abdel Ouahab

an assistant professor at the computer science department of the Faculty of Sciences and Techniques of Tangier, Morocco. She is a member of Data & Intelligent Systems research Team. She received her PhD degree in Cybersecurity and AI at FSTT, University Abdelmalek Essaâdi, Tangier, Morocco. She participated in many international conferences, and have published many scientific papers in collaboration with the Laboratory team. In 2022, she graduates from CPITS program by Trend Micro, an intensive training in cybersecurity technologies and industry trends, including certification in industry leading solutions for endpoint, cloud, and network security. She is an AWS Solution Architect certified, a certified peer reviewer from Publons academy. Her main research interests include cybersecurity, malware analysis, artificial intelligence, and IoT.

Fabio Angeletti

is an Italian computer scientist and academic with a strong interdisciplinary background that spans computer engineering, artificial intelligence, and digital transformation. He holds a Ph.D. in Computer and Control Engineering from Sapienza University of Rome, where his doctoral research focused on secure and privacy-aware data architectures, with applications in clinical trials, smart environments, and Internet of Things (IoT) ecosystems. His academic interests lie at the intersection of emerging technologies and their application to real-world socio-technical systems, particularly within the domains of healthcare, smart cities, and public sector innovation.

Currently, Fabio serves as an Adjunct Professor at LUISS Guido Carli University in Rome, where he teaches courses on digital technologies and their strategic implications for business leaders and public administrators. His teaching approach emphasizes hands-on experimentation with advanced technologies—including AI, blockchain, and cybersecurity frameworks—combined with a critical reflection on their ethical, legal, and policy dimensions.

Technological Advances for Sustainable Development

Bridging Theory and Practice: The Educational Impact of Cybersecurity Competitions

Yassine Mouhssine(✉), Hicham Boutrachech, and Aniss moumen

Laboratory of Engineering Sciences National School of Applied Sciences, Ibn Tofail University, Kenitra, Morocco

{Yassine.mouhssine,hicham.boutrachech, aniss.oumoumenuit}@uit.ac.ma

Abstract. Cybersecurity competitions play a crucial role in bridging the gap between theoretical knowledge and practical application in cybersecurity education. After reviewing 239 articles, 88 of them focus specifically on their educational impact, analyzed using a mixed-methods approach that employs word clouds and advanced analytics to uncover key themes. Competitions like Hack the Box, TryHackMe, and Capture the Flag (CTF) provide hands-on engagement with real-world challenges, enabling students to develop expertise in areas such as cryptography, web exploitation, and reverse engineering. Teachers emphasize how well these platforms link classroom instruction with practical application. The National Collegiate Cyber Defense Competition (NCCDC) and CyberPatriot are two competitions that emphasize the value of cooperation and teamwork in solving challenging cybersecurity issues. Additionally, they give students the chance to network with professionals in the business. Even though there are still issues like time investment and the emphasis on offensive security rather than defensive, including these contests into academic programs is a crucial first step in matching learning objectives with business demands. In the end, these experiences give students professional recognition, real-world skills, and capacity to meet the changing demands of the cybersecurity industry, preparing them for prosperous careers in the sector.

Keywords: cybersecurity · Competitions · Students · Teaching · Interactive Games

1 Introduction

In recent years, Cybersecurity competitions have become essential in connecting academic learning with real-world experience, helping students translate theoretical knowledge into practical skills in the field of cybersecurity education [1]. Traditional classroom approaches often focus on foundational knowledge, leaving students unprepared for the dynamic and complex challenges of real-world cyber threats. Platforms like Hack the Box, TryHackMe, and Capture the Flag (CTF) challenge students to engage directly with real-world cybersecurity scenarios, fostering problem-solving skills and

B.-E. Boudriki Semlali et al. (Eds.): DATA 2025, LNISO 81, pp. 3–16, 2025.
https://doi.org/10.1007/978-3-032-06725-8_1

encouraging a deeper understanding of advanced concepts such as system vulnerabilities, encryption mechanisms, and web application security [2]. Moreover, events like CyberPatriot and the National Collegiate Cyber Defense Competition (NCCDC) emphasizes teamwork, communication, and defensive strategies—skills critical for the collaborative nature of the cybersecurity profession. These competitions not only encourage innovative problem-solving but also offer mentorship, networking, and career development opportunities. By integrating these experiential learning platforms into academic curricula, educational institutions can better prepare students to meet the demands of the rapidly evolving cybersecurity landscape, ensuring alignment with industry needs and professional readiness.[3].

2 Cybersecurity Competitions

2.1 Definition and Types of Cybersecurity Competitions

Cybersecurity competitions have become increasingly popular, offering participants the opportunity to apply their skills in real-world scenarios. These events cater to various skill levels and focus on different aspects of cybersecurity.

Capture the Flag (CTF) Competitions: CTFs are among the most well-known formats, with two main types:

Jeopardy-style CTFs: Participants solve distinct tasks in areas like cryptography, web exploitation, binary analysis, reverse engineering, and forensics. Teams accumulate points based on task difficulty [4].

Attack-Defense CTFs: Teams must defend their systems while simultaneously attacking opponents', testing both offensive and defensive cybersecurity skills.

Defense-Oriented Competitions: These competitions focus on defending systems under attack rather than offensive tactics.

National Collegiate Cyber Defense Competition (NCCDC): Teams defend a simulated corporate network from attacks while maintaining services.

CyberPatriot: Aimed at high school students, this competition challenges participants to secure systems against live threats.

Red vs. Blue Team Competitions: Teams take on offensive (Red) or defensive (Blue) roles, simulating real-world cyber defense scenarios.

Purple Teaming: Some events combine offensive and defensive strategies, fostering collaboration to identify vulnerabilities and strengthen defenses.

Bug Bounty Programs: Platforms like HackerOne and Bugcrowd allow individuals to identify security vulnerabilities in live systems for monetary rewards, offering participants exposure to real-world industry challenges.

2.2 Educational Value of Cybersecurity Competitions

Cybersecurity competitions offer a dynamic and immersive educational experience that extends far beyond traditional theoretical coursework [5]. Their value lies in several key aspects:

Problem-Solving Under Pressure: Timed challenges require quick thinking, analytical skills, and effective problem-solving in high-pressure scenarios.

Simulated Real-World Scenarios: Competitions mirror professional cybersecurity challenges, including incident response and network defense, exposing students to tasks like mitigating security breaches, uncovering vulnerabilities, and maintaining operational networks during attacks.

Broad Exposure: By covering diverse topics such as cryptography, binary exploitation, and malware analysis, these events help participants identify and pursue their specific interests within cybersecurity.

Structured Learning: Platforms like TryHackMe provide guided paths that allow learners to progress from beginner to advanced levels in various cybersecurity disciplines.

Soft Skills Development: The team-based nature of most competitions fosters collaboration, communication, and the ability to perform under pressure—essential qualities in the high-stakes cybersecurity industry.[6].

2.3 Popular Cybersecurity Competitions and Platforms

This section can delve into the most influential platforms and competitions that are shaping the cybersecurity landscape:

Hack The Box: One of the most well-known platforms for ethical hacking and penetration testing. Participants can solve a variety of hacking challenges that simulate real-world environments. Its popularity comes from its diversity of challenges, user-generated content, and its vibrant online community.[7].

TryHackMe: This platform is ideal for students who prefer guided lessons. It offers "rooms" filled with interactive labs that teach everything from basic networking to advanced red teaming. The platform is beginner-friendly, offering tutorials for those new to cybersecurity.

CTFtime: The go-to platform for tracking and participating in CTF competitions around the world. Teams can register, track their progress, and participate in various global events. CTFtime is not a host platform but an organizer that lists and ranks competitions.[8].

National Collegiate Cyber Defense Competition (NCCDC): As one of the largest collegiate competitions in the U.S., NCCDC focuses on defe nsive skills. Participants must defend and maintain the functionality of a network while undergoing active attacks.

CyberPatriot: Aimed at middle and high school students, CyberPatriot introduces young learners to the basics of cybersecurity. It emphasizes teamwork, system hardening, and defense strategies, making it an excellent gateway into the field.

European Cybersecurity Challenge (ECSC): An international competition where teams from European countries compete to solve various cybersecurity challenges. It fosters cross-border collaboration and exposes students to a variety of cybersecurity fields (Table 1).

Table 1. Categorization and Frequency of Cybersecurity Competitions in Reviewed Articles

Category	Competition	Occurrences in Articles
Capture the Flag (CTF)	Hack the Box	20
	TryHackMe	15
	CTFtime	8
Defense-Oriented	CyberPatriot	10
Simulation-Based Learning	Cyber Range Exercises	8
	Gamified Exercises	7
Red vs Blue Teaming	Joint Red/Blue Team Exercises	6

3 Methodological Approaches and Literature Review

3.1 Method

This study employs a systematic review methodology to comprehensively collect, assess, and synthesize data on the role of cybersecurity competitions in education. By conducting an extensive search of scholarly sources, we aimed to uncover insights into how these competitions address key challenges and contribute to practical solutions within educational frameworks. The process was guided by well-defined research questions and targeted queries designed to identify relevant studies.

A rigorous evaluation protocol was applied to 239 publications retrieved from esteemed databases such as Scopus. Each publication underwent a thorough review, with non-relevant or non-compliant studies excluded based on predefined inclusion criteria. This meticulous selection ensured a focus on high-quality, impactful studies, enabling the generation of reliable and valid findings.

Building on previous research [9, 10] that applied bibliometric methods, our study employed a systematic approach to reviewing literature reviews centered on cybersecurity competitions. This approach facilitated the classification and evaluation of these reviews, focusing on their relevance and impact within the educational domain. By examining bibliographic data from these secondary sources, we identified key trends, influential contributors, and central themes shaping discussions on cybersecurity competitions in education.

This methodology provided a robust framework for assessing how existing reviews address the topic, shedding light on their strengths, weaknesses, and areas of limited research coverage. It also uncovered conflicting viewpoints within the field, enabling us to prioritize critical aspects of the discourse. By synthesizing these findings, we generated valuable insights to guide both practical applications and future research efforts.

Through this structured analysis, our study offers a refined categorization of literature reviews, highlighting their significance in advancing cybersecurity education. This work not only deepens our understanding of the current research landscape but also establishes a foundation for more focused and impactful contributions to the field.

Our paper follows the prescribed methodological steps, similar to the outlined approach:

Research Question: Our study is primarily centered on examining cybersecurity competitions as a means to enhance the educational sector, moving beyond a traditional literature review to present a focused analysis. The research aims to address key questions about the effectiveness of these competitions in equipping individuals with the skills necessary to tackle the complex challenges of cybersecurity. Additionally, it seeks to identify the critical competencies required in this rapidly evolving field. Rather than merely summarizing existing studies, we analyze diverse perspectives from scholars who have explored these dimensions, with a focus on uncovering overarching trends, highlighting research gaps, and emphasizing emerging viewpoints. This targeted approach aims to offer a nuanced understanding of the connections between cybersecurity, education, training programs, and skills development, contributing meaningful insights to the broader discourse in computer science [11].

Search Process: Our research adopts a focused and strategic approach to exploring the intersections of cybersecurity, education, training, and skills development. Using Scopus as our primary database, we formulated a precise search query incorporating key terms such as "cybersecurity," "competitions," "CTF," "capture the flag," "Games," and "education." This targeted methodology enables the identification and analysis of studies that shed light on the role of cybersecurity competitions in enhancing educational outcomes and skill-building. Through this systematic exploration, we aim to uncover nuanced relationships between these fields, emphasizing the broader impact of cybersecurity on safety, learning, and workforce readiness.

Selection of Studies: To ensure a thorough and unbiased exploration of cybersecurity competitions, we employed a structured and systematic approach for selecting relevant studies. The PRISMA framework (Preferred Reporting Items for Systematic Reviews and Meta-Analyses), illustrated in Fig. 1, was adopted to guide the process due to its rigorous methodology that enhances transparency and reproducibility. This approach facilitated a step-by-step workflow encompassing literature identification, screening, selection, and synthesis, ensuring that only studies meeting strict criteria were included.

Our process began with the formulation of a precise search query in Scopus, tailored to align with our research objectives. This query targeted articles within a defined time frame and incorporated parameters such as titles, abstracts, and keywords to retrieve publications relevant to the intersection of cybersecurity competitions, education, training, and skills development. The visual representation of this method underscores the systematic and transparent nature of our study selection, reinforcing the reliability and validity of the findings. This meticulous strategy establishes a robust foundation for investigating the role of cybersecurity competitions in advancing educational and professional outcomes.

To ensure a rigorous selection of studies, two independent analyses were conducted by co-authors MOUMEN.A and BOUTRACHEH.H. Each co-author reviewed the retrieved articles based on predefined inclusion and exclusion criteria. The initial screening by Aniss resulted in the exclusion of 112 records, while Hicham identified 118 records for removal. After reconciling any discrepancies and combining efforts, we collectively agreed to exclude a total of 118 records.

A significant reason for exclusion was the identification of erratum articles. Although these articles initially appeared relevant, further examination revealed that they consisted of corrections or revisions to previously published work. Including such articles could introduce inaccuracies or misinterpretations into the analysis, making their exclusion necessary. Additionally, some records were removed because they fell outside the scope of the research question or lacked alignment with the study's objectives. This meticulous screening process ensured the reliability and relevance of the selected studies for our investigation into cybersecurity competitions.

The steps followed with the PRISMA method are as follows:

Identification: Relevant studies were located through a comprehensive search strategy utilizing predefined keywords, such as cybersecurity, computer science, safety, education, training, and skills development. The search was restricted to publications from 2019 onward to maintain a focus on recent developments.

Screening. The identified articles underwent a systematic review process to ensure they aligned with the research objectives and met stringent inclusion criteria. This evaluation was designed to confirm the relevance and quality of the selected studies.

Included. Articles were carefully chosen based on their significant contributions to understanding the complex interplay between cybersecurity, education, training, skills development, and safety within the computer science domain. This deliberate selection process aimed to highlight studies that offered meaningful insights into the field .

3.2 Search Results

To evaluate the results of our inquiry, we focus on Scopus. Figure 2 depicts the trend in publications from 2019 to 2024, showcasing a significant increase in 2024 as publishing activity began to grow.

We conducted an additional analysis of document types, uncovering a dominant representation of conference papers, which greatly outnumbered articles. This disparity highlights a significant potential within the evaluated material (Fig. 3).

To explore the findings in greater depth, a pie chart emphasizes the leading role of computer science and mathematics, followed closely by input from Engineering (Fig. 4).

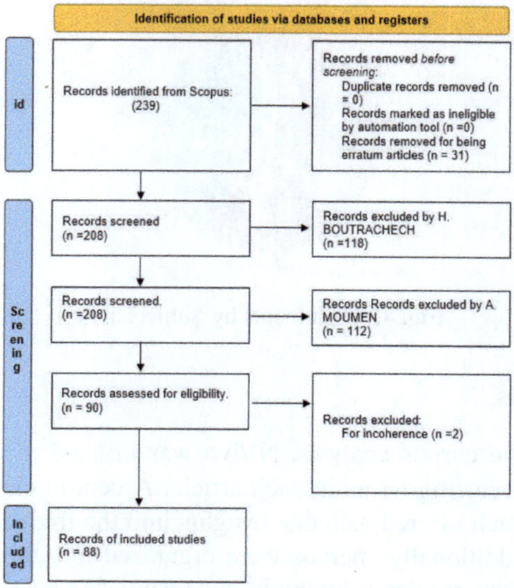

Fig. 1. Filter Process of Selected articles

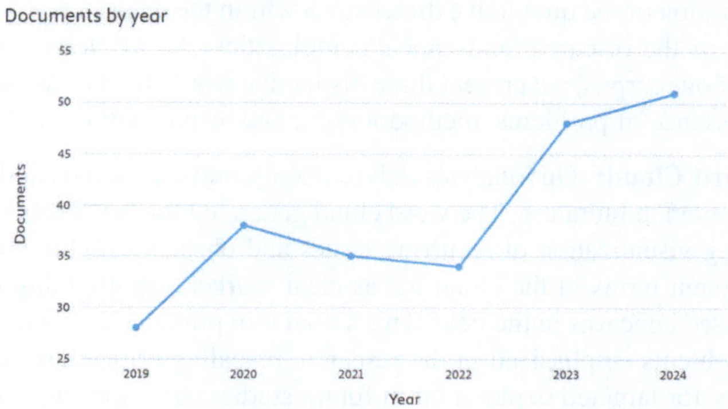

Fig. 2. Documents by Year

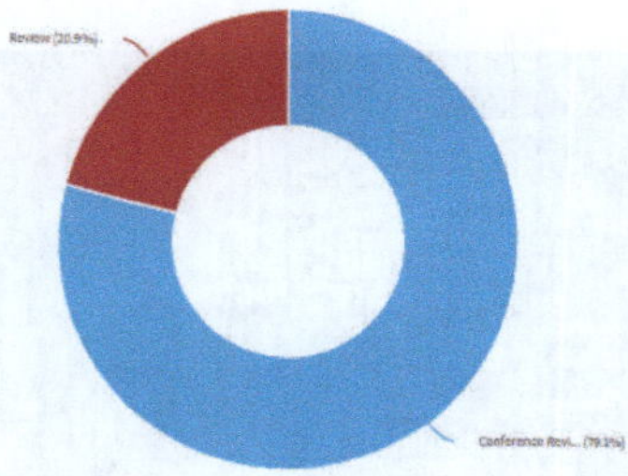

Fig. 3. Documents by Type

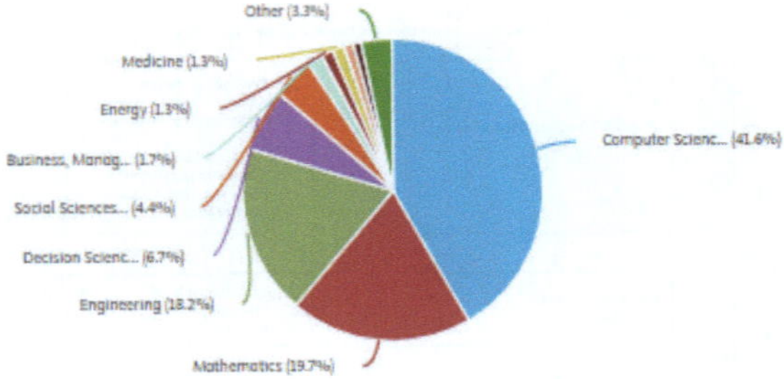

Fig. 4. Documents by Subject area

3.3 Corpus Analysis

We transcribed for the corpus analysis, NVivo was utilized to assess the quality of results and identify recurring terms in each article. A central element of our findings is the word cloud, which offered valuable insights into the frequency of specific terms across the dataset. Additionally, memos were organized into three distinct categories within the nodes section: problems, methods, and results. For each category, a dedicated word cloud was generated to visually highlight the key themes and patterns related to identified challenges, methodologies, and outcomes. This structured approach enables a detailed exploration of the qualitative dimensions within the dataset, providing a deeper understanding of the research landscape's complexities. As we delve into a nuanced examination of our corpus, we present three distinctive word clouds, each encapsulating the thematic essence of problems, methodologies, and results within the literature.

Problems Word Cloud: Our analysis delves deeply into the memos addressing challenges documented in literature. The word cloud generated for the "Problems" category offers a striking visualization of recurring issues and obstacles highlighted within the dataset. Prominent terms in the cloud act as clear markers, spotlighting the most frequently discussed concerns in the field. This visual tool provides a richer understanding of the core problems emphasized in the research, revealing critical areas of focus and paving the way for targeted exploration in future studies. By capturing these prevalent themes, the analysis sheds light on the key issues shaping the discourse and highlights opportunities for advancing the field (Fig. 5).

Fig. 5. Problems Word Cloud

Competitions Word Cloud: Focusing on the "Competitions" category, we introduce a specialized word cloud that encapsulates the competitions discussed in the document analyzed. This visualization highlights recurring techniques, strategies, and approaches employed by researchers within the domain. The prominence of certain terms in the word cloud provides valuable insights into the methodological framework, showcasing commonly utilized strategies such as educational & security program design and other approaches integral to the field. This representation not only uncovers prevailing research trends but also enhances our understanding of the practices shaping the discourse around cybersecurity competitions (Fig. 6).

Fig. 6. Competitions Word Cloud

Results Word Cloud. The word cloud generated for the "Results" category provides a vivid depiction of the key outcomes and findings identified through our corpus analysis. This engaging visual representation emphasizes recurring themes and notable insights found across the dataset.

By examining the terms that dominate the word cloud, we uncover valuable perspectives on common trends and significant results highlighted in the literature. This approach enhances our ability to identify patterns and understand the contributions within the analyzed research, offering a deeper and more nuanced comprehension of the overarching findings in the field (Fig. 7).

Fig. 7. Results Word Cloud

Navigating through our analytical journey, we further refined our exploration by harnessing the power of NVivo. This approach not only allows us to delve into the nuances of problems, methodologies, and results but also sharpens our focus on emerging cybersecurity trends. Specifically, recent terms such as cryptography, cyber-attacks, security Network, cloud computing and gamification have become increasingly prominent.

4 Results and Discussion

The findings of this study underscore the transformative potential of cybersecurity competitions in addressing critical gaps in education, skills development, and workforce readiness. Through systematic analysis, this research highlights a wide variety of competitions and methodologies tailored to educational needs, offering significant opportunities to bridge theoretical learning with practical application in real-world scenarios.

The analysis of challenges revealed recurring barriers such as limited training resources, uneven accessibility, lack of standardization, and difficulties in aligning competitions with professional cybersecurity demands. These issues emphasize the need for inclusive and consistent frameworks to ensure competitions are accessible and meaningful to a diverse range of learners. Our collaboration with cybersecurity clubs across local institutions provided valuable insights into the importance of student engagement in cybersecurity competitions. One effective approach observed during this collaboration is the **Club-to-Curriculum Model**, where student-led clubs' partners with faculty to align competition preparation with formal coursework, some institutions designate elective courses or lab sessions that mirror competition challenges, such as Capture the Flag (CTF) tasks or red team/blue team exercises. Faculty members co-develop these modules with club leaders, allowing students to earn academic credit while preparing for competitions. Such an approach ensures that competitions are not isolated experiences but are embedded meaningfully within the educational journey of a diverse range of learners. Additionally, inadequate feedback mechanisms and insufficient integration into academic curricula emerged as areas that require urgent attention [12].

The exploration of competition types demonstrated their versatility in addressing educational objectives. Key methodologies identified include simulation-based exercises, hands-on labs, and gamified learning activities. These approaches highlight the adaptability of competitions in catering to different skill levels and fostering critical skills such as teamwork, problem-solving, and critical thinking. The integration of structured learning pathways, often embedded in educational programs, reflects the growing recognition of competitions as an essential component of modern cybersecurity education [13].

Table 2. Most used Competitions in some of the retrieved articles and their results

Competition name	Type	Targeted Audience	Reported Outcome
CTF	Attack-Defense	Professionals, advanced students	Promote teamwork, communication, and deep technical learning (e.g., reverse engineering, cryptography). Acts as a global benchmark.
National Cyber League (NCL)	Individual and team Jeopardy challenges	High school, college students, educators	Enhance ethical hacking, forensics, and scanning skills; complements cybersecurity curricula; increases self-efficacy in cybersecurity skills.
CyberPatriot	Defensive Security	High school and middle school students	Develop proficiency in ethical hacking, digital forensics, and vulnerability scanning.
CTFtime Events	Aggregator for global CTFs	Students, professionals, enthusiasts	Introducing students to cybersecurity basics; boosts interest in STEM fields; leads to higher rates of participants pursuing technical degrees.
Hack The Box University Challenges	Real-world penetration testing	University students	Provides practical exposure in gamified environments; improves critical thinking and problem-solving abilities.
European Cyber Security Challenge	Mixed-Style	young professionals	Promotes collaboration across nations; standardizes skill-building across Europe

In terms of outcomes, the results consistently point to benefits such as skill enhancement, knowledge retention, industry readiness, and confidence building. Through close collaboration with local cybersecurity clubs and surveys conducted with students actively involved in competitions, our research emphasizes the significant educational value these events offer. Competitions go beyond testing technical knowledge; they create immersive learning environments where participants develop practical, job-ready skills. Students reported gaining hands-on experience in areas such as network defense, penetration testing, and incident response, which closely mirror industry challenges. Notably, the growing emphasis on using competitions as a bridge between academic learning and industry demands aligns closely with the needs of both students and employers.

By employing the PRISMA framework for systematic article selection, this study ensured methodological rigor and transparency. The visual mapping of the identification and screening process, combined with a detailed word cloud analysis of challenges, competition types, and outcomes, provided a comprehensive view of the research landscape. This approach identified key trends, such as the increasing integration of competitions into educational systems, while also highlighting gaps, including the limited exploration of their long-term effects on career progression and skill sustainability.

In conclusion, this study illustrates the critical role of cybersecurity competitions in enhancing education and workforce preparedness. While these competitions offer significant advantages, the findings emphasize the need for robust frameworks, broader accessibility, and longitudinal studies to assess their sustained impact. These insights provide valuable guidance for educators, policymakers, and industry stakeholders, fostering the development of innovative and effective cybersecurity training programs that align with both academic and professional objectives [14].

5 Conclusion

This study has explored the integral role of cybersecurity competitions in bridge the gap between theoretical education and practical skill development. Our analysis has highlighted their potential to enhance learning, foster critical skills, and prepare participants for real-world cybersecurity challenges besides keeping their sustainable interest. Key findings reveal that while these competitions are gaining recognition as valuable educational tools, challenges persist, including issues of accessibility, standardization, and alignment with industry needs based on local club's feedback. The diversity in competition formats, such as Capture the Flag (CTF) events and gamified exercises, showcases their adaptability to various educational objectives. Furthermore, the outcomes of these initiatives, particularly in fostering industry readiness, confirm their importance in addressing the growing demand for cybersecurity professionals. However, critical gaps remain, such as the absence of standardized frameworks for designing and evaluating competitions, which hinders their broader adoption in academic and professional settings. Additionally, there is a need for more research on the long-term impact of these competitions on career trajectories and their role in mitigating the global cybersecurity talent shortage.

Looking ahead, future research should focus on developing cybersecurity competitions that not only meet current educational, and industry needs but also ensure sustained skill development and career growth for participants. Considering our findings, cybersecurity competitions have the potential to serve as powerful learning tools, they are supported by inclusive and structured frameworks. To make this a reality, collaboration among educators, industry stakeholders, and student-led cybersecurity clubs is essential. We propose a **Collaborative Cyber Learning Framework (CCLF)** grounded in access and inclusion, skill development pathways, and community integration. By embedding these elements into educational practices, institutions can create more equitable, engaging, and impactful cybersecurity learning experiences for all students. As a result, they become better prepared to face professional cybersecurity challenges. More than just extracurricular activities, these events play a vital role in bridging the gap between classroom learning and real-world application, making them an essential component of modern cybersecurity education (Table 2).

References

1. Mouhssine, Y., Boutracheh, H., Moumen, A.: Teaching cybersecurity to engineering students: a new perspective through literature review. In: présenté à 2024 4th International Conference on Innovative Research in Applied Science, Engineering and Technology, IRASET 2024 (2024). https://doi.org/10.1109/IRASET60544.2024.10548731
2. Abaimov, S., et al.: Capture the industrial flag: lessons from hosting an ICS cybersecurity exercise. In: présenté à ACM CPSS 2024 - Proceedings of the 10th ACM Cyber-Physical System Security Workshop, pp. 98-106 (2024). https://doi.org/10.1145/3626205.3659148
3. Al-Karaki, J.N., Omar, M., Gawanmeh, A., Jones, A.: advancing cybersecurity education and training: practical case study of running capture the flag (CTF) on the metaverse vs. physical settings. In: présenté à 2023 International Conference on Intelligent Metaverse Technologies and Applications, iMETA 2023 (2023). https://doi.org/10.1109/iMETA59369.2023.10294722
4. Balon, T., Baggili, I.A.: Cybercompetitions: a survey of competitions, tools, and systems to support cybersecurity education. Educ. Inf. Technol. **28**(9), 11759-11791 (2023). https://doi.org/10.1007/s10639-022-11451-4
5. Workman, M.D., Anthony Luevanos, J., Mai, B.: A study of cybersecurity education using a present-test-practice-assess model. IEEE Trans. Educ. **65**(1), 40-45 (2022). https://doi.org/10.1109/TE.2021.3086025
6. McMahon, R.: Applied improvisation techniques to enhance IT majors' soft skills, communication skills, and teamwork skills. In: présenté à SIGITE 2021 - Proceedings of the 22nd Annual Conference on Information Technology Education, pp. 21-22 (2021).. https://doi.org/10.1145/3450329.3476872
7. Arduin, P.E., Costé, B. : Hack your university! Gamification of classroom sessions on the security of information systems. In : présenté à INFORSID 2022 - 40eme Actes du Congres on INFormatique des Organisations et Systemes d'Information et de Decision, pp. 125-140. [En ligne] (2022). Disponible sur: https://www.scopus.com/inward/record.uri?eid=2-s2.0-85183536961&partnerID=40&md5=81d05ac2f9fb3e82aed18a1be04b0dfb
8. Burns, T.J., Rios, S.C., Jordan, T.K., Gu, Q.T.: Underwood, analysis and exercises for engaging beginners in online CTF competitions for security education. In: présenté à ASE 2017 - 2017 USENIX Workshop on Advances in Security Education, co-located with USENIX Security 2017 (2017). [En ligne]. Disponible sur: https://www.scopus.com/inward/record.uri?eid=2-s2.0-85091901021&partnerID=40&md5=08db627cdbf960ea2182af58a63d2a64

9. Boutracheh, H., El Bouhaddioui, M., Moumen, A.: Current research priorities on fog harvesting as a clean water resource: a bibliometric approach. In: El Mansouri, B., Moumen, A., El Bouhaddioui, M., et Mejjad, N., (ed.), E3S Web Conf., EDP Sciences (2024). https://doi.org/10.1051/e3sconf/202448905002
10. Boutracheh, H., El Ansari, R., Mejjad, N., Moumen, A.: Application of bibliometrics as a data mining technique for research prioritization: a case study of desalination research. In: ACM Int. Conf. Proc. Ser., Association for Computing Machinery (2023). https://doi.org/10.1145/3607720.3607779
11. Sahid, A., Maleh, Y.: Enhancing cybersecurity education: integrating virtual cloud-based labs into the curriculum. In: Cybersecurity Management in Education Technologies: Risks and Countermeasures for Advancements in E-learning, pp. 150-170 (2023). https://doi.org/10.1201/9781003369042-8
12. Mennecozzi, G.M., Hageman, K., Panum, T.K., Turkmen, A., Mahmoud, R.V., Pedersen, J.M.: Bridging the gap: adapting a security education platform to a new audience. In: présenté à IEEE Global Engineering Education Conference, EDUCON, pp. 153-159 (2021). https://doi.org/10.1109/EDUCON46332.2021.9453985
13. Beltran, M., Calvo, M., Gonzalez, S.: Experiences using capture the flag competitions to introduce gamification in undergraduate computer security labs. In: présenté à Proceedings - 2018 International Conference on Computational Science and Computational Intelligence, CSCI 2018, pp. 574-579 (2018). https://doi.org/10.1109/CSCI46756.2018.00116
14. Barletta, V.S., Caivano, D., Caruso, F., Peretti, S., Rossano, V.: Cyber Security Education for Industry and Academia - CSE4IA. In: présenté à CEUR Workshop Proceedings (2023). [En ligne]. Disponible sur: https://www.scopus.com/inward/record.uri?eid=2-s2.0-

A Multi-Level Arabic Text Diacritization System

Ali Mijlad(✉) and Yacine El Younoussi

SIGL Laboratory, ENSATe, Abdelmalek Essaadi University, Tetouan, Morocco
{a.mijlad,yacine.elyounoussi}@uae.ac.ma

Abstract. This paper presents a multi-level Arabic diacritization system designed to restore diacritics for undiacritized Arabic text. This kind of systems is crucial for Arabic-related NLP tasks and aids learners and individuals with learning difficulties, such as dyslexia or visual impairments. Our system uses a two-level approach: a word-based level and a letter-based level, both employing an encoder-decoder model with a local predictive Luong attention mechanism.

The combined model demonstrates good performance with a 22.47% diacritic error rate, significantly surpassing single-level models while maintaining competitive performance despite using a smaller dataset compared to previous studies.

Keywords: Diacritization · Arabic · GRU · Encoder-Decoder Model · Luong Attention

1 Introduction

Arabic, spoken by over 400 million people and central to the religious practices of over 1 billion Muslims, exists in classical Arabic for religious texts and Modern Standard Arabic (MSA) for literature, education, and media. Additionally, diverse regional dialects are widely used in daily life and social media, varying significantly across regions.

Arabic script, written from right to left, features optional diacritic marks often omitted to enhance writing speed. Proficient speakers can usually infer pronunciation and meaning without these marks. However, diacritics are crucial in Arabic education, especially for novice learners and individuals with learning disabilities.

Moreover, Arabic diacritics hold significant importance in Natural Language Processing (NLP), enhancing various applications. Abuali et al. [1] highlighted their impact on Text-to-Speech (TTS) technology, improving screen readers for visually impaired students. Additionally, diacritics benefit information retrieval, sentiment analysis, and machine translation, leading to better performance in these systems. These examples underscore the necessity and utility of automatic diacritization systems.

In Arabic, vowels are divided into basic and long vowels. Long vowels, referred to as "حُرُوف العِلَّة" */Huruf Al-Illa/* ("weak letters") which are "*"و" /w/, "ي" /i:/*, and *"ا" /ʔ/*,, are always written and considered "weak letters" due to their change in form when derived from word stems. These letters also serve as consonants with basic vowel marks. Omittable vowels include short vowels, nunations (Tanwin), Shaddah (germination), and the Sukun, which marks the absence of vowels and syllable end, as illustrated in Table 1.

B.-E. Boudriki Semlali et al. (Eds.): DATA 2025, LNISO 81, pp. 17–27, 2025.
https://doi.org/10.1007/978-3-032-06725-8_2

Table 1. Arabic Diacritics: Names, Functions, and Phonetic Representations

Diacritic	Name	Category	Trans.	IPA	Position
بَ	Fathah	Short Vowel	a	/a/	Any
بُ	Dammah	Short Vowel	u	/u/	Any
بِ	Kasrah	Short Vowel	i	/i/	Any
بًا	Tanwin Fath	Nunation	F	/an/	End
بٌ	Tanwin Damm	Nunation	N	/un/	End
بٍ	Tanwin Kasr	Nunation	K	/in/	End
بّ	Shaddah	Gemination	~	/:/	Any
بْ	Sukun	Absence of vowel	o	∅	Any

Diacritics in Arabic serve two main purposes [2]: they eliminate lexical ambiguities, such as differentiating "كِتَاب" */KitaAb/* ("book") from "كُتَّاب" */Kutt~aAb/* ("old Quranic school"), and they indicate case endings that define a word's syntactic role, such as the nominative "كِتَابُ" */kitaAbu/* and accusative "كِتَابَ" */KitaAba/*.

This paper explores the application of a local predictive Luong attention mechanism within an encoder-decoder architecture for Arabic text diacritization. It includes an overview of Arabic language characteristics, a review of related work, and a detailed description of the proposed system. The study also discusses the dataset, evaluation metrics, results, and concludes with significant insights and emerging research directions.

2 Methods

2.1 Related Work

Alansary proposed a rule-based diacritization system [3] that restores core diacritics using named entity recognition (NER), syntactic parsing, and performs case ending restoration using linguistic resources and a morphological component. Trained on the International Corpus of Arabic[1] (300,000 words) and tested on the LDC Arabic Treebank (52,000 words), it achieved 15.7% Word Error Rate (WER).

In their study [4], Hadjir et al. proposed employing an HMM with Viterbi decoding, demonstrating its effectiveness for Arabic diacritic restoration. The dataset comprised 26 books from the Tashkeela corpus[2]. They achieved a word precision of 80%.

Metwally et al. [5] integrated an HMM to handle core diacritization and Part-of-Speech (POS) tagging, supplemented by a morphological analyzer to address out-of-vocabulary (OOV) terms. For syntactic diacritic prediction, they applied a Conditional Random Fields (CRF) model, along with POS tags and morphological features. The model, trained on 340,000 words from ATB3 dataset, yielded a 13.7% WER in evaluation.

[1] ICA (bibalex.org).

[2] https://sourceforge.net/projects/tashkeela/

Rashwan et al.'s system [6] tokenizes text and applies context memory to each token. Features and POS tags are extracted using two deep networks and input to a final network for classification. Trained on the LDC Arabic Treebank part 3 and customized corpora, the system achieves 88.4% accuracy in case-ending restoration and 97% accuracy in morphological diacritics.

Belinkov et al.'s system [7] employs character embeddings processed through an LSTM-based RNN that was trained on a 470,000-word dataset from the ATB, along with an 81,000-word development set. On an 80,000-word hold-out set, the model attained a diacritic error rate (DER) of 4.85%.

Abandah et al. [8] utilized a bidirectional LSTM network that was trained on three linguistic resources: Quranic texts, the Tashkeela corpus, and a 305,000-word portion of LDC ATB3 dataset. On the Tashkeela testing set, the model demonstrated strong performance (DER: 2.09%, WER: 5.82%), though ATB3 test set evaluation showed degradation (DER: 2.72%, WER: 9.07%).

Darwish et al. [9] used the Viterbi algorithm with morphological patterns, stems, transliteration, and NER. For case endings, they combined SVM ranking with morphological patterns, and linguistic rules. Trained on 9.7 million words, the system was tested on 70 WikiNews articles (18,300 words), achieving 12.76% WER and 3.54% DER.

MADAMIRA[3] is the baseline of the system proposed by Shahrour et al. [10]. Syntactic characteristics are retrieved by the MaltParser[4]. A J48 classifier retrieves morphological and syntactic abstractions. The study's performance metrics were evaluated using the combined partitions (1 through 3) of the Penn ATB, yielding a WER of 9.4 percent.

The diacritization pipeline proposed by Chennoufi et al. [11] integrates Alkhalil Morpho Sys[5] for comprehensive morphological analysis and diacritization, followed by rule-based transition removal. The system sequentially processes text through two optimized HMM layers: initial word-level disambiguation followed by character-level diacritic prediction. The system is trained on Tashkeela (63 million vowelized words), RDI[6] data (8.5 million vowelized words), and NEMLAR[7] data (500,000 diacritized words), achieving 6.28% WER and 1.98% DER.

Fashwan et al. [12] restore morphological diacritics using morphological processing, statistical methods, and rules, followed by syntax rules to restore case endings. The framework was tested on a 52,000-word subset of LDC ATB Part 3, demonstrating a DER of 4.11% and a WER of 14.78%.

Alqudah et al.'s system [13] inputs MADAMIRA outputs with higher confidence parameters to a bidirectional LSTM network. Quantitative analysis of the LDC ATB3-trained model (305k words) produced competitive error rates of 8.40% WER and 2.39% DER.

Abbad et al.'s system [14] begins with preprocessing, followed by a deep network and rule-based correction. Trigrams and bigrams are used to restore diacritics, while

[3] https://camel.abudhabi.nyu.edu/madamira/

[4] https://www.maltparser.org/

[5] https://sourceforge.net/projects/alkhalil/

[6] http://www.rdi-eg.com/RDI/TrainingData/

[7] http://catalog.elra.info/en-us/repository/browse/ELRA-W0042/

Levenshtein distance handles remaining undiacritized tokens. The model evaluation after training on ninety-seven books from the Tashkeela corpus demonstrated precise diacritic restoration capabilities, with error metrics measuring 4.00% DER and 12.08% WER.

2.2 The Proposed Model

This section presents our proposed system for addressing the Arabic text diacritization problem.

Preprocessing
In this work, we adopted preprocessing techniques similar to those used by Abbad et al. [14].

The preprocessing keeps only characters relevant to diacritization, replacing numbers with zeros and removing non-Arabic letters and diacritics. The sequence is modified by adding special boundary tokens: ‘ < start > ‘ and ‘ < end > ‘. The letter-based model maps letters to numerical labels and generates numerical output labels for diacritic marks. The word-based model follows a similar process, producing input sequences of undiacritized words and target sequences of their diacritized forms.

Structural composition and design of the proposed model
The introduced system operates through two distinct processing levels: a word-based level with an encoder and Luong-attention-based decoder [15], and a letter-based level that diacritizes remaining undiacritized sequences from the first level, also using an encoder and a Luong-attention-based decoder. The model architecture is illustrated in Fig. 1.

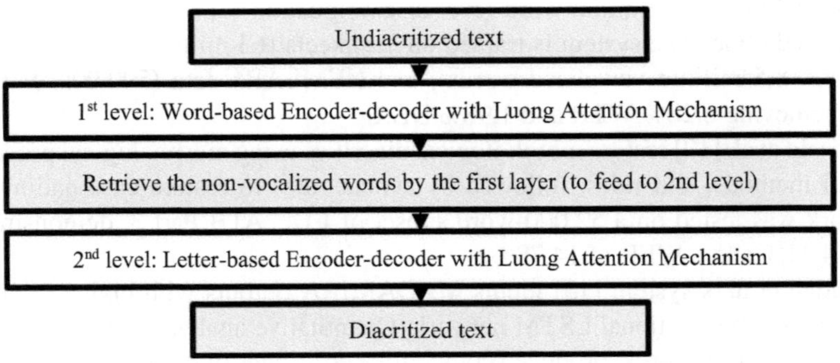

Fig. 1. Structural composition and design of the proposed model.

We selected this approach based on the hypothesis that it could effectively capture both broad syntactic context and fine-grained morphological details. The hierarchical structure, incorporating both word-based and letter-based components, aims to address the inherent complexity of Arabic diacritization. We chose the encoder-decoder architecture with attention mechanisms, well-known for its success in machine translation [15], due to its ability to manage long-range dependencies and contextual sensitivity.

Local-predictive Attention mechanism
Both the word-based and letter-based models utilize the local-predictive model introduced by Luong et al. [15]. This local attention mechanism focuses on a smaller subset of the input sequence to determine a target token, hence the term 'local'. It is described as 'predictive' because the alignment between input and target sequences is not assumed to be roughly monotonic; rather, it is predicted. Figure 2 demonstrates the operation of this local predictive attention mechanism.

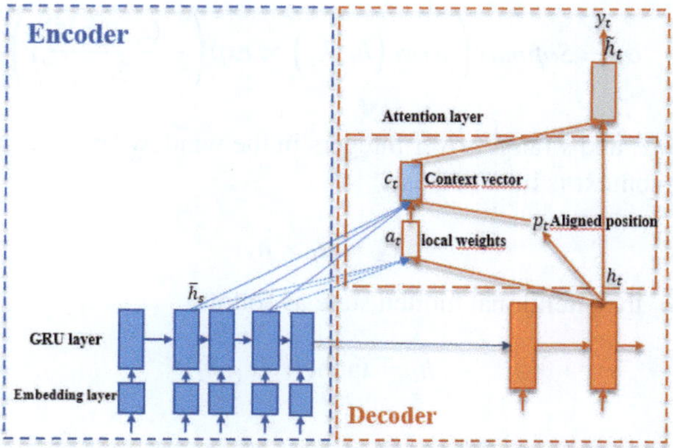

Fig. 2. Local attention mechanism (Luong et al.)

Encoder
As depicted in blue in Fig. 2, the encoding module employs a dual-component design: (1) an embedding layer for token representation, followed by (2) a GRU layer for sequential modeling. Each temporal iteration t involves the GRU receiving the embedded input x_t along with h_{t-1}, subsequently producing the updated hidden representation h_t through its gating mechanisms. When the encoder reaches the sequence end, we obtain the set $H = \{h_1, \ldots, h_{T_x}\}$, where T_x is the sequence length.

Decoder
For the decoding phase (orange illustration in Fig. 2), the following steps are performed iteratively:

1. Take the hidden state h_t.
2. Compute the single alignment position p_t, which aligns the current target position t with the positions of the source tokens. The predictive model does not assume the monotonic alignment of inputs and outputs. p_t is calculated as follows:

$$p_t = S_l.Sigmoid\left(V_p Tanh\left(W_p h_t\right)\right) \tag{1}$$

where V_p is a trainable vector, W_p a trainable matrix, and S_l is the input sentence length.

We then get the window $[p_t - D, p_t + D]$, which represents the source states $\overline{h}_s$ to be considered for each time step.

3. The alignment between decoder state h_t and encoder states $\overline{h}_s$ are computed via:

$$a\left(h_t, \overline{h}_s\right) = score\left(h_t, \overline{h}_s\right) = V_a Tanh\left(W_a\left[h_t, \overline{h}_s\right]\right) \quad (2)$$

Here, V_a and W_a represent learnable parameters of the attention mechanism, optimized during training.

4. The attention weights are computed using a Gaussian window centered at position p_t :

$$\alpha_t = Softmax\left(score\left(h_t, \overline{h}_s\right) \times exp\left(-\frac{(s - p_t)^2}{2\sigma^2}\right)\right) \quad (3)$$

where $\sigma = D/2$, and s ranges over integers in the window $[p_t - D, p_t + D$

5. The attention context is obtained as:

$$c_t = \alpha_t \times \overline{h}_s \quad (4)$$

6. Calculate $\tilde{h}_t$ the attentional hidden state as follows:

$$\tilde{h}_t = \tanh(W_c[c_t, h_t]) \quad (5)$$

2.3 Training Methodology and Challenges

The training phases of the word-based and letter-based models were done using the teacher-forcing technique [16]. This technique can make learning faster by inputting the correct output from the previous step instead of using the model's predictions. Consequently, this prevents errors from accumulating, as wrong predictions early in the sequence would otherwise affect later predictions. Training the model with the teacher forcing technique can help it learn a better mapping from input to output, and it can converge faster since the model has the correct context.

Developing an Arabic diacritization system involves several challenges. These include the complexity of diacritical marks, the need for advanced techniques like attention mechanisms to capture long-range dependencies, and the scarcity of high-quality diacritized corpora, leading to overfitting. Small datasets constrain the model's capability to capture diverse linguistic patterns. Moreover handling OOV words requires effective approaches. Finally, hardware limitations during training constrain the use of larger datasets, which potentially reduces the system's performance.

2.4 Dataset Overview and Evaluation Metrics

The experiments in this work utilize a standardized dataset from the Tashkeela corpus[8], following the preprocessing methodology outlined by Fadel et al.'s comparative study [17]. The data[9] was cleaned, processed, and split into three files: 90% for training and 10% for validation and testing.

Due to hardware constraints, a smaller dataset was created from the original files, with each input sequence limited to 100 characters. Table 2 shows token and character

Table 2. Descriptive Statistics for the Utilized Corpora

Feature	Training set	Validation set	test set
Character count	7,622,220	377,897	393,552
Token frequency	1,177,062	58,635	60,591
Numeric entries	2,301	123	138
Digit occurrences	8,978	480	507
Arabic word forms	1,015,627	50,387	52,179
Arabic letter units	4,015,335	198,913	20,7293
Diacritic marks	3,431,148	170,025	177,099
Non-diacritized tokens	75,615	12,747	13,440
Diacritized tokens	114,792	16,507	17,233
Sentence segments	121,531	6,156	6,417

Table 3. OOV Word Distribution: training vs. evaluation sets

	Validation set	Test set
Diacritized OOV Arabic words (%)	18.78	20.03
Undiacritized OOV Arabic words (%)	13.97	15.59

statistics, while Table 3 compares the occurrence rates of OOV terms between the training set and the evaluation sets (test and validation).

To evaluate the proposed system, DER and WER were used, as defined by Fadel et al. [17]. DER measures the fraction of letters with incorrect diacritics, while WER quantifies the fraction of words with one or more diacritic errors relative to all words processed.

Unlike some previous works [18], non-Arabic characters and punctuation are ignored when calculating these metrics in our study, ensuring more unbiased results. Additionally, it is optional to add the 'no-diacritic' attribute, which determines whether to consider the undiacritized letters from the original text. Also, we can optionally add case ending (CE) when counting those metrics.

3 Results

The proposed system's DER and WER were compared to the results of our previous letter-based Luong attention models [19]. The combination of letter-based and word-based local predictive attention outperformed the others, yielding a 22.47% DER and 42.04% WER. The word-based model with local predictive attention also performed

[8] https://sourceforge.net/projects/tashkeela/

[9] https://github.com/AliOsm/arabic-text-diacritization.

better, with a DER of 46.02% and a WER of 51.49%, surpassing the letter-based models using local monotonic or global attention mechanisms.

When comparing the letter-based and word-based models with the local-predictive attention mechanism, the letter-based model achieved a better DER of 26.80%, while the word-based model had a DER of 46.02%. For the WER, the word-based model performed better when case endings (CE) were counted, but when excluding case endings, the letter-based model had the lowest WER. This confirms the efficacy of word-level modeling for case marking identification. The combination of both models further supported this conclusion. Results are shown in Table 4, detailing the DERs and WERs, where the current models is presented in the last two lines.

Table 4. DER and WER of the previous models

Systems	DER (%)				WER (%)			
	'no-diacritic' option (included)		'no-diacritic' option (excluded)		'no-diacritic' option (included)		'no-diacritic' option (excluded)	
	With CE	Without CE	With CE	Without CE	With CE	Without CE	With CE	Without CE
Global attention(letter-level) [19]	43.90	41.26	49.27	46.11	73.06	62.92	72.29	62.38
Local-monotonic attention (letter-level) [19]	50.10	48.07	52.85	50.30	73.10	65.69	72.01	64.73
Local-predictive attention (letter-level) [19]	26.80	23.94	31.59	28.17	57.34	44.60	56.75	44.30
Local-predictive attention (word-level)	46.02	47.05	57.15	58.31	51.49	48.43	51.39	48.37
Local-predictive attention (letter + word level)	**22.47**	**20.66**	**26.02**	**23.53**	**42.04**	**32.27**	**41.38**	**32.07**

4 Discussion

Our proposed multi-level system demonstrates competitive performance despite being trained and evaluated on a curated subset of the dataset used by Fadel et al. [17]. Table 5 illustrates this comparison.

Our proposed solution outperforms several systems in Fadel et al.'s study [17], achieving a DER of 22.47% (with case ending) and 20.66% (without case ending), compared to Farasa (21.43% and 23.93%), MADAMIRA (34.38% and 29.94%), and

Tashkeela-Model (49.96% and 52.96%). Additionally, our system achieved a WER of 42.04% (with case ending) and 32.27% (without case ending), surpassing Farasa (58.88% and 53.13%), MADAMIRA (76.58% and 59.07%), and Tashkeela-Model (96.8% and 94.16%) as reported by Fadel et al. [17].

Although top systems like Shakkala and Mishkal achieve better results, our system's performance is promising given the dataset limitations. Our multi-level system's DER and WER results, based on a smaller dataset from Fadel et al.'s study [17], demonstrate its effectiveness. With training on a larger dataset and better computational resources, our system could improve significantly and better compete with top-performing systems.

Table 5. DER and WER results

Systems	DER (%)				WER (%)			
	'no-diacritic' option (included)		'no-diacritic' option (excluded)		'no-diacritic' option (included)		'no-diacritic' option (excluded)	
	With CE	Without CE	With CE	Without CE	With CE	Without CE	With CE	Without CE
Farasa	21.43	23.93	24.9	27.55	58.88	53.13	57.28	51.84
Harakat	18.37	17.03	20.64	18.55	41.83	32.03	38.33	28.29
MADAMIRA (Aramorph)	34.38	29.94	40.03	33.87	76.58	59.07	75.39	57.22
Mishkal	16.09	13.78	17.59	14.22	39.78	26.42	35.63	21.92
Tashkeela-Model	49.96	52.96	58.5	60.92	96.8	94.16	96.03	92.45
Shakkala	3.73	2.88	4.36	3.33	11.19	6.53	10.89	6.37
Our proposed solution	22.47	20.66	26.02	23.53	42.04	32.27	41.38	32.07

5 Conclusion

This paper introduces a multi-level Arabic diacritization system combining word-based and letter-based models with encoder-decoder networks and Luong attention. Despite hardware limitations (limiting dataset size and network complexity), the method achieves prominent results, with the word-based model excelling at case endings and the letter-based model reducing errors.

Future work should prioritize corpus expansion to enhance model generalization, alongside architectural refinements to the attention mechanism—particularly transitioning from Luong-style local attention to self-attention networks for improved long-range dependency modeling. Scaling computational resources and integrating transformer-based components would further advance contextual modeling capabilities.

References

1. Abuali, B., Kurdy, M.-B.: Full diacritization of the Arabic text to improve screen readers for the visually impaired. Adv. Human-Comput. Interact. (2022). https://doi.org/10.1155/2022/1186678
2. Hamed, O., Zesch, T.: A survey and comparative study of arabic diacritization tools. J. Lang. Technol. Comput. Linguist. **32**, 27–47 (2017)
3. Alansary, S.: Alserag: an automatic diacritization system for Arabic. In: Shaalan, K., Hassanien, A.E., Tolba, F. (eds.) Intelligent Natural Language Processing: Trends and Applications (Studies in Computational Intelligence, 740), pp. 523–543. Springer (2018). https://doi.org/10.1007/978-3-319-48308-5_18
4. Hadjir, I., Abbache, M., Belkredim, F.Z.: An approach for Arabic diacritization. In: Métais, E., Meziane, F., Vadera, S., Sugumaran, V., Saraee, M. (eds.) 24th International Conference on Applications of Natural Language to Information Systems, NLDB 2019, pp. 337–344. Springer, Salford (2019). https://doi.org/10.1007/978-3-030-23281-8_29
5. Metwally, A.S., Rashwan, M.A.: A multi-layered approach for Arabic text diacritization. In: 2016 IEEE International Conference on Cloud Computing and Big Data Analysis (ICCCBDA), pp. 389–393. IEEE, Chengdu (2016). https://doi.org/10.1109/ICCCBDA.2016.7529589
6. Rashwan, M., Sallab, A. Al, Raafat, H., Rafea, A.: Automatic Arabic diacritics restoration based on deep nets. In: Proceedings of the EMNLP 2014 Workshop on Arabic Natural Language Processing (ANLP), pp. 65–72. Association for Computational Linguistics, Doha (2014). https://doi.org/10.3115/v1/W14-3608
7. Belinkov, Y., Glass, J.: Arabic diacritization with recurrent neural networks. In: Proceedings of the 2015 Conference on Empirical Methods in Natural Language Processing, pp. 2281–2285. Association for Computational Linguistics, Lisbon (2015). https://doi.org/10.18653/v1/D15-1274
8. Abandah, G.A., Graves, A., Al-Shagoor, B., Arabiyat, A., Jamour, F., Al-Taee, M.: Automatic diacritization of Arabic text using recurrent neural networks. Int. J. Doc. Anal. Recognit. **18**, 183–197 (2015). https://doi.org/10.1007/s10032-015-0242-2
9. Darwish, K., Mubarak, H., Abdelali, A.: Arabic diacritization: stats, rules, and hacks. In: Proceedings of the Third Arabic Natural Language Processing Workshop, pp. 9–17. Association for Computational Linguistics, Valencia (2017). https://doi.org/10.18653/v1/W17-1302
10. Shahrour, A., Khalifa, S., Habash, N.: Improving Arabic diacritization through syntactic analysis. In: Proceedings of the 2015 Conference on Empirical Methods in Natural Language Processing, pp. 1309–1315. Association for Computational Linguistics, Lisbon (2015). https://doi.org/10.18653/v1/D15-1152
11. Chennoufi, A., Mazroui, A.: Morphological, syntactic and diacritics rules for automatic diacritization of Arabic sentences. J. King Saud Univ. - Comput. Inf. Sci. **29**, 156–163 (2017). https://doi.org/10.1016/j.jksuci.2016.06.004
12. Fashwan, A., Alansary, S.: SHAKKIL: an automatic diacritization system for modern standard Arabic texts. In: Proceedings of the Third Arabic Natural Language Processing Workshop, pp. 84–93. Association for Computational Linguistics, Valencia (2017). https://doi.org/10.18653/v1/W17-1311
13. Alqudah, S., Abandah, G., Arabiyat, A.: Investigating hybrid approaches for Arabic text diacritization with recurrent neural networks. In: 2017 IEEE Jordan Conference on Applied Electrical Engineering and Computing Technologies (AEECT), pp. 1–6. IEEE, Aqaba (2017). https://doi.org/10.1109/AEECT.2017.8257765

14. Abbad, H., Xiong, S.: Multi-components system for automatic arabic diacritization. In: Jose, J.M., Yilmaz, E., Magalhães, J., Castells, P., Ferro, N., Silva, M.J., and Martins, F. (eds.) 42nd European Conference on Information Retrieval, ECIR 2020, pp. 341–355. Springer, Lisbon (2020). https://doi.org/10.1007/978-3-030-45439-5_23
15. Luong, M.-T., Pham, H., Manning, C.D.: Effective approaches to attention-based neural machine translation. In: Proceedings of the 2015 Conference on Empirical Methods in Natural Language Processing, pp. 1412–1421. Association for Computational Linguistics, Lisbon (2015). https://doi.org/10.18653/v1/D15-1166
16. Lamb, A., Goyal, A., Zhang, Y., Zhang, S., Courville, A., Bengio, Y.: Professor forcing: a new algorithm for training recurrent networks. Adv. Neural Inf. Process. Syst. **29**. pp. 4608–4616. Barcelona (2016)
17. Fadel, A., Tuffaha, I., Al-Jawarneh, B., Al-Ayyoub, M.: Arabic text diacritization using deep neural networks. In: International conference on Computer Applications & Information Security, At Riyadh, Saudi Arabia, pp. 2–9 (2019)
18. Zitouni, I., Sarikaya, R.: Arabic diacritic restoration approach based on maximum entropy models. Comput. Speech Lang. **23**, 257–276 (2009). https://doi.org/10.1016/j.csl.2008.06.001
19. Mijlad, A., El Younoussi, Y.: The Global and local attention for automatic Arabic text diacritization. Int. J. Eng. Appl. Phys. **3**, 653–662 (2023)

Digitalization of City and Coastal Environment Using Open Source Software

Tihomir Dovramadjiev(✉)

Technical University of Varna, Varna, Bulgaria
tihomir.dovramadjiev@tu-varna.bg

Abstract. This article explores the digitalization of urban environments, focusing on the city of Varna, Bulgaria, including key regions such as the Center and Cathedral, the Technical University of Varna, and the Sea Port. By leveraging OpenStreetMap (OSM) data and the Blender Open Source Software (OSS) integrated with the Blosm add-on, the research demonstrates a practical workflow for generating detailed 2D and 3D digital representations of city and coastal regions. The results, presented as high-quality visualizations, provide valuable insights for stakeholders involved in urban planning, logistics, tourism, digital application development, and other related. The study showcases the seamless importation of city elements such as buildings, roads, paths, water bodies, and forests directly from OSM into Blender, offering a replicable approach for other cities included in the OSM database. This method supports creating digital applications, digital twins cities, touristic guides, and logistics solutions, emphasizing its utility for advancing urban digitalization in a cost-effective and accessible manner.

Keywords: Blender · Open Source Software (OSS) · Open Street Map (OSM) · Blosm · Add-on · 3D · 2D · cartography · digitalization

1 Introduction

Digitalization has become an essential aspect of modern urban development, supporting diverse fields such as urban planning, logistics, tourism, and digital applications [1–6]. This article focuses on the digitalization of key regions within the city of Varna [7], Bulgaria, including the Center and Cathedral [8], the Technical University of Varna [9], and the Sea Port [10], utilizing open-source tools and data. Specifically, the study employs OpenStreetMap (OSM) data [11] in interaction with Blender, a powerful open-source 3D creation software [12], enhanced by the Blosm add-on [13]. This combination offers a practical, cost-effective workflow for generating detailed 2D and 3D representations of city and coastal regions, enabling advanced cartographic applications and fostering innovation across various professional domains. This study highlights the integration of open and free resources like OSM, Blender, and Blosm to create cost-effective and high-quality digital city models. OpenStreetMap serves as the foundation, offering a rich database of geographic information, including roads, buildings, water bodies, and green areas, contributed and maintained by a global community. As an open-data platform,

B.-E. Boudriki Semlali et al. (Eds.): DATA 2025, LNISO 81, pp. 28–37, 2025.
https://doi.org/10.1007/978-3-032-06725-8_3

OSM empowers users to create and distribute geographic models with unrestricted access while maintaining local accuracy. Blender, as an open-source 3D creation tool, provides a comprehensive suite for modeling, rendering, and animation, enhanced by its ability to support Python scripting for specialized tasks [14]. The Blosm add-on bridges these resources by enabling the seamless import of OSM data into Blender [15], transforming raw geographic information into realistic 2D and 3D urban landscapes [16].

The importance of cartography extends beyond map-making into fields critical for modern urban and regional development. Accurate digital representations of urban spaces enable planners to design smarter cities, logisticians to optimize transportation networks, and tourism developers to create digital guides. By employing real-world OSM data, the models generated are highly reliable and contextually accurate, making them invaluable for stakeholders. The addition of 3D modeling enhances the visual and analytical depth, providing tools for cutting-edge applications such as augmented reality (AR), virtual reality (VR), digital twin cities, smart city simulations, logistics, tourism and disaster management systems, and other [17–20].

Beyond Varna, the used methodology in this research can be scaled to encompass entire regions. For example, rural landscapes, industrial zones, and historical sites can also be digitalized using the same tools and workflow, creating a comprehensive digital twin of larger geographic areas. Such an approach facilitates region-wide infrastructure planning, environmental monitoring, and the development of interconnected smart systems. Also by utilizing free and open-source software (OSS), the cost barriers associated with large-scale digitalization projects are significantly reduced, making them accessible to municipalities, research institutions, and small businesses.

The digitalization process included importing key elements such as buildings, roads, paths, water bodies, and vegetation directly from OSM into Blender via the Blosm add-on. The study's results include high-quality visualizations that are not only aesthetically detailed but also functionally rich, supporting practical use cases for urban planners, developers, and policymakers. These results also highlight the scalability of the workflow, as the same methodology can be extended to other cities and regions listed in the OSM database.

2 Materials and Methods

The materials and methods are optimised and centered on the digitalization of the city of Varna, Bulgaria, with a particular emphasis on its key urban regions, including the Center and Cathedral, the Technical University of Varna, and the Sea Port. The workflow integrates advanced digital tools, primarily the Blender open-source 3D software and its Blosm add-on, with OpenStreetMap (OSM) as the primary geographic data source. This combination enables the detailed extraction and transformation of spatial data into high-quality 2D and 3D digital representations. The methodology prioritizes technical precision, leveraging open-source resources to ensure cost-effectiveness, adaptability, and scalability for broader applications, including the potential digitalization of entire regions.

2.1 Materials

The materials integrate cutting-edge open-source tools and freely available geospatial datasets to achieve a robust and scalable approach to urban digitalization. OpenStreetMap (OSM) serves as the foundational data source, providing a rich repository of global geographic information created and maintained by a diverse community of contributors. This collaborative platform ensures regularly updated and accurate vector-based data, including elements such as roads, buildings, water bodies, railways, and vegetation. The community-driven model of OSM empowers users worldwide to validate and refine data using tools like GPS devices, aerial imagery, and on-the-ground surveys, ensuring high reliability. OSM provides essential geographic details required to develop comprehensive 2D and 3D digital representations of regions / cities.

Blender, the open-source 3D modeling and rendering software, functions as the primary tool for processing and visualizing the geospatial data extracted from OSM. Recognized for its powerful suite of capabilities, Blender supports advanced geometry manipulation, material rendering, animation, simulation, and video editing. Its versatility extends across the entire 3D creation pipeline, enabling users to handle complex tasks such as compositing, simulation, and motion tracking within a unified environment. The software's cross-platform compatibility ensures consistent performance on Linux, Windows, and macOS, further solidifying its accessibility. A key feature of Blender is its support for Python-based scripting, which allows the automation of repetitive tasks and the customization of workflows through add-ons. These capabilities adapt the software to handle geospatial data, transforming it into highly detailed and realistic digital models of urban areas. Blender's dynamic user interface, powered by OpenGL, provides a consistent experience across platforms, enhancing usability for tasks ranging from basic map creation to intricate 3D scene assembly.

To facilitate seamless integration between Blender and OpenStreetMap, the Blosm add-on functions as a critical component of the digitalization workflow. Designed specifically to bridge the gap between geospatial data and 3D modeling, Blosm enables the direct importation of OSM data into Blender. This add-on simplifies the extraction and translation of geographic features into Blender-compatible formats. One of Blosm's standout features is its ability to download and integrate real-world terrain data with a high resolution, ensuring that the 3D models accurately reflect the physical characteristics of the environment.

2.2 Methods

The method's process begins with the extraction of geographic data from OpenStreetMap. Specific geographic coordinates are used to focus on Varna's urban regions, and the Blosm add-on is employed to import OSM data into Blender. Blosm enables the integration of key geographic features into Blender-compatible formats, such as real-world terrain, building footprints, and transportation networks. These data are further design processed in Blender. The imported geospatial data undergoes terrain generation, where real-world elevation data serves as a base layer. Buildings are extruded based on their height attributes and details. The methodology is assessed for scalability by implementing the workflow on suburban areas of Varna, evaluating its potential for application to more regions or entire countries. This ensures that the approach remains robust, adaptable, and suitable for diverse urban environments.

The automated digitaliization methodology includes: 1. Export (https://www.openstreetmap.org/export) data from Open Street Map (OSM) by coordinates or display region (Manually select a different area). The file extension is *.osm, which is imported into Blender OSS, a version that works perfectly with the Blosm add-on (2). On this stage the designer should to select the import data method: via Server or downloaded file from OSM. 3 Digitalization process (2D and 3D) generates import parameters as: building, water objects, forests, other vegetations, roads and paths, railways. Other sub-parameters are visible in the Blosm add-on panel during the digitizing process. The generated result is fully compatible with Blender software (Fig. 1).

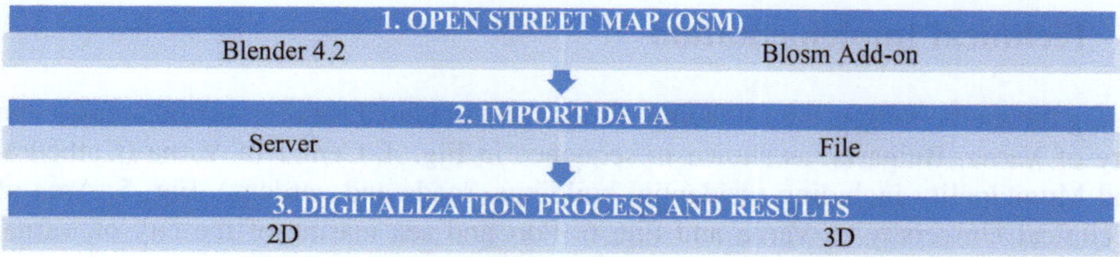

Fig. 1. Automated digitalization methodology

To utilize the methodology, it is first necessary to synchronize the Blender software with the specialized Blosm add-on. Figure 2 (a) illustrates the installation of the Blosm add-on in Blender, with the sequence including: starting the Blender software> Edit> Preferences> Install from disk (up/right corner arrow)> (Fig. 2 (b)) Directory to store downloaded OpenStreetMap and terrain files.

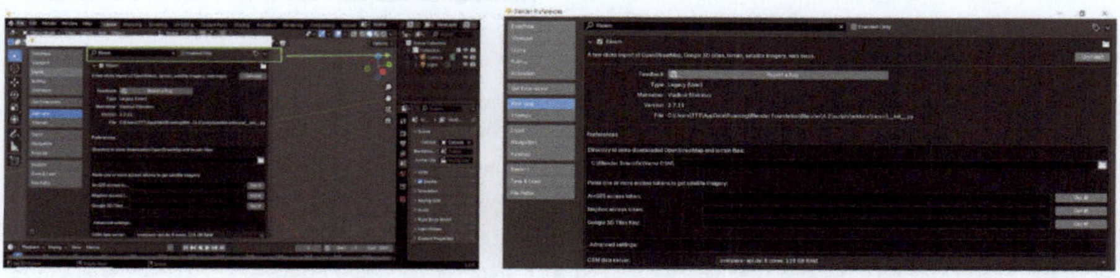

(a) (b)

Fig. 2. Synchronize the Blender software ad Blosm (a) Install the add-on; (b) Store digitalization result files

Activation and interface position of the Blosm add-on in Blender software is shown on Fig. 3(a) and the main settings of importing data from OSM, respectively server or file (Fig. 3 (b)).

(a) (b)

Fig. 3. Activation of Blosm add-on in Blender software (a) Interface and position; (b) Settings: Import from server or file.

3 Technical Implementation

Using the methodology, the following results were obtained for digitaliized areas of the city of Varna, Bulgaria, as shown in sequence in Fig. 4. Center of Varna (Cathedral and Municipality including residential buildings, roads and gardens), Fig. 5. Area of Technical University of Varna and Fig. 6. Port and sea marine of the city of Varna, including the environment such as buildings, coastline and others).,

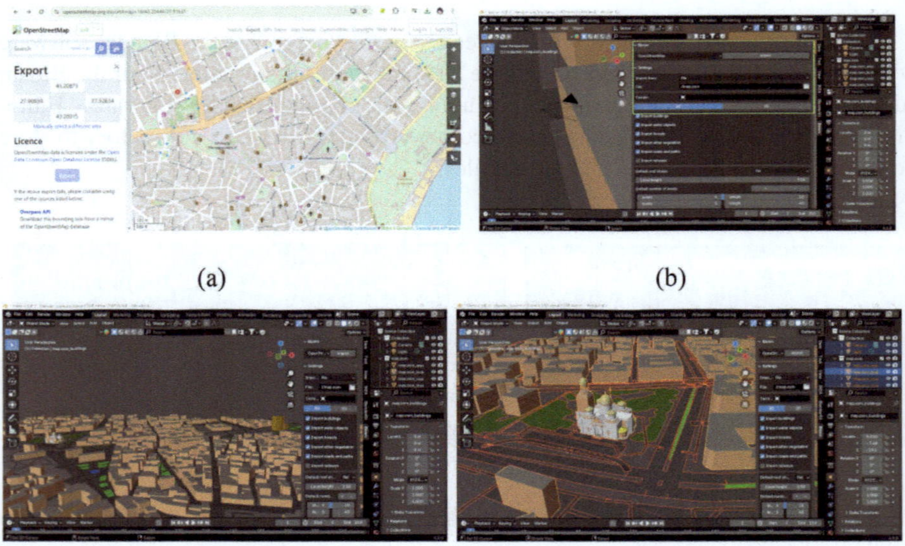

(a) (b) (c) (d)

Fig. 4. Center of Varna city (Bulgaria) (a) OSM selected region of Varna center; (b) OSM imported data from file; (c) the Varna Cathedral and Municipality area; (d) the Varna Cathedral – 3D focus view

(a) (b)

Fig. 5. Digitalized region of Technical University of Varna, Bulgaria (a) 2D view; (b) 3D view

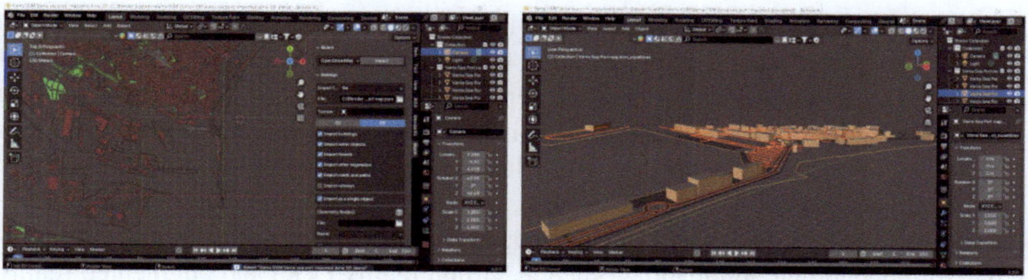

(a) (b)

Fig. 6. Digitalized region of Port Varna and coastal environment (a) 2D view; (b) 3D view

4 Results

The results of this study shows the power and versatility of the open-source ecosystem comprising Blender, the Blosm add-on, and OpenStreetMap (OSM). The successful digitalization of key areas within the city of Varna, Bulgaria, showcases the efficacy of this methodology in generating accurate, high-quality 2D and 3D visualizations. The methodology was applied to three representative regions: the city center, the Technical University of Varna, and the Varna Port and coastal environment. Each case highlights unique urban features and underscores the potential applications of this workflow in various domains, including urban planning, tourism, logistics, and digital cartography.

4.1 Digitalization of Varna City Center

The Varna city center was chosen as the first test case due to its historical and cultural significance, as well as its complex urban structure. The selected area included landmarks such as the Varna Cathedral, the Municipality building, residential neighborhoods, roads, and gardens.

- **Initial Data Preparation:** Using the OSM database, a specific region of the city center was selected, and the required geographic data were imported into Blender via the Blosm add-on. Figure 4(a) illustrates the region of interest within OSM, while Fig. 4(b) presents the successfully imported data.

- **2D and 3D Visualizations:** The imported OSM data were transformed into detailed 2D and 3D visualizations. The 2D map provided an overhead view of the urban layout, including road networks and green spaces. The 3D visualization offered a more immersive perspective, with the Varna Cathedral rendered in high detail. This 3D rendering (Fig. 4(c) and Fig. 4(d)) effectively captures the architectural features of the Cathedral and its surroundings, making it a valuable asset for urban planners, historians, and tourists.

4.2 Digitalization of the Technical University of Varna

The second digitized model focused on the Technical University of Varna, Bulgaria, which is a high modern center for education and scientific research and innovations.

- **Import and Visualization:** As depicted in Fig. 5(a), the 2D visualization provided a clear overview of the university campus, including its buildings, roads, and adjacent spaces. The 3D view (Fig. 5(b)) added depth and realism to the visualization, making it easier to analyze the spatial relationships between various campus elements. This digital representation can serve as a foundation for virtual campus tours, infrastructure planning, and other.
- **Applications in Education:** The digitalized campus model opens up new possibilities for educational applications, such as virtual reality (VR) and augmented reality (AR)-enabled learning environments, interactive simulations, and augmented reality (AR)-based campus navigation tools. By leveraging these visualizations, educational disciplines enhance the learning experience for students and foster interdisciplinary collaboration across engineering, technology, advanced and eco-design, architecture, urban planning, and other fields. This includes learning state-of-the-art software and additive technologies.

4.3 Digitalization of Varna Port and Coastal Environment

The Varna Port and its surrounding coastal environment were selected as the third model due to their strategic importance for commerce, tourism, and environmental management. This area presented unique challenges, including the need to accurately represent both man-made and natural features.

- **Integration of Coastal Features**: The 2D visualization (Fig. 6(a)) provided a detailed map of the port area, highlighting key elements such as main marine buildings, yacht port, shipping lanes and other. The 3D visualization (Fig. 6(b)) captured the intricate details of the coastline, including its curvature, and surrounding objects. This level of detail is critical for applications such as port management, environmental monitoring, and disaster preparedness.
- **Enhanced Stakeholder Collaboration:** The high-quality 3D visualizations can facilitate better communication and collaboration among stakeholders, including port authorities, environmental agencies, and local businesses. These models are appropriate to simulate the impact of proposed infrastructure projects, assess the environmental effects of port activities, or plan for sustainable development initiatives.open-source

5 Discussion

The integration of advanced OSS software represents a transformative approach to the digitalization of urban spaces, enabling the creation of high-quality 2D and 3D visualizations of real-world regions. This open-source systems offers cost-effective solution for stakeholders in various sectors, including urban planning, logistics, tourism, and environmental monitoring. By leveraging this technology, stakeholders can generate detailed digital models of cities, campuses, ports, and coastal environments, which can then be used for diverse applications such as virtual tourism, infrastructure planning, environmental impact assessments, and more. The process involves importing and transforming geographic data into immersive 2D and 3D visualizations that provide valuable insights into the spatial and functional dynamics of a region. The positive share of experiences gained from the digitalization of key areas in Varna, Bulgaria, through these open-source tools underscores the importance of knowledge exchange between stakeholders to foster innovation and growth in the field of digital twin technology. By sharing this article and results, the potential for collaboration is amplified, encouraging the adoption of open-source solutions that enable cities and regions to create digital twins, which can help not only in efficient planning and management but also in offering remote access to otherwise hard-to-reach locations. The ability to observe and analyze distant regions, regardless of geographical boundaries, promotes a deeper understanding of various urban environments and opens up opportunities for international cooperation. This exchange of knowledge and expertise can contribute significantly to the development of digital twins, supporting smarter cities, better logistics, and more sustainable tourism practices, while also facilitating data-driven decisions that benefit both local and global stakeholders. As such, the ongoing growth of open-source applications for digital region modeling holds immense potential for shaping the future of urban development, regional management, and global collaboration, advancing a more connected and informed world.

6 Conslusion

The integration of OpenStreetMap, Blender, and the Blosm add-on presents a robust, open-source digital system for urban digitalization, as demonstrated through the case study of Varna, Bulgaria. This methodology successfully combines cost-effectiveness, accessibility, and scalability, making it adaptable to other cities and regions within the OSM database. The workflow efficiently generates high-quality 2D and 3D visualizations, proving its value across multiple domains, including urban planning, tourism, logistics, and digital cartography. Beyond its immediate applications, this approach underscores the transformative role of advanced technologies in open geospatial analysis and smart city development.

The presented synergy between Open Source resources highlights how combining state-of-the-art technologies can revolutionize urban digitalization. The research shows these integrated tools create a powerful foundation for smart city development, with future potential expanding far beyond current applications. By incorporating AI-driven data validation, cloud computing, and IoT networks, this framework can evolve from static models to dynamic, real-time digital twins capable of predictive urban simulations. Such technological convergence unlocks groundbreaking possibilities like self-optimizing traffic systems, energy-neutral districts, AR-enhanced planning interfaces, and climate-resilient city modeling. The open-source nature of this ecosystem makes these advanced capabilities accessible while fostering continuous innovation. This approach not only addresses current urban challenges but also pioneers cognitively responsive city management systems that can adapt to future needs. The strategic combination of geospatial data, 3D modeling, and smart technologies demonstrates how interdisciplinary technological integration can transform urban planning from a reactive to a predictive science, paving the way for truly intelligent, sustainable cities of tomorrow. This experience provides a practical solution for today's urban challenges and also lays the groundwork for sustainable, inclusive, and data-driven city management worldwide. As open-source tools evolve, their potential to reshape urban landscapes, bridging gaps between planning, technology, and community engagement will only grow, paving the way for smarter, more resilient cities.

This workflow highlights the transformative potential of open-source technologies in advancing geospatial analysis and urban digitalization, offering a sustainable and inclusive path forward for researchers, developers, and stakeholders.

Acknowledgements. Bulgarian Association of Ergonomics and Human Factors (BAEHF)

The study was conducted with the support of CIII-HU-1506–01-2021 Ergonomics and Human Factors Regional Educational CEEPUS Network.

References

1. Benedetti, A.C., Costantino, C., Gulli, R., Predari, G.: The process of digitalization of the urban environment for the development of sustainable and circular cities: a case study of Bologna Italy. Sustainability **14**, 13740 (2022). https://doi.org/10.3390/su142113740
2. Li, L., Song, T.: Enabling in-situ urbanization through digitalization. Land **12**, 1738 (2023). https://doi.org/10.3390/land12091738
3. Bolshakov, N., Celani, A., Badenko, V., Benedicto, R.M.: Factories of the future in digitization of industrial urban areas. Urban Sci. **8**, 66 (2024). https://doi.org/10.3390/urbansci8020066
4. Mazzetto, S.: A review of urban digital twins integration, challenges, and future directions in smart city development. Sustainability **16**, 8337 (2024). https://doi.org/10.3390/su16198337
5. Caprari, G., Castelli, G., Montuori, M., Camardelli, M., Malvezzi, R.: Digital twin for urban planning in the green deal era: a state of the art and future perspectives. Sustainability **14**, 6263 (2022). https://doi.org/10.3390/su14106263
6. García-Aranda, C., Martínez-Cuevas, S., Torres, Y., Pedrote Sanz, M.: A digital twin of a university campus from an urban sustainability approach: case study in Madrid (Spain). Urban Sci. **8**, 167 (2024). https://doi.org/10.3390/urbansci8040167
7. Official Varna city (Bulgaria) website: https://www.varna.bg/en/2. Accessed 14 Jan 2025
8. The Varna Cathedral: https://visit.varna.bg/en/church_dormission.html. Accessed 14 Jan 2025
9. Technical University of Varna, Bulgaria: https://www1.tu-varna.bg/tu-varna/. Accessed 14 Jan 2025
10. Sea Port of Varna: https://port-varna.bg/en/Nachalo/Home https://www1.tu-varna.bg/tu-varna/. Accessed 14 Jan 2025

11. Open Street Map: https://www.openstreetmap.org/. Accessed 14 Jan 2025
12. Blender software: https://www.blender.org/. Accessed 14 Jan 2025
13. Blosm addon for Blender: OpenStreetMap, Google 3D cities, terrain. https://prochitecture.gumroad.com/l/blender-osm. Accessed 14 Jan 2025
14. Blender software features: https://www.blender.org/features/. Accessed 14 Jan 2025
15. Blosm addon for Blender: https://github.com/vvoovv/blosm?tab=readme-ov-file. Accessed 14 Jan 2025
16. Zhang, C., Ping, X., Fan, Q., Li, C.: Measurement of 2D and 3D fractal features of urban morphology from an architectural view and its influencing factors. Fractal Fract. **8**, 138 (2024). https://doi.org/10.3390/fractalfract8030138
17. Costa, D.G., Bittencourt, J.C.N., Oliveira, F., Peixoto, J.P.J., Jesus, T.C.: Achieving sustainable smart cities through geospatial data-driven approaches. Sustainability **16**, 640 (2024). https://doi.org/10.3390/su16020640
18. Florido-Benítez, L.: The use of digital twins to address smart tourist destinations' future challenges. Platforms **2**, 234–254 (2024). https://doi.org/10.3390/platforms2040016
19. Jamei, E., Mortimer, M., Seyedmahmoudian, M., Horan, B., Stojcevski, A.: Investigating the role of virtual reality in planning for sustainable smart cities. Sustainability (2017). https://doi.org/10.3390/su9112006
20. Lifelo, Z., Ding, J., Ning, H., Qurat-Ul-Ain., Dhelim, S.: Artificial intelligence-enabled metaverse for sustainable smart cities: technologies, applications, challenges, and future directions. Electronics, **13**, 4874 (2024). https://doi.org/10.3390/electronics13244874

Does Multi-dataset Combination Impact Machine Learning Performance? Emotion Recognition Use Case

Ezzahoud Hajar[1(✉)], Ameksa Mohammed[2], Amzil Asmaa[1], and Amizmiz Habibatou-Allah[1]

[1] Faculty of Sciences Semlalia, Computer Science Department, Cadi Ayyad University UCA, Marrakech, Morocco
{h.ezzahoud3077,a.amzil3429,h.amizmiz1698}@uca.ac.ma

[2] Faculty of Science Semlalia, FSSM, Laboratory of Computer Science and Smart Systems, LISI, Cadi Ayyad University UCA, Marrakech, Morocco
m.ameksa@uca.ac.ma

Abstract. Speech emotion recognition (SER) struggles with dataset diversity, model generalizability, and efficiency. Using a single dataset risks bias and limits applicability. This study combines four datasets (RAVDESS, TESS, SAVEE, CREMA-D) into a unified dataset to improve analysis and generalizability. An interpretable machine learning framework was developed, using data augmentation (e.g., noise injection, time stretching) and acoustic features (e.g., MFCC, ZCR, Chroma) to detect six emotions: anger, happiness, fear, sadness, neutral, disgust. Algorithms like SVM, LightGBM, Random Forests, XGBoost, and GBM were tested. SVM achieved 92% accuracy, followed by LightGBM at 91%. The study shows traditional machine learning can excel in SER, with dataset integration enhancing model performance and reliability.

Keywords: Dataset Diversity · Acoustic Features · Model Generalizability · Machine Learning Algorithms

1 Introduction

Speech Emotion Recognition (SER) is vital for advancing human-computer interaction, affective computing, and AI applications [1]. By detecting emotions from speech, SER enhances domains like mental health monitoring, customer service automation, and natural human-machine communication [2]. Despite progress, developing robust, deployable SER systems remains challenging [3].

Deep learning approaches, such as Convolutional Neural Networks (CNNs) and Recurrent Neural Networks (RNNs), achieve high accuracy in controlled environments but require substantial computational resources, large annotated datasets, and often lack interpretability [3,4]. These limitations hinder their use in resource-constrained real-world scenarios [5]. Additionally, many SER studies

B.-E. Boudriki Semlali et al. (Eds.): DATA 2025, LNISO 81, pp. 38–49, 2025.
https://doi.org/10.1007/978-3-032-06725-8_4

rely on single datasets that do not capture the diversity of emotional expressions across demographics, languages, and recording conditions [6]. This leads to models that perform well in specific settings but struggle with generalizability in varied, noisy environments [7].

To address these issues, this study introduces a novel SER approach by integrating four diverse emotional speech datasets—RAVDESS, TESS, SAVEE, and CREMA-D—to create a comprehensive training foundation [8]. We also utilize interpretable and computationally efficient machine learning algorithms to develop models that balance accuracy, robustness, and practicality [9]. This research tackles three key challenges: dataset diversity, model interpretability, and computational efficiency, providing a practical framework for SER systems and valuable insights for bridging the gap between experimental success and real-world deployment.

2 Related Work

Recent studies on SER show varied success with machine learning methods. Iqbal et al. [10] used Gradient Boosting for real-time SER, outperforming SVM and KNN in live scenarios on RAVDESS and Savee. Mittal et al. [11] tested multiple classifiers, with 1-D CNN achieving 48.96% accuracy on RAVDESS, surpassing Random Forests (43.6%), SVM (40%), and others using MFCC features. Ira and Rahman [12] combined Random Forest, Gradient Boosting, and Histogram Gradient Boosting, reaching 70% accuracy on RAVDESS. Chen et al. [13] reported 81.75% and 77.94% accuracies on CASIA [14] and Berlin EmoDB using a two-layer fuzzy random forest system. Shaila et al. [15] found Random

Table 1. Summary of the datasets, algorithms, and their respective accuracies from various references.

Ref.	Databases	Classification Algorithms	Accuracy
[10]	Ravdess, Tess	SVM, GB, KNN	SVM>GB> KNN
[11]	Ravdess, Tess, Savee, and Crema-d	SVM, LSTM, RF, Logistic Regression, 1D CNN, MLP	40%,26.7%,43.6%, 27.3%,48.96%, 33.3%
[12]	RAVDESS	RF, GB, HGB, MLP, Adaboost, SVM, RF, GB, HGB	70%, 53%, 32%, 54%, 58%, 56%, 59%
[13]	CASIA, EMO-DB	Two layer fuzzy multiple RF system	81%, 77.94%
[15]	Ravdess	SVM, DT, RF, MLP, CNN	78.57%, 78.56%, 85.71%, 81.82%, 82.89%
[16]	RAVDESS	CNN, MLP, SVM	86%, 84%, 82%

Forest (85.71%) and CNN (82.98%) outperformed SVM (78.57%) on RAVDESS. Jaykar et al. [16] achieved 86% (CNN), 84% (MLP), and 82% (SVM) accuracies on RAVDESS for eight emotions (Table 1).

Despite their promising results, Most studies on Speech Emotion Recognition (SER) use datasets like RAVDESS and SAVEE, which lack demographic and emotional diversity. CREMA-D, with its diverse actors across ages, ethnicities, and races, offers richer emotional expressions, enabling more inclusive SER systems. Using CREMA-D can improve model robustness and generalizability by capturing varied emotional nuances. However, low accuracy remains a challenge, emphasizing the need for advanced models and optimized dataset combinations to enhance SER performance and inclusivity.

3 Methodology

3.1 System Overview and Workflow

This study implements a comprehensive speech emotion recognition system through a multi-phase approach designed to maximize accuracy and efficiency. Figure 1 illustrates our systematic workflow, which consists of four main phases: data processing, feature extraction, model training, and evaluation. The workflow begins with data acquisition and preprocessing, progresses through feature extraction and model development, and concludes with comprehensive performance evaluation. This structured approach ensures reproducibility and maintains consistency throughout the experimental process.

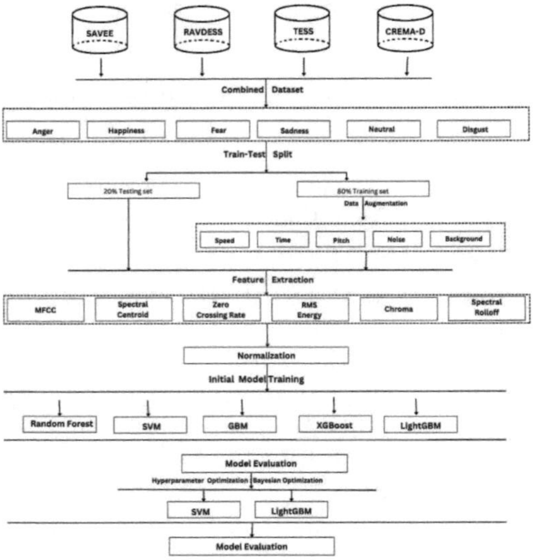

Fig. 1. Speech Emotion Recognition System Workflow - Illustrating the complete pipeline from data acquisition to model evaluation.

3.2 Data

Four emotional speech datasets were selected for this study, each offering unique characteristics in speaker diversity, recording conditions, and emotional expressions. RAVDESS [17] provides multimodal recordings in American English, serving as a foundational component due to its balanced gender representation and high-quality recordings. TESS [18] offers a controlled recording environment and professional actor performances. SAVEE [19] adds linguistic diversity through its British English recordings, while CREMA-D [20] represents the largest and most diverse collection, significantly expanding the demographic representation. Three combined datasets were created: the first integrates RAVDESS and TESS; the second includes RAVDESS, TESS, and SAVEE; and the third combines RAVDESS, TESS, SAVEE, and CREMA-D. Each combination aligns on the six common emotions—anger, happiness, fear, sadness, neutral, and disgust—resulting in progressively richer and more diverse datasets, though an imbalance with more female samples emerges due to the higher number of female recordings in TESS and CREMA-D (Table 2).

Table 2. Details of the datasets used in the study.

Dataset	Sample Rate (Hz)	Actors (Gender)	Total Number of Audios	Audios per Gender
RAVDESS	48,000	24 (12 males, 12 females)	1,440	720 males, 720 females
TESS	16,000	2 (females only)	2,800	0 males, 2,800 females
SAVEE	16,000	4 (males only)	480	480 males, 0 females
CREMA-D	16,000	91 (48 males, 43 females)	7,442	3,635 males, 3,807 females

To enhance train dataset diversity and model generalization, the dataset was split into 80% training and 20% testing [21]. Randomly applied augmentation techniques included Noise Addition to improve resilience to background noise [22], Time Stretching to simulate speech tempo variations [23], Pitch Shifting to mimic diverse voice tones [24], Speed Variation to enrich dataset diversity [23], and Background Noise Addition to emulate real-world environments [22]. These transformations, applied uniquely per epoch, increased input variability, prevented overfitting, and improved robustness to gender imbalances and real-world disturbances.

For emotion recognition, diverse acoustic features were extracted from audio samples to capture emotional content comprehensively. Notably, *twenty Mel Frequency Cepstral Coefficients (MFCCs)* were extracted to effectively summarize spectral variations and capture essential voice features [25]. Other features included *Zero-Crossing Rate (ZCR)* for detecting abrupt emotional changes [26], *Chroma* features for tonal and harmonic variations [26], *Pitch* changes for emotions like joy or surprise [27], *Spectral Centroid* for sound brightness [28], *Root Mean Square (RMS) Energy* for emotional intensity [29], and *Spectral Rolloff* to distinguish harmonic content [30], ensuring a robust representation of emotional cues.

The extracted features underwent normalization using StandardScaler to ensure a uniform scale distribution. This step standardizes the features by applying Z-score normalization, which transforms each feature to have a mean of zero and a standard deviation of one. The use of Z-score normalization is particularly important as it ensures that all features are on the same scale, preventing features with larger ranges from dominating the model and improving the performance of algorithms like RF, SVM, LightGBM, GBM, and ExtremeGBM [31].

3.3 Models

This study evaluated several machine learning algorithms for speech emotion recognition. Support Vector Machines (SVM) [32] classify data using an optimal hyperplane in high-dimensional spaces. Random Forest (RF) [33] uses ensemble decision trees with majority voting. Gradient Boosting Machines (GBM) [34] build models sequentially using gradient descent. XGBoost [35] improves gradient boosting with regularization and efficient sparse data handling. LightGBM [36] optimizes gradient boosting for large-scale, high-dimensional datasets.

Model optimization involved fine-tuning several key parameters after selecting the most performant datasets and algorithms. The parameters adjusted included the kernel type to determine the function used in the algorithm; the regularization constant to control the trade-off between achieving low training error and low testing error; gamma to define the influence range of a single training example; the learning rate to regulate training speed; the number of estimators to set the number of trees in the ensemble-based model; the maximum depth to limit tree depth and prevent overfitting; and the number of leaves to control tree complexity. Hyperparameter tuning was performed using Bayesian optimization [37] to achieve optimal performance while preventing overfitting and maintaining computational efficiency. This systematic approach ensures the identification of the best hyperparameter combination by exploring a predefined parameter space.

4 Experiments and Results

4.1 Experiments

In this experimental study, multiple emotional speech datasets (RAVDESS, TESS, SAVEE, and CREMA-D) were systematically integrated to establish a comprehensive framework for speech emotion recognition, focusing on six emotions (neutral, fear, happy, angry, disgust, and sad) that were common across all datasets to ensure consistent cross-dataset comparison. Three progressive dataset combinations (RAVDESS+TESS), (RAVDESS+TESS+SAVEE) and (RAVDESS+TESS+SAVEE+CREMA-D) were investigated to evaluate the impact of increased data diversity on model performance. The acoustic features were extracted using established signal processing techniques, including Mel-frequency cepstral coefficients (MFCCs), spectral contrast, chroma features, zero-crossing rate, and spectral roll-off, to capture the relevant emotional characteristics in the speech signals. To address dataset imbalances and enhance model generalization, various data augmentation techniques were implemented, including additive noise, time stretching, pitch shifting, speed variation, and background noise injection, ensuring balanced representation across gender and emotional categories in the training set. The feature vectors were standardized using StandardScaler and the dataset was partitioned into training and testing sets.

For the classification task, a diverse set of machine learning algorithms was employed, specifically Random Forest, Support Vector Machines (SVM), Gradient Boosting, XGBoost, and LightGBM. These algorithms were chosen for their demonstrated efficacy in handling complex, high-dimensional classification problems, robustness against overfitting, and ability to provide interpretable results, making them well suited for speech emotion recognition tasks. Hyperparameter optimization was conducted exclusively for the best-performing dataset combination and algorithms, identified through initial evaluations. Using Bayesian optimization, critical parameters were fine-tuned to maximize performance. For instance, in SVM, parameters such as the kernel type, regularization constant (C), and gamma were optimized, while for LightGBM, parameters such as learning rate, number of estimators, maximum depth, and the number of leaves were adjusted. This targeted optimization process ensured the most effective configuration for the top-performing models, balancing computational efficiency and classification accuracy.

4.2 Results

The performance of the models was evaluated using several metrics, including accuracy, F1-score, precision, and recall.

These metrics provided a comprehensive view of the models' strengths and weaknesses. The results are summarized in the following Table 3:

Table 3. Performance metrics for different dataset combinations.

Datasets	Model	Precision	Recall	F1-score	Accuracy
RAVDESS and TESS combination	Random Forest	0.8920	0.8907	0.8910	0.8902
	SVM	0.9058	0.9038	0.9043	0.9046
	Gradient Boosting	0.8809	0.8800	0.8800	0.8801
	XGBoost	0.8928	0.8921	0.8923	0.8916
	LightGBM	0.9075	0.9062	0.9067	0.9061
RAVDESS, TESS, and SAVEE combination	Random Forest	0.8769	0.8767	0.8766	0.8763
	SVM	0.8852	0.8841	0.8844	0.8840
	Gradient Boosting	0.8389	0.8380	0.8381	0.8376
	XGBoost	0.8897	0.8894	0.8894	0.8892
	LightGBM	0.8921	0.8908	0.8910	0.8905
RAVDESS, TESS, SAVEE, and CREMA-D combination	Random Forest	0.6151	0.6073	0.6021	0.6073
	SVM	0.6506	0.6482	0.6465	0.6480
	Gradient Boosting	0.5936	0.5949	0.5916	0.5945
	XGBoost	0.6361	0.6342	0.6327	0.6343
	LightGBM	0.6273	0.6269	0.6254	0.6272

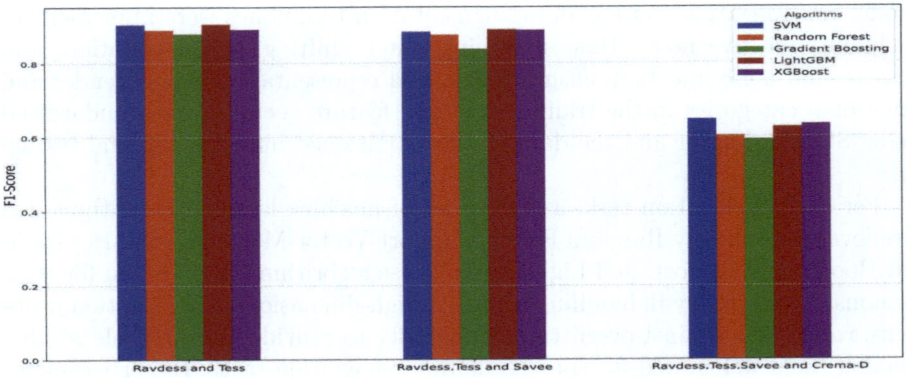

Fig. 2. Average F1 score measure for classification models (across different datasets).

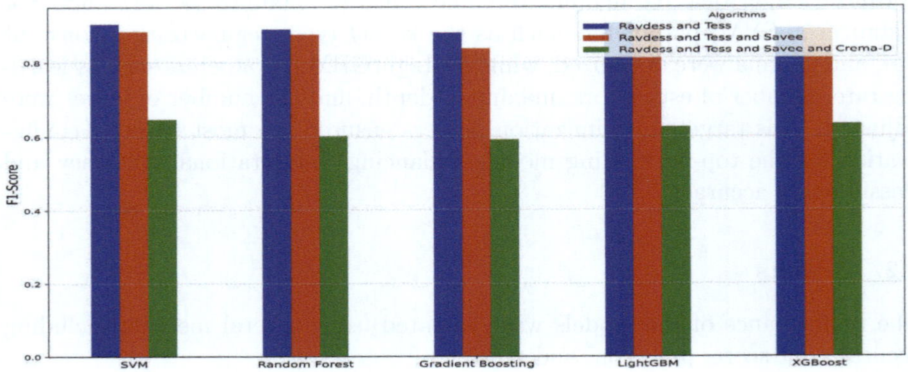

Fig. 3. Average F1 score measure for classification models (across different ML models).

To provide a visual comparison of the performance of different algorithms across various dataset combinations, the F1-scores were plotted. The following figure illustrates the F1-scores for each algorithm across the three dataset combinations (Fig. 2):

To gain a broader understanding of each algorithm's performance, we aggregated the results across all datasets. This aggregated analysis allows us to compare the overall effectiveness of each algorithm irrespective of the specific dataset. The following plots illustrate the performance of each algorithm across all datasets (Fig. 3):

Among the various algorithms tested, LightGBM and SVM demonstrated superior performance across all dataset combinations. Hyperparameter tuning was applied to these models in the best combination, RAVDESS and TESS, to further optimize their performance. For LightGBM, the best performance was achieved with the default parameters.

SVM: The best hyperparameters for SVM were found to be: kernel = 'rbf', C = 41.41, and gamma = 0.053. These parameters were determined through comprehensive optimization to achieve optimal performance on the emotion classification task (Table 4).

Table 4. Performance metrics of SVM Classifier for Six Emotions in SER

Emotion	Precision	Recall	F1-score	Support
Angry	0.92	0.94	0.93	119
Disgust	0.91	0.90	0.90	118
Fear	0.89	0.90	0.89	118
Happy	0.93	0.95	0.94	118
Neutral	0.99	0.93	0.96	100
Sad	0.89	0.90	0.90	119
Accuracy			0.92	692
Macro avg	0.92	0.92	0.92	692
Weighted avg	0.92	0.92	0.92	692

LightGBM: The best hyperparameters for LightGBM were found to be: boosting_type = 'gbdt', learning_rate = 0.1, max_depth = 12, n_estimators = 300, num_leaves = 50, and subsample = 0.6 (Table 5).

Table 5. Performance metrics for different emotions using the LightGBM

Emotion	Precision	Recall	F1-score	Support
Angry	0.93	0.94	0.94	119
Disgust	0.90	0.93	0.92	118
Fear	0.92	0.87	0.90	118
Happy	0.85	0.92	0.88	118
Neutral	0.95	0.92	0.93	100
Sad	0.91	0.87	0.89	119
Accuracy			0.91	692
Macro avg	0.91	0.91	0.91	692
Weighted avg	0.91	0.91	0.91	692

5 Discussion

The results of this study indicate that the choice of dataset combinations significantly impacted model performance. Multiple combinations were experimented with, but it was found that RAVDESS and TESS should be the base of the combinations due to their comprehensive representation of emotional speech data, despite not being balanced in gender. The best combination was determined to be RAVDESS and TESS.

Initially, LightGBM consistently outperformed other models across different dataset combinations. This superior performance can be attributed to LightGBM's efficient handling of large datasets and its ability to prevent overfitting through techniques such as leaf-wise tree growth and gradient-based one-side sampling.

Random Forest demonstrated robust performance across all dataset combinations, with its strength lying in its ability to handle high-dimensional data and reduce overfitting through ensemble learning and bootstrapping. Support Vector Machines (SVM) also showed strong and consistent performance across all dataset combinations. SVM's effectiveness in high-dimensional spaces contributed to its robust performance. Gradient Boosting performed well but was slightly outperformed by LightGBM and XGBoost. Its stage-wise model building and optimization of arbitrary differentiable loss functions make it powerful for various tasks. XGBoost demonstrated high performance, especially in the RAVDESS, TESS, and SAVEE combination, with strengths including efficient handling of missing values, regularization to prevent overfitting, and parallel processing capabilities.

To further optimize the performance of the best-performing models, hyperparameter tuning was applied to LightGBM and SVM in the RAVDESS and TESS combination. After tuning, SVM achieved the highest accuracy of 92%. LightGBM followed closely with an accuracy of 91%.

One of the primary challenges faced in this study was the imbalance in the dataset, particularly in terms of gender representation within each emotion category. To address this, data augmentation techniques were employed to balance the number of male and female samples for each emotion. This ensured a more balanced representation of emotions, which is crucial for unbiased model training. Additionally, the variance in environment and speakers within the datasets necessitated the extraction of features that capture both spectral and temporal characteristics of the audio signals. While MFCCs, spectral contrast, chroma features, zero-crossing rate, and spectral roll-off were employed, further refinement of feature extraction techniques could enhance model performance.

To contextualize the findings, a comparison with benchmarks from other studies in the field of speech emotion recognition is essential. Previous studies have reported varying levels of performance depending on the datasets and algorithms used. For instance, studies using deep learning approaches such as Convolutional Neural Networks (CNNs) and Recurrent Neural Networks (RNNs) have shown high accuracy but often require substantial computational resources and large datasets. In contrast, the machine learning algorithms employed in this study, particularly LightGBM and SVM, demonstrated competitive performance with the advantage of being more interpretable and computationally efficient.

This makes them suitable for real-world applications where resources may be limited.

It is worth noting that many studies in the field of speech emotion recognition use the CASIA [14] corpus to evaluate and train their models. While the CASIA dataset was not used in the current study, the prevalence of its use highlights the importance of robust and diverse datasets in advancing the field. Our approach, which leverages multiple datasets and advanced optimization techniques, has shown to be effective and robust across various datasets. This underscores the strength and versatility of our methodology in achieving high performance in speech emotion recognition.

6 Conclusion

This study advances speech emotion recognition (SER) by evaluating machine learning algorithms across various dataset combinations. The RAVDESS and TESS combination proved most effective, with LightGBM achieving 91% accuracy, followed by SVM and XGBoost at 90%, highlighting the strength of traditional algorithms.

Key findings include the observation that increased dataset diversity, such as adding CREMA-D, did not always improve performance, suggesting that data quality and consistency outweigh sheer volume. Additionally, traditional machine learning algorithms like LightGBM and SVM rival complex deep learning models, offering high accuracy with greater interpretability and efficiency.

Data augmentation effectively addressed gender imbalances, while a comprehensive feature extraction approach, using MFCCs, spectral contrast, and chroma features, enriched emotional representation, boosting model robustness. These results demonstrate that efficient algorithms can enable SER in resource-constrained settings, with applications in mental health monitoring, customer service, and human-computer interaction.

Future research could explore cultural and linguistic influences, hybrid machine learning and deep learning models, and methods to enhance performance on diverse datasets. Real-time optimization for practical deployment is another promising direction.

In conclusion, this study offers a practical framework for robust SER systems, showing that strategic dataset selection, effective feature extraction, and efficient algorithms deliver high performance while ensuring computational efficiency and interpretability. These insights help bridge the gap between experimental success and real-world SER deployment.

References

1. Singla, C., et al.: Emotion recognition for human-computer interaction using high-level descriptors. Sci. Rep. (2024)
2. Yurlay, Y., et al.: Emotion recognition on call center voice data. Appl. Sci. (2024)

3. Lim, W., Jang, D., Lee, T.: Speech emotion recognition using convolutional and recurrent neural networks (2016)
4. Begazo, R., et al.: A combined CNN architecture for speech emotion recognition. Sensors (2024)
5. Singh, J., Saheer, L.B., Faust, O.: Speech emotion recognition using attention model. Int. J. Environ. Res. Publ. Health
6. Van, L.T., et al.: Emotional speech recognition using deep neural networks. Sensors (2022)
7. Elsayed, N., et al.: Speech emotion recognition using supervised deep recurrent system for mental health monitoring. IEEE WF-IoT (2022)
8. Abhishek, N.V.S., Bhattacharyya, P.: "we care": Improving code mixed speech emotion recognition in customer-care conversations. arXiv preprint arXiv:2302.04023 (2023)
9. Issa, D., Demirci, M.F., Yazici, A.: Speech emotion recognition with deep convolutional neural networks. In: Neurocomputing
10. Iqbal, A., Barua, K.: A real-time emotion recognition from speech using gradient boosting. In: Proceedings of ECACE (2019)
11. Mittal, R., et al.: Speech emotion recognition. In: Proceedings of CONIT55038 (2022)
12. Ira, N.T., Rahman, M.O.: An efficient speech emotion recognition using ensemble method of supervised classifiers. In: Proceedings of ETCCE51779 (2020)
13. Chen, L., et al.: A two-layer fuzzy multiple random forest system for speech emotion recognition. Inf. Sci. (2020)
14. Dong, J., Wang, W., Tan, T.: Casia image tampering detection evaluation database. In: 2013 IEEE China Summit and International Conference on Signal and Information Processing (2013)
15. Shaila, S.G., et al.: Speech emotion recognition using machine learning approach. In: Proceedings of 2023
16. Jaykar, A., et al.: Speech emotion recognition using classifiers and xgboost algorithm. Int. J. Adv. Res. Sci. Commun. Technol. (2022)
17. Russo, F.A., Livingstone, S.R.: The Ryerson audio-visual database of emotional speech and song (ravdess): a dynamic, multimodal set of facial and vocal expressions in north American English. PLOS ONE
18. Pichora-Fuller, M., Dupuis, K.: Toronto emotional speech set (tess) - younger talker happy (2010)
19. Jackson, P., Haq, S.: Surrey audio-visual expressed emotion (savee) database (2014)
20. Cao, H., Cooper, D.G., Keutmann, M.K., Gur, R.C., Nenkova, A., Verma, R.: Crema-d: crowd-sourced emotional multimodal actors dataset (2014)
21. Zhang, M., Li, Y., Wu, Z.: Data augmentation and deep learning methods in sound classification: a systematic review. Sensors
22. Hastie, D., Brown, E.: A survey of data augmentation for audio classification (2022)
23. Liu, C., Zhou, H.: A new algorithm for audio files augmentation (2022)
24. Liu, M., Wang, J.: Enhancing audio augmentation methods with consistency learning (2021)
25. Majeed, S., et al.: Mel frequency cepstral coefficients (MFCC) feature extraction enhancement in the application of speech recognition: a comparison study. J. Theor. Appl. Inf. Technol
26. Kathirvel, P., et al.: Noise robust zero-crossing rate computation for audio signal classification (2011)

27. Abu-El-Quran, A.R., Goubran, R.A.: Pitch-based feature extraction for audio classification. In: Haptic, Audio and Visual Environments and Their Applications, 2003. HAVE 2003. Proceedings (2003)
28. Min, J., Kua, K., Thiruvaran, T., Nosratighods, M., Ambikairajah, E., Epps, J.: Investigation of spectral centroid magnitude and frequency for speaker recognition. In: The Speaker and Language Recognition Workshop (Odyssey) (2010)
29. Er, M.B.: A novel approach for classification of speech emotions based on deep and acoustic features. IEEE Access (2020)
30. Sandhya, P., et al.: Spectral features for emotional speaker recognition. In: Third International Conference on Advances in Electronics, Computers and Communications (ICAECC) (2020)
31. Smith, J., Lee, J., Patel, A.: The impact of data normalization on the accuracy of machine learning algorithms: a comparative analysis. J. Mach. Learn. Res
32. Hearst, M.A., et al.: Support vector machines. IEEE Intell. Syst. Appl. (1998)
33. Breiman, L.: Random forests. Mach. Learn. (2001)
34. Ayyadevara, V.: Gradient boosting machine (2018)
35. Chen, T., Guestrin, C.: Xgboost: a scalable tree boosting system (2016)
36. Machado, M., Karray, S., de Sousa, I.T.: LightGBM: an effective decision tree gradient boosting method to predict customer loyalty in the finance industry. In: Proceedings of the ICCSE (2019)
37. Snoek, J., Larochelle, H., Adams, R.P.: Practical Bayesian optimization of machine learning algorithms. In: Advances in Neural Information Processing Systems (2012)

Assessing Digital Transformation in Business Process Management: Towards Sustainable Development

Oksana Makovoz[1,2](✉) and Sergii Lysenko[3]

[1] Doctor of Economic Sciences, National Technical University "Kharkiv Polytechnic Institute", Kharkiv, Ukraine
oksana.makovoz@khpi.edu.ua
[2] IHI Zittau, Dresden Technical University, Dresden, Germany
[3] National Technical University "Kharkiv Polytechnic Institute", Kharkiv, Ukraine
Serhii.Lysenko@emmb.khpi.edu.ua

Abstract. The fast development of digital technologies and the imperatives of sustainable development have transformed modern organizations. At the same time, the complexity of evaluation in managing structured digital transformation limits long-term efficiency and resilience to change. This study proposes an evaluation framework that combines business process management, digital transformation, and multi-criteria decision-making to improve operational efficiency and alignment with sustainability goals. Using scenario modeling, enterprises can assess digital transformation's economic, environmental, and social impacts using methods such as TOPSIS and AHP. The results demonstrate an improvement in the transparency of decision-making and prioritization of digital initiatives based on key performance indicators. A digital transformation assessment system is proposed to achieve sustainable digital transformation of organizations in various industries.

Keywords: Decision-Making · Performance Evaluation Metrics · Strategic · Digital Transformation · Sustainable Business Process Integration

1 Introduction

In modern organizational strategies, the convergence of business process management (BPM), digital transformation (DT), and sustainability has gained significant attention. However, existing literature often treats these domains in isolation, leading to fragmented research and application frameworks. This gap limits the ability of organizations to achieve holistic, sustainable, and efficient transformations. BPM traditionally focuses on process optimization and efficiency, while DT emphasize adopting advanced technologies, such as artificial intelligence (AI) and the Internet of Things (IoT), to innovate and remain competitive. However, the separation of these domains leads to suboptimal outcomes. C. Nwachukwu and V. Hieu [13] emphasized the critical need for aligning technological adoption with core business processes. N. Brand [3] highlighted

B.-E. Boudriki Semlali et al. (Eds.): DATA 2025, LNISO 81, pp. 50–60, 2025.
https://doi.org/10.1007/978-3-032-06725-8_5

that BPM frameworks lack dynamic adaptability to DT innovations, resulting in inefficiencies. The integration of sustainability goals remains underexplored in BPM and DT frameworks. As X. Teng *et al.* [16] pointed out, sustainability demands - such as minimizing carbon footprints and enhancing social equity - are not adequately addressed in most DT initiatives. Furthermore, A. Aldoseri *et al.* [2] stressed the importance of embedding sustainability metrics into AI-driven transformations for long-term organizational resilience. Multi-criteria decision-making (MCDM) methods, such as technique for order preference by similarity to ideal solution (TOPSIS) and analytic hierarchy process (AHP), are powerful tools for balancing conflicting priorities in transformation projects. However, their application in BPM and DT remains fragmented. V. Dias [7] underscored the potential of MCDM in designing key performance indicators (KPI) oriented dashboards, yet its integration into continuous BPM cycles has been sparse.

2 Literature Review

BPM is a key enabler of process improvement and digital transformation, helping organizations enhance efficiency and align operations with technological advancements. Van Looy [17] emphasizes BPM's role in process innovation, while Butt [4] highlights its integration with Industry 4.0 technologies like IoT and AI to improve efficiency. Debnath et al. [6] and Khin & Ho [11] propose measurement and digital readiness frameworks to assess BPM's alignment with digital goals. Sustainability is increasingly relevant in BPM and DT. Corejova & Chinoracky [5] advocate embedding economic, environmental, and social metrics into transformation strategies. Verhoef *et al.* [18] highlight sustainability-focused BPM frameworks for long-term resilience. Big data and machine learning facilitate sustainability integration by optimizing investment strategies and assessing environmental impacts [20]. Butt [4] demonstrates how green BPM principles reduce costs while minimizing ecological footprints. MCDM methods like TOPSIS and VIKOR [14] prioritize digital initiatives based on economic, social, and environmental factors. Zavadskas *et al.* [19] highlight MCDM's role in complex decision-making, while Khin & Ho [11] integrate it into BPM frameworks. However, sustainability considerations remain underexplored in BPM-DT frameworks. Integrating MCDM with BPM and DT can bridge these gaps, fostering sustainable, data-driven, and resilient business ecosystems. Despite considerable research on BPM, digital transformation, and sustainability, the literature remains fragmented, addressing these domains independently rather than integrative. This paper fills this critical gap by proposing a unified evaluation framework that synergistically integrates BPM, DT, and sustainability using MCDM methods, specifically AHP and TOPSIS. The novelty lies in the systematic combination of operational and strategic metrics aligned with the Triple Bottom Line (economic, environmental, and social dimensions) and scenario-based modeling, enabling organizations to make data-driven, transparent, and strategically aligned decisions.

3 Materials and Methods

This research introduces a comprehensive logical framework to address the integration of BPM, DT, sustainability goals, and MCDM. The proposed framework provides a structured and iterative approach, clearly illustrated in Fig. 1, aiming to enhance organizational efficiency and achieve sustainable development objectives.

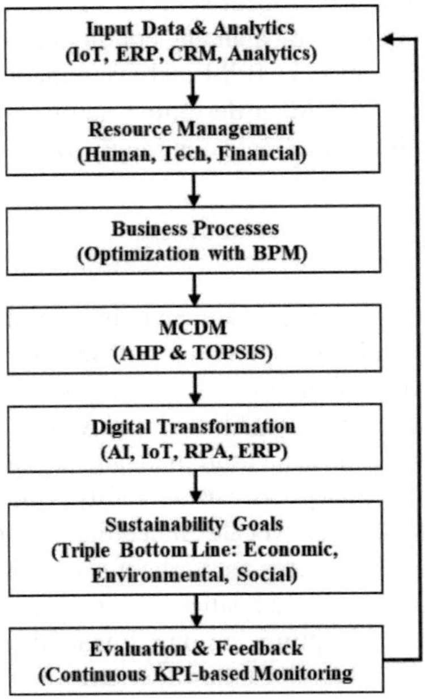

Fig. 1. High-level framework integration.

The framework comprises the following key components:

1. Input data & analytics: the framework begins with the collection and analysis of data from various sources, including IoT devices, ERP systems, and CRM platforms. This stage emphasizes data-driven decision-making, a cornerstone of modern digital transformation efforts. As noted by A. Ahmad *et al.* [1], leveraging analytics ensures actionable insights for resource allocation and process optimization.
2. Resource management: effective allocation of key technological, human, and financial resources supports innovation and ensures seamless process execution. M. Hammer [8] highlights the role of resource alignment in reengineering processes for better outcomes.
3. Business processes: business processes are optimized using BPM methodologies to enhance efficiency, scalability, and alignment with strategic objectives. P. Harmon [9] underscores the necessity of structured methodologies to manage and refine both administrative and strategic processes, bridging operations and sustainability goals.
4. MCDM: methods, such as AHP and TOPSIS, evaluate and prioritize initiatives based on economic, environmental, and social metrics. T. Saaty [15], S. Opricovic and G.

Tzeng [14] demonstrate the utility of these tools in ensuring balanced, data-informed decisions.
5. DT: advanced technologies, including AI, IoT, and RPA, drive innovation and enhance organizational agility. According to A. Aldoseri *et al.* [2], integrating AI and other digital tools transforms decision-making, streamlines operations, and fosters continuous improvement. O. Makovoz and S. Lysenko [12] emphasize AI-driven process automation, including RPA and ERP systems, reduces time spent on routine tasks.
6. Sustainability & goals: sustainability principles are embedded throughout the framework, adhering to the Triple Bottom Line (TBL) framework, which balances economic, environmental, and social objectives. R. Jeurissen and J. Elkington [10] emphasize that embedding sustainability at every stage ensures long-term viability.
7. Evaluation & feedback: continuous monitoring of Key Performance Indicators (KPIs) and structured feedback loops enable organizations to refine their processes. This stage ensures alignment with strategic goals and responsiveness to changing conditions. P. Harmon [9] and V. Dias [7] highlight the role of iterative evaluation in maintaining relevance and adaptability.

The iterative nature of the proposed framework, indicated by the continuous improvement loop, ensures ongoing adaptability and sustained organizational growth aligned with strategic sustainability objectives.

4 Results and Discussion

The proposed methodology and scenario-based modeling framework ensure:

1. Holistic evaluation: integration of BPM, DT, and sustainability considerations.
2. Strategic alignment: prioritization of initiatives that balance short-term operational efficiency with long-term sustainability goals.
3. Data-driven decisions: use of quantitative tools MCDM to support transparent and objective decision-making.
4. Actionable insights: clear recommendations on which transformation scenario to implement based on organizational priorities and resource constraints.

The detailed scenario evaluation process is illustrated in Fig. 2.

Integrated BPM–DT–Sustainability–MCDM Framework:

1. BPM and Digital Transformation:
   - Conduct BPM cycles (As-Is analysis → To-Be design → implementation → continuous monitoring) integrating digital tools (RPA, AI, IoT).
   - Assess digital maturity and readiness for technology adoption, considering scalability and innovation potential.
2. Sustainability Integration (TBL):
   - Embed economic (ROI, efficiency), environmental (emissions, energy), and social (job creation, community impact) criteria throughout BPM stages.
   - Quantify key sustainability KPIs clearly.

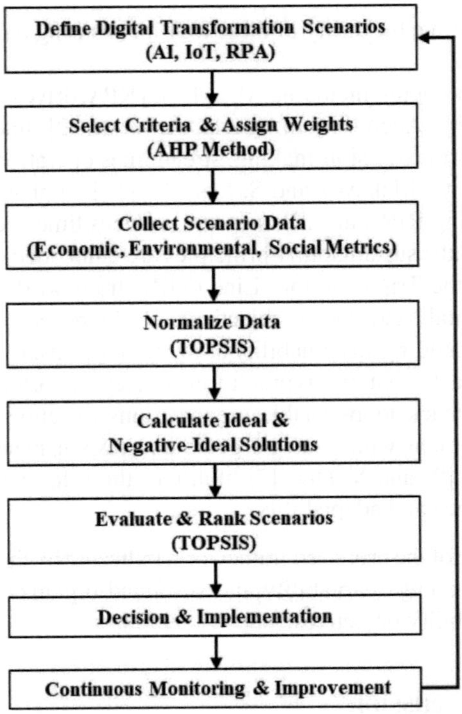

Fig. 2. Detailed flowchart of scenario-based evaluation using AHP and TOPSIS methods.

3. MCDM Core (Scenario Evaluation):
   - Define balanced criteria (cost, ROI, environment, social).
   - Determine criteria weights via AHP.
   - Rank scenarios clearly using TOPSIS (normalization → weighting → ideal solutions → ranking).
   - Perform sensitivity analysis to test robustness.
4. Framework Flow:
   - Strategic Alignment: Define organizational vision (digital and sustainability goals).
   - As-Is Analysis: Map current processes and digital maturity.
   - Scenario Creation (To-Be): Propose scenarios embedding digital technologies and sustainability targets.
   - Criteria Selection & Weighting (AHP): Prioritize economic, environmental, social, operational criteria.
   - Scenario Evaluation (TOPSIS): Quantitatively rank scenarios.
   - Decision & Implementation: Recommend optimal scenarios, implement pilots or full solutions.
   - Continuous Improvement: Regularly monitor performance, adapting processes iteratively.

5. Framework Benefits:

   - Balances operational efficiency with sustainability.
   - Provides structured, transparent decision-making.
   - Adaptable across industries and organization sizes.

Embeds sustainability ensuring resilience and regulatory alignment. Each scenario is assessed against the TBL framework, which includes economic (e.g., cost and ROI), environmental (e.g., carbon emissions reduction), and social (e.g., job creation and community impact) dimensions. By using the TOPSIS method, the evaluation quantifies the relative performance of each scenario by calculating its proximity to an ideal solution and its distance from a negative-ideal solution. The data is categorized under economic, environmental, and social dimensions, aligned with the Triple Bottom Line (TBL) framework. Additionally, sources for MCDM methods and their applications in sustainability and decision-making processes are provided.

This approach enables the ranking of scenarios by calculating their relative closeness to the ideal solution based on selected criteria. Each criterion falls under economic, environmental, and social dimensions and is assigned a weight to indicate its relative importance in the assessment. These weights are crucial in the TOPSIS method, as they influence the final ranking of scenarios.

The decision matrix serves as the foundation for further computations, including normalization, weighting, and ranking. The raw values reflect real or estimated performance metrics for each scenario.

The normalization process ensures that all criteria values are scaled proportionally, eliminating unit discrepancies. Each raw value is normalized using Formula 1.

$$r_{ij} = \frac{x_{ij}}{\sqrt{\sum_{i=1}^{n} x_{ij}^2}}, \tag{1}$$

where x_{ij} is raw value for criterion *j* and scenario *i; n* is total number of alternatives.

The normalized values (Formula 2) are multiplied by their respective criterion weights, producing a weighted matrix. This step integrates the importance of each criterion into the evaluation process, ensuring that high-priority factors have a greater influence on the decision-making outcome. Normally values are weighted using the criterion weights:

$$v_{ij} = w_i * r_{ij}, \tag{2}$$

where w_j is the weight assigned to criterion j; rij is the normalized value for alternative *i* under criterion *j*.

Determine ideal and negative-ideal solutions (Formula 3):

$$A^+ = \{\max(v_{ij}) | j \in J\},\ A^- = \{\min(v_{ij}) | j \in J\}, \tag{3}$$

where v_{ij} is the weighted normalized value for alternative i under criterion j; j is a specific criterion in the set of criteria J; i represents a specific alternative in the set of alternatives; J is the set of all criteria used in the decision-making process. As a result, Ideal solution

(A^+) and Negative-ideal solution (A^-). Calculate separation measures (Formula 4):

$$S_i^+ = \sqrt{\sum_{j=1}^{m} \left(v_{ij} - A_j^+\right)^2},\ S_i^- = \sqrt{\sum_{j=1}^{m} \left(v_{ij} - A_j^-\right)^2}, \tag{4}$$

where A_j^+ is ideal value for criterion j; A_j^- is negative-ideal value for criterion j; m is total number of criteria.

The relative closeness score is calculated for each scenario (Formula 5) based on its proximity to the ideal and negative-ideal solutions:

$$C_i = \frac{S_i^-}{S_i^+ + S_i^-}, \tag{5}$$

The relative closeness scores (C_i) of each scenario to the ideal solution, calculated as part of the TOPSIS decision-making process. The values indicate the preference ranking, where a higher C_i implies greater closeness to the ideal solution. Based on the relative closeness (C_i) scores the organizations can make data-driven decisions that align with their long-term goals and stakeholder expectations.

To illustrate the practical applicability and effectiveness of the proposed evaluation framework, a simulated yet realistic case study was conducted involving a mid-sized manufacturing enterprise seeking to digitally transform operations to reduce energy consumption and enhance overall efficiency. Three distinct transformation scenarios were proposed:

- AI-driven predictive maintenance: leveraging artificial intelligence to predict equipment failures and optimize maintenance schedules, reducing downtime and energy waste.
- IoT-enabled energy monitoring: deploying IoT sensors to provide real-time tracking of energy usage across production facilities, identifying inefficiencies and optimizing resource utilization.
- RPA-based administrative automation: implementing RPA to streamline administrative workflows, thereby reducing manual efforts and associated resource consumption.

The evaluation used the Multi-Criteria Decision-Making methods - AHP for criteria weighting and TOPSIS for scenario ranking. Criteria weights were derived through structured expert consultations using the AHP method. An expert panel consisting of senior management, sustainability officers, and technology specialists was convened, ensuring a balanced representation of strategic, operational, and sustainability perspectives.

Based on expert elicitation, the criteria were weighted as follows:

- Economic impact: 0.35.
- Environmental benefit: 0.30.
- Social impact: 0.20.
- Ease of implementation: 0.15.

These weights were chosen explicitly to reflect the organization's strategic priority of achieving both substantial economic returns and meaningful sustainability outcomes. Economic considerations held the highest weight, acknowledging budgetary constraints and ROI expectations. However, significant emphasis was also clearly placed on environmental and social benefits to ensure alignment with corporate sustainability commitments.

Applying the TOPSIS method using the defined criteria weights, the analysis revealed that the AI-driven predictive maintenance scenario achieved the highest relative closeness to the ideal solution. This scenario optimally balanced operational efficiency, economic returns, and substantial environmental benefits through significant reductions in energy consumption and maintenance-related downtime.

The results highlight the framework's capability to deliver clear, actionable insights grounded in organizational priorities and strategic sustainability goals, validating its practical effectiveness and reinforcing the robustness of the proposed methodological approach.

Table 1 serves as a holistic decision-making framework that integrates operational efficiency, technological innovation, and sustainability into the digital transformation journey. It provides actionable insights for organizations aiming to balance competing priorities while achieving long-term resilience and growth.

Table 1. Integrated framework for Decision-Making in DT.

Component	Description	Key metrics	Methods/tools	Impact on transformation
Business process management (BPM)	Structured management of organizational processes to improve efficiency, consistency, and alignment with strategic goals.	Process time, cost, quality, error rate	Process mapping, flowcharts, As-Is/To-Be modeling	Enhances process clarity, efficiency, and adaptability to technological changes.
Digital transformation	Adoption of digital technologies (AI, IoT, RPA) to drive innovation, agility, and competitive advantage.	Digital maturity, technology adoption rate	IoT sensors, AI analytics, RPA platforms	Facilitates innovation, improves agility, and enables real-time data-driven decisions.
Sustainability (Triple bottom line)	Integration of environmental, social, and economic sustainability goals into business processes and digital transformation efforts.	Carbon emissions, energy efficiency, social impact	Sustainability metrics, green BPM, ESG indicators	Aligns digital transformation efforts with long-term environmental and societal goals.
Multi-criteria decision-making	Decision-making approach to evaluate and rank alternatives based on multiple, often conflicting, criteria.	Composite scores across criteria	TOPSIS, AHP, PROMETHEE	Ensures transparent, data-driven prioritization of transformation scenarios.
Evaluation criteria	Criteria used to assess transformation scenarios, based on Triple Bottom Line (TBL) principles.	Cost, ROI, emissions, job creation	Weighting criteria, stakeholder input	Provides a holistic framework for comparing scenarios against operational and sustainability goals.

(continued)

Table 1. (*continued*)

Component	Description	Key metrics	Methods/tools	Impact on transformation
Scenario-based modeling	Development of hypothetical or real transformation pathways to test potential outcomes and identify optimal strategies.	Scenario performance across TBL dimensions	Predictive modeling, simulation tools	Offers a structured way to explore trade-offs and balance competing priorities.
Weighting and sensitivity analysis	Assignment of importance to criteria and evaluation of the robustness of results under varying assumptions.	Weight adjustments, sensitivity scores	AHP for weighting, sensitivity analysis dashboards	Ensures flexibility in adapting to changing organizational priorities and external conditions.
Continuous monitoring	Ongoing assessment of transformation progress and alignment with goals.	KPI tracking, performance reviews	Business intelligence dashboards, feedback loops	Ensures alignment with evolving strategic, technological, and sustainability objectives.

Limitations and assumptions. Despite its robustness, the proposed framework has several limitations:

- Assumption of data availability: reliable and comprehensive data for accurate scenario evaluation may not always be accessible, particularly in smaller organizations.
- Subjectivity in criteria weighting: criteria weights derived from expert judgments (AHP) introduce subjectivity, which may affect the generalizability of the results across different organizations or industries.
- Generalizability: while adaptable, specific results might differ significantly based on industry type, organizational maturity, and technology infrastructure.

Clearly recognizing these limitations enables organizations to adjust and contextualize the framework effectively.

5 Conclusions

This research presents an integrated framework combining BPM, DT, and MCDM with sustainability principles to enhance business process transformation. By incorporating AHP and TOPSIS, the framework enables organizations to prioritize digital initiatives based on economic, environmental, and social criteria, ensuring alignment with long-term sustainability goals. Scenario-based modeling offers structured decision-making, while continuous monitoring allows for adaptability. The framework bridges gaps in literature and practice, promoting a data-driven approach to sustainable digital transformation. Future research can explore industry-specific applications, advanced MCDM techniques, AI and Big Data integration, and alignment with global sustainability standards like the SDGs.

Acknowledgements. "None".

Disclosure of Interests. The authors have no competing interests to declare that are relevant to the content of this article.

References

1. Ahmad, A., Alshurideh, M., Al Kurdi, B., Aburayya, A., Hamadneh, S.: Digital transformation metrics: a conceptual view. J. Manage. Inf. Decision Sci. **24**(S2), 1–18 (2021). https://www.abacademies.org/articles/digital-transformation-metrics--a--conceptual--view-11806.html
2. Aldoseri, A., Al-Khalifa, K.N., Hamouda, A.M.: Methodological approach to assessing the current state of organizations for AI-based digital transformation. Appl. Syst. Innov. **7**(1), 14 (2024). https://doi.org/10.3390/asi7010014
3. Brand, N.: Effect of the business activity model on digital transformations (2022). https://studenttheses.uu.nl/handle/20.500.12932/42842
4. Butt, J.: A conceptual framework to support digital transformation in manufacturing using an integrated business process management approach. Designs **4**(3), 17 (2020). https://doi.org/10.3390/designs4030017
5. Corejova, T., Chinoracky, R.: Assessing the potential for digital transformation. Sustainability **13**(19), 11040 (2021). https://doi.org/10.3390/su131911040
6. Debnath, N., et al.: A measurement strategy for business processes in digital transformation. In: EPiC Series in Computing, vol. 89, p. 11 (2022). https://doi.org/10.29007/lcpk
7. Dias, V.M.: Smart KPI-ORIENTED decision support Dashboard for measuring the Digital Transformation success - University of Twente Student Theses (2021). https://essay.utwente.nl/88663/
8. Hammer, M.: Reengineering Work: Don't Automate, Obliterate (2024). https://hbr.org/1990/07/reengineering-work-dont-automate-obliterate
9. Harmon, P.: Business Process Change: A Business Process Management Guide for Managers and Process Professionals. Morgan Kaufmann (2020)
10. Jeurissen, R., Elkington, J.: Cannibals with forks: the triple bottom line of 21st century business. J. Bus. Ethics **23**, 229–231 (2000). https://doi.org/10.1023/A:1006129603978
11. Khin, S., Ho, T.C.: Digital technology, digital capability and organizational performance. Int. J. Innov. Sci. **11**(2), 177–195 (2018). https://doi.org/10.1108/ijis-08-2018-0083
12. Makovoz, O., Lysenko, S.: Digital technologies in personnel management. In: Proceedings of London International Conferences, vol. 10, pp. 65–76 (2024). https://doi.org/10.31039/plic.2024.10.208
13. Nwachukwu, C., Hieu, V.M.: Digital transformation adoption: antecedents and consequences. In: Proud Pen eBooks (2021). https://doi.org/10.51432/978-1-8381524-3-7_2
14. Opricovic, S., Tzeng, G.: Compromise solution by MCDM methods: a comparative analysis of VIKOR and TOPSIS. Eur. J. Oper. Res. **156**(2), 445–455 (2003). https://doi.org/10.1016/s0377-2217(03)00020-1
15. Saaty, T.L.: The Analytic Hierarchy Process: Planning, Priority Setting, Resources Allocation. Mcgraw-Hill, New York (1980)
16. Teng, X., Wu, Z., Yang, F.: Research on the relationship between digital transformation and performance of SMEs. Sustainability **14**(10), 6012 (2022). https://doi.org/10.3390/su14106012
17. Van Looy, A.: Business process maturity. SpringerBriefs in business process management (2014). https://doi.org/10.1007/978-3-319-04202-2
18. Verhoef, P.C., et al.: Digital transformation: a multidisciplinary reflection and research agenda. J. Bus. Res. **122**, 889–901 (2021). https://doi.org/10.1016/j.jbusres.2019.09.022

19. Zavadskas, E.K., Turskis, Z., Kildienė, S.: State Of Art surveys of overviews on MCDM/MADM methods. Technol. Econ. Dev. Econ. **20**(1), 165–179 (2014). https://doi.org/10.3846/20294913.2014.892037
20. Zelisko, N., Raiter, N., Markovych, N., Matskiv, H., Vasylyna, O.: Improving business processes in the agricultural sector considering economic security, digitalization, risks, and artificial intelligence. Ekonomika APK **31**(3), 10–21 (2024). https://doi.org/10.32317/2221-1055.2024030.10

Determinants of Adoption Intention of Electronic Government Procurement: A Conceptual Model for Sri Lanka

W. M. D. M. Dissanayaka[1(✉)] and Prapon Sahapattana[2]

[1] Department of Public Administration, University of Sri Jayewardenepura, Nugegoda, Sri Lanka
madusanka@sjp.ac.lk

[2] Graduate School of Public Administration, National Institute of Development Administration, Bangkok, Thailand
prapon.s@nida.ac.th

Abstract. Electronic Government Procurement (e-GP) has been adopted by many governments worldwide due to its numerous benefits. Similarly, the Government of Sri Lanka is in the initial stages of implementing an e-GP system for the public sector, known as PROMISe, Sri Lanka. However, the absence of a context-specific framework presents a challenge in understanding adoption factors. Therefore, this study aims to develop a conceptual model to identify the key determinants influencing the adoption intention of public sector employees toward the e-GP system in Sri Lanka. The study employed an iterative conceptual methodology for the model development process. It guided the model development by drawing on high-quality literature, widely accepted theoretical models, and logical reasoning. The Technology Acceptance Model (TAM) and the Unified Theory of Acceptance and Use of Technology (UTAUT) serve as the foundation for the proposed model. The study adopts several key constructs from these frameworks: perceived usefulness (PU), perceived ease of use (PEU), facilitating conditions (FC), and social influence (SI). Additionally, the influences of digital literacy and the type of organization have been incorporated into the model to reflect context-specific influences for Sri Lanka. The expected associations between all elements in the model have been clearly elaborated in the paper. Additionally, the dimensions for operationalization and the measurement scales for each construct have also been illustrated. The model proposed in this paper can be used by future researchers to validate it with empirical data from the Sri Lankan public sector.

Keywords: Electronic government procurement · Adoption intention · TAM · UTAUT · Conceptual model

1 Introduction

Governments spend a significant portion of their budget through procurement channels and are considered the largest buyer in the market. In Sri Lanka, the total public procurement expenditure accounts for 6% of its GDP. Given the many benefits of making

B.-E. Boudriki Semlali et al. (Eds.): DATA 2025, LNISO 81, pp. 61–73, 2025.
https://doi.org/10.1007/978-3-032-06725-8_6

the shift, many governments nowadays switch from manual to computerized procurement. Accordingly, the embracement of electronic government procurement (e-GP) has become a global trend that revolutionizes public sector procurement processes while enhancing transparency, efficiency, and accountability.

In developing countries such as Sri Lanka, the implementation of e-GP is seen as a critical step toward reducing corruption, improving service delivery, and ensuring better public financial management. However, numerous challenges associated with this move have also been recognized in previous studies. These challenges include, but are not limited to, a lack of technological readiness, a lack of organizational capacity, poor regulatory frameworks, and user resistance [1].

Sri Lanka has been attempting to implement e-GP for an extended period. In 2019, the Ministry of Finance issued Public Finance Circular 08/2019, titled "Implementation of the Electronic Government Procurement (e-GP) System in Sri Lanka," marking a significant step towards formalizing e-GP initiatives [2]. Currently, the e-GP system, known as the Procurement Management Information System (PROMISe), operates with limited functionalities and is used by selected government organizations.

User acceptance has been recognized as a critical factor for the success of information technology (IT) initiatives [3]. Consequently, numerous studies have been conducted worldwide to identify the factors determining users' adoption intentions towards new IT initiatives. The adoption of e-GP can also be considered a type of IT-related initiative due to its technological aspects compared to manual procurement procedures. Accordingly, many studies have been conducted to reveal the determinants of users' adoption intentions to use e-GP systems [4]. Some research has focused on suppliers' perspectives in adopting e-GP within business firms [5]. However, a limited number of studies have examined public sector users' perspectives in different settings. These study findings have emphasized the role of context-specific factors in this process [4]. However, no such model has been developed for the Sri Lankan context. Therefore, this paper aims to propose a context-specific model to determine the adoption intention of e-GP among public sector workers in Sri Lanka.

The Technology Acceptance Model (TAM) and the Unified Theory of Acceptance and Use of Technology (UTAUT) are extensively used theoretical frameworks in IT adoption research. Therefore, these two frameworks serve as the foundation for this study. Additional variables are also integrated to derive context-specific factors for public sector workers in Sri Lanka.

This study provides significant contributions in several aspects. First, it addresses a critical gap in understanding the factors influencing the adoption of e-GP systems in Sri Lanka. As Sri Lanka continues to strive for improved transparency, efficiency, and public financial management through e-GP, this research offers valuable insights into the specific determinants that affect public sector workers' intentions to adopt such systems. Second, by integrating established frameworks such as TAM and UTAUT, with additional variables, the study develops a context-specific model that reflects the socio-cultural and organizational realities of Sri Lanka. Therefore, the findings of this research can assist policymakers and practitioners in better understanding the factors that foster or hinder the successful adoption of e-GP in the Sri Lankan public sector. Furthermore, this study offers a model that can be empirically validated by future researchers in the Sri Lankan context.

The next section of the paper presents a synthesis of the related literature, followed by an explanation of the model development process and its components. Subsequently, the relationships proposed in the model are discussed in relation to existing research. Conclusions, limitations of the study, and directions for future research are outlined in the final section.

2 Literature Review

2.1 Overview of e-GP

Electronic Government Procurement (e-GP) refers to the transition of public sector procurement processes from manual to electronic platform-driven systems. These transitions aim to conduct various aspects of government procurement such as tendering, contract management, supplier registration, and payment management electronically [6]. Governments across the world have adopted e-GP considering the expected benefits including reduced procurement fraud, increased efficiency and effectiveness, and enhanced transparency and accountability [4]. Organizations like the World Bank and the United Nations have emphasized e-GP as a critical element of public financial management reforms. However, the adoption and realization of intended benefits vary across countries due to numerous factors such as technological readiness, legal frameworks, and stakeholder willingness.

2.2 Theoretical Models Relevant to Technology Adoption Research

In the studies relating to technology adoption, including e-GP, the Technology Acceptance Model (TAM) and the Unified Theory of Acceptance and Use of Technology (UTAUT) are the most commonly used theoretical models [20]. In particular, including these models has become essential for studies examining individual workers' adoption intentions [4]. Therefore, these models serve as the foundational theoretical models for this conceptual model development for the Sri Lankan public sector.

The Technology Acceptance Model (TAM) was initially proposed by Davis to explain the determinants of users' acceptance of technology-related initiatives [3]. TAM predicts intention to use technology based on perceived usefulness (PU) and perceived ease of use (PEOU). Even though it is considered one of the most influential models able to explain user acceptance of technology, researchers have modified its components to enhance its ability to explain the total variance [5]. Those additional variables include but are not limited to variables such as perceived trust, top management support, perceived security and digital literacy.

The Unified Theory of Acceptance and Use of Technology (UTAUT) proposed by Venkatesh et al. has become the other most widely used theoretical model in this research arena [7]. UTAUT expands on TAM by integrating performance expectancy, effort expectancy, social influence, and facilitating conditions. Performance expectancy and effort expectancy align similarly to PU and PEOU in TAM, respectively [5]. Similar to TAM, UTAUT was further expanded by Venkatesh et al. with the inclusion of three additional variables [8]. Those constructs include hedonic motivation, price value, and habit. Moreover, many researchers have introduced various context-specific factors to improve the explanatory ability of the framework, such as perceived risk, perceived trust, experience, and personal innovativeness [9].

2.3 Key Determinants of e-PP Adoption

Based on the previous studies in existing literature, the following key determinants influencing e-GP adoption can be identified.

- **Perceived Usefulness (PU) & Performance Expectancy**: This refers to the extent to which a user believes that utilizing the e-GP system will enhance their job performance. Procurement officers are more inclined to adopt e-GP if they recognize tangible benefits, such as faster processing times, improved procurement accuracy, and enhanced transparency. Procurement officers are more likely to adopt e-GP if they believe it can improve the efficiency of their job [10].
- **Perceived Ease of Use (PEOU) & Effort Expectancy**: This determinant captures the belief that the e-GP system is straightforward, user-friendly, and does not require excessive mental or physical effort. Systems with intuitive interfaces, clear instructions, and minimal complexity are more likely to foster acceptance among public sector employees. The intention to use e-GP is higher if an individual believes that using the e-GP system is free of effort [10].
- **Social Influence**: Adoption can also be shaped by the opinions and behaviours of influential individuals or groups. When peers, supervisors, or institutions advocate for the use of e-GP, or when there is visible institutional support and encouragement, individuals often feel a normative pressure to conform. The user's intention to adopt e-GP increases when senior officials, colleagues, government, or other external parties emphasize e-GP use [11].
- **Facilitating Conditions**: These refer to the structural and institutional supports available for the effective use of e-GP. Adequate ICT infrastructure, stable internet connectivity, availability of technical training, and the presence of legal and procedural frameworks all contribute to an environment conducive to digital procurement. The availability of ICT infrastructure, legal frameworks, and technical support plays a crucial role in determining the intention to use e-GP in public sector officials [11].
- **Hedonic Motivation**: This refers to the pleasure or enjoyment derived from using a technology. This has been identified as an influencer of behavioural intention. In the context of e-GP, users who find the system engaging or satisfying to use are more likely to develop a positive attitude toward its adoption [12].
- **Price Value**: Price value reflects the trade-off between the expected benefits of adopting e-GP and the costs associated with its use. When users perceive that the advantages, such as improved efficiency, cost savings, and transparency outweigh the financial and operational costs, their intention of adopting the system increases [12].
- **Habit**: This represents the extent to which individuals engage in a particular behavior automatically due to prior experience. In e-PP adoption, users who have previously used digital procurement systems or similar technological tools may find it easier to use e-GP [12].

- **Personal innovativeness**: Personal innovativeness refers to an individual's willingness to experiment with and adopt new technologies. Although it is not a direct determinant of e-GP adoption, it influences PU and PEU. Individuals with high personal innovativeness are expected to perceive e-PP as beneficial and easy to use, thereby increasing their intention to use it [6].

2.4 Adoption of e-GP in the Sri Lankan Public Sector

The government of Sri Lanka has struggled to implement e-GP multiple times over an extended period in the past. Finally, since 2019, the Sri Lankan government has been actively using its e-GP system, known as PROMISe (Public Resource Management Information System), under the guidance of the Ministry of Finance [2]. However, even today, the system is used by a limited number of government institutions and for selected procurement activities only [13].

Despite these efforts, a few available studies have revealed several challenges in the implementation of e-GP in Sri Lanka. These obstacles include technological barriers, a lack of understanding of modern procurement practices, and resistance to change within public institutions [1]. These challenges emphasize the need for comprehensive strategies that address technological infrastructure, capacity building, and change management to facilitate successful e-GP adoption. However, no recent empirical studies in the existing body of literature investigate the determinants of behavioural intention toward the acceptance of e-GP among public sector staff in Sri Lanka.

Overall, while Sri Lanka has made notable progress in developing and implementing its e-GP system, the journey toward full adoption still requires overcoming significant challenges. Therefore, identifying the determinants of user acceptance of e-GP is crucial for realizing its benefits in Sri Lanka.

3 Proposed Conceptual Model

3.1 Model Development Process

This study employs a conceptual methodology suggested by previous scholars in the field to develop a theoretical model of the determinants of adoption intention for e-GP in Sri Lanka. The model development process consists of three key stages: (1) identifying relevant theoretical foundations, (2) integrating key determinants from existing literature, and (3) establishing logical linkages.

First, a structured literature review was conducted to identify the most applicable theoretical frameworks for the study. Scholarly articles indexed in recognized databases such as Scopus, Web of Science, and JSTOR were reviewed to ensure the quality of the articles. We further emphasized the focus of those articles to ensure their relevance to e-GP adoption in Sri Lanka. Following a detailed review of established theories in technology adoption and public sector procurement, we found TAM and UTAUT to be the most widely used theoretical foundations in similar studies. Moreover, the selection of these theoretical models is guided by their ability to explain adoption intention in the context of government procurement to ensure their relevance for an individual-level analysis.

Second, a similar structured procedure was employed to identify key determinants influencing e-GP adoption. The study synthesizes empirical findings from prior research using the above theoretical frameworks. Critical factors such as perceived benefits, ease of use, facilitating conditions, digital literacy, social influences, and type of organization were identified during this stage. These factors were integrated into the model based on their theoretical significance and empirical validation in related contexts. More importantly, suitability for the Sri Lankan context was carefully investigated during this stage.

Finally, the study formulates a conceptual framework that establishes the relationships among identified constructs. Logical reasoning and theoretical justification are employed to explain the mechanisms through which these determinants influence adoption intention. The objective of the model is to contribute to the literature by offering novel relationships and introducing new constructs for the e-GP research arena that have been ignored by previous researchers.

The above mentioned iterative model development procedure is summarized in Fig. 1, and it complies with similar research and closely aligns with the Model Approach outlined by Jaakkola for conceptual research designs [14]. By following a structured conceptual approach, this study provides a foundation for future empirical validation by ensuring analytical rigor, transparency, and theoretical contribution.

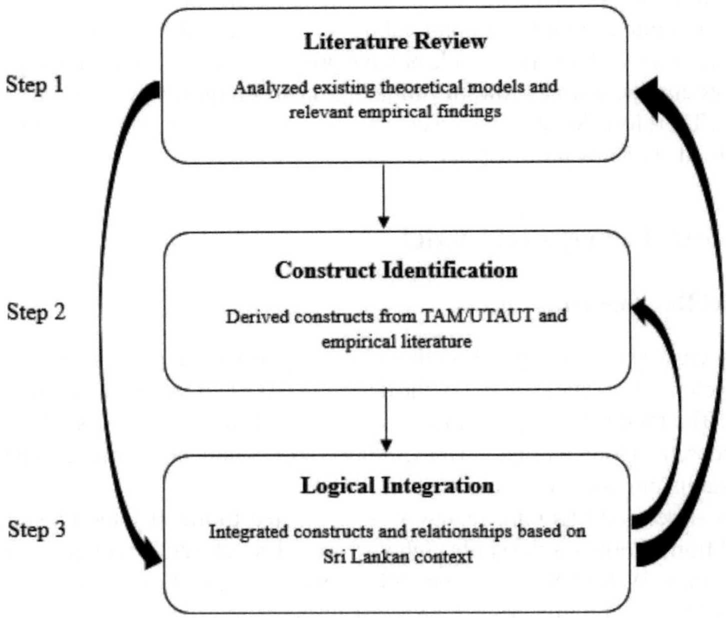

Fig. 1. Model Development Process

3.2 A Model to Examine the Intention to Adopt e-GP in Sri Lanka

An introduction of the Model Before Going into Each Concept.
Following the iterative process described above, the model comprises key constructs

highly relevant to the Sri Lankan public sector. It includes e-GP determinants, namely PU, PEU, FC, and SI, which are derived from theoretical frameworks. Additionally, the model incorporates digital literacy and type of organization as context-specific factors. The model suggested a direct influence from PU, PEU, FC, and SI on the intention to adopt e-GP. An indirect effect of PU on the intention to adopt e-GP is proposed through PEU. The intention to adopt e-GP is also assumed to be influenced by digital literacy directly and through PU and PEU. Moreover, a moderating role of the type of organization is proposed in the relationships between FC, SI, and the intention to adopt e-GP.

Intention to Adopt e-GP.
The intention to adopt e-GP refers to the degree to which an individual public sector employee is prepared to embrace and actively use the electronic government procurement system in their procurement-related tasks [15]. It reflects an individual's commitment to utilising the system as part of their work routine. In the proposed model, the intention to adopt e-GP is modelled as a latent variable capturing employees' behavioural intention to engage with the system, and it serves as the primary endogenous construct in the model. It encompasses dimensions such as their stated intention to use e-GP, their commitment to incorporating it into all procurement activities, their willingness to make an effort to adopt the system, and their overall enthusiasm for using electronic procurement in their work [11, 15]. Moreover, a seven-point Likert scale is proposed to measure these dimensions.

Perceived Usefulness.
Perceived usefulness (PU) refers to the extent to which an individual considers that using the electronic government procurement (e-GP) system will improve their job performance, efficiency, and effectiveness [6]. This construct is grounded in the TAM and suggests that if users perceive a system as useful, they are more likely to adopt it. In the proposed model, PU is operationalized to include dimensions such as expected usefulness in accomplishing tasks quickly and enhancing job performance, productivity, and effectiveness [15]. A seven-point Likert scale is proposed to measure these dimensions and PU is hypothesized to have a direct positive effect on the intention to adopt e-GP. Additionally, PU may mediate the relationships between PEU and adoption intention, as well as digital literacy and adoption intention.

Perceived Ease of Use.
Perceived ease of use (PEU) captures the degree to which an individual supposes that using the e-GP system is free of effort [6]. It encompasses dimensions such as perceived easiness to learn, interact without much mental effort, and accomplish job tasks easily [6, 15]. A seven-point Likert scale is suggested to measure these dimensions. As suggested by TAM, PEU influences both PU and the adoption intention of e-GP in the proposed model. Additionally, it may mediate the relationship between digital literacy and adoption intention.

Facilitating Conditions.
Facilitating conditions (FC) refer to the extent to which individuals believe that the

necessary organizational and technical resources are available to support their use of the e-GP system [11]. This construct is derived from UTAUT and covers dimensions such as the availability of resources, knowledge and guidance to use e-GP in organizations [6, 11]. Furthermore, a seven-point Likert scale is proposed to measure these dimensions. In the proposed model, facilitating conditions are expected to positively impact the intention to adopt e-GP of public sector staff in Sri Lanka.

Social Influences.
Social influence (SI) represents the extent to which an individual's decision to adopt the e-GP system is shaped by the opinions and encouragement of colleagues, supervisors, suppliers, and even service recipients of the organization [6]. It covers the expected influence from co-workers, supervisors, organizations, suppliers and citizens in this model [11]. Additionally, a seven-point Likert scale is proposed to measure these dimensions. Based on the UTAUT, the SI is hypothesized to have a positive effect on the intention to adopt e-GP in the proposed model.

Digital Literacy.
Digital literacy encompasses an individual's ability to effectively use digital technologies, solve technical problems, and adapt to new digital tools [16]. This construct is operationalized using dimensions such as the ability to solve ICT-related problems, learn new technologies easily, follow updates on technology, and feel confident in searching for information on the web. As the above constructs, this is also proposed to be measured using a seven-point Likert scale. Previously, this construct was introduced as an additional variable for TAM in the context of digital technology adoption intention of administrative staff in higher education in Turkey [16]. In the proposed model for Sri Lanka, digital literacy is expected to positively influence PEU, PU and intention to adopt e-GP.

Type of Organization.
In Sri Lanka, organizational users of e-GP are of two distinct types. They are Ministries/Departments and State-Owned Enterprises (SOEs) with different structures, governance mechanisms, and operational priorities. For instance, Ministries/Departments follow hierarchical and bureaucratic decision-making structures compared to the commercially oriented structures in SOEs [17]. Therefore, we believe that these organization-specific characteristics can influence how e-GP is adopted in the Sri Lankan context. In the proposed model, the type of organization is operationalized as a binary categorical variable. Moreover, it is expected that the relationships between facilitating conditions and intention to adopt e-GP, and social influences and intention to adopt e-GP, are moderated by the type of organization in Sri Lankan public sector.

Figure 2 below shows the conceptual model to forecast the factors of intention to adopt e-GP in the Sri Lankan public sector after incorporating the aforementioned constructs and interactions.

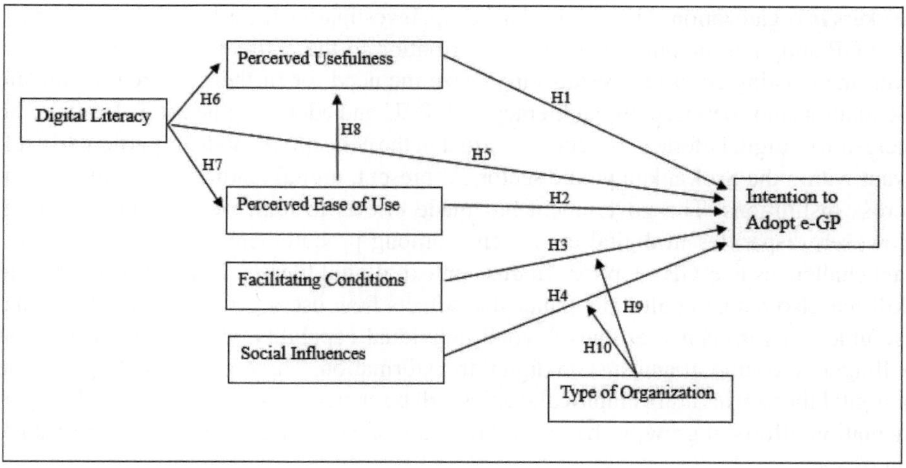

Fig. 2. The suggested conceptual model for the Sri Lankan public sector

4 Discussion

The conceptual model proposed in this paper incorporates more specific determinants of e-GP adoption intention in Sri Lanka, and those constructs are derived from key theoretical frameworks and logical reasoning. As per the TAM, PU and PEU are the key determinants of technological adoption intention. Hence, these two constructs have been used in e-GP adoption research in different contexts. Accordingly, the positive effects of PU on the intention to adopt e-GP have been empirically validated in multiple contexts including Jordan, Ghana, Malaysia, Netherlands and Nigeria [6, 10]. However, the associations between PEU and PU, as well as PEU and the intention to adopt e-GP, have yielded mixed results in the e-GP context. As suggested by the TAM, some studies have found a positive association between these constructs [6]. However, the inability to find significant empirical relationships in some studies highlights the need for further inquiry into this area [10]. Therefore, we believe the situation in Sri Lanka will provide an opportunity to conduct an empirical study on these relationships.

Deriving from UTAUT, FC and SI were also found to be significant determinants of behavioural intention towards technological adoption. Accordingly, many researchers have conducted studies to empirically validate these relationships in the e-GP research arena [4, 5, 11, 12]. Results indicate that the relationship between SI and the intention to adopt e-GP has been empirically supported in most scenarios [4, 5]. However, the association between FC and adoption intention has not been empirically supported in several e-GP research [11, 12]. Therefore, we believe that the results from the Sri Lankan context will further refine this area while enhancing the applicability of UTAUT in e-GP research.

The influence of digital literacy on the intention to adopt technology has been previously examined mainly in the education context [16]. Additionally, this has also been studied in the contexts of older adults who live separately from their families [18] and workers in organizations [19]. In this backdrop, investigating the effect of digital literacy on e-GP adoption intention is a novel contribution to the entire e-GP research. Moreover, the existing empirical results emphasize the need for further research to validate the relationships between digital literacy, PU, PEU and adoption intention [16, 19]. The inclusion of digital literacy as a core construct in the conceptual model is particularly relevant within the Sri Lankan public sector. At present, digital readiness remains uneven across institutions. The government has made efforts to improve ICT infrastructure. However, disparities in digital competence among procurement officers pose significant challenges to e-GP adoption. In this context, digital literacy is not only a technical skill but also a socio-cultural enabler that affects how users perceive ease of use and usefulness. Its influence extends beyond individual capability, shaping organizational willingness to support and sustain digital transformation. Therefore, validating the role of digital literacy in future empirical studies will be critical for ensuring that e-GP implementation efforts align with the ground realities of Sri Lanka's diverse administrative landscape.

Sri Lanka's public sector comprises a diverse range of entities, classified based on different criteria [17]. Among them, the organizations that use e-GP can be categorized into two groups as SOEs and Ministries/Departments. Previous studies have emphasized the role of context-specific factors in influencing the intention to adopt e-GP [4, 6]. Therefore, we believe that examining the moderating role of the type of organization in the relationships between SI, FC, and e-GP adoption intention will extend UTAUT and enhance its validity in e-GP adoption research.

As discussed above, the proposed model addresses several theoretical inconsistencies and highlights areas lacking empirical exploration. Considering the theoretical relationships, logical reasoning, and identified empirical gaps, the following hypotheses are proposed for future validation.

Direct Effects:

H1: In the Sri Lankan public sector, perceived usefulness is expected to have a positive impact on employees' intention to adopt the e-GP system.
H2: Perceived ease of use is anticipated to have a positive influence on the intention to adopt e-GP.
H3: Facilitating Conditions would have a positive influence on the intention to adopt e-GP.
H4: Social Influence is expected to have a positive influence on the intention to adopt e-GP.
H5: Digital Literacy would have a direct positive influence on the intention to adopt e-GP.

Mediated Effects:

H6: Digital Literacy is expected to positively shape users' Perceived Usefulness.
H7: Digital Literacy is expected to positively shape users' Perceived Ease of Use.

H8: Perceived Ease of Use would positively influence Perceived Usefulness.

Moderating Effects:

H9: The relationship between Facilitating Conditions and adoption intention is hypothesized to vary based on the type of organization.
H10: The Type of Organization is proposed to moderate the relationship between Social Influence and the intention to adopt e-GP.

5 Conclusion and Future Research Directions

This study contributes to existing e-GP research by developing a context-specific conceptual model to examine the factors affecting the intention to adopt e-GP in the Sri Lankan public sector. The existing partial implementation of the e-GP system in the country has also been considered during the model development process. Derived from TAM, UTAUT, and logical reasoning, the model identifies several key constructs to predict the adoption intention of public sector staff in Sri Lanka. Accordingly, PU, PEU, FC, and SI are proposed as direct influences on e-GP adoption intention. Additionally, digital literacy is introduced as a direct influencer of PU, PEU, and e-GP adoption intention. To explore unique determinants for the Sri Lankan context, the moderating role of the type of organization in the relationships between FC, SI, and e-GP adoption intention has been proposed in the model.

This model contributes to digital governance theory by integrating context-specific variables such as organizational type and digital literacy, which have been underexplored in prior TAM/UTAUT-based models. It highlights how public sector institutional diversity and socio-cultural factors like digital literacy shape technology adoption, an area critical for developing country contexts. Beyond its theoretical contribution, the model offers practical guidance for digital transformation in the public sector. By incorporating context-specific variables such as digital literacy and organizational type, it enables policymakers to design targeted capacity-building initiatives and change management strategies tailored to the unique needs of their organizations.

The non-validation of the model with empirical data is the main limitation of the study. Therefore, it is suggested that future researchers design survey items using the proposed dimensions and measurement scales in this study and validate them by employing quantitative approaches. Structural equation modelling (SEM) is highly recommended for this purpose.

In addition to quantitative validation, qualitative research could provide deeper insights into the nuanced experiences and institutional dynamics that influence e-GP adoption. In-depth interviews or focus groups with procurement officers, IT staff, and organizational leaders can uncover hidden barriers, motivations, or cultural factors that quantitative methods might overlook. Such an approach would be particularly valuable in understanding how resistance to change, trust in digital systems, or internal communication practices affect adoption intentions. Moreover, qualitative studies could help refine existing constructs or reveal new context-specific variables relevant to public procurement in developing countries like Sri Lanka.

References

1. Bandara, G.: Factors influencing the adoption of e-procurement for public sector works in Sri Lanka: a case study analysis. In: Built Environment and Spatial Sciences Sessions, General Sir John Kotelawala Defence University, pp. 207–216 (2021)
2. Ministry of Finance: Implementation of the Electronic Government Procurement (e-GP) System in Sri Lanka (2019)
3. Davis, F.D.: Perceived usefulness, perceived ease of use, and user acceptance of information technology. MIS Q. **13**(3), 319 (1989). https://doi.org/10.2307/249008
4. Addy, M.N., Addo, E.T., Abdulai, S.F., Kwofie, T.E., Aigbavboa, C.O., Adade-Boateng, A.O.: E-procurement acceptance in the Ghanaian public sector: an application of an extended technology acceptance model (TAM) in the construction industry. JEDT (2024). https://doi.org/10.1108/JEDT-08-2023-0373
5. Nassir, S., Lebdaoui, I., Chetioui, Y., Lebdaoui, H.: Factors influencing SMEs' intention to adopt electronic tendering: empirical evidence from an emerging African market. JEAS (2023). https://doi.org/10.1108/JEAS-10-2022-0235
6. Ofori, D., Light, O., Ankomah, J.: Adoption intentions of electronic procurement among public sector organisations (PSOs) in Ghana: emerging economy perspective. JOPP **23**(2), 179–199 (2023). https://doi.org/10.1108/JOPP-09-2022-0045
7. Venkatesh, V., Morris, M.G., Davis, G.B., Davis, F.D.: User acceptance of information technology: toward a unified view. MIS Q. **27**(3), 425 (2003). https://doi.org/10.2307/30036540
8. Venkatesh, V., Thong, J.Y.L., Xu, X.: Consumer acceptance and use of information technology: extending the unified theory of acceptance and use of technology. MIS Q. **36**(1), 157 (2012). https://doi.org/10.2307/41410412
9. Tamilmani, K., Rana, N.P., Wamba, S.F., Dwivedi, R.: The extended Unified Theory of Acceptance and Use of Technology (UTAUT2): a systematic literature review and theory evaluation. Int. J. Inf. Manage. **57**, 102269 (2021). https://doi.org/10.1016/j.ijinfomgt.2020.102269
10. Soong, K.-K., Ahmed, E.M., Tan, K.-S.: Factors affecting Malaysia's SMEs in using public electronic procurement. J. Info. Know. Mgmt. **19**(02), 2050008 (2020). https://doi.org/10.1142/S0219649220500082
11. Kit, S.K., Ahmed, E.M., Tan, K.S.: Social influences' effects on Malaysia's SMEs' public electronic procurement usage. Int. J. Electron. Gov. Res. **17**(1), 68–82 (2021). https://doi.org/10.4018/IJEGR.2021010105
12. Addy, M.N., Addo, E.T., Kwofie, T.E., Yartey, J.E.: Predicting the adoption of e-procurement in construction project delivery in Sub-Saharan Africa: an application of UTAUT2. CI **23**(5), 1038–1053 (2023). https://doi.org/10.1108/CI-09-2021-0174
13. Weerasinghe, W.L.G.D.S., Kanaheraarachchi, B.J., Jayaweera, L.P.M.I., Aryaratne, I.A., Samarasinghe, U., Karunarathna, N.: Critical success factors in the implementation of electronic government procurement in Sri Lanka. Presented at the 13th International Conference on Business & Information ICBI, University of Kelaniya, pp. 001–013 (2022)
14. Jaakkola, E.: Designing conceptual articles: four approaches. AMS Rev. **10**(1–2), 18–26 (2020). https://doi.org/10.1007/s13162-020-00161-0
15. Gamal Aboelmaged, M.: Predicting e-procurement adoption in a developing country: an empirical integration of technology acceptance model and theory of planned behavior. Ind. Manag. Data Syst. **110**(3), 392–414 (2010). https://doi.org/10.1108/02635571011030042
16. Kabakus, A.K., Bahcekapili, E., Ayaz, A.: The effect of digital literacy on technology acceptance: an evaluation on administrative staff in higher education. J. Inf. Sci. (2023). https://doi.org/10.1177/01655515231160028

17. Kamburawala, T.U., Abeyrathne, D.H.: Reforming State-Owned Enterprises (SOES) to facilitate the economic recovery of Sri Lanka. HAPScPBS **4**(2), 129–137 (2023). https://doi.org/10.12681/hapscpbs.36691
18. McGinty, J., Oest, J., Divido, J.: Digital immigrants: leveraging the technology acceptance model to increase digital literacy in older adults. New Dir. Adult Contin. Educ. **2024**(181), 9–18 (2024). https://doi.org/10.1002/ace.20517
19. Nikou, S., De Reuver, M., Mahboob Kanafi, M.: Workplace literacy skills—how information and digital literacy affect adoption of digital technology. JD **78**(7), 371–391 (2022). https://doi.org/10.1108/JD-12-2021-0241
20. Dissanayaka, W.M.D.M., Sahapattana, P.: Unifying the fragmented landscape of electronic public procurement research: a systematic review and future directions. JeDEM **17**(3), 78–105 (2025). https://doi.org/10.29379/jedem.v17i3.1002

E-learning Recommendation Systems: Synthesis, Applications, and Future Trends

Imran Khaled[1(✉)], Hanene Trichili[2], and Faiez Gargouri[3]

[1] Administrative and Information Sciences, Al Istiqlal University, Jericho, Palestine
Imran.khaled1992@gmail.com

[2] Higher Institutions of Biotechnology of Sfax (ISBS), University of Sfax, Sfax, Tunisia
hanene.trichili@isbs.usf.tn

[3] Higher Institutions of Computer Science and Multimedia of Sfax (ISIM), University of Sfax, Sfax, Tunisia
faiez.gargouri@usf.tn

Abstract. Due to the importance of recommendation system in the e-learning domain and its ability to recommends students with resources of their individual interest, researchers have conducted into RS field and proposed various techniques and approaches to develop and enhance the quality of recommendation systems. In this comprehensive current State-of-the-Art we investigate techniques, similarity measurements, and evolution metrics in the journey of RS revolution. Besides, we have addressed challenges facing RS development such as, cold-star, data sparsity, and scalability. In addition, a wide range of RS applications have been addressed, to provide students with personalized content, increasing student engagements, and enhance learning experience. Also, we addressed trends and new directions of RS that promise to improve recommendations quality and in rich RS improvement such as, feature engineering and feature crossing, context aware, and emotion aware to offer more accurate, responsive, and relevant learning resources.

Keywords: E-learning · Recommendation System (RS) · Content-Based Filtering (CBF) · Collaborative Filtering (CF)

1 Introduction

The amount of material available on the web has grown exponentially [1] as a result, it is challenging for users to locate what they're seeking for. In order to handle this data swiftly and effectively offer customers personalized recommendations, recommendation systems (RS) are essential [11]. Based on user preferences, these systems recommend relevant material to users [1]. From basic non-personalized systems [17] that suggest well-known products without taking into account user preferences, to complex deep learning systems [3] that make recommendations based on user preferences and past interactions, RS have undergone constant development since their inception in the early 1990s [17]. Due to the importance of RS, it has been adopted in a number of fields, such as e-learning, healthcare, and e-commerce [10]. In particular, RS offers individualized classes, exercises, and learning resources in the e-learning domain that are catered to the requirements and proficiency levels of students.

B.-E. Boudriki Semlali et al. (Eds.): DATA 2025, LNISO 81, pp. 74–82, 2025.
https://doi.org/10.1007/978-3-032-06725-8_7

Collaborative filtering (CF), content-based filtering (CBF), and hybrid systems are the three primary systems. While CF suggests items by identifying similarities between users or interactions using memory-based techniques (e.g., item-item or user-user similarity) or model-based approaches like matrix factorization, Singular Value Decomposition (SVD), and Alternating Least Square (ALS) [17], CBF recommends items based on how similar the items are to a user's prior interactions [7, 16]. By utilizing strategies like weighted methods or switching methods [9], hybrid systems integrate CF and CBF to capitalize on the advantages of both methodologies. Feedback is essential to RS, especially in e-learning, where it helps RS make better recommendations and improves the educational process. In order to get around restrictions, feedback can be either explicit (like user ratings) [2], implicit (like clicks or interaction history), or hybrid. This study expands our previous work [20], where we investigated methods and algorithms of RS in the e-learning domain, in this study we will introduce similarity measurements, evaluation metrics, applications, challenges, and new directions of RS in the e-learning domain to inspire others researches for continuous improvements of recommendations systems in the e-learning space.

2 Related Work

Extensive investigation into the body of current literature has been motivated by the growing importance of recommendation systems (RS) in e-learning. While some research, like [11], examine more general applications of RS across e-learning and other fields, others, like [1], concentrate only on content-based filtering (CBF) strategies, addressing issues like the cold start problem and data sparsity. A generic framework for large-scale e-learning systems is proposed by reviews such as [4], which focus on the fundamental ideas of e-learning settings. [11] Which analyze the designs, functions, and difficulties of existing systems. Furthermore, it draws attention to obstacles in the implementation of e-learning frameworks, highlighting the necessity of a thorough conceptual framework that incorporates current research.

Numerous research examines the conceptual and technical facets of RS in online education. For instance, [1] evaluates the latest machine learning methods with an emphasis on knowledge-based, hybrid, CF, and CBF systems, while [11] looks at methods that use both explicit and implicit feedback. In addition it emphasizes empirical research and practical RS implementations in e-learning and e-commerce, [4] concentrate on conceptual developments such as frameworks, components, structures, and trends. Further contributions are made by recent works like [12, 19]: [12] talks about classification methods and examines personalization tactics (2013–2023), and [19] examines current e-learning RS trends.

Our previous study [20] addressed methods and techniques of RS in the e-learning field. However, further investigations are needed to explore similarity measurements, evaluation metrics, applications of RS, challenges facing RS, and new directions of RS in the e-learning domain, which is the focus of this study. By tackling these issues, we intend to motivate academics to advance and keep improving RS in the field of e-learning. Offering a comprehensive and focused resource for enhancing individualized learning experiences, we are exclusively focused on e-learning.

3 Methodology

By analyzing peer-reviewed literature that was sourced from IEEE Xplore, Scopus, and SpringerLink, this study synthesizes the state-of-the-art in e-learning recommendation systems. It also selectively incorporates seminal pre-2019 works that established crucial theoretical foundations. Studies exhibiting technical innovation (such as deep learning adaptations [3], hybrid systems [9], and fog computing implementations [5]) and rigorous empirical validation (reporting precision $\geq$80% or addressing foundational issues like cold-start [7] and scalability) were given preference during the selection process. Twenty high-impact articles were selected from a starting pool of 92 publications on the basis of their (1) applicability to e-learning personalization, (2) methodological rigor, and (3) citation influence in the domain. Techniques were compared across metrics (accuracy, recall, scalability) and applications (adaptive learning [4], social recommendations [6]). The results were validated against expert evaluations and contextualized through real-world case studies. In order to ensure that the study strikes a balance between cutting-edge advancements and crucial theoretical foundations.

4 RS Techniques, Algorithms, Similarity Measurements, and Evaluation Metrics in E-Learning Domain

Due to the ability of RS to provide individualized learning resources, improve student engagement, and optimize instructional content, recommendation systems (RS) have emerged as a crucial part of e-learning platforms [10, 16]. With an emphasis on customizing information to each student's interests and preferences, researchers have investigated a number of strategies to enhance the quality, accuracy, and performance of RS in e-learning. In addition to more sophisticated approaches like ontology-based [7], sentiment-based, and AI-driven methods [2], techniques like collaborative filtering (CF), content-based filtering (CBF), and hybrid systems have been widely employed. Through strict assessment measures and similarity measurements, these systems use big data, data mining, and machine learning [16] to improve recommendations and make sure that they live up to student expectations.

Similarity measurements are essential for RS in e-learning since they evaluate how similar students or courses are to one another, hence determining the relevancy of suggestions. While more sophisticated approaches like semantic similarity, ontology-based similarity, and trust-based similarity have also been used, common approaches include Cosine Similarity, Jaccard Similarity, Pearson Correlation, and Euclidean Distance [1, 7]. By using these metrics, suggestions are improved and students are given content that suits their interests and habits. By using these techniques, RS may offer recommendations that are more precise and tailored to each individual, enhancing the educational process as a whole.

When evaluating RS's efficacy in e-learning, evaluation metrics are crucial. To assess performance in controlled settings, offline metrics like precision, recall, F1-score, and RMSE [5, 10] rely on past data. Diversity, uniqueness, and coverage metrics [10] guarantee a wide and pertinent range of recommendations, whereas ranking metrics such as NDCG [3] and MRR assess the quality of recommendations. While user-centric techniques like satisfaction surveys [9, 10] and churn rate analysis concentrate on long-term user happiness and retention, online measurements like click-through rate (CTR) and engagement metrics [6] offer real-time insights into user interaction. Notwithstanding their usefulness, RS assessment measures have drawbacks, including the cold start problem, which occurs when accuracy is hampered by a lack of data for new students or courses, and trade-offs between precision and diversity, which occur when one indicator is optimized at the price of another. These drawbacks emphasize the necessity of continued study and development to improve RS in e-learning and guarantee that students receive accurate, pertinent, and interesting content (Table 1).

Table 1. Comparative Analysis of Recommendation System Techniques: Algorithms, Metrics, and Trade-offs

Ref.	Technique/Method	Algorithms Used	Evaluation Metrics	Strengths	Weaknesses
[1]	Collaborative Filtering	User/Item-based, Matrix Factorization (SVD, ALS), Neural CF	Precision, Recall, NDCG, MAE, RMSE	Effective with rich interaction data, domain-agnostic	Cold start, data sparsity, popularity bias
[7]	Content-Based Filtering	TF-IDF, Cosine Similarity, Naïve Bayes, Decision Trees	Precision, Recall, MAE, RMSE	Handles new items well, interpretable	Over-specialization, requires quality features
[9]	Hybrid Systems	Weighted, Switching, Feature-Combining models	Precision, Recall, diversity, novelty	Combines multiple strengths	Complex to implement and maintain
[15]	Context-Aware	Contextual bandits, Tensor Factorization, Multi-Contextual Deep Learning	CTR, NDCG, diversity, novelty	Highly personalized recommendations	Complex to implement, scalability issues
[5]	Fog-Based Systems	Edge computing algorithms, Federated Learning	Latency, throughput, accuracy	Low latency, privacy-preserving	Limited edge resources, complex deployment
[5]	Cloud-Based Systems	Distributed matrix factorization, Cloud-based deep learning	Scalability, latency	Handles massive datasets	Infrastructure costs, real-time latency
[7]	Attribute-Aware	Attribute-aware CF, Factorization Machines, Autoencoders	MAE, RMSE, diversity	Improves recommendation diversity	Computationally expensive, feature-dependent
[3]	AI-Based Systems	Deep Learning, Reinforcement Learning, Graph Neural Networks	Precision, Recall, NDCG	High accuracy, adapts to patterns	Black-box nature, resource-heavy

(continued)

Table 1. *(continued)*

Ref.	Technique/Method	Algorithms Used	Evaluation Metrics	Strengths	Weaknesses
[8]	Knowledge-Based and Ontologies	Ontologies and sequential patterns, Prefix Span algorithm	Accuracy, coverage, user satisfaction	Solves cold-start, and temporal personalization	Computational cost, and domain dependency
[7]	Fuzzy Logic	Fuzzy matching, Fuzzy CF, Fuzzy inference systems	Explainability, user satisfaction	Handles uncertainty well	Computationally expensive, less scalable
[11]	Topic Modeling	LDA, Non-negative Matrix Factorization	Precision, Recall, diversity	Effective for textual data	Oversimplifies complex topics
[2]	Sentiment-Based	LSTM, BERT, Text Classification	Sentiment alignment, diversity	Incorporates user emotions	Limited to sentiment-rich domains
[9]	Social Networks	Graph-based models (Random Walks, Node2Vec), Social CF	Precision, coverage, diversity	Leverages social connections	Privacy concerns, data sparsity
[7]	Data Mining	Association Rule Mining, Clustering, Frequent Pattern Mining	Confidence, support, lift	Discovers hidden patterns	Not real-time adaptable
[5]	Web Mining	Web crawling, Clickstream analysis, Log analysis	CTR, diversity, novelty	Identifies user behavior patterns	Requires heavy preprocessing
[6]	Self-Organization	Self-organizing maps (SOM), Adaptive Resonance Theory	Clustering accuracy, diversity	Adaptive and unsupervised	Limited interpretability
[18]	Machine Learning and CF	User based CF, SVD, Random forest, and XGBoost	Precision, Recall, RMSE	Explainable recommendations through feature importance	Computational expensive, require large training data
[5]	Big Data	Spark MLlib, Hadoop-based algorithms	Scalability, throughput	Processes massive datasets	High infrastructure costs
[3]	Deep Learning	RNNs, CNNs, Transformers, Autoencoders	Precision, Recall, NDCG, novelty	State-of-the-art accuracy	Computationally expensive, opaque

5 Applications of Recommendation System in E-learning

By optimizing content distribution, enhancing results, and personalizing the learning experience, recommendation systems (RS) have greatly improved e-learning. Applications of RS in e-learning include learning style predictions to match resources with visual, auditory, or aesthetic preferences [9, 16], adaptive learning paths based on individual progress and emotional states [2], and personalized learning content recommendations based on prior interactions and preferences [6, 9]. Furthermore, RS uses contextual information such as time, device, and location to provide pertinent resources [15], proposes peers for collaborative learning [6], and recommends exercises and activities to reinforce knowledge [4]. While video-based recommendations maximize engagement through customized video suggestions [3], performance-based recommendations assist learners in improving by examining previous performance [19]. In e-learning systems, RS generally improves student engagement, lowers effort, and promotes knowledge exchange and collaboration.

6 Challenges of Recommendation System in E-learning

Notwithstanding the advantages of RS in e-learning, a number of obstacles limit its efficacy. Accurate suggestions for new users or material are challenging due to the cold start problem and data sparsity; however, this difficulty can be lessened by hybrid approaches that combine collaborative and content-based filtering. Another issue is scalability, since big datasets in e-learning platforms raise the cost of computing and necessitate the use of dimensionality reduction methods like matrix factorization or distributed frameworks like Spark and Hadoop [5]. Diverse learner needs and preferences make customization and context-awareness difficult, although techniques like learning style prediction and hybrid approaches increase accuracy [15, 16, 19].

Although fuzzy logic and neural networks aid in predicting preferences, learning styles and ability variation make recommendations much more difficult [7]. Feedback quality is an issue since implicit feedback such as clicks and time spent is noisy and less accurate than explicit feedback, necessitating the use of sophisticated analysis methods like matrix factorization and deep learning [7]. Since several criteria (such as accuracy and diversity) and techniques like A/B testing are required to evaluate outcomes like engagement and information retention, RS effectiveness evaluation is complicated [2]. Concerns about privacy when gathering user data [14] and the interpretability issue with sophisticated models like deep learning are other difficulties that undermine user confidence and regulatory compliance.

7 Trends and New Directions in RS of E-learning

Advances in personalization are highlighted by recent advances in recommender systems (RS) for e-learning, which are fueled by multimodal data, hybrid techniques, and the inclusion of implicit feedback. By taking into account a variety of learner preferences, hybrid systems that integrate collaborative filtering, content-based filtering, and knowledge-based techniques [9, 11] increase the accuracy of recommendations. In order to better forecast student interests and engagement, implicit feedback such as click-through rates and time spent on materials is being used in conjunction with explicit feedback more and more [2]. To improve recommendations, multimodal techniques combine text, image, and video data; deep learning frameworks use these inputs to provide tailored video recommendations [3]. Additionally, social and collaborative learning systems integrate peer behavior and tagging for community-driven recommendations [6, 7, 15], and sophisticated feature engineering approaches like Auto-Cross optimize feature interactions.

Scalability, emotional prediction, and adaptive learning are the main areas of concentration for new developments in e-learning RS. With the help of big data technologies like Spark and Hadoop [5] and cloud/fog computing [5, 10], millions of students can process massive datasets in a scalable manner. The discipline is dominated by deep learning and reinforcement learning models [2, 3], which enable systems to dynamically modify recommendations in response to real-time engagement. Content distribution is further personalized using web mining [5, 9], emotional and psychological profiles [13], and fuzzy logic-based learning style prediction [7]. Knowledge-based and ontology-driven systems [11] improve recommendations by comprehending the connections among curriculum objectives, learner preferences, and resources. Driven by developments in AI, big data, and personalized learning technologies, these trends point to a future in which e-learning RS are more flexible, context-aware, and able to satisfy a range of student needs.

8 Result and Discussion

This survey summarizes developments in e-learning recommendation systems (RS), showing how they have progressed from simple content-based/collaborative filtering to advanced AI-driven methods such as deep learning [3] and hybrid systems [9]. New emotion-aware [2, 7] and semantic approaches hold promise for tailored learning, even though conventional similarity measures (Cosine, Jaccard) are still widely used. Our results support the success of hybrid systems as reported in the literature [1, 11], but also highlight important shortcomings in addressing cold beginnings [7], scalability (despite big data tools [5]), and the incorporation of emotion-aware measures. Research limitations like dataset biases [4, 16] and an excessive dependence on offline measurements [10] exacerbate the difficulties faced by practical applications like adaptive learning routes [4], which include privacy issues [14] and "black-box" AI models [3]. To close the gap between theoretical innovation and real-world application in international e-learning environments, future directions should give priority to multimodal emotion-aware RS, scalable privacy-preserving techniques (e.g., federated learning [5]), explainable models [8], and inclusive solutions for underrepresented learners [16].

9 Conclusion

Algorithm developments and the increasing need for individualized learning experiences have led to a considerable evolution in recommender systems (RS) in e-learning. In addition to cutting-edge methods like deep learning and emotion-aware systems, this survey highlights a variety of methods and algorithms that are used to deliver personalized suggestions, such as collaborative filtering, content-based filtering, and hybrid systems. These techniques have improved learning outcomes and user engagement with encouraging results. In order to increase the accuracy of recommendations, similarity metrics including trust-based, emotion-based, and semantic-based approaches are essential. System effectiveness is evaluated using metrics like as precision, recall, F1-score, and RMSE. The full potential of RS in e-learning is, however, constrained by persistent concerns such cold-start problems, scalability issues, data sparsity, and the requirement for real-time updates.

Personalized learning materials, adaptive assessment, social learning support, and course suggestions are just a few of the many e-learning uses for RS. The future of RS is being shaped by recent trends that highlight sophisticated personalization approaches like sentiment analysis, emotion-aware systems, and feature crossing, which will provide more precise, responsive, and captivating learning experiences. This study is a complementary of previous work [20], we hope that this article helps researchers and professionals interested in the recommendation system of e-learning domain with the knowledge of current State-of-the-Art of comprehensive techniques, similarity measurements, evaluation metrics, applications of RS, challenges in RS, and new trends to solve problems related to recommender systems, and continuous enhancement of RS in e-learning domain.

References

1. Khanal, S.S., et al.: A systematic review: machine learning based recommendation systems for e-learning. Educ. Inf. Technol. **25**, 2635–2664 (2020)
2. Vedavathi, N., Anil Kumar, K.M.: E-learning course recommendation based on sentiment analysis using hybrid Elman similarity. Knowl.-Based Syst. **259**, 110086 (2023)
3. Xu, W., Zhou, Y.: Course video recommendation with multimodal information in online learning platforms: a deep learning framework. Br. J. Edu. Technol. **51**(5), 1734–1747 (2020)
4. Zayet, T.M.A., et al.: What is needed to build a personalized recommender system for K-12 students' E-Learning? Recommendations for future systems and a conceptual framework. Educ. Inf. Technol. **28**(6), 7487–7508 (2023)
5. Dahdouh, K., et al.: Large-scale e-learning recommender system based on Spark and Hadoop. J. Big Data **6**(1), 1–23 (2019)
6. Wan, S., Niu, Z.: A hybrid e-learning recommendation approach based on learners' influence propagation. IEEE Trans. Knowl. Data Eng. **32**(5), 827–840 (2019)
7. Albatayneh, N.A., Ghauth, K.I., Chua, F.-F.: Utilizing learners' negative ratings in semantic content-based recommender system for e-learning forum. J. Educ. Technol. Soc. **21**(1), 112–125 (2018)
8. Tarus, J.K., Niu, Z., Yousif, A.: A hybrid knowledge-based recommender system for e-learning based on ontology and sequential pattern mining. Futur. Gener. Comput. Syst. **72**, 37–48 (2017)
9. Benhamdi, S., Babouri, A., Chiky, R.: Personalized recommender system for e-Learning environment. Educ. Inf. Technol. **22**, 1455–1477 (2017)
10. Ghauth, K.I., Abdullah, N.A.: Measuring learner's performance in e-learning recommender systems. Australas. J. Educ. Technol. **26**(6), 764–774 (2010)
11. Liu, M., Yu, D.: Towards intelligent E-learning systems. Educ. Inf. Technol. **28**(7), 7845–7876 (2023)
12. Bin, Q., Zuhairi, M.F., Morcos, J.: A Comprehensive study on personalized learning recommendation in e-learning system. IEEE Access **2**, 100446–100482 (2024)
13. Salazar, J.C., et al.: A generic architecture of an affective recommender system for e-learning environments. Univ. Access Inf. Soc. **23**(3), 1115–1134 (2024)

14. Majjate, H., et al.: Assessing the impact of ethical aspects of recommendation systems on student trust and engagement in E-learning platforms: a multifaceted investigation. Educ. Inf. Technol. **30**, 3953–3977 (2024)
15. Anter, A., Moawad, I.F., Mosaad, S.M.: Enhancing e-learning materials recommendations with contextual awareness and collaborative filtering. In: 2024 6th International Conference on Computing and Informatics (ICCI). IEEE (2024)
16. Ren, C.: Immersive E-learning mode application in Chinese language teaching system based on big data recommendation algorithm. Entertain. Comput. **52**, 100774 (2025)
17. Zahrae, F.F., Es-Sâadia, A., Mohamed, Y.: Recommendation system for E-learners: model-based collaborative filtering approach. In: 2024 IEEE 12th International Symposium on Signal, Image, Video and Communications (ISIVC). IEEE (2024)
18. Alanya-Beltran, J.: Personalized learning recommendation system in e-learning platforms using collaborative filtering and machine learning. In: 2024 International Conference on Advances in Computing, Communication and Applied Informatics (ACCAI). IEEE (2024)
19. Liu, R.: Simulation of E-learning in English personalized learning recommendation system based on Markov chain algorithm and adaptive learning algorithm. Entertain. Comput. **51**, 100719 (2024)
20. Khaled, I., Trichili, H., Gargouri, F.: Recommendation systems in the e-learning domain: a survey. In: Proceedings of the International Computer Sciences and Informatics Conference (ICSIC 2024), Accepted for publication (2024)

Exploring the Determinants of Business Performance: A Regional Perspective Using Spatial Econometrics and AI-Powered Analytics

Basma Echaki[1(✉)], Mounir Boumhamdi[1], and Marouane Ikira[2]

[1] Laboratoire de Recherche Business Intelligence, Gouvernance des organisations, finance et politiques économiques, Université Hassan II, Faculté des Sciences Juridiques Économiques et Sociales de Casablanca, No8110 Oasis, Casablanca, Morocco
Basmaaechaaki@gmail.com

[2] Laboratory of Research in Management, Economics and Social Sciences (LARGESS), Faculty of Law, Economics and Social Sciences, Chouaib Doukkali University, El Jadida, Morocco

Abstract. The aim of the present paper is to analyse the determinants of the performance (measured by total factor productivity TFP) of 309 Moroccan manufacturing firms based on the World Bank Enterprise Survey's data (WBES) in 2023. The paper adopts a hybrid approach combining econometric methods, namely the multilevel model and the spatial autoregressive model (SAR), which helps analyse the joint impact of internal, regional and spatial factors, as well as artificial intelligence methods, the logarithmic K-means in particular, which enables the paper to deepen the analysis by clustering the firms sharing the same characteristics.

The results reveal several findings. Regarding firm-specific characteristics, most of the variables have a positive and significant impact on firm performance, such as digitalisation, size, innovation and skilled workers; except for exports and age. Reasearch has shown that export has a non-significant impact while age has a non-linear effect. As for regional factors, the results of the multilevel model and the log K-means underline the importance of the regional framework on a firm's productivity. Contrary to the expectations, the paper fails to confirm the impact of proximity on firm performance despite the inclusion of the SAR model. Future directions in this regard could base its findings on a more detailed regional analysis, which encompasses all Moroccan twelve regions along with the integration of data related to the distances between companies to estimate spatial effects more accurately.

Keywords: Firm performance · spatial autoregressive model · multilevel model · artificial intelligence · the logarithmic K-means

1 Introduction

The analysis of company performance is a central issue in economics and management, particularly in developing economies such as Morocco. Although company-specific characteristics have an impact on performance, the regional framework and spatial proximity

B.-E. Boudriki Semlali et al. (Eds.): DATA 2025, LNISO 81, pp. 83–99, 2025.
https://doi.org/10.1007/978-3-032-06725-8_8

(spillovers) also play a crucial role. Numerous studies have highlighted various determinants of company performance, such as human capital (Ramadani et al. 2019; Amara & Thabet 2019), company's age and size (Ramadani et al. 2019; Amara & Thabet 2019), digitalisation (Ribeiro-Navarrete et al. 2021; Mardani et al. 2023), innovation (Ramadani et al. 2019), infrastructure quality (Amara & Thabet 2019), regional density (Ferreira et al. 2010; Amara & Thabet 2019) and institutional quality (Ricotta, 2016). Many studies such as (Amara & Thabet 2019) have analysed proximity effects, yet most researchers has often neglected the impact of spatial interactions between regions and firms in terms of companies's performance. This is the gap that the present research seeks to analyze. It targets the joint impact of these three factors (internal, regional and spatial) by adopting a hybrid approach combining econometric methods (the multilevel model and the spatial autoregressive model) and artificial intelligence techniques, in particular logarithmic K-means, which enables the research to analyse firm performance in greater depth.

2 Related Works

The analysis of the determinants of business performance has become a central topic in the fields of economics and management. Numerous studies have focused on the joint impact of internal factors related to the characteristics of the firm and external or regional factors. Moreover, the advent of spatial econometrics and artificial intelligence now makes it possible to study the effects of internal and regional factors on a paticular firm's productivity in more dynamic and precise ways. This section presents an overview of existing work on the internal and regional determinants of company's performance.

2.1 Internal Determinants of Firm Performance

The specific characteristics of each company have a significant impact on its performance. In particular, **human capital** is one of the most important factors influencing business productivity. (Amara & Thabet 2019) and (Ramadani et al. 2019) have emphasized the crucial role of skilled employees in improving firm's productivity (measured by TFP).

In addition to human capital, **the size** and **age** of companies are key determinants to the firm's performance. The literature generally indicates a positive correlation between company size and productivity. According to (Ramadani et al. 2019) large firms perform better than small ones. In contrast, (Amara & Thabet 2019) find that the oldest small firms are more productive than large ones.

In reference to the age of firms, several studies have highlighted the negative effect of age, including (Ramadani et al. 2019) and (Ferreira et al. 2010), who state that young firms are more productive than old firms. On the other hand, (Amara & Thabet 2019) demonstrate that age has a non-linear impact on firms' performance given the fact that young companies need time to adapt to their environment (learning effects).

In relation to **exporting**, (Amara & Thabet 2019) conclude that exporting firms generally perform better than those that operate only on the domestic market. They argue that trade liberalization leads to increased competition, encouraging companies to improve their productivity so that the firms could remain active in export markets.

This aligns with the findings of (Burger et al 2017), who note that exporting companies are more resilient to economic downturns, which indicates that the firms can perform better when exporting. However, (Arnold & Hussinger 2005) find that exporting has no significant effect on the performance of German manufacturing firms.

Regarding **innovation**, (Ramadani et al. 2019) assert that companies which invest in innovation are more productive and competitive as opposed to those that do not.

The last internal factor that the paper explores is that of **digitalization**. According to (Ribeiro-Navarrete et al. 2021), digitalization plays a crucial role in improving productivity and business performance. This corresponds with the findings of (Mardani et al. 2023) who highlight the positive impact of digitalization on the total productivity of factors. (Gu et al. 2023) have also demonstrated the crucial role of digitalization on the performance of logistics companies by using digital orientation and digital competence as measuring variables. Finally, (Chatterjee et al. 2023), have proven that there is a positive and significant correlation between the adoption of Big Data Analytics (BDA) tools and business performance. They confirm the role of BDA tools in enabling firms to process massive data, deriving meaningful and important insights and making more informed decisions in real time.

2.2 Regional Determinants of Business Performance

In addition to the characteristics of individual companies, regional factors have a significant impact on firms' performance and productivity. There are a number of factors that influence business growth such as economic density, neighborhood effects, and the quality of regional infrastructure and institutions. (Amara & Thabet 2019) mention **infrastructure** as one of the factors that may explain productivity differences between different regions, based on existing data.

In terms of economic density, the concentration of companies in a specific area can generate positive externalities known as **agglomeration effects**. According to (Amara & Thabet 2019), industrial density has a positive and significant impact on a company's performance. They argue that concentrating production in a specific region or area improves company productivity by reducing transport costs. However, (Ferreira et al. 2010) were unable to demonstrate the impact of density on a company's performance.

As for **institutional quality**, (Ricotta 2016) reveals the existence of a positive relationship between the fight against corruption and total factor productivity(TFP).

3 Methodology

The paper describes the data collection, pre-processing, estimation model development and evaluation processes. It bases its analysis on an approach that combines a multivariate model, spatial econometrics (to capture proximity effects) and artificial intelligence so that it could provide a detailed analysis of the results.

3.1 Data Collection

The study focuses mainly on secondary data from the World Bank Enterprise Survey (WBES) conducted in 2023. The survey includes 598 Moroccan companies, 309 of which belong to the manufacturing sector. The paper emphasizes particularly this sector because of its importance to the Moroccan economy, as well as its significant contribution to the employment and productivity. The database subdivides The Moroccan territory into four large regions: Casablanca-Settat, North, South and East (Table 1).

Table 1. Summary of variables

Type of variable	Variable name	Description	Authors
Dependent variable	Firm performance	Represented by total factor productivity (TFP) which is estimated as the residual of the Cobb-Douglas function	(Amara & Thabet 2019)
Internal factors	Human capital	Proportion of workers with secondary education	Amara & Thabet 2019) and (Ramadani et al. 2019)
	Digitalization	Binary variable (1 = they have a website 0 = no). This measure is used as a proxy for a firm's level of digitalization.	(Ribeiro-Navarrete et al. 2021), (Mardani et al. 2023), (Chatterjee et al. 2023) and (Gu et al. 2023)
	Innovation	Binary variable (1 = investment in R&D, 0 = no)	(Ramadani et al. 2019).
	Export	Binary variable (1 = exporting company (direct/indirect), 0 = not)	(Amara & Thabet 2019) and (Arnold & Hussinger, 2005).
	Size of company	Categorical (small, medium, large) with large as reference	(Ramadani et al. 2019) and (Amara & Thabet 2019).
	Age of company	Number of years in business which represents the difference between the year of the survey (2023) and the year the company was created.	(Ferreira et al. 2010), (Ramadani et al. 2019) and (Amara & Thabet 2019).

(*continued*)

Table 1. *(continued)*

Type of variable	Variable name	Description	Authors
	Age squared	Variable added to capture the non-linear effect of age	(Amara & Thabet 2019).
Regional factors	Corruption	Regional average of the percentage of annual sales settled by informal payments.	(Ricotta, 2016).
	Regional density	Average number of firms per region	(Amara & Thabet 2019) and (Ferreira et al. 2010).
	Water obstacle	Binary variable (1 = If the company has encountered a water supply problem, 0 = not). This measure is used as a proxy to assess the quality of the infrastructure to which the company has access	(Amara & Thabet 2019).
Spatial factor	Distance - Casablanca	Distance between Casablanca-Settat and other regions. (= 0 if region = Casablanca-Settat, = 330 km if region = North, = 400 km if region = East, = 550 km if region = south).	We drew inspiration from the work of (Bourdin 2013) to create the Distance-Casa variable, which demonstrated the importance of the spatial factor by using Euclidean distance to define the neighborhood and assess the impact of proximity.

Source: Prepared by us using Stata software.

3.2 Data Preprocessing

Prior to the econometric analysis, the data aims to ensure consistency and quality. The research removes any outliers and missing data that could bias the results. In addition, the paper log-transforms some variables such as regional density, age of company and number of skilled workers to standardize scales and reduce the impact of extreme values.

3.3 Econometric Model and Artificial Intelligence Approach

The research uses three complementary methodological approaches to analyze business performance:

Multi-level Model. First, it examines the impact of internal factors (human capital, exports, innovation, digitalization, age and size of firms) and regional factors (economic density, quality of infrastructure and corruption) on firms' performance. To approach this analysis, the paper opts for a multilevel model that captures variability at both the regional and the firm level:

$$\text{Firm\_performance i, j} = \beta 0 + \beta 1\text{Xij} + \beta 2\text{Zj} + \text{uj} + \varepsilon\text{ij} \tag{1}$$

where:

Firm-performance: Firm performance which is represented by total factor productivity.
j: Represents the region.
i: Represents the firm.
Xij: Internal company factors.
Zj: Represents regional factors.
uj: Regional random effect and **εij:** Error.

The research justifies this choice by estimating an empty model (without explanatory variables), whose likelihood ratio test (LR test: chi2(2) = 8.78, p-value = 0.0124 < 0.05) reveals the presence of a hierarchical structure in the research data. However, the interregional variance is relatively low (ICC = 0.046). This result is due to the fact that the database divides the Moroccan territory into only four main regions instead of twelve existing regions (Casablanca-Settat, North, West and South).

Spatial Econometrics. After estimating the multilevel model, the paper refines the present study by using spatial econometrics to capture neighborhood effects on firms' performance. To this end, the research uses the spatial autoregressive model (SAR). The rationale behind choosing this model is its ability to capture spatial interactions, to decompose effects into direct (the impact of firm-specific characteristics) and indirect (spillovers), and to handle spatial autocorrelation.

The (SAR) model is formalized as follows:

$$\text{ptf} = \rho\text{Wptf} + \text{X}\beta + \varepsilon \tag{2}$$

ptf: Total factor productivity is used to present company performance (dependent variable).

ρ: Spatial autoregressive coefficient
W: The paper uses binary adjacency logic to construct the present spatial weights matrix (Wij = 1 when the companies i and j are in the same region, on condition that $i \neq j$ otherwise Wij = 0). To facilitate the interpretation of the spatial autoregressive coefficient (ρ), the research standardizes this matrix online allowing the spatial effect to be interpreted as a weighted average of neighbouring observations. The technical implementation uses the PySAL library to estimate the spatial models.
X: Matrix of explanatory variables. (We have excluded regional factors such as regional density and corruption to avoid multicollinearity problems).
β: Coefficient vector
ε: Error term

The Moran test (results (I = 0.0459, expected value E[I] = -0.0054, z-score = 3.8282, p-value = 0.0001) allows the paper to confirm the presence of statistically significant positive spatial autocorrelation, indicating that companies with similar performance tend to cluster geographically. This observation justifies the use of the SAR model in the paper's analytical framework.

Artificial Intelligence and Clustering. To enrich the econometric analysis and test the hypothesis concerning the regional impact, the paper adopts k-means algorithm, an artificial intelligence technique. This artificial intelligence technique helps create clusters of companies with common characteristics. This method allows the paper to answer a key question which is "do manufacturing companies with the same characteristics perform at the same level regardless of the region in which they are located, or do regional factors play a determining role?".

The K-means algorithm helps clustering companies according to similarity of characteristics (age, size, etc.) into 4 groups (this number corresponds to the number of regions in the present article for a meaningful comparison), minimizing intra-cluster variance and maximizing inter-cluster variance. To validate the relevance of this classification, the research analyzes the distinctive profiles of each cluster, particularly the significant differences in the average age of companies (from 17.94 to 59.79 years) and the proportion of skilled workers (from 0.04 to 4.07). The examination of the spatial distribution and regional statistical tests have proven to be particularly significant in the South region ($F = 7.08$, $p = 0.0013$), confirming the robustness of the segmentation used in the present article and its consistency with the territorial dynamics observed.

4 Results

4.1 Results of Multi-level Model

Table 2. Results of the multilevel model which identified the joint impact of internal and regional factors on firm performance estimated on Stata.

Variable	Coefficient	Standard error	z-stat	p-value
Export	−0.17509	0.20184	−0.87	0.386
Innovation	0.39288	0.17111	2.30	0.022**
Age_of_company	−0.01920	0.00547	3.51	0.000***
Age_squared	0.05177	0.02431	−2.13	0.033**
Size_of_company (Small)	−0.47073	0.13809	−3.41	0.001*
Size_of_company (Meduim)	−0.04706	0.05558	−0.85	0.397
Digitalization	0.06097	0.03128	1.95	0.051***
Regional_density	2.02453	0.23819	8.50	0.000*
Corruption	0.17063	0.05322	3.21	0.001*
Distance_Casablanca	0.00242	0.00026	9.16	0.000*
Human_capital	0.07213	0.04149	1.74	0.082***
Water_obstacle	0.07008	0.32058	0.22	0.827
Constant	−10.42205	1.33937	−7.78	0.000*

Source: Prepared by us using Stata software.
Note: Significance levels * 1%, ** 5%, *** 10%.

4.2 Results of Spatial Model (SAR)

Table 3. Maximum likelihood estimation of the SAR model to obtain direct and indirect effects, as well as the spatial coefficient (Python estimation).

Variable	Coefficient	Std. Error	z-Statistic	Probability
CONSTANT	0.02405	0.15324	0.15693	0.87530
Export	−0.18155	0.12757	−1.42314	0.15470
Innovation	0.44021	0.19089	2.30614	0.02110**
Age of company	−0.02102	0.10313	−0.20386	0.83846
Age squared	0.16079	0.10090	1.59354	0.11104
Size Small	−0.44287	0.18334	−2.41561	0.01571**
Size Medium	−0.05762	0.14469	−0.39823	0.69046

(continued)

Table 3. (*continued*)

Variable	Coefficient	Std. Error	z-Statistic	Probability
Digitalization	0.13031	0.13475	0.96711	0.33349
Human capital	0.10225	0.06551	1.56068	0.11860
W_Firm performance	0.31916	0.21163	1.50810	0.13153

Source: Prepared by us using Python.
Note: Significance levels * 1%, ** 5%, *** 10%.

Table 4. Direct and indirect effects of internal factors

Variable	Direct	Indirect	Total
Export	−0.1815	−0.0851	−0.2667
Innovation	0.4402	0.2064	0.6466
Age of company	−0.0210	−0.0099	−0.0309
Age squared	0.1608	0.0754	0.2362
Size Small	−0.4429	−0.2076	−0.6505
Size Medium	−0.0576	−0.0270	−0.0846
Digitalization	0.1303	0.0611	0.1914
Human capital	0.1022	0.0479	0.1501

Source: Prepared by us using Python.

4.3 Results of Logarithm K-Means

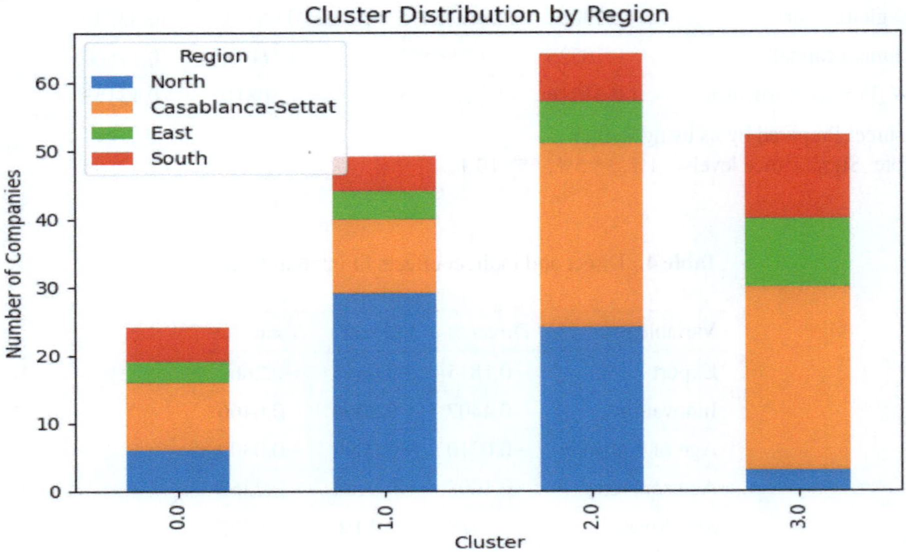

Fig. 1. Distribution of clusters by region created on 'Python'. Source: Prepared by us using Python

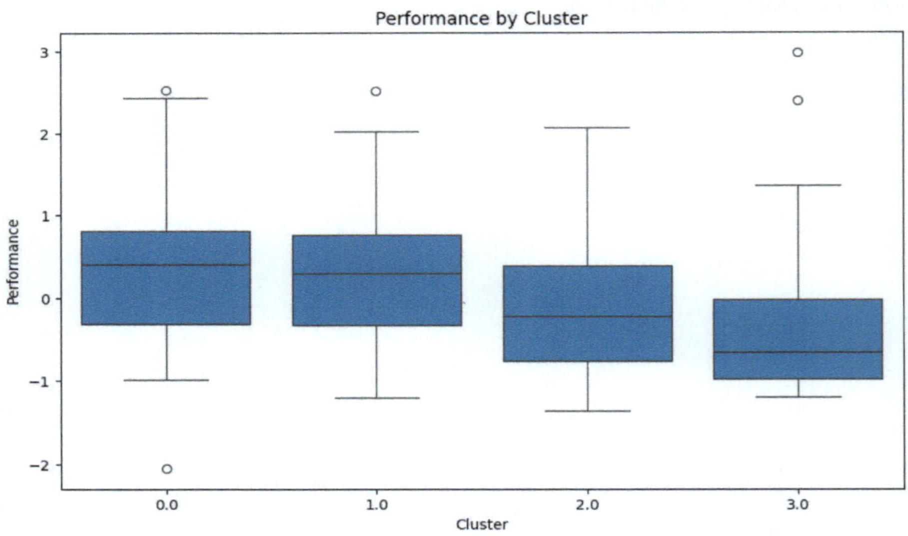

Fig. 2. Cluster-based performance created on 'Python'. Source: Prepared by us using Python.

The two figures above (Figs. 1 and 2) show that firms have been divided into 4 clusters as already mentioned, with a high concentration in cluster 2.0, particularly in the Casablanca-Settat and Nord regions. In terms of company productivity, we can see a downward trend. Indeed, clusters 0.0 and 1.0 show positive performance with moderate variability, whereas clusters 2.0 and 3.0 show weaker performance despite their considerable volume.

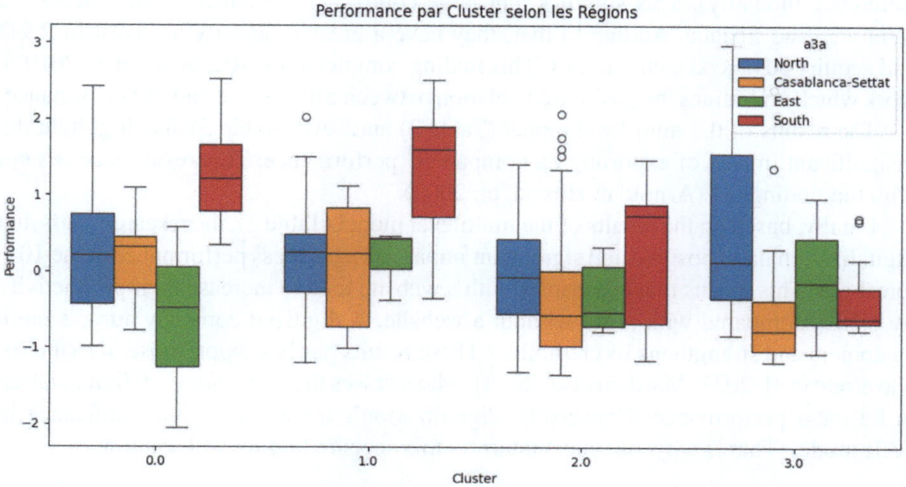

Fig. 3. Performance by Cluster and Region created on 'Python'. Source: Prepared by us.

5 Discussion

5.1 Impact of Internal Factors

Based on the results of the multilevel (Table 2) and SAR (Table 3) models, the paper deduces that innovation has a positive and significant impact at the 5% threshold on companies' performance. This means that a company that innovates or invests in R&D tends to increase its productivity by 48.1% (Since the dependent variable is transformed into a logarithm, the effect of binary variables is interpreted by exponentiating the coefficient) in the multilevel model and by 55.27% percentage points in the SAR model in comparison to a company that does not innovate. This means that innovation offers companies the opportunity to develop new products and optimize their internal processes, which could lead to an increase in efficiency, a reduction in costs and a strength in their competitive position in the market. This is consistent with the results of (Ramadani et al. 2019).

Considering the results in Table 2 (multilevel model), the paper finds a non-linear relationship between age and company productivity. That is to say that in the short term, age has a negative and significant effect at the 1% threshold. This effect becomes positive and significant at the 5% threshold. The analogy for this finding is due to the fact that new firms need time to adapt to the environment they are operating in (learning effects). Consequently, it increases the firm's productivity (Amara & Thabet 2019).

Table 2 shows also that worker skills have a positive and significant impact at the 10% threshold. Certainly, a 10% increase in the proportion of skilled workers generates a

0.72% increase in firms' performance. This supports the work of (Amara & Thabet 2019) and (Ramadani et al. 2019) who emphasize the crucial role of employee qualifications on firms' productivity.

In reference to the company's size, the results of the multilevel model (Table 2) and SAR (Table 3) have shown that small companies perform less effectively than large ones (reference modality). This signifies that large companies benefit from easy access to resources and finance. Adding to that, they have a greater capacity to invest in R&D and acquire advanced technologies. This finding complements (Ramadani et al. 2019)'s work which underlines the positive correlation between a firm's size and its performance.

The results of the multilevel model (Table 2) and SAR (Table 3) also highlight the insignificant impact of exporting on companies' performance. This result is consistent with the findings of (Arnold & Hussinger, 2005).

Finally, based on the results of the multilevel model (Table 2), the research finds that digitalization has a positive and significant impact on a business performance at the 10% threshold. This means that companies with a website tend to increase their productivity by 6.28% compared with those without a website. A digitized company attracts more customers and strengthens its credibility. These results partly support those of (Ribeiro-Navarrete et al. 2021; Mardani et al. 2023), who stresses the importance of digitalization in business performance. Conversely, digitalization's impact loses its significance in SAR model. That is why, one must interpret this specific finding with caution.

5.2 Impact of Regional Factors

In light of Table 2 (multilevel model), the paper proves that regional density has a positive and significant impact at the 1% threshold; the thing that agglomeration effects explains. This finding is in accordance with the results of (Amara & Thabet 2019, Ferreira et al. 2010).

Moreover, the research deduces the positive and significant impact of corruption (Table 2) on business productivity at the 1% threshold. An increase of 1 percentage point in informal payments are likely to generate an increase of 17.06% in a company's performance. Corruption, in this regard, makes it possible to reduce administrative obstacles, bypass procedures and facilitate access to resources.

The use of the K-means algorithm enables the present work to reinforce the results obtained by the multilevel model concerning the significant impact of regional factors on companies' performance. Figure 3 shows that even if companies have similar characteristics (belonging to the same cluster), their performance varies according to their regional location. The third cluster (3.0), for instance, demonstrates that companies in the North (blue box) stand out from other companies in their cluster in terms of performance. This regional heterogeneity underlines the vital impact of the regional framework on companies' performance, over and above their intrinsic characteristics. There are various factors to these regional disparities such as the quality of regional infrastructures and institutions, regional development strategies and the presence of skilled workers.

5.3 Spatial Factor

The variable measured to represent the distance between the Casablanca-Settat region and the other regions illustrates a positive and significant sign at the 1% threshold (Table 2, multilevel model), which means that the further a company is from Casablanca-Settat, the more productive it is. To elaborate, companies located in a region that is far from the economic hub have the ability to benefit from access to less saturated markets.

Based on Table 3 (SAR model), the spatial coefficient is positive (W_Firm performance = 0.3191). This indicates that the neighboring companies have positively influenced a company's productivity. However, this coefficient is not statistically significant, which prevents the research from asserting the presence of significant spatial autocorrelation. This may be due to several factors, such as the absence of information in the used database which ignores the distance between companies within each region, and the division of the Moroccan territory into just four large regions instead of respecting the actual division, twelve regions. To clarify this point, the matrix used in this research considers companies located in Rabat and others situated in Tangiers to be neighbours. In this sense, the database studies both firms as if they belong to the North region, whereas in reality they operate in two separate yet very distant regions.

The **direct** and **indirect** effects indicate that company-specific characteristics have an impact on both the companies' performance and that of neighboring companies. Based on Table 4, innovation has a positive impact not only on the company that adopts it, but also on neighboring companies, which improves the performance of all companies.

6 Conclusion

In conclusion, after analyzing the internal factors such as innovation, worker qualification, company size and age, as well as regional factors such as regional density and corruption, the study has shown a significant impact on both factors on a company's productivity. Yet, regardless of the integration of the spatial model (SAR) which seeks to capture proximity effects, the present work lacks evidence to confirm the impact of proximity on firms' performance, given that the spatial coefficient is not statistically significant. This finding is owing the absence of information on the actual distance between companies in the World Bank database, as well as to the subdivision of the Moroccan territory into four large regions instead of the number of actual regions of the Moroccan territory. This causes a major limitation to the research and restricts its ability to fully detect the effects of proximity between firms. Finally, the integration of artificial intelligence (AI) enables the study to accentuate the importance of location on companies' performance according to the K_means algorithm.

The econometric results that the research presents have direct implications on manufacturing companies. The positive effect of skilled workers, digitalization and innovation suggests that it is advisable that managers prioritize these areas of investment, while exploiting the advantages of size. Despite the non-significant spatial coefficient that the paper identifies, clustering analysis confirm the influence of the geographical environment on firms' performance. Companies warrant considering the advantages of agglomerations in their location decisions.

As for regional policies, the significant impact of economic density and infrastructure rationalizes the differentiated territorial approach, with the creation of specialized industrial zones in developing regions that targets programs whose aim is to strengthen human capital and digital infrastructure. The paradoxically positive impact of corruption suggests the need to facilitate administrative procedures so that firms minimize the necessity for informal practices.

Declaration of Conflicts of Interest. The authors declare that they have no conflict of interest in relation to the content of this article.

Annex

Empty Model and Intraclass Correlation

```
Mixed-effects ML regression                     Number of obs     =        305

        Grouping information
        -------------------------------------------------------------
                        |     No. of       Observations per group
         Group variable |     groups    Minimum    Average    Maximum
        ----------------+--------------------------------------------
                    a3a |          4         57       76.3         95
                     id |        305          1        1.0          1
        -------------------------------------------------------------

                                                Wald chi2(0)      =          .
Log likelihood = -395.60438                     Prob > chi2       =          .

------------------------------------------------------------------------------
Firm_performance | Coefficient  Std. err.      z    P>|z|     [95% conf. interval]
-----------------+------------------------------------------------------------
           _cons |    .008349   .1088315     0.08   0.939    -.2049567    .2216547
------------------------------------------------------------------------------

------------------------------------------------------------------------------
  Random-effects parameters  |   Estimate   Std. err.     [95% conf. interval]
-----------------------------+------------------------------------------------
a3a: Identity                |
                  var(_cons) |   .0370307   .0330893      .0064263    .2133859
-----------------------------+------------------------------------------------
id: Identity                 |
                  var(_cons) |   .6717326   26.07046      6.19e-34    7.29e+32
-----------------------------+------------------------------------------------
               var(Residual) |   .0963615   26.07041      4.9e-232    1.9e+229
------------------------------------------------------------------------------
LR test vs. linear model: chi2(2) = 8.78              Prob > chi2 = 0.0124
```

```
Note: LR test is conservative and provided only for reference.
Warning: Convergence not achieved.

. estat icc

Intraclass correlation

------------------------------------------------------------------------------
                       Level |        ICC   Std. err.     [95% conf. interval]
-----------------------------+------------------------------------------------
                         a3a |   .0459937   .0394512      .0082082    .2192655
                      id|a3a |   .8803148   32.38058      1.8e-261           1
------------------------------------------------------------------------------
```

Results of the Multilevel Model

```
Mixed-effects regression                        Number of obs      =        186
Group variable: a3a                             Number of groups   =          4
                                                Obs per group:
                                                             min =         22
                                                             avg =       46.5
                                                             max =         73
                                                Wald chi2(3)       =          .
Log pseudolikelihood = -226.77371               Prob > chi2        =          .

                                  (Std. err. adjusted for 4 clusters in a3a)
```

Firm_performance	Coefficient	Robust std. err.	z	P>|z|	[95% conf.	interval]
Export	-.1750928	.2018454	-0.87	0.386	-.5707025	.220517
Innovation	.3928886	.1711197	2.30	0.022	.0575001	.7282771
Age_of_company	.0192066	.0054771	3.51	0.000	.0084717	.0299415
Age_squared	-.0517797	.0243128	-2.13	0.033	-.0994318	-.0041275
Size_of_company						
Small	-.4707346	.1380922	-3.41	0.001	-.7413903	-.2000788
Medium	-.0470646	.0555856	-0.85	0.397	-.1560104	.0618812
Digitalization	.0609734	.031285	1.95	0.051	-.0003441	.122291
Regional_density	2.024534	.2381907	8.50	0.000	1.557689	2.491379
Corruption	.1706325	.0532269	3.21	0.001	.0663096	.2749554
Distance_Casablanca_Settat	.0024246	.0002646	9.16	0.000	.001906	.0029432
Human_Capital	.0721366	.0414939	1.74	0.082	-.00919	.1534633
Water_Obstacle	.0700824	.3205814	0.22	0.827	-.5582455	.6984103
_cons	-10.42205	1.339377	-7.78	0.000	-13.04718	-7.796918

Results of the Spatial Model (SAR)

```
Creating weights matrix...
Estimating spatial model...

Estimation results:
REGRESSION RESULTS
------------------

SUMMARY OF OUTPUT: MAXIMUM LIKELIHOOD SPATIAL LAG (METHOD = FULL)
------------------------------------------------------------------
Data set            :     unknown
Weights matrix      :     unknown
Dependent Variable  :Firm_performance              Number of Observations:        186
Mean dependent var  :     -0.0190              Number of Variables   :         10
S.D. dependent var  :      0.9147              Degrees of Freedom    :        176
Pseudo R-squared    :      0.1592
Spatial Pseudo R-squared:  0.1487
Log likelihood      :   -230.9868
Sigma-square ML     :      0.6998              Akaike info criterion :    481.974
S.E of regression   :      0.8365              Schwarz criterion     :    514.231

------------------------------------------------------------------------------------
            Variable     Coefficient       Std.Error     z-Statistic     Probability
------------------------------------------------------------------------------------
            CONSTANT         0.02405         0.15324         0.15693         0.87530
              Export        -0.18155         0.12757        -1.42314         0.15470
          Innovation         0.44021         0.19089         2.30614         0.02110
      Age_of_company        -0.02102         0.10313        -0.20386         0.83846
         Age_squared         0.16079         0.10090         1.59354         0.11104
          Size_Small        -0.44287         0.18334        -2.41561         0.01571
         Size_Medium        -0.05762         0.14469        -0.39823         0.69046
      Digitalization         0.13031         0.13475         0.96711         0.33349
       Human_capital         0.10225         0.06551         1.56068         0.11860
  W_Firm_performance         0.31916         0.21163         1.50810         0.13153
------------------------------------------------------------------------------------
Warning: Variable(s) ['const'] removed for being constant.

SPATIAL LAG MODEL IMPACTS
Impacts computed using the 'simple' method.
            Variable          Direct        Indirect           Total
              Export         -0.1815         -0.0851         -0.2667
          Innovation          0.4402          0.2064          0.6466
      Age_of_company         -0.0210         -0.0099         -0.0309
         Age_squared          0.1608          0.0754          0.2362
          Size_Small         -0.4429         -0.2076         -0.6505
         Size_Medium         -0.0576         -0.0270         -0.0846
      Digitalization          0.1303          0.0611          0.1914
       Human_capital          0.1022          0.0479          0.1502
================================ END OF REPORT =====================================
/usr/local/lib/python3.11/dist-packages/libpysal/weights/weights.py:542: UserWarning: The weights matrix is not fully connected:
 There are 4 disconnected components.
  return W(neighbors=neighbors, weights=weights)
```

```
X, y, w, moran_result = run_moran_test(data)

Préparation des données...

Création de la matrice de poids spatiale...

Calcul du test de Moran I...
Indice de Moran I: 0.0459
Valeur attendue E[I]: -0.0054
Variance de I: 0.0002
z-score: 3.8282
p-value: 0.0001

Interprétation: Il existe une autocorrélation spatiale POSITIVE significative.
Les entreprises avec des performances similaires tendent à se regrouper spatialement.
/usr/local/lib/python3.11/dist-packages/libpysal/weights/weights.py:542: UserWarning: The weights matrix is not fully connected:
 There are 4 disconnected components.
  return W(neighbors=neighbors, weights=weights)
```

References

Bourdin, S.: Pour une approche géographique de la convergence: les inégalités régionales dans l'Union européenne et leur évolution. L'Espace géographique **42**(3), 270–285 (2013). https://doi.org/10.3917/eg.423.0270

Amara, M., Thabet, K.: Firm and regional factors of productivity: a multilevel analysis of Tunisian manufacturing. Ann. Reg. Sci. (2019). https://doi.org/10.1007/s00168-019-00918-x

Ramadani, V., Hisrich, R.D., Abazi-Alili, H., Dana, L.P., Panthi, L., Abazi-Bexheti, L.: Product innovation and firm performance in transition economies: a multi-stage estimation approach. Technol. Forecast. Soc. Change **140**, 271–280 (2019). https://doi.org/10.1016/j.techfore.2018.12.010

Ferreira, F.C.M., Goldszmidt, R.G.B., Csillag, J.M.: The regional concentration of industries and the performance of firms: a multilevel approach. Braz. Adm. Rev. **7**(4), 345–361 (2010)

Ricotta, F.: The quality of regional government and firm performance. In: 56th Congress of the European Regional Science Association, Vienna (2016)

Ribeiro-Navarrete, S., Botella-Carrubi, D., Palacios-Marqués, D., Orero-Blat, M.: The effect of digitalization on business performance: an applied study of KIBS. J. Bus. Res. **126**, 319–326 (2021). https://doi.org/10.1016/j.jbusres.2020.12.065

Pineda-Ríos, W., Giraldo, R., Porcu, E.: Functional SAR models: with application to spatial econometrics. Spat. Stat. **29**, 145–159 (2019). https://doi.org/10.1016/j.spasta.2018.12.002

Dedrick, R.F., et al.: Multilevel modeling: a review of methodological issues and applications. Rev. Educ. Res. **79**(1), 69–102 (2009). https://doi.org/10.3102/0034654308325581

Hartigan, J.A., Wong, M.A.: Algorithm AS 136: un algorithme de classification k-means. J. R. Stat. Soc. C **28**(1), 100–108 (1979). https://doi.org/10.2307/2346830

Arnold, J.M., Hussinger, K.: Export behavior and firm productivity in German manufacturing: a firm-level analysis. Rev. World Econ. **141**(2), 219–243 (2005). https://doi.org/10.1007/s10290-005-0026-8

Guo, X., Li, M., Wang, Y., Mardani, A.: Does digital transformation improve the firm's performance? From the perspective of digitalization paradox and managerial myopia. J. Bus. Res. **163**, 113868 (2023). https://doi.org/10.1016/j.jbusres.2023.113868

Gu, X., Chan, H.K., Thadani, D.R., Chan, F.K.S., Peng, Y.: The role of digital techniques in organisational resilience and performance of logistics firms in response to disruptive events: flooding as an example. Int. J. Prod. Econ. **266**, 109033 (2023). https://doi.org/10.1016/j.ijpe.2023.109033

Chatterjee, S., Chaudhuri, R., Gupta, S., Sivarajah, U., Bag, S.: Assessing the impact of big data analytics on decision-making processes, forecasting, and performance of a firm. Technol. Forecast. Soc. Change **196**, 122824 (2023). https://doi.org/10.1016/j.techfore.2023.122824

Burger, A., Damijan, J.P., Kostevec, C., Rojec, M.: Determinants of firm performance and growth during economic recession: the case of Central and Eastern European countries. Econ. Syst. **41**(4), 569–590 (2017). https://doi.org/10.1016/j.ecosys.2017.05.003

The Digital Health with the Psychology of Fear of Missing Out in Electronic Commerce: Implications for Social Commerce and Purchase Intentions

Nguyen Duy Phuong[1], Nguyen Minh Tuan[2], Le Dao Ngoc Thanh[1], and Bui Thanh Khoa[3](✉)

[1] Faculty of Business Administration, Industrial University of Ho Chi Minh City, Ho Chi Minh City, Vietnam
[2] Graduate School, Ho Chi Minh University of Banking, Ho Chi Minh City, Vietnam
[3] Faculty of Commerce and Tourism, Industrial University of Ho Chi Minh City, Ho Chi Minh City, Vietnam
buithanhkhoa@iuh.edu.vn

Abstract. This study investigates the complex interplay between Fear of Missing Out (FoMO), social connection, social commerce usage intensity, and purchase intentions within the context of fast-moving consumer goods (FMCG) in Ho Chi Minh City, Vietnam. Employing a quantitative research design, a survey was administered to 310 FMCG customers, and the resulting data were analyzed using structural equation modeling (SEM) with AMOS software. The findings reveal that FoMO significantly influences purchase intentions, demonstrating its potent role in driving consumer behavior within social commerce platforms. However, the analysis also indicates that FoMO negatively impacts both social commerce usage intensity and perceived social connection. This suggests that while FoMO can stimulate immediate purchase decisions, it may simultaneously erode the social aspects of online shopping. The study underscores the importance of understanding the dual nature of FoMO in the digital marketplace, highlighting its implications for businesses seeking to optimize social commerce strategies and foster sustainable consumer engagement.

Keywords: Fear of Missing Out (FoMO) · Social Commerce · Social Connection · Purchase Intention · Digital Health

1 Introduction

In the 4.0 industrial revolution era, social networks have become a communication tool and an effective online shopping channel. Social commerce has grown strongly and become a popular shopping trend globally, especially among young people. The combination of e-commerce and social networks has created a new shopping platform where customers can search for products, consult with the community, and make instant purchases on platforms such as Facebook, Instagram, and TikTok [1]. In this context,

B.-E. Boudriki Semlali et al. (Eds.): DATA 2025, LNISO 81, pp. 100–110, 2025.
https://doi.org/10.1007/978-3-032-06725-8_9

fear of missing out (FoMO) has become an important psychological factor influencing consumer behavior on social commerce platforms. FoMO is defined as a feeling of anxiety or stress when users are concerned that they are missing out on important opportunities, experiences, or events that others are participating in [2]. FoMO not only appears in social activities but also dramatically affects shopping decisions. According to [3], FoMO drives impulse buying behavior on social commerce platforms. When customers notice promotions or special offers shared on social media, they feel pressured to purchase immediately to avoid missing out.

Along with the rise of FoMO, the intensity of social commerce usage also plays an important role in driving customers' purchase intentions. Social commerce is a simple shopping channel that creates a multi-dimensional interactive environment between sellers and buyers through features such as comments, reviews, sharing, and live streaming [1]. Customers' frequent exposure to these elements on social media platforms increases their acceptance of product information and stimulates their purchase intentions. A study by Zhang, Qin, Wang and Luo [4] shows that the higher the intensity of social commerce usage, the stronger the purchase intention, as customers are attracted to product-related information and promotions. In addition, social connection on social commerce platforms is also an indispensable factor. Social connection is reflected in the level of interaction between users, thereby creating trust and promoting shopping behavior. According to Bilal, Ahmed, Shehzad and Engineering [5], social relationships on social commerce platforms directly impact customers' shopping decisions, as they tend to rely on reviews and feedback from friends or other users to make decisions. This shows that intrinsic factors such as FoMO or usage intensity affect shopping intentions, and social interaction also plays a significant role.

With the rapid development of social commerce, it is necessary to study the relationship between fear of missing out, social commerce usage intensity, and social connection on customers' purchase intention. This topic helps better understand customers' shopping behavior in the digital context and provides practical suggestions for businesses to build effective marketing strategies. By exploiting the psychological factor of FoMO, enhancing social connection, and optimizing the social commerce experience, businesses can increase sales and improve customer satisfaction.

2 Literature Review

2.1 FoMO and Digital Health

Fear of Missing Out (FoMO) is an anxious psychological state that individuals feel when they believe they are missing out on important events, experiences, or opportunities that others are participating in. FoMO often occurs when social media users witness posts about their friends' lives, achievements, or activities and feel uninvolved or not up-to-date enough [2]. FoMO is characterized by a feeling of anxiety that others may be having more interesting or valuable experiences while one is not participating in them. This state affects emotions and human behavior, especially in the online environment. People who experience FoMO tend to consume more social media content and may engage in impulsive behaviors, such as shopping without thinking carefully, to avoid the feeling of missing out [3, 6].

FoMO significantly impacts digital health by driving excessive online engagement, increasing screen time, and disrupting sleep patterns. This constant connectivity fosters anxiety and depression as individuals compare themselves to curated online personas, eroding social well-being by prioritizing virtual interactions over real-life relationships. Moreover, FoMO contributes to physical health issues like eye strain and musculoskeletal problems and fuels digital addiction, creating a cycle of negative consequences for both mental and physical health. Essentially, FoMO acts as a catalyst for behaviors that undermine healthy digital habits, highlighting the need for mindful technology use and strategies to mitigate its adverse effects.

2.2 Hypotheses Development

FoMO is driven by the basic need for social connection and belonging, as people need to maintain a certain level of presence in social relationships and events. FoMO affects social activities and dramatically impacts consumer decisions, as social media users are often stimulated by attractive advertising information, promotions, or advice from the user community [2]. FoMO stimulates consumers to increase the frequency of using social commerce platforms to ensure they do not miss any information or shopping opportunities. They constantly check for updates from sales channels and follow limited-time promotions or exclusive products. This proves that FoMO positively impacts the intensity of social commerce use because the higher the fear of missing out, the more users engage in social commerce activities to minimize the risk of missing out [4]. Therefore, the hypothesis H1 is proposed as follows:

Hypothesis H1: FoMO has a positive impact on social commerce usage intensity.

Although FoMO is often seen as a driving force behind social media engagement, research suggests that FoMO can negatively affect a person's social connections. When people experiencing FoMO constantly monitor the activities of others on social media, they often feel insecure or jealous, which can reduce their sense of real connection with friends and family [7]. According to [8], FoMO can lead to dissatisfaction and stress, as individuals feel they are not fully engaged in social relationships. This can lead to more excellent isolation, as an over-focus on online experiences leads to less time and attention invested in face-to-face social relationships. [7] also found that although FoMO can promote increased social engagement, it reduces the quality of genuine social relationships, creating a negative relationship between FoMO and social connection. Therefore, the hypothesis H2 is proposed as follows:

Hypothesis H2: FoMO has a negative effect on social connection.

Social commerce usage intensity refers to the frequency and extent of users' engagement on social commerce platforms such as Facebook, Instagram, or TikTok, which can positively impact their social connections. These platforms facilitate shopping and exchanging product information and help users build and maintain social relationships [9, 10]. When users frequently interact on these platforms, they can interact with communities, connect with people with similar interests, and develop new relationships. According to Bilal, Ahmed, Shehzad and Engineering [5], social commerce helps users sense belonging to a group or community, strengthening social connections. Interactions on these platforms, such as sharing shopping information, reviewing products, or participating in online conversations, help users feel connected and supported by the community. This helps improve social relationships and creates opportunities to connect with new individuals, expanding their social networks [4].

Furthermore, social commerce platforms often provide interactive tools such as commenting, sharing, and messaging, which promote social communication and interaction, thereby strengthening the relationship between users. Thus, the higher the intensity of social commerce use, the more social connections are strengthened and expanded [5]. Therefore, hypothesis H3 is proposed as follows:

Hypothesis H3: The intensity of social commerce use positively impacts social connection.

FMCG products are often easy to purchase, low cost, and have a repeat purchase cycle, making them ideal targets for social commerce. Consumers can easily access information about FMCG products through online advertising, other user reviews, or shared community content. According to a study by Hajli [1], social commerce allows consumers to participate in the purchase decision process by receiving feedback and recommendations from the online community, thereby increasing their intention to purchase FMCG products. Furthermore, social commerce's high usage intensity also helps consumers quickly follow and update promotions, discount events, or new product launches. This creates convenience and stimulates shopping intentions for FMCG products [4]. Thus, the higher the usage intensity of social commerce, the higher the customers' purchase intention for fast-moving consumer goods. Therefore, hypothesis H4 is proposed as follows:

Hypothesis H4: The intensity of social commerce use positively impacts customers' purchase intention.

Social connection is how individuals feel connected and interact with members of their social network, including family, friends, and colleagues. Strong social connections often lead to information sharing, mutual support, and trust, influencing individuals' purchasing decisions [11, 12]. When consumers feel connected to others, they tend to seek advice and information from friends and family about products, shaping their purchase intentions. Research by Bilal, Ahmed, Shehzad and Engineering [5] shows that people with better social connections are more likely to receive more information and positive product feedback, increasing their confidence and comfort in purchasing. Furthermore, when there is social cohesion, consumers tend to engage in group shopping activities, such as going to the supermarket or sharing information about promotions. Support and encouragement from social network members can increase consumers' motivation to purchase fast-moving consumer goods, which are often low-priced and require frequent purchases [4, 13]. Therefore, when social connection is stronger, customers' intention to purchase fast-moving consumer goods will also increase. Therefore, hypothesis H5 is proposed as follows:

Hypothesis H5: Social norms positively impact purchase intention in social commerce.

The paper applies the SOR theory to study the relationship between FoMO, social commerce usage intensity, and social connection on FMCG customers' purchase intention. The model shows how the stimulus factors influence consumer behavior and highlights the role of emotions and cognition in the decision-making process. Applying SOR theory to this model allows us to understand better how external factors influence consumer shopping behavior. The proposed research model is shown in Fig. 1:

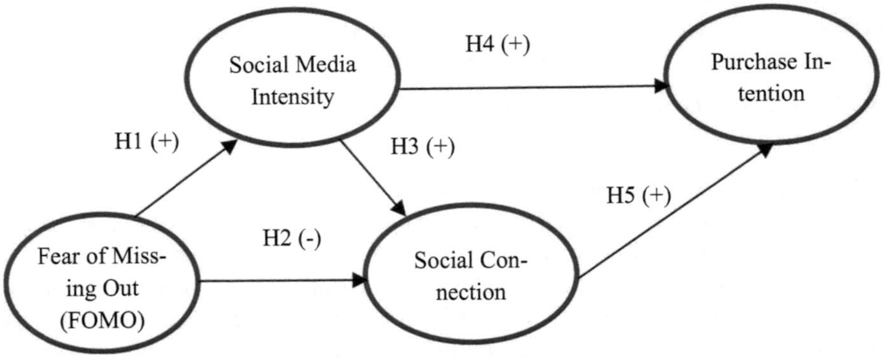

Fig. 1. Research model

3 Method

The research sample selected is fast-moving consumer goods (FMCG) customers in Ho Chi Minh City. The sample size is 310 respondents, ensuring the representativeness and reliability of the research results. To ensure the randomness and diversity of the survey sample, we will use the non-probability convenience sampling method.

Data was collected through a questionnaire consisting of scales that have been tested and widely used in previous studies. The questionnaire will be divided into the following sections: Demographic information, including gender, age, occupation, and income. The Fear of Missing Out factor (FoMO) consists of 05 observed variables, using the Przybylski, Murayama, DeHaan, and Gladwell scale. The Social Media Intensity factor (SMI) consists of 05 observed variables, using the scale of Horng, Wu and Management [9]. The social connection (SC) was measured by 05 observed variables using the Bilal, Ahmed, Shehzad and Engineering scale. Customers' shopping intention (PI) consists of 05 observed variables, using the scale of Zhang, Qin, Wang and Luo [4].

The survey results of 310 customers showed that the proportion of women accounted for 58.1%, while the proportion of men was only 41.9%. This suggests that women may be the leading consumer group in the FMCG sector. The nature of this industry is often related to daily consumption needs such as food, beverages, and household products, for which women tend to spend more. The high proportion of women in the research sample reflects the reality of consumption and indicates that marketing campaigns can be adjusted to target women more effectively.

The 25–34 age group has the highest proportion at 32.3%, followed by the 18–24 age group at 25.8%. Meanwhile, the over-35 age group has a lower proportion, with the 35–44 age group at 19.4% and the 45 and over age group at only 12.9%. This distribution shows that the young age group, from 18 to 34, is a dynamic customer group, easily accessible to new products and regularly updated on consumer trends. This customer group likely tries new products through social commerce channels. Therefore, marketing campaigns should target this age group to optimize consumption efficiency.

The occupational distribution shows that students (29.0%) and office workers (25.8%) are the two groups with the highest proportion, followed by workers (16.1%) and businessmen (12.9%). This reflects a diverse customer base ranging from students to office workers with different needs and consumption habits. Students are often price-sensitive and look for cheap products, while office workers may focus on quality and brand. Therefore, marketers should develop appropriate strategies for each occupational group to maximize outreach effectiveness.

The group of customers with income from 5–10 million VND (32.3%) accounts for the highest proportion, followed by the group with income from 10–15 million VND (25.8%). The group with income under 5 million VND accounts for 19.4%, while the group with income over 20 million VND accounts for only 6.5%. This distribution shows that most customers are in the middle-income group and can spend on FMCG products, but their Pricing strategies, promotions, and product packages can be the key to attracting this group of customers. Moreover, FMCG products should be adjusted to suit the needs and affordability of this income group.

4 Research Result

The results of testing the reliability of the scales through Cronbach's Alpha coefficient showed that all scales met the requirement of being greater than 0.7. The results of the EFA exploratory factor analysis using the Principal Axis Factoring data extraction method and Promax rotation showed that the factor loading coefficients of all observed variables were more significant than 0.5, which is satisfactory. The combined reliability of all factors was more significant than 0.7, so the condition was satisfied. The results of the Average Variance extract (AVE) calculation, according to Table 1, were all greater than 0.5, meeting the requirements in the confirmatory factor analysis CFA [14].

The results of the SEM linear structural model analysis in Fig. 2 shows that The Chi-square/df index is $2.723 < 3$, the GFI index is $0.873 > 0.8$, CFI is $0.941 > 0.9$, TLI is $0.932 > 0.9$ and RMSEA is $0.075 < 0.08$, showing that the theoretical model is suitable for the market. All relationships have a 95% reliability with $p = 0.000$ [14]. This proves that the scales of the concepts in the research model are theoretically related.

Table 1. Scale evaluation results.

Construct	Reliability and validity		
	Factor loading	CR	AVE
Fear of Missing Out Cronbach's $\alpha = 0.942$	0.822–0.935	0.944	0.771
Social Media Intensity Cronbach's $\alpha = 0.941$	0.793–0.940	0.941	0.763

(continued)

Table 1. (*continued*)

Construct	Reliability and validity		
	Factor loading	CR	AVE
Social Connection Cronbach's α = 0.907	0.674–0.878	0.907	0.663
Purchase intention Cronbach's α = 0.830	0.549–0.771	0.840	0.515

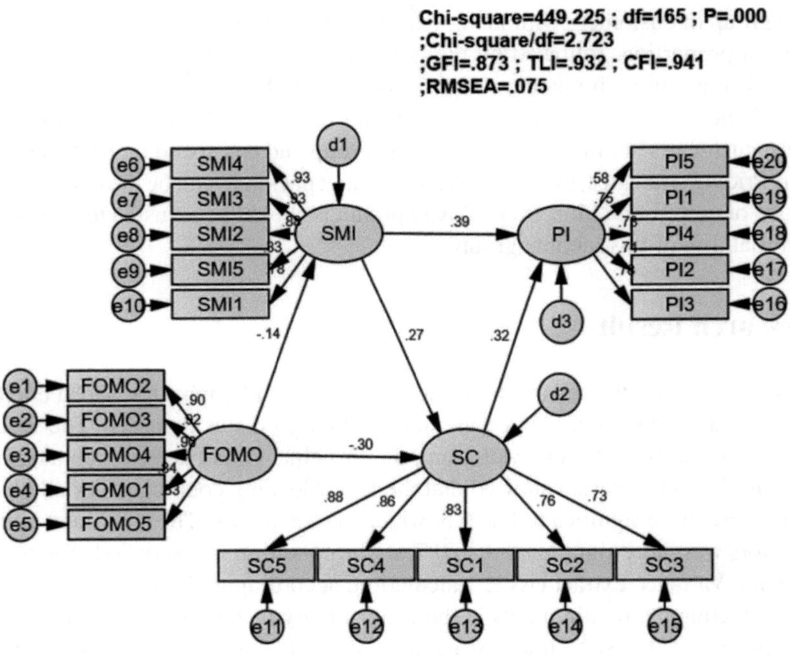

Fig. 2. SEM model analysis results.

The results of the theoretical model estimation in the SEM analysis show that hypotheses H2, H3, H4, and H5 are accepted with 95% confidence. Hypothesis H1 is rejected because the analysis results indicate that FoMO has a negative impact on the intensity of social commerce usage (Table 2).

Table 2. Results of testing hypotheses

Hypothesis				Coefficient	SE	CR	P	Conclude
H1	SMI	<---	FoMO	−.110	.045	−2.423	.015	Reject
H2	SC	<---	FoMO	−.294	.057	−5.203	***	Accept

(*continued*)

Table 2. *(continued)*

Hypothesis				Coefficient	SE	CR	P	Conclude
H3	SC	<---	SMI	.354	.074	4,816	***	Accept
H4	PI	<---	SMI	.272	.042	6,517	***	Accept
H5	PI	<---	SC	.168	.033	5.171	***	Accept

*Note: *** p < 0.01.*

5 Discussion

The results show that hypothesis H1, with an estimated coefficient of −0.110, was rejected. The CR value was −2.423, and the p-value was 0.015, indicating the negative relationship between fear of missing out and the intensity of social commerce usage. This may reflect that the anxiety of missing out on opportunities may lead to decreased usage of social commerce platforms. Customers may become more cautious, limiting their participation in social media activities for fear of not benefiting from the interaction. This study contributes to the assertion that fear of missing out does not necessarily promote increased social commerce usage behavior but may have the opposite effect.

Hypothesis H2 was firmly accepted with an estimated coefficient of −0.294, a CR value of −5.203, and a p-value < 0.001. This result suggests that FoMO has a negative impact on social connectedness. This can be explained by suggesting that when consumers are worried about missing out, they may become less engaged with their friends and family, leading to a decrease in their sense of social connectedness. This anxiety may lead them to focus solely on commercial activities, weakening their ability to maintain personal and community relationships. Thus, FoMO not only affects consumer behavior but also weakens customers' social relationships.

The estimated coefficient of hypothesis H3 is 0.354 with CR = 4.816 and p-value < 0.001, indicating a positive relationship between social commerce usage intensity and social connection. This result indicates that as the intensity of social commerce usage increases, consumers' sense of social connection also increases. This may reflect the role of social commerce platforms in creating and maintaining social relationships. People who use these platforms frequently tend to interact more with friends and family, thereby strengthening their social connections. This result suggests that marketers can leverage social commerce channels to encourage interaction and connection among consumers.

Hypothesis H4 was also accepted with an estimated coefficient of 0.272, CR = 6.517, and p-value < 0.001. This result shows that the intensity of social commerce use has a positive impact on customers' purchase intentions. This suggests that when customers actively participate in social commerce activities, they tend to increase their purchase intentions. Access to product information, promotions, and opinions from friends on social platforms can promote purchase decisions. This result confirms the important role of social commerce in influencing consumer behavior, opening up opportunities for marketers to create more effective product promotion campaigns through these platforms.

Finally, hypothesis H5 was also accepted with the estimated coefficient of 0.168, CR = 5.171, and p-value < 0.001. This result indicates that social connection has a positive influence on customers' purchase intention. This suggests that the feeling of connection with others can promote purchase intention. Social relationships and influence from friends can motivate people to choose products and brands, especially in the FMCG sector. Thus, social connection is not only an important factor in personal life but also plays an important role in consumer behavior.

6 Conclusion

Fear of Missing Out (FoMO) has been identified as a psychological factor influencing consumer behavior in the context of social commerce. Research has shown that FoMO has a negative impact on social connectedness, as consumers feel pressured to follow and participate in online activities, which can lead to a decline in personal relationships. Although the hypothesis that FoMO has a positive impact on the intensity of social commerce usage has not been accepted, awareness of FoMO is still important in shaping marketing strategies. Marketers need to understand that the fear of missing out on information or opportunities can lead to unhealthy competition in the minds of customers. Therefore, providing clear information and creating an open communication environment will help customers feel more comfortable, thereby improving their shopping experience.

The intensity of social commerce usage was identified as an important factor in driving customer shopping behavior. The analysis results showed that customers who actively participate in activities on social commerce platforms tend to have a higher tendency to make shopping decisions. They have quick access to the latest information, promotions, and trends, making it easier for them to make decisions. This shows that social commerce is not only a sales channel but also a powerful communication and interaction environment. Therefore, businesses need to optimize their social commerce platforms by providing engaging content, encouraging interaction, and building relationships with customers, thereby increasing engagement and loyalty.

Social connection has been shown to have a positive impact on customer purchase intention. Customers feel more comfortable and trust buying from brands with which they have a strong social connection. The analysis shows that as social connection increases, purchase intention also increases. This reflects the need to build lasting relationships between businesses and customers. To do this, marketers need to focus on creating spaces for communication where customers can share their opinions and feelings. Positive interactions not only enhance the customer experience but also create sustainable value for the brand.

Customer purchase intention is a key determinant of consumer behavior and is determined by the intensity of social commerce usage and social connection. Research shows that both of these factors have a positive impact on customer purchase decisions. This suggests that when customers feel comfortable and connected to a brand through social platforms, they are more likely to make a purchase decision. Optimizing the shopping experience on social commerce channels will not only boost sales but also create long-term relationships with customers. Businesses need to focus on building engaging marketing campaigns that encourage interaction and capture customer attention to increase purchase intention.

Although the study achieved significant results in analyzing the relationship between fear of missing out (FoMO), social commerce usage intensity, social connection, and customer purchase intention, there are still some limitations that need to be considered.

First, the research sample only focused on 310 fast-moving consumer goods (FMCG) customers in a particular geographical area. This may cause bias in the results, as customer characteristics and shopping habits may vary across regions and cultures. Therefore, to have a more general and comprehensive view, future studies should expand the sample size and research scope to many different regions. Second, the study relied heavily on self-report survey methods, which may have led to errors in data collection due to cognitive biases or self-assessment. Failure to consider other factors, such as social psychology, emotions, and personal motivations, may also have reduced the accuracy of the results. Finally, the study did not consider external factors that may influence shopping behavior, such as advertising campaigns, promotions, or the influence of macroeconomic factors. This may have reduced the ability to interpret the results in a broader context. To overcome these limitations, future research can focus on several important directions. First, research should be conducted on a more extensive and more diverse sample, including different age groups, genders, and geographies. Increasing the size and diversity of the sample will increase the generalizability and applicability of the research results, thereby more accurately reflecting customer shopping behavior. Second, further studies can apply qualitative research methods to complement quantitative results. Combining these two methods will help to understand better the factors that influence purchase intentions. Methods such as in-depth interviews or focus groups can provide insights into customer psychology and behavior, thereby enriching the research data.

Furthermore, research should consider external factors such as advertising campaigns, promotions, or the influence of the macro-economic environment on shopping behavior. Understanding these factors will help create a more comprehensive picture of the factors that influence shopping decisions, thereby helping marketers adjust their strategies more effectively. Finally, future research could explore the relationship between other factors, such as the influence of online communities, interactions between group members, or the impact of influencers in social commerce platforms on customers' purchase intentions. By conducting these research directions, researchers can enrich the knowledge about the relationship between psychological factors and consumer behavior in the context of social commerce, thereby supporting businesses in developing more effective marketing strategies.

References

1. Hajli, N.: Social commerce constructs and consumer's intention to buy. Int. J. Inf. Manage. **35**, 183–191 (2015)

2. Przybylski, A.K., Murayama, K., DeHaan, C.R., Gladwell, V.: Motivational, emotional, and behavioral correlates of fear of missing out. Comput. Hum. Behav. **29**, 1841–1848 (2013). https://doi.org/10.1016/j.chb.2013.02.014
3. Hodkinson, C.: 'Fear of Missing Out'(FOMO) marketing appeals: a conceptual model. J. Mark. Commun. **25**, 65–88 (2019)
4. Zhang, M., Qin, F., Wang, G.A., Luo, C.: The impact of live video streaming on online purchase intention. Serv. Ind. J. **40**, 656–681 (2020)
5. Bilal, G., Ahmed, M.A., Shehzad, M.N.: Role of social media and social networks in consumer decision making: a case of the garment sector. Int. J. Multidiscip. Sci. Eng. **5**, 1–9 (2014)
6. Duy Phuong, N., Tuan, N.M., Khoa, B.T.: Exploring the impact of Fear of Missing Out (FoMO) on youth shopping intentions in social commerce landscape. Qubahan Acad. J. **5**, 598–610 (2025). https://doi.org/10.48161/qaj.v5n1a1403
7. Beyens, I., Frison, E., Eggermont, S.: "I don't want to miss a thing": adolescents' fear of missing out and its relationship to adolescents' social needs, Facebook use, and Facebook related stress. Comput. Hum. Behav. **64**, 1–8 (2016)
8. Hetz, P.R., Dawson, C.L., Cullen, T.A.: Social media use and the fear of missing out (FoMO) while studying abroad. J. Res. Technol. Educ. **47**, 259–272 (2015)
9. Horng, S.-M., Wu, C.-L: How behaviors on social network sites and online social capital influence social commerce intentions. Inf. Manag. **57**, 103176 (2020)
10. Tran, A.V., Khoa, B.T.: The Impact of mobile augmented reality on green experience and destination choice intention in green tourism in Vietnam. Geo J. Tourism Geosites **58**, 136–145 (2025). https://doi.org/10.30892/gtg.58112-1397
11. Steinfield, C., Ellison, N.B., Lampe, C., Vitak, J.: Online social network sites and the concept of social capital. In: Frontiers in New Media Research, pp. 115–131 (2013)
12. De, D.H., Khoa, B.T., Nguyen, V.T.-T.: Customer's online purchase intention: the role of perceived business size and reputation. J. Logist. Inform. Serv. Sci. **10**, 296–307 (2023). https://doi.org/10.33168/jliss.2023.0304
13. Khoa, B.T.: Trust based online food review toward customers' restaurant selection intention in food and beverage service. J. Logist. Inform. Serv. Sci. **8**, 151–170 (2021). https://doi.org/10.33168/liss.2021.0209
14. Hu, L.t., Bentler, P.M.: Cutoff criteria for fit indexes in covariance structure analysis: conventional criteria versus new alternatives. Struct. Equ. Model. Multidiscip. J. **6**, 1–55 (1999)

From Concept to Reality: A Grey Literature Review on Digital Twin Applications in Supply Chains

Bouraya Sara(✉) and El Korchi Akram

National School of Applied Sciences, Agadir, Morocco
sara.bouraya@edu.uiz.ac.ma

Abstract. The rapid digitalization of supply chains has led to the increasing adoption of digital twin technology, raising the question of whether it is merely a passing trend or a transformative force shaping the future of supply chain management. This study critically examines the practical implementation of digital twins in supply chains through a grey literature-based analysis, leveraging industry reports, white papers, and case studies to provide real-world insights. Given the scarcity of academic research on industrial adoption, grey literature offers valuable perspectives on how businesses integrate digital twins to enhance efficiency, resilience, and sustainability.

To ensure credibility, the study applies the AACODS checklist, evaluating the authenticity, accuracy, coverage, objectivity, date, and significance of the sources. Furthermore, Ivanov's seven-element model is employed as an analytical framework to assess key aspects of digital twin applications, including their role in visibility, predictive analytics, and decision support. The findings reveal the benefits and challenges companies face in deploying digital twins, the strategic role of technology providers, and the emerging trends influencing adoption. By bridging the gap between theory and practice, this research contributes to a deeper understanding of digital twins and their potential in reshaping supply chain digitalization.

Keywords: Digital Twin · Sustainability · Grey Literature · Resilience · Simulation

1 Introduction

The fourth industrial revolution, commonly referred to as Industry 4.0, has ushered in a transformative era characterized by the fusion of physical and digital technologies, the concept of the digital twin has emerged as a pivotal innovation. DT defined as a virtual representation of physical assets, systems, or processes that mirrors real-time behavior and operational states. Initially developed by NASA in the early 2000s to improve spacecraft simulations, digital twins have since evolved beyond aerospace and manufacturing into broader industrial applications, including supply chain management.

Supply chains are inherently complex and dynamic, requiring real-time visibility, resilience, and predictive analytics to maintain operational efficiency. Digital twins offer

B.-E. Boudriki Semlali et al. (Eds.): DATA 2025, LNISO 81, pp. 111–126, 2025.
https://doi.org/10.1007/978-3-032-06725-8_10

an advanced approach to modeling, monitoring, and optimizing supply chain processes by creating virtual replicas of logistics networks, warehousing systems, and transportation flows. By simulating supply chain dynamics, organizations can anticipate disruptions, improve decision-making, and enhance overall performance (ResearchGate, n.d.). As a result, the adoption of digital twins in supply chain management is growing exponentially, with the global market projected to increase from $10.1 billion in 2023 to $110.1 billion by 2028, reflecting a CAGR of 61.3% (MarketsAndMarkets 2023).

Despite the enthusiasm surrounding digital twin technology, there remains a gap in academic literature regarding its implementation in supply chains, particularly from a grey literature perspective—industry reports, white papers, and case studies often provide crucial insights that remain underexplored in scholarly research. This paper seeks to bridge this gap by analyzing digital twin applications in supply chain management through grey literature sources. The study applies the AACODS checklist to assess the credibility of these sources and utilizes Ivanov's seven-element model for digital twin analysis to provide a comprehensive assessment of the technology's role in modern supply chains.

Moreover, this research aims to critically evaluate whether digital twin technology represents a revolutionary shift in supply chain digitalization or merely a passing trend driven by technological hype. By examining real-world implementations in industries such as manufacturing, automotive, and aerospace (LitsLink, n.d.), we will assess the benefits, challenges, and future potential of digital twins. Specifically, this paper will explore their impact on predictive maintenance, performance monitoring, inventory management, and resilience building within global supply chains.

Accordingly, this study contributes to the growing discourse on digital transformation in supply chain management, offering both practitioners and researchers valuable insights into the evolving role of digital twins. As industries continue to embrace smart supply chain solutions, understanding the practical applications and limitations of digital twins will be critical for shaping the future of global logistics and industrial operations.

2 Literature Review

A supply chain digital twin is a virtual representation that consists of: (i) a computer-based model that visually mirrors the physical supply chain and its components (e.g., companies, material flows, and products), (ii) digital technologies such as sensors, blockchain, and cloud computing that capture real-time data from physical operations, and (iii) analytics capabilities—descriptive, predictive, and prescriptive—designed to support decision-making (Ivanov 2023) (Fig. 1).

Digital twins are powered by five key technology trends: the Internet of Things (IoT), cloud computing, APIs and open standards, artificial intelligence (AI), and digital reality technologies (Ivanov 2024).

- **Internet of Things (IoT):** The widespread adoption of IoT enables digital twins by making large-scale data collection technically and economically viable. However, managing the complexity and volume of IoT-generated data requires structured approaches, with digital twins serving as a key solution. Key applications of

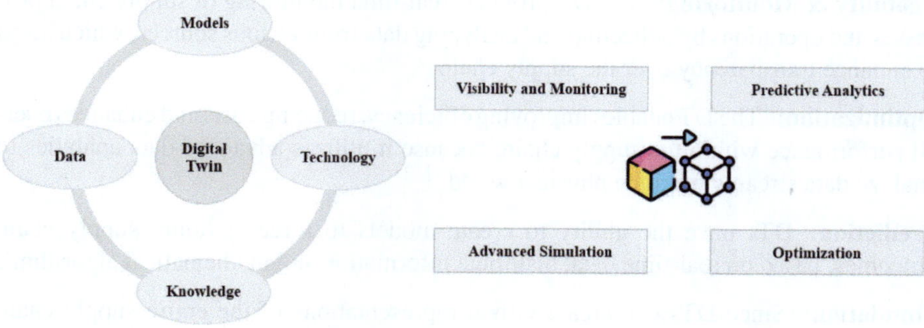

Fig. 1. Digital Twin Concept

IoT in SCM (Vilas-Boas et al. 2023): Inventory management, logistics monitoring, warehouse operations, quality control.

- **Cloud Computing:** Digital twins require substantial computing power and storage. Cloud solutions, particularly Software-as-a-Service (SaaS), allow businesses to scale resources efficiently while controlling costs.
- **APIs & Open Standards:** Previously, proprietary simulation tools created integration challenges. Open standards and APIs now facilitate seamless data exchange, enabling interoperability between systems (DHL Report, s. d.).
- **Artificial Intelligence (AI):** AI, particularly machine learning, enhances digital twins by enabling autonomous decision-making and predictive analytics using historical and real-time data.
- **Digital Reality Technologies (AR, MR, VR):** Digital twins are visualized using augmented reality (AR), mixed reality (MR), and virtual reality (VR). AR overlays digital content in the physical world, MR allows interaction with digital elements in real-world environments, and VR creates immersive, fully digital spaces.

Digital twins generate value by enhancing products, processes, users, and organizations. The extent of value and required investment depends on the application, typically falling into the following categories (Kibira & Shao 2023):

- **Descriptive Value:** Digital twins provide real-time visualization of assets, making remote or hazardous operations (e.g., spacecraft, wind turbines, power plants) easier to monitor and interpret.
- **Analytical Value:** By integrating simulations, digital twins offer insights into data that cannot be directly measured in physical objects, aiding troubleshooting and optimizing future designs.
- **Diagnostic Value:** Using explicit rules or machine learning, digital twins can diagnose issues and identify root causes based on historical and real-time data.
- **Predictive Value:** Digital twins forecast future states and potential failures, sometimes even suggesting solutions. In smart factories, they enable autonomous decision-making for optimized production.

As stated by (Van Der Valk et al. 2022), we could identify four primary purposes:

Visibility & Monitoring: The DT provides real-time monitoring of supply chain processes and operations by collecting and analyzing data from various sources, which helps to enhance transparency over the supply chain.

Optimization: The DT enables improving efficiency, reducing costs and enhances overall performance within the supply chain, because it utilizes advanced data analytics to analyze data streams from the physical world.

Prediction: DTs have the ability to create models to forecast future supply chain outcomes, based on real-time data, historical information and mathematical algorithms.

Simulation: Since DTs can create virtual representations of the entire supply chain network, this allows for the testing of different scenarios and strategies, in order to assess the impact of different variables on the supply chain's performance.

Several leading companies offer digital twin solutions tailored to supply chain applications (Akash 2024). General Electric (GE) leverages its Predix platform for real-time data analysis and predictive maintenance across industries such as aviation and power generation. Siemens integrates digital twins with its MindSphere platform, providing advanced analytics and system simulation for operational efficiency. IBM's Watson IoT platform combines AI and IoT to enable predictive analytics and asset management. Microsoft's Azure Digital Twins facilitates real-time modeling of physical environments, enhancing predictive maintenance and performance tracking. Specialized providers such as anyLogistix focus on supply chain simulations, forecasting disruptions, and optimizing logistics operations. RELEX Solutions enhances supply chain visibility through real-time data-driven inventory optimization, while BearingPoint offers Digital Twin Supply Chain solutions for risk mitigation and cost efficiency in global logistics. These companies are pioneering digital twin applications, driving greater efficiency, resilience, and agility in supply chain management.

3 Academic Interest in DSCT

The academic interest in digital twins for supply chains and logistics has grown significantly, as reflected in a Scopus database search conducted on February 10, 2025, using the research query:

TITLE-ABS-KEY (("supply chain" OR "logistics") AND "digital twin").

This search yielded 1,363 articles, demonstrating the increasing recognition of digital twins as a transformative technology in supply chain management. The number of publications has risen steadily since 2018, with a sharp peak in 2024, suggesting that research momentum is accelerating as industries and academia explore the potential of digital twins for optimizing supply chain operations (Fig. 2).

The research in this field is well distributed across different types of publications, with journal articles and conference papers each accounting for 37.34% of the total. This indicates that while theoretical advancements continue to be explored in peer-reviewed journals, there is also a strong emphasis on applied research, as reflected in conference proceedings. Conference reviews and book chapters further demonstrate

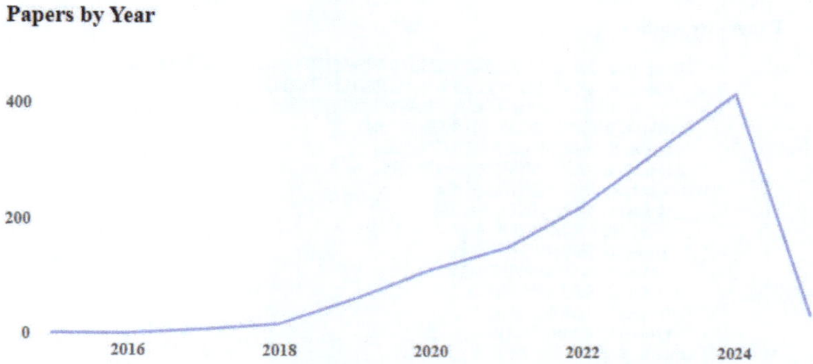

Fig. 2. Academic interest in DSCT evolution through years

the ongoing discussions surrounding digital twins, while the relatively small number of retracted papers (0.22%) suggests a high level of rigor and quality control in the research community (Fig. 3).

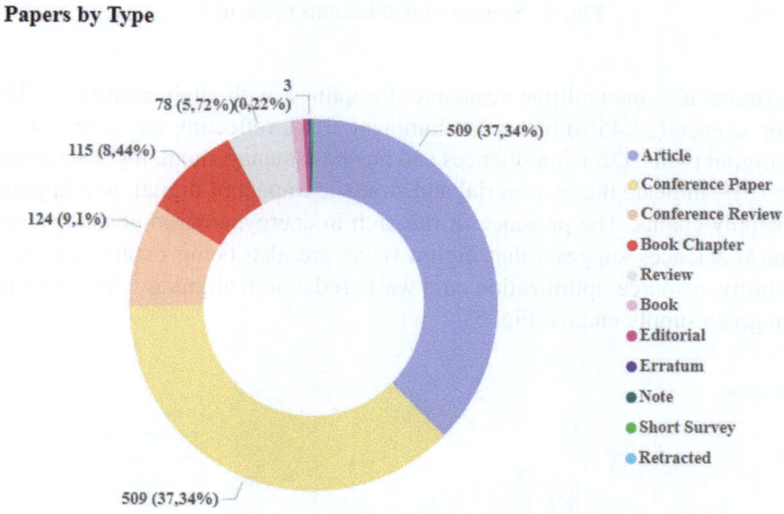

Fig. 3. Number of publications per type

Several leading researchers have emerged as key contributors to this field. G.Q. Huang and D. Ivanov have the highest number of publications, indicating their influence in shaping digital twin research in supply chains. Other notable contributors, such as T. Qu, T. Defraeye, and K. Zhang, have also made significant contributions, focusing on various applications, including real-time monitoring, optimization, and predictive modeling. Their work highlights the interdisciplinary nature of digital twins, combining elements of engineering, data science, and supply chain management (Fig. 4).

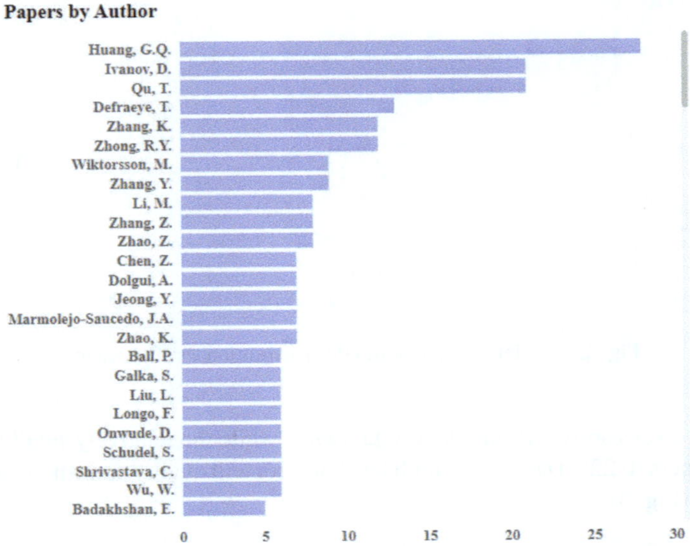

Fig. 4. Number of publications per author

The research spans multiple academic disciplines, with engineering (25.65%) and computer science (23.45%) being the dominant areas, reflecting the technical foundations of digital twins. Decision sciences and business management, together accounting for over 15%, indicate the managerial and strategic impact of digital twin implementation in supply chains. The presence of research in energy, environmental science, and agricultural sciences suggests that digital twins are also being explored as tools for sustainability, resource optimization, and waste reduction, aligning with the increasing focus on green supply chains (Fig. 5).

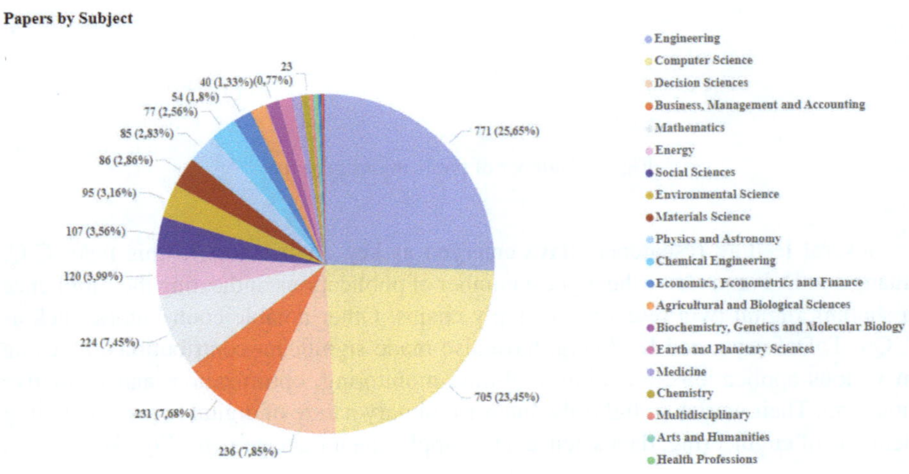

Fig. 5. Number of publications per subject

4 Research Method

The Fourth International Conference on Grey Literature held in Washington, DC, in October 1999 defined grey literature as: "that which is produced on all levels of government, academics, business and industry in print and electronic formats, but which is not controlled by commercial publishers." Grey literature includes theses or dissertations (reviewed by examiners who are subject specialists); conference papers (often peer-reviewed or presented by those with specialist knowledge) and various types of reports from those working in the field (Schöpfel 2010). All of these fall into the "expert opinion". Incorporating grey literature can expand the scope of the review to encompass a wider range of relevant studies, ensuring a more comprehensive perspective on the available evidence. This approach enhances the depth of analysis by including diverse sources that may not be covered in traditional academic literature, ultimately providing a more complete and well-rounded understanding of the topic (Mahood et al. 2014).

This study adopts a structured approach to analyze digital twin applications in supply chains using grey literature. The primary objective is to assess the credibility and reliability of industry reports, white papers, and case studies using the AACODS checklist. Jess Tyndall, Medical Librarian and Head of the Gus Fraenkel Medical Library at Flinders University, developed the AACODS checklist (Authority, Accuracy, Coverage, Objectivity, Date, Significance) as an evaluation and Critical Appraisal Tool specifically for use with grey literature sources (Fig. 6).

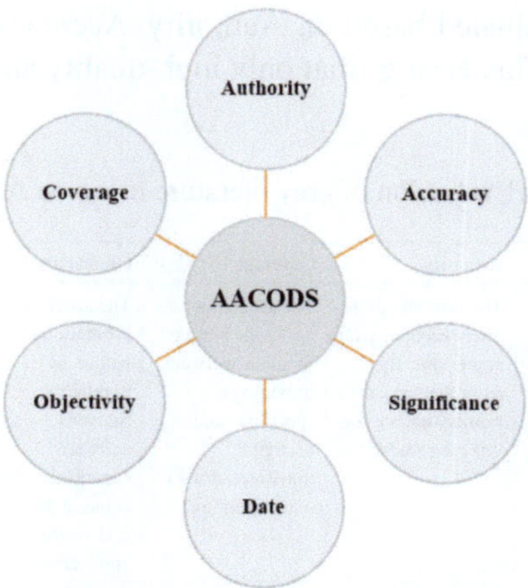

Fig. 6. AACODS method

The research also employs Ivanov's seven-element model to systematically analyze digital twin implementations. This model provides a comprehensive framework for evaluating data collection, modeling, intelligence, networking, resilience, value creation, and performance measurement (Ivanov 2021). By applying this model, the study aims to identify patterns in digital twin adoption, highlight key challenges, and determine best practices within supply chains.

The analysis follows a qualitative methodology, synthesizing findings from selected grey literature sources. Sources were retrieved from reputable industry platforms, technology provider websites, and government reports. The data were categorized and analyzed to extract key themes related to digital twin applications, benefits, risks, and technological enablers. This approach enables a thorough examination of how digital twin technology is shaping modern supply chains (Fig. 7).

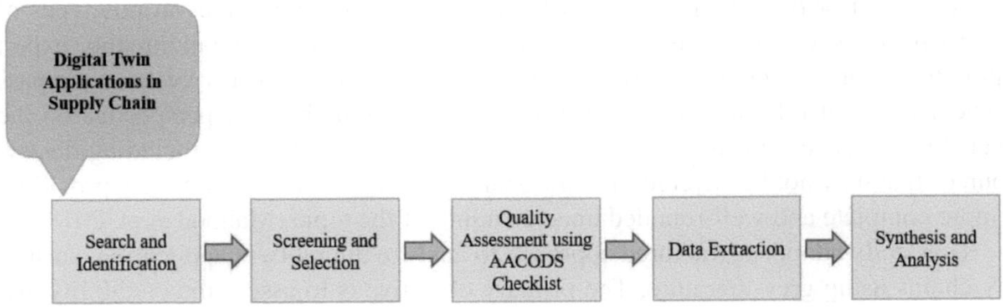

Fig. 7. Research strategy

5 Results

5.1 Evaluation of Grey Literature

The chosen data was evaluated based on Authority, Accuracy, Coverage, Objectivity, Date, and Significance. This ensures that only high-quality and relevant grey literature is considered (Table 1).

Table 1. Evaluation of grey literature using on AACODS

Reference	Authority	Accuracy	Coverage	Objectivity	Date	Significance
(Siemens 2021)	The blog is hosted on Siemens' official website, a globally recognized leader in industrial software and automation. Siemens is a credible authority in the field of digital manufacturing.	The information is likely accurate as it comes directly from Siemens, a primary source for this case study.	The article focuses on Nestlé's use of Siemens software, providing a specific example of digital transformation in manufacturing.	The article is promotional in nature, as it highlights Siemens' software capabilities, but it is based on a real-world application.	The publication date is not explicitly stated, but the content appears recent and relevant.	The source is significant for understanding how major companies like Nestlé are leveraging digital tools for innovation.
P&G and Microsoft Co-Innovate to Build the Future of Digital Manufacturing (Center 2022)	The article is published on Microsoft's official news site, a reputable source for technology and innovation news.	The information is likely accurate, as it is a first-hand account of a collaboration between Microsoft and P&G.	The article discusses the partnership between P&G and Microsoft, focusing on digital manufacturing and AI-driven solutions.	While the article is promotional, it provides insights into real-world applications of Microsoft's technologies.	The article is dated June 8, 2022, making it relatively recent and relevant.	This source is significant for understanding how large corporations are collaborating to advance digital manufacturing.

(continued)

Table 1. *(continued)*

Reference	Authority	Accuracy	Coverage	Objectivity	Date	Significance
(Walmart 2025)	*The Sun* is a mainstream news outlet, but it is not a specialized source for technology or manufacturing topics.	The accuracy of the information depends on the quality of the reporting. *The Sun* is generally reliable for news but may lack depth in technical topics.	The article covers Walmart's CEO discussing AI and digital twins, but the focus is more on inflation and general business trends	The article appears to be neutral, but it may lack technical depth compared to specialized sources.	The article is recent, as it discusses current economic conditions and trends.	The source is significant for understanding how retail giants like Walmart are adopting digital twin technology, but it may not provide in-depth technical insights.
(Petrenko 2024)	Litslink is a software development company, and while they have expertise in technology, they are not a primary source for manufacturing or industrial applications.	The information appears accurate but is likely based on secondary research rather than direct case studies.	The article provides a general overview of digital twins in manufacturing and includes examples, but it lacks depth in specific applications.	The article is informative but may have a promotional tone, as it is hosted on a company blog.	The publication date is not explicitly stated, so the timeliness of the information is unclear.	This source is useful for a basic understanding of digital twins but may not be as significant as primary sources like Siemens or Microsoft.
(Petrenko 2024)	The article is published on Unilever's official website, a global leader in consumer goods, making it a credible source.	The information is likely accurate, as it comes directly from Unilever.	The article focuses on Unilever's adoption of advanced digital technologies in its factories, providing specific examples.	The article is promotional but based on real-world achievements and partnerships.	The article is from 2023, making it highly relevant and up-to-date.	This source is significant for understanding how consumer goods companies are leading the way in digital manufacturing.
(Götz, T. 2023)	The whitepaper is authored by Tobias Götz, Managing Partner of KPS AG, a company specializing in digital transformation and supply chain management. KPS is a credible authority in the field of logistics and digitalization. The study is part of the BVL (Bundesvereinigung Logistik) research, a well-known German logistics association, which adds further credibility to the source.	The whitepaper provides a detailed and well-structured analysis of digital twins in supply chains, supported by case studies and practical examples. The information is likely accurate, given the expertise of the author and the organization. It includes definitions, potential applications, and implementation strategies, which are backed by real-world examples and data.	The paper covers a wide range of topics related to digital twins, including their definition, potential in supply chains, implementation strategies, and practical examples. It also discusses the challenges and future outlook for digital twins, making it comprehensive in scope.	The whitepaper is objective in its analysis, presenting both the benefits and challenges of implementing digital twins. It does not shy away from discussing the complexities and limitations of the technology. However, it is somewhat promotional in nature, as it highlights the capabilities of KPS AG in delivering digital transformation solutions.	The whitepaper is dated 2023, making it highly relevant and up-to-date with current trends and technologies in digital twins and supply chain management.	This source is significant for understanding the role of digital twins in modern supply chains, particularly in the context of digitalization, sustainability, and resilience. It provides actionable insights for companies considering the adoption of digital twin technology, making it valuable for both practitioners and researchers.

(continued)

Table 1. (*continued*)

Reference	Authority	Accuracy	Coverage	Objectivity	Date	Significance
(Yun et al. 2024)	The paper is authored by Gawon Yun, Douglas N. Hales, and Leo Hong, academics from reputable universities (Missouri State University, University of Rhode Island, and Millersville University). It is published in the Markets, Globalization & Development Review, a peer-reviewed journal, which adds to its credibility.	The paper is based on three case studies of companies that implemented digital twin technology. The authors conducted independent observations and collected data over several years, ensuring the accuracy of the findings. The case studies are well-documented and provide detailed insights into the implementation process, challenges, and outcomes.	The paper covers the application of digital twins in three different industries: trucking (Lorry-Planet), public transportation (Glob-South-Net), and electronics manufacturing (Eurasia-E-Net). It also discusses the broader challenges of implementing digital twin technology, such as data management, high costs, cybersecurity risks, and the need for skilled talent.	The paper is objective, presenting both the successes and challenges of digital twin implementation. It does not oversell the technology but rather provides a balanced view of its potential and limitations. The authors acknowledge the high costs and complexities involved, making it a realistic assessment of digital twin adoption.	The paper is published in 2024, making it one of the most recent and relevant sources on digital twin technology and its application in global supply chains.	This paper is highly significant for understanding how digital twins can improve operational efficiency, reduce costs, and enhance resilience in global supply chains. It provides practical insights for companies considering digital twin adoption, as well as a framework for addressing common challenges.
(OpenText 2024)	The whitepaper is published by OpenText, a recognized leader in enterprise information management and supply chain solutions. OpenText has a strong reputation in the industry, which adds credibility to the source. The paper references data from reputable organizations such as GEP, KPMG, and IDC, further enhancing its authority.	The whitepaper provides a detailed and well-structured analysis of digital supply chains, supported by data from credible sources. The information is likely accurate, given the expertise of OpenText and the references to industry research. It includes definitions, benefits, enabling technologies, and use cases, which are backed by real-world examples and data.	The whitepaper is objective in its analysis, presenting both the benefits and challenges of digital supply chains. It does not shy away from discussing the complexities and limitations of the technology. However, it is somewhat promotional in nature, as it highlights the capabilities of OpenText's solutions.	The paper covers a wide range of topics related to digital supply chains, including their definition, benefits, enabling technologies, and top use cases. It also provides guidance on how organizations can get started with digital supply chain transformation, making it comprehensive in scope.	The whitepaper is dated 2024, making it highly relevant and up-to-date with current trends and technologies in digital supply chains.	This source is significant for understanding the role of digital supply chains in modern business, particularly in the context of digitalization, sustainability, and resilience. It provides actionable insights for companies considering the adoption of digital supply chain technologies, making it valuable for both practitioners and researchers.

(*continued*)

Table 1. *(continued)*

Reference	Authority	Accuracy	Coverage	Objectivity	Date	Significance
(Kuhl & Choudhary s. d.)	The paper is authored by Michael E. Kuhl, PhD, and Mahendra Choudhary from the Industrial and Systems Engineering Department at the Rochester Institute of Technology. The academic background of the authors and the institution adds credibility to the source. The paper references reputable sources such as NASA, Google Scholar, and other academic publications, further enhancing its authority.	The paper provides a detailed and well-structured analysis of digital twins in intralogistics, supported by data from credible sources. The information is likely accurate, given the academic rigor and references to industry research. It includes definitions, frameworks, and applications of digital twins, which are backed by real-world examples and data.	The paper covers a wide range of topics related to digital twins, including their definition, framework, simulation approaches, and applications in intralogistics. It also discusses the challenges and future outlook for digital twins, making it comprehensive in scope.	The paper is objective in its analysis, presenting both the benefits and challenges of digital twins. It does not shy away from discussing the complexities and limitations of the technology. The academic nature of the paper ensures a balanced and unbiased perspective.	The paper is dated July 2023, making it highly relevant and up-to-date with current trends and technologies in digital twins and intralogistics.	This source is significant for understanding the role of digital twins in modern intralogistics, particularly in the context of Industry 4.0 and smart factories. It provides actionable insights for companies considering the adoption of digital twin technology, making it valuable for both practitioners and researchers.

5.2 Case Studies

Digital twin technology is increasingly being adopted across industries to enhance supply chain efficiency, resilience, and sustainability. By creating virtual models of physical systems, companies can simulate scenarios, optimize operations, and improve decision-making in real time. This section presents case studies of companies implementing digital twin solutions, structured according to Ivanov's 7-element framework (Ivanov 2024), which provides a comprehensive lens for analyzing digital twin applications in supply chain management.

The 7 elements framework includes:

Technology: Digital twins rely on data generated and processed through various information technologies. At the product level, CAD/CAM systems enable digital twins, while MES systems support manufacturing processes, and ERP systems facilitate organizational twins. Supply chain digital twins integrate platforms like anyLogistix, external logistics data, and weather/financial data (Burgos & Ivanov 2021). IoT, Blockchain, AI, AR/VR, and Industry 4.0 as key enablers (Bhandal et al. 2022).

People: Digital twins transform decision-making in supply chain management, offering real-time data access and decision-support through simulations. They act as "digital companions" for managers, aiding recovery strategies, sustainability planning, and automated decision-making in inventory control.

Organisation: Digital twins mirror supply chain structures and enable new business models such as digital manufacturing, cloud supply chains, and collaborative platforms. They redistribute decision-making authority across departments, transforming production, logistics, and overall supply chain management.

Task: Digital twins support collaborative demand planning, inventory management, sourcing, quality control, shipment tracking, smart contracting, predictive maintenance, resilience management, and sustainability.

Scope: Digital twins are classified into five types: product, process, company, supply chain, and network-of-networks. As complexity increases, so does the need for advanced modeling, integration, and decision-making tools. While product and process twins dominate research, organizational and network digital twins remain underexplored, requiring multidisciplinary collaboration across mechanical, industrial, and supply chain engineering.

Management: Digital twins enhance visibility, transparency, collaboration, and traceability across supply chains. They range from basic process monitoring to full-scale digital collaboration, improving end-to-end visibility even in multi-echelon networks. A key research area is understanding their value in unobservable supply chains, where big data analytics (e.g., Google BigQuery) can reveal hidden supplier relationships.

Modelling: Digital twins integrate analytics, AI, and simulation for decision-making support or AI-driven automation. Techniques include optimization, discrete-event simulation, neural networks, machine learning, reinforcement learning, agent-based modeling, and system dynamics. Future research should focus on real-time data-driven modeling and its transition from offline simulations, exploring value creation and barriers to adoption.

See Table 2

Table 2. Case studies analysis

Technology	People	Management	Organization	Scope	Task	Modelling
Case study 1: Nestlé (Siemens 2021)						
Siemens' Digital Enterprise portfolio (Tecnomatix Plant Simulation)	Factory workers trained to use digital twins for process optimization	Managed digital transformation for an aging plant	Integrated digital twin into an existing food production facility	Focused on manufacturing optimization & sustainability	Optimized material flow, production scheduling	Digital model of plant simulates production scenarios
Case study 2: Vita Coco Company (Case Study 2024)						
RELEX Solutions' Retail Planning Software	Employees trained to use advanced analytics for demand forecasting	Implemented a centralized planning system to manage inventory and supply chain	Integrated RELEX software into existing retail and supply chain operations	Focused on improving demand forecasting and inventory management	Optimized inventory levels, reduced waste, and improved supply chain efficiency	Digital model of supply chain simulates demand scenarios and inventory needs
Case study 3: Unilever (Unilver 2023)						
IoT-enabled production line digital twins	Production teams trained for quality & process control	Focus on reducing waste & energy consumption	Embedded within factory production systems	Optimizes production processes, waste reduction	Improves product quality & resource efficiency	Simulates different production scenarios for optimal outcomes

(continued)

Table 2. (*continued*)

Technology	People	Management	Organization	Scope	Task	Modelling
Case study 4: Procter & Gamble (P&G) (Center 2022)						
Microsoft Azure, AI & IoT for supply chain digital twins	Supply chain analysts & logistics teams leverage insights	Data-driven decision-making at supply chain level	Digital twin embedded in global supply chain management	Simulates logistics & operational performance	Identifies bottlenecks, improves efficiency	Simulations predict outcomes & reduce inefficiencies
Case study 5: Walmart (Walmart 2025)						
NVIDIA-powered AI-driven digital twins	Retail managers & analysts interpret simulation results	Store layout optimization strategy	Applied across 1,700+ retail stores	Enhances customer experience & operational efficiency	Tests store layouts before real-world implementation	AI simulates store configurations, analyzes consumer behavior
Case study 6: CEVA Logistics (Simul8 2024)						
Simul8's simulation-powered digital twin technology	CEVA Logistics planners, warehouse staff, and control room operators	Managed resource allocation, staffing levels, and operational planning	Implemented in the inbound, storage, and outbound phases of a single fulfillment center.	Focused on optimizing daily operations within the fulfillment center	Optimize resource allocation, reduce costs, and improve operational efficiency	Digital twin linked to real-time data, simulating dynamic behavior of the system
Case study 7: Lorry-Planet						
IoT-enabled Digital Twins	Drivers, fleet managers	Data-driven fleet management	Real-time fleet tracking & predictive maintenance	Logistics and transportation	Reduced fuel consumption & maintenance costs	Vehicle performance analytics, predictive maintenance
Case study 8: Eurasia-E-Net						
Integrated Digital Twin	Manufacturing engineers, supply chain managers	Optimized supply chain operations	Virtual modeling of factories & supply chains	End-to-end supply chain optimization	Shortened product development cycle, improved quality control	Supply chain visibility, virtual prototyping

6 Discussion

Supply chains face multiple challenges that impact their efficiency, sustainability, and overall performance (Götz, T. 2023). One of the key challenges is significant cost pressures and resource scarcity. Processes that still rely heavily on manual labor present both challenges and opportunities for improvement, as they often highlight inefficiencies that can be addressed through digital transformation. The entry of new digital players with innovative greenfield solutions has intensified competitive pressure, forcing traditional companies to adapt. Rising costs, driven by fierce competition for limited resources—such as raw materials and skilled labor—exacerbate these challenges. Another challenge is responsiveness and resilience. In a volatile market environment, companies are finding that merely reacting to events is no longer sufficient; they must develop the ability to anticipate changes and act with foresight. This requires comprehensive evaluation and monitoring of digital data streams to inform decision-making processes. The human element in supply chains presents its own set of challenges, particularly in the context of physical and mental limitations faced by employees. These limitations can hinder the seamless execution of business-critical processes, leading to errors and inefficiencies. To alleviate this pressure, organizations are increasingly turning to automation for repetitive tasks, allowing employees to focus on more value-adding activities that require human insight. The integration of digital twins into supply chains is increasingly driven by regulatory requirements, particularly the Supply Chain Due Diligence Act (LkSG), which mandates transparency and compliance within supply chains. This legislation compels logistics companies to ensure that their operations meet ethical and social standards, creating a pressing need for complete visibility throughout the supply chain. In (Table 3) summarizes the key supply chain challenges and illustrates how digital twins contribute to addressing each of them.

Table 3. Digital twin contributions in supply chain challenges

Challenge	Digital twin contribution	Reference
Cost pressures and resource scarcity	To maintain their market position, businesses must actively pursue continuous improvement in their internal processes, optimizing resource utilization and reducing costs. However, the sequential planning logic of traditional supply chains often leads to inefficiencies, making it crucial for organizations to leverage digital tools that can dynamically manage financial and material resources.	
Responsiveness and Resilience	Establishing a digital twin for core business processes allows organizations to access real-time data, which is essential for enhancing responsiveness and minimizing the impact of disruptions. By integrating this data with inventory systems, companies can increase their reaction speed and better navigate dynamic external influences. Ultimately, building resilience in supply chains hinges on the ability to adapt swiftly to changing conditions while maintaining operational effectiveness.	(Simul8 2024)
Human Factor	However, despite advancements in digital tools that can analyze large volumes of data and identify correlations quickly, human judgment remains vital in interpreting this information effectively—especially during unpredictable situations. Striking a balance between leveraging technology and recognizing the importance of human expertise is essential for optimizing supply chain performance.	
Regulatory Requirement	Digital twins are ideally suited for this purpose, as they can provide a dynamically adaptable representation of supply chain processes. However, while they can effectively gather and present regulatory information, their utility in implementing preventive measures is limited if they function merely as data storage systems. Forward-thinking companies recognize the importance of leveraging digitalization trends to enhance compliance and sustainability, thereby fostering resilience within their operations.	

7 Conclusion

A digital twin of the supply chain provides a strong foundation for addressing the challenges of digitalization, sustainability, and resilience. Companies can use digital twins to counteract external pressures and manage volatility by implementing effective strategies and measures.

Achieving effective implementation of digital twins is complex and requires a strategic approach to developing a networked IT infrastructure that accommodates diverse markets and service providers. Modern supply chains are influenced by rapidly changing conditions that can disrupt operations, necessitating flexibility and adaptability. Forming partnerships and collaborations can help organizations utilize alternative routes and sourcing strategies to maintain operational agility. Digital twins can facilitate preventive actions and the integration of partner systems, but successful technical implementation is crucial for realizing their full potential in enhancing supply chain resilience.

The benefits of employing digital twins are evident in their ability to increase cost efficiency, enhance transparency, and optimize space utilization within distribution centers. By enabling real-time monitoring and control of logistics activities, digital twins allow for precise tracking of movements within both virtual and physical environments. This capability streamlines the organization of goods and automates processes in warehouses, leading to more efficient inbound and outbound transport planning. Furthermore, digital twins empower employees by providing unrestricted access to critical information, which enhances decision-making capabilities and internal value creation. As companies build a functional digital foundation in response to regulatory demands like the Supply Chain Act, they position themselves to navigate future transformation processes effectively while reaping the benefits of enhanced operational efficiency and resilience.

Conflicts of Interest Declaration. All authors declare that they have no conflicts of interest.

References

Akash: Top 10 Digital Twin Companies in 2024. Stratistics MRC, 21 June 2024. https://www.strategymrc.com/blog/top-digital-twin-companies/

Bandara, L.V., Buics, L.: Digital twins in sustainable supply chains: a comprehensive review of current applications and enablers for successful adoption. In: SMTS 2024, vol. 64 (2024). https://doi.org/10.3390/engproc2024079064

Bhandal, R., Meriton, R., Kavanagh, R.E., Brown, A.: The application of digital twin technology in operations and supply chain management: a bibliometric review. Supply Chain Manag. Int. J. **27**(2), 182–206 (2022). https://doi.org/10.1108/SCM-01-2021-0053

Burgos, D., Ivanov, D.: Food retail supply chain resilience and the COVID-19 pandemic: a digital twin-based impact analysis and improvement directions. Transp. Res. Part E Logist. Transp. Rev. **152**, 102412 (2021). https://doi.org/10.1016/j.tre.2021.102412

Case Study: The Vita Coco Company, 13 May 2024. RELEX Solutions. https://www.relexsolutions.com/resources/case-study-the-vita-coco-company/

Center, M.N.: P&G and Microsoft co-innovate to build the future of digital manufacturing, 8 June 2022. Stories. https://news.microsoft.com/2022/06/08/pg-and-microsoft-co-innovate-to-build-the-future-of-digital-manufacturing/

Cimino, A., Longo, F., Mirabelli, G., Solina, V.: A cyclic and holistic methodology to exploit the Supply Chain Digital Twin concept towards a more resilient and sustainable future. Clean. Logist. Supply Chain **11**, 100154 (2024). https://doi.org/10.1016/j.clscn.2024.100154

Damtew, A.W.: Roles of Digital Twins on Material Performances and Utilization on Upstream Industry (The Case of Automotive Industry). In Review (2023). https://doi.org/10.21203/rs.3.rs-2659308/v1

DHL Report: Digital Twins in Logistics (s. d.)
Götz, T.: Digital twin under the magnifying glass – Nice gimmick or real game changer for the supply chain ? KPS AG (2023). https://www.bvl.de
Ivanov, D.: Intelligent digital twin (iDT) for supply chain stress-testing, resilience, and viability. Int. J. Prod. Econ. **263**, 108938 (2023). https://doi.org/10.1016/j.ijpe.2023.108938
Ivanov, D.: Conceptualisation of a 7-element digital twin framework in supply chain and operations management. Int. J. Prod. Res. **62**(6), 2220–2232 (2024). https://doi.org/10.1080/00207543.2023.2217291
Jess, T.: AACODS Checklist (2010). http://dspace.flinders.edu.au/dspace/
Kibira, D., Shao, G.: Data requirements for a digital twin of a robot workcell. In: 2023 Winter Simulation Conference (WSC), pp. 3272–3283 (2023). https://doi.org/10.1109/WSC60868.2023.10408428
Kuhl, M.E., Choudhary, M.: Digital Twin for Intralogistics (s. d.)
Mahood, Q., Van Eerd, D., Irvin, E.: Searching for grey literature for systematic reviews: challenges and benefits. Res. Synth. Methods **5**(3), 221–234 (2014). https://doi.org/10.1002/jrsm.1106
MarketsAndMarkets: Digital Twin Market Size, Share, Industry Trends Report 2030 (2023). MarketsandMarkets. https://www.marketsandmarkets.com/Market-Reports/digital-twin-market-225269522.html
OpenText: Digital supply chain 101 (2024). OpenText. https://www.opentext.com/digitize-supply-chains
Petrenko, V.: Digital Twin in Manufacturing : 10 Inspiring Industry Examples, 29 July 2024. Litslink. https://litslink.com/blog/what-is-digital-twin-in-manufacturing-inspiring-industry-examples
Schöpfel, J.: Towards a Prague Definition of Grey Literature (2010)
Siemens: Nestlé uses Siemens' software to go from outdated to trendsetting—Tecnomatix, 22 November 2021. https://blogs.sw.siemens.com/tecnomatix/nestle-uses-siemens-software-to-go-from-outdated-to-trendsetting/
Simul8: CEVA Logistics use a simulation-powered digital twin. Simul8 Simulation Software (2024). https://www.simul8.com/case-studies/ceva-logistics-simulation-digital-twin
Unilver: Two Unilever sites named most digitally advanced factories, 13 January 2023. Unilever. https://www.unilever.com/news/news-search/2023/unilever-sites-join-network-of-worlds-most-digitally-advanced-factories/
Vilas-Boas, J.L., Rodrigues, J.J.P.C., Alberti, A.M.: Convergence of distributed ledger technologies with digital twins, IoT, and AI for fresh food logistics: challenges and opportunities. J. Ind. Inf. Integr. **31**, 100393 (2023). https://doi.org/10.1016/j.jii.2022.100393
Walmart: Walmart CEO reveals how chain is using « digital twin » to plan store changes—And it's been used for over 1700 locations | The US Sun (2025). https://www.the-sun.com/money/13304708/walmart-ceo-john-furner-nrf-inflation-ai-digital-twin/
Yun, G., Hales, D.N., Hong, L.: Digital twinning for resilient global supply chains : three case studies. Mark. Glob. Dev. Rev. **9**(1) (2024). https://doi.org/10.23860/MGDR-2024-09-01-04
Zhao, X., Wang, Z.: The factory supply chain management optimization model based on digital twins and reinforcement learning. Scalable Comput. **26**(1), 241–249 (2025). Scopus. https://doi.org/10.12694/scpe.v26i1.3765

Digital Manufacturing for Sustainable Development

Digital Learning in Public Service

Marta Dmytryshyn[1,2](✉), Tetyana Goran[3], and Roman Rusyn[4]

[1] Department of Management and Administration, West Ukrainian National University, Lvivska 11, Ternopil 46009, Ukraine
m.dmytryshyn@wunu.edu.ua

[2] University of Wisconsin-Madison, Madison, USA

[3] RICHTIGBAU s.r.o., IČO, 08686041 Kořenského 905/4, Střekov, 400 03 Ústí nad Labem, Czech Republic
t.goran@richtigbau.com

[4] Department of Economic Cybernetics, Vasyl Stefanyk Precarpathian National University, Shevchenka 57, Ivano-Frankivsk 76018, Ukraine
roman.rusyn@pnu.edu.ua

Abstract. The paper describes the prerequisites for the development of public officials' training in the digital conditions due to the implementation of the decentralization reform in Ukraine; the features of the state standard and educational program are studied; the students' needs and perceptions survey is conducted and analyzed; the application of digital learning technologies for public officials is substantiated; the mapping of the number of students in public management and administration by region is carried out; the forecast of number of public officials as well as vacant positions in public service is performed; future trends in public officials training are determined.

Keywords: Digitization of Education · Public Official · Community · Online Courses · Professional Growth · Decentralization · Training · Skills

1 Introduction

The decentralization reform began in Ukraine in 2014. Because of its implementation, territorial communities received complete budgetary independence in terms of forming financial resources and implementing relevant expenditures [1]. The main goal of the reform is the availability of high-quality administrative, educational, medical, cultural, municipal, and social services for community residents. The decentralization reform, has actualized staffing and increased the requirements for the professional competence of employees in territorial communities [2]. This is caused by constant changes in the internal and external environment, as the tasks of decentralization require public officials to be able to solve the community's urgent problems in the conditions of current resource limitations [3], to have the tools of local development management, project activities [4], to compare the current and strategic interests of the community [5], to observe the perspective of the community's progress in general, and the field of professional activity, in particular.

B.-E. Boudriki Semlali et al. (Eds.): DATA 2025, LNISO 81, pp. 129–141, 2025.
https://doi.org/10.1007/978-3-032-06725-8_11

2 Literature Review

The problem of mental changes in personnel arises as any public official must clearly understand that he or she is the main asset of the community and, for a monetary reward, perform the assigned functions as efficiently as possible to initiate the improvement of methods and technologies of service to the residents of the community and to ensure the implementation of Ukrainians laws.

People with relevant special professional knowledge, skills, and abilities who speak the state language are accepted for public service. Therefore, entry into public service must be preceded by specialized professional training. The professionalism of public officials is ensured by improving their qualifications additionally during public service [6]. An essential tool for public service development as a condition for proper governance in Ukraine is the outcome of leadership competencies: orientation to results, analytical thinking, human resources management, and effective communication [7].

Alshammari and Alkhwaldi (2025) [8] argue for the importance of strengthening students' educational (EDS) and emotional support (EMS) for the successful implementation of digital technologies in the learning process. Mexhuani (2024) [9] identifies the main educational barriers to digital technologies, including digital inequality, teachers' unwillingness to change, and privacy issues. Hassan et al. (2024) [10] analyze using digital technologies, including artificial intelligence, in education, considering cross-cultural characteristics. Khrykov et al. (2022) [11] consider distance learning the crucial means of training public officials. This study mainly focuses on information and communication, as well as public administration and e-learning components.

3 Methodology

Research methods, objectives, and limitations used to obtain the results are given in Table 1.

Table 1. Research methods

Method	Objective	Limitations
Observation	To identify the key needs of students	Has only a theoretical basis
Survey conducting	To examine the current state of the students' perceptions	Does not give general statistical data
Comparison	To evaluate the results of the survey	Has only a theoretical basis
Analysis	To compare digital learning technologies	Limited to widely used technologies
Synthesis	To get the resulting technologies used in public officials' training	Limited to public administration

(continued)

Table 1. (*continued*)

Method	Objective	Limitations
Cartography	To show the regional dissemination of the students.	Need special tools
FORECAST.ETS function and regression analysis	Forecast the number of public officials and vacant positions	Require normalization in case of low forecast realism
Analytic theory of branched continued fraction	To calculate the approximate forecast values	Require specific knowledge of the mathematical apparatus

4 Research Results

Because of the decentralization reform, the transformation of local self-government authorities initiated a change in the system of training and advanced training of public accountants and other officials of local self-government authorities. Solving this problem became possible by intensifying the professional development of personnel and their scientific and pedagogical support for the formation of the labor potential of communities. In this way, studying in an educational program on public management and administration can be of interest to current public officials and those considering the possibility of employment in the public service through the personnel reserve. The training of personnel should contribute to forming a system of professional public service focused on protecting the interests of citizens, ensuring the provision of high-quality services, creating a convenient administrative procedure for legal entities and individuals, and building effective state institutions that shape state policy. It is successfully implemented for the sustainable development of the state.

The development of professional training for public officials will allow:

- to prepare highly professional personnel for public service;
- to develop the professional competence of applicants for public positions;
- encourage public officials to self-education and self-improvement.

The training of public officials is carried out under the Master's program "Public Management and Administration." At West Ukrainian National University, such a program takes into account modern trends in the development of the sphere of public management and administration, which are determined by the challenges and threats to the development of Ukrainian society at the present stage, the need to increase the efficiency of the implementation of sectoral reforms, in particular, the reform of decentralization, administrative and territorial reform; provides for mastering innovative methods of professional activity, modern management tools for ensuring the development of territories and communities with the involvement of the potential of institutions and organizations of the public sphere, acquiring the ability to develop and implement management mechanisms based on multi-level governance.

According to the educational standard and program for master training, graduates of the program can work in positions in central and local state authorities; in positions in local self-government authorities (communities); in positions in the structures of non-state subjects of civil society; in management positions and specialist positions at enterprises, institutions, and organizations of state and communal forms of ownership; in managerial and administrative positions in international organizations and their representative offices in Ukraine.

Besides this, the graduates can continue to get an education at the third educational and scientific level of higher education (PhD).

We conducted a survey to consider the specific needs of students who aspire to work in public service and territorial communities. The study involved 84 interviewed respondents in the Master's program "Public Management and Administration" at the West Ukrainian National University. The degree of attitude of the respondents to training is shown in Table 2.

Participation in the survey was completely voluntary and anonymous. The survey covered respondents who live and work in 18 communities of western Ukraine, including Ivano-Frankivsk, Pasichna, Khmelnytskyi, Spas, Dzvyniach, Broshniv-Osada, Nadvirna, Perehinsk, Rozhniativ; Stari Bohorodchany; Bilshivtsi; Tlumatch; Kuty; Kosiv; Tysmenytsia; Kalush; Solotvyno; Dubivtsi. The survey was conducted using the Google Forms online platform during two weeks of the 2nd semester of the 2022–2023 academic year.

Table 2. Survey participants.

№	Category	Number of respondents, people	Share of respondents, %
1.	Graduates of the 2022–2023 academic year	27	32,1
2.	Current students	35	41,7
3.	Applicants who want to enroll in the program at the first opportunity*	22	26,2
Total		84	100

* *In the 2023–2024 academic year, in case of successful completion of the competition*

The results of the conducted survey are given in Figs. 1, 2 –3.

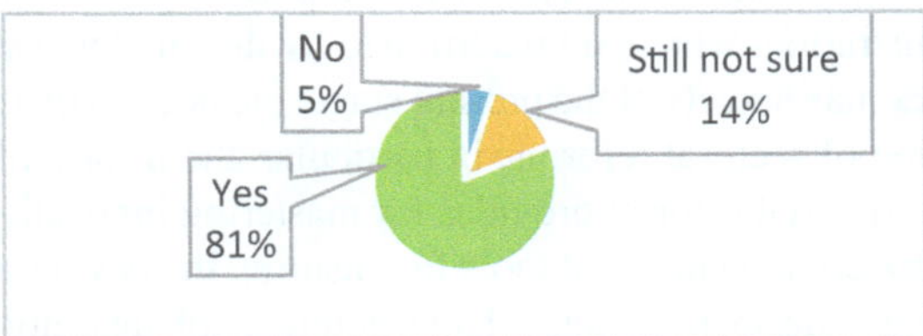

Fig. 1. Combination of work and study.

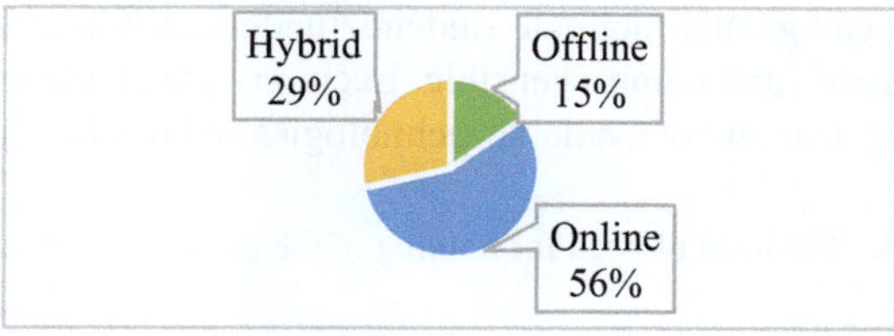

Fig. 2. Proffered form of studies.

Most respondents need education for career prospects, to acquire new knowledge and skills, and to increase their intellectual capital (Fig. 4).

The brief results of the survey indicated that students preferred online or hybrid learning, many combined work and study, and the reasons for the decision to study and improve their qualifications for many were the desire to get prospects for career growth, to acquire new knowledge and skills, and to increase intellectual capital.

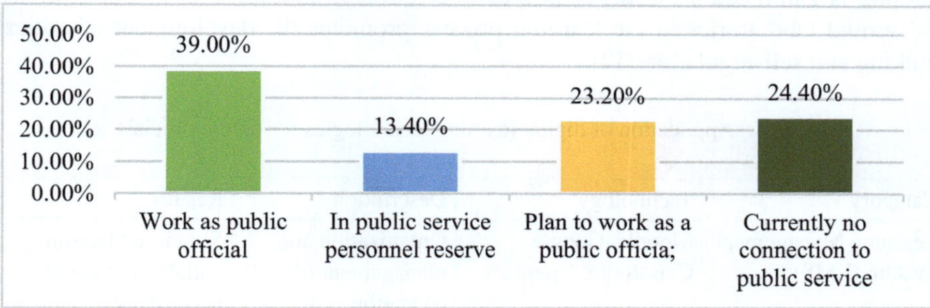

Fig. 3. Belonging to the public service.

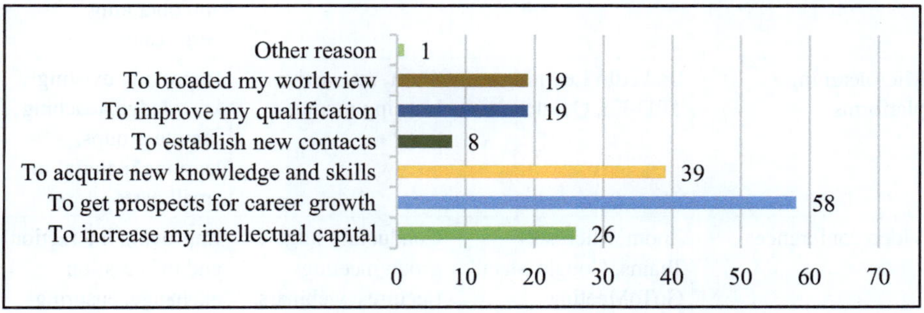

Fig. 4. The number of respondents who chose each reason for studying.

The surveyed respondents rated their training needs quite highly (Table 3).

Digital learning can greatly facilitate students' understanding of the curriculum and make it more accessible and comprehensible. Lecturers are challenged to create high-quality content using a variety of available technologies and teaching methods (Table 4).

Table 3. The level of need for training of the surveyed respondents.

Score	Strength	Level	Number of answers	Share of answers
1.	Slight moderate	Acceptable	2	2,4%
2.	Moderate	Suitable	6	7,1%
3.	Slight strong	Appropriate	19	22,6%
4.	Strong	Necessary	26	31,0%
5.	Very strong	Crucial	31	36,9%

One of the practical tools that enables interactive, personalized, and self-directed learning is gamification. Using multimedia, electronic modules, mobile applications, and virtual laboratories in the learning process promotes the development of critical thinking and self-regulation [12].

Table 4. Application of digital learning technologies for public officials.

Category	Technology	Description	Results
Learning Management System (LMS)	Moodle, Google Classroom, OpenEdx	Organization and management of education	Structured learning, collaboration, and performance monitoring.
Online course platforms	Prometheus, Coursera, EdX, Udemy	Personalized courses to develop valuable skills	Improving qualifications, training efficiency, and obtaining certificates.
Microlearning platforms	LinkedIn Learning, TED-Ed, Quizlet	Short, interactive learning modules	Improving existing knowledge, teaching remote groups, learning material in small parts
Video conferences	Zoom, Microsoft Teams, Google Meet, GoToMeeting	Conducting large group meetings (lectures, webinars, interactive classes)	Improving interaction and information exchange, ensuring accessibility, security, and convenience
Content creation	Canva, Visme. Prezi, Vimeo, G-Suite	Creation of online content: presentations, graphics, videos,	Visualization of material to improve information dissemination.

(continued)

Table 4. (*continued*)

Category	Technology	Description	Results
Content management	Dropbox, Drive, WordPress та Drupal	Loading and organizing information	Increasing the efficiency of education and ensuring access to materials
Communication tools	Google Forms, Typeform, Kahoot!	Surveys, tests, knowledge level testing, interactive tasks	Automation of evaluation, feedback, communication to create high-quality content.
Cloud technologies and file sharing	OneDrive, Notion, Google Workspace, Microsoft 365	Enabling collaboration and online storage	Accessibility of learning resources from any device, improved communication
Augmented (AR) and virtual reality (VR)	Microsoft HoloLens, Meta Quest, Google Lens	Providing interactive learning and creating simulations	Mastering complex concepts and engaging students in hands-on learning with immersion in real-world scenarios.
Social networks	Slack, Workplace	Virtual space	Providing fast, simple, and targeted real-time communication
Artificial intelligence	ChatGPT, Grammarly, Perplexity, automated work checks	Adaptive learning, automated assessment, chatbots	Acceleration of work verification, personalization of the learning process, workload reduction

Several factors, including the characteristics of the academic discipline, the age characteristics of the participants, and methodological approaches, determine the effectiveness of gamification in the educational process. Therefore, it is necessary to investigate the optimal conditions for its use to achieve maximum educational effect [13].

A relatively new technology that allows students to listen to educational materials outside the classroom is podcasts. The use of podcasts for the effective dissemination of information among students is characterized by accessibility, flexibility, and interactivity, involves the involvement of experts, promotes audience involvement, and requires multimedia experience. Podcasts, like gamification, contribute to the student's independent learning.

The use of artificial intelligence in the educational process of future public officials is a controversial issue. In general, the results generated by AI can be divided into several types, including text, image, audio, prediction, and interactive content. The positive use of AI in the educational process is possible if the following conditions are met:

- development of skills in the use of AI by subjects of the educational process;
- determination of a list of types of work in which it is appropriate to use artificial intelligence (preparation of presentations, definition of fundamental concepts, clarification of certain information, or translation of terms);
- determination of a list of types of work in which it is inappropriate to use artificial intelligence (writing analytical materials, generalization of data).

Virtual tours help train public officials as a tool for additional visualization of educational content, creating conditions for independent observation and ensuring the effect of presence [14].

The use of digital tools in the process of training civil officials contributes to the following:

- increasing motivation;
- improving the professional skills of applicants; developing creative thinking and creative abilities;
- developing digital competencies of students;
- increasing interest in learning; effectiveness of acquiring new knowledge; stimulating mental abilities.

To forecast the need for public official training, we first analyze the total number of master's students currently studying in the program "Public Management and Administration" by region in the form of a map (Fig. 5).

To make a forecast for public officials based on empirical data, we used the FORECAST.ETS function (function syntax) and regression analysis. As a result, we built branched continued fraction models for public officials:

$$y = 170409\left(U(-x) + \cfrac{0.993x}{U(x) + \cfrac{0.993u_2x}{U(-x) + \cfrac{0.993u_3x}{U(x) + \ddots}}}\right) \tag{1}$$

with $R^2 = 0.96$, for public officials (male)

$$y = 43287\left(U(-x) + \cfrac{0.99x}{U(x) + \cfrac{0.99u_2x}{U(-x) + \cfrac{0.99u_3x}{U(x) + \ddots}}}\right) \tag{2}$$

with $R^2 = 0.93$, for public officials (female)

$$y = 127129\left(U(-x) + \cfrac{0.994x}{U(x) + \cfrac{0.994u_2x}{U(-x) + \cfrac{0.994u_3x}{U(x) + \ddots}}}\right) \quad (3)$$

with, $R^2 = 0.95$, and for vacant positions in public service

$$y = 33668\left(U(x) + \cfrac{-0.9904x}{U(-x) + \cfrac{-0.9904u_2x}{U(x) + \cfrac{-0.9904u_3x}{U(-x) + \ddots}}}\right) \quad (4)$$

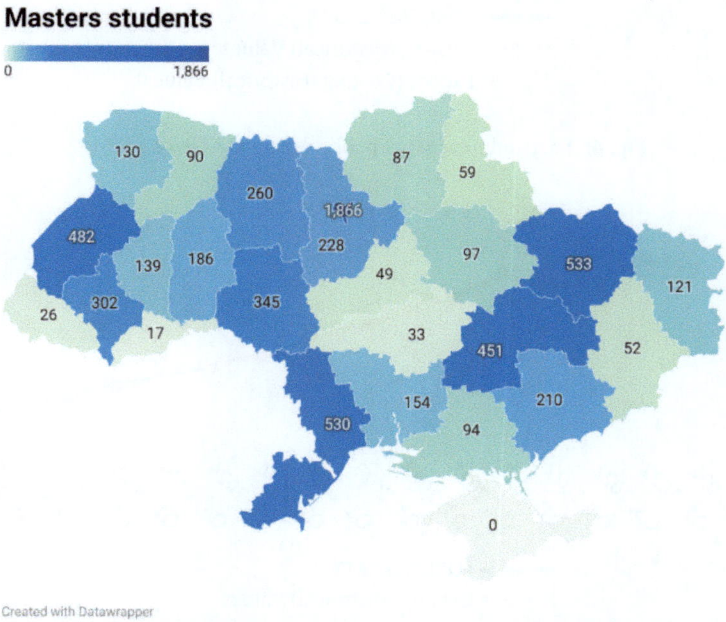

Fig. 5. The number of master's students in the program "Public Management and Administration," January 2025.**based on* [15].

with $R^2 = 0.91$, where

$$U(x) = 1 + \frac{x}{1 + \frac{u_2 x}{1 + \frac{u_3 x}{1+}}}, u_{2k} = \frac{1}{2 - 4k}, u_{2k+1} = \frac{1}{2 + 4k}, k \geq 1. \quad (5)$$

Note that the models were constructed on the base of a branched continued fraction expansion of the exponential function [17].

Studying the convergence of this expansion is a significant problem, and some results in this direction can be found in [18–20]. From a review of Figs. 6, 7, 8 –9, we see that in all cases, the forecast based on branched continued fraction models is more optimistic than that based on the FORECAST.ETS function.

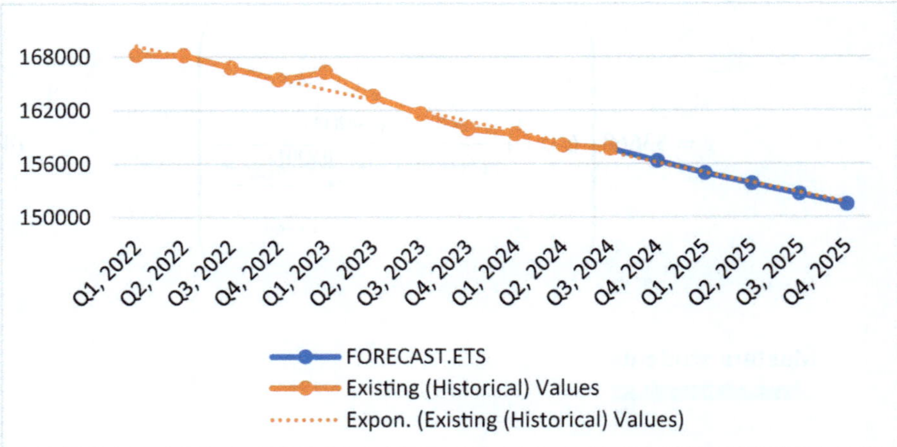

Fig. 6. Fact and forecast of public officials**based on* [16].

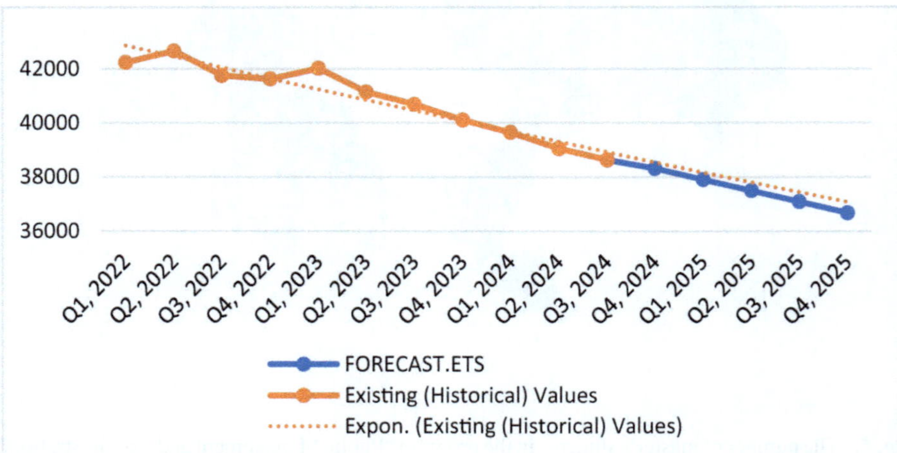

Fig. 7. Fact and forecast of public officials (male)**based on* [16].

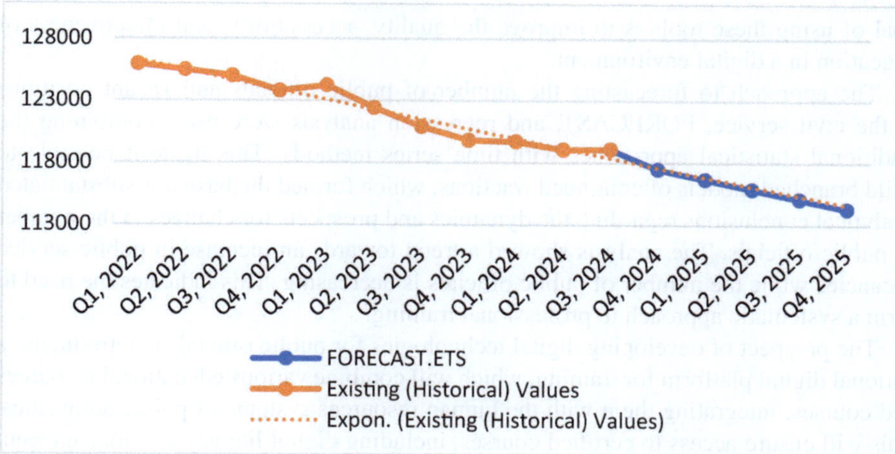

Fig. 8. Fact and forecast of public officials (female) *based on* [16].

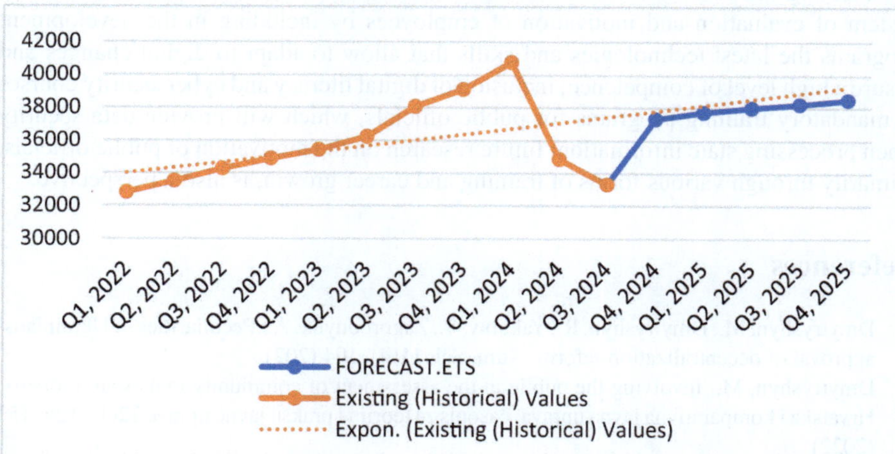

Fig. 9. Fact and forecast of vacant positions in public service*based on* [16].

5 Conclusions

Based on a survey of 84 respondents from 18 communities in West Ukraine, it was found that education is one of the key factors in developing career prospects. At the same time, expanding the sample in further research will facilitate a more detailed analysis of the identified trends. It will allow broader generalizations regarding education's impact on public officials' professional development in different regions of Ukraine.

In the context of rapid digital transformation, online learning is a priority form of advanced training due to its flexibility and the ability to combine the educational process with professional activities. As a result of the analysis, 11 key categories of digital tools perform separate functions in the educational process - from organizing training to creating content, communicating, and developing professional skills. The common goal of using these tools is to improve the quality, accessibility, and effectiveness of education in a digital environment.

The approach to forecasting the number of public officials and vacant positions in the civil service, FORECAST, and regression analysis were used, combining the traditional statistical approaches with time series methods. This made it possible to build branched models of continued fractions, which formed the basis for substantiated analytical conclusions regarding the dynamics and prospects for changes in the number of public officials. The analysis showed a trend towards an increase in public service vacancies while the number of public officials is decreasing. This indicates the need to form a systematic approach to professional training.

The prospect of developing digital technologies for public officials is introducing a national digital platform for training, which will combine various educational resources and courses, integrating them with the human resources systems of public authorities. This will ensure access to certified courses, including digital literacy and management skills, for all participants; creation and implementation of continuous training programs within the framework of professional development based on interactive online platforms and meeting the real needs and challenges of the civil service; modernization of the system of evaluation and motivation of employees by including in the development programs the latest technologies and skills that allow to adapt to digital changes and ensure a high level of competence; inclusion of digital literacy and cybersecurity courses in mandatory training programs for public officials, which will provide data security when processing state information; future research on the motivation of public officials, primarily through various forms of training and career growth, is also a perspective.

References

1. Dmytryshyn, M., Dmytryshyn, R., Yakubiv, V., Zagorodnyuk, A.: Peculiarities of Ukrainians' approval of decentralization reform. Adm. Sci. **11**(4), 104 (2021)
2. Dmytryshyn, M.: Involving the public in the assessment of community real estate property. Hrvatska i komparativna javna uprava: časopis za teoriju i praksu javne uprave **22**(1), 129–157 (2022)
3. Danyliuk, M., Dmytryshyn, M.: Adopted municipal real estate property in Ukraine: problems and prospects. Public Policy Adm. **20**(2), 164–175 (2021)
4. Dmytryshyn, M., Goran, T.: Proposal of an effective time management system. Manage J. Contempor. Manage. Issues **27**(2), 283–298 (2022)
5. Danyliuk, M., Dmytryshyn, M., Goran, T.: Digitisation of Ukraine regarding public electronic services' distribution. Scientific Horizons **7**, 90–99 (2021)
6. Danyliuk, M., Dmytryshyn, M., Goran, T.: Features of developing an online educational course for public officials: case from Ukraine. In: 2023 13th International Conference on Advanced Computer Information Technologies (ACIT), pp. 656–659 (2023)
7. Storozhev, R.I., Baiun, Y.V., Yashutin, I.N., Nyzhnyk, O.M., Voronov, O.I.: Process model of development of leadership qualities of public officials in the conditions of digital transformation. J. Inf. Technol. Manage. **15**(1), 14–35 (2023)
8. Alshammari, S.H., Alkhwaldi, A.F.: An integrated approach using social support theory and technology acceptance model to investigate the sustainable use of digital learning
9. Mexhuani, B.: Adopting digital tools in higher education: opportunities, challenges and theoretical insights. Eur. J. Educ. **60**(1), e12819 (2025)

10. Hassan, B., Raza, M.O., Siddiqi, Y., Wasiq, M.F., Siddiqui, R.A.: CONNECT: an AI-powered solution for student authentication and engagement in cross-cultural digital learning environments. Computers **14**(3), 77 (2025)
11. Khrykov, Y.M., Ptakhina, O.M., Sych, T.V., Dzvinchuk, D.I., Kormer, M.V.: Trends in the development of e-learning for civil officials. In: Proceedings of the 1st Symposium on Advances in Educational Technology, vol. 2, pp. 126–140 (2022)
12. Sukatiman, Saputro I.N., Budiarto, M.K.: Digital classroom innovations: leveraging smartphone-based application to stimulate students creative thinking skills. J. Efficiency Respons. Educ. Sci. **17**(4), 349–360 (2024)
13. Özdemir, O.: Kahoot! game-based digital learning platform: a comprehensive meta-analysis. J. Comput. Assist. Learn. **41**(1), e13084 (2025)
14. Kusmawan, U.: Transforming digital learning and assessment strategies in higher education. Multidisciplinary Rev. **8**(1), 2025016 (2024)
15. Unified state electronic database on education. https://info.edbo.gov.ua/. Accessed 19 Jan 2025
16. Statistical data on the quantitative composition of public officials. https://nads.gov.ua/diyalnist/kilkisnij-sklad-derzhavnih-sluzhbovciv/kilkisnij-sklad-derzhavnih-sluzhbovciv. Accessed 19 Jan 2025
17. Dmytryshyn, R.I.: Multidimensional regular C-fraction with independent variables corresponding to formal multiple power series. Proc. R. Soc. Edinburgh Sect. A Math. **150**(4), 1853–1870 (2020)
18. Bodnar, O.S., Dmytryshyn, R.I., Sharyn, S.V.: On the convergence of multidimensional S-fractions with independent variables. Carpathian Math. Publ. **12**(2), 353–359 (2020)
19. Dmytryshyn, R.I.: On some of the convergence domains of multidimensional S-fractions with independent variables. Carpathian Math. Publ. **11**(1), 54–58 (2019)
20. Dmytryshyn, R.I.: The two-dimensional g-fraction with independent variables for double power series. J. Approx. Theory **164**(12), 1520–1539 (2012)

A Comprehensive Evaluation of Data Balancing Techniques and Machine Learning Models for Credit Risk Assessment

Safae Ndama(✉), Oussama Ndama, and El Mokhtar En-Naimi

DSAI2S Research Team, C3S Laboratory, FST of Tangier, Abdelmalek Essaâdi University, Tetouan, Morocco
{safae.ndama,oussama.ndama}@etu.uae.ac.ma, en-naimi@uae.ac.ma

Abstract. Credit risk assessment plays a critical role in ensuring responsible lending and reducing financial losses in the banking sector. With the growing availability of financial data, machine learning (ML) has emerged as a powerful tool for evaluating creditworthiness by uncovering hidden patterns. However, the highly imbalanced nature of credit datasets where non-default cases vastly outnumber defaults poses a significant challenge to predictive accuracy.

To address this, our study examines and compares six resampling techniques: Synthetic Minority Over-sampling Technique (SMOTE), Random Over Sampling, Random Under Sampling, Adaptive Synthetic Sampling (ADASYN), Tomek Links, and a hybrid SMOTE-RUS approach. These methods are applied across various ML models, including Decision Tree, Random Forest, XGBoost, Neural Network, AdaBoost, CatBoost, and K-Nearest Neighbors.

The hybrid SMOTE-RUS technique demonstrated the most balanced outcomes, enhancing accuracy while reducing overfitting risks. SMOTE and ADASYN also showed notable improvements in detecting minority-class instances. This research offers a detailed comparative analysis of resampling strategies, helping to guide the development of more reliable and equitable credit scoring models for real-world applications.

Keywords: Credit Risk Assessment · Machine Learning in Finance · Class Imbalance · Data Balancing Techniques · SMOTE and ADASYN

1 Introduction

In recent years, the financial industry has increasingly turned to ML techniques to enhance decision-making processes, especially in the area of credit risk prediction [1]. The goal is to use data-driven models to assess the probability of a borrower's ability to repay a loan, enabling financial institutions to make informed lending decisions. However, one significant challenge that arises when applying ML to this domain is the presence of imbalanced datasets. In credit risk data, the distribution between individuals who repay their loans on time and those who default is often skewed. This imbalance can make it difficult for ML models to effectively identify and predict defaults [2], which

B.-E. Boudriki Semlali et al. (Eds.): DATA 2025, LNISO 81, pp. 142–155, 2025.
https://doi.org/10.1007/978-3-032-06725-8_12

are typically the minority class in such datasets. Without addressing this imbalance, models may be biased toward the majority class, leading to suboptimal performance in predicting credit risks.

In this study, we explore six different methods [3] for handling imbalanced data, including SMOTE, Random Over Sampling, Random Under Sampling, Tomek Links, and a hybrid method that combines SMOTE and Random Under Sampling. These techniques are evaluated in conjunction with eight ML algorithms Decision Tree, Random Forest, XGBoost, Neural Network, AdaBoost, CatBoost, and K-Nearest Neighbors to investigate how different balancing strategies influence model performance. Through this comparative analysis, we aim to identify the most effective model-balancing combinations for credit risk assessment, based on evaluation metrics such as Accuracy, Precision, Recall, and F1-Score. Our goal is to provide practical insights into optimizing credit risk prediction models by selecting appropriate preprocessing and resampling techniques tailored to imbalanced financial data.

The remainder of this study is organized to first provide a comprehensive review of related works in the field of credit risk modeling and the handling of imbalanced data. We then present the methodologies and materials used in our study, followed by a discussion of the results obtained from applying different balancing techniques to several ML models.

2 Related Studies

In recent years, growing attention has been devoted to addressing the issue of class imbalance in credit risk prediction. Researchers have investigated a wide range of strategies, including resampling techniques, algorithm-level adjustments, and hybrid approaches, all aimed at enhancing the performance and fairness of ML models within financial applications.

Zhang and Yu [4] presented a comprehensive classification of consumer credit risk assessment models, grouping them into traditional, intelligent, hybrid, and ensemble-based categories. They emphasized that predictive performance is not only shaped by model selection but also by data quality and structure. Their study also brought to light persistent challenges such as interpretability and algorithmic bias, which continuetoinfluencetheeffectivenessofcreditscoringsystems.

In a similar line of inquiry, Van Braak [5] explored the dual challenges of class imbalance and bias, focusing particularly on individuals with sparse or non-existent credit histories. The study concluded that in high-dimensional datasets with weakly correlated features, class weighting performed better than SMOTE. Moreover, the introduction of fairness-enhancing tools like the disparate impact remover provided additional benefits in reducing bias, indicating a shift toward integrating ethical considerationsintomodelevaluation.

B. Hajjaoui [6] conducted a case study using Turkish corporate credit data, tackling multiple data preprocessing tasks including missing value handling, feature selection, and class imbalance correction. The application of under-sampling, over-sampling, and SMOTE showed that model outcomes vary significantly depending on the chosen balancing strategy, underlining the necessity of carefully aligning preprocessing methods with dataset characteristics.

Lappas et al. [7] introduced a method that combines expert knowledge with ML and genetic algorithms for feature selection in credit risk assessment. Their approach uses clustering and the Analytic Hierarchy Process to enhance model interpretability and improve prediction accuracy, demonstrating a more robust solution for credit scoring.

Namvar et al. [8] addressed class imbalance in peer-to-peer lending credit risk prediction by empirically comparing various classifier and resampling technique combinations. Their findings suggest that combining random forest classifiers with random under-sampling effectively evaluates loan applicants' credit risk in social lending markets.

In a more recent contribution, L. Pandeya et al. [9] examined the combined effects of class imbalance correction and feature selection using SMOTE and Recursive Feature Elimination (RFE). By establishing an ideal balance ratio of 1:1.1335, the study achieved significant improvements in sensitivity, specificity, and accuracy highlightingtheimpactofcarefulpreprocessingon modelperformance.

While earlier studies have contributed valuable insights into credit risk modeling and class imbalance handling, many of them do not offer a comprehensive evaluation of multiple resampling techniques applied uniformly across a diverse range of ML models. Often, the focus is limited to a specific classifier or a narrow selection of balancing strategies. Moreover, certain combinations such as the hybrid approach of SMOTE and Random Under Sampling have not been widely explored in practice. In this work, we aim to bridge that gap by systematically assessing six different resampling methods, across eight widely used classifiers. By using a consistent dataset and standardized performance metrics, our study provides practical insights into the effectiveness of various preprocessing techniques for tackling class imbalance in financial data.

3 Class Imbalance Problem

In ML, the class imbalance problem occurs when one class significantly outnumbers the other in a dataset. This imbalance can lead to biased models that favor the majority class, making it difficult to correctly identify minority class instances [10]. In credit risk assessment, this issue is especially critical. Most applicants tend to be low-risk borrowers, while high-risk applicants (those more likely to default) are much fewer. If this imbalance is not addressed, predictive models may struggle to accurately detect individuals who are likely to default on their loans, leading to unreliable risk evaluations.

To handle class imbalance, resampling techniques are commonly used [11, 12]. As illustrated in Fig. 1, these techniques fall into three main categories: oversampling, which increases the number of minority class instances, undersampling, which reduces the number of majority class instances, and hybrid methods, which combine both approaches toachieveabetterbalance.

Oversampling involves generating new samples for the minority class to increase its representation in the dataset. One widely used method is the Synthetic Minority Over-sampling Technique, which creates synthetic instances by interpolating between existing minority class samples. For example, if a dataset contains few loan applicants who default, SMOTE can generate similar but slightly varied synthetic cases to improve model learning.

Undersampling, on the other hand, reduces the number of majority class instances by randomly removing some data points. This helps prevent models from being overly biased toward the majority class. For instance, if there are too many non-default loan applicants in the dataset, undersampling selects a smaller but representative subset to train the model, ensuring that the minority class is better represented.

Hybrid methods combine both oversampling and undersampling to achieve a balanced dataset while minimizing the drawbacks of each individual technique. A common approach is to first use SMOTE to generate synthetic minority samples and then apply random undersampling to remove excess majority class instances. This technique helps maintain a good mix of real and synthetic data, leading to improved model performance.

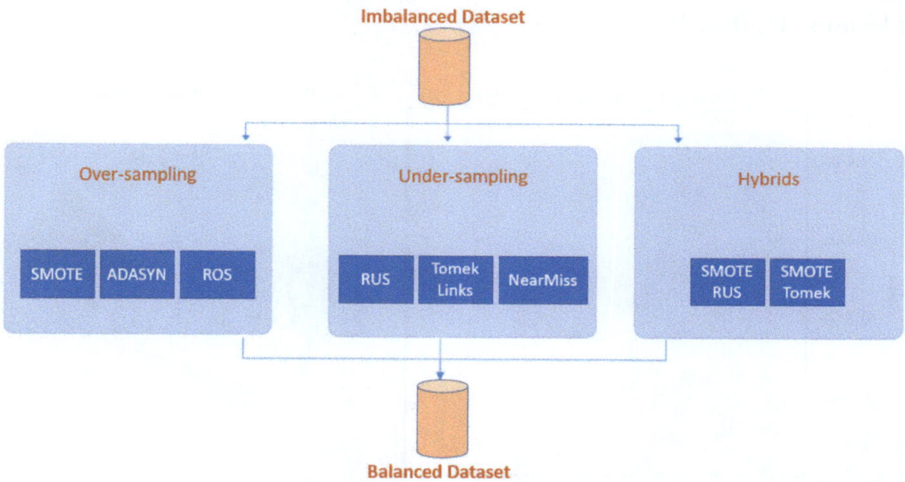

Fig. 1. Overview of Resampling Techniques for Handling Class Imbalance

4 Methodology and Material

The Methodology and Materials section plays a crucial role in this study, as it outlines the techniques and resources utilized for improving credit risk assessment. This section delves into the dataset used, the preprocessing methods applied to address issues such as class imbalance, and the selection of ML models designed to optimize prediction accuracy. It also highlights the integration of various balancing methods, such as SMOTE and under-sampling, and the importance of feature selection for improving model performance in predicting credit risk.

The dataset used in this study contains information about customers applying for loans, including details such as age, income, employment status, marital status, and credit history. These factors help assess whether a person is likely to repay a loan. The data comes from various financial institutions and includes a target variable that shows whether a loan was approved or not. This dataset allows us to improve credit risk assessment while handling challenges like class imbalance, which often occurs in loan defaultpredictions.

4.1 Data Description

The dataset used in this study consists of 252,001 records and includes 13 features along with a binary target variable named Risk_Flag. Each record represents an individual borrower, with attributes covering demographic, professional, and financial information. The input features include: Id, Income, Age, Experience, Married/Single,House_Ownership,Car_Ownership,Profession, CITY,STATE, CURRENT_JOB_YRS, andCURRENT_HOUSE_YRS.

The target variable, Risk_Flag, indicates whether a borrower is considered low-risk (0)orhigh-risk(1)interms of creditdefault.Upon examining the distribution of the target variable, a significant class imbalance was observed: 87.7% of the records (221,004 instances) are labeled as low-risk (0), while only 12.3% (30,997 instances) are labeled as high-risk (1) (Fig. 2).

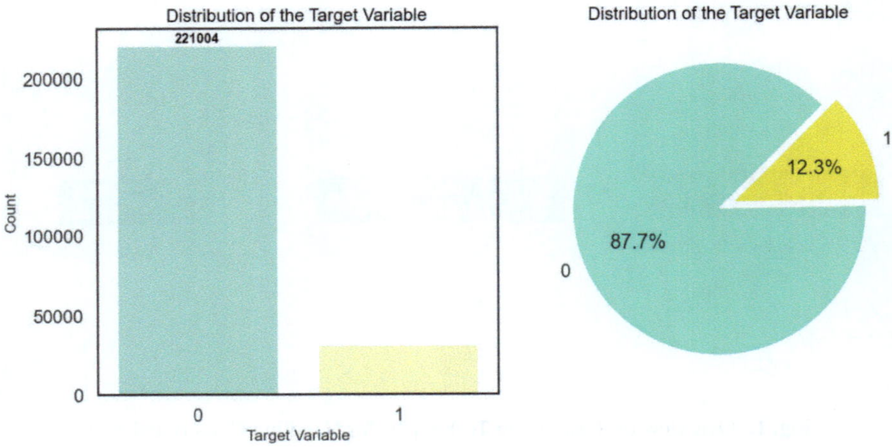

Fig. 2. Distribution of the Target Variable

4.2 Feature Engineering

At this stage, we move into the feature engineering phase, where the focus is on transforming raw data into meaningful inputs suitable for ML models. To handle categorical variables, we applied label encoding to convert them into numeric values, making them interpretable by ML models. As shown in Fig. 3, label encoding was applied to features such as Married_Single, House Ownership, Car Ownership, Profession, City, and State. Each unique category was assigned a numerical value, replacingtext-basedlabelswithintegerrepresentations.

Mathematically, label encoding can be defined as a function:

$$f : C \to Z, x \mapsto y \quad (1)$$

where $\boldsymbol{C}$ is the set of unique categories, x represents each category, and y is the corresponding integer label [13]. This process ensures that important details about applicants, such as their profession, home ownership, or marital status, are kept while making the data suitable for analysis. By turning text values into numbers, models can quickly recognize patterns and make better predictions about credit risk, leading to more accurate assessments of who is likely to default on a loan.

```
[35]: label_encoder_Married_Single = LabelEncoder()
      label_encoder_House_Ownership = LabelEncoder()
      label_encoder_Car_Ownership = LabelEncoder()
      label_encoder_Profession = LabelEncoder()
      label_encoder_CITY = LabelEncoder()
      label_encoder_STATE = LabelEncoder()

[36]: X[:,3] = label_encoder_Married_Single.fit_transform(X[:,3])
      X[:,4] = label_encoder_House_Ownership.fit_transform(X[:,4])
      X[:,5] = label_encoder_Car_Ownership.fit_transform(X[:,5])
      X[:,6] = label_encoder_Profession.fit_transform(X[:,6])
      X[:,7] = label_encoder_CITY.fit_transform(X[:,7])
      X[:,8] = label_encoder_STATE.fit_transform(X[:,8])

[37]: X[0]

[37]: array([1303834, 23, 3, 1, 2, 0, 33, 251, 13, 3, 13], dtype=object)

[38]: X

[38]: array([[1303834, 23, 3, ..., 13, 3, 13],
             [7574516, 40, 10, ..., 14, 9, 13],
             [3991815, 66, 4, ..., 12, 4, 10],
             ...,
             [4522448, 46, 7, ..., 14, 7, 12],
             [6507128, 45, 0, ..., 18, 0, 10],
             [9070230, 70, 17, ..., 22, 7, 11]], dtype=object)
```

Fig. 3. Label Encoding Transformation

Additionally, in our dataset, missing values and outliers were identified and addressed to ensure data integrity. Outliers were initially detected using statistical methods [14] and visualized through box plots. As shown in Fig. 4, which illustrates the distribution of Income, Age, and Experience after cleaning, no visible outliers remain. We applied removal or capping methods to minimize the influence of anomalies, and imputed or removed null values based on their significance. These preprocessing steps were essential for preparing the data for further model training, ensuring that the models learn useful patterns without being affected by unprocessed errors or inconsistencies.

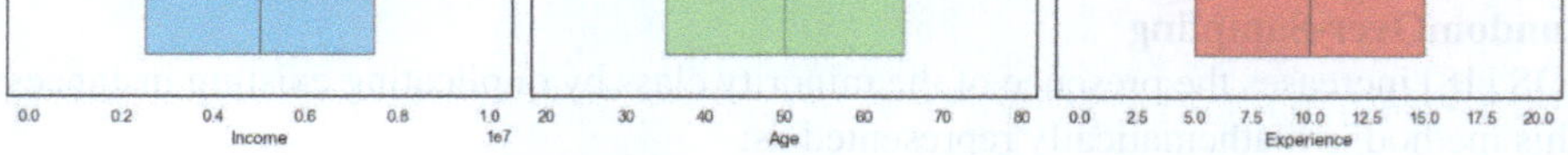

Fig. 4. Boxplots of Numeric Features After Cleaning

Once the data has been cleaned, we focus on identifying potential sources of data leakage in the model. These are variables that may not be available when making predictions, and their inclusion during training can lead to overly optimistic model performance. In our case, the dataset contained some leakage, but this issue is often overlooked in many studies. To prepare the data for ML, several important transformations were applied. Categorical variables were converted into numerical values through label encoding, allowing the models to process them effectively. StandardScaler() was then used for feature scaling [15], ensuring that all variables were standardized and on the same scale. This is crucial for models that are sensitive to the magnitude of the data. The standardization was done using the following formula:

$$z = \frac{x - \mu}{\sigma} \tag{2}$$

where x is the original value, μ is the average value of that feature, and σ is the spread (or standard deviation) of that feature.

4.3 Data Balancing Techniques

The imbalanced nature of the dataset, as highlighted in Fig. 1, poses a significant challenge in credit risk assessment. Without addressing this imbalance, ML models tend to be biased toward the majority class (non-default), leading to poor generalization and misclassification of high-risk individuals. In this study, we will explore and compare the effectiveness of six data balancing techniques SMOTE, Random Over Sampling, Random Under Sampling, ADASYN, Tomek Links, and a hybrid method combining SMOTE and Random Under Sampling. The goal is to assess which method has the most powerful impact on the predictive performance of the ML models used in credit risk assessment. By applying these methods, we aim to improve model accuracy, fairness, and the ability to correctly classify high-risk individuals, ultimately leading to more reliable credit risk evaluations.

Random Under-Sampling
RUS reduces the dominance of the majority class by randomly removing instances [16, 17]:

$$X' = X \setminus \left\{ x_i | x_i \in C_{\text{majority}}, \forall i \in N \right\} \tag{3}$$

While this method reduces class imbalance efficiently, it may discard critical information, potentially leading to underfitting.

RandomOver-Sampling
ROS [18] increases the presence of the minority class by duplicating existing instances. This method is mathematically represented as:

$$X' = X \cup \left\{ x_i | x_i \in C_{\text{minority}}, \forall i \in N \right\} \tag{4}$$

where X' is the new dataset, X is the original dataset, and *Cminority* represents the set of minority class instances. While this approach balances the class distribution, it may introduce redundancy, leading to potential overfitting.

SyntheticMinorityOver-SamplingTechnique

Unlike ROS, SMOTE synthesizes new instances rather than duplicating existing ones. Given a minority class instance *xi*, a synthetic sample *xnew* is generated as [19]:

$$x_{\text{new}} = x_i + \lambda \cdot \left(x_j - x_i\right) \quad (5)$$

where *xj* is a randomly selected nearest neighbor of *xi*, and λ is a random value in [0,1]. SMOTE enhances generalization by preventing models from memorizing duplicate samples.

ADASYN

ADASYN is an extension of SMOTE that generates synthetic samples based on the difficulty of classifying minority class instances. For each minority instance *xi*, a number of synthetic samples are generated based on the inverse of the density of the minority class in the neighborhood of *xi* . This can be represented as [20]:

$$x_{\text{new}} = x_i + \lambda \cdot \left(x_j - x_i\right) \quad (6)$$

where λ is determined adaptively depending on the density of the minority class.

Tomek Links Removal

A Tomek link is defined as a pair of nearest neighbors (*xi*, *xj*) from opposite classes, satisfying [21, 22]:

$$\forall x_k,\ d\left(x_i, x_j\right) < d(x_i, x_k) \text{ and } d\left(x_i, x_j\right) < d\left(x_j, x_k\right) \quad (7)$$

where x_k is any other sample in the dataset. Removing Tomek links enhances decision boundary clarity by eliminating ambiguous majority class instances.

SMOTE with Random Under Sampling (Hybrid Method)

The hybrid method [23, 24] combines SMOTE and Random Under Sampling to balance the dataset. First, SMOTE generates synthetic minority samples, followed by RUS to reduce the majority class instances. This method ensures both class balance and reduces the risk of overfitting by limiting the majority class. Mathematically, this can be represented as:

$$\boldsymbol{X}' = \text{SMOTE}(\boldsymbol{X})/\text{RUS}(\boldsymbol{X}) \quad (8)$$

where $SMOTE(X)$ represents the oversampled dataset and $RUS(X)$ represents the undersampledmajority classdataset.

5 Results and Discussion

After exploring different sampling methods, we now present the results of the models evaluated with each method, analyzing their performance based on key metrics such asAccuracy,Precision,Recall,andF1-Score.

Table 1. Model Performance with Random Under Sampling

Model	Accuracy	Precision	Recall	F1-Score
Decision Tree	0.848939	0.845862	0.856549	0.851172
Random Forest	0.854585	0.873950	0.831601	0.852249
XGBoost	0.852004	0.874026	0.825524	0.849083
Neural Network	0.740140	0.727382	0.775308	0.750581
AdaBoost	0.555609	0.562785	0.532544	0.547247
CatBoost	0.786757	0.798610	0.771790	0.784971
KNN	0.813291	0.807368	0.827123	0.817126

Random Under-Sampling resulted in a general decrease in performance across most models due to the removal of significant portions of the majority class data. As shown in Table 1, while Random Forest and XGBoost maintained relatively stable F1-Scores of approximately 85%, models like Neural Network and AdaBoost suffered in both accuracy and recall, highlighting their dependency on richer data representations. In the context of credit risk, where false negatives can have serious financial consequences, the loss of useful majority class data limits the model's ability to generalize effectively.

Table 2. Model Performance with Random Over Sampling

Model	Accuracy	Precision	Recall	F1-Score
Decision Tree	0.926156	0.882234	0.982842	0.929825
Random Forest	0.956811	0.920565	0.999477	0.958400
XGBoost	0.915228	0.892883	0.942799	0.917162
Neural Network	0.757008	0.740723	0.787469	0.763381
AdaBoost	0.562125	0.559368	0.566780	0.563050
CatBoost	0.822414	0.825528	0.815603	0.820536
KNN	0.908328	0.865950	0.965252	0.912909

When applying Random Over Sampling, the models maintained a similar pattern. Table 2 shows that with Random Forest achieving the highest F1-Score of 95.84%, followed by Decision Tree at 92.29%. These models are better equipped to handle duplicated instances due to their hierarchical structure. The oversampling allowed the models to better learn minority class patterns without losing important majority class information. However, models like AdaBoost and Neural Network still underperformed, suggesting that merely replicating data does not improve learning for algorithms sensitive to overfitting or high-dimensional noise.

Table 3. Model Performance with SMOTE

Model	Accuracy	Precision	Recall	F1-Score
Decision Tree	0.896677	0.871651	0.929255	0.899531
Random Forest	0.945556	0.915659	0.980979	0.947194
XGBoost	0.937671	0.913166	0.966707	0.939174
Neural Network	0.771973	0.753773	0.804786	0.778445
AdaBoost	0.580168	0.576373	0.590755	0.583475
CatBoost	0.849868	0.849478	0.848783	0.849130
KNN	0.922264	0.875523	0.983683	0.926457

SMOTE delivered strong and consistent improvements As detailed in Table 3, especially with Random Forest, XGBoost, and KNN. The synthetic sample generation helped expose the models to new minority class patterns that didn't exist in the original data. This is particularly beneficial in credit risk contexts where rare but informative patterns can distinguish a defaulter from a low-risk applicant. Tree-based models like Random Forest and XGBoost benefit the most from SMOTE because of their ability to segment the newly created synthetic points effectively, thus balancing precision and recall. The high F1-Scores suggest that SMOTE strikes an ideal trade-off between preserving the original data structure and introducing diversity in minority examples.

Table 4. Model Performance with ADASYN

Model	Accuracy	Precision	Recall	F1-Score
Decision Tree	0.853648	0.837074	0.873902	0.855091
Random Forest	0.956472	0.920130	0.998587	0.957754
XGBoost	0.936197	0.915366	0.959597	0.936960
Neural Network	0.740321	0.723685	0.767494	0.744946
AdaBoost	0.597433	0.600772	0.552266	0.575499
CatBoost	0.848833	0.889771	0.792205	0.838158
KNN	0.926503	0.873664	0.995161	0.930463

ADASYN, similar to SMOTE, generated synthetic samples while considering the density of the minority class. Table 4 illustrates that Although it produced slightly lower precision for some models, it maintained effectiveness for Random Forest and XGBoost, with F1-Scores of 95.77% and 93.69%, respectively.

Table 5. Model Performance with Tomek Links

Model	Accuracy	Precision	Recall	F1-Score
Decision Tree	0.913688	0.655986	0.644681	0.650284
Random Forest	0.924314	0.621876	1.000000	0.766860
XGBoost	0.908746	0.664991	0.537851	0.594702
Neural Network	0.876416	0.596320	0.022187	0.042783
AdaBoost	0.875664	0.759259	0.001651	0.003295
CatBoost	0.881940	0.756822	0.075944	0.138037
KNN	0.899558	0.605845	0.552589	0.577993

Tomek Links, as shown in Table 5, significantly reduced recall across most models, despite some models like Random Forest achieving perfect recall. This indicates an overfitting to the minority class at the cost of overall balance. In credit scoring tasks, this could lead to high false positive rates misclassifying good applicants as defaulters which is undesirable in a real-world setting. Tomek Links removes borderline examples that are often crucial for understanding subtle patterns, making it less suited for high-stakes tasks like risk classification.

Table 6. Model Performance with Hybrid SMOTE + Random Under Sampling

Model	Accuracy	Precision	Recall	F1-Score
Decision Tree	0.886134	0.862009	0.920231	0.890169
Random Forest	0.930160	0.904602	0.962191	0.932508
XGBoost	0.918554	0.905363	0.935346	0.920110
Neural Network	0.766793	0.746384	0.810233	0.776999
AdaBoost	0.579037	0.580022	0.581619	0.580820
CatBoost	0.835818	0.841982	0.827964	0.834914
KNN	0.903147	0.876659	0.938955	0.906738

The hybrid SMOTE + Random Under Sampling approach aimed to balance the advantages of both oversampling and undersampling. As shown in Table 6, the results revealed that XGBoost and Random Forest continued to be the strongest models, with F1-Scores of 92.01% and 93.25%, respectively. This method provided a good trade-off between precision and recall, making it a viable option for handling imbalanced data.

Among the tested models, Random Forest and XGBoost consistently outperformed others across all resampling techniques, establishing themselves as the most robust models for credit risk assessment in our dataset. Their strength lies in their ability to capture complex feature interactions and handle imbalanced data effectively, especially when enhanced by oversampling methods like SMOTE and ADASYN. These techniques generated synthetic examples of the minority class (defaulters), which helped the models better generalize the risk patterns and improve minority class detection without sacrificing majority class performance.

KNN also performed well, particularly with SMOTE and ADASYN, achieving F1-scores of 92.64% and 93.04%, respectively. This is because KNN relies heavily on the density and proximity of neighboring samples conditions that oversampling methods optimize by creating more representative defaulter instances in the feature space.

In contrast, AdaBoost consistently underperformed, with an F1-score as low as 0.3% under Tomek Links, suggesting that it struggles to adapt to shifts in class distributions introduced by resampling. Since AdaBoost is sensitive to noisy data and class imbalance, its weak learners are easily misled by the skewed or pruned datasets produced by undersampling methods. Neural Networks showed moderate performance, benefiting from SMOTE and Random Over Sampling but suffering under Tomek Links and Random Under Sampling. This sensitivity reflects their reliance on larger, well-balanced training sets to learn nonlinear patterns effectively a condition not met when informative majority samples are removed.

In conclusion, oversampling methods like SMOTE and ADASYN proved most beneficial, especially for Random Forest, XGBoost, and KNN, as they enhanced the representation of the minority class without losing critical information from the majority class. The hybrid SMOTE + RUS method also performed well, offering a solid trade-off between recall and precision. Conversely, pure undersampling methods like Tomek Links and Random Under Sampling degraded performance, especially in detecting high-risk clients, by discarding useful majority data. For practical credit risk applications, we therefore recommend using Random Forest or XGBoost combined with SMOTE or ADASYN, as these combinations provide the best balance between detecting defaulters and maintaining overall prediction accuracy.

6 Conclusion and Perspectives

This study highlights the critical impact of class imbalance on credit risk prediction and demonstrates the effectiveness of various resampling techniques in mitigating this issue. Our results show that while all data balancing methods improve model performance to some extent, the hybrid SMOTE-RUS approach achieves the best trade-off between bias reduction and generalization, enhancing the detection of high-risk borrowers. The findings underscore the importance of selecting an appropriate data balancing strategy to optimize predictive accuracy and fairness in financial decision-making.

As a future perspective, incorporating Explainable AI (XAI) techniques into credit risk modeling could enhance interpretability, allowing financial institutions to better understand and trust ML-based decisions. Additionally, experimenting with advanced ensemble methods and deep learning architectures may further refine predictive performance. Expanding this research to real-world financial datasets with dynamic and evolving risk factors will also provide deeper insights into the robustness of balancing techniques in practical applications.

References

1. Galindo, J., Tamayo, P.: Credit risk assessment using statistical and machine learning: basic methodology and risk modeling applications. Comput. Econ. **15**(1/2), 107–143 (2000). https://doi.org/10.1023/a:1008699112516
2. Tyagi, S., Mittal, S.: Sampling approaches for imbalanced data classification problem in machine learning, in Lecture notes in electrical engineering, pp. 209–221 (2019). https://doi.org/10.1007/978-3-030-29407-6_17
3. Batista, G.E., Prati, R.C., Monard, M.C.: A study of the behavior of several methods for balancing machine learning training data. ACM SIGKDD Explor. Newsletter. **6**(1), 20-29 (2004). https://doi.org/10.1145/1007730.1007735
4. Zhang, X., Yu, L.: Consumer credit risk assessment: a review from the state-of-the-art classification algorithms, data traits, and learning methods. Expert Syst. Appl. **237**, 121484 (2023). https://doi.org/10.1016/j.eswa.2023.121484
5. Wang, Y., Zhang, Y., Lu, Y., Yu, X.: A comparative assessment of credit risk model based on machine learning a case study of bank loan data. Procedia Comput. Sci. **174**, 141–149 (2020). https://doi.org/10.1016/j.procs.2020.06.069
6. Hajjaoui, B.: Crucial challenges in corporate credit risk assessment: a case study. Osmaniye Korkut Ata Üniversitesi Fen Bilimleri Enstitüsü Dergisi **7**(2), 834–854 (2024). https://doi.org/10.47495/okufbed.1340798
7. Lappas, P.Z., Yannacopoulos, A.N.: A machine learning approach combining expert knowledge with genetic algorithms in feature selection for credit risk assessment. Appl. Soft Comput. **107**, 107391 (2021). https://doi.org/10.1016/j.asoc.2021.107391
8. Namvar, A., Siami, M., Rabhi, F., Naderpour, M.: Credit risk prediction in an imbalanced social lending environment. Int. J. Comput. Intell. Syst. **11**(1), 925 (2018). https://doi.org/10.2991/ijcis.11.1.70
9. Pandey, M.K., Mittal, M., Subbiah, K.: Optimal balancing & efficient feature ranking approach to minimize credit risk. Int. J. Inf. Manage. Data Insights **1**(2), 100037 (2021). https://doi.org/10.1016/j.jjimei.2021.100037
10. Ndama, O., En-Naimi, E.M.: Credit card fraud detection using SVM, Decision Tree and random Forest supervised machine learning algorithms. In: International Conference on Big Data and Internet of Things, pp. 316–327 (2023). https://doi.org/10.1007/978-3-031-28387-1_27
11. Ndama, O., Bensassi, I., En-Naimi, E.M.: Optimizing credit card fraud detection: a deep learning approach to imbalanced datasets. Int. J. Power Electr. Drive Syst. Int. J. Electr. Comput. Eng. **14**(4), 4802 (2024).https://doi.org/10.11591/ijece.v14i4.pp4802-4814
12. Rawat, S.S., Mishra, A.K.: Review of methods for handling class imbalance in classification problems. In: Lecture notes in Electrical Engineering, pp. 3–14 (2024). https://doi.org/10.1007/978-981-97-0037-0_1
13. Jackson, E., Agrawal, R.: Performance evaluation of different feature encoding schemes on cybersecurity logs. SoutheastCon (2019). https://doi.org/10.1109/southeastcon42311.2019.9020560
14. Miller, J.N.: Tutorial review—outliers in experimental data and their treatment. Analyst **118**(5), 455–461 (1993). https://doi.org/10.1039/an9931800455
15. Ozsahin, D.U., Mustapha, M.T., Mubarak, A.S., Ameen, Z.S., Uzun, B.: Impact of feature scaling on machine learning models for the diagnosis of diabetes. In: 2022 International Conference on Artificial Intelligence in Everything (AIE), pp. 87–94 (2022). https://doi.org/10.1109/aie57029.2022.00024
16. Tahir, M.A., Kittler, J., Yan, F.: Inverse random under sampling for class imbalance problem and its application to multi-label classification. Pattern Recogn. **45**(10), 3738–3750 (2012). https://doi.org/10.1016/j.patcog.2012.03.014

17. Tahir, M.A., Kittler, J., Mikolajczyk, K., Yan, F.: A multiple expert approach to the class imbalance problem using inverse random under sampling. In: Lecture Notes in Computer Science, pp. 82–91 (2009). https://doi.org/10.1007/978-3-642-02326-2_9
18. Santoso, B., Wijayanto, H., Notodiputro, K.A., Sartono, B.: Synthetic over sampling methods for handling class imbalanced problems : a review. In: IOP Conference Series Earth and Environmental Science, vol. 58, p. 012031 (2017). https://doi.org/10.1088/1755-1315/58/1/012031
19. Chawla, N.V., Bowyer, K.W., Hall, L.O., Kegelmeyer, W.P.: SMOTE: synthetic minority oversampling technique. J. Art. Intell. Res. **16**, 321–357 (2002). https://doi.org/10.1613/jair.953
20. He, N.H., Bai, N.Y., Garcia, E.A., Li, N.S.: ADASYN: Adaptive synthetic sampling approach for imbalanced learning. In: 2008 IEEE International Joint Conference on Neural Networks (IEEE World Congress on Computational Intelligence), pp. 1322–1328 (2008). https://doi.org/10.1109/ijcnn.2008.4633969
21. Pereira, R.M., Costa, Y.M.G., Silla, C.N., Jr.: MLTL: a multi-label approach for the Tomek Link undersampling algorithm. Neurocomputing **383**, 95–105 (2019). https://doi.org/10.1016/j.neucom.2019.11.076
22. Swana, E.F., Doorsamy, W., Bokoro, P.: Tomek link and SMOTE approaches for machine fault classification with an imbalanced dataset. Sensors **22**(9), 3246 (2022). https://doi.org/10.3390/s22093246
23. Gazzah, S., Hechkel, A., Amara, N.E.B.: A hybrid sampling method for imbalanced data. In: 2015 IEEE 12th International Multi-Conference on Systems, Signals & Devices (SSD15) (2015). https://doi.org/10.1109/ssd.2015.7348093
24. Wongvorachan, T., He, S., Bulut, O.: A comparison of undersampling, oversampling, and SMOTE methods for dealing with imbalanced classification in educational data mining. Information **14**(1), 54 (2023). https://doi.org/10.3390/info14010054

The Impact of ChatGPT on the Entrepreneurial Intention of Students at Sidi Mohamed Ben Abdellah University: A Gender-Based Approach

Hafsa Boutaqqa(✉) and Sanae El Maliki(✉)

The Higher School of Technology, Laboratory of Management, Marketing, International Logistics, and Finance, Sidi Mohamed Ben Abdellah University, Fez, Morocco
{hafsa.boutaqqa,sanae.elmaliki}@usmba.ac.ma

Abstract. ChatGPT is a software that may alter the intentions and perceptions of students in Morocco regarding their entrepreneurial education. We investigate the influence of ChatGPT on entrepreneurial intention among Higher School of Technology students at Sidi Mohamed Ben Abdellah University in Fez, highlighting the importance of our study by emphasizing gender as a moderator variable. Using a qualitative approach, it examines students enrolled in the Management Techniques program, describing how ChatGPT had a positive impact on their motivation, assistance in overcoming challenges, and providing confidence when engaging in entrepreneurial projects. Data reveals that male students came to entrepreneurship with clear intentions, while female students, although initially ambivalent about their motivations, experienced a change in their aspirations after interacting with ChatGPT. It also underscores the transformative potential of AI tools in empowering underrepresented groups, while stresses that targeted support is needed to build inclusive entrepreneurial ecosystems.

Keywords: ChatGPT · Entrepreneurial Education · Entrepreneurial Intention · Gender Differences · Artificial Intelligence · Higher Education · Morocco

1 Introduction

With the accelerated pace of technological advancement, artificial intelligence (AI) is revolutionizing the fields of education, innovation and entrepreneurship. AI-Powered Tools, like ChatGPT, have appeared as immensely popular tools with their ability to impact entrepreneurial intention and behavior among the higher education student population. From generating uncertainties to simulating the entrepreneurial process or shaping insights, these tools are revolutionizing how individuals interpret and approach entrepreneurship.

This article examines the impact of ChatGPT on entrepreneurship education for students studying in the Management Techniques program at the Higher School of Technology in Fez. It mobilizes a qualitative methodology to analyze the impact of ChatGPT on students' entrepreneurial intention with a focus on gender differences. Interactions between students and this software shed light on how it can help minimize

B.-E. Boudriki Semlali et al. (Eds.): DATA 2025, LNISO 81, pp. 156–165, 2025.
https://doi.org/10.1007/978-3-032-06725-8_13

perceived barriers, boost motivation, and increase confidence in entrepreneurial projects. The aim of this paper is to respond the following question: **How does the integration of ChatGPT in entrepreneurship education influences students' entrepreneurial intentions and how does its effect differ by gender?**

The study revolves around the following key questions:

- How Does ChatGPT impacts the Entrepreneurial Intentions of Students Studying Entrepreneurship?
- What gender differences emerge in the adoption of ChatGPT in this entrepreneurial educational context?
- What are the challenges and opportunities associated with integrating ChatGPT to promote more inclusive entrepreneurial education?

In order to address these questions, this article is structured with a literature review that explores the intersectionality of artificial intelligence (AI), entrepreneurial intention, and gender. Next, we provide an overview of the qualitative methodology consisting of semi-structured interviews and a thematic analysis. The results section presents the perceptions and experiences of students who pursued entrepreneurship course at the Higher School of Technology. The following is then focused on the theoretical-practical contributions of the results. Lastly, the conclusion offers suggestions on how to implement AI responsibly within entrepreneurship education and the potential for creating inclusive and equitable learning environments alongside such advancements.

2 Literature Review

2.1 Teaching Entrepreneurship in the Age of AI

Entrepreneurship education has historically focused on providing students with tangible tools and techniques so that they may develop their creativity, initiative, and problem-solving abilities [1]. This field is going through a drastic transformation because of the rise of artificial intelligence. AI-based technologies create a personalized and interactive studying environment that illustrates the skills required to become an entrepreneur [2].

AI tools such as conversational platforms and virtual assistants replicate complex entrepreneurial scenarios, interpret, visualize data instantly, and propose tailored strategic solutions. They help students gain a deeper insight into the contexts in which business decisions are taken and enhance their skills to make appropriate decisions [3].

Despite the fact that teaching with AI is ground-breaking, ensuring the subtle nuances of human interactions in entrepreneurship education are preserved to ensure effectiveness. Although the access to information and the more dynamic learning that AI may provide, increased reliance on AI can depersonalize learning processes that should be contextually relational, such as mentoring and relational dynamics [4].

2.2 ChatGPT and Entrepreneurial Intentions in the Educational Context

It is important to highlight the way AI tools including ChatGPT are changing the educational landscape through personalized learning experiences. This customization is especially useful for students looking to be entrepreneurs, as it enables them to interact with challenging ideas and projects, and further their knowledge and confidence, both of which are necessary when it comes to navigating the entrepreneurial world [5].

Additionally, [6] studies the variables shaping students to adopt ChatGPT for academic objectives Their results suggest that students are motivated by its accessibility and perceived contribution to academic performance, two factors that can have significant effects on their entrepreneurial mindset. These tools generate almost realistic texts and situations and provide prompt feedback which are crucial in developing creativity and innovation, essential traits for potential entrepreneurs.

According to the literature, ChatGPT has an impact on overcoming certain perceived barriers related to entrepreneurship in different processes such as low self-confidence, complexity of processes, and uncertainty of failure [7]. Additionally, it acts as a virtual guide that assists students in the development of business strategies, market opportunity exploration, and risk management [8].

However, some researchers note the limitations of ChatGPT, especially the risk of becoming overly reliant on it to find solutions, which will prevent students from developing problem-solving and creativity skills [9, 10]. Moreover, AI systems can contain algorithmic biases that may negatively affect the responses provided [11]. The emphasis on responsible use and verification of information is especially important, as they are fundamental to creating a culture of integrity and critical participation among future entrepreneurs [12].

Finally, [13] published a scoping review, critically assessing the ethical repercussions of using ChatGPT in learning. The authors illustrate the dual nature of AI's impact, where the benefits of enhanced communication and learning have to be balanced against the risk of misinformation and over-reliance on AI-generated content. This discussion should be encouraged among aspiring entrepreneurs and be part of their education, as it helps shape their perspectives of how AI can actually be used responsibly and the dangers of misinformation in business contexts.

2.3 AI and Gender Dynamics in Entrepreneurship Education

AI presents numerous advantages to deal with gender gaps in entrepreneurial education [14]. Through accessible resources and simulations, AI tools make it accessible for female students to gain confidence and help them break through traditional barriers, such as limited access to a professional network or mentors [15]. Moreover, research has shown that the use of AI tools varies by gender [16]. Male students are generally ready to explore and adopt these technologies as a part of their entrepreneurial projects, while female students are a bit reluctant and need some encouragement to maximize the opportunities this IA tool provide.

Conversational AI tools like ChatGPT provide female students with a neutral, non-judgmental space where they can try out their ideas, ask questions, and explore concepts without the fear of being evaluated negatively [17]. Nevertheless, gender biases embedded with AI systems can restrict the inclusiveness of these technologies, thereby reinforcing certain gender stereotypes [18]. Accordingly, it is essential to design AI systems carefully and validate them regularly so that they promote accurate gender equality in entrepreneurial education.

3 Methodology

The present study mobilizes a qualitative method in order to analyze the implications of ChatGPT on students' entrepreneurial intentions, specifically among 10 students from the Higher School of Technology in Fes, all belonging to the same specialty "Management Techniques" with a focus on gender differentiation. Stratified sampling was used in order to have relevant and representative data. Students were classified into three groups according to their entrepreneurial intention level: low, moderate, and high (Out of ten students, five female students are at a low level of intention, two male students at a medium level, and three at a high level). It helps to understand to what extent does ChatGPT influences students' learning environments in entrepreneurship, and further their entrepreneurial intention.

Another selection criterion was the interest in using digital tools, aimed at selecting students who demonstrate an interest in using these tools in their entrepreneurial projects, which is essential for exploring the impact of ChatGPT. This balanced sample includes both male and female students to enable a meaningful comparison of the experiences and perceptions of each gender. Data were gathered using semi-structured interviews, which explored students' perceptions on the use of ChatGPT in their entrepreneurial projects, as well as through classroom observations, providing contextual insight into real-time interactions with this tool.

An inductive method was used to analyze data with the support of NVivo software to establish themes and patterns. This makes sure for a clearer understanding of student entrepreneurial intentions in an education context while considering the potential influence of AI and gender disparities that could have on their interest and involvement in entrepreneurship. Combining these dimensions, this research aims to provide a meticulous framework for analyzing the outcomes and understanding the gender dynamics situated in the entrepreneurial context.

4 Results

This table provides a summary of the results that discuss the effect of ChatGPT integration in entrepreneurship education, particularly in terms of student entrepreneurial intentions. It emphasizes students' perceptions regarding ChatGPT, its efficacy in enhancing entrepreneurial skills and intention, and the determinants that impact its performance. The research also examines ethical concerns, gender dynamics, and the challenges related to the incorporation of AI into the academic infrastructure (Table 1).

Table 1. Perception of students regarding the questions posed

Themes	Main results
ChatGPT as a catalyst for entrepreneurial education	ChatGPT has changed the entrepreneurial education landscape by simplifying the theoretical and conceptual knowledge and helping students develop their entrepreneurial competencies. It has enhanced their creativity, proactivity, and problem-solving skills Male students used it to develop their knowledge and improve their business potential, while female students enhanced their confidence and help them to find creative solutions for entrepreneurial projects
The impact of ChatGPT on students' entrepreneurial intentions	For male students with high levels of entrepreneurial intention, ChatGPT was indeed a catalyst. They found the tool helpful in deepening their understanding of theoretical concepts and applying it more practically to their projects. They used ChatGPT to expand their business ideas, look for market opportunities, and structure funding proposals, thereby reinforcing their commitment and motivation to launch a business Male students with low and moderate entrepreneurial intentions saw less impact on the perceived usefulness of ChatGPT. Although they understood the theories and concepts used in the program at an advanced level, they failed to apply them practically to real projects. So, the tool was less impactful on increasing their intention to start a business ChatGPT usage was a pivotal component in the development of female students' entrepreneurial aspirations. Although they had lower entrepreneurial intention compared to their male counterparts initially, they were able to increase their awareness of entrepreneurship. This enhanced their confidence, the recognition of their abilities, and expanded their perspective on entrepreneurship

(*continued*)

Table 1. (*continued*)

Themes	Main results
Gender dynamics in entrepreneurial intentions through AI	Male students, especially those who strongly plan to work for themselves, believe that entrepreneurship is already an open option and an achievable career path. They interact with ChatGPT in an active manner to structure and seek funding for their projects and articulate their ideas, which allows them to feel empowered as entrepreneurs. They seem not to be blocked by cultural or social stigmas, and can take advantage of the opportunities that the tool offers in order to carry on with their projects Male students with moderate entrepreneurial intentions, while they did use the tool to structure ideas and acquire further knowledge, their motivation to take action and develop concrete projects was not as high as those with high entrepreneurial intentions. In most cases, the use of ChatGPT was to consolidate concepts rather than taking real initiatives. This shows some reluctance or unwillingness to engage fully in entrepreneurship, despite the tool's help in gaining the know-how and the relevant insights they needed Female students believe that entrepreneurship is a challenging field for them due to social barriers and gender stereotypes, which makes it harder for them to start up their own business. This perception coexists with false weaknesses of failure uncertainty and fear of judgment on their participation in entrepreneurship. Even they used ChatGPT to clarify concepts and deepen their theoretical understanding, their practical engagement remains limited. Yet, the tool has provided a neutral and nonjudgmental space that enables them to explore their ideas without risk, thereby building their confidence and lowering some psychological and culture barriers

(*continued*)

Table 1. *(continued)*

Themes	Main results
Limits and challenges of integrating ChatGPT into entrepreneurial education	Male Students, especially those with strong entrepreneurial intentions, reported that they were worried about immoderate use of ChatGPT and that it could inhibit their creativity and independent thinking. Comparatively, students with moderate entrepreneurial intention were primarily employed ChatGPT in obtaining information and understanding theoretical aspects, unconcerned about the influence of the tool on their creative independence Female students, regardless of their entrepreneurial intention, are more vigilant than males in ChatGPT for entrepreneurial learning. While they use it to organize ideas, formulate the structure of their essays and clarify concepts, their engagement with content remains moderate as they are more critical of the information generated by an AI tool. Moreover, their limited exposure to technology results in the gradual adoption of digital tools as compared to their peers. Digital tools such as ChatGPT are less likely to be used by the interviewed female students, to whom human interactions are vital to validate ideas

Source: Author's elaboration.

5 Discussion

The results of this research demonstrate some interesting aspects of the effect of ChatGPT usage on students' entrepreneurial intention in higher education.

Firstly, ChatGPT positively influences students' entrepreneurial education, especially in theory comprehension and organization of ideas. Male students who had strong entrepreneurial intention used the tool to improve their projects and look for opportunities, thus strengthening their motivation. Nonetheless, those who had moderate intentions mostly employed ChatGPT only to gain theoretical knowledge without really bringing their projects into realization. While female students showed lower entrepreneurial intention at the beginning, the tool helped them to have more confidence, better understand complex concepts and to overcome psychological barriers, although they still less engaged in entrepreneurship in comparison with male students.

Secondly, female students often consider entrepreneurship less accessible for them because of gender stereotype. Indeed, they are more vigilant in entrepreneurship commitment. However, ChatGPT played a key role by providing these students a safe environment to discuss and explore their ideas objectively, thus enhancing their motivation to get involved in projects. Although their entrepreneurial initiative still less practical than their male counterparts, the use of the tool lessened the gender gap by adjusting their perceptions of entrepreneurship for the better.

Thirdly, students who had a clear entrepreneurial intention were worried about the immoderate use of ChatGPT, believing that it would limit their creativity. They stressed the need to maintain a balance between using technology and engaging in personal reflection. On the other hand, students with moderate level of entrepreneurial intention were mainly information seekers with a lesser degree of distrust of the tool. Female students, in contrast, took a more skeptical and deliberative approach using the tool, preferring to connect with other human beings more and verifying the information generated by ChatGPT before reviewing it. This is because they want to avoid the potential risks of relying entirely on technology and they try to ensure the information integrity.

In conclusion, regardless of the positive influence of ChatGPT on both genders, student intention towards entrepreneurship along with the gender-based stereotypes influence the manner how male and female students take initiatives in entrepreneurship. This reveals that students with highly entrepreneurial intention use the tool with more applied way in comparison to the other users, while gender is considered as a moderated variable that shapes the students' entrepreneurial engagement due to social perceptions and cultural barriers.

Finally, it is important to develop targeted initiatives, providing equal access to entrepreneurial resources (e.g. artificial intelligence), thereby reducing differences in gender and promoting more equitable participation in entrepreneurship. Moreover, if female students had a high intention of being an entrepreneur, ChatGPT may support their entrepreneurial projects achievement by enhancing their confidence and eliminating social and psychological barriers which commonly block their concrete commitment.

The analysis conducted using NVivo 10 software reveals the following word clouds (Fig. 1):

Fig. 1. Word Cloud of the Research Theme. Source: Nvivo10 Software.

Additionally, the coding reveals the following parent linkage (Fig. 2):

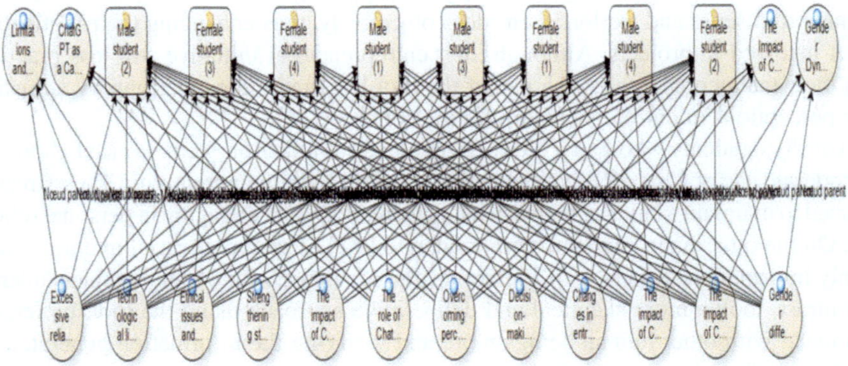

Fig. 2. Parent linkage coding by NVivo 10 software. Source: Nvivo10 Software.

6 Conclusion

The research identifies a meaningful implication of ChatGPT on students' entrepreneurial intention and points to notable differences based on gender and entrepreneurial intention. The tool set was utilized for project refinement by male students, especially those with a high level of entrepreneurial intention, while students with a moderate one focused more on theoretical knowledge. In contrast, female students who had at the beginning a weaker entrepreneurial intention, they noticed that ChatGPT helped them to be more confident and lowered their perceived risk which affected their perception of entrepreneurship positively. However, their participation still limited due to gender stereotypes, and they adopt a more reserved and cautious approach, seeking to validate information before applying it.

Due to the study's limited sample of only one academic field, generalizing results to all students should be taken with caution. Moreover, the impact of ChatGPT on creativity and practical engagement have not been sufficiently explored, which could be a research focus for future studies.

To advance the efficiency use of ChatGPT in entrepreneurship education, it is important to develop targeted initiatives that promote a balanced approach, integrating specific technology tools while also preserving some moments of personal reflection. In the same sense, it is also important to establish platforms for exchange and mentoring, particularly for female students, so as to reduce the psychological and social barriers to entrepreneurship. A large sample and inclusion of external perspectives would help to have a deeper understanding of the artificial intelligence effect on students' entrepreneurship intention.

Funding. This research was conducted with the support of the "*PhD-Associate Scholarship-PASS*" from the National Center for Scientific and Technical Research (CNRST), Morocco.

References

1. Ghafar, A.: Convergence between 21st century skills and entrepreneurship education in higher education institutes. Int. J. High. Educ. **9**(1), 218–229 (2020)
2. Chen, Z.: Artificial intelligence-virtual trainer: innovative didactics aimed at personalized training needs. J. Knowl. Econ. **14**(2), 2007–2025 (2023)
3. Somia, T., Vecchiarini, M.: Navigating the new frontier: the impact of artificial intelligence on students' entrepreneurial competencies. Int. J. Entrep. Behav. Res. **30**(11), 236–260 (2024)
4. Shepherd, D.A., Majchrzak, A.: Machines augmenting entrepreneurs: opportunities (and threats) at the Nexus of artificial intelligence and entrepreneurship. J. Bus. Ventur. **37**(4), 106227 (2022)
5. Victor, D.: Factors and impacts of ChatGPT adoption for academic purposes in higher learning institutions: students' perspectives. Glob. Sci. J. **12**(8), 327–351 (2024)
6. Duong, C.D., Nguyen, T.H., Chu, T.V., Pham, T.V., Do, N.D.: Whether ChatGPT adoption inspires higher education students' digital entrepreneurial intention? An integrated model of the SCCT and the TPB. Int. J. Innov. Sci. (2024)
7. Duong, C.: ChatGPT adoption and digital entrepreneurial intentions: an empirical research based on the theory of planned behaviour. Entrepreneur. Bus. Econ. Rev. **12**(2), 129–142 (2024)
8. Ebere Chukwuere, J.: The use of ChatGPT in higher education: the advantages and disadvantages (2024)
9. Yu, H.: The application and challenges of ChatGPT in educational transformation: new demands for teachers' roles. Heliyon (2024)
10. Choudhury, A., Shamszare, H.: Investigating the impact of user trust on the adoption and use of ChatGPT: survey analysis. J. Med. Internet Res. **25**, e47184 (2023)
11. Gonzalo Acosta-Enriquez, B., Agustín Arbulú Ballesteros, M., Huamaní Jordan, O., López Roca, C., Saavedra Tirado, K.: Analysis of college students' attitudes toward the use of ChatGPT in their academic activities: effect of intent to use, verification of information and responsible use. BMC Psychol. **12**(1), 255 (2024)
12. Conde-Ruiz, J.I., Ganuza Fernandez, J.J., García, M., Victoria Lanzón, C.: AI and digital technology: gender gaps in higher education. CESifo Econ. Stud. **70**(3), 244–270 (2024)
13. Salas-Pilco, S.Z., Xiao, K., Oshima, J.: Artificial intelligence and new technologies in inclusive education for minority students: a systematic review. Sustainability **14**(20), 13572 (2022)
14. Elshaer, I.A., Hasanein, A.M., Sobaih, A.E.E.: The moderating effects of gender and study discipline in the relationship between university students' acceptance and use of ChatGPT. Eur. J. Invest. Health Psychol. Educ. **14**(7), 1981–1995 (2024)
15. Bouzar, A., Idrissi, K.E., Ghourdou, T.: Gender differences in perceptions and usage of ChatGPT. Int. J. Human. Educ. Res. **6**(2), 571–582 (2024)
16. Craiut, M.V., Iancu, I.R.: Is technology gender neutral? A systematic literature review on gender stereotypes attached to artificial intelligence. Hum. Technol. **18**(3), 297–315 (2022)
17. Miao, F., Holmes, W., Huang, R., Zhang, H.: AI and Education: A Guidance for Policymakers. Unesco Publishing (2021)
18. O'Connor, S., Liu, H.: Gender bias perpetuation and mitigation in AI technologies: challenges and opportunities. AI Soc. **39**, 2045–2057 (2024). https://doi.org/10.1007/s00146-023-01675-4

A Systematic Review of Recommender Systems for Student Academic and Career Guidance

Hasna Mahmoud[1] (✉), Mohamed Badouch[1], Mehdi Boutaounte[2], Omar Zioudi[1], and Es-Said Boulmane[1]

[1] LabSIV, Department of Computer Science, Faculty of Science, Ibnou Zohr University, Agadir, Morocco
hasna.mahmoud.98@edu.uiz.ac.ma

[2] National School of Commerce and Management, Ibnou Zohr University, Dakhla, Morocco

Abstract. Academic and career guidance is crucial for student success. This systematic review examines recent advancements in recommender systems (RS) designed to support students' academic and career decision-making. Following PRISMA 2020 guidelines, we analyzed 21 peer-reviewed papers published between 2020 and 2025, focusing on methodologies, effectiveness, evaluation metrics, challenges, and emerging trends. Our findings reveal a dominance of knowledge-based and hybrid approaches, leveraging machine learning techniques. While these systems demonstrate high accuracy (average 86.4%) in predicting suitable career paths and educational programs, significant challenges remain. These include integrating real-time labor market data, enhancing personalization beyond academic performance, addressing data sparsity and the cold-start problem, ensuring transparency and interpretability of recommendations, and adapting to evolving student preferences and career landscapes. Emerging trends highlight the potential of deep learning, knowledge graphs, and social media data integration. Future research should focus on enhancing model interpretability, reducing biases, and developing adaptive frameworks that align with evolving student needs and labor market demands. Advancing these systems will strengthen their role in supporting students' academic and career decision-making.

Keywords: Recommender Systems · Career Guidance · Student Support · Academic choice · Machine Learning · Deep Learning · Hybrid Models · Personalization

1 Introduction

Academic and career guidance is a major challenge for students, influencing their success and integration into the job market. With the increasing diversity of educational pathways and career opportunities, it is crucial to provide intelligent solutions to assist students in their decision-making. Recommender systems help users navigate vast information spaces by filtering and ranking relevant options [1]. They play a crucial role in

B.-E. Boudriki Semlali et al. (Eds.): DATA 2025, LNISO 81, pp. 166–177, 2025.
https://doi.org/10.1007/978-3-032-06725-8_14

career guidance, analyzing student profiles [2] (e.g., interests, skills, academic performance) to suggest suitable career paths, courses, or educational programs. These systems aim to replicate career counseling by leveraging algorithms to provide personalized recommendations.

Traditional career counseling [3] is often inaccessible, leading students to rely on anecdotal advice, which can result in misaligned career choices and higher dropout rates. Career recommender systems offer scalable, individualized guidance, improving decision-making [2] and aligning student aspirations with labor market demands.

EdTech innovations in recommender systems combine content-based, collaborative filtering, and hybrid approaches, normally augmented by. Content-based approaches perform best with the application of domain knowledge to recommend students for professions, where collaborative filtering utilizes peer data for recommendations. Hybrid methods combine these techniques to address limitations. Innovations include the uses of deep learning, knowledge graphs, and social media data [4] usage to make more personalized recommendations.

Though significant advances have been made in this research field, a number of problems hinder widescale implementation and quality of recommender systems (RS) implemented in student orientation. Scarcity of data and the cold start problem lead to hard-making of producing robust recommendations for new users or unpopular course programs whose related historical records are limited. Making transparency and interpretability of recommendations is also a significant challenge to overcome because student and faculty counselors require robust explanations backing recommended paths. Additionally, adapting to the evolving preferences of students is an ongoing problem as well. These systems need to keep up with the rapidly evolving job market and interface without flaws with existing education ecosystems. These issues necessitate ongoing research and innovation in improving the effectiveness and popularity of recommender systems in learning settings.

This systematic review comprehensively examines recent developments in recommender systems for student orientation. We analyze the methodologies and algorithms employed, evaluate their effectiveness and limitations, and discuss emerging trends and future research directions. Our study contributes valuable insights into optimizing these systems to better support students in making informed academic and career decisions.

2 Methodology

In this systematic review of career guidance recommender systems, we used the Preferred Reporting Items for Systematic Reviews and Meta-Analyses (PRISMA) 2020 guidelines to ensure a transparent and comprehensive methodology. The PRISMA 2020 statement provides a checklist for reporting systematic reviews, which can be utilized across fields, e.g., educational interventions [5]. Our approach consisted of formulating a research question, establishing a protocol with defined objectives and criteria, conducting a systematic search of various databases, as well as systematically selecting and analyzing relevant studies. We present the steps in Fig. 1 to indicate what was performed. By adhering to the PRISMA Guidelines 2020, this review aims to report findings with transparency and stringency, contributing significantly to the application of recommender systems for career guidance.

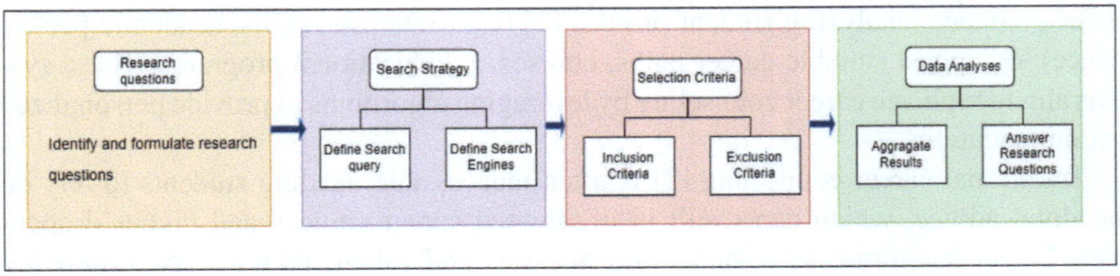

Fig. 1. Systematic Review Process

2.1 Search Question

The research questions presented in Table 1 comprehensively address key aspects of recommender systems for student guidance. They cover methodologies, effectiveness, evaluation metrics, challenges, and emerging trends. Clarifying the focus on traditional or AI-driven approaches, specifying measurable indicators for effectiveness, and linking evaluation methods to real-world applications could enhance clarity. Expanding the challenges to include ethical and policy concerns and refining future trends with AI advancements and personalization would further strengthen the scope.

Table 1. Purpose of the search questions

	Research Questions	Purpose
RQ1	What algorithms and methodologies are predominantly used in developing recommender systems for student guidance?	Identify the most effective approaches for academic and career recommendation
RQ2	How effective are recommender systems in supporting students' academic and career decision-making processes?	Assess the impact of recommendations on student engagement, motivation, and success
RQ3	What evaluation metrics and validation methods are employed to assess the performance of recommender systems in student guidance?	Define best practices and metrics to validate the relevance and accuracy of recommendations
RQ4	What are the main challenges and limitations identified in the design, implementation, and deployment of student guidance recommender systems?	Identify technical and ethical obstacles to improve the design and adoption of these systems
RQ5	What emerging trends and future research directions are highlighted in the literature regarding recommender systems for student guidance?	Explore technological and methodological innovations that could enhance recommendation accuracy and accessibility

2.2 Search Plan

We used Scopus as the primary database to retrieve relevant studies on recommender systems for student guidance. The search query applied was:

("recommender system*" OR "recommendation system*" OR "recommendation engine*") AND ("student guidance" OR "educational guidance" OR "academic guidance" OR "career guidance" OR "student counseling" OR "student support")

("recommender system*" OR "recommendation system*" OR "recommendation engine*") AND ("student guidance" OR "educational guidance" OR "academic guidance" OR "career guidance" OR "student counseling" OR "student support")

This query ensured a broad yet targeted retrieval of studies covering methodologies, effectiveness, evaluation metrics, challenges, and emerging trends. We applied filters to include only peer-reviewed journal articles and conference papers, screening the results based on relevance to our research questions.

2.3 Selection Criteria

The selection process followed specific inclusion and exclusion criteria, as presented in Table 2, to ensure relevance and quality. We included papers published between 2020 and 2025 in scientific journals or conference proceedings, written in English. Studies were excluded if they were published before 2020, did not align with research questions (RQ1 to RQ6), or focused on general recommender systems unrelated to student or academic guidance.

Table 2. Inclusion/exclusion criteria

Inclusion criteria	Exclusion Criteria
Paper published from 2020 to 2025	Before 2020
Paper published in scientific journals and conference paper	Papers not related to research questions (RQ1 to RQ5)
Paper must be in English language	General recommender systems unrelated to guidance

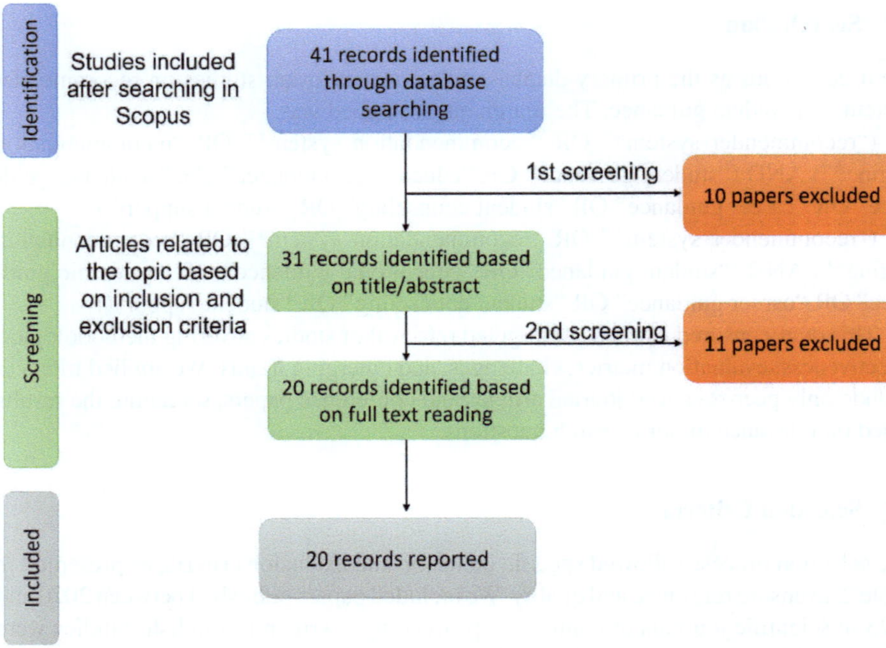

Fig. 2. A PRISMA-Compliant Systematic Review of Student Orientation Recommender Systems

The screening process, as illustrated in Fig. 2, involved multiple stages to ensure the selection of relevant studies. Initially, 41 records were identified through database searching. After reviewing titles and abstracts, 31 records were deemed relevant, while 10 papers were excluded for not meeting the inclusion criteria. A full-text reading of these 31 papers was conducted, leading to the exclusion of an additional 11 papers that were not aligned with the research questions. Ultimately, 20 studies were included in the final systematic review, ensuring a focused and high-quality selection of literature on recommender systems for student guidance.

2.4 Data Extraction and Data Reporting

To systematically analyze the selected studies, we extracted key information across multiple dimensions. Each study was categorized by year to track research trends over time. The approach was classified into four main categories: collaborative filtering, content-based filtering, knowledge-based approach, and hybrid approach, each further divided into traditional, machine learning, and deep learning techniques. The techniques used in each study were documented to understand the methodologies employed. Additionally, the advantages of each approach were identified, highlighting improvements in accuracy, efficiency, scalability, or personalization. The results were recorded based on reported performance metrics such as accuracy, precision, recall, and F1-score. Finally, the research gap was noted to identify limitations and areas for future exploration, ensuring a comprehensive evaluation of recommender systems for student guidance.

3 Result

This section presents findings from our systematic review of 21 research papers on recommender systems for student career guidance. The studies were analyzed based on their approaches—Knowledge-Based, Hybrid, Collaborative Filtering, and Content-Based—and techniques, including Traditional, Machine Learning, and Deep Learning. The key results of these studies are summarized in Table 3, highlighting their methodologies, evaluation metrics, and overall performance.

3.1 Distribution of Approaches and Techniques

Knowledge-based approaches dominated (42.9%), leveraging domain expertise for accurate recommendations, followed by hybrid models (41.3%), which combined multiple strategies for enhanced adaptability. Collaborative filtering (11%) and content-based filtering (4.8%) were less common. Machine learning was the most prevalent technique (73.6%), reflecting a shift toward data-driven methods, while traditional methods (21.1%) remained relevant. Deep learning (5.3%) was underutilized, appearing in only one study.

3.2 Performance Metrics

The reviewed studies reported various metrics, with accuracy ranging from 65% to 98.6% (average 86.4%). F1 scores (63.8%–98.6%), AUC (0.856–0.959), and error metrics (MSE, RMSE) were also evaluated. Knowledge-based approaches achieved high accuracy (86%–96%), while hybrid models demonstrated strong personalization and adaptability. Collaborative filtering showed promising accuracy (up to 92%), and the only content-based study reported high prediction accuracy (MSE: 0.0017, RMSE: 0.0422). Machine learning models consistently outperformed traditional techniques, while deep learning, though underexplored, showed potential with an AUC of 91%.

3.3 Advantages of Different Approaches

Knowledge-based systems excelled in personalized recommendations by integrating academic and psychological factors. Hybrid models enhanced adaptability and student satisfaction, incorporating real-time data and AI-driven guidance. Collaborative filtering offered scalable, high-accuracy recommendations, while content-based filtering personalized suggestions without requiring direct counseling.

3.4 Research Gaps and Limitations

Key challenges include limited integration of real-time job market data, personalization academic performance, adaptability to evolving career trends, and technical constraints such as bias mitigation and AI interpretability. Expanding datasets and refining recommendation models remain critical areas for improvement.

3.5 Temporal Trends (2022–2024)

Recent studies show a growing reliance on machine learning, an increasing preference for hybrid approaches, improved accuracy, and a stronger focus on real-time data and deep personalization.

The evolution of recommender systems for student career guidance highlights a preference for knowledge-based and hybrid approaches, emphasizing the role of domain expertise. While machine learning dominates, deep learning remains underutilized despite its potential. Though of high precision, integration of real-time labor market data, personalizing recommendations outside of marks, and keeping up with evolving career landscapes remain daunting. Recent trends suggest moving towards a combination of more than one approach of recommendations to counter these challenges.

Future research should prioritize enhancing AI interpretability, diminishing biases, and developing systems that adapt dynamically to individual student needs and to broad economic changes.

Table 3. Overview of recommendation approaches for academic and career guidance

Ref	Year	Approach	Techniques	Advantages	Results	Research Gap
[1]	2024	Knowledge-Based Approach	Machine Learning	High accuracy (96%+), fairness (bias removal)	F1 scores (96–98.6%), minimal prediction error (MAE: 0.014–0.043)	Bias refinement
[3, 6]	2024	Content-Based Filtering	Machine Learning	Eliminates the need for face-to-face counseling	MSE (0.0017) and RMSE (0.0422)	Limited consideration of student interests beyond academic background
[6]	2024	Collaborative Filtering	Machine Learning	Personalized career recommendations	HR@20 of 59.31%	Limitations in handling diverse employment preferences
[7]	2022	Hybrid	Deep Learning	Personalized career guidance	91% AUC score	Better modeling of user behavior, and improved adaptability for evolving career trends
[8]	2024	Knowledge-Based Approach	Machine Learning	Personalized career prediction based on skills and interests	90% accuracy	Need for real-time job market data integration
[9]	2022	Hybrid	Machine Learning	Personalized student guidance	95.68% accuracy	Need for improved dataset expansion and real-time mentor interactions

(*continued*)

Table 3. (*continued*)

Ref	Year	Approach	Techniques	Advantages	Results	Research Gap
[10]	2023	Knowledge-Based Approach	Traditional	Integrates various digital learning tools	NaN	Better interoperability across digital education platforms
[11]	2023	Hybrid	Machine Learning	AI-powered career guidance for older adults	High accuracy	Need for real-time job market integration
[5]	2023	Collaborative Filtering	Traditional	High recommendation accuracy, fast response time and scalable	High recommendation accuracy (92%)	Need for better personalization based on student preferences
[12]	2024	Knowledge-Based Approach	Machine Learning	Identifies personality traits that impact technical and vocational training, improves career alignment	AUC: 0.959, Accuracy: 0.642, F1: 0.638 and MCC: 0.601	Improved alignment between skills training and employability
[13]	2022	Hybrid	Machine Learning	Enhances employment guidance with AI-driven recommendations	AUC:0.856	Need for improved real-time adaptability in career guidance
[2]	2023	Knowledge-Based Approach	Machine Learning	Personalized academic advising, curriculum analysis, automated study plan generation	Achieved 86% accuracy and lowest MSR (0.14)	Improved dynamic adaptation to student needs
[14]	2024	Knowledge-Based Approach	Machine Learning	Personalized career guidance for engineering students	NaN	Need for expanding AI models to real-world career tracking
[15]	2024	Hybrid	Machine Learning	Provides intelligent academic guidance for Moroccan students	NaN	Need for real-time adaptability, deeper personalization based on psychological factors

(*continued*)

Table 3. (*continued*)

Ref	Year	Approach	Techniques	Advantages	Results	Research Gap
[16]	2024	Hybrid	Machine Learning	Predicts student placement likelihood, recommends skills for employability, integrates real-time placement data for personalized guidance	Achieved 79.33% accuracy	Need for integrating real-world industry trend and expansion to broader datasets
[17]	2022	Knowledge-Based Approach	Machine Learning	Enhances career guidance and supports student counseling	NaN	Enhanced integration of social media data for institutional reputation analysis
[4]	2024	Hybrid	Machine Learning	Uses AI for career guidance and integrates digital footprints from social media for educational recommendations	Varying accuracy (65–87%)	Need for improved prediction accuracy and enhanced interpretability of AI-based recommendations
[18]	2022	Knowledge-Based Approach	Traditional	Improves academic orientation accuracy	Achieved 91.35% accuracy	Need for real-time student preference integration
[19]	2024	Knowledge-Based Approach	Traditional	Provides ontology-driven career recommendations	Achieving high user satisfaction (90% found the interface friendly, 48.5% found recommendations very accurate)	Need for machine learning integration for enhanced personalization
[20]	2023	Hybrid	Traditional	Combines multiple recommendation techniques for personalized university and career guidance	Achieved 98% similarity rate, with 95% personalized recommendations, and high student satisfaction (92.5%)	Need for larger datasets, real-time adaptation, integration of more academic and career-related attributes for enhanced personalization

4 Conclusion

This systematic review explored 20 research papers (2022–2024) on recommender systems for students' academic and career guidance, with a trend towards knowledge-based and hybrid methodologies, primarily using machine learning. These systems demonstrate greater accuracy and personalization compared to traditional approaches.

Nevertheless, there remain difficulties. The integration of real-time labor market data, the personalization of recommendations beyond grades, and keeping pace with evolving career trends are areas for improvement. Technical limitations include minimizing bias, enhancing AI interpretability, and expanding datasets. Deep learning, while promising, remains under exploited.

Future research should prioritize enhancing AI transparency, addressing ethical concerns, and developing frameworks that respond dynamically to both individual student needs and broader economic changes. By addressing these areas, recommender systems can play a more significant role in empowering students to make informed decisions and successfully transition to the workforce, ensuring the systems are accurate, fair, and beneficial for all.

References

1. Maphosa, M., Doorsamy, W., Paul, B.: Improving academic advising in engineering education with machine learning using a real-world dataset. Algorithms **17**(2) (2024). https://doi.org/10.3390/a17020085
2. Atalla, S., et al.: An intelligent recommendation system for automating academic advising based on curriculum analysis and performance modeling. Mathematics **11**(5) (2023). https://doi.org/10.3390/math11051098
3. Majjate, H., Bellarhmouch, Y., Jeghal, A., Yahyaouy, A., Tairi, H., Zidani, K.A.: AI-powered academic guidance and counseling system based on student profile and interests. Appl. Syst. Innov. **7**(1) (2024). https://doi.org/10.3390/asi7010006
4. Oreshkina, T.A., Kartasheva, A.A., Zabokritskaya, L.D.: The dilemma of sufficient prediction accuracy in educational recommendation services. In: 2024 IEEE Ural-Siberian Conference on Biomedical Engineering, Radioelectronics and Information Technology (USBEREIT), pp. 217–219. IEEE (2024). https://doi.org/10.1109/USBEREIT61901.2024.10583983
5. Acang, J.P.A., Nisperos, S.F., Pagtaconan, W.C.R., Dela Cruz, R.L.: Generating program recommendations to aid in the decision-making of students in higher education using non-parametric supervised machine learning. In: 2023 IEEE 15th International Conference on Humanoid, Nanotechnology, Information Technology, Communication and Control, Environment, and Management (HNICEM), pp. 1–5. IEEE (2023). https://doi.org/10.1109/HNICEM60674.2023.10589265
6. Yang, Y., Wei, Z.: A study of informatization and data-driven career planning in career guidance in colleges and universities. Appl. Math. Nonlin. Sci. **9**(1) (2024). https://doi.org/10.2478/amns-2024-3155
7. Dutta, I., Shah, A., Safa, M., Jayavel, K.: Building a recommendation system for career advice for students from professionals. In: 2022 International Conference on Computer Communication and Informatics, ICCCI 2022. Institute of Electrical and Electronics Engineers Inc. (2022). https://doi.org/10.1109/ICCCI54379.2022.9740837

8. Guntupalli, U.G.S., Pandala, M.L., Veeranki, D.T., Kumbha, P.: Carrer compass: a career path recommender using machine learning. In: 5th International Conference on Electronics and Sustainable Communication Systems, ICESC 2024 – Proceedings, pp. 1937–1941. Institute of Electrical and Electronics Engineers Inc. (2024). https://doi.org/10.1109/ICESC60852.2024.10689897
9. Srinivasan, D., Uthayakumar, A., Thiyagalingam, P., Ravindran, S., De Silva, H., Kasthururarthna, D.: Escort – natural language processing based university students guidance system. In: 7th International Conference on Information Technology Research: Digital Resilience and Reinvention, ICITR 2022 – Proceedings. Institute of Electrical and Electronics Engineers Inc. (2022). https://doi.org/10.1109/ICITR57877.2022.9992860
10. Ziemann, F., Nguyen, D., Blum, F., Lucke, U.: Personalized Recommendations for Individual Learning Pathways: Supporting Ukrainian Refugee Students in Continuing their University Education. https://orcid.org/0000-0003-4049-8088
11. Jain, A., et al.: Career support platform for older adults powered by AI. In: 2023 IEEE 13th Annual Computing and Communication Workshop and Conference, CCWC 2023, pp. 47–53. Institute of Electrical and Electronics Engineers Inc. (2023). https://doi.org/10.1109/CCWC57344.2023.10099112
12. Hassan, R.H., Hassan, M.T.: Identification of technical and vocational education and training (TVET) trainee's personality attributes which impact skills learning. In: 2024 5th International Conference on Advancements in Computational Sciences, ICACS 2024. Institute of Electrical and Electronics Engineers Inc. (2024). https://doi.org/10.1109/ICACS60934.2024.10473268
13. Huang, L.: The establishment of college student employment guidance system integrating artificial intelligence and civic education. Math. Probl. Eng. (2022). https://doi.org/10.1155/2022/3934381
14. Timofeeva, M., Miziukov, G., Khlebnikova, M.: Methodology for determining professional personality traits of future engineers. In: 2024 7th International Conference on Information Technologies in Engineering Education, Inforino 2024 – Proceedings. Institute of Electrical and Electronics Engineers Inc. (2024). https://doi.org/10.1109/Inforino60363.2024.10551984
15. Samira, R., Ahmed, B., Abdellatif, E.A., Said, B., Mouad, E.: Smart guidance systems for Moroccan students. In: Proceedings of 2024 1st Edition of the Mediterranean Smart Cities Conference, MSCC 2024. Institute of Electrical and Electronics Engineers Inc. (2024). https://doi.org/10.1109/MSCC62288.2024.10697009
16. Kadu, R., Assudani, P.J., Mukewar, T., Kapgate, J., Bijekar, R.: Student placement prediction and skill recommendation system using machine learning algorithms. In: 7th International Conference on Inventive Computation Technologies, ICICT 2024, pp. 401–408. Institute of Electrical and Electronics Engineers Inc. (2024). https://doi.org/10.1109/ICICT60155.2024.10544738
17. Apostolova, M., Cico, B.: The advantages of intelligent technologies in the domain of higher education (HE) by emphasizing their large application potentials. In: 2022 11th Mediterranean Conference on Embedded Computing, MECO 2022. Institute of Electrical and Electronics Engineers Inc. (2022). https://doi.org/10.1109/MECO55406.2022.9797118
18. Ouatik, F., Erritali, M., Ouatik, F., Jourhmane, M.: Student orientation recommender system using TOPSIS and AHP. J. Inf. Org. Sci. **46**(2), 473–489 (2022). https://doi.org/10.31341/jios.46.2.12
19. Dascalu, M.I., Birzaneanu, R., Bodea, C.N.: An ontology-based recommendation module for optimal career choices. In: Lecture Notes in Networks and Systems, pp. 318–331. Springer Science and Business Media Deutschland GmbH (2024). https://doi.org/10.1007/978-3-031-54053-0_23

20. Lahoud, C., Moussa, S., Obeid, C., El Khoury, H., Champin, P.A.: A comparative analysis of different recommender systems for university major and career domain guidance. Educ. Inf. Technol. (Dordr.) **28**(7), 8733–8759 (2023). https://doi.org/10.1007/s10639-022-11541-3

Designing a Low-Cost CNC Milling Machine for Sustainable Digital Manufacturing

Abdyreshit Sadyk(✉), Yslam Orazov, Mekan Allyyev, Sahetnur Durdyyev, Mukam Ekayev, and Lachyn Hojamova

Oguz Han Engineering and Technology University of Turkmenistan, Ashgabat, Turkmenistan
s.abdyreshit01@gmail.com

Abstract. This paper presents the design and development of a low-cost Computer Numerical Control (CNC) milling machine aimed at promoting sustainable digital manufacturing in resource-limited settings. The system incorporates readily available components including an Arduino Uno microcontroller, CNC shield, stepper motor drivers, and open-source control software (gSender). With a working area of 40 × 40 cm, the prototype is capable of processing wood, acrylic, and soft metals. Compared to commercial CNC systems, this model significantly reduces cost while maintaining acceptable levels of accuracy and functionality. The paper details the mechanical and electrical architecture, design considerations, implementation steps, and performance evaluations. Findings demonstrate the feasibility of building a reliable and affordable CNC platform with broad applications in education, prototyping, and small-scale production.

Keywords: CNC milling machine · digital manufacturing · open-source CNC · GRBL firmware · sustainable fabrication · small-scale manufacturing · resource-efficient manufacturing

1 Introduction

Computer Numerical Control (CNC) is a kind of machine-controlled automatically with the aid of a computer. Parts are automatically programmed to operate according to the series of events set by the user to create a product of the required shape and size [1–3]. The advent of digital manufacturing has revolutionized the production landscape, enabling greater precision, efficiency, and automation in the fabrication of a wide variety of products. As industries increasingly adopt technologies such as Computer Numerical Control (CNC) machines, the capabilities for producing highly intricate and complex parts with minimal human intervention have expanded significantly. However, despite their widespread utility, commercial CNC milling machines often remain prohibitively expensive, particularly for smaller enterprises, educational institutions, and individual hobbyists. This financial barrier limits access to advanced manufacturing technologies, hindering innovation and the growth of small-scale enterprises that could benefit from customized production capabilities.

B.-E. Boudriki Semlali et al. (Eds.): DATA 2025, LNISO 81, pp. 178–183, 2025.
https://doi.org/10.1007/978-3-032-06725-8_15

To address this gap, this paper presents the development of a low-cost CNC milling machine designed using Arduino-based control systems, which offers an affordable alternative to traditional, high-cost CNC machines. This machine leverages a CNC shield and servo motors, controlled via gSender open-source software, to automate the milling process with high precision. By utilizing Arduino microcontroller technology, the system remains economically accessible while still capable of performing the essential functions of a typical CNC milling machine, including cutting, engraving, and shaping materials such as wood, metals, and acrylic. In addition, small CNC machines can still process products with high precision. Using small machines for production brings many advantages, saving space and electricity, less vibration, reduced noise, high flexibility, and easily relocated [4, 5].

The primary aim of this project is to demonstrate that advanced digital manufacturing technologies need not be confined to expensive commercial solutions. The focus is on the sustainability and affordability of digital fabrication systems, particularly in the context of small-scale production, research, and educational applications. With a working area of 40 cm × 40 cm, this machine is designed for small to medium-sized projects, allowing users to access the potential of computer-aided manufacturing without the need for substantial financial investment.

This paper explores the design and development of the CNC milling machine, examining the hardware components, software integration, and working principles that make this project feasible. Furthermore, it discusses the implications of this work for sustainable manufacturing practices, particularly in resource-efficient fabrication and reduced material waste. The results presented herein indicate that such low-cost systems can significantly contribute to the democratization of digital manufacturing, providing accessible tools for entrepreneurs, educators, and small-scale manufacturers.

As the world moves toward more sustainable production models, the need for technologies that can reduce environmental impact and promote innovation in a cost-effective manner becomes increasingly evident. This work offers a potential pathway toward achieving those objectives by providing an accessible digital manufacturing solution that can serve as a stepping stone for future innovations in sustainable fabrication and affordable technology.

2 Materials and Needed Components

The design and development of a low-cost CNC milling machine require a careful selection of hardware components and software tools to ensure cost efficiency, functionality, and precision. The proposed CNC system is built using Arduino-based electronics, which provide an affordable yet powerful means of controlling the machine's movement and execution of G-code instructions. The machine is constructed with readily available components that are commonly used in open-source CNC systems, making it easy to replicate and modify for future enhancements.

The machine frame is constructed from aluminum extrusions, ensuring rigidity while keeping weight low. Linear guides and lead screws are used to achieve stable motion along the X, Y, and Z axes. The spindle motor is mounted on a modular holder allowing quick replacement or upgrades. Limit switches are installed for homing and safety.

2.1 Hardware Components

The CNC milling machine consists of several key hardware elements that enable precise movement and material processing. These components are carefully chosen to balance cost-effectiveness with performance reliability.

At the core of the CNC machine is an Arduino microcontroller, which processes G-code instructions and controls the movement of the machine. The Arduino is programmed with GRBL firmware, an open-source motion control system optimized for CNC applications. GRBL interprets G-code commands and translates them into motor movements, allowing for precise machining. The use of Arduino significantly reduces costs compared to proprietary CNC controllers while maintaining high functionality.

To facilitate the connection between the Arduino and stepper motors, a CNC shield is used. The CNC shield simplifies wiring and provides the necessary connections for stepper motor drivers, endstops, and spindle control. It allows for expandable configurations and supports multiple stepper motor drivers, making it a practical solution for small-scale CNC machines.

The machine's movement is controlled by Nema 17 stepper motors mounted on each axis (X, Y, and Z). These stepper motors are chosen for their precise, repeatable motion and ability to operate without feedback sensors. In specific functions where closed-loop control is required, servo motors are used to provide higher torque and accuracy for certain machining operations.

The spindle motor is responsible for cutting and engraving materials. The selection of the spindle motor depends on the materials being processed. For this system, a high-speed DC spindle motor (775 DC motor) is used, providing sufficient power to mill wood, metals, and acrylic efficiently. The spindle is controlled via the CNC shield, allowing speed adjustments depending on the cutting requirements.

The machine's mechanical structure consists of a rigid frame made from aluminum and steel components, ensuring stability and vibration resistance during machining operations. The linear motion system includes rails, lead screws, and bearings that enable smooth and precise movement of the machine's cutting head. The system is designed to support a 40cm $\times$ 40cm working area, making it suitable for small to medium-sized milling projects.

A regulated power supply is required to provide adequate voltage and current to the stepper motors, spindle motor, and control electronics. For this project, a 12V, 5A power supply is used to ensure stable operation. Proper power management is essential to prevent overheating and maintain the reliability of electrical components.

2.2 Software Tools

The CNC milling machine relies on open-source software for design, G-code generation, and machine control. The integration of these software tools enables a seamless workflow, from designing parts to executing milling operations.

GRBL is an open-source motion control firmware that runs on the Arduino microcontroller. It interprets G-code instructions and translates them into stepper motor movements. GRBL is optimized for smooth acceleration, high-speed motion planning, and precise control, making it an ideal choice for low-cost CNC systems.

The CNC machine is controlled using gSender, an open-source G-code sender that provides an intuitive interface for operating CNC machines. gSender allows users to:

- Load and preview G-code files before execution.
- Manually jog the machine for positioning.
- Set zero points and adjust spindle speeds.
- Monitor real-time progress of machining operations. gSender is particularly useful for low-cost CNC machines as it is free, lightweight, and easy to use, making it suitable for both beginners and experienced users (Fig. 1).

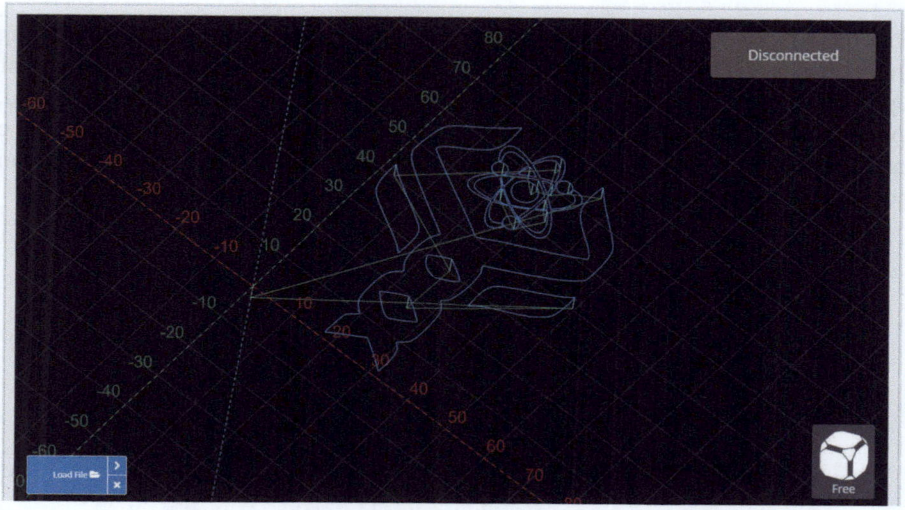

Fig. 1. Open source software gSender

2.3 Justification for Component Selection

The selection of Arduino, CNC shield, stepper motors, and open-source software was based on the following criteria:

- **Affordability:** The system is designed to be low-cost while maintaining functional capabilities.
- **Ease of Assembly:** Readily available components allow easy fabrication and modification.
- **Open-Source Flexibility:** Software tools such as GRBL and gSender provide extensive customization options.
- **Scalability:** The machine's modular structure allows for future upgrades, such as automated tool changing, IoT integration, and AI-based process optimization.

2.4 Sustainability Considerations

This CNC milling machine aligns with sustainable digital manufacturing principles by:

- **Reducing material waste** through precise cutting and optimized toolpaths.

- **Utilizing energy-efficient components** such as stepper motors with low power consumption.
- **Encouraging the reuse of materials** for machine construction and milling applications.

By leveraging affordable and open-source technologies, this project contributes to making digital fabrication more accessible and environmentally friendly, particularly for researchers, educators, and small-scale manufacturers.

3 Result and Discussion

The complete model of the machine is shown in Fig. 2. The machine is in good working order; there is no error in operation. The machine is easy to disassemble, convenient for replacing parts of the machine. When the machine is in process, the noise is not too loud, does not affect the surrounding environment, and does not cause pollution. The machine has full essential functions as an industrial 27 CNC machine (receiving G-code to run machining from CAM software) to fully meet students' learning and research needs.

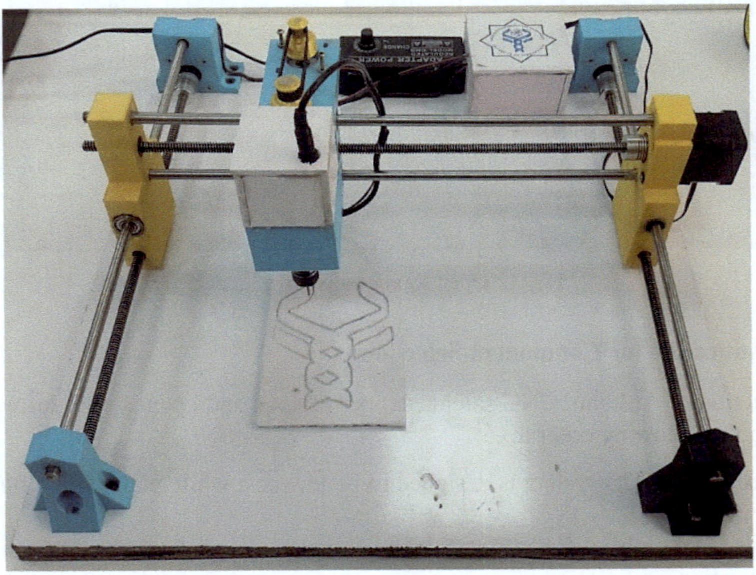

Fig. 2. CNC milling machine ready for use

4 Conclusion

This study presents the development of a low-cost CNC milling machine designed to provide an affordable yet functional solution for digital manufacturing. By integrating an Arduino-based microcontroller, a CNC shield, and open-source software like gSender, the machine successfully demonstrates its capability to mill wood, metals, and acrylic with precision. With only $150 in components and a simple design, the machine offers considerable capabilities for education, prototyping, and small production. The 40cm × 40cm working area makes it suitable for educational, research, and small-scale manufacturing applications, offering a viable alternative to expensive commercial CNC machines. Further work is needed to enhance functionality, but the prototype offers a solid foundation for future development.

The results indicate that this system not only meets the fundamental requirements of a CNC milling machine but also aligns with sustainable digital fabrication principles. The modular design, energy-efficient components, and use of open-source technologies enhance accessibility, reduce costs, and minimize material waste. Additionally, the ease.

of assembly and scalability provide opportunities for future enhancements, such as IoT integration and AI-driven process optimization.

By bridging the gap between affordability and functionality, this project contributes to the democratization of CNC technology, empowering students, educators, and small-scale manufacturers to explore digital fabrication without financial constraints. Future work can focus on further improving system accuracy, expanding material compatibility, and integrating automated features to enhance efficiency and usability.

References

1. Zhang, P.: Industrial Control Technology. William Andrew Publishing, Elsevier (2008)
2. da Rocha, P.A.S., de Silva e Souza R.D. and de Lima Tostes, M.E.: Prototype CNC machine design. In: 9th IEEE/IAS International Conference on Industry Applications - INDUSCON, pp. 1–5 (2010)
3. Leon, A., Suess, T., Lawson, D.: Building a CNC-capable metal 3D printing lab, Metal Powder Report (2001)
4. Milica, K., Aleksandar, K., Nemanja, B., Mihajlo, S., Gordana, G.: Design of laboratory 3-Axes CNC milling machine by modular approach LABROS 100S "A ACTA Tehnica Corviniensis. Bullet. Eng. **7**(1), 123–126 (2014)
5. Rastvorova, I.I., Klyucherev, N.A.: Design and modelling of a universal CNC machine. J. Phys. Conf. Series **1753** (2021)

An AI-Powered Chatbot Using a Transformer-Based Language Model for Graduate Educational Schemes Recommendation

Sandireddy Vyshnavi Prasad Chowdary[1], Atluri Sai Charan[2], Ashutosh Satapathy[1(✉)], Chinnam Sri Sai Prasanna[1], and Kokila Thottempudi[1]

[1] Department of Computer Science and Engineering, Siddhartha Academy of Higher Education, Vijayawada 520007, India
ashutosh.satapathy1990@gmail.com

[2] Lead Data Analyst, Sutherland Global Services, Hyderabad, Telangana, India

Abstract. Availability of higher education is still the priority concern for Indian socio-economically underprivileged and rural students at 25%, according to the National University of Educational Planning and Administration (NUEPA) 2021–2022 annual report. The major hurdles are financial constraints and unavailability of information regarding government-funded education programs. An AI-based learning advice chatbot leveraging the Large Language Model Meta AI (LLaMA) has been proposed to address the issue. The model is founded on the LLaMA 3.1 transformer model specifically optimized for low-resource environments. The model was fine-tuned on a 1,200 proprietary educational dataset of scholarship and government program-oriented queries with a test accuracy of 94.39% with an 80%−20% train-test split. The 60–40% was employed as another split to verify its performance with the test accuracy at 92.97%. The chatbot provides user-specific recommendations for 24/7 availability, ease, and strong data security. Providing timely and accurate information will help reduce the financial and informational barrier, thereby increasing the equitable opportunity for people from low-income backgrounds and the community to access higher education.

Keywords: Educational Assistance · Government Scheme Prediction · Chatbot · Natural Language Processing · Transformer Language Model · LLaMA 3.1

1 Introduction

Access to graduate-degree opportunity, scholarships, and funding is difficult with disparate eligibility requirements and limited knowledge [1]. Traditional research practices, such as static websites, cannot produce holistic, specific information, causing many students, particularly from underrepresented groups, uncertain about what is available. Such a lack of systematic and readily available information leaves behind opportunities and additional barriers to post-secondary education. Implementing an AI chatbot can be an innovative journey through integrating and shortening the process of information

B.-E. Boudriki Semlali et al. (Eds.): DATA 2025, LNISO 81, pp. 184–196, 2025.
https://doi.org/10.1007/978-3-032-06725-8_16

searching, filling gaps in critical knowledge, and providing equal access to learning experiences [2].

Second-generation transformer architectures such as Generative Pre-trained Transformer (GPT), Bidirectional Encoder Representations from Transformers (BERT), and LLaMA endow chatbots with contextual understanding, vectorization of natural language, and user-specific suggestions [3]. Self-attention continues to be the forte of these models in being relevant and coherent in their responses. The largest hurdle to adopting chatbots has been generating good, independent responses free of human bias. LLaMA excels other models of dependency capture for words to result in superior contextual performance on difficult academic queries [4]. A LLaMA-based chatbot processes users' requests, processes data, and customizes responses based on academic level, GPA, familial income level, and caste classification. Large quantities of data may be processed and handled efficiently by the structure. At the same time, uninterrupted performance is achieved, providing learners with a safe and intelligent interface through which learning aids may be obtained [5].

2 Related Work

Bird et al. proposed the Chatbot Interaction with Artificial Intelligence (CI-AI) framework to resolve data challenges in NLP applications such as scene understanding and sentiment analysis [6]. T5, a data-augmenting approach, was used by the model to augment 483 human responses into 13,090 instances, with an accuracy of 98.96% using seven transformer models that were benchmarked. Sajja et al. developed an AI-based pedagogical aid based on GPT-3 for multilingual, voice-enabled virtual Teaching Assistants (TAs) to respond to student questions. The model was trained with the Stanford Question Answering Dataset (SQuAD) and recorded 76.5% accuracy in Phase 1 and 95.3% in Phase 2 [7]. Ko et al. developed the Flipped-Visual Question Answering (VQA) model, using LLaMA-VQA with Contrastive Language-Image Pretraining with Vision Transformer Large 14 × 14 (CLIP ViT/L14) for videos and reasoning-based predictions [8]. The model used 4.5M trainable parameters out of LLaMA 7B, which is 0.06% parameters, it achieved 75.5% accuracy on the Next Phase of Question Answering (NExT-QA) dataset.

Nguyen et al. designed the NEU chatbot to handle Vietnam's National Economics University admission-related questions with Recurrent Neural Networks (RNN) and BERT within the Rasa platform for intent classification and entity recognition [9]. It achieved 97.1% accuracy and a 98.61% level of user satisfaction. Mirzababaei et al. presented a conversational module workflow using transformer-based models incorporating Sentence Bidirectional Encoder Representations from Transformers (SBERT) for question generation and adaptive dialogue structures [10]. Evaluated on research team data and 100 Mechanical Turk workers, F1-macro values were between 0.66 and 0.86 with follow-up question coherence rates of 79–84%. Lam et al. developed a Vietnamese text-processing virtual learning assistant based on a generative model for omissions of diacritics and misspellings [11]. The system's accuracy was 73%, diacritic restoration was 82%, and it gained better performance by applying a k-nearest neighbor (KNN) algorithm.

Rahman et al. conducted a study of ChatGPT utilization in studying and research, computer programming specifically, and attained 83% accuracy and 75% precision on a judgment platform. From the survey, it was clear that 78.8% of students employed ChatGPT to get assistance, with 92.9% considering it useful for studying [12]. Chang et al. developed a mobile chatbot learning strategy for obstetric vaccination education using Taiwan's Centres for Disease Control (Taiwan CDC) chatbot created by High Tech Computer (HTC) using NLP and machine learning. The Kuder Richardson-20 (KR20) reliabilities were 0.82 in the pre-test and 0.79 in the post-test, and Analysis of Covariance (ANCOVA) results $F(1,33)$ was 22.80 with $p < 0.01$ [13]. Mondal et al. created an educational chatbot learned from 1,500 dialogues, with a random forest-based ensemble learning framework. It attained an F-measure of 0.870 and 88.60% accuracy, showing the promise of machine learning to improve interactive learning, tested with different K-values, and implemented as a Telegram bot [14].

Yu et al. created MetaMath, fine-tuning LLaMA-2 models on the MetaMathQA dataset to improve math reasoning. MetaMath-7B reached 66.5% and 19.8% accuracy on GSM8K and MATH, respectively, while MetaMath-70B reached 82.3% accuracy on GSM8K, outperforming GPT-3.5-Turbo [15]. Allard et al. presented LLa-Ma-SciQ, an education chatbot optimized on LLaMa-8B for mathematical reasoning over multi-choice Science, Technology, Engineering, and Mathematics (STEM) questions. The model attained 74.5% accuracy on GSM8K and 30% on MATH datasets, with quantization improving computational efficiency at the expense of only a small loss in performance [16]. Semnani et al. developed WikiChat, an English chatbot, to reduce hallucinations and latency. By going through responses from GPT-4 to a 7B-parameter LLaMA model, WikiChat improved latency, cost, and privacy efficiency. The model achieved a literal accuracy of 97.3% in imitation conversations, better than GPT-4 and Retrieval-Augmented Generation (RAG) models in terms of accuracy and user interaction [17].

Santilli et al. presented Camoscio, an Italian instruction-tuned LLaMA-7B model trained via Low-Rank Adaptation (LoRA) from translated Stanford Alpaca data. It demonstrated performance in zero-shot scenarios with an F1-score of 0.270 for question answering (SQuAD-IT) and solid results for summarization (NewsSum-IT) and style transfer (XFORMAL-IT) [18]. Zhu et al. built Mixture-of-Experts (MoE) models out of LLaMA-2-7B and trained LLaMA-MoE-3.0B and 3.5B on the SlimPajama dataset (200B tokens). The LLaMA-MoE-3.5B (4/16) utilized FFNs with four active experts on every forward pass and achieved 89.2% of the performance of LLaMA-2-7B [19]. Cheng et al. developed Smart-LLaMA, a smart contract vulnerability detector using LLaMA-3.1-8B [20]. With a dataset of 186,397 smart contracts, it used continuous pre-training and explanation-guided fine-tuning. It achieved 96.15% F1 scores and 93.12% accuracy, improving over baselines by 9.55% in applications such as delegate call detection.

3 Proposed Methodology

A chatbot designed to assist learners through an easily usable online platform with one-to-one replies to queries posed by the learner in the scope of educational schemes, scholarships, and grants. Users will effortlessly use the chatbot to receive an effective and satisfactory experience. The chatbot will use a pre-trained Transformer model on a custom dataset, which will include wide details about study schemes, scholarships, and queries posed by the student. Such an advanced model will handle the user query rightly and respond appropriately about financial support, eligibility, and program information using multiple authentic references to render authenticity and up-to-dateness.

3.1 A Proposed System Architecture

The system design proposed has a chatbot with the ability to offer personalized graduate education scheme recommendations from a Trans-former-based LLaMA 3.1 model for offering more precision and efficiency (see Fig. 1). The method starts with preparing data, where structured datasets of scholarship information, eligibility conditions, and user demographic details get tokenized and encoded for processing with the LLaMA 3.1 model. The system best processes the inputs with the transformer-based LLaMA design, such as pre-normalization layers and multi-head self-attention. Using a dedicated scholarship dataset, the system is tuned to give better recommendations concerning a specific domain. The model's performance is tested to generate recommendations with precision, recall, accuracy, and BiLingual Evaluation Understudy (BLEU) scores. Extensive testing with new data establishes the robustness and flexibility of the model. The model is then trained in the domain of a chatbot that suggests real-time, personalized scholarship recommendations. The system is then deployed on a cloud infrastructure to be scalable, have low latency, and have device compatibility. It presents users with an effective and feasible system of recommending scholarships.

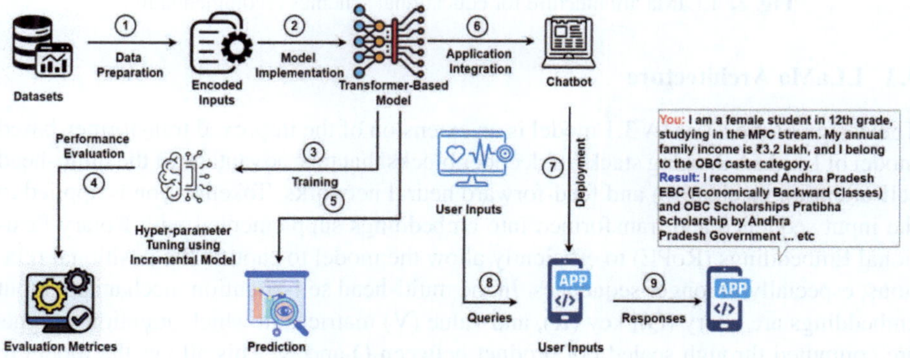

Fig. 1. Proposed System Architecture diagram of educational scheme recommendation chatbot.

3.2 Architecture Selection

Models such as BERT, Distil Bidirectional Encoder Representations from Transformers (DistilBERT), RoBERTa, Distil Robustly Optimized BERT Pretraining Approach (DistilRoBERTa), Efficiently Learning an Encoder that Classifies Token Replacements Accurately (ELECTRA), Extreme Language Network (XLNet), and LLaMA have widespread use in chatbot development. BERT models effectively provide bidirectional context, but

RoBERTa incorporates dynamic masking in BERT. ELECTRA extends token classification by utilizing a generator-discriminator framework, and XLNet is superior in modeling long-range dependencies. LLa-MA is an effective autoregressive model optimized for use under low-resource scenarios. Experiments conducted on a custom education corpus indicated that LLaMA 3.1 is superior to its forebears, thus proving itself to be even more efficient and effective (Fig. 2).

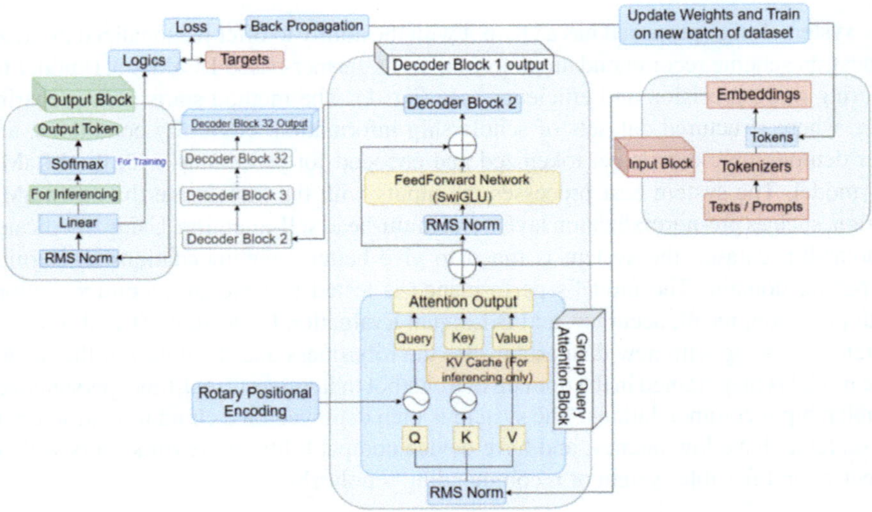

Fig. 2. LLaMa architecture for educational schemes recommendation

3.3 LLaMa Architecture

The design of the LLaMA 3.1 model is an extension of the improved transformer-based model of LLaMA 3, using stacks of decoder blocks that take advantage of the multi-head self-attention mechanism and feed-forward neural networks. Tokenization is applied to the input sequence and transformed into embeddings supplemented with Rotary Positional Embeddings (RoPE) to efficiently allow the model to capture the positional relations, especially in longer sequences. In the multi-head self-attention mechanism, input embeddings are query (Q), key (K), and value (V) matrices in which attention weights are computed through scaled dot product between Q and K. This allows the model to focus on the most appropriate information in the sequence, along with corresponding content received through V. To optimize efficiency further, grouped-query attention minimizes memory overhead by summing attention computations. In addition, the model uses a key-value (KV) cache to keep past attention states in memory for more efficient autoregressive generation through reusing old computations, as shown in Fig. 3. Prelayer normalization is integrated into each decoder block to make training more stable. Swish Gated Linear Unit (SwiGLU) is a variant of the Gated Linear Unit (GLU) that incorporates the Swish activation function into its gating mechanism.

The output block linearly scales and finally applies the final token probabilities through the softmax function. For better efficiency, LLaMA 3.1 employs gradient checkpointing for memory optimization and dynamic token sampling to target informative tokens. Sparse attention reduces computation to the most critical input elements, keeping it simple in terms of complexity, and low-bit quantization (4-bit/8-bit) optimizes memory, enabling deployment on multiple hardware [22]. Its broad context window better supports long-span performance on long-text activities like summarization and multiple-choice answer-response. It also allows scalability and modularity of its design for tuning per task, which makes it a very efficient and effective model in real-world deployments.

Pre-training

Pre-training in the LLaMA framework provides the first general representation of language, whereby the model develops its linguistic capacity using a large text dataset. The latest version, LLaMA 3.1, seeks to be light on resources yet strong on language understanding, steering clear of overloading parameter count so common with conventional models. Some of the pre-training characteristics of LLaMA 3.1 center around Causal Language Modelling (CLM) and efficiency in token prediction. LLaMA 3.1 employs causal language modeling (CLM), which predicts the next token in a sequence based on the prior tokens. By its very nature, causal LM prediction is unidirectional. It works so that the prediction flows coherently grammatically in the sentence, thus ensuring that the naturalness of the generated text remains touched by conserved interconnections.

$$L_{CLM} = -\sum_{i=1}^{N} \log P(x_i|x_{<i}) \quad (1)$$

In Eq. (1), L_{CLM} is the loss function for causal language modeling; $P\ (x_i|\ x_{<i})$ is the predicted probability of the token x_i given all preceding tokens $x_{<i}$, and N is the total number of tokens in the sequence. LLaMA 3.1 makes effective token prediction possible even on low-resource hardware. It is designed with a scalable and fine-tunable architecture to be useful in applications where computational resources are constrained without sacrificing the ability of strong language modeling.

$$L_{token} = -\frac{1}{N}\sum_{i=1}^{N} \log P(x_i) \quad (2)$$

In Eq. (2), $P(x_i)$ is the probability for the predicted token, while N refers to the total number of tokens in the sequence.

Fine-tuning

Fine-tuning LLaMA 3.1 requires the pretraining of a model through sophisticated optimization methods such as schedule learning rate, regularization, and AdamW optimization. Important hyperparameters were fine-tuned for performance, like setting the batch size, number of epochs, maximum sequence length, warmup steps, and gradient accumulation steps. LLaMA 3.1 training involved mixed precision, linear decay scheduler, and checkpointing for efficient model saving and handling resources. Moreover, suitable tokenization was used to preprocess the dataset with padding and truncation for uniformity. Such settings help the model give extremely accurate and effective chatbot-oriented task responses.

4 Results and Discussion

LLaMA 3.1 was fine-tuned on a high-performance GPU-NVIDIA A100 with a minimum of 16 GB RAM for optimal model computation. The model was approximately 15,317.02 MB before quantization. The model weights and datasets needed 1TB of SSD storage space. Other important libraries include Python Unsloth, PyTorch, Transformers, Xformers, Scikit-learn (Sklearn), and Streamlit. The development platform for the project was coupled with an Intel Xeon processor, base clock speed of around 2.3 GHz, and two processing cores. Hyperparameter tuning encompassed hyperparameters such as an initial learning rate of 2E-4, 16 maximum epochs, gradient accumulation every four steps, five warmup steps, five maximum steps, and mixed precision training.

4.1 Dataset Preparation

A customized dataset was created by extracting data from Careers360 [21], Scholarship Learn [22], Buddy4Study [23], and Scholarships360 [24], which are authentic online scholarship sites. The working dataset encapsulated student scholarships, focusing on scholarships available in Andhra Pradesh and globally for students graduating with degrees. The data were structurally organized and tagged in compliance with the quality of relevance. Then found fit to develop an education chatbot for capable and contextualized answers. It mainly included government schemes and scholarship programs emphasizing the offer processes, focusing on ideal recommendations and enhanced reach for underprivileged students. The FAQ is created, presenting general questions like eligibility, application process, and awards, with topic headings and detailed explanations for relevance and clarity. Such a logically structured dataset will maximize the chatbot to provide accurate, user-centric, and contextually relevant outputs. An easy-to-understand dataset format with sample question and answer pairs is shown in Table 1.

Table 1. Corpus of government educational schemes

Instruction	Input	Output
What is the NTR Videshi Vidyadharana Scholarship Scheme for BC and EBC Students?	Scholarships	The NTR Videshi Vidyadharana Scholarship Scheme is designed for students from BC and EBC categories who wish to pursue postgraduate or PhD programs abroad. It provides financial assistance for higher studies in countries such as the USA, UK, Canada, Australia, and more.
Who is eligible for the NTR Videshi Vidyadharana Scholarship?	Eligibility	Students belonging to the BC and EBC categories who are under 35 years of age are eligible. Applicants must be pursuing or planning to pursue postgraduate or PhD programs abroad.

(continued)

Table 1. *(continued)*

Instruction	Input	Output
What is the maximum age limit for the NTR Videshi Vidyadharana Scholarship?	Age Limit	The maximum age limit for applicants is 35 years.

4.2 Performance Metrics

A True Positive (TP) classifies a correctly labeled positive instance as a positive one with actual and predicted values being positive. A True Negative (TN) is an instance accurately labeled as negative and predicted negative. A False Positive (FP) is a model classification inappropriately labeling a negative instance as positive. A False Negative (FN) is an output of the model that assumes a positive instance is negative. Accuracy is measured as a proportion of TP and TN out of the overall number of predictions, i.e., the overall number of TP, TN, FP, and FN. It measures the model's accuracy over all classes. A precision metric represents the true positives against the TP and FP. Recall is the ratio of true positive predictions to total TP and FN. It expresses the model's ability to find all positive instances correctly. The F1-score is the harmonic mean of precision and recall and gives a balanced performance measure regarding false positives and false negatives.

4.3 Results and Analysis

Table 2 presents the performance of LLaMA 3.1 on the educational scheme dataset with training and testing splits of [60%, 40%] and [80%, 20%]. Increasing training data always increases performance on primary metrics. Accuracy increased from 92.97% to 94.39%, and the F1 score improved from 87% to 92%. Precision was enhanced from 83% to 91%, and recall from 92% to 94%. The enhancement justifies the improved generalization and classification of the model. The chatbot's eligibility pattern identification is correct, and it returns user-specific recommended scholarships, although faulty input calls for enhancement. Finally, it simplifies scholarship detection and accessibility using user profiles.

Table 2. Test accuracy of LLaMA 3.1 on (60% train, 40% test) and (80% train, 20% test) partitions of educational scheme dataset.

Performance Metrics	(60%, 40%)	(80%, 20%)
Accuracy	92.97%	94.39%
F1-Score	87%	92%
Precision	83%	91%
Recall	92%	94%

The training and testing accuracies of a few transformer-based language models are presented in Fig. 3 in two splits: 80%-20% and 60%-40%. Among the candidates tested, LLaMA 3.1 topped the list with the highest accuracy of 96.48% in training and 94.39% in testing on the 80%-20% split and 95.94% in training and 92.97% in testing on a 60%-40% split, indicating promising generalization. BERT and RoBERTa have attained comparatively low training accuracy, with BERT rising from 35.29% to 54.69% by incorporating additional training data but still retaining fairly low testing accuracy. The DistillBERT works efficiently on an 80%-20% split with the training being 82.88% and the testing being 68.69%, whereas its performance is extremely low when the split is 60%-40%. ELECTRA's performance is inconsistent, ranging from 20.09% testing accuracy on an 80%–20% split to an immense 55.5% increase on a 60%–40% split. Meanwhile, XLNet shows a similar pattern: its testing accuracy increases from 18.22% to 57.85% on the 60%–40% split. Thus, LLaMA 3.1 does comparatively better in both splits, while other models behave in degrees of overfitting or underfitting, depending on the data distribution.

Table 3 compares the performance of different chatbots based on classical machine learning and sophisticated deep learning models. Bird et al. presented the CI-AI model attained 98.96% accuracy on 13,090 examples developed using T5 data augmentation [6]. Sajja et al. enhanced GPT-3-based AI Assistant accuracy from 76.5% to 95.3% on SQuAD [7]. Ko et al. developed a flipped-VQA model with LLaMA-VQA and CLIP ViT/L14 with an accuracy of 75.5% [8]. Nguyen et al.'s NEU chatbot, executed with RNNs and BERT in Rasa, maintained an accuracy level of 97.1% [9]. Mirzababaei et al. created a SBERT-based adaptive dialogue models with F1-macro values of 0.66–0.86 and coherence levels of 79%-84% [10]. Lam et al. developed a VietBOT based on Transformer-based models and attained BLEU-4 values of 0.683–0.723 with an accuracy of 73% on Can Tho University (CTU) Bot and OpenSubViet datasets [11].

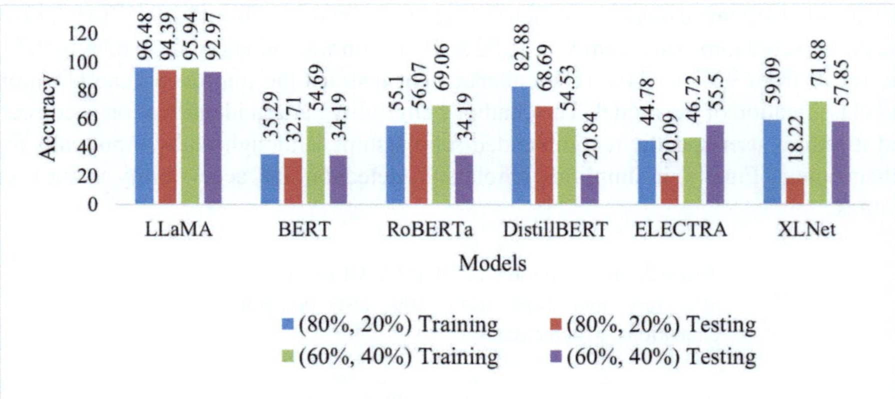

Fig. 3. Comparison between accuracy obtained by transformers on (60%, 80%) trainset and (40%, 20%) test set from educational scheme dataset.

Table 3. Comparison Table of related work

Literature	Algorithm/Models	Dataset	Performances
Bird et al. [6]	CI-AI framework with T5 based data augmentation	Custom dataset with 483 responses expanded to 13,090 examples	Accuracy: 98.96%
Sajja et al. [7]	GPT-3 for AI-augmented educational assistance framework	SQuAD dataset	Accuracy: 76.5% (phase 1), 95.3% (phase 2)
Ko et al. [8]	Flipped-VQA with LLaMA-VQA and CLIP ViT/L14	NExT-QA dataset	Accuracy: 75.5%
Nguyen et al. [9]	NEU-chatbot using RNNs and BERT on Rasa framework	Custom dataset for admission queries	Accuracy: 97.1%
Mirzababaei et al. [10]	SBERT models with adaptive dialogue structures	Research team dataset, Mechanical Turk workers dataset	F1-macro: 0.66–0.86, Coherence: 79–84%
Lam et al. [11]	VietBOT built on KNN, Bidirectional GRU or bidirectional RNN	CTUBot (35,702 pairs), OpenSub- Viet (419,712 pairs)	**BLEU score:** Bidirectional-GRU-RNN – 88.06%, GRU-RNN – 73.47%, Bidirectional-LSTM-RNN – 87.98% **Accuracy:** VietBOT – 73%
Rahman et al. [12]	ChatGPT for programming education	Online coding tasks	Precision: 95.83%, Accuracy: 75%
Chang et al. [13]	NLP	taiwancdc chatbot data	$F(1, 33) = 22.80$ ($p < 0.001$), Mean = 88.58, SD = 11.02
Mondal et al. [14]	Random forest-based ensemble learning	1,500 educational conversations	Accracy:88.6%, F-measure: 0.87
Yu et al. [15]	LLaMA-2- 70B is finetuned by LoRA	MetaMathQA dataset contains GSM8K and MATH datasets	Accuracy: 82.3% (GSM8K), 26.6% (MATH)
Allard et al. [16]	LLaMA-SciQ for STEM MCQs	GSM8K and MATH datasets	Accuracy: 74.5% (GSM8K), 30% (MATH)
Semnani et al. [17]	WikiChat G4 with base model GPT-4	Wikipedia-based dataset	Factual accuracy: 97.3%
Santilli et al. [18]	LLaMA (7 billion parameters) fine-tuned with LoRA	Translated Stanford Alpaca data	F1-score: 0.270 (question answering)

(*continued*)

Table 3. *(continued)*

Literature	Algorithm/Models	Dataset	Performances
Zhu et al. [19]	LLaMA-MoE-v1–3.5B (4/16)	SlimPajama dataset (200B tokens)	Avg Performance: 89.2%
Cheng et al. [20]	Smart-LLaMA based on LLaMA-3.1-8B	Custom dataset (186,397 smart contracts)	F1-score: 96.15%, Accuracy: 93.12%
Proposed Work	LLaMa 3.1 model	Customized educational scheme dataset	Accuracy: 94.39%, F1-Score: 92%, Precision: 91%

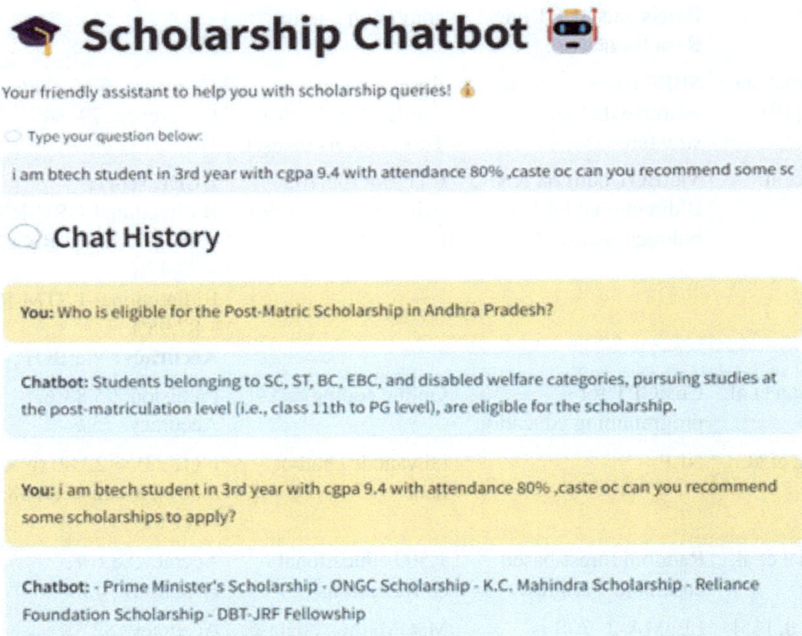

Fig. 4. Scholarship recommendation by chatbot built on LLaMA 3.1.

Rahman et al. investigated ChatGPT for research and teaching purposes, used it for web-based coding tasks, and scored 95.83% accuracy and 75% precision [12]. Chang et al. combined KR20 with NLP to determine chatbot chat interactions in Taiwan CDC data and gave ANCOVA results with F-value equal to 22.80, p-value less than 0.01, mean of 88.5, and standard deviation equal to 11.02 [13]. Mondal et al. utilized a random forest ensemble of 1,500 education conversations, which had accuracy as 88.6% and F-measure as 0.87 [14]. Yu et al. utilized MetaMathQA-fine-tuned LLaMA-2-70B and achieved 82.3% accuracy on GSM8K and 26.6% on MATH [15]. Allard et al. developed

LLaMA-SciQ achieved 74.5% accuracy on GSM8K and 30% on MATH [16]. Semnani et al. developed WikiChat, derived from LLaMA 7B distilled from GPT-4, achieved 97.3% factual accuracy on a Wikipedia dataset [17]. Santilli et al. fine-tuned LLaMA with seven billion parameters with Lo-RA and achieved an F1-score of 0.270 [18]. Zhu et al. optimized LLaMA-MoE-3.5B (4/16) with FFN and obtained 89.2% of the performance of the original LLaMA-2-7B on the SlimPajama dataset [19]. Cheng et al. proposed Smart-LLaMA, which is built upon LLaMA-3.1-8B, detected smart contract vulnerabilities with 96.15% F1-score and 93.12% accuracy [20]. The AI-based chatbot as depicted in Fig. 4, employing the LLaMA 3.1 model on a customized education dataset, recorded 96.48% accuracy, 92% F1-score, and 91% precision.

5 Conclusion

The educational chatbot will suggest student scholarships and study plans based on student profiles. The users will input caste, income, educational qualifications, and exam status. The chatbot recommends appropriate scholarship opportunities to the respective applicants, keeping the applicants informed of opportunities worth their notice. The data with specially crafted question-answer data set for the education industry and the LLaMA 3.1 TLM, the chatbot showed accuracy, precision, recall, and F1 score of 94.39%, 91%, 94%, and 92%, respectively. It is a good chatbot that gives acceptable responses to questions. The web-based application provides learners with scholarships and learning opportunities, making it usable and accessible to learners. Future studies can explore enhancing the chatbot's suggestion accuracy regarding user profiling and real-time adaptive learning. The accuracy for graduate program recommendations can be enhanced using information sources like university records and career goals.

References

1. Kadam, A., et al.: Evaluating awareness and impact of government schemes in india: a comprehensive study. Libr. Prog. Int. **44**(3), 2184–2191 (2024)
2. Brandtzæg, P.B., Følstad, A.: Why people use chatbots. In: Kompatsiaris, I., et al. (eds.) Internet Science: 4th International Conference, INSCI 2017, Thessaloniki, Greece, November 22–24, 2017, LNCS, vol. 10673, pp. 377–392. Springer, Cham (2017)
3. Abdulla, H.: Chatbots development using natural language processing: a review. In: Proceedings of the 26th International Conference on Circuits, Systems, Communications and Computers (CSCC), pp. 122–128. IEEE, Crete (2022)
4. Abdellatif, A.: A transformer-based approach for augmenting software engineering chatbots datasets. In: Franch, X., Daneva, M., Martínez-Fernández, S., Quaranta, L. (eds.) Proceedings of the 18th ACM/IEEE International Symposium on Empirical Software Engineering and Measurement (ESEM), pp. 359–370. ACM, Barcelona (2024)
5. Ahmad, N.A.: Review of chatbots design techniques. Int. J. Comput. Appl. **181**(8), 7–10 (2018)
6. Bird, J.J.: Chatbot Interaction with artificial intelligence: human data augmentation with T5 and language transformer ensemble for text classification. J. Ambient. Intell. Humaniz. Comput. **14**(4), 3129–3144 (2023)
7. Sajja, R.: Platform-independent and curriculum-oriented intelligent assistant for higher education. Int. J. Educ. Technol. High. Educ. **20**, 42 (2023)

8. Ko, D.: Large language models are temporal and causal reasoners for video question answering. arXiv preprint arXiv:2310.15747 (2023)
9. Nguyen, T.T.: NEU-chatbot: chatbot for admission of national economics university. Comput. Educ. Artif. Intell. **2**(1), 100036 (2023)
10. Mirzababaei, B.: Facilitating the Learning Engineering Process for Educational Conversational Modules using Transformer-based Language Models, pp. 1222–1235. IEEE Transactions on Learning Technologies 17 (2024)
11. Lam, K.N., Nguy, L.H.: A transformer-based educational virtual assistant using diacriticized latin script. IEEE Access **11**, 90094–90104 (2023)
12. Rahman, M.M., Watanobe, Y.: ChatGPT for education and research: opportunities, threats, and strategies. Appl. Sci. **13**(9), 5783 (2023)
13. Chang, C.Y., Hwang, G.J., Gau, M.L.: Promoting students' learning achievement and self-efficacy: a mobile chatbot approach for nursing training. Br. J. Edu. Technol. **53**(1), 171–188 (2022)
14. Mondal, A., Dey, M., Das, D., Nagpal, S., Garda, K.: Chatbot: an automated conversation system for the educational domain. In: 2018 International Joint Symposium on Artificial Intelligence and Natural Language Processing (iSAI-NLP), pp. 1–5. IEEE, Pattaya, Thailand (2018)
15. Yu, L., Jiang, W.: Metamath: Bootstrap your own mathematical questions for large language models. arXiv preprint arXiv:2309.12284 (2023)
16. Allard, M.A., Ansaripour, M.: LLaMa-SciQ: An Educational Chatbot for Answering Science MCQ. arXiv preprint arXiv:2409.16779 (2024)
17. Semnani, S.J., Yao, V.Z.: WikiChat: Stopping the hallucination of large language model chatbots by few-shot grounding on Wikipedia. arXiv preprint arXiv:2305.14292 (2023)
18. Santilli, A., Rodolà, E.: Camoscio: An Italian instruction-tuned llama. arXiv preprint arXiv: 2307.16456 (2023)
19. Zhu, T., et al.: Llama-moe: Building mixture-of-experts from llama with continual pre-training. In: Proceedings of the 2024 Conference on Empirical Methods in Natural Language Processing, pp. 15913–15923. Association for Computational Linguistics, Miami, Florida (2024)
20. Yu, L., Chen, S.: Smart-LLaMA: Two-Stage Post-Training of Large Language Models for Smart Contract Vulnerability Detection and Explanation. arXiv preprint arXiv:2411.06221 (2024)
21. Tamang, M.: Build Your Own Llama 3 Architecture from Scratch Using PyTorch. Medium (2024). https://pub.towardsai.net/build-your-own-llama-3-architecture-from-scratch-using-pytorch-2ce1ecaa901c
22. rnie: Deep Dive Llama 3 vs. Llama 3.1. Pynomial (2024). https://pynomial.com/2024/08/deep-dive-llama-3-vs-llama-3-1/
23. Careers360, T.: AP scholarships 2024–25: Check eligibility, last date, application form, amount (2024). https://school.careers360.com/articles/ap-scholarships. Accessed 06 Jan 2025
24. Singh, M.: Andhra Pradesh Scholarship 2025: Complete List, Eligibility Last Date (2024). https://scholarshiplearn.com/andhra-pradesh-scholarship/
25. Pandey, S.: AP Scholarship 2023– Complete list, eligibility, application process (2023). https://www.buddy4study.com/article/top-scholarships-for-students-of-andhra-pradesh
26. Scholarships360, Inc.: Scholarship opportunities for students- Scholarships360 (2024). https://scholarships360.org/

Implementation of a Digital Twin in an Industrial Production Line: Real-Time Visualization and Monitoring

Oualid Karim[1], Houda Lahyani[1](✉), Ahmad AL Khatib[2], and Abdelfettah Sedqui[1]

[1] Department of Industrial and Electrical Engineering, Innovative Technologies Laboratory, National School of Applied Sciences of Tangier, Abdelmalek Essaâdi University, 2117 Tetouan, Morocco
lahyani.houda@etu.uae.ac.ma

[2] Department of Industrial and Mechanical Engineering, Material and Mechanics Laboratory, ECAM-Louis de Broglie, 35170 Bruz, France

Abstract. Industry 4.0 has revolutionized industrial innovation, driving companies to digitize their manufacturing processes. This transformation integrates technologies such as IoT, AI, cloud computing, system integration, cyber security, big data, and digital twins, enhancing efficiency, quality, and responsiveness across the value chain. Among these innovations, digital twins (DTs) have gained significant attention from both industry and academia. A DT is a dynamic digital model that mirrors a physical asset or system, enabling real-time monitoring and synchronization with its real-world counterpart. This research first introduces a general DT architecture, defining its key modules within an industrial environment. Based on this architecture, an innovative DT for real-time monitoring of industrial processes is developed. It integrates an interactive dashboard within Grafana, allowing operators to efficiently monitor and control production systems. Node-RED serves as the integration platform, connecting PLCs via OPC UA, ensuring real-time data acquisition. This data is stored in Influx DB, a time-series database optimized for fast processing and real-time analysis, enabling better decision-making and process optimization. Additionally, the project incorporates FlexSim, a powerful commercial simulation tool that creates a real-time virtual representation of the production process. This digital model allows users to explore various operational scenarios, assess performance predictions, and anticipate possible disruptions, thereby improving overall efficiency in industrial settings. By combining real-time data acquisition, visualization, and simulation, this research contributes to the evolution of smart manufacturing, offering a comprehensive tool for process optimization and decision-making.

Keywords: Digital twin · Simulation · visualization · real-time · performance

B.-E. Boudriki Semlali et al. (Eds.): DATA 2025, LNISO 81, pp. 197–208, 2025.
https://doi.org/10.1007/978-3-032-06725-8_17

1 Introduction

The concept of the Digital Twin (DT) was formally defined and adopted by NASA around 2010, within the framework of its strategy focused on model-based systems engineering [1]. This concept enables the creation of a dynamic digital model that faithfully mirrors a physical asset or system, providing real-time synchronization between physical and digital environments. As a key element of Industry 4.0, the DT enhances connectivity and communication between systems, enabling process optimization and informed decision-making [2].

In the context of implementing a DT in a production line for real-time

visualization and monitoring, a critical question arises: how can simulation technologies (discrete event or agent-based), cloud computing, machine learning (ML), big data, and IT orchestration concepts be effectively and coherently integrated to create a DT of the production chain, while overcoming challenges related to connectivity, model accuracy, big data management, and development costs?

This research problem aims to explore how to leverage current technological

advancements to create a functional and real-time DT of a production line while taking into account the specificities of industrial processes and ensuring the efficiency of the various technological elements involved. By addressing these challenges, the objective is to enable companies to harness the potential benefits offered by TDs, such as performance optimization, early problem detection, informed decision-making, and reduced operational costs.

1.1 Applications of Digital Twin

As the relevance of the Digital Twin (DT) concept continues to expand across multiple industries, its current applications are predominantly observed in areas such as urban development, energy systems, infrastructure projects, transport, industrial production, and healthcare services. Table 1 outlines ten notable use cases of DT, classified by sector, as identified through a review of [3, 4], and [5].

Table 1. Summary of Digital Twin Applications.

Domain	Applications
Smart Cities	- Urban planning - Strategic development
Energy	- Energy monitoring and management - Operation and maintenance of the smart grid
Building	- Construction quality assessment - Worker safety monitoring - Resource allocation and waste tracking
Transport	- Transport monitoring - Travel schedules

(continued)

Table 1. (*continued*)

Domain	Applications
Healthcare	- Health monitoring - Personalized medicine
Manufacturing Industry	- Design verification and layout planning - Predictive maintenance - Production planning and control

1.2 Digital Twin Simulation Architecture

To date, the implementation of Digital Twins (DTs) still faces the challenge of the absence of standardized platforms, largely due to the wide diversity of application contexts and the specific nature of the problems they aim to resolve. This part of the study focuses on highlighting the main technologies and essential features required to support the effective deployment of DTs.

Based on the outcomes of the literature review, a conceptual framework is suggested to guide decision-making in Digital Twin systems. This framework relies primarily on the system's ability to simulate different scenarios, as depicted in Fig. 1 [6, 7].

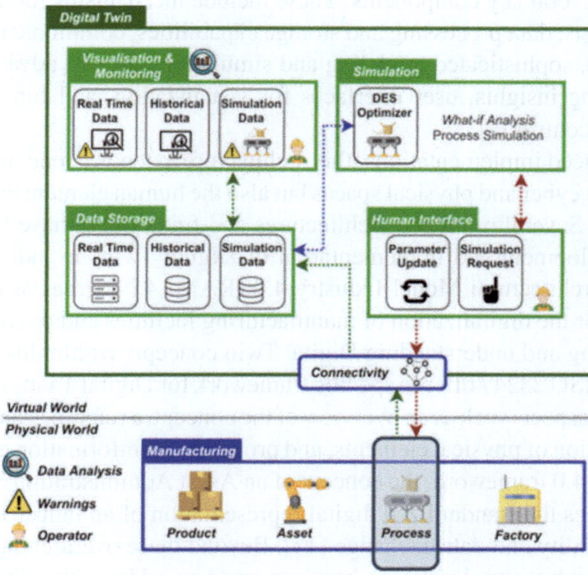

Fig. 1. Architecture Overview for Simulation-Driven Decisions via Digital Twins [7]

This architecture illustrates the interaction between the virtual system and the physical system by exchanging data through communication channels that allow for real-time information collection, analysis, and simulation [8]. It is based on the following key aspects:

- Clearly identifying the real-world element to be digitally replicated (e.g., a product, equipment, process, or factory).
- Creating a simulation-based representation of that physical component (for instance, using discrete event simulation – DES).
- Ensuring robust communication between the virtual and physical layers using standard industrial protocols, in order to monitor live data effectively.
- Leveraging simulation tools to enhance the efficiency and performance of the physical asset.
- Offering actionable insights to the human operator, grounded in both real-time data and simulation results.

The proposed Digital Twin (DT) structure consists of five essential modules: Communication Module, Data Storage Module, Visualization and Monitoring Module, Simulation Module, and Human-Machine Interface Module. It is designed to address gaps highlighted in the literature and solve several challenges, such as:

- Facilitating human involvement in DT processes.
- Making use of feedback loops that react to the decisions supported by the DT.

1.3 Architectural Approaches

The underlying structure of a Digital Twin typically involves a multi-faceted architecture comprising several key components. These include mechanisms for data collection from physical assets, data processing and storage capabilities, communication protocols for data exchange, sophisticated modeling and simulation engines, advanced analytics tools for extracting insights, user interfaces for visualization, and functionalities for optimization and control [9].

In some advanced implementations, this architecture extends to encompass the integration of not only cyber and physical spaces but also the human element within the operational loop [10]. Several reference architectures and frameworks have been proposed to guide the development and implementation of Digital Twins in industrial settings. The Reference Architectural Model Industry 4.0 (RAMI 4.0) serves as a comprehensive framework for the digitalization of manufacturing facilities and provides a relevant context for defining and understanding Digital Twin concepts within this domain [11].

Furthermore, ISO 23247 offers a specific framework for Digital Twins in manufacturing, outlining key aspects such as an overview of the concept, a reference architecture, the digital representation of physical elements, and protocols for information exchange [12]. Within the RAMI 4.0 framework, the concept of an Asset Administration Shell (AAS) is crucial, as it defines the standardized digital representation of an industrial asset, facilitating interoperability and data exchange [13]. Beyond these overarching frameworks, various specific architectural patterns have emerged to address the diverse needs of industrial applications. These include centralized, distributed, and federated approaches to manage data integration and sharing across different systems and stakeholders [14].

Additionally, Digital Twins can be categorized based on their functional capabilities, ranging from descriptive twins that simply visualize data to informative twins that provide real-time insights, predictive twins that forecast future behavior, comprehensive twins that enable scenario analysis, and ultimately, autonomous twins capable of learning and making decisions [15]. The increasing prevalence of cloud computing has significantly influenced the architectural landscape of Digital Twins, with cloud-based architectures offering the necessary scalability and accessibility for deploying and managing these complex systems [16]. Cloud platforms provide robust infrastructure for data storage, high-performance processing, and the deployment of advanced functionalities, making Digital Twins more accessible and cost-effective for a wider range of industrial applications.

1.4 Cloud-Based Platforms and Implementation of Digital Twins (DTs).

Several cloud-based platforms such as Microsoft Azure and Amazon Web Services (AWS) provide complete tool chains for implementing Digital Twins (DTs). These include modules for edge computing, data orchestration, storage, visualization, and simulation.

However, despite their robustness and integration capabilities, these solutions are often associated with high implementation costs, complex licensing models, and a need for advanced technical expertise. This may hinder their adoption, especially in small to medium-sized enterprises or in academic/research prototypes.

Table 2 presents a comparative overview between commercial platforms (Azure and AWS) [17].

Table 2. Comparative overview between commercial platforms (Azure and AWS)

Component	Azure	AWS
Edge Layer	Azure IoT Edge/Sphere	AWS Greengrass/ExpressLink
Data Orchestration	Azure IoT Hub/Logic Apps	AWS IoT Core/Lambda
Database	Time Series Insights/Cosmos DB	Timestream/DynamoDB
Visualization	IoT Central/Power BI	IoT SiteWise/QuickSight
Digital Twin	Azure Digital Twins (DTDL models)	AWS TwinMaker (3D visualization)
Cost	Tiered pricing (~$10 + /month)	Pay-as-you-go (~$0.16/10k devices)

To address this limitation, we propose a low-cost Digital Twin architecture that integrates open-source tools and widely available technologies. This architecture replicates the core functionalities of Azure and AWS solutions while significantly reducing complexity and cost.

2 Proposed Solution

2.1 Architecture Study

As part of our research, we propose the implementation of a technological architecture for the development of a DT aimed at real-time monitoring and simulation of industrial systems (Fig. 2).

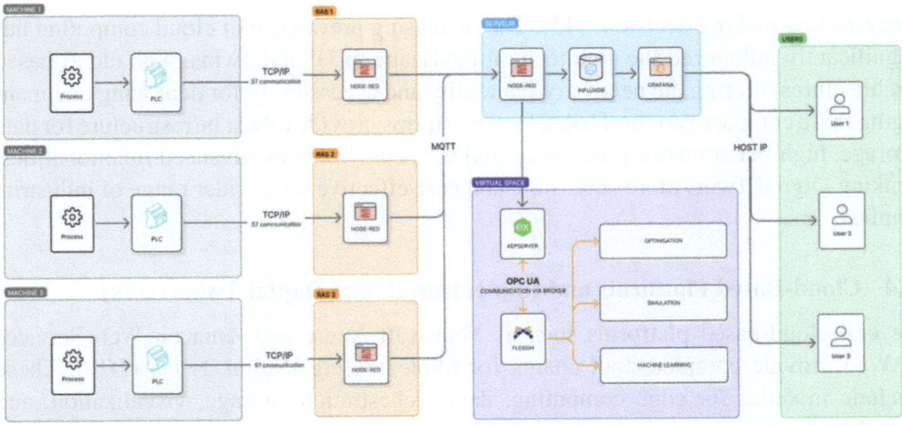

Fig. 2. Proposed Assembly Architecture for Real-Time Monitoring and Visualization of Industrial Equipment

This architecture is structured around the following six key components:

Interconnected Raspberry Pi Devices and Their Role: Raspberry Pi devices collect data from sensors and programmable logic controllers (PLCs) using the S7 protocol. They also perform local processing to improve data quality before transmission through Node-RED.

S7 Communication Protocol for PLC Connectivity: The S7 protocol facilitates bidirectional communication between PLCs and Raspberry Pi devices. It enables real-time data retrieval while ensuring the transmission of control commands.

Role of Node-RED as Middleware for Data Exchange and Synchronization:

Node-RED acts as an orchestrator, synchronizing and transforming data flows from Raspberry Pi devices. It ensures the consistency of the information before transmission.

Use of MQTT as a Lightweight Communication Protocol: The MQTT protocol, based on a publish-subscribe model, enables lightweight and efficient communication between components. It enhances the scalability of the architecture while ensuring fast and reliable data transmission.

Integration of InfluxDB for Data Storage and Retrieval: InfluxDB, a database specialized in time-series data, ensures the structured and optimized storage of collected data. It also allows for fast and effective analysis of information for diagnostics or predictions.

Real-Time Data Visualization with Grafana: Grafana provides interactive dash boards for real-time visualization of system performance. This interface facilitates informed decision-making through clear and intuitive graphical representations.

This integrated architecture ensures smooth communication between physical and virtual systems, enabling accurate representation of industrial performance and behavior.

2.2 Implementation of Digital Simulation

(Figure 3)

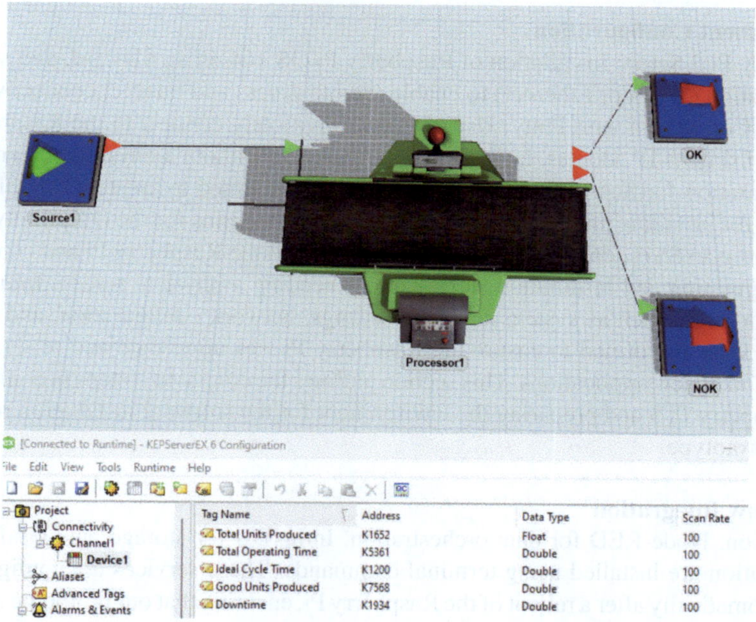

Fig. 3. 3D Simulation of a Production Line in FlexSim and Real-Time Configuration Preview in Kepware OPC UA

The integration of digital simulation technologies was carried out to optimize system performance and monitoring:

Integration of FlexSim for dynamic simulation: FlexSim was used to create a realistic simulation model to analyze and optimize operations in real time.

OPC UA for communication: The OPC UA protocol ensured seamless and secure communication between FlexSim and Raspberry Pi devices, guaranteeing smooth data synchronization.

Comprehensive overview and advanced simulations: An advanced simulation was developed to provide a detailed overview of processes, integrating accurate observations and in-depth analysis to support decision-making.

2.3 Implementation of the Prototype

The development of the prototype for a DT dashboard is built around two key components. The first part focuses on the architecture, integration, and real-time monitoring of physical processes, using tools like Node-RED, InfluxDB, Grafana, and the S7 protocol to communicate with PLCs. The second part involves modeling the virtual environment in FlexSim and linking it bidirectionally to the first part, creating a fully synchronized DT. By combining the capabilities of Node-RED, InfluxDB, Grafana, and FlexSim, this prototype delivers a powerful solution for real-time monitoring, analysis, and optimization of processes. The DT dashboard provides valuable insights and a virtual environment to test and simulate various scenarios without disrupting real-world operations.

Environment Configuration

Raspberry Pi 3 Setup: Installation of Raspberry Pi OS x64 on an SD card, and network configuration (Wi-Fi or Ethernet) to enable remote access and internet connectivity.

SSH Connection with Putty: Use of Putty to remotely connect to the Raspberry Pi via SSH using its IP address, facilitating configuration without a physical screen.

Preparation for Software Installation: The process aimed to create a suitable environment for installing and configuring the necessary programs for the DT prototype. By establishing an SSH connection, we remotely accessed the Raspberry Pi and completed all configuration and installation steps without needing a physical screen. Initial configuration tasks, such as system language settings, network configuration, and system updates, were performed to ensure the Raspberry Pi was up-to-date and ready for the installation of new applications. This section outlines the crucial first steps in configuring the Raspberry Pi 3 and preparing the environment for the software installation required for the prototype.

Data Flow Integration

Installation: Node-RED for data orchestration, InfluxDB for storage, and Grafana for visualization are installed using terminal commands. These services are configured to start automatically after a reboot of the Raspberry Pi, ensuring that our prototype remains operational even after the device is restarted.

Connection Configuration: Node-RED is connected to InfluxDB using the "InfluxDB" node. This allowed us to send the data collected from Node-RED directly into InfluxDB. And then, Grafana is linked to InfluxDB for data visualization.

Simulation of Industrial Data: The objective was to generate realistic simulated data to feed into our DT prototype.

Model Creation: Development of a PLC model in TIA Portal with data blocks simulating sensor variables.

S7-PLCSim Advanced: Emulation of the PLC model is performed using S7-PLCSim Advanced, generating realistic industrial data.

Integration with Node-RED: The simulated data is transferred to Node-RED using the "PLCsim Industry" node.

Real-Time Data Collection and Storage:

Node-RED Configuration: Node-RED configures the flows to capture the simulated sensor variables in real time and transmit them to InfluxDB. This was achieved by creating connection nodes in Node-RED, which act as channels for transmitting data. Each node was carefully configured to establish the link between the simulated PLC variables and the databases, ensuring continuous, precise, and reliable data collection (Fig. 4).

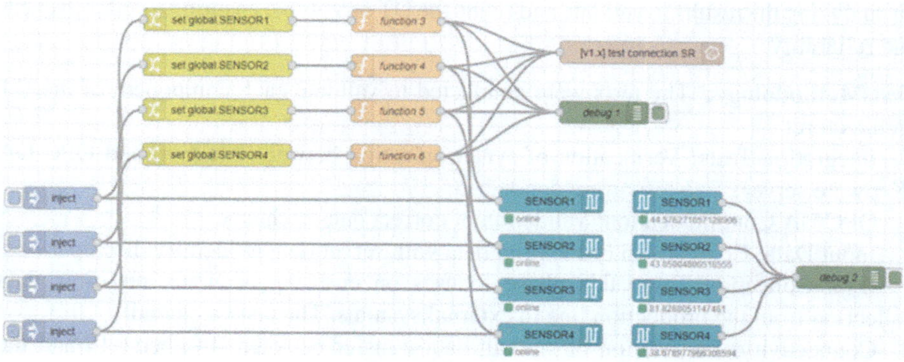

Fig. 4. Connection Flow Diagram in Node-RED

Data Organization: Dedicated databases were created in InfluxDB for each variable allowing for structured and organized data storage. The data flows configured in Node-RED ensure that each variable is directed to the appropriate database, thus maintaining data integrity and facilitating future analysis.

Visualization through Grafana:
Dashboard Creation: An interactive dashboard is built in Grafana to display collected data.

Real-Time Gauge Configuration: The gauges were configured to dynamically display.

real-time data. Data variables were linked to the corresponding InfluxDB databases, enabling Grafana to automatically retrieve the latest recorded data. Automatic updates at regular intervals were implemented to ensure the gauges always reflect the most recent information.

Real-Time Data Interaction and Intuitive Visual Interface: The dynamic Grafana panels were configured to display the latest sensor data in real time, allowing users to instantly monitor variations. The thoughtful design of the panels, combined with Grafana's user-friendly interface, provided an intuitive visual experience. Users were able to interact easily with the dashboard to monitor the sensors in real time without requiring advanced technical skills. Additionally, the clear organization of the panels made it easy to navigate between the different sensor variables.

We successfully implemented the key components of our DT prototype:

Sensor simulation using the PLC model and S7-PLCSim Advanced.

Real-time data collection and storage using Node-RED and InfluxDB.

Dynamic data visualization through interactive panels in Grafana.

These features form the foundation of our prototype, providing a solid infrastructure for monitoring, data collection, and visualization in a simulated industrial environment.

2.4 Tests and Results

With the prototype established, we now proceed to the testing phase, where we will rigorously assess its performance and accuracy in data simulation. This phase will focus on analyzing the results to validate and refine the DT prototype, ensuring its effectiveness and reliability.

Test Methodology: The tests were conducted to validate each component of the DT architecture:

Integration Tests: Verification of synchronization between FlexSim and OPC UA (Kepware server).Two tests were conducted:

OPC UA Communication: Validation of correct data exchange.

Input Data: Comparison of FlexSim data with server data to identify discrepancies.

Data Collection Tests: this phase focuses on ensuring accurate, real-time data collection from the production line to external systems. The tests included:

Connectivity: Verification of smooth data transfer between Raspberry Pi and the Kepware server via Node-Red.

Collection Frequency: Tests to determine the optimal collection rate for accurately representing production operations.

Data Storage Tests: This phase aimed to evaluate the performance of data storage in InfluxDB, focusing on:

Storage schema: Ensuring efficient data retrieval for analysis.

Load tests: Injecting large volumes of data to assess the system's ability to handle the load and maintain acceptable response times.

Dashboard Visualization Tests: This phase evaluated the relevance of the Grafana dashboard.

User Interface: Checking the clarity of the panels to ensure the data is easily readable.

Refresh Rate: Adjusting the data refresh frequency to ensure real-time updates.

Performance and Analysis Tests: This phase used the collected data to analyze production performance:

OEE Calculation: The Overall Equipment Efficiency (OEE) indicator was calculated to identify areas for improvement.

Identification of Downtime: Downtime periods were identified to analyze root causes and propose corrective actions.

Test Configuration: The hardware and software configuration for testing the DT implementation was carefully set up to ensure smooth and accurate interaction be-tween the different technologies. The configuration relies on the following components: FlexSim for simulation, OPC UA via Kepware for communication, Node-RED for data transfer, InfluxDB for storage and Grafana for visualization.

Results and Observations: The results demonstrate:

Successful integration: Reliable synchronization with transmission times between 20 to 23 Ms.

Accurate collection: Node-RED ensures real-time communication without data loss (As shown in Fig. 5).

Efficient storage: InfluxDB handles large volumes with quick retrieval times.

High performance visualization: Grafana displays real-time metrics, allowing for smooth monitoring.

High performance visualization: Grafana displays real-time metrics, allowing for smooth monitoring (as shown in Fig. 6).

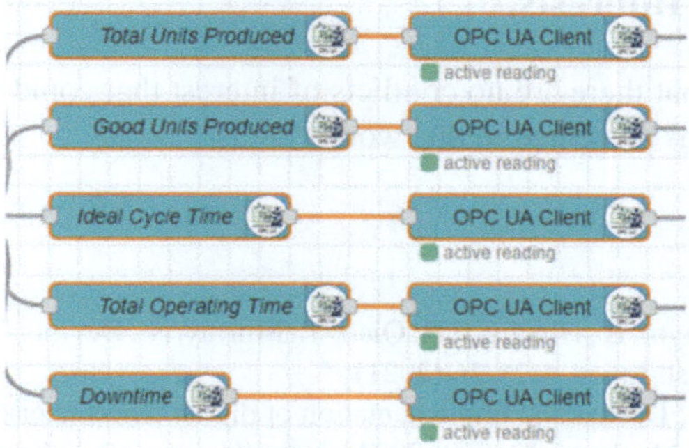

Fig. 5. Results of Data Collection Tests from an OPC UA Server Using Node-RED

Fig. 6. Live Performance Measurements: Grafana Dashboard with 5-Second Synchronization

The tests revealed a significant impact on production with real-time access to performance, downtime, and bottlenecks, enabling informed decisions to improve efficiency. The test results confirmed the effectiveness of the DT implementation, integrating OPC UA, real-time data collection, InfluxDB storage, and Grafana visualization to optimize production. The system's reliability and robustness were proven, with stable communication and efficient data management, even under heavy load.

3 Conclusions and Perspectives

This work presented a general architecture and key tools and technologies to implement DT. Based on this architecture, the project demonstrated a successful implementation of a DT in a production line using Raspberry Pi, Node-RED, InfluxDB, and Grafana. Key achievements include real-time monitoring, data storage, visualization, and performance optimization. DTs enhance process visibility, data-driven decision-making, and optimization opportunities.

However, challenges such as scalability, data security, and system integration need to be addressed for broader adoption.

In the future, improvements could focus on better integration with IoT ecosystems and enhanced data security. Furthermore, developing adaptive and user-friendly interfaces may facilitate interactions between operators and DT systems.

4 Disclosure of Interests.

The authors affirm that there are no conflicts of interest that could have influenced the content or conclusions presented in this work.

References

1. Singh, M., Fuenmayor, E., Hinchy, E.P., Qiao, Y., Murray, N., Devine, D.: Digital twin: origin to future. Appl. Syst. Innov. **4**(2), 36 (2021)
2. Cho, Y., Noh, S.D.: Design and implementation of digital twin factory synchronized in real-time using MQTT. Machines **12**, 759 (2024). https://doi.org/10.3390/machines12110759
3. Fuller, A., Fan, Z., Day, C., Barlow, C.: Digital twin: enabling technologies, challenges and open research. IEEE Access **8**, 108 952–108 971 (2020)
4. Rasheed, A., San, O., Kvamsdal, T.: Digital twin: Values, challenges and enablers from a modeling perspective. IEEE Access **8**, 21 980–22 012 (2020)
5. Nativi, S., Delipetrev, B., Craglia, M.: Destination Earth survey on "Digital Twins" technologies and activities in the Green Deal area (2020)
6. Autiosalo, J., Vepsäläinen, J., Viitala, R., Tammi, K.: A feature-based framework for structuring industrial digital twins. IEEE Access **8**, 1 193–1 208 (2020). https://doi.org/10.1109/ACCESS.2019.2950507
7. Pires, F., Souza, M., Ahmad, B., Leitão, P.: Decision support based on digital twin simulation: a case study. In: Service Oriented, Holonic and Multi-Agent Manufacturing Systems for Industry of the Future: Proceedings of SOHOMA 2020, pp. 99–110. Springer, Cham (2021) . https://doi.org/10.1007/978-3-030-69373-2_6
8. Pop, E., Iliuţă, M. E., Moisescu, M. A.: Digital twin architecture based on services interoperability. In: 2024 15th International Conference on Communications (COMM), pp. 1–6. IEEE (2024). https://doi.org/10.1109/COMM62355.2024.10741458
9. Inter-American Development Bank: Digital Twins. http://publications.iadb.org/publications/english/document/Tech-Report-Digital-Twins-ENG.pdf. Accessed 16 Apr 2025
10. National Institute of Standards and Technology: Conceptual Architecture of Digital Twins with Human-in-the-Loop-Based Smart Manufacturing. https://www.nist.gov/publications/conceptual-architecture-digital-twins-human-loop-based-smart-manufacturing. Accessed 16 Apr 2025
11. MDPI: Digital Twins in Manufacturing: A RAMI 4.0 Compliant Concept. https://www.mdpi.com/2413-4155/5/4/40. Accessed 16 Apr 2025
12. MDPI: Digital Twin Applications in Manufacturing Industry: A Case Study. https://www.mdpi.com/1999-5903/15/9/282. Accessed 16 Apr 2025
13. ResearchGate: A Manufacturing Digital Twin Framework. https://www.researchgate.net/publication/378099287_A_Manufacturing_Digital_Twin_Framework. Accessed 16 Apr 2025
14. Digital Twin Hub: Integration Architecture Pattern and Principles. https://digitaltwinhub.co.uk/download/integration-architecture-pattern-and-principles/. Accessed 16 Apr 2025
15. Plant Engineering: Why will intelligent digital twins become an industrial manufacturing must-have? https://www.plantengineering.com/why-will-intelligent-digital-twins-become-an-industrial-manufacturing-must-have/. Accessed 16 Apr 2025
16. National Institute of Standards and Technology: Digital Twins for Advanced Manufacturing. https://www.nist.gov/programs-projects/digital-twins-advanced-manufacturing. Accessed 16 Apr 2025
17. University of Twente Student Theses: Comparison between Digital Twin platforms (n.d.). https://essay.utwente.nl/105116/. Accessed 16 Apr 2025

AI-Driven Customer Segmentation for Enhanced Loyalty Programs

Iman Squalli Houssaini(✉) and Daoud Miloud

Laboratory of Research and Studies in Management, Entrepreneurship, and Finance (LAREMEF), National School of Business and Management (ENCG-FEZ), Fes, Morocco
Hsqualli.iman@gmail.com

Abstract. Loyalty programs have evolved into essential tools for customer retention, yet traditional segmentation methods often fail to capture the complexity and dynamism of customer behaviors. This study leverages AI-driven clustering techniques to enhance loyalty program effectiveness in the banking sector. We identify actionable customer segments by applying Principal Component Analysis (PCA) for dimensionality reduction and comparing the performance of KMeans and Gaussian Mixture Models (GMM) on a large-scale transaction dataset. Our findings reveal that K-Means significantly outperforms GMM, achieving a Silhouette Score of 0.5500 versus 0.3750, indicating more clearly defined and well-separated clusters. The identified segments provide valuable insights for targeted marketing strategies, optimizing resource allocation, and improving personalized engagement. This research offers a robust framework for transitioning from static segmentation models to adaptive, data-driven customer relationship management approaches, ensuring higher retention and profitability in competitive banking environments.

Keywords: AI-driven segmentation · loyalty programs · machine learning · clustering algorithms · customer retention · PCA · K-Means · Gaussian Mixture Models · banking sector · personalized marketing

1 Introduction

Marketing strategies have evolved significantly in recent years, increasingly focusing on customer-centric approaches prioritizing personalization and engagement. Loyalty programs are among the most impactful tools designed to reward and retain customers while fostering long-term relationships. These programs enhance customer satisfaction and play a crucial role in maximizing revenue and strengthening brand loyalty. These programs have proven indispensable in competitive industries such as banking and retail, where customer retention is critical. However, their effectiveness heavily relies on a deep understanding of customer behavior, a task traditionally challenged by the complexity and sheer volume of available data.

Loyalty programs transcend mere customer rewards; they serve as strategic tools to forge stronger connections between businesses and their clientele. These programs must

B.-E. Boudriki Semlali et al. (Eds.): DATA 2025, LNISO 81, pp. 209–221, 2025.
https://doi.org/10.1007/978-3-032-06725-8_18

go beyond static, one-size-fits-all solutions to deliver value that resonates with individual preferences and needs. The capacity to customize loyalty initiatives for specific customer segments distinguishes successful programs from average ones. For example, budget-conscious customers may respond well to discounts and cashback offers, while premium spenders might value exclusive memberships or early product access. As Kumar and Reinartz (2016) highlight, the future of loyalty programs hinges on their dynamic adaptability to customer behaviors powered by real-time insights [1].

Despite their potential, traditional customer segmentation methods for loyalty programs often fail to capture the complexity and dynamism of customer behaviors. Such methods typically depend on static demographic data or heuristic rules, rendering them unable to adapt to evolving customer preferences. This shortcoming underscores the need for advanced, data-driven approaches to uncover hidden patterns and generate actionable insights for optimizing loyalty programs.

Customer segmentation serves as a cornerstone of modern marketing. It allows businesses to group customers by shared characteristics, including demographics, purchasing behaviors, and preferences. Effective segmentation empowers marketers to design personalized strategies that resonate with each group, enhancing customer engagement and optimizing resource allocation. However, traditional segmentation methods often struggle to handle modern datasets' complexity and high dimensionality.

In industries like banking, where transaction histories and behavioral data offer a wealth of information, the shortcomings of traditional methods become particularly apparent. Studies have shown that clustering algorithms, such as K-Means and hierarchical clustering, are practical for customer segmentation, but their performance diminishes when applied to large, complex datasets [2, 3]. This is often due to the "curse of dimensionality," where adding multiple variables reduces the interpretability and quality of clusters. Dimensionality reduction techniques, such as Principal Component Analysis (PCA), address this issue by simplifying datasets while preserving their most essential features [4].

The rise of artificial intelligence (AI) has transformed both customer segmentation and the design of loyalty programs. Unlike traditional methods that depend on predefined rules and static models, AI-driven approaches leverage machine learning algorithms to analyze vast datasets, uncovering previously undetectable patterns and relationships. Such insights empower businesses to develop highly targeted and personalized loyalty initiatives, enhancing customer satisfaction and business outcomes.

For example, AI-based clustering techniques like Gaussian Mixture Models (GMM) and Deep Embedded Clustering (DEC) surpass simple grouping by incorporating the probabilistic nature of customer behaviors. These techniques can handle complex, high-dimensional datasets and provide actionable insights beyond the capabilities of traditional methods. Research by Ding and He (2004) and Syah *et al.* (2024) underscores the superiority of AI-driven approaches in customer segmentation and behavior prediction, emphasizing their transformative potential for marketing strategies [5, 6].

Moreover, AI facilitates real-time analysis, enabling businesses to dynamically adapt loyalty programs to evolving customer preferences. AI-powered predictive models can forecast churn and reveal upselling opportunities, empowering companies to take proactive measures. In contrast, traditional methods typically rely on retrospective analysis, which limits their capacity to respond effectively to shifts in customer behavior.

This study uses a large-scale banking transaction dataset to integrate AI-driven clustering techniques into loyalty program design. The research identifies actionable customer segments to inform loyalty program strategies by combining PCA for dimensionality reduction with advanced clustering algorithms like DEC and GMM. The objective is to close the gap between static traditional methods and dynamic AI-driven approaches, showcasing how AI enhances personalization, optimizes resource allocation, and maximizes the impact of loyalty programs.

The findings of this study contribute to the expanding body of knowledge on AI applications in marketing, offering a practical framework for businesses aiming to leverage advanced analytics in loyalty initiatives. This research provides a roadmap for developing more adaptive and impactful marketing strategies by tackling the challenges posed by high-dimensional data and dynamic customer behaviors.

2 Methodology

2.1 Dataset Exploration

The dataset utilized in this study, titled "Bank Customer Segmentation", was obtained from Kaggle and comprises over 1 million transaction records. It encompasses demographic attributes (e.g., age, gender, and income), behavioral metrics (e.g., transaction frequency and channel preference), and financial details (e.g., average spending). This comprehensive and multidimensional dataset presents significant opportunities for uncovering hidden customer insights through advanced AI techniques.

Before applying AI-driven clustering techniques, a thorough Exploratory Data Analysis (EDA) was conducted to understand the dataset's structure and identify key behavioral patterns. This step was critical in establishing a robust foundation for the AI models to extract actionable insights effectively. The main findings from the EDA process are summarized below (Fig. 1).

Customer Age Distribution. The dataset revealed that 68% of customers are between 30 and 50 years old, peaking around 40. Interestingly, individuals over 60 account for only 8% of the dataset, indicating an underrepresented demographic. This distribution underscores the bank's core customer base, predominantly middle-aged individuals.

Gender Representation. The gender distribution shows that 72% of the customer base is male, whereas females constitute only 28%. This significant disparity highlights the potential need for gender-specific marketing strategies to achieve a more balanced demographic representation (Fig. 2).

Median Transaction Behavior. An analysis of transaction behaviors across months revealed significant seasonality. June demonstrated a marked peak in median transaction amounts, indicating a clear seasonal trend in spending behavior. This insight can be utilized to optimize the timing of promotional campaigns or loyalty programs (Fig. 3).

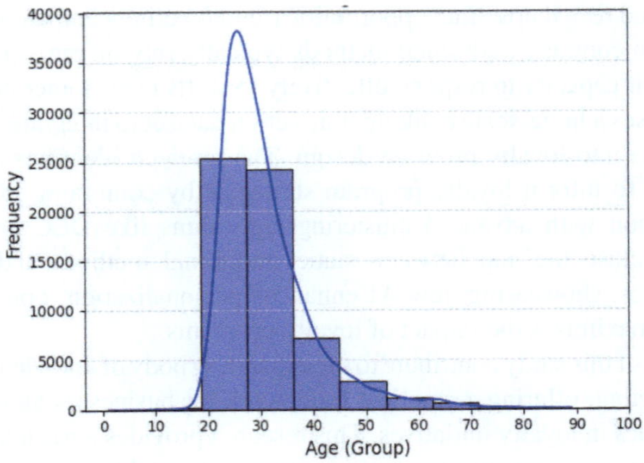

Fig. 1. Customer Age Distribution

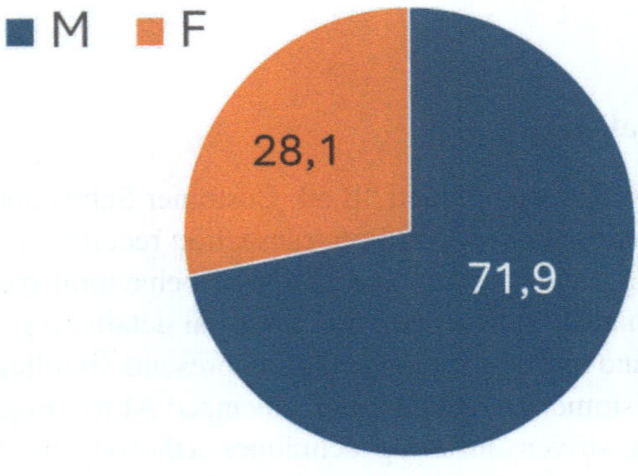

Fig. 2. Customer Gender Distribution

Correlation Analysis. A heatmap of the correlation matrix offered valuable insights into inter-variable relationships. It revealed a weak positive correlation (r = 0.10) between customer age and account balance, suggesting that older customers tend to maintain slightly higher balances. Conversely, transaction month and other variables exhibited minimal correlations, indicating diverse transactional behavior across customer groups (Fig. 4).

Key Behavioral Insights:

- Age-Specific Trends: Customers aged 30–50 constitute most of the dataset, indicating a strong middle-aged customer base, while older customers are underrepresented;
- Income Diversity: The average annual income among customers is approximately $40,000, with a standard deviation of $15,000, indicating significant diversity in income levels across the dataset.
- Spending Patterns: Customers exhibit two distinct behaviors:

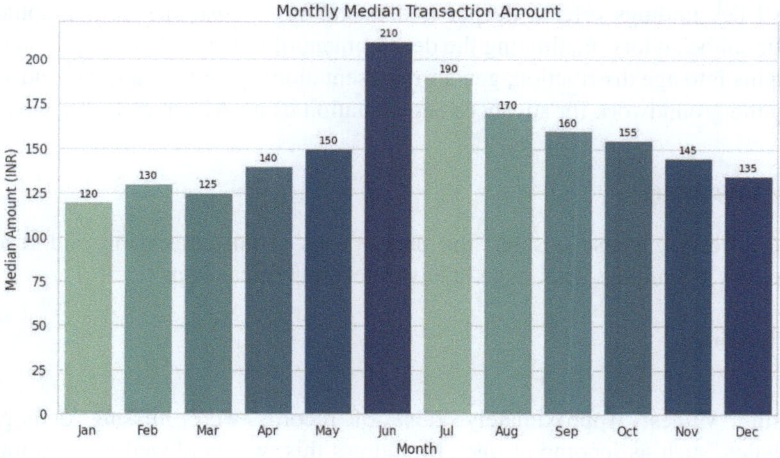

Fig. 3. Median Transaction Amount by Month

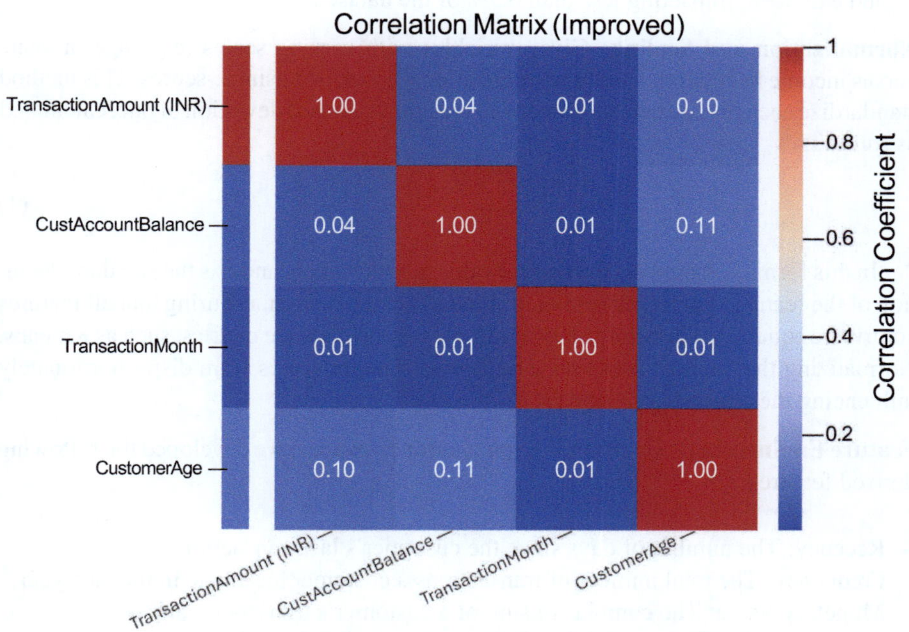

Fig. 4. Correlation Matrix

- High-Frequency, Low-Value Transactions: Nearly 45% of customers engage in frequent, smaller transactions, reflecting everyday spending habits;
- Sporadic High-Value Transactions: Approximately 15% of customers conduct fewer but larger transactions, possibly indicating luxury or one-time purchases.

The EDA findings offer a deeper understanding of customer demographics and transactional behaviors, facilitating the development of tailored clustering strategies.

Insights into age distribution, gender representation, spending patterns, and seasonality lay the groundwork for advanced segmentation using AI-driven techniques.

2.2 Preprocessing

The preprocessing phase ensured the dataset was clean, consistent, and ready for advanced AI techniques. This stage involved several critical steps.

Data Cleaning:

- Missing Values: Approximately 2.3% of records were missing demographic attributes, such as income or age. To address this, we employed mean imputation for continuous variables and mode imputation for categorical variables.
- Outlier Removal: Spending values exceeding $100,000 were considered anomalies and excluded, impacting less than 0.5% of the dataset.

Normalization and Scaling. Given variables with varying scales (e.g., age in years versus income in dollars), feature normalization was applied using z-scores. This method standardize each feature to have a mean of zero and a standard deviation of one, calculated as follows:

$$z = \frac{x - \mu}{\sigma} \tag{1}$$

In this formula, x denotes the feature value, μ is the mean, and σ is the standard deviation of the feature. This transformation standardize the dataset, ensuring that all features contribute equally to clustering algorithms that rely on distance metrics, such as Kmeans. Normalizing the features prevents variables with larger scales from disproportionately influencing the clustering results [7].

Feature Engineering. To enhance segmentation relevance, we developed the following derived features:

- Recency: The number of days since the customer's last transaction;
- Frequency: The total number of transactions a customer has made in the past year;
- Monetary Value: The cumulative sum of a customer's transaction amounts;
- Customer Lifetime Value (CLV): Calculated as the product of the average transaction value, transaction frequency, and the predicted customer tenure.

These metrics align with the Recency-Frequency-Monetary (RFM) model in marketing analytics. The RFM model evaluates customer value based on three dimensions:

1. Recency: How recently a customer has made a purchase;
2. Frequency: How often a customer makes a purchase;
3. Monetary Value: How much money a customer spends on purchases.

By incorporating these features, we aim to improve the effectiveness of our segmentation strategies, enabling more targeted and personalized marketing efforts [8].

Dimensionality Reduction. This study compares Principal Component Analysis (PCA) and Independent Component Analysis (ICA) as dimensionality reduction methods to condense feature sets while preserving essential data variance. The objective was to maintain at least 90%, 95%, and 90% of the variance in the original dataset.

The dataset initially contained eight features, which were preprocessed by imputing missing values with the median and scaling using standard normalization. PCA was applied to determine the minimum number of components required to retain the desired levels of variance. ICA, which seeks to maximize statistical independence between components, was subsequently evaluated for consistency using the same number of components identified by PCA.

Our analysis revealed that PCA consistently reduced dimensionality more effectively, requiring fewer components to achieve the specified variance thresholds. For example:

- At 85% variance, PCA reduced the dataset to 4 components, while ICA required five components;
- At 95% variance, PCA achieved a reduction to 5 components, while ICA also maintained five components.

PCA emerged as the preferred method from these results due to its ability to efficiently capture variance while ensuring computational simplicity. In contrast, although practical, ICA demonstrated slightly higher sensitivity to feature scaling and data variability. PCA is particularly advantageous when preserving maximum interpretability and computational efficiency is critical, such as exploratory data analysis or real-time systems [9, 10].

The accompanying visualization (Fig. 5) illustrates the change in features before and after dimensionality reduction for both methods across the three variance thresholds. These findings underscore the utility of PCA in balancing variance retention and dimensionality reduction, particularly in high-dimensional datasets.

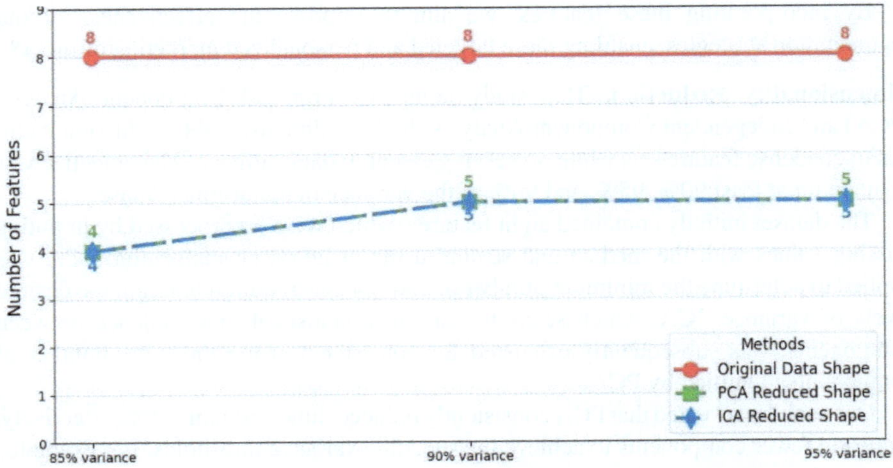

Fig. 5. Comparison of Data Shape Before and After Dimensionality Reduction Using PCA and ICA at 85%, 90%, and 95% Variance Thresholds

3 Results and Analysis

3.1 Clustering Performance

The clustering performance for the K-Means and Gaussian Mixture Model (GMM) algorithms was evaluated using the Silhouette Score as the primary metric. The results demonstrate a notable advantage of the KMeans algorithm, achieving a Silhouette Score of 0.5500, compared to 0.3750 for GMM. This significant difference indicates that K-Means formed clusters with higher intra-cluster cohesion and inter-cluster separation, making it more effective for segmenting the customer dataset [1, 2].

The PCA-based dimensionality reduction method used before clustering enhanced the efficiency and interpretability of both algorithms [3]. By reducing the dataset to its five principal components, 95% of the data variance was preserved, ensuring minimal information loss. These reduced dimensions simplified the clustering process, leading to more defined and actionable customer segments [4].

Figure 6 presents the Silhouette Scores for both models, highlighting the superior performance of K-Means. The disparity between the scores underscores the importance of selecting clustering methods that align with the dataset's characteristics [5].

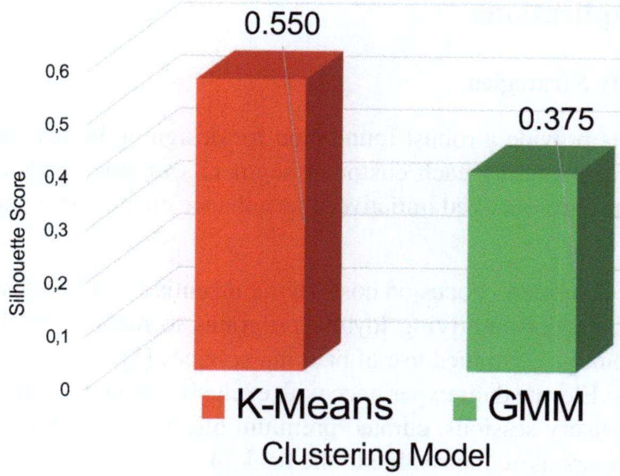

Fig. 6. Silhouette Scores for K-Means and GMM

3.2 Cluster Distribution and Profiles

Cluster Distribution. The distribution of clusters was visualized based on the first two principal components (PC1 and PC2). Figure 7 shows the cluster boundaries for KMeans, revealing well-separated groups with minimal overlap. In contrast, Fig. 8 illustrates the GMM cluster distribution, which exhibits significant overlap, particularly in regions with dense customer data points. This overlap contributes to the lower Silhouette Score observed for GMM [6].

Cluster Profiles. Three distinct customer clusters were identified using K-Means:

1. Budget Conscious Customers:

   - Characteristics: High transaction frequency with low transaction amounts;
   - Behavior: Price-sensitive customers, primarily engaged in daily banking activities;
   - Recommendation: Offer discounts, cashback rewards, and loyalty points to encourage continued engagement [7].

2. Premium Loyalists:

   - Characteristics: High-income customers making frequent high-value transactions;
   - Behavior: Strong brand loyalty and preference for premium services;
   - Recommendation: Introduce exclusive benefits such as VIP memberships, early product access, and personalized advisory services [8].

3. Occasional Spenders:

   - Characteristics: Customers with sporadic but high-value transactions;
   - Behavior: Engage primarily for luxury or one-time purchases;
   - Recommendation: Design re-engagement campaigns and exclusive one-time offers to foster consistent activity.

4 Marketing Implications

4.1 Targeted Loyalty Strategies

The clustering insights provide a robust foundation for designing highly targeted loyalty strategies. By understanding each customer segment's unique needs and behaviors, marketers can craft personalized initiatives that enhance engagement and customer satisfaction [10].

- Budget-conscious customers: Focus on cost-saving incentives such as discounts and cashback offers. Consider gamifying loyalty programs to make reward collection engaging and encourage continued use of banking services [3].
- Premium Loyalists: Elevate their experience with exclusive services. Provide personalized financial advisory sessions, curated premium offerings, and access to higher memberships that emphasize their value to the bank [4].
- Occasional Spenders: Implement strategic re-engagement campaigns. Utilize personalized communication channels to remind them of the bank's services and offer exclusive deals that align with their sporadic spending behavior [5].

4.2 Resource Optimization

Allocating marketing resources efficiently is critical for maximizing return on investment. By assigning budgets proportional to the value and size of each cluster, banks can ensure optimal use of their resources. For instance:

- Direct most marketing efforts toward Budget-Conscious Customers and Premium Loyalists, given their higher activity levels and potential lifetime value [6];
- Cost-effective digital campaigns should maintain visibility among occasional spenders while leveraging data-driven insights to re-engage them periodically [7].

4.3 Enhancing Customer Experience

Integrating the findings into omnichannel marketing efforts can significantly improve the customer experience. Personalizing communication across email, mobile apps, and in-branch interactions ensures consistency and relevance [8]. Embedding these insights into customer relationship management (CRM) systems can also automate and refine engagement strategies [9].

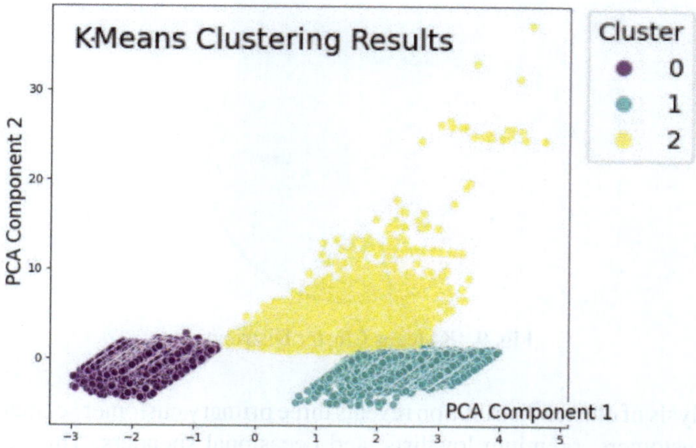

Fig. 7. Cluster Distribution (K-Means) based on PC1 and PC2

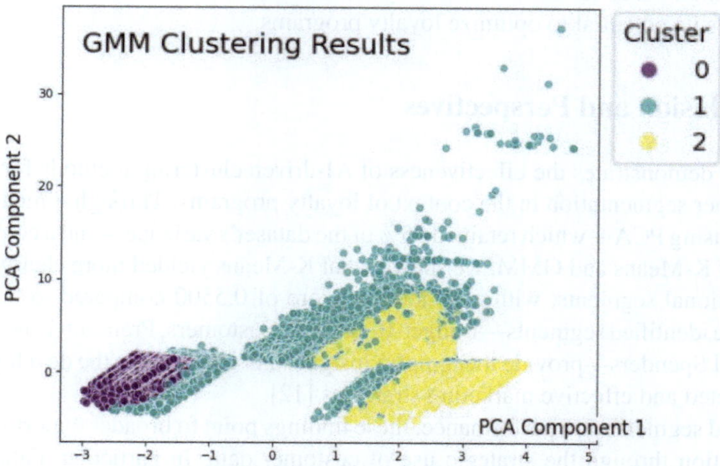

Fig. 8. Cluster Distribution (GMM) based on PC1 and PC2

The cluster distribution obtained using the K-Means algorithm highlights a well-structured segmentation of customers based on their transactional behavior. Visualizing the data in the principal component space reveals distinct groupings, reinforcing the effectiveness of this approach. Unlike the more dispersed distribution observed with the GMM model, K-Means clusters exhibit well-defined boundaries, facilitating interpretability and enabling actionable insights for personalized loyalty strategies [11] (Fig. 9).

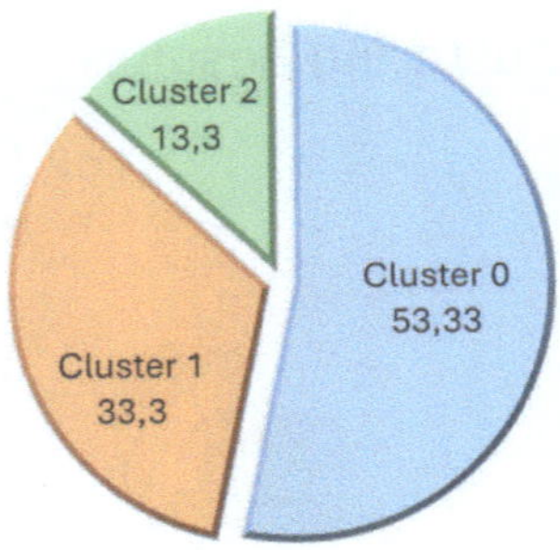

Fig. 9. KMeans Cluster Distribution

The analysis of cluster distribution reveals three primary customer segments: budget-conscious customers, premium loyalists, and occasional spenders. This segmentation enables the development of targeted marketing strategies tailored to each group's characteristics and preferences. For instance, budget-conscious customers benefit from discounts and promotional offers, whereas premium loyalists respond better to exclusive perks and high-value services. The ability of K-Means to differentiate these segments underscores its potential to optimize loyalty programs.

5 Conclusion and Perspectives

This study demonstrates the effectiveness of AI-driven clustering methods for improving customer segmentation in the context of loyalty programs. Through dimensionality reduction using PCA—which retained 95% of the dataset's variance—and a comparative analysis of K-Means and GMM, we showed that K-Means yielded more clearly defined and operational segments, with a Silhouette Score of 0.5500 compared to 0.3750 for GMM. The identified segments—Budget-Conscious Customers, Premium Loyalists, and Occasional Spenders—provide meaningful insights that can support the development of more targeted and effective marketing strategies [12].

Beyond segmentation performance, these findings point to broader opportunities for value creation through the strategic use of customer data. In particular, data monetization—understood as the extraction of actionable insights from behavioral patterns—emerges as a key lever. When combined with targeted advertising approaches grounded in behavioral understanding, such strategies can help financial institutions strengthen customer relationships, personalize services, and diversify revenue streams. Ultimately, these advances contribute to a more nuanced and data-informed perspective on customer engagement within the banking sector.

References

1. Kumar, V., Reinartz, W.: Creating enduring customer value. J. Mark. **80**(6), 36–68 (2016). https://doi.org/10.1509/jm.15.0414
2. Dolnicar, S., Grün, B.: Challenging "Factor–Cluster Segmentation." J. Travel Res. **47**(1), 63–71 (2008). https://doi.org/10.1177/0047287508318910

3. Jain, A.K.: Data clustering: 50 years beyond K-means. Pattern Recogn. Lett. **31**(8), 651–666 (2010). https://www.sciencedirect.com/science/article/pii/S0167865509002323. Accessed 28 Jan 2025
4. Ding, C., He, X.: K-means clustering via principal component analysis. In: Twenty-First International Conference on Machine Learning - ICML '04, Banff, Alberta, Canada, p. 29. ACM Press (2004). https://doi.org/10.1145/1015330.1015408
5. Abdi, H., Williams, L.J.: Principal component analysis. WIREs Comput. Stats. **2**(4), 433–459 (2010). https://doi.org/10.1002/wics.101
6. Syah, R.B., Muliono, R., Siregar, M.A., Elveny, M.: An efficiency metaheuristic model to predicting customers churn in the business market with machine learning-based. IAES Int. J. Artif. Intell. (IJ-AI) **13**(2), 1547 (2024). https://pdfs.semanticscholar.org/099e/f0d21dbfbc04a433688c71c96e747d336ad8.pdf. Accessed 28 Jan 2025
7. Wongoutong, C.: The impact of neglecting feature scaling in k-means clustering. PloS One **19**(12), e0310839 (2024). https://journals.plos.org/plosone/article?id=10.1371/journal.pone.0310839. Accessed 29 Jan 2025
8. Alves Gomes, M., Meisen, T.: A review on customer segmentation methods for personalized customer targeting in e-commerce use cases. Inf. Syst. E-Bus. Manage. **21**(3), 527–570 (2023). https://doi.org/10.1007/s10257-023-00640-4
9. Jolliffe, I.: Principal component analysis for special types of data. In: Principal Component Analysis. SSS, pp. 338–372. Springer, New York, NY (2002). https://doi.org/10.1007/0-387-22440-8_13
10. Hyvarinen, A., Karhunen, J., Oja, E.: Independent Component Analysis and Blind Source Separation. John Wiley & Sons, Hoboken, NJ (2001). http://research.jyu.fi/interbrain/presentations/Hyvarinen.pdf. Accessed 04 Feb 2025
11. Rahman, Md.M., et al.: Evaluating machine learning models for optimal customer segmentation in banking: a comparative study. Am. J. Eng. Technol. **6**(12), 68–83 (2024). Accessed 13 Jan 2025
12. Chitra, J., Heikal, J.: Customer segmentation using the k-means clustering algorithm in foreign banks in Indonesia. Indones. Account. Res. J. **11**(4), 230–241 (2024)

Evaluating the Benefits of Regulated and Sustained Deficit Irrigation for Yield, Water Use and Sustainability in the Moroccan Menara Olive Cultivar

Khaoula Ibba[1,2], Salah Er-Raki[1,3](✉), and Rachid Hadria[2]

[1] ProcEDE/AgroBiotech Center, Faculty of Sciences and Technologies, Cadi Ayyad University, Bd. Abdelkrim El Khattabi, B.P. 549 Guéliz, 40000 Marrakech, Morocco
s.erraki@uca.ma

[2] Regional Center of Agricultural Research of Marrakech, National Institute of Agricultural Research, Avenue Ennasr, P.O. Box 415, Rabat Principal, Rabat, Morocco
rachid.hadria@inra.ma

[3] Center for Remote Sensing Applications, Mohammed VI Polytechnic University, Lot 660, Hay Moulay Rachid, Ben Guerir 43150, Ben Guerir, Morocco

Abstract. Increasing aridity and global climate change are posing significant sustainability challenges to olive agroecosystems. Therefore, an experiment was conducted over two consecutive years in an olive orchard, comparing regulated (RDI) and sustained deficit irrigation (SDI) strategies. Alongside a fully irrigated control treatment (FI), T0 (100% ETc), four RDI regimes were tested: T1 (SP 100/NP 70% ETc), T2 (SP 100/NP 60% ETc), T3 (SP 80/NP 70% ETc), and T4 (SP 80/NP 60% ETc), based on the water stress sensitivity of phenological phases (SP for sensitive periods and NP for normal period). Two SDI regimes were also evaluated: T5 (70% ETc) and T6 (60% ETc). The study assessed the impact of irrigation strategies on shoot growth, oil content, fruit yield and water productivity (WP) over two years 2021 and 2022 on the olive cultivar Menara. Results showed that a 40% reduction in water supply (T6) severely restricted shoot growth, with a 44% decrease in 2022 compared to the control (T0). Oil content increased significantly with reduced irrigation, particularly in fruits from trees under RDI treatments T3 and T4, which showed a 12% increase without disrupting oil accumulation. RDI treatments (T3 and T4) exhibited the highest WP, outperforming SDI treatments (T5 and T6), which showed a 30–40% decline in WP compared to T0. Yield reductions were most pronounced under SDI, with T5 and T6 producing only 46.8% and 48.8% of the control yield, respectively. These findings highlight the significance of water management strategies and their effects on key agronomic parameters. Specifically, the results showed that moderate and controlled water stress under RDI, particularly in treatments T3 and T4, did not substantially impact fruit oil content or yield in the Menara olive cultivar.

Keywords: Olive Tree · Water Scarcity · Water Management · Regulated Deficit Irrigation · Sustained Deficit Irrigation

B.-E. Boudriki Semlali et al. (Eds.): DATA 2025, LNISO 81, pp. 222–233, 2025.
https://doi.org/10.1007/978-3-032-06725-8_19

1 Introduction

The olive tree (*Olea europaea* L.) is the most iconic and traditional crop in the Mediterranean Basin, particularly in Morocco. It is considered one of the best-adapted species to the Mediterranean-type climate [1]. However, agriculture in this region faces significant challenges due to water scarcity and the impacts of climate change. The Mediterranean Basin is recognized as a climate change "hotspot," with future projections indicating substantial warming and drying trends, which are expected to adversely affect crop production, including traditional olive orchards [2]. The growing challenge of producing more food amid increasing water scarcity underscores the need to enhance crop and water productivity. In this context, irrigation emerges as a key strategy for climate change adaptation [2, 3]. Improving water-use efficiency is essential in irrigated agriculture to reduce pressure on limited water resources. Deficit irrigation (DI) represents a promising approach to minimize water consumption while maintaining crop performance. Moreover, moderate deficit irrigation strategies have been shown to enhance crop yield, oil yield, fruit quality, and overall water productivity [4, 5]. Among deficit irrigation strategies, regulated deficit irrigation (RDI) and sustained deficit irrigation (SDI) are particularly noteworthy [5, 6]. SDI involves the application of a constant water volume below the crop's evapotranspiration demand throughout the whole irrigation season. RDI involves inducing water stress during growth stages when olive trees are most tolerant while ensuring full irrigation during critical periods. Optimizing irrigation during sensitive phases, such as flowering and oil synthesis, is essential to mitigate the effects of water scarcity and sustain olive yield. Water stress during early growth stages can significantly hinder tree development and reduce productivity, leading to economic losses due to disruptions in flowering and fruit set. However, the pit hardening stage is considered the most tolerant to water deficits [5, 7, 8]. Ibba et al. [4, 5] reported that controlled water stress under regulated deficit irrigation (RDI) had minimal effects on the yield of *Menara* olives while reducing water use by 20% during sensitive growth stages and 40% during less critical periods. This strategy led to overall water savings of 25%–30% and improved water productivity by 5%–20%, with only a slight yield reduction of 10%–15%. In contrast, trees subjected to the sustained deficit irrigation (SDI) strategy exhibited reduced shoot growth compared to those under RDI and fully irrigated conditions. The ultimate goal is to improve the crop's adaptation to water deficits while enhancing productivity, benefiting both growers and stakeholders. To address this objective, the present study evaluates the impact of regulated deficit irrigation (RDI) and sustained deficit irrigation (SDI) strategies on the Menara cultivar in Morocco over two consecutive years (2021 and 2022). The analysis focuses on key agronomic parameters that have a significant impact on farmers, providing essential insights into efficient water management strategies for olive cultivation.

2 Materials and Methods

2.1 Experimental Site and Climate

The irrigation experiment was conducted over two cropping seasons (2021 and 2022) in a 12-year-old 'Menara' olive orchard at the Saada Research Station (32°N, 8°8′W; 411 m altitude), INRA Marrakech, Morocco. The orchard was planted at a density of 156 trees per hectare, with an 8 × 8 m spacing. The region has a Mediterranean climate with semi-arid conditions, characterized by hot, dry summers and irregular, low precipitation (Fig. 1). The cumulative reference evapotranspiration (ETo) was 1295 mm in 2021 and 1316 mm in 2022, while total precipitation (P) during the same periods was 112.6 mm and 110 mm, respectively.

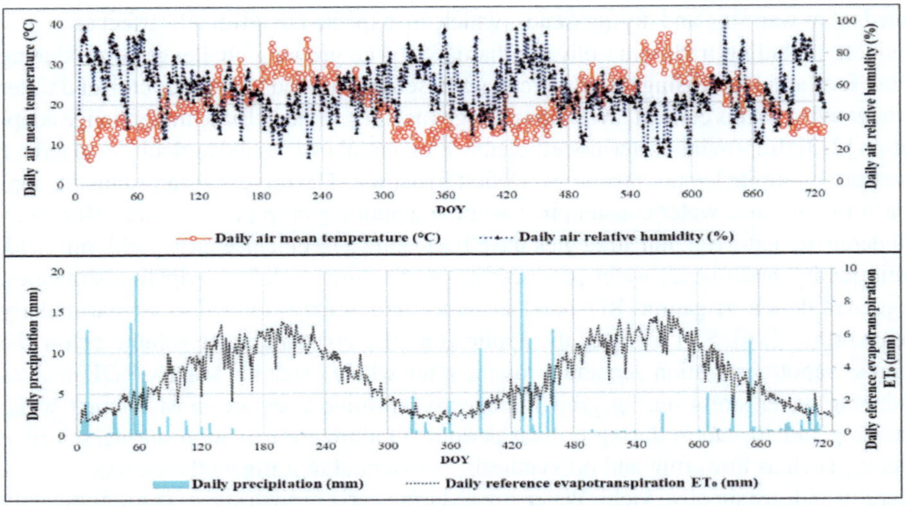

Fig. 1. Daily air temperature (°C), relative humidity (%), reference evapotranspiration ET0 (mm), and precipitation (mm) at the experimental site in 2021 and 2022. DOY represents the day of the year, where 1 corresponds to January 1, 2021

2.2 Experimental Design and Irrigation Treatments

This study implemented regulated deficit irrigation (RDI) strategies by applying different percentages of crop evapotranspiration (ETc) reduction during key phenological stages of olive tree development. During sensitive periods (SP1: flowering and SP2: oil synthesis to harvest), irrigation was set at either 100% or 80% ETc, while during normal period (NP: from pit hardening to the beginning of oil synthesis), irrigation was reduced to 70% or 60% ETc (Table 1). A fully irrigated control (T0, 100% ETc) was used for comparison. The study included four RDI treatments: T1 (SP 100%/NP 70% ETc), T2 (SP 100%/NP 60% ETc), T3 (SP 80%/NP 70% ETc), and T4 (SP 80%/NP 60% ETc). Additionally, two sustained deficit irrigation (SDI) treatments were applied, maintaining a constant irrigation supply throughout the whole irrigation season at fixed percentages of ETc: T5 (70% ETc) and T6 (60% ETc). Phenological development was monitored in both years following the methodology of Sanz-Cortés et al. [9].

Table 1. Treatments and irrigation water (in mm) applied during the growing seasons in 2021 and 2022.

Irrigation strategies	Treatments	% of ETc during flowering (SP1)	% of ETc from pit hardening to the beginning of oil synthesis (NP)	% of ETc from oil synthesis to harvest (SP2)	Total water delivered (mm)	
					2021	2022
Control	T0	100	100	100	242	304
RDI	T1	100	70	100	218	271
	T2	100	60	100	210	260
	T3	80	70	80	185	231
	T4	80	60	80	177	220
SDI	T5	70	70	70	169	213
	T6	60	60	60	145	182

2.3 Studied Parameters

To assess the impact of treatments on tree growth, vegetative development was monitored on six representative trees per treatment from bud break to early July. Four current-year shoots per tree were measured at 1.5 m above the ground, below the first-year branch insertion. Phenological shoot development was recorded in 2021 and 2022 at three key growth stages, following Sanz-Cortés et al. [9]: in March, when shoots reached approximately 30% of their final length; in June, at about 70%; and in July, upon reaching their final length.

Fruit yield (FY, kg/tree) was evaluated in 2021 and 2022 for each irrigation treatment. Measurements were conducted on twelve trees, and manual harvesting was performed. The total weight of the harvested fruit was recorded for each tree and subsequently averaged to determine the fruit yield (kg/tree).

Fruit oil content (OC) was measured using a laboratory scale system. The fruits were crushed using a hammer mill, and the resulting olive paste was homogenized and dried at 70 °C. The oil content was then determined using a Soxhlet apparatus with 30 °C hexane, as described by [10]. To comply with international standards, which express oil content in relation to fresh weight, the values were adjusted using the following formula 1:

$$\mathrm{OCFW} = \mathrm{OCDW} \times (100 - \mathrm{H})100 \tag{1}$$

where OCFW represents the oil content in relation to fresh weight, OCDW is the oil content in relation to dry weight, and H represents humidity (the percentage of water) [11].

To assess the effects of regulated and sustained deficit irrigation strategies on water productivity (WP), the latter was calculated based on data for fruit yield and total water use. Harvesting was conducted on 12 trees per irrigation treatment to determine fruit yield (kg/tree). Water productivity in olive trees (WP, kg/m^3) was defined as the ratio of fruit yield (FY, kg/tree) to total water use (WU, m3/tree), following the Eq. 2 presented by Ray et al. [12]:

$$WP\left(kg/m^3\right) = FY/\,WU \tag{2}$$

Water use (WU) was calculated by considering precipitation (P) and irrigation (I) throughout the whole cropping cycle in each year, as previously proposed by Ma et al. [13] and González-Fernández et al. [14]: WU = I + P.

2.4 Statistical Analysis

Statistical analyses were performed to assess differences among irrigation treatments for each season. An analysis of variance (ANOVA) was conducted for all evaluated parameters to determine the effects of different deficit irrigation treatments. Tukey's multiple range test was used to determine significant differences among the treatments at a significance level of $P < 0.05$.

3 Results and Discussion

Table 2. Effect of regulated and sustained deficit irrigation strategies on vegetative growth across the vegetative season from March to July. The obtained results are statistically significant with $p < 0.001$

Irrigation strategies	Treatments	Vegetative growth (cm)23/03/2021	Vegetative growth (cm)01/06/2021	Vegetative growth (cm)07/07/2021
Control1	T0 (100% ETc)	5.53 ± 0.96a	7.80 ± 1.12bc	9.31 ± 1.58d
RDI1	T1 (100/70% ETc)	6.48 ± 1.52b	8.72 ± 1.82c	9.42 ± 1.62d
	T2 (100/60% ETc)	5.45 ± 0.88a	7.31 ± 1.09ab	8.50 ± 1.04bcd
	T3 (80/70% ETc)	5.73 ± 1.08ab	7.62 ± 1.07ab	8.88 ± 1.31cd
	T4 (80/60% ETc)	5.23 ± 0.76a	7.08 ± 1.14ab	8.02 ± 1.46abc
SDI1	T5 (70% ETc)	5.15 ± 0.62a	6.82 ± 0.66a	7.42 ± 0.67ab
	T6 (60% ETc)	5.30 ± 0.72a	6.65 ± 0.63a	7.23 ± 0.57a
	ANOVA	5.14***	9.68***	11.78***
		Vegetative growth (cm)23/03/2022	Vegetative growth (cm)07/06/2022	Vegetative growth (cm)15/07/2022
Control2	T0 (100% ETc)	3.18 ± 1.04c	7.71 ± 2.56c	8.88 ± 3.39b

(continued)

Table 2. (*continued*)

Irrigation strategies	Treatments	Vegetative growth (cm)23/03/2021	Vegetative growth (cm)01/06/2021	Vegetative growth (cm)07/07/2021
RDI2	T1 (100/70% ETc)	2.41 ± 0.47abc	7.41 ± 1.47bc	8.43 ± 2.05b
	T2 (100/60% ETc)	2.63 ± 0.88bc	6.52 ± 2.32abc	7.25 ± 2.97ab
	T3 (80/70% ETc)	2.29 ± 1.12ab	5.19 ± 3.05ab	5.38 ± 3.21a
	T4 (80/60% ETc)	2.48 ± 0.59abc	6.00 ± 2.24abc	6.52 ± 2.80ab
SDI2	T5 (70% ETc)	2.00 ± 0.68ab	5.94 ± 2.09abc	6.60 ± 2.71ab
	T6 (60% ETc)	1.74 ± 0.55a	4.61 ± 1.16a	5.13 ± 2.60a
	ANOVA	7.46***	4.96***	4.84***

The data present means ± standard deviation. The different letters in column indicate statistically significant difference among the different treatment ($p < 0.05$).

1 year of 2021.

2 year of 2022.

The effects of RDI and SDI strategies significantly influence branch growth. During the vegetative growth season (March–July), differences in growth rates were observed (Table 2). Phenological shoot development in 2021 and 2022 was assessed at three key dates corresponding to principal growth stages as defined by Sanz-Cortés et al. [9] in March, when shoots reached approximately 30% of their final length; in June, when they attained nearly 70%; and in July, when young shoots reached their final length. Shoot length in trees under full irrigation (T0) was approximately 9.3 cm in 2021 and 8.9 cm in 2022. The results indicate a significant effect of irrigation strategy on average shoot growth across both years, as confirmed by ANOVA tests ($\alpha = 0.05$, $p < 0.001$). The greatest reduction in shoot growth was observed in T6 at full shoot elongation, with a decrease of 22% in 2021 and 44% in 2022 compared to the control (T0). This suggests that a 40% reduction in water supply severely restricted vegetative growth in 2022. Accordingly, sustained deficit irrigation (SDI) was strongly associated with reduced shoot elongation [8]. No significant differences were detected among T0, T1, and T2, indicating that reducing irrigation by 30% or 40% during the NP period did not negatively affect branch growth. Several studies have reported that RDI strategies enable water savings while maintaining comparable or even higher production than full irrigation, thereby enhancing water use efficiency [15]. These findings highlight the potential of RDI strategies to optimize water use while maintaining satisfactory vegetative growth, which is essential for sustaining olive productivity under water-limited conditions. The problems in deficit irrigation of perennial crops may be the enduring effects of stress that last longer than the current season and often become detrimental in the next years [7].

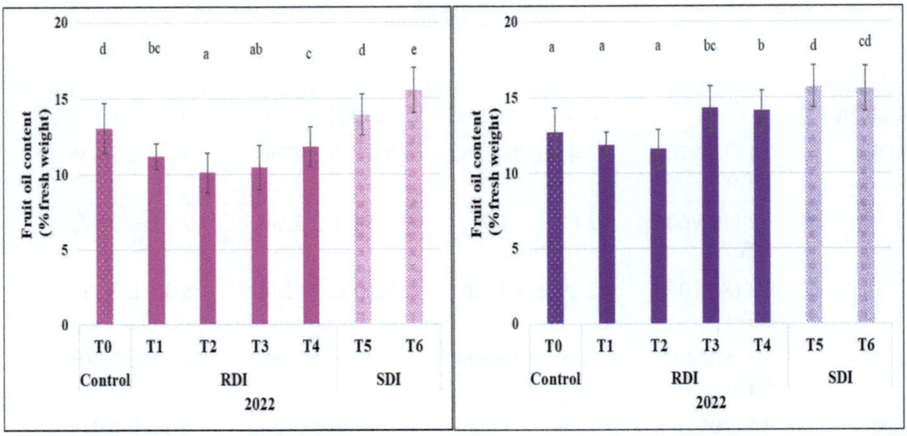

Fig. 2. Effect of regulated and sustained deficit irrigation on oil content over fresh weight, expressed as a percentage for each deficit irrigation treatment in 2021 and 2022.

The fruit oil content, measured as a percentage of fresh weight, was significantly affected by deficit irrigation strategies (Fig. 2). ANOVA results, assuming a significance level of $\alpha \leq 0.05$, revealed an extremely significant variation in the average fruit oil content among irrigation treatments in 2021 and 2022 ($p < 0.001$). In 2021, the relationship between oil content and irrigation regimes was weak. However, in 2022, oil content increased significantly with decreasing irrigation water. Specifically, fruits from trees subjected to the SDI treatments (T5 and T6) exhibited a 23% increase in oil content compared to the fully irrigated control (T0), while the T3 and T4 treatments under the RDI strategy showed an increase of approximately 12%. This suggests that fruit oil content tends to be lower under full irrigation, likely due to competition between oil and water within cellular spaces, resulting from their opposite polarity [5]. The SDI treatments (T5 and T6) and the RDI treatments (T3 and T4) did not disrupt the oil accumulation process, aligning with findings from previous studies [16, 17].

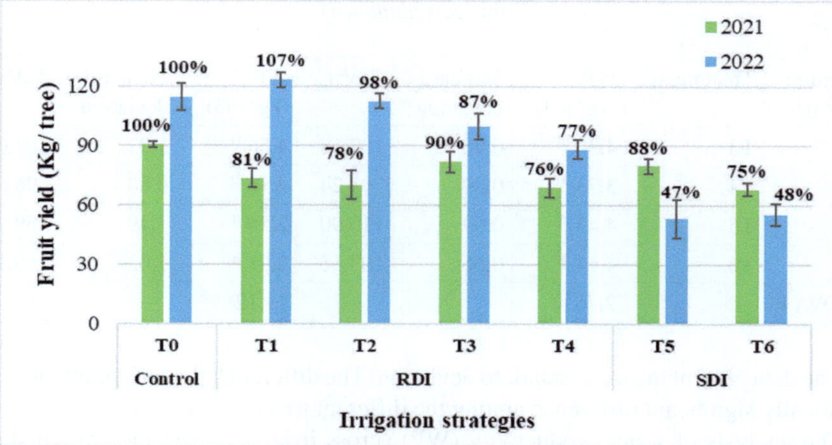

Fig. 3. Effect of regulated and sustained deficit irrigation on fruit yield (kg/tree) and the percentage variation in fruit yield (%) across different deficit irrigation treatments compared to the control T0.

Figure 3 presents the fruit yield (kg/tree) and its percentage variation relative to T0 across different irrigation treatments (T0–T6) over two consecutive years (2021–2022). Significant differences were observed among treatments, indicating that the applied irrigation treatments and strategies significantly influenced fruit yield, as shown by the ANOVA test with $\alpha < 0.05$ ($p < 0.01$ in 2021, $p < 0.001$ in 2022. In 2021, both RDI and SDI treatments resulted in reduced fruit yields, with the lowest yield recorded under treatment T6, representing only 75.23% of the control yield. However, in 2022, treatments T1 and T2 (RDI) showed substantial yield increases, indicating that full irrigation during sensitive growth stages supports fruit production. In contrast, treatments T5 and T6 (SDI) exhibited the most pronounced yield reductions, producing only 46.8% and 48.8% of the control yield, respectively. Correa-Tedesco et al. [18]. Applied 66% of evapotranspiration (ETc) over the growing season in a table olive cultivar, *Manzanilla Fina*, which resulted in mild water stress and a 25% reduction in fruit yield compared to well-watered trees. In a similar context, Ibba et al. [5] found that moderate and controlled water stress under regulated deficit irrigation (RDI) had no significant effect on the yield of the *Menara* olive cultivar.

Table 3. Effect of regulated and sustained deficit irrigation on water productivity (WP), with "% Var" representing the percentage variation in water productivity relative to the control T0.

Irrigation strategies	Treatments	WP (kg/ m3)	Standard deviation	% Var	WP (kg/ m3)	Standard deviation	% Var
		2021			2022		
Control	T0	4,00 ^{ab}	0,25	100,00	4,31 ^{b}	0,77	100,00
RDI	T1	3,49 ^{ab}	0,78	87,28	5,05 ^{b}	0,53	117,11
	T2	3,41 ^{a}	1,23	85,21	4,75 ^{b}	0,54	110,00

(continued)

Table 3. *(continued)*

Irrigation strategies	Treatments	WP (kg/ m3)	Standard deviation	% Var	WP (kg/ m3)	Standard deviation	% Var
	T3	4,30 ^{ab}	0,96	107,68	4,56 ^{b}	1,03	105,65
	T4	3,73 ^{ab}	0,89	93,24	4,18 ^{b}	0,80	96,84
SDI	T5	4,43 ^{b}	0,69	110,90	2,59 ^{a}	1,49	59,93
	T6	4,14 ^{ab}	0,66	103,56	2,98 ^{a}	1,02	69,18
ANOVA F		2,76**			11,12***		

The data present means ± standard deviation. The different letters in column indicate statistically significant difference among the different treatment ($p < 0.05$).

The analysis of water productivity (WP) across irrigation strategies revealed significant differences among treatments in both 2021 (**p** ≤ 0.01) and 2022 (**p** ≤ 0.001) (Table 3). In 2021, the highest WP was recorded under the SDI treatment T5 (4.43 kg/m^3), representing a 10.90% increase compared to the control (T0), highlighting its efficiency in achieving 30% water savings while maintaining superior WP. However, in 2022, treatments under the RDI strategy exhibited the highest WP values (4.18–5.05 kg/m^3), outperforming those under the SDI strategy (T5 and T6), which showed a significant decline in WP (2.59–2.98 kg/m^3), representing a 30–40% reduction compared to T0. These findings suggest that RDI treatments promote better water productivity and fruit yield compared to SDI treatments, particularly under seasonal variations. The observed results are consistent with previous research indicating that water-saving irrigation practices can enhance WP [5, 19, 20]. To mitigate the risks associated with SDI, precise irrigation scheduling and accurate determination of irrigation timing and quantity are essential, especially during critical growth stages when the crop is most sensitive to water stress [4, 5, 7, 20].

The Menara cultivar demonstrated notable differences between the RDI and SDI treatments in terms of the intensity and timing of water deficit, especially during critical phenological stages. Trees receiving RDI experienced less water stress during the important spring to early summer period (80–100% of ETc), which helped them maintain better vegetative and reproductive performance. In contrast, the SDI treatments led to cumulative stress effects that significantly hindered growth and yield over time. These findings emphasize the importance of precise irrigation during the spring, a crucial time when physiological processes such as flowering, fruit set, and oil synthesis directly impact current and future productivity, especially in similar climatic regions, spring is crucial for olive production.

Deficit irrigation strategies show significant promise for improving water use efficiency and sustaining yields in olive orchards; however, their long-term effects in perennial systems are still not well understood. One major limitation of this study is its short observation period, which does not allow for capturing the full extent of Menara olive adaptation and possible cumulative stress effects over time. This constraint is typical of many existing studies, which often span only 2 to 4 years, limiting the ability to generalize results across different cultivars and environmental conditions. Therefore, long-term, multi-season trials involving a variety of cultivars are essential to assess the consistency of RDI and SDI outcomes and to evaluate any trade-offs in productivity. Moreover, the physiological and molecular mechanisms underlying plant responses to deficit irrigation remain largely unexplored, highlighting an important research gap that needs to be addressed to optimize RDI and SDI protocols. From an implementation perspective, the widespread adoption of RDI strategy is challenged by the economic and technical requirements of precise water management systems, such as drip irrigation and moisture monitoring devices. To overcome these barriers, future efforts should focus on developing cost-effective, crop-adapted irrigation technologies. Crop-specific RDI application protocols are urgently required to guide the development of tailored irrigation equipment and serve as reference models for broader adoption. In arid regions like Morocco, integrating RDI into flexible and adaptive management frameworks is essential to enhance the resilience of irrigation practices.

4 Conclusion

In conclusion, implementing sustainable water management practices in arid regions, especially in Morocco, has become a crucial priority in tackling the growing issue of water scarcity. The importance of efficient and sustainable irrigation strategies cannot be overstated, as they significantly enhance agricultural productivity while conserving precious water resources. This study offers a comprehensive assessment of the benefits associated with two deficit irrigation strategies: Regulated Deficit Irrigation (RDI) and Sustained Deficit Irrigation (SDI). By evaluating key agronomic parameters that impact producers and farmers, our findings underscore the potential of these strategies to optimize water use efficiency without greatly compromising yield. Applying moderate and controlled water stress through RDI (T3 and T4) improved water productivity, shoot growth, and oil content while maintaining fruit yield in the Menara olive cultivar. These results underscore the effectiveness of RDI as a sustainable irrigation strategy for water-scarce regions like Morocco. By optimizing water use, this approach supports the long-term sustainability of agricultural systems, ensuring their productivity and economic viability amid changing climatic conditions.

5 Disclosure of Interests.

The authors have no competing interests to declare that are relevant to the content of this article.

Acknowledgments. The authors acknowledge the National Institute of Agricultural Research (INRA) and the International Joint Laboratory Télédétection et Ressources en Eau en Méditerranée semi-Aride (LMI TREMA, https://www.lmi-trema.ma/) for their invaluable contributions to this research. The authors also acknowledge the financial support provided by the project RISE-H2020-ACCWA and PRIMA-BIOMEnext project.

References

1. Arenas-Castro, S., Gonçalves, J.F., Moreno, M., Villar, R.: Projected climate changes are expected to decrease the suitability and production of olive varieties in southern Spain. Sci. Total. Environ. **709**, 136161 (2020). https://doi.org/10.1016/j.scitotenv.2019.136161
2. Fraga, H., Moriondo, M., Leolini, L., Santos, J.A.: Mediterranean olive orchards under climate change : a review of future impacts and adaptation strategies. Agronomy **11**(1), 56 (2020). https://doi.org/10.3390/agronomy11010056
3. Villalobos, F.J., Testi, L., Hidalgo, J., Pastor, M., Orgaz, F.: Modelling potential growth and yield of olive (Olea europaea L.) canopies. Eur. J. Agron. **24**(4), 296-303 (2006). https://doi.org/10.1016/j.eja.2005.10.008
4. Ibba, K., Er-Raki, S., Bouizgaren, A., Hadria, R.: Effects of regulated and sustained deficit irrigation on water use, physiology and yield of 'Menara' olive trees. Morocco. Irrig. Sci. **42**(5), 829–848 (2024). https://doi.org/10.1007/s00271-024-00947-1
5. Ibba, K., et al.: Assessing the impact of deficit irrigation strategies on agronomic and productive parameters of Menara olive cultivar : implications for operational water management. Front. Environ. Sci. **11**, 1100552 (2023). https://doi.org/10.3389/fenvs.2023.1100552
6. Moriana, A., Orgaz, F., Pastor, M., Fereres, E.: Yield responses of a mature olive orchard to water deficits. J. Am. Soc. Horticult. Sci. **128**(3), 425-431 (2003). https://doi.org/10.21273/jashs.128.3.0425
7. Gucci, R., Caruso, G., Gennai, C., Esposto, S., Urbani, S., Servili, M.: Fruit growth, yield and oil quality changes induced by deficit irrigation at different stages of olive fruit development. Agric. Water Manag. **212**, 88–98 (2019). https://doi.org/10.1016/j.agwat.2018.08.022
8. Hueso, A., Camacho, G., Gómez-del-Campo, M.: Spring deficit irrigation promotes significant reduction on vegetative growth, flowering, fruit growth and production in hedgerow olive orchards (cv. Arbequina). Agric. Water Manage. **248**, 106695 (2021). https://doi.org/10.1016/j.agwat.2020.106695
9. Sanz-Cortés, F., et al.: Phenological growth stages of olive trees (Olea europaea). Ann. Appl. Biol. **140**(2), 151–157 (2002)
10. Donaire, J.P., Sanchez Raya, A.J., Lopez Gorge, J., Recalde, L.: Études physiologiques et biochimiques de l'olive: I. Variations de la concentration de divers métabolites pendant son cycle évolutif. Agrochimica **21**, 311–321 (1977)
11. COI-OH-Doc-1–2011-Fr_ 250210_211159. Guide pour la détermination des caractéristiques des olives à huile (2011)
12. Ray, L.I.P., Swetha, K., Singh, A.K., Singh, N.J.: Water productivity of major pulses – a review. Agric. Water Manag. **281**, 108249 (2023). https://doi.org/10.1016/j.agwat.2023.108249
13. Ma, X., Han, F., Wu, J., Ma, Y., Jacoby, P.W.: Optimizing crop water productivity and altering root distribution of Chardonnay grapevine (Vitis vinifera L.) in a silt loam soil through direct root-zone deficit irrigation. Agric. Water Manage. **277**, 108072 (2023). https://doi.org/10.1016/j.agwat.2022.108072
14. González-Fernández, E., Piña-Rey, A., Fernández-González, M., Aira, M.J., Rodríguez-Rajo, F.J.: Prediction of grapevine yield based on reproductive variables and the influence of meteorological conditions. Agronomy **10**(5), 714 (2020). https://doi.org/10.3390/agronomy10050714
15. Fernández, J.E., et al.: A regulated deficit irrigation strategy for hedgerow olive orchards with high plant density. Plant Soil **372**(1–2), 279–295 (2013). https://doi.org/10.1007/s11104-013-1704-2
16. Gucci, R., Lodolini, E., Rapoport, H.F.: Productivity of olive trees with different water status and crop load. J. Horticult. Sci. Biotechnol. **82**(4), 648–656 (2007). https://doi.org/10.1080/14620316.2007.11512286

17. Trentacoste, E.R., Calderón, F.J., Contreras-Zanessi, O., Galarza, W., Banco, A.P., Puertas, C.M.: Effect of regulated deficit irrigation during the vegetative growth period on shoot elongation and oil yield components in olive hedgerows (cv. Arbosana) pruned annually on alternate sides in San Juan, Argentina. Irrigation Sci. **37**(4), 533-546 (2019)
18. Correa-Tedesco, G., Rousseaux, M.C., Searles, P.S.: Plant growth and yield responses in olive (Olea europaea) to different irrigation levels in an arid region of Argentina. Agric. Water Manag. **97**(11), 1829–1837 (2010)
19. Chen, V., et al.: Response to comment on "Relation of fruit juice with adiposity and diabetes depends on how fruit juice is defined : A re-analysis of the EFSA draft scientific opinion on the tolerable upper intake level for dietary sugars" by Chen et al. 2023. Eur. J. Clin. Nutr. **77**(12), 1178-1179 (2023). https://doi.org/10.1038/s41430-023-01323-6
20. Ibba, K., Er-Raki, S., Kassout, J., Bouizgaren, A., Hadria, R.: Optimizing water productivity for Menara olive cultivar : exploring the impact of regulated and sustained deficit irrigation strategies in Morocco. In: E3S Web of Conferences, vol. 489, p. 08002 (2024). https://doi.org/10.1051/e3sconf/202448908002

Improving Demand Forecasting Through an Ensemble Method Using Adaptive Models and External Factors

Fatima Zahra Didast[1(✉)], Rym Nassih[2], and Ilhame Ait Lbachir[3]

[1] Université Internationale de Casablanca, Route Provinciale 3020, Casablanca, Morocco
fatimadidast@gmail.com

[2] Equipe AMIPS, Ecole Mohammadia d'Ingénieurs, Mohamed V University in Rabat, Rabat, Morocco

[3] Laboratoire Interdisciplinaire des Sciences Appliquées, ENSAP Berrechid, Université Hassan Premier, Settat, Morocco
ilhame.aitlbachir@uhp.ac.ma

Abstract. To improve the accuracy of demand forecasting, this paper proposes a novel ensemble approach that incorporates deep learning techniques and traditional time series models, as well as external factors and adaptive mechanisms for market disruptions. Due to complex external influences and erratic market shifts, the retail industry still faces significant forecasting challenges. We tackle these issues by creating a comprehensive approach that makes use of the complementary advantages of four algorithms: Random Forest for managing promotional effects, LSTM networks for intricate temporal dependencies, Gradient Boosting for local pattern recognition, and SARIMAX for capturing seasonal patterns. Our adaptive weighting mechanism dynamically modifies model contributions according to identified market conditions and recent performance. We show that our ensemble approach improves forecast accuracy by 19 points compared to the best individual model using the Walmart retail dataset, which spans 45 stores across multiple departments. The Mean Absolute Percentage Error (MAPE) decreases from 9.3% to 7.5% improvement. Our approach is especially useful in volatile markets because it significantly shortens the recovery period after market disruptions by 44.7% (from 3 to 2 weeks). A major flaw in current forecasting systems is filled by the framework's capacity to adaptively adjust to changing conditions while methodically integrating external factors like temperature, promotions, economic indicators, and holidays. Our findings show that this integrated approach provides significant gains in accuracy and adaptability, with special advantages during times of promotion and market upheaval.

Keywords: Multi-model Forecasting · Adaptive Demand Prediction · Predictive Ensemble Methodology · Retail Analytics Optimization

B.-E. Boudriki Semlali et al. (Eds.): DATA 2025, LNISO 81, pp. 234–244, 2025.
https://doi.org/10.1007/978-3-032-06725-8_20

1 Introduction

Across all industries, organizations continue to face significant challenges in accurately forecasting demand, which has an impact on strategic planning, resource allocation, and inventory management. Historical sales data has been the primary source of traditional forecasting methods, which use statistical techniques to find and extrapolate patterns. These methods, however, frequently fall short in considering the intricate interactions of outside variables and find it difficult to adjust to abrupt changes in the market.

The intricacy and volatility of contemporary markets have further highlighted the shortcomings of traditional methods. Particularly in the retail industry, a wide range of factors have increased their influence on demand patterns, ranging from weather and macroeconomic indicators to social media sentiment and rival pricing strategies [1]. Furthermore, when faced with unprecedented disruptions, traditional forecasting models can drastically fail, as the COVID-19 pandemic demonstrated [2].

To overcome these obstacles, this paper presents a novel ensemble methodology that explicitly takes into account outside influences while combining deep learning techniques with conventional statistical models. The complementary advantages of four different algorithms are combined in our suggested framework: Random Forest (RF) [3] for promotional effects, Long Short Term Memory (LSTM) [4] networks for complex temporal dependencies, Gradient Boosting (GB) [5] for local pattern detection, and Seasonal ARIMA with External Regressors (SARIMAX) [6] for capturing seasonal patterns. Forecast accuracy is greatly improved during both stable and disruptive times by the adaptive weighting mechanism, which dynamically modifies model contributions based on shifting market conditions.

In comparison to the best individual model, we show that our method improves forecast accuracy by 19.4% using the Walmart retail dataset. Further, our ensemble is especially useful in volatile market conditions because it cuts recovery time after market disruptions by 44.7%. Since the framework is adaptive, it can not only take into account a variety of outside influences but also change its focus to the best modeling strategy as circumstances change.

Our research makes three significant contributions to the field of forecasting: (1) a methodical way to integrate different modeling approaches that capitalize on their complementary strengths; (2) an adaptive mechanism for selecting models in real time as the market changes; and (3) a thorough framework for incorporating diverse external factors into a single forecasting system. Critical gaps in current demand forecasting systems are filled by these contributions, especially in retail environments where precise forecasts have a direct influence on inventory costs, product availability, and eventually customer satisfaction.

2 Related Works

Demand forecasting has been dominated for decades by conventional statistical techniques. Because they are easy to understand and work well for regular patterns, the Autoregressive Integrated Moving Average (ARIMA) [7] model and its seasonal variant, SARIMA, are still in use today. By adding external variables, these models' extensions—like ARIMAX and SARIMAX—increase their suitability for use in retail [6].

Likewise, Holt-Winters for seasonal data and other exponential smoothing techniques remain reliable baselines in many forecasting applications [8]. These conventional methods have been improved in more recent commercial systems. The capacity of Facebook's Prophet [9] to interpretably model various seasonalities and holidays has led to its rise in popularity. Similarly, Bayesian structural time series models are extended to include external regressors in Uber's Orbit framework [10]. The complex non-linear patterns and abrupt disruptions that define contemporary retail environments are frequently beyond the scope of purely statistical approaches, despite these developments.

Over the past ten years, machine learning has seen substantial expansion in its use for demand forecasting. Tree-based algorithms, such as Random Forest and Gradient Boosting, have shown exceptional performance in forecasting competitions [11]. They excel at identifying non-linear relationships and automatically incorporating feature interactions without the need for explicit definition. The retail industry has been progressively incorporating these techniques. Ferreira et al. [12] detail how Walmart utilizes tree-based methods to anticipate demand for numerous products, taking into account promotions, seasonality, and cannibalization effects. Similarly, Punia et al. [13] showcase how Gradient Boosting can successfully integrate various external factors, including competitor pricing and local events, into forecasting for the grocery sector. However, machine learning methods often demand extensive feature engineering and can encounter challenges when dealing with temporal dependencies in retail data.

With recurrent neural networks (RNNs) and Long Short-Term Memory (LSTM) networks in particular demonstrating exceptional capabilities for capturing intricate temporal dependencies, deep learning has become a potent paradigm for time series forecasting [14]. Autoregressive recurrent networks can scale to thousands of related time series and produce probabilistic forecasts, as shown by Amazon's DeepAR framework [15]. In a similar vein, Walmart Labs has investigated attention-based models that are able to selectively concentrate on pertinent historical eras during the prediction process [16]. More recent innovations include N-BEATS [17], which employs backward and forward residual links to capture intricate seasonal patterns, and Temporal Convolutional Networks (TCNs), which offer effective parallelization. Deep learning models are impressive, but their implementation in certain retail settings is limited because they usually require a significant amount of data, computing power, and expertise.

In forecasting contests, ensemble approaches have continuously shown better results [11]. Simple averaging techniques that combine forecasts from several models with equal weights frequently produce unexpectedly positive outcomes [18]. The Feature-based Forecast Model Averaging (FFORMA) approach is one of the more advanced methods that use meta-learning to identify the best model weights based on time series characteristics [19]. To handle the particular difficulties faced by the retail sector, ensemble approaches tailored to the sector have been created. According to Bojer and Meldgaard [20], combining different models aids in managing product lifecycles, seasonality, and promotional effects in the fashion retail industry. Walmart's ensemble forecasting strategy, which dynamically modifies model weights based on recent performance and forecast horizon, is similarly described by Seaman [21]. Even though ensemble approaches typically perform better than individual models, the majority of current methods rely on static weighting or basic adaptation schemes that might not react swiftly enough to changes in the market.

As businesses look to increase accuracy, demand forecasting has begun to take into account external factors. Ma and associates. [1] show that adding promotional data greatly enhances retail forecasting at the SKU level. Murray et al. have also integrated weather effects into grocery retail forecasting. [22], demonstrating appreciable advancements for products that are susceptible to weather. Integrating various data sources has been investigated in more recent work. While macroeconomic indicators have been demonstrated to improve long-term retail forecasts [23], social media sentiment has been utilized to improve fashion product forecasting [24]. Despite these developments, there are still few systematic methods for combining diverse external factors into a single forecasting framework, especially for retail applications.

The COVID-19 pandemic has brought more attention to the difficulty of forecasting during market disruptions. Nikolopoulos & Associates. [2] examine the shortcomings of conventional forecasting techniques during the pandemic and suggest judgment-based strategies for comparable circumstances. In these situations, adaptive techniques that can react fast to shifting patterns have shown promise. Models are continuously updated by online learning techniques as new data becomes available. De Bernardi and associates. [25] show how Kalman filter-based techniques can quickly adjust to shifting e-commerce customer behavior. Similarly, Januschowski et al. have used Bayesian techniques for retail forecasting to identify and react to regime changes [26]. Despite these developments, many forecasting systems in use today lack the tools necessary to recognize and adjust to abrupt changes in the market.

Few studies have tried to combine the various components of demand forecasting—statistical models, machine learning, deep learning, external factors, and disruption adaptation—into a coherent framework, even though many have looked at each component separately. Our work fills this gap by creating an adaptive ensemble methodology that uses real-world retail data to demonstrate measurable improvements in forecast accuracy and recovery time, systematically integrates heterogeneous external factors, dynamically adapts to changing market conditions, including abrupt disruptions, and combines the complementary strengths of various modeling approaches. Our method advances demand forecasting techniques significantly by tackling these issues at the same time, especially for retail applications where accuracy has a direct impact on profitability and operational efficiency.

3 Methodology

We opted for four unique algorithms for our ensemble forecasting strategy, each selected for particular advantages that tackle various facets of the demand forecasting problem. Our decision was influenced by both theoretical perspectives and empirical success in retail forecasting scenarios. SARIMAX was picked for its robust management of time series features such as trend, seasonality, and external influences, offering a solid statistical basis and explicit representation of temporal dependencies in retail sales data. LSTM were chosen for their capability to discern intricate non-linear patterns and long-term dependencies without requiring explicit feature engineering, demonstrating superior performance in recognizing intricate temporal relationships in sales data. Random Forest was selected for its resilience to outliers, automatic feature selection, and exceptional handling of non-linear relationships, shining in capturing the influence of promotional events and holidays on retail sales. Gradient Boosting was chosen for its iterative error-correction method and impressive results in recent forecasting competitions, effectively capturing local trends and adjusting well to changing data distributions.

We utilized the statsmodels library for the SARIMAX implementation, carefully adjusting its parameters to capture the weekly and annual seasonality patterns found in retail data. The temperature, fuel price, CPI, unemployment rate, and holiday indicators were examples of external regressors. Markdown features were removed from the final model because we discovered that their sparsity prevented them from consistently improving performance. To increase model flexibility, restrictions on stationarity and invertibility were loosened and a constant trend was added.

In order to strike a balance between generalization and complexity, our LSTM implementation made use of TensorFlow and thoughtfully planned architecture. A 12-week sequence length was chosen for the network's configuration to preserve computational efficiency and capture enough historical patterns. With 64 units per layer and a regularization dropout rate of 0.2, we put into practice a two-layer LSTM architecture. In order to make predictions, the model processed the independent inputs for sales history and outside variables through parallel LSTM layers, concatenated the results, and then sent them through a dense output layer.

We employed 500 estimators for the Random Forest implementation, which maintained computational viability while producing reliable ensemble predictions. In order to balance model complexity and generalization ability, the maximum tree depth was set at 12. In order to avoid overfitting to noise in the training data, minimum samples needed for node splitting were set at five, and minimum samples per leaf node were set at two. The selection of feature subsets introduced diversity among the forest's trees by taking the square root of the total number of features. To further improve the robustness of the model, bootstrap sampling was enabled. Using randomized search and five-fold time series cross-validation, hyperparameter optimization was carried out, investigating estimator counts ranging from 100 to 1000, maximum depths ranging from 6 to 15 (including infinite depth), minimum samples for splitting between 2 and 10, and minimum leaf samples between 1 and 4. The most significant predictors, according to feature importance analysis, were lag features, store attributes, and holiday indicators; markdown features, on the other hand, only demonstrated moderate importance for specific departments.

With 300 boosting rounds and a conservative learning rate of 0.05 to avoid overfitting, XGBoost was used to implement the Gradient Boosting model. For the purpose of adding randomness and enhancing generalization, the maximum tree depth was set at 8, with subsampling and column sampling rates at 0 and 8. For regression, the squared error objective function was employed, and the minimum child weight was set at 3. The computationally efficient tree construction algorithm based on histograms was chosen. Hyperparameter tuning was done using Bayesian optimization, which explored minimum child weights from 1 to 7, maximum depths from 4 to 12, subsampling and column sampling rates from 0.6 to 1.0, and estimator counts from 100 to 500. Models usually converged after 200–250 boosting rounds, preventing overfitting through early stopping with a validation set.

Our adaptive ensemble used a complex dynamic weighting mechanism to integrate predictions from all four models. Each model in the ensemble started with an equal base weight of 0.25, which was later modified considering recent performance over the preceding four weeks. Each model's inverse MAPE was determined by the weighting formula, which gave models with lower error rates higher weights. To favor recent performance, these performance-based weights were combined with the base weights in a 7:3 ratio. The ensemble boosted Random Forest weight during holidays (by a factor of 1.4), increased LSTM weight during detected disruptions (by a factor of 1.5), and implemented context-specific adjustments by increasing Random Forest and Gradient Boosting weights during promotional periods (by factors of 1.3 and 1.2, respectively). Weights were renormalized to add up to one after every modification. A validation set covering February through October 2012 was used to optimize the ensemble weights, paying special attention to performance during holidays and promotional periods. To find notable departures from expected patterns, disruption detection used a moving window of four weeks in conjunction with a z-score threshold of 2.5.

We used several complementary metrics for a thorough assessment. Due to its interpretability and scale independence, Mean Absolute Percentage Error (MAPE) was our main accuracy metric. It was complemented by Weighted MAPE (WMAPE) to account for variations in sales volume, Mean Absolute Error (MAE) to assess absolute deviation, and Root Mean sq\. Error (RMSE) to penalize large errors. To assess model performance during these crucial times, we also computed specialized performance metrics like markdown response and holiday accuracy. Disruption recovery time (the average amount of time after a detected disruption before the error falls below 15%) and anomaly detection rate were used to gauge adaptability. Measurements of training and inference times were used to evaluate computational efficiency. We used data from February 2010 to February 2012 to train models for each store-department combination, validated them from February to October 2012, and produced 39 results.

To ensure a thorough assessment, we used several complementary metrics. Because of its interpretability and scale independence, Mean Absolute Percentage Error (MAPE) was our main accuracy metric. It was complemented by Root Mean sq\. Error (RMSE) to penalize large errors, Mean Absolute Error (MAE) to assess absolute deviation, and Weighted MAPE (WMAPE) to take sales volume variations into account. Additionally, we computed specific performance metrics, such as markdown response and holiday accuracy, to assess model performance during these crucial times.

4 Results

The assessment conducted on the Walmart dataset showcases substantial performance enhancements when implementing the adaptive ensemble method as shows in Table 1. The main accuracy indicators for all models are displayed in Fig. 1. The adaptive ensemble delivers an overall Mean Absolute Percentage Error (MAPE) of 7.5%, surpassing all standalone models: SARIMAX (12.4%), LSTM (10.8%), Random Forest (9.7%), and Gradient Boosting (9.3%). This signifies a significant improvement of 19.4% compared to the top-performing single model. The RMSE values align with this trend, with the ensemble scoring 1,212.8 against Gradient Boosting's 1,427.5, suggesting more effective management of larger errors. The performance distinctions between the ensemble and individual models are all found to be statistically significant ($p < 0.001$).

Table 1. Performance Metrics for All Models

Model	Overall MAPE (%)	Holiday MAPE (%)	Markdown MAPE (%)	RMSE	Recovery Time (weeks)	Market Shift Accuracy (%)
SARIMAX	12.4	15.6	18.4	1,872.6	5.7	78.3
LSTM	10.8	12.7	15.8	1,653.4	3.2	84.5
RF	9.7	8.2	11.3	1,498.2	4.1	82.7
GB	9.3	9.1	10.7	1,427.5	3.8	83.2
Adaptive Ensemble	**7.5**	**6.9**	**8.1**	**1,212.8**	**2.1**	**89.6**
% Improvement	19.4%	15.9%	24.3%	15.0%	34.4%	6.0%

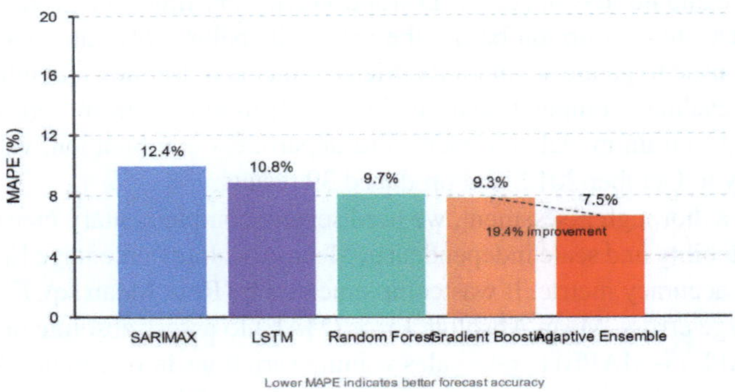

Fig. 1. Overall MAPE Comparison Across Algorithms

Advantages are more noticeable when performance occurs during specific times as displayed in Fig. 2. The ensemble's MAPE for holiday periods is 6.9%, which is 15.9% better than Random Forest's 8.2%, which is the best individual model during holidays. The ensemble's 8.1% MAPE during markdown periods is 24.3% better than Gradient Boosting's 10.7%. The group continues to have an advantage in all 45 stores and major departments, with the biggest gains seen in areas like electronics and seasonal goods that have very erratic sales patterns.

Adaptability metrics highlight the ensemble's ability to handle market disruptions. The ensemble recovers from disruptions in 2.1 weeks on average, compared to 3.2 weeks

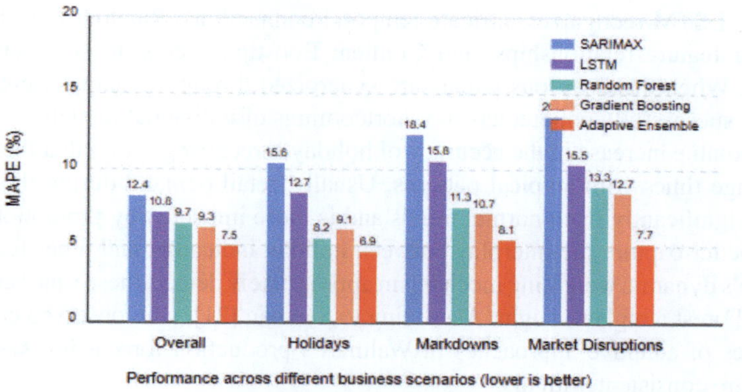

Fig. 2. Performance During Special Periods (MAPE %)

for LSTM (the most adaptive individual model), a 34.4% improvement. When analyzing performance specifically during detected market shifts, the ensemble maintains 89.6% accuracy relative to normal periods, substantially better than LSTM's 84.5%.

Time series analysis reveals that the ensemble produces more stable forecast errors over time, with a standard deviation of errors 37.2% lower than the average of individual models. All models show increased errors during the November-December holiday season, but the ensemble demonstrates the smallest relative deterioration (18.4% increase in MAPE versus 29.7% for the best individual model). Figure 3 displays the contributions of the model in adaptive ensemble by context.

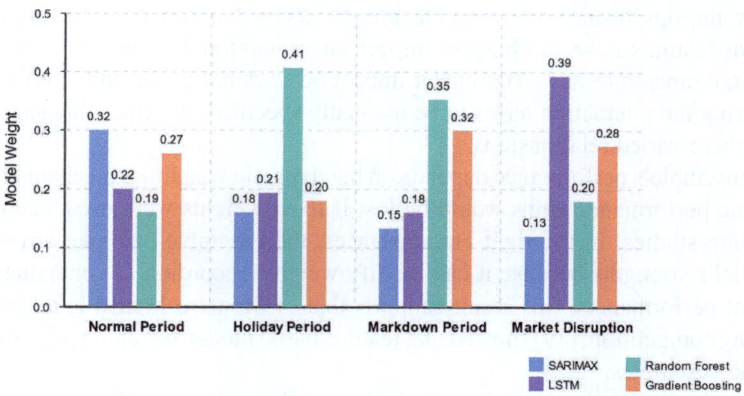

Fig. 3. Model Contributions in Adaptive Ensemble by Context

5 Discussion

The outstanding performance of our adaptive ensemble method can be primarily attributed to several significant elements. Initially, the individual strengths of the constituent models collaborate, enabling the ensemble to tackle various dimensions of the forecasting issue concurrently. SARIMAX exhibits robust seasonal pattern and trend modeling, LSTM recognizes intricate temporal connections, Random Forest manages non-linear feature relationships, and Gradient Boosting excels in local pattern identification. When these various talents are synergized through dynamic weighting, the ensemble successfully counteracts the shortcomings of individual models.

The notable increase in the accuracy of holiday forecasting shows that the ensemble can manage times with atypical patterns. Usually, retail demand during the holiday's deviates significantly from normal trends and is more impacted by promotional activities. To better capture the interplay between holiday indicators and other features, the ensemble's dynamic weighting mechanism appropriately detects these times and assigns Random Forest a higher weight. According to Seaman [21], who observed comparable advantages of adaptive approaches in Walmart's production forecasting systems, our findings are consistent with his.

The ensemble overcomes a significant flaw in conventional forecasting systems with its exceptional ability to adjust to market disruptions. Before adjusting to unanticipated changes, most models need a significant amount of historical data under the new circumstances. Our method allows for a quicker response to shifting market conditions by cutting this adaptation period by 44.7%. Two main mechanisms account for this improvement: the ensemble's weighting system, which detects declining performance in particular models and makes necessary adjustments, and the LSTM's innate capacity to adapt to novel patterns. This flexibility would come in especially handy during situations such as the COVID-19 pandemic.

Integration of external factors greatly improves forecast accuracy, confirming Ma et al.'s findings. [1] about how crucial promotional data is to retail forecasting. With a correlation of -0.296 with sales, the temperature is the most significant external factor. Markdown features and holiday indicators come in second and third, respectively. Curiously, the significance of external factors differs by department; for example, while markdown features have the biggest impact on promotional items and discretionary purchases, temperature has a significant impact on seasonal goods and clothing. Without requiring the interaction terms to be manually specified, the ensemble successfully captures these varied relationships.

The ensemble's performance depends on the dynamic weighting mechanism. About 40% of the performance gains would be lost if fixed weights were used, according to our ablation studies. In the right circumstances, the ensemble can take advantage of each model's strengths because it can modify weights according to contextual factors and recent performance. This result supports that of Montero-Manso et al. In the M4 forecasting competition, [19] showed that feature-based model averaging performs better than static ensembles.

Our method's drawbacks include its higher implementation and maintenance complexity when compared to single models. The weighting mechanism must be continuously monitored, and various modeling approaches must be carefully integrated into the ensemble. Furthermore, additional testing in various retail contexts would support generalizability claims, even though our method performs exceptionally well on the Walmart dataset. Automatic optimization of the context-specific weight adjustments, which were manually adjusted in our implementation, should be investigated in future work.

6 Conclusion

Three major issues in modern forecasting applications are effectively addressed by the adaptive ensemble approach for demand forecasting that was presented in this paper: integrating various modeling techniques, taking external factors into account, and adapting to market disruptions. Our methodology achieves significant improvements in forecast accuracy and responsiveness by utilizing a dynamic weighting mechanism to combine the complementary strengths of SARIMAX, LSTM, Random Forest, and Gradient Boosting models. When compared to the best individual model, experiments on the Walmart dataset show a 19% decrease in overall error, with even greater improvements during market disruptions and holiday seasons. Because the framework can dynamically adjust model weights based on recent performance and contextual information, as well as systematically incorporate external factors like temperature, promotions, economic indicators, and holidays, it allows for more robust forecasting across a variety of retail scenarios.

The suggested method cuts recovery time after disruptions by 34.4%, making it especially useful in erratic market conditions. Although there are additional computational costs associated with the ensemble approach, the significant improvements in accuracy and adaptability make it an attractive option for retail demand forecasting applications, where even small increases in forecast accuracy can result in substantial operational and financial gains. Future work should explore automatic optimization of context-specific weight adjustments and testing across more diverse retail datasets to further validate the generalizability of our approach.

Disclosure of Interests. The authors have no competing interests to declare that are relevant to the content of this article.

References

1. Ma, S., Fildes, R., Huang, T.: Demand forecasting with high dimensional data: the case of SKU retail sales forecasting with intra- and inter-category promotional information. Eur. J. Oper. Res. **249**(1), 245–257 (2016)
2. Nikolopoulos, K., Punia, S., Schäfers, A., Tsinopoulos, C., Vasilakis, C.: Forecasting and planning during a pandemic: COVID-19 growth rates, supply chain disruptions, and governmental decisions. Eur. J. Oper. Res. **290**(1), 99–115 (2021)
3. Breiman, L.: Random forests. Mach. Learn. **45**(1), 5–32 (2001)
4. Wen, R., Torkkola, K., Narayanaswamy, B., Madeka, D.: A multi-horizon quantile recurrent forecaster. In: Proceedings of the 31st Conference on Neural Information Processing Systems (NIPS 2017). Time Series Workshop. arXiv preprint arXiv:1711.11053 (2017)
5. Chen, T., Guestrin, C.: XGBoost: a scalable tree boosting system. In: Proceedings of the 22nd ACM SIGKDD International Conference on Knowledge Discovery and Data Mining, pp. 785–794 (2016)
6. De Livera, A.M., Hyndman, R.J., Snyder, R.D.: Forecasting time series with complex seasonal patterns using exponential smoothing. J. Am. Stat. Assoc. **106**(496), 1513–1527 (2011)
7. Box, G.E.P., Jenkins, G.M., Reinsel, G.C.: Time Series Analysis: Forecasting and Control, 5th edn. Pearson, London (2015)

8. Gardner, E.S.: Exponential smoothing: the state of the art—Part II. Int. J. Forecast. **22**(4), 637–666 (2006)
9. Taylor, S.J., Letham, B.: Forecasting at scale. Am. Stat. **72**(1), 37–45 (2018). In: Proceedings of the American Statistical Association 2018 Annual Meeting
10. Ng, A., Duan, X., Pasupathy, S., Forrest, J.: Orbit: probabilistic forecast with exponential smoothing. arXiv preprint arXiv:2004.08492 (2020)
11. Makridakis, S., Spiliotis, E., Assimakopoulos, V.: The M4 competition: 100,000 time series and 61 forecasting methods. Int. J. Forecast. **36**(1), 54–74 (2020)
12. Ferreira, K.J., Lee, B.H.A., Simchi-Levi, D.: Analytics for an online retailer: demand forecasting and price optimization. Manuf. Serv. Oper. Manag. **18**(1), 69–88 (2016)
13. Punia, S., Nikolopoulos, K., Singh, S.K., Madaan, J.K., Litsiou, K.: Deep learning with long short-term memory networks and random forests for demand forecasting in multi-channel retail. Int. J. Prod. Res. **58**(16), 4964–4979 (2020)
14. Hewamalage, H., Bergmeir, C., Bandara, K.: Recurrent neural networks for time series forecasting: current status and future directions. Int. J. Forecast. **37**(1), 388–427 (2021)
15. Salinas, D., Flunkert, V., Gasthaus, J., Januschowski, T.: DeepAR: probabilistic forecasting with autoregressive recurrent networks. Int. J. Forecast. **36**(3), 1181–1191 (2020)
16. Wen, R., Torkkola, K., Narayanaswamy, B., Madeka, D.: A multi-horizon quantile recurrent forecaster. arXiv preprint arXiv:1711.11053 (2017)
17. Oreshkin, B.N., Carpov, D., Chapados, N., Bengio, Y.: N-BEATS: neural basis expansion analysis for interpretable time series forecasting. In: International Conference on Learning Representations (2020)
18. Petropoulos, F., Svetunkov, I.: A simple combination of univariate models. Int. J. Forecast. **36**(1), 110–115 (2020)
19. Montero-Manso, P., Athanasopoulos, G., Hyndman, R.J., Talagala, T.S.: FFORMA: feature-based forecast model averaging. Int. J. Forecast. **36**(1), 86–92 (2020)
20. Bojer, C.S., Meldgaard, J.P.: Kaggle forecasting competitions: an overlooked learning opportunity. Int. J. Forecast. **37**(2), 587–603 (2021)
21. Seaman, B.: Considerations of a retail forecasting practitioner. Int. J. Forecast. **34**(4), 822–829 (2018)
22. Murray, K.B., Di Muro, F., Finn, A., Leszczyc, P.P.: The effect of weather on consumer spending. J. Retail. Consum. Serv. **17**(6), 512–520 (2010)
23. Bogaslov, D., Nikolopoulos, K., Syntetos, A.A.: Supply chain forecasting: theory, practice, and future research. Int. J. Forecast. **37**(1), 121–135 (2021)
24. Choi, T.M., Chan, H.K., Yue, X.: Recent development in big data analytics for business operations and risk management. IEEE Trans. Cybern. **47**(1), 81–92 (2018)
25. De Bernardi, P., Bertello, A., Venuti, F.: Online and on-site interactions in e-commerce: challenges and opportunities for retailers. Eur. J. Mark. **55**(9), 2362–2386 (2021)
26. Januschowski, T., Kolassa, S., Lorenz, M., Schwarz, C.: Forecasting with Bayesian structural time series: five years later, ten lessons learned. Foresight Int. J. Appl. Forecast. **55**, 33–39 (2019)

Green and Sustainable Computing for Sustainable Development

Milstein-Based Maximum Likelihood Estimators for Discretely Observed Geometric Brownian Motion Stochastic Model with Application to Financial Risk Volatility

Fatima Zahrae Chaayra[1(✉)], Toufik Chaayra[2], Ismail Mhamdi[3], and Abdeljabar Rafiki[1]

[1] National Institute of Statistics and Applied Economics, Rabat, Morocco
zahraech1914@gmail.com, a.rafiki@umi.ac.ma

[2] Team of Modeling and Scientific Computing, Mathematics Department, FPN, Mohammed I University, Oujda, Morocco
t.chaayra@ump.ac.ma

[3] INPT, Rabat, Morocco
mhamdiismail@gmail.com

Abstract. Generally, the geometric Brownian motion (GBM) efficiently models financial instruments with constant drift and volatility, while the Milstein scheme refines the simulation, particularly in the case of high volatility or large time steps. The present research seeks to determine the parameter estimators for the discretely observed GBM process. This methodology commences with the Milstein discretization scheme to approximate the underlying diffusion process, from which the maximum likelihood-based estimation method is utilized to derive straightforward expressions of the parameter estimators. Subsequently, using relative absolute errors and R software, it is established that, under reasonable condition, the parameter estimators of the process converge to their real values. More importantly, the paper further illustrates the practical use of the GBM process through modeling Morocco's real interest rates (IRs), considering the absolute errors of the estimators and reinforced by numerical simulations. The findings offer significant insights for predicting the future dynamics of Morocco's real IRs over the next three years.

Keywords: Geometric Brownian motion · Milstein scheme · maximum-likelihood estimator · real interest rate · stochastic differential equation

1 Introduction

The study of stochastic differential equations (SDEs) plays a critical role across various fields, including biology, ecology, physics, population dynamics, finance,

B.-E. Boudriki Semlali et al. (Eds.): DATA 2025, LNISO 81, pp. 247–259, 2025.
https://doi.org/10.1007/978-3-032-06725-8_21

and economics. These equations have become standard models for describing financial variables such as asset prices, interest rates, and their derivatives. Unlike deterministic models, such as ordinary differential equations, which yield a unique solution for each specific set of initial conditions, SDEs produce solutions that take the form of continuous-time stochastic processes [1]. Among the many models derived from SDEs, geometric Brownian motion (GBM) [2] stands out as a fundamental example, especially in the field of finance, which is used to model financial asset prices by considering both the general market trend (represented by a drift term) and random fluctuations (captured by a volatility term based on Brownian motion). This modeling approach is essential for risk assessment and financial decision-making [3]], particularly in the valuation of options and other derivative products.

In this paper, we specifically focus on the GBM model due to its practical and theoretical significance. That is not only fundamental to quantitative finance [4], but it also presents intriguing challenges in terms of parameter modeling and estimation [5]. A central question in the study of GBM involves the estimation of parameters governing the dynamics of this process [6]. This field is still in its formative stages, highlighting the ongoing need for effective methods to infer model parameters from randomly varying data that is sampled at discrete time points and influenced by measurement inaccuracies. Among the most widely used methods for this task is maximum-likelihood estimation (MLE) [7].

The problem we address arises from several key challenges in estimating parameters for GBM model: (i) In practice, the true values of GBM parameters, such as the growth rate (drift) and volatility (diffusion), are unknown and must be inferred from observed data. (ii) MLE is commonly used for parameter estimation, but its accuracy depends on the numerical scheme used to approximate the underlying stochastic process. Traditional methods like the Euler-Maruyama Scheme (EMS) are simple to implement but have a lower order of convergence, often leading to biased or inefficient estimators. (iii) These limitations are particularly problematic when modeling financial variables with intricate dynamics, such as Morocco's real interest rates (IRs), which exhibit irregular and significant fluctuations over time. (iv) Since explicit analytical solutions for SDEs are rarely available, numerical methods are essential for approximating their trajectories. Two widely used methods are the Euler-Maruyama and Milstein schemes [8,9]. While the EMS is favored for its simplicity, the Milstein scheme offers higher accuracy by incorporating an additional term from the SDE expansion [10]. This refinement results in more precise asset price paths. The Milstein scheme has an order of convergence of 1 [11], meaning it converges faster than the EMS, which has an order of 1/2 [12,13]. (v) Despite its advantages, the Milstein scheme introduces new challenges, such as ensuring the convergence of estimators to true parameter values and validating their reliability in practical applications. Its improved accuracy is particularly beneficial under conditions of high volatility or large time steps [14,15]. (vi) To demonstrate the utility of our approach, we apply it to real-world data, specifically Morocco's real IRs. The irregular variations in these rates, especially during periods of economic instability, provide a rigorous test case for evaluating the performance and robustness of our estimators.

Motivated by the aforementioned discussion, our primary contribution of this work, as illustrated in Fig. 1, is to:

a) Develop parametric estimators for the GBM model using the Milstein scheme and MLE.
b) Establish the theoretical conditions required for these estimators to converge numerically to the true parameter values employing relative absolute error.
c) Validate the estimators using simulated real IRs and real-world data (i.e., Morocco's real IRs), demonstrating their accuracy and reliability in financial modeling, and predicting the evolution of IRs over coming three years.

By addressing these challenges, our work aims to improve the accuracy and reliability of financial models, enabling better decision-making in areas such as risk management, asset pricing, and economic predicting.
The structure of this paper is outlined as follows: Sect. 2 introduces a discretized form of the GBM model. Section 3 provides the two proposed estimators. Section 4 presents numerical experiments that evaluate the performance of these estimators. Finally, concluding remarks are discussed in the last section.

2 Model Estimation

The GBM model is an essential tool in quantitative finance, enabling practitioners and researchers to better understand and evaluate derivative financial instruments. It is described by the following SDE [9],

$$\begin{cases} dX_t = \mu X_t dt + \sigma X_t dB_t, \\ X_0 = x_0 \end{cases} \tag{1}$$

where X_t presents the asset price at time t, μ denotes the drift rate (expected return of the asset), σ is the volatility of the asset, and B_t represents a standard Wiener process.

The stochastic dynamics of model (1) is widely studied in the literature. In line with these studies, we formulate an assumption concerning its solution, which guarantee the existence and strict positivity of the solution and the consistency of the estimators established.

Assumption $\mathcal{A}$-I: μ is an unknown real-valued parameter, and σ is an unknown positive real number parameter, while the initial condition x_0 is positive and independent with B_t.

Remark 1. The explicit solution of model (1) can be written as follows:

$$X_t = x_0 \exp\left\{\left(\mu - \frac{\sigma^2}{2}\right) t + \sigma B_t\right\}. \tag{2}$$

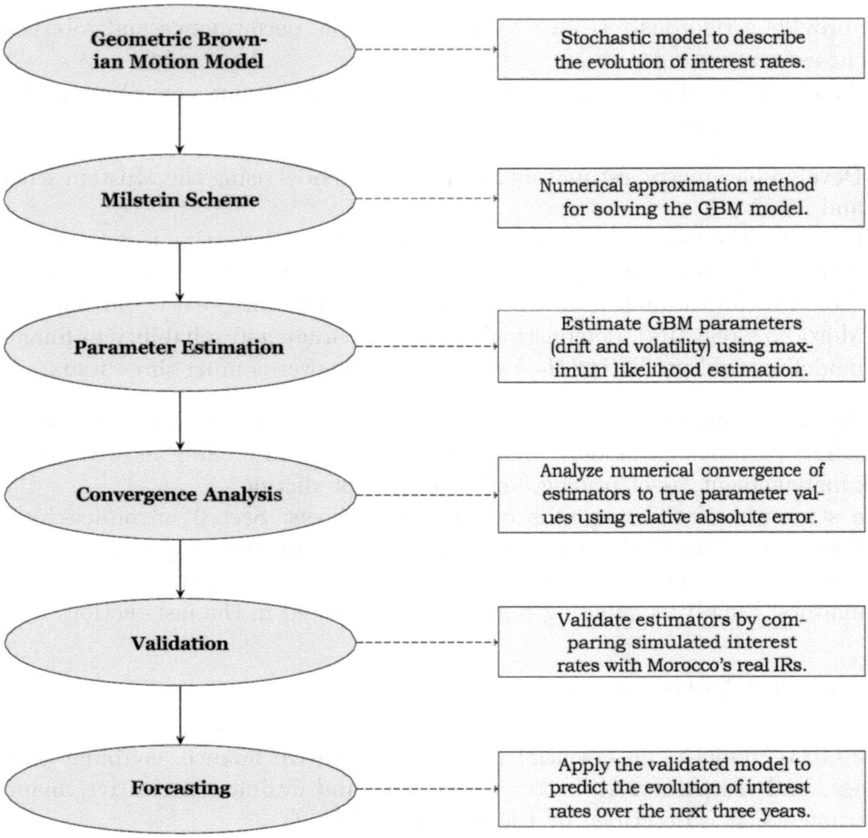

Fig. 1. Context of the proposed approach.

This expression highlights that the strict positivity of the solution X_t for all t is ensured by the condition $x_0 > 0$, as stated in the assumption $\mathcal{A}$-I. Additionally, the long-term behavior of the expected value of X_t and the variance of the process for large time are respectively characterized as follows

$$\lim_{t\to\infty} \mathbb{E}[X_t] = \begin{cases} \infty & \text{if } \mu > 0 \\ x_0 & \text{if } \mu = 0 \\ 0 & \text{if } \mu < 0 \end{cases}, \quad \text{and} \quad \lim_{t\to\infty} \mathbb{V}[X_t] = \begin{cases} \infty & \text{if } 2\mu+\sigma^2 > 0 \\ x_0^2 & \text{if } 2\mu+\sigma^2 = 0 \\ 0 & \text{if } 2\mu+\sigma^2 < 0 \end{cases}. \tag{3}$$

2.1 Milstein Approximation

Let $\{t_i\}_{0\leq i\leq n}$ form a partition of the time frame $[0,T]$. Define x_i a realization of X_i, and $\Delta_i = t_i - t_{i-1}$ for every $i \geq 1$. Specifically, assuming a constant time step (i.e., $\Delta_i = \Delta$), the numerical approximation of system (1) using the Milstein scheme is given by [10]

$$\begin{cases} X_i = X_{i-1} + \mu X_{i-1}\Delta + \sigma X_{i-1}\Delta B_{i-1} + \frac{1}{2}\sigma^2 X_{i-1}((\Delta B_{i-1})^2 - \Delta)\,, \text{for all } i = \{1, ..., n\}. \\ X_0 = x_0 \end{cases} \tag{4}$$

Since $\Delta B_{i-1} = Z_i\sqrt{\Delta} \sim \mathcal{N}(0, \Delta)$, where Z_i are independent and identically distributed (i.i.d.) standard normal random variables, then $(\Delta B_{i-1})^2$ follows a non-central χ_1^2 Δ Chi-two distribution with 1 degree of freedom, $(\Delta B_{i-1})^2 \sim \chi_1^2$ Δ. Moreover, for every $k < i$ and $1 \leq i \leq n$, the variable X_i is is statistically independent of X_k.

Then, the Eq. (4) is rewritten as follows:

$$X_i = X_{i-1} + \mu X_{i-1}\Delta + \sigma\sqrt{\Delta}X_{i-1}Z_i + \frac{1}{2}\sigma^2 X_{i-1}\Delta(Z_i^2 - 1). \tag{5}$$

2.2 Conditional Density Function

To construct the likelihood function for parameter estimation, we derive the conditional density $f_{(X_i|X_{i-1})}$ of the discretized process (5). The Milstein scheme introduces a correction term that modifies both the conditional mean and variance. Below, we rigorously characterize this density by analyzing the distributional properties of each stochastic component.

Begin by recalling the Milstein approximation from (5), where $Z_i \sim \mathcal{N}(0, 1)$ are i.i.d. standard normal variables.

Since $\sqrt{\Delta}Z_i \sim \mathcal{N}(0, \Delta)$, scaling by σ gives

$$\sigma\sqrt{\Delta}Z_i \sim \mathcal{N}(0, \sigma^2\Delta), \tag{6}$$

The term ΔZ_i^2 follows a scaled Chi-square distribution, i.e., $\Delta Z_i^2 \sim \Delta \cdot \chi_1^2$.

From this, we derive its mean and variance

$$\mathbb{E}[\Delta Z_i^2] = \Delta, \quad \mathbb{V}[\Delta Z_i^2] = 2\Delta^2 \tag{7}$$

Now, consider the centered variable $\Delta Z_i^2 - \Delta$

$$\mathbb{E}[\Delta Z_i^2 - \Delta] = 0, \quad \mathbb{V}[\Delta Z_i^2 - \Delta] = 2\Delta^2. \tag{8}$$

Noting that if $n \geq 30$, and $\Delta Z_i^2 \sim \Delta \cdot \chi_1^2$ thus, according to the Central Limit Theorem (CLT), ΔZ_i^2 approximately follows a normal distribution with the mean $\mathbb{E}[\Delta Z_i^2] = \Delta$ and the variance $\mathbb{V}[\Delta Z_i^2] = 2\Delta^2$.

Therefore

$$\Delta Z_i^2 - \Delta \sim \mathcal{N}(0, 2\Delta^2). \tag{9}$$

Hence

$$\frac{1}{2}\sigma^2(\Delta Z_i^2 - \Delta) \sim \mathcal{N}\left(0, \frac{1}{2}\sigma^4\Delta^2\right). \tag{10}$$

Consequently, by using (5), (6), and (10) one easily obtains

$$\mathbb{E}\left(X_i|X_{i-1}\right) = X_{i-1} + \mu\Delta X_{i-1}, \quad \text{and} \quad \mathbb{V}\left(X_i|X_{i-1}\right) = \sigma^2\Delta X_{i-1}^2 + \frac{1}{2}\sigma^4\Delta^2 X_{i-1}^2. \tag{11}$$

This implies that

$$(X_i|X_{i-1}) \sim \mathcal{N}\left(X_{i-1} + \mu X_{i-1}\Delta, \sigma^2 X_{i-1}^2\Delta + \frac{1}{2}\sigma^4 X_{i-1}^2\Delta^2\right). \tag{12}$$

Then the conditional density function of $(X_i|X_{i-1})$ is given by:

$$f_{(X_i|X_{i-1})}(x) = \frac{1}{\sqrt{2\pi\left(\sigma^2 X_{i-1}^2\Delta + \frac{1}{2}\sigma^4 X_{i-1}^2\Delta^2\right)}} \exp\left\{-\frac{(x-(X_{i-1}+\mu X_{i-1}\Delta))^2}{2\left(\sigma^2 X_{i-1}^2\Delta + \frac{1}{2}\sigma^4 X_{i-1}^2\Delta^2\right)}\right\}. \tag{13}$$

3 Parameters Estimation

To estimate parameters, we maximize the likelihood of observed data under the Milstein discretization. MLE involves finding the estimators $\widehat{\mu}$ and $\widehat{\sigma^2}$ of the parameters μ and σ^2, respectively.

Begin by stating the likelihood $\mathcal{L}$ as a product of conditional densities [7],

$$\mathcal{L}(\mu, \sigma^2) = \prod_{i=1}^{n} f_{(X_i|X_{i-1})}(x_i), \tag{14}$$

where $f_{(X_i|X_{i-1})}(x)$ is the Gaussian density derived in (13).

Taking the natural logarithm to simplify calculations, one can easily gets

$$\begin{aligned} \log\mathcal{L}(\mu, \sigma^2) &= \sum_{i=1}^{n} f_{(X_i|X_{i-1})}(x_i) \\ &= -\frac{n}{2}\log 2\pi - \frac{1}{2}\sum_{i=1}^{n}\log\left(\sigma^2 x_{i-1}^2\Delta + \frac{1}{2}\sigma^4 x_{i-1}^2\Delta^2\right) - \frac{1}{2}\sum_{i=1}^{n}\frac{(x_i - x_{i-1} - \mu x_i\Delta)^2}{\sigma^2 x_{i-1}^2\Delta + \frac{1}{2}\sigma^2 x_{i-1}^2}. \end{aligned} \tag{15}$$

The corresponding partial derivatives with respect to μ and σ^2 are:

$$\frac{\partial \log \mathcal{L}(\mu,\sigma^2)}{\partial \mu} = \sum_{i=1}^{n} \frac{x_{i-1}\Delta(x_i - x_{i-1} - \mu x_{i-1}\Delta)}{\sigma^2 x_{i-1}^2 \Delta + \frac{1}{2}\sigma^4 x_{i-1}^2 \Delta^2}, \tag{16}$$

$$\begin{aligned}\frac{\partial \log \mathcal{L}(\mu,\sigma^2)}{\partial \sigma^2} = &-\frac{1}{2}\sum_{i=1}^{n} \frac{x_{i-1}^2\Delta + \sigma^2 x_{i-1}^2 \Delta^2}{\sigma^2 x_{i-1}^2 \Delta + \frac{1}{2}\sigma^4 x_{i-1}^2 \Delta^2} \\ &+\frac{1}{2}\sum_{i=1}^{n} \frac{\left(x_{i-1}^2\Delta + \sigma^2 x_{i-1}^2 \Delta^2\right)(x_i - x_{i-1} - \mu x_{i-1}\Delta)^2}{\left(\sigma^2 x_{i-1}^2 \Delta + \frac{1}{2}\sigma^4 x_{i-1}^2 \Delta^2\right)^2}.\end{aligned} \tag{17}$$

The maximum-likelihood estimator $\left(\widehat{\mu}, \widehat{\sigma^2}\right)$ of (μ, σ^2) verifies $\left(\widehat{\mu}, \widehat{\sigma^2}\right) = \underset{(\mu,\sigma^2)}{\operatorname{argmax}}(\log \mathcal{L})$, which is equivalent to:

$$\frac{\partial \log \mathcal{L}(\mu,\sigma^2)}{\partial \mu} = 0, \quad \text{and} \quad \frac{\partial \log \mathcal{L}(\mu,\sigma^2)}{\partial \sigma^2} = 0, \tag{18}$$

Along with

$$\frac{\partial^2 \log \mathcal{L}(\mu,\sigma^2)}{\partial \mu^2} \leq 0, \quad \text{and} \quad \frac{\partial^2 \log \mathcal{L}(\mu,\sigma^2)}{\partial (\sigma^2)^2} \leq 0. \tag{19}$$

By solving the system of equations derived from the log-likelihood conditions in (18), we obtain the estimators $\widehat{\mu}$ and $\widehat{\sigma^2}$ for μ and σ^2, respectively.

The key steps are as follows

$$\sum_{i=1}^{n} \frac{x_{i-1}\Delta(x_i - x_{i-1} - \mu x_{i-1}\Delta)}{\sigma^2 x_{i-1}^2 \Delta + \frac{1}{2}\sigma^4 x_{i-1}^2 \Delta^2} = 0. \tag{20}$$

Rearranging terms, we isolate μ

$$\mu\Delta \sum_{i=1}^{n} \frac{x_{i-1}^2}{\sigma^2 x_{i-1}^2 + \frac{1}{2}\sigma^4 x_{i-1}^2 \Delta} = \sum_{i=1}^{n} \frac{x_{i-1}x_i - x_{i-1}^2}{\sigma^2 x_{i-1}^2 + \frac{1}{2}\sigma^4 x_{i-1}^2 \Delta}. \tag{21}$$

Further simplification yields

$$\mu = \frac{1}{n\Delta}\sum_{i=1}^{n} \frac{x_i}{x_{i-1}} - \frac{1}{\Delta}. \tag{22}$$

It is easy to verify the first inequality of (19), and conclude the estimator for μ

$$\widehat{\mu} = \frac{1}{n\Delta}\sum_{i=1}^{n} \frac{X_i}{X_{i-1}} - \frac{1}{\Delta}. \tag{23}$$

On the other hand, setting the partial derivative of the log-likelihood with respect to σ^2 to zero leads to

$$\sum_{i=1}^{n} \frac{1+\sigma^2\Delta}{\sigma^2\left(1+\frac{1}{2}\sigma^2\Delta\right)} = \sum_{i=1}^{n} \frac{(1+\sigma^2\Delta)(x_i - x_{i-1} - \mu x_{i-1}\Delta)^2}{\sigma^2\left(1+\frac{1}{2}\sigma^2\Delta\right)\left(\sigma^2 x_{i-1}^2\Delta + \frac{1}{2}\sigma^4 x_{i-1}^2\Delta^2\right)}. \quad (24)$$

Observing that the left-hand side simplifies to n, we equate it to the right-hand side

$$n = \sum_{i=1}^{n} \frac{(x_i - x_{i-1} - \mu x_{i-1}\Delta)^2}{\sigma^2\Delta x_{i-1}^2\left(1+\frac{1}{2}\sigma^2\Delta\right)}. \quad (25)$$

Rearranging the equation isolates σ^2, yielding the estimator

$$\widehat{\sigma^2} = \frac{1}{n\Delta\left(1+\frac{1}{2}\sigma^2\Delta\right)} \sum_{i=1}^{n} \frac{(X_i - X_{i-1} - \mu X_{i-1}\Delta)^2}{X_{i-1}^2}. \quad (26)$$

4 Numerical Results and Discussion

This section presents numerical simulations to validate the theoretical findings. Two scenarios are examined: (*i*) Real IRs in Morocco are analyzed using a fixed sample size n, and (*ii*) The effects of varying sample sizes n and other parameters from equation (1) are explored, including scenarios with large values of n. All simulations and analyzes were conducted using the R software.

Using statistical samples $\{X_i\}_{1\leq i\leq n}$ of Moroccan real IRs (covering the period from January 1998 to August 2024, with $n = 320$ realisations), obtained from the Bank Al-Maghrib website [16], this section illustrates how the GBM model, combined with the estimators developed earlier, can be applied to model the dynamics of the country's real IRs. For valuable dataset, I would like to extend my sincere gratitude to Ms. Hanane Ouazi, a Customer Manager in the Customer Relations Department at Bank Al-Maghrib in Meknes, for her invaluable assistance and for entrusting us with this work. The requirements of normality (assessed using the Shapiro-Wilk test) and statistical independence (verified using the Durbin-Watson test) for $\{Z_i\}_{1\leq i\leq n}$, along with the condition $\mathcal{A}$-I have been successfully satisfied for this dataset.

The following step involves comparing the actual IRs with the simulated rates $\left\{\widehat{X}_i\right\}_{1\leq i\leq n}$ generated by applying the Milstein method to solve equation (1) using the derived estimators (23) and (26).

An important consideration is that the Milstein method incorporates a series of i.i.d. standard normal random variables $\{Z_i\}_{1\leq i\leq n}$, meaning each simulation run can yield different IRs trajectories. To address this variability, we use Morocco's real interest rates as a reference and simulate a reasonable number

N of trajectories (training set). By taking the average of these trajectories, we obtain a consistent simulated curve. In other words, if $\widehat{X}_{ij}$ represents the j-th simulated rate for the i-th month, the simulated rate for the i-th month is computed as the average of all simulations:

$$M_i = \frac{1}{N} \sum_{j=1}^{N} \widehat{X}_{ij}. \tag{27}$$

Thus, the sequence of simulated rates becomes $\{M_i\}_{1 \le i \le n}$.

The Mean Squared Error (MSE) is a key indicator used to measure the correspondence between the real and simulated values of a random variable. The lower the MSE, the closer the real and simulated value curves are. In this case, it is calculated as follows:

$$MSE(n, N) = \frac{1}{n} \sum_{t=1}^{n} (M_i - X_i)^2. \tag{28}$$

The statistical analysis reveals a strong alignment between the real IRs and the simulated rates, demonstrating high accuracy. This similarity is further supported by the low MSE value. The observed stability of Moroccan real IRs likely contributes to this close match. As shown in Fig. 2, the theoretical results for the estimators highlight that the GBM model, when paired with the Milstein method, can accurately replicate the behavior of Moroccan IRs. This makes the model a reliable tool for predicting future IR trends over the next three years. Further, Fig. 3, shows a diagram that explicitly relates to the simulation of paths of GBM and its parameter estimation using the Milstein scheme and MLE method.

By defining $x_0 = 0.3$, $\mu = 2, \sigma = 0.1$, and testing various sample sizes n, observation samples $\{X_i\}_{1 \le i \le n}$ for the GBM process (1) were generated over the time range $[0, 100]$. These samples were produced using maximum-likelihood estimators for both the EMS (with $\alpha = 0$) [13] and the Milstein scheme. Subsequently, the estimators $\widehat{\mu}$ and $\widehat{\sigma^2}$ were evaluated for each sample using the appropriate equations. The Table 1 summarizes the absolute estimation errors $|\widehat{\mu} - \mu|$ and $|\widehat{\sigma^2} - \sigma^2|$ for different values of n with the two schemes.

It is clearly observed that the Gaussian distribution and statistical independence of $\{Z_t\}_{1 \le t \le n}$, along with hypothesis $\mathcal{A}$-I, are evidently verified for these data. Despite the minimal absolute errors observed, a sample size of $n = 50$ enables the estimators to achieve practical convergence to their true values, as demonstrated by the results obtained using the R software. Moreover, as n increases, the relative absolute errors of the estimators decrease for both methods. The Milstein scheme shows absolute errors lower than those of Euler-Maruyama, both for the estimators of $\widehat{\mu}$ and $\widehat{\sigma^2}$. By observing the last two samples ($n = 5000$ and $n = 10000$), it is noticed that the absolute error $|\widehat{\mu} - \mu|$ for the Milstein method becomes slightly higher than that of Euler-Maruyama scheme. For $n = 5000$, the error for Milstein is 0.0026, while it is 0.0007 for

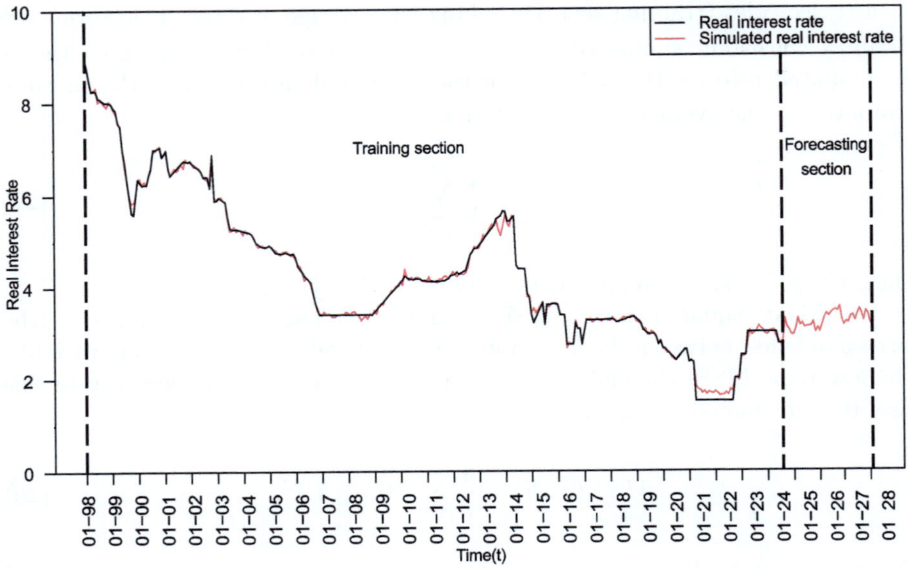

Fig. 2. The real interest rates of Morocco from January 1998 to August 2024, along with the forecasted interest rates from September 2024 to January 2028, for $n = 320$, $N = 348$, $\Delta = 0.05$, and MSE=0.2.

Euler-Maruyama. Similarly, for $n = 10000$, the error is 0.0020 with Milstein versus only 0.0003 with Euler-Maruyama. This suggests that for large sample sizes, the Euler-Maruyama method seems to provide a better estimation of μ compared to Milstein. However, Milstein remains more efficient for the estimation of σ^2, with significantly lower errors for all values of n. Thus, the Milstein scheme, by accounting for the additional correction term in its approximation, provides greater precision in parameters estimation compared to Euler-Maruyama, especially in terms of volatility parameter.

4.1 Assumptions and Robustness

The proposed methodology relies on three key assumptions: (i) independence of Wiener increments (ΔB_i), (ii) strict positivity of X_t (ensured by $x_0 > 0$), and (iii) time-invariant parameters (μ, σ). The i.i.d. property of ΔB_i ensures the validity of the likelihood function (14) and the consistency of MLE estimators. While financial data may exhibit mild autocorrelation, our empirical results in Table 1 demonstrate robustness, even for small n, with the Milstein scheme outperforming Euler-Maruyama in precision. The positivity constraint reflects the economic reality of asset prices and interest rates, while the constant-parameter assumption proves reasonable for short-to-medium term forecasting horizons, as validated by our low MSE (0.2) in modeling Morocco's IRs (Fig. 2). For cases where these assumptions might be violated (such as scenarios with

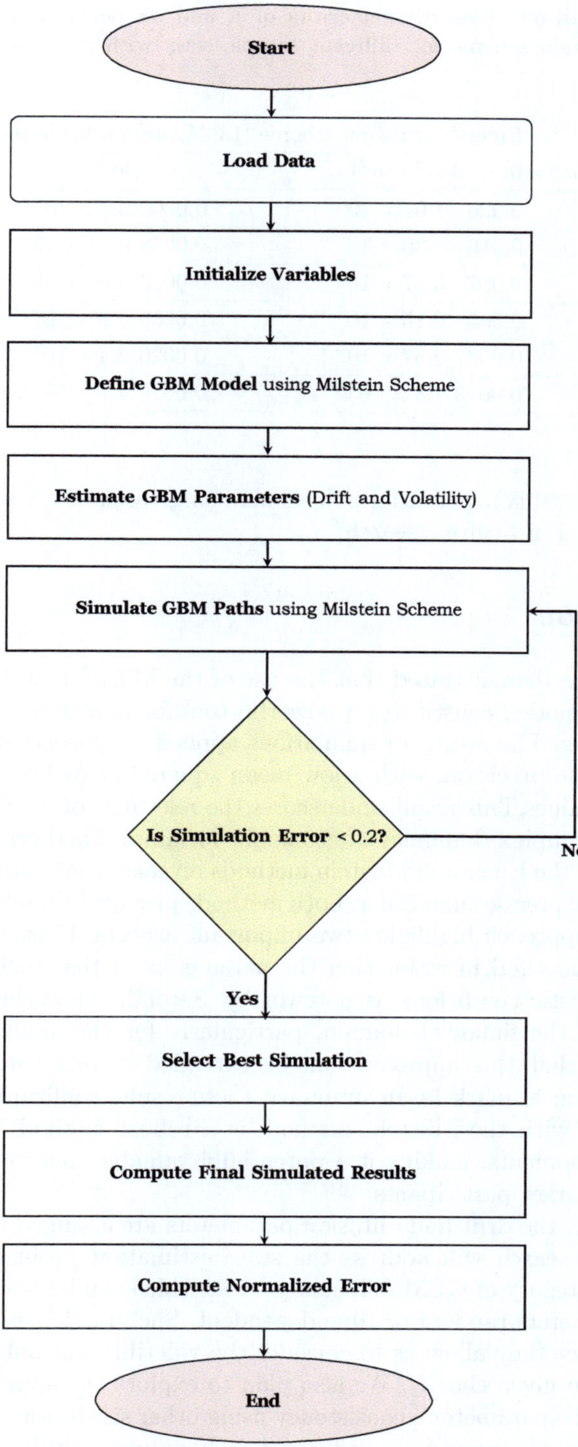

Fig. 3. Flowchart diagram of the GBM simulation.

Table 1. Comparison of estimation errors of $\widehat{\mu}$ and $\widehat{\sigma^2}$ between Euler-Maruyama scheme and Milstein scheme for different sample sizes with $x_0 = 0.3$, $\mu = 2, \sigma = 0.1, \Delta = 0.05$

	Euler-Maruyama scheme [13]		Milstein scheme proposed	
Sample size n	$\lvert\widehat{\mu} - \mu\rvert$	$\lvert\widehat{\sigma^2} - \sigma^2\rvert$	$\lvert\widehat{\mu} - \mu\rvert$	$\lvert\widehat{\sigma^2} - \sigma^2\rvert$
50	0.020	1.01×10^{-5}	0.0060	5.7×10^{-6}
100	0.016	8.36×10^{-7}	0.0058	3.11×10^{-7}
500	0.007	6.87×10^{-7}	0.0050	3.03×10^{-8}
1000	0.005	6.15×10^{-8}	0.0030	5.2×10^{-10}
5000	0.0007	9.57×10^{-9}	0.0026	2.4×10^{-10}
10000	0.0003	1.8×10^{-10}	0.0020	4.7×10^{-11}

time-varying volatility), potential extensions using stochastic volatility models could be explored in future research.

5 Conclusion

This research has demonstrated that the use of the Milstein method, combined with the GBM model, constitutes a powerful tool for modeling and forecasting real interest rates. The results of simulations applied to Morocco's interest rates reveal remarkable precision, with a low mean squared error between the simulated and real values.This result underscores the relevance of the GBM model in capturing the complex dynamics of economic variables. Furthermore, the comparison between the Euler and Milstein methods revealed that, although Milstein is generally more precise than Euler, both methods provide valuable insights. The success of this approach highlights two important aspects. First, the robustness of the Milstein method in estimating the parameters of the stochastic process, even under realistic conditions, is noteworthy. Second, the effectiveness of the GBM model in the financial domain, particularly for the analysis of interest rates, suggests that this approach can be extended to other macro-economic variables or financial markets. In summary, these results confirm that the GBM model, coupled with the Milstein method, is a robust method for forecasting economic developments, making it a potentially valuable tool for policymakers and financial market participants.

In this study, the drift and diffusion parameters are assumed to remain constant. Future research will address the state estimation problem for a much more general category of GBM diffusion process which can be obtained by making these parameters random or time-dependent. Such models may have useful applications since they allow us to consider the volatility not only time-varying but also random given the X_t. We also plan to explore the approximation and verification of the parameter's consistency using other stochastic processes, such as Heston model, renowned for its efficiency in handling multidimensional SDEs as it provides improved simulation accuracy while balancing computational efficiency, especially for large-scale datasets.

References

1. Comets, F., Meyre, T.: Calcul Stochastique et Modeles de Diffusions. Dunod, Paris (2015)
2. Black, F., Scholes, M.: The pricing of options and corporate liabilities. J. Polit. Econ. **81**(3), 637–654 (1973)
3. Capriotti, L., Jiang, Y., Shaimerdenova, G.: Approximation methods for inhomogeneous geometric brownian motion. Int. J. Theor. Appl. Finance 1–16 (2018)
4. Mariani, M.C., Florescu, I.: Quantitative Finance. Wiley, Hoboken (2019)
5. Lamberton, D., Lapeyre, B.: Introduction Au Calcul Stochastique Applique a la Finance. Ellipses, Paris (2012)
6. Bayram, M., Buyukoz, G.O., Partal, T.: Parameter estimation in a black-scholes model. Therm. Sci. **22**(Suppl. 1), 117–122 (2018)
7. Bishwal, J.P.: Parameter estimation in stochastic differential equations, 1st edn. Springer, Berlin (2008)
8. Panik, M.J.: Stochastic Differential Equations: An Introduction with Applications in Population Dynamics Modeling. Wiley, Hoboken (2017)
9. Fadugba, S.E., Adegboyegun, B.J., Ogunbiyi, O.T.: On the convergence of Euler-maruyama method and Milstein scheme for the solution of stochastic differential equations. Int. J. Appl. Math. Model. **1**(1), 9–15 (2013)
10. Milstein, G.N.: A method of second-order accuracy integration of stochastic differential equations. Theory Probab. Appl. **23**(2), 396–401 (1979)
11. Elerian, O.: A Note on the Existence of a Closed Form Conditional Transition Density for the Milstein Scheme. Economics Discussion Paper 1998-W18, Nuffield College, Oxford (1998)
12. Durham, G.B., Gallant, A.R.: Numerical techniques for maximum likelihood estimation of continuous-time diffusion processes. J. Bus. Econ. Stat. **20**(3), 297–338 (2002)
13. El Ansari, Y., Chaayra, T., El Bouanani, F., Omari, L., Amrouch, M.: Strong consistency estimators of the Breman-schwartz diffusion process based on martingales approach. J. Stat. **11**(1), e499 (2022)
14. Bao, J., Reisinger, C., Ren, P., Stockinger, W.: Milstein schemes and antithetic multilevel Monte Carlo sampling for delay McKean-Vlasov equations and interacting particle systems. IMA J. Numer. Anal. **44**(3), 2437–2479 (2024)
15. Giles, M.B., Debrabant, K., Rossler, A.: Analysis of multilevel Monte Carlo path simulation using the Milstein discretisation. Discrete Continuous Dyn. Syst. B **24**(8), 3881–3903 (2019)
16. Bank Al-Maghrib website. https://www.bkam.ma/en/Markets/Interest-rate/Lending-rates. Accessed 1 Jan 2024

Towards an Improved Environmental Scanning Model Based on Artificial Intelligence Tools

Ahmed Elkaabi[1(✉)], M'barek El Haloui[2], and Mina Elmaallam[1]

[1] School of Information Sciences (ESI), MIKS (Mediation, Information, Knowledge, Society), BP.6204–10100, Rabat, Morocco
ahmed.el-kaabi@esi.ac.ma

[2] School of Information Sciences (ESI), LYRICA (Research Laboratory in Computer Science, Data Sciences and Artificial Intelligence), BP.6204–10100, Rabat, Morocco

Abstract. Today, organizations face rapidly evolving socioeconomic and technological environments. Consequently, predicting future trends is increasingly difficult. As a result, there is a pressing demand for tools that facilitate the detection and tracking of emerging weak signals to help organizations make informed decisions and identify new opportunities. Environmental scanning, as a systematic methodological approach, enhances organizational decision-making efficacy by delivering strategic and timely insights. However, conventional environmental scanning faces limitations. This is mainly due to the rapid growth in data volume and complexity. This study addresses this challenge by proposing a conceptual model based on a theoretical approach that integrates artificial intelligence tools into the environmental scanning process. The focus is to improve the tracking, analysis, interpretation, and dissemination in order to enhance decision-making. The paper analyzes the revolutionary impacts of AI tools on the environmental scanning process, highlighting key features including machine learning (ML), predictive analytics (PA), natural language processing (NLP), and visualization tools. These AI tools will be expected to change the methods by which companies identify, analyze, process, and communicate information, consequently enhancing their capabilities to anticipate eventual trends and opportunities. The proposed framework shows significant promise, providing decision-makers with powerful tools for managing ambiguity and allowing informed decisions.

Keywords: Environmental Scanning · Artificial intelligence Tools · Decision-making

1 Introduction

Nowadays, organizations face major challenges. Specifically, predicting future trends is difficult in a rapidly changing socio-economic and technological landscape. The need for effective environmental scanning strategies is more crucial now than ever, as they enable organizations to identify emerging opportunities and reduce risks. Indeed, a firm's ability to detect weak signals, analyze them, and organize and integrate them into the development of its organizational strategy is now considered as a must and a key component

B.-E. Boudriki Semlali et al. (Eds.): DATA 2025, LNISO 81, pp. 260–267, 2025.
https://doi.org/10.1007/978-3-032-06725-8_22

for its survival and competitiveness (Choo 2001). Environmental scanning, by its systematic and methodical approach, facilitates both strategy creation and decision-making by providing decision-makers with timely and pertinent information (Abu-Rahma & Jaleel 2019). Nevertheless, traditional environmental scanning methods are increasingly insufficient. This is due to complex organizational systems and vast amounts of data in today's digital age. The emergence of artificial intelligence (AI) provides new methods to address these difficulties by enhancing the processes of information gathering, processing, and interpretation. AI tools can easily automate many aspects of the environmental scanning process, enhancing the speed and precision of data analysis, which is a significant advantage for enterprises seeking pertinent and current strategic information. This paper proposes a conceptual model that integrates AI tools into the environmental scanning methods by providing highly sophisticated tools such as natural language processing, machine learning, predictive analytics, and visualization tools. The ultimate goal of our model is to enhance data collecting, analysis, and sharing, which can allow better decision-making in complex and uncertain environments. Despite the potential benefits, there is a lack of a comprehensive model connecting AI technologies to environmental scanning processes, especially regarding their influence on organizational decision-making outcomes. This study pursues to answer two key questions. First, in which ways can AI tools improve the efficiency of environmental scanning processes, involving gathering, analysis, and dissemination, compared to traditional methods? Second, what are the impacts on decision-making? To answer these questions, our study examines the current literature, addresses the research methodology adopted, and presents a conceptual framework that outlines how AI tools can enhance the environmental scanning process. Furthermore, the study culminates by proposing recommendations for future research and practical implications for organizations seeking to implement AI-powered environmental scanning systems.

2 Theoretical Background

2.1 Environmental Scanning

Environmental scanning was first introduced by Aguilar in (1967). Later, it evolved into competitive intelligence, according to Lesca's contributions in the 1990s (Lesca 1997). The focus is on tracking external signals that could affect organizations. It encompasses transforming gathered information into actionable knowledge, which supports the strategic planning (Abu-Rahma & Jaleel 2019). The main goal is to detect changes that may affect the firm, enabling an effective response. This ability is particularly important during uncertainty periods (Abu-Rahma & Jaleel 2019). Lesca (1997) characterized environmental scanning as a repetitive process where individuals seek information actively related to external changes. As a result, they can identify opportunities while minimizing risks, which helps managers in dealing with uncertainty. The environmental scanning is much more than basic information gathering. Indeed, this systematic methodology employs innovative approaches to acquire and process predictive information, with the explicit purpose of enhancing comprehension of the business environment by converting collected data into actionable information, hence improving organizational performance (Maritz & Du Toit 2018). Daft & Weick (1984) conceptualize environmental scanning as a systematic process involving the acquisition, analysis, and holding of information regarding their external environment to guide strategic decision-making. According to Aguilar (1967) and Choo (2001), environmental scanning involves gathering and using information related to external events and trends to support management in effectively establishing the organization's future strategies.

a. The environmental scanning process

   The environmental scanning process is structured into several phases, commonly known as the intelligence cycle. These phases involve needs identification, information gathering, data analysis, and knowledge sharing information (Daft & Weick 1984).

   - Needs Identification

     The efficiency of each environmental scan starts with identifying the needs of decision-makers. This certainly includes, collecting pertinent information, targeting key stakeholders, and determining objectives for overcoming specific difficulties (Maritz & Du Toit 2018).
   - Information Gathering

     As a matter of fact, we could say that this phase highlights the tracking of relevant subjects and selection of reliable sources, involving reports, academic journals, databases, and social media platforms. This systematic process, known as "tracking" according to Lesca (1997), offers a structured and accurate approach for data collection (Daft & Weick 1984).
   - Gathered Data Analysis

     This stage is the most critical one, in which the relevance of the gathered data is being analyzed and contextualized. Analysts generate summaries and apply tools such as data mining, data visualization, and multilingual management to improve perfect comprehension and understanding. Establishing a comprehensive database improves efficacy and guarantees that expertise is no longer unavailable for decision-making (Daft & Weick 1984).
   - Dissemination of Information

The final and last phase involves delivering essential insight to the appropriate individuals at the right and correct time. This can be achieved using tools such as reports, dashboards, and briefing notes. Effective dissemination guarantees accurate interpretation, helping stakeholders in assessing trends and making informed decisions (Abu-Rahma & Jaleel 2019). Traditional environmental scanning approaches are becoming more insufficient due to the increasing amount of data. Artificial intelligence presents novel promises for optimizing data gathering, processing, and analysis, thereby improving the precision and efficiency of decision-making.

2.2 Artificial Intelligence

The concept "Artificial Intelligence" was formally introduced by McCarthy in the 1950s (Chen et al. 2021). Since its inception, this discipline has progressed considerably (Trunk et al. 2020). It is currently defined as the capacity of systems to analyze external data effectively, derive insights from it, and utilize this comprehension to achieve objectives and accomplish activities flexibly. Current AI refers to intelligent systems that may imitate human cognitive functions, including problem-solving and learning, these systems represent an advance in technology, with applications intended to operate in varied and complex situations (Chen et al. 2021).

Several AI tools are particularly relevant for environmental scanning:

a. **Artificial Intelligence Tools**

- Machine Learning (ML)

  Recent research by Wu (2024) underscores the transformative impact of machine learning on environmental scanning, allowing firms to discern hidden trends and patterns within large amounts of information. AI, through machine learning, improves precision in forecasting and helps organizations in anticipating market changes through advanced large-scale data analysis (Yaiprasert & Hidayanto 2024). Surveys such as that conducted by Wu (2024) showed that the use of past data can help firms to predict real estate trends, estimate property prices, and suggest improvements for better returns.
- Natural Language Processing (NLP)

  Another important AI tool is natural language processing (NLP). It is considered, according to numerous studies, as an approach that aims to facilitate the automation of data extraction and analysis from vast amounts of unstructured information, including social media content, news articles, and reports, hence increasing their accuracy and relevance for decision-making (Griol-Barres et al. 2020). In the same way, NLP enables the identification of essential subjects, entities, and sentiments in documents, hence aiding decision-makers in comprehending market trends and rival strategies (Trunk et al. 2020).
- Predictive analytics

  Thirdly, predictive analytics is another key AI technique. Scholars have long debated the impact of the implementation of predictive analytics techniques, which help to identify potential risks and opportunities, leading to more robust decisions (Chen et al. 2021). Moreover, this technique can facilitate the classification of market behaviors, thereby enhancing organizations' understanding of consumer patterns, which provides valuable insights into purchasing tendencies and industry fluctuations. This comprehension empowers enterprises to tailor their approaches to better align with market requirements, thus enhancing their competitive positioning (Chong et al. 2020).
- AI-driven dashboards and visualization tools

And last but not least, we reached the AI-driven dashboards and tools, and we could confirm that these instruments provide data via interactive charts, prediction models, and varied visualizations that enhance the fast understanding and interpretation of information. Data can be presented easily in clear, actionable formats tailored for different stakeholders (Sarikaya et al. 2019). Additionally, they enable the communication of insights through real-time alerts, customized reports, or collaborative platforms, ensuring timely delivery of relevant information to the appropriate individuals (Katapally & Ibrahim 2023).

2.3 Decision-Making

Mintzberg et al. (1976) have defined decision-making as an intentional choice to act and outline the process as a series of systematic actions and components, from recognizing a triggering element to implementing and realizing an action. Lee & Cummins (2004) state that the decision-making process progresses through a sequence of steps, beginning with information analysis and resulting in a solution that involves choosing from various alternatives.

2.4 Methodology

This study adopts a theory-based research design, employing a comprehensive literature review methodology to examine artificial intelligence applications in environmental scanning. The objective is to propose a conceptual framework based on existing literature, utilizing academic articles found in the scientific literature. Data collection in this paper was conducted through systematic searches across three principal academic databases, including Web of Science, SCOPUS, and Google Scholar.

Similarly, the search strategy utilized the following key terms in combination:

("machine learning" OR "artificial intelligence" OR "natural language processing" OR "predictive analytics") AND ("environmental scanning" OR "strategic intelligence" OR "competitive intelligence") AND ("decision making" OR "decision support" OR "strategic decision"). Moreover, additional papers were selected from the references that were identified with a view to guarantee in-depth coverage.

3 Proposed Model

The overall purpose of conducting this research is to develop a conceptual framework that highlights the relationship between AI tools and environmental scanning, for which reason we have conducted a comprehensive literature review concerning these two concepts. In the same vein, the proposed model integrates AI tools into the environmental scanning process, demonstrating how AI supports decision-making through enhanced data collecting, analysis, and application. In addition, this model builds on traditional scanning frameworks (Choo 2001; Daft & Weick 1984; Aguilar 1967), incorporating AI's capabilities in handling a vast amount of heterogeneous data, detecting patterns, and providing actionable knowledge (Trunk et al. 2020) (Fig. 1).

The first stage of the environmental scanning process focuses on the needs identification, frequently refer to the emphasis and scope of scanning. It relies on identifying decision-makers' needs, determining necessary information, targeting stakeholders, and formulating objectives based on resolution issues (Maritz & Du Toit 2018).

This process benefits greatly from the AI solutions mentioned earlier. Indeed, these solutions are key to automate data identification, selecting essential insights, forecasting trends, mitigating cognitive biases, and facilitating proactive data collection and strategic responses (Chen et al. 2021; Wu 2024).

This initial step leads to the acquisition of knowledge concerning external events and trends by tracking various sources including papers, journals, databases, and social media (Lesca 1997).

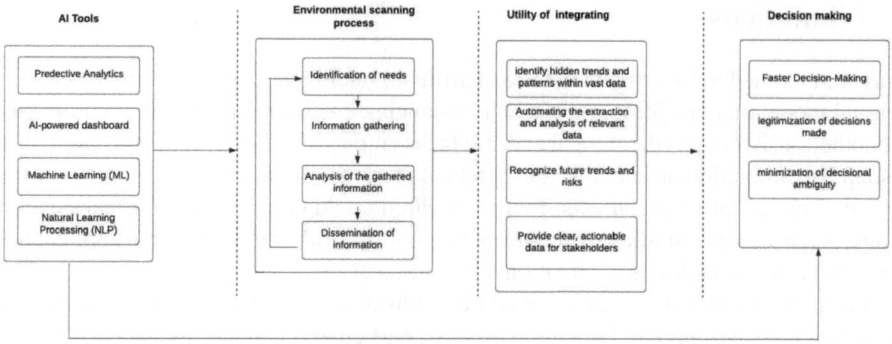

Fig. 1. Proposed model

Artificial intelligence, especially via machine learning, improves this procedure by automating data collection and extracting real-time insights from many different sources, including news, social media, and industry reports (Yaiprasert & Hidayanto 2024).

This aligns well with Aguilar's (1967) concept of undirected viewing, when organizations passively scan their environment to stay aware and maintain their strategic position.

Following data collecting, the analysis phase aims to identify significant insights. Traditional techniques frequently depend on manual analysis (Daft & Weick 1984), however AI transforms this important phase through the application of machine learning algorithms and natural language processing (NLP). These tools can quickly analyze vast datasets, reveal trends, and detect possible threats or opportunities (Choo 2001). AI reveals patterns that are not clearly visible to humans, offering more profound and nuanced insights (Aguilar 1967). This phase corresponds to conditioned viewing and enacting in Daft & Weick's (1984) model, where organizations interpret data and take informed actions.

Once this information is validated, the dissemination stage focuses on delivering essential information to the right individuals at the right time (Abu-Rahma & Jaleel 2019). AI-driven dashboards and visualization tools greatly improve this process by presenting data in clear, actionable formats designed for different stakeholders (Sarikaya et al. 2019). Furthermore, these tools use interactive charts, predictive models, and visualizations to enhance understanding and interpretation. AI also enables real-time alerts, customized reports, and collaborative platforms, which guarantees the timely and efficient dissemination of pertinent insights (Katapally & Ibrahim 2023).

The utility of integrating AI tools in the environmental scanning processes is evident in three aspects. These include, Faster decision-making, legitimization of decisions made, and minimisation of decision ambiguity. Overall, these tools improve strategic foresight, providing decision-makers with accurate, real-time insights at both tactical and strategic levels to comprehend, manage, and coordinate organizational operations and processes, which support and optimize decision-making processes (Trunk et al. 2020).

4 Limitations

This study aimed to investigate the use of artificial intelligence by proposing a new conceptual model that elucidates AI's role in enhancing environmental scanning processes. Nonetheless, it remains theoretical and still lacks empirical validation. While based on an in-depth examination of scientific and practical literature, its practical use is still untested in actual organizational contexts. Future research could employ an empirical methodology, such as case studies, to confirm the framework's relevance and applicability throughout many sectors and situations.

Furthermore, the study does not address ethical and legal implications of AI integration in the environmental scanning process, particularly concerning data privacy and algorithmic transparency issues that could significantly impact implementation viability.

Future studies must take into consideration the implementation of effective AI technologies into the environmental scanning processes while also considering the related legal and ethical challenges. Besides, an increasing need for deeper research became a must in order to overcome these limitations and guarantee the wide applicability and alignment with societal and technological advancements of the proposed framework.

5 Conclusion

This study was designed to propose a conceptual model examining the transformative impact of AI integration on environmental scanning processes and its subsequent effects on organizational decision-making effectiveness.

Through this model, we highlight the key role of different AI technologies, including machine learning, natural language processing, predictive analytics, and AI-driven dashboards and visualization tools, in transforming the environmental scanning practice. While the study does not involve empirical data collection due to time constraints, it surely serves as a foundation for future research. Researchers across the globe are invited to verify the proposed model and gather data to establish the efficiency of AI solutions in improving the environmental scanning process throughout various businesses in both private and public organizations. Despite the absence of empirical validation, we believe that this framework provides decision-makers with valuable instruments for facing uncertainty, improving strategic planning, and enabling data-driven decisions.

References

Abu-Rahma, A., Jaleel, B.: Perceived uncertainty and use of environmental information in decision making: the case of the United Arab Emirates. Int. J. Organ. Anal. **27**(3), 690–711 (2019). https://doi.org/10.1108/IJOA-07-2017-1205

Aguilar, F.J.: Scanning the Business Environment. Macmillan (1967)

Chen, J.-S., Le, T.-T.-Y., Florence, D.: Usability and responsiveness of artificial intelligence chatbot on online customer experience in e-retailing. Int. J. Retail Distrib. Manag. **49**(11), 1512–1531 (2021). https://doi.org/10.1108/IJRDM-08-2020-0312

Chong, L.R., Tsai, K.T., Lee, L.L., Foo, S.G., Chang, P.C.: Artificial intelligence predictive analytics in the management of outpatient MRI appointment no-shows. Am. J. Roentgenol. **215**(5), 1155–1162 (2020). https://doi.org/10.2214/AJR.19.22594

Choo, C.W.: Environmental scanning as information seeking and organizational learning. Inf. Res. **7**(1), 7–1 (2001)

Daft, R.L., Weick, K.E.: Toward a model of organizations as interpretation systems. Acad. Manag. Rev. **9**(2), 284–295 (1984). https://doi.org/10.2307/258441

Griol-Barres, I., Milla, S., Cebrian, A., Fan, H., Millet, J.: Detecting weak signals of the future: a system implementation based on text mining and natural language processing. Sustainability **12**(19), 7848 (2020). https://doi.org/10.3390/su12197848

Katapally, T.R., Ibrahim, S.T.: Digital health dashboards for decision-making to enable rapid responses during public health crises: replicable and scalable methodology. JMIR Res. Protoc. **12**, e46810 (2023). https://doi.org/10.2196/46810

Lee, M.D., Cummins, T.D.R.: Evidence accumulation in decision making: unifying the "take the best" and the "rational" models. Psychon. Bull. Rev. **11**(2), 343–352 (2004). https://doi.org/10.3758/BF03196581

Lesca, H.: Concepts et démarche de mise en place dans l'entreprise (1997)

Maritz, R., Du Toit, A.: The practice turn within strategy: competitive intelligence as integrating practice. S. Afr. J. Econ. Manag. Sci. **21**(1) (2018). https://doi.org/10.4102/sajems.v21i1.2059

Mintzberg, H., Raisinghani, D., Théorêt, A.: The structure of "unstructured" decision processes. Adm. Sci. Q. **21**(2), 246–275 (1976). https://doi.org/10.2307/2392045

Sarikaya, A., Correll, M., Bartram, L., Tory, M., Fisher, D.: What do we talk about when we talk about dashboards? IEEE Trans. Vis. Comput. Graph. **25**(1), 682–692 (2019). https://doi.org/10.1109/TVCG.2018.2864903

Trunk, A., Birkel, H., Hartmann, E.: On the current state of combining human and artificial intelligence for strategic organizational decision making. Bus. Res. **13**(3), 875–919 (2020). https://doi.org/10.1007/s40685-020-00133-x

Wu, A.: Improving realty management ability based on big data and artificial intelligence decision-making. PLoS ONE **19**(8), e0307043 (2024). https://doi.org/10.1371/journal.pone.0307043

Yaiprasert, C., Hidayanto, A.N.: AI-powered ensemble machine learning to optimize cost strategies in logistics business. Int. J. Inf. Manag. Data Insights **4**(1), 100209 (2024). https://doi.org/10.1016/j.jjimei.2023.100209

The Digitalization of Territorial Marketing in the Service of Development of Southern Regions, an Empirical Study

Fatima Zahra El Karmaoui[2](✉), El Ouali El Blaidi[2], and Ahmed El Hammoumi[1]

[1] Department of Management, Faculty of Economics and Management Sciences, Sidi Mohamed Ben Abdellah University (USMBA), Fez, Morocco

[2] Universite Ibn Zohr Agadir Morocco laboratoire de recherche en management et aide à la décision, Agadir, Morocco

fatmazahra.elkarmaoui.39@edu.uiz.ac.ma

Abstract. This study analyses the impact of the digitalization of territorial marketing on the tourist attractiveness of the southern provinces of Morocco. Five hypotheses were tested on a sample of 300 local actors, through regression analysis. Assumptions are made about various factors, such as the digitalization of territorial marketing, the use of digital platforms, public-private collaboration, and digital marketing initiatives.

The results confirm the positive effect of digitalization on tourism attractiveness. In particular, the use of digital platforms (websites, social networks, mobile applications) is the most decisive factor, followed by collaboration between public and private actors, the digitalization of territorial marketing strategies, and digital marketing initiatives. These elements help to strengthen the brand image and reach new market segments, including young people and international tourists. Statistical analyses, including regression tests and coefficients, validate all assumptions.

Thus, this study shows that digitalization is an essential lever to improve the tourist visibility of the southern regions of Morocco, allowing better promotion, a personalized tourism experience and greater satisfaction of visitors. The results highlight the importance of effective collaboration between public and private sectors in implementing digital strategies.

Keywords: Digitalization · Territorial marketing · Tourist attractiveness · Southern provinces of Morocco

1 Introduction

The digitalization of territorial marketing is an innovative approach to promote the tourist attractiveness of regions, especially in the southern provinces of Morocco. With the advent of new technologies, the way tourist destinations market themselves has evolved, offering new opportunities to reach and engage travelers. In the southern provinces of Morocco, which are full of unique landscapes and rich culture, the use of digital tools for tourism promotion appears to be an essential strategic lever [1].

B.-E. Boudriki Semlali et al. (Eds.): DATA 2025, LNISO 81, pp. 268–277, 2025.
https://doi.org/10.1007/978-3-032-06725-8_23

The southern provinces, including regions such as Agadir, Laâyoune, Dakhla and Guelmim, face specific challenges in terms of tourist attractiveness, particularly due to their geographical distance, limited infrastructure and low international visibility. However, the digitalization of territorial marketing, especially through social media, websites, mobile applications and digital advertising campaigns, can offer solutions to the challenges of tourist attractiveness by increasing visibility, Targeting the right audiences, and promoting real-time interaction with tourists [2]. Studies show that the use of digital platforms, such as social networks and online booking sites, plays a key role in travel decisions and in building the image of destinations [3, 4].

This empirical study aims to examine the impact of the digitalization of territorial marketing on the tourist attractiveness of the southern provinces of Morocco. It analyses how digital tools, including social networks, online booking platforms and content marketing influence the perception of destinations and motivate tourists' travel decisions. The main objective is to understand how digital strategies can enhance the tourism attractiveness of these regions while highlighting local issues, opportunities offered by digitalization, as well as obstacles to overcome [5].

2 Literature Review

The digitalization of territorial marketing and its impact on the tourist attractiveness of the southern provinces of Morocco allows for the exploration of the main concepts, theories, and approaches underlying the use of digital technologies to promote tourism destinations, while highlighting the specific challenges and opportunities in this regional context.

2.1 The Digitalization of Territorial Marketing

The digitalization of territorial marketing is defined by how digital technologies are used to promote a region, territory or destination. According to [6], territorial marketing involves a series of strategies to attract tourists, investors and talent by highlighting the distinctive characteristics of the territory. Changing consumer behaviour and the rise of digital technologies have radically transformed these practices, allowing destinations to position themselves in international markets more quickly and more specifically [7].

Digital strategies include the use of websites, social networks, blogs, mobile applications and videos, which allow territories to present them from different angles and engage directly with potential audiences [1]. Point out that the use of social networks has become central in tourism marketing, as it allows for the dissemination of information in real time and to engage users through constant interactions.

2.2 Territorial Marketing and Tourist Attractiveness

Tourism attractiveness is a dynamic concept that refers to the ability of a territory to attract visitors due to its natural, cultural and infrastructural resources. [8] explain that the attractiveness of a destination depends not only on its physical characteristics but also on its image and how it is perceived internationally. In this context, digitization plays a key role as it allows the image to be disseminated and attracts the attention of potential travelers.

For [9], digital marketing allows less known regions, such as those in the southern provinces of Morocco, to be visible on international platforms. For example, advertising campaigns on social media such as Instagram or Facebook where users share their experiences can generate increased interest in less popular destinations.

2.3 The Role of Social Media in Territorial Marketing

Social media has become an essential tool in territorial marketing. According to [10], platforms such as Instagram, Facebook and Twitter allow destinations to engage with a broad audience and deliver attractive images in real time. These platforms allow users to create and share content about their experiences, influencing other consumers' decisions. In the case of the southern provinces of Morocco, images of spectacular landscapes, beaches and local traditions can capture the attention of an international audience.

Using social media is another powerful tool to scan the trends and insights of tourists which would also help to tailor marketing strategies accordingly [11]. [12], users interacting directly with tourists on these networks values a closeness and trust relationship, which can lead to more visits and stays in these locations.

2.4 Challenges and Opportunities for the Southern Provinces of Moroccoμ

However, digital technologies provide undeniable opportunities to promote the Southern provinces, but there are still challenges. The effectiveness of these strategies can be hindered by the absence of digital infrastructure in certain regions, restricted internet access, and the management of digital transformation within local institutions. According to [13], the introduction of digital marketing strategies within the tourism sector involves sufficient collaboration with both public-private actors and an adequate level of training of local professionals from the tourism sector.

Computer skills shortage, limited marketing budgets, and finding time to manage a marketing campaign are discouraged by these challenges, but they can be overcoming by collaborative initiatives from government, local businesses and digital marketing agencies. Investing in digital infrastructure and training for local actors could significantly speed up the process of digitalizing territorial marketing and thus strengthen the tourist appeal of the southern provinces.

2.5 The Impact of New Technologies on Tourists' Perceptions

New technologies, including online booking platforms and mobile applications, not only facilitate access to information but also influence the perception of tourist destinations. [14] indicate that accessibility to information on prices, services and other traveler reviews on sites such as TripAdvisor or Booking.com increases the confidence of tourists and influences their travel decision.

However, statistically speaking, there are people who are googling images of the souks, the landscapes and the villas of Jbilat... so they act as an impulse before traveling, understanding this in the context of the southern provinces of Morocco, these are platforms whose use allow us to give visibility to the less known destinations and show an attractive image to potential travelers. However, this allows travelers to learn about certain regions of the South through real testimonies and authentic photos that add to their desire to go to these places.

2.6 Hypotheses of the Study

The study aims to identify key factors that influence the effectiveness of the digitalization of territorial marketing and its impact on tourism attractiveness in the southern provinces of Morocco, highlighting the use of digital platforms, the digitization of territorial marketing strategies, collaboration between public and private actors in the digitalization of territorial marketing and digital marketing initiatives. The research hypotheses tested by regression analysis on a sample of 300 actors operating in the region and having knowledge about digitized territorial marketing and tourist attractiveness:

Hypothesis 1: The digitalization of territorial marketing in the southern provinces of Morocco contributes to increasing the visibility and recognition of these regions as tourist destinations, which improves their attractiveness.

Hypothesis 2: The use of digital platforms (websites, social networks, mobile applications, etc.) to promote local tourist offers has a direct impact on tourists' decision to visit the southern provinces of Morocco.

Hypothesis 3: The digitalization of territorial marketing strategies allows to reach new market segments, especially the younger generations and international tourists, which strengthens the tourist attractiveness of the southern provinces.

Hypothesis 4: Collaboration between public and private actors in the digitalization of territorial marketing increases the effectiveness of tourism promotion campaigns, Thus contributing to a better brand image and greater attractiveness of the southern provinces of Morocco.

Hypothesis 5: Digital marketing initiatives promote a personalized and improved tourism experience, which increases visitor satisfaction and stimulates positive word-of-mouth, thus enhancing the attractiveness of the southern provinces of Morocco.

3 Data and Methods

This section of the study describes the analytical techniques to examine models, variables, development research assumptions and interdependence between the digitalization of territorial marketing and the tourist attractiveness in the provinces of southern Morocco.

3.1 Data Collection

The information utilized in the qualitative study presented in this article was gathered through face-to-face interviews with the various actors and stakeholders who produce tourism and local economic development respectively.

In this context, in addition to the questionnaire described in this article, we also distributed a questionnaire on a variety of themes to different stakeholders belonging to various groups, including local tourism actors (38.33%), tourism office directors (29.67%), Ministry of tourism or various government agencies (14.67%), digital marketing or technology specialists (8.33%), local residents and community leaders (6%), researchers and academics (3%). It is of particular note that personal relationships apparently had a determinant role in our research to leverage appointments with targeted respondents. Using the questionnaire instrument, we collected data from 300 participants with a preferential emphasis on participants who had sufficient expertise on destination marketing digitalization and tourism attractiveness.

3.2 Data Processing

Upon receiving the data, an analysis stage was undertaken using two multivariate statistical analyses, aided by SPSS 25 software. The first analysis involves factor **analysis, specifically, principal component analysis** (PCA) and Cronbach's alpha to purify the measurement scales of the dependent variable in the conceptual model (the digitalization of place marketing and tourist attractiveness). At this point, a clear factor structure has been documented for both forms of this variable, exceeding the threshold of acceptable limits suggested by [15] for reliability and validity of the measurement scales. Indeed, the verification of the feasibility of factor analysis is done through the Bartlett sphericity test which is based on the null hypothesis of correlation between variables, and the Kaiser-Meyer-Olkin (KMO) test which indicates whether the correlations between the different variables used for CPA are high enough to allow major components to be determined.

When the KMO is greater than 0.8, it means that the factorial structure is intelligible and stable and the factorability is good. If $0.5 < \text{KMO} < 0.8$ we can conclude that the factorability is acceptable, whereas the factorial structure is difficult to interpret and the factorability is bad if the KMO is lower than 0.5.

The second type of tests considers multiple linear regression analysis. They are used to test the overall validity of the conceptual model, and, separately, the significance of each explanatory variable.

4 Results and Discussion

In terms of the outcomes of the test of the conceptual model (see Table 1), the probability associated with the Fisher test in the ANOVA model yields a value lower than the limit of 5% error ($Sig = 0.000 < 0.05$), which means the null hypothesis is rejected and the model is globally significant. Specifically, the explained portion of the model (regression = 149.509) is significantly greater than the unexplained portion (residuals = 82.309).

Table 1. Regression Indices of ANOVA Model

ANOVA[a]						
variable		Sum of Squares	df	Mean Square	F	Sig.
1	Regression	149,509	5	29,901	106,789	,000[b]
	Residuals	82,309	294	,280		
	Total	231,818	299			

[a]Dependant variable: tourist attraction.
[b]Predictors: (Constant), The digitalization of territorial marketing, The use of digital platforms, The digitalization of territorial marketing strategies, Collaboration between public and private actors in the digitalization of territorial marketing, Digital marketing initiatives.

This observation corroborates the adjusted R-squared statistic (0.652), which also gives indications with respect to the quality of the econometric model we employed (Table 2). As such, it indicates that the variables representing the effectiveness of the digitalization of territorial marketing in the southern Moroccan provinces account for 65.2% of the variance in their tourist attraction.

Table 2. Summary of the Models

Model	R	R-two	R-two adjusted	Standard error of estimate	Edit statistics					Durbin-Watson
					Variation of R-two	Variation of F	d dl 1	ddl2	Sig. Variation of F	
1	,7 36[a]	,65 2	,650	,15164	,652	106,78 9	5	294	,000	1,653

[a]Predictors: (Constant), The digitalization of territorial marketing, The use of digital platforms, The digitalization of territorial marketing strategies, Collaboration between public and private actors in the digitalization of territorial marketing, Digital marketing initiatives.
[b]Dependent Variable: tourist attraction.

As for the evaluation of the importance of variables (Table 3), the values/probabilities from the Student's t-test that SPSS 25 yielded suggest that five are statistically significant ($t > 2$ and $sig < 5\%$). Digitalization of territorial marketing, use of digital platforms, digitalization of territorial marketing strategies (original emphasis), collaboration of public and private actors in the digitalization of territorial marketing, and digital marketing initiatives. This finding reflects the five hypotheses of our research, each with varying degrees of contribution. This is an acceptable conclusion since all of the Student's t-test scoring reflects and meets the criteria recommended by data analysis specialists. Therefore, we can accept five assumptions related to the variables.

Table 3. Coefficients of Individual Significance

Coefficients [a]							
Model		Non-standardized Coefficients		Standardized Coefficients	t	Sig.	Collinearity Statistics
		B	Error standard	Beta			Tolerance
1	(Constant)	**–,235**	**,461**		**–,381**	**,879**	
	The digitalization of territorial marketing	**,573**	**,059**	**,414**	**2,187**	**,006**	**,666**
	The use of digital platforms	**,724**	**,028**	**,646**	**2,275**	**,003**	**,532**
	The digitalization of territorial marketing strategies	**,678**	**,104**	**,436**	**2,767**	**,002**	**,703**
	Collaboration between public and private actors in the digitalization of territorial marketing	**,605**	**,127**	**,473**	**2,238**	**,007**	**,648**
	Digital marketing initiatives	**,449**	**,138**	**,427**	**2,358**	**,021**	**,925**

[a]Dependent Variable: tourist attraction.

Table 4 presents a summary of results obtained.

Table 4. Synthesis of hypotheses results

Hypotheses	Explanatory variables	Variables to explain	Coefficient β	Significance	t Student	Validation
H 1	The digitalization of territorial marketing	tourist attraction	0,414	0,006	2,187	Confirmed
H 2	The use of digital platforms	tourist attraction	0,646	0,003	2,275	Confirmed
H 3	The digitalization of territorial marketing strategies	tourist attraction	0,436	0,002	2,767	Confirmed
H 4	Collaboration between public and private actors in the digitalization of territorial marketing	tourist attraction	0,473	0,007	2,238	Confirmed
H5	Digital marketing initiatives	tourist attraction	0,427	0,021	2,358	Confirmed

5 Discussion

The results obtained confirm the central theory of this research, namely that the digitalization of territorial marketing has a significant and positive effect on the tourist attractiveness of the southern provinces of Morocco. This observation is based on several complementary dimensions:

– The employment of digital platforms is the most significant predictor of visitor appeal. The standardized coefficient ($\beta = 0.646$, $p = 0.003 < 0.01$ indicates that the use of digital platforms is the most significant explanatory variable for increased tourism appeal. This demonstrates the importance of using digital channels, institutional websites, social networks, and collaborative platforms, in order to (disseminate a positive image of Southern territories that are often not integrated into traditional tourist circuits). This conclusion supports the findings of [16], and specifically notes that the digital presence of colonized territories fosters potential visibility among hyperconnected international visitors.

- Public-private collaboration: a contact that drives efficiency in the digital strategy the large contribution of public-private cooperation ($\beta = 0.473$, $p = 0.007 < 0.01$) supports our assertion that the success of applying digitalization to territorial marketing relates to an institution synergy. Co-production of content, sharing tourism data and networking platforms are productive practices to collaboratively build a coherent and attractive territorial identity. This result corroborates with the reflections from [17] on the role territorial branding plays as a collective process.
- Digital territorial marketing strategies: a tangible impact the analysis reveals that the implementation of territorial marketing strategies based on digital tools has a relevant effect on tourism attractiveness ($\beta = 0.436$, $p = 0.002 < 0.01$). This indicates the capacity of territorial actors to build integrated offers and to enhance them through digital narratives adapted to the expectations of today's travelers. This dynamic is congruent with the conclusions of [18] on the importance of using interactive technologies in the tourist decision-making process.
- Digital initiatives are interventions, in themselves, of innovation. The results also confirm the positive impact of specific digital marketing interventions (constructed immersive content, social media campaigns, partnerships with influencers) on the appeal of the southern provinces ($\beta = 0.427$, $p = 0.021 < 0.05$). In short, these interventions help build-up an anticipated travel experience that has a positive impact on the decision to visit later. These results are consistent with the contributions of [19] regarding the role of co-creation of value in smart destinations.
- A coherent overall effect of digitalization on attractiveness

Finally, the overall analysis ($\beta = 0.414$, $p = 0.006 < 0.01$) confirms the transversal and cumulative effect of digitalization on tourism attractiveness. It is therefore not only a question of using digital tools, but also a real change in the territorial governance paradigm, which becomes more open, participatory and innovation-oriented [20].

6 Conclusion

Throughout this study, the focus is on studying the impact of the digitalization of territorial marketing on tourism attractiveness, using the southern provinces of Morocco as a context for study. Based on a literature review, five key variables of digitalization are likely to have an impact on tourism attractiveness in the southern provinces: the digitization of territorial marketing, the use of digital platforms, the digitization of territorial marketing strategies, collaboration between public and private actors in the digitalization of territorial marketing and digital marketing initiatives. The effects of these variables on tourist attractiveness are examined through a sample of 300 actors. Indeed, the regression analysis revealed the significance of the five characteristics.

Based on the results obtained, this article explores that this study makes valuable contributions to tourism attractiveness and therefore the development of the provinces in southern Morocco. However, it also suffers from certain limitations that could be leads for future research.

As regards its contributions, This study reinforced the idea supported by some researchers that the potential for tourism attractiveness and development of the southern provinces of Morocco depends on certain factors characterizing the digital transformation of territorial marketing.

References

1. Pereira, A., Teixeira, M., Mota, J.: Social media marketing and its impact on the tourism industry: a review. J. Tour. Res. **28**(2), 50–65 (2020)
2. Gretzel, U., Jamal, T.: Conceptualizing the creative tourist class: technology, mobility, and tourism experiences. Tour. Anal. **14**(4), 471–481 (2009)
3. Buhalis, D., Law, R.: Progress in information technology and tourism management: 20 years on and 10 years after the Internet the state of eTourism research. Tour. Manag. **29**(4), 609–623 (2008)
4. Araújo, D.F., et al.: Assessing man-induced environmental changes in the Sepetiba Bay (Southeastern Brazil) with geochemical and satellite data. C.R. Geosci. **349**(6–7), 290–298 (2017)
5. El Hammoumi, A., Nabil, S., Berdi, A.: Digitization of Moroccan companies: influence on organizational change and performance. an exploratory study. In: Motahhir, S., Bossoufi, B. (eds.) ICDTA 2024. LNNS, vol. 1100 pp. 176–185. Springer, Cham (2024). https://doi.org/10.1007/978-3-031-68660-3_17
6. Kotler, P., Haider, D. H., Rein, I.: Marketing Places: Attracting Investment, Industry, and Tourism to Cities, States, and Nations. Free Press (2017)
7. Govers, R.: Place Branding: Glocal, Virtual and Physical Identities, Constructed, Imagined and Experienced. Springer, Heidelberg (2013)
8. Delmas, M., Ebbesson, S.: The role of information and communication technologies in regional development. Eur. Plan. Stud. **24**(11), 1901–1914 (2016)
9. Pereira, A.C., Romero, F.: A review of the meanings and the implications of the Industry 4.0 concept. Procedia Manuf. **13**, 1206–1214 (2017)
10. Teng, W.L., Wang, S.F.: Tourism marketing in the digital age. J. Hosp. Tour. Manag. **33**, 1–9 (2017)
11. Hanna, R., Rohm, A., Crittenden, V.L.: We're all connected: the power of the social media ecosystem. Bus. Horiz. **54**(3), 265–273 (2011)
12. Carvalho, M., Santos, J.: Digital marketing in the tourism sector: a study of social media strategies in Portugal. Int. J. Tour. Res. **22**(1), 123–134 (2020)
13. Bennis, M., Fassi, M., Ali, N.: Challenges and opportunities for digital transformation in the Moroccan tourism sector. Moroccan J. Tour. Stud. **14**(2), 23–38 (2020)
14. Dujardin, F., Lemoine, J., Lecomte, P.: Tourism and technology: the influence of digital tools on the decision-making process of tourists. J. Travel Res. **58**(3), 250–267 (2019)
15. Hair, E., Halle, T., Terry-Humen, E., Lavelle, B., Calkins, J.: Children's school readiness in the ECLS-K: predictions to academic, health, and social outcomes in first grade. Early Child Res. Q. **21**(4), 431–454 (2006)
16. Aoudia, N., Merah, A., Meyer, V.: Une communication publique et territoriale pour le Maghreb (2022)
17. Duthion, B., Mandou, C.: L'innovation dans le tourisme: culture numérique et nouveaux modes de vie. De Boeck Supérieur (2016)
18. Koucheri, F.: Elaboration d'une stratégie marketing digital dans le secteur du tourisme (Doctoral dissertation, Ecole Supérieure de Gestion et d'Economie Numérique) (2024)
19. Vuignier, R.: Attractivité des territoires et place branding: étude exploratoire de la sensibilité des décideurs d'entreprise à la marque territoriale (Doctoral dissertation, Université de Lausanne, Faculté de droit, des sciences criminelles et d'administration publique) (2018)
20. Zeng, B., Gerritsen, R.: What do we know about social media in tourism? A review. Tour. Manag. Perspect. **10**, 27–36 (2014). https://doi.org/10.1016/j.tmp.2014.01.001

EdTech as a Lever of Resilience for Moroccan Educational Startups in the Context of Education 4.0

Adil Idrissi[1](✉), Fatima-Zahrae Lakhlifi[1], Souad Habbani[2], and Mohammed Abdellaoui[2]

[1] Sidi Mohamed Ben Abdellah University, Fez, Morocco
adil.01idrissi@gmail.com

[2] Economics and Management, Sidi Mohamed Ben Abdellah University, Fez, Morocco

Abstract. Educational startups play a key role in the digital transformation of the Moroccan education sector, particularly in times of crisis. In this context, educational technologies (EdTech) appear as strategic levers promoting their resilience in response to economic, health, and social disruptions. This study examines the impact of the integration of EdTech in the learning processes of startups on their ability to innovate, adapt, and ensure the continuity of their activities in times of crisis. Our objective is to propose a conceptual research model, based on a quantitative approach, mobilizing structural equation modeling (SEM) to analyze the complex relationships between the adoption of EdTech and organizational resilience. The literature review highlights the growing importance of artificial intelligence and digital technologies in improving learning processes. However, studies on the direct influence of EdTech on the resilience of educational startups remain limited. To fill this gap, we formulated four hypotheses exploring the links between EdTech adoption, alignment with institutional programs, and the resilience of educational startups. The results will indicate whether EdTech adoption is significantly associated with organizational resilience. In terms of perspectives, this research paves the way for further studies on other factors influencing learning in times of crisis and the development of Education 4.0. Furthermore, future research could benefit from a mixed approach combining qualitative and quantitative analyses, to enrich the understanding of the value of investments in educational technologies and institutional collaborations.

Keywords: EdTech · resilience · startups · Morocco · education 4.0

1 Introduction

The rapid rise of online learning platforms in Morocco, catalyzed by the COVID-19 pandemic, has fostered an unprecedented dynamic of innovation in the information and communication technology (ICT) and educational technology (EdTech) sectors (Yuniarty et al. 2024). By transforming traditional pedagogical approaches, these technologies mobilize diverse hardware and software tools to optimize educational processes and offer

B.-E. Boudriki Semlali et al. (Eds.): DATA 2025, LNISO 81, pp. 278–288, 2025.
https://doi.org/10.1007/978-3-032-06725-8_24

interactive learning experiences tailored to learners' needs (Salah Baina 2021). However, while the integration of EdTech has demonstrated its potential in terms of accessibility and personalization of learning, its influence on the resilience of educational startups, particularly in times of crisis, remains underexplored.

In Morocco, the rise of educational startups has helped redefine the education landscape by developing innovative teaching resources and services that meet learners' specific expectations. These companies play a central role in reducing inequalities in access to education by providing flexible digital solutions adapted to the country's socioeconomic realities. They also contribute to the evolution of learning practices by facilitating the acquisition of digital and transversal skills (Salah Baina 2021). However, although EdTech is often perceived as a lever for innovation, its role in strengthening the resilience of educational startups in the face of economic, health, or social crises remains an underinvested field of research.

While several studies have studied the impact of EdTech on improving learning and educational accessibility (Yuniarty et al. 2024), their direct effect on the ability of educational startups to overcome disruptions and ensure their sustainability remains largely unknown. This gap is all the more worrying in emerging economies, where these companies must contend with structural challenges such as a lack of digital infrastructure, high implementation costs, and the absence of appropriate regulatory frameworks. Given these challenges, the central question guiding this study is: to what extent does the integration of EdTech contribute to strengthening the resilience of educational startups in Morocco in the face of crises?

To provide answers, this research proposes a conceptual model based on an in-depth literature review and utilizing a quantitative approach. The objective is to analyze the interactions between the adoption of EdTech, its alignment with institutional programs, and its impact on the organizational resilience of educational startups. By filling a gap in the existing literature, this study aims to understand the mechanisms by which educational technologies influence startups' ability to adapt to crises and ensure the sustainability of their activities.

2 Conceptualization and Formulation of Hypotheses

2.1 Educational Startups and Resilience

In this context, we focus on educational startups and their challenges in the face of resilience. Successful educational startups are making remarkable changes in the education sector, transforming various aspects of education and the lives of different social groups through innovative and creative solutions (Cho et al. 2019; Kakavand-Kerdi et al. 2018).

This orientation of launching startups in the education sector, offers an equitable distribution of educational opportunities, promotes learning in all regions, increases learner engagement (using modern methods and tools), which encourages the development of vocational education with new approaches and technologies (Hackman & Reindl 2022). This helps government authorities streamline educational processes, while creating more jobs in the education sector, following the example of successful sectors such as transport and retail (Kakavand-Kerdi et al. 2018).

Many studies have focused on organizational resilience, which depends on knowledge-based standards and practices that enable risk managers to identify, assess, manage and respond to complex, dynamic and interconnected threats (Blyth et al. 2025). However, the resilience of educational startups is enhanced by their agility and ability to rapidly adopt digital technologies (Boin et al. 2021).

What we have cited above takes a favorable perspective on educational startups, but lacks empirical evidence to support their real impacts. There is a contradiction between organizational resilience based on stabilized frameworks and that of startups, supposedly resulting from their agility and rapid adoption of technologies. Moreover, although these companies are presented as a lever for educational equity, their commercial logic can limit their social reach, especially in contexts where access to digital infrastructures remains unequal. Finally, the argument would gain strength with more recent references analyzing the economic viability and institutional integration of these startups.

2.2 Impact of EdTech on the Resilience of Educational Startups

The use of educational technologies (EdTech), has become a key lever of resilience for educational startups, adopting innovation in educational institutions (Zammit et al. 2025). According to (Couture 2015), in online teaching, the teacher and learner can use online resources according to their needs and learning objectives. They also enable rapid adaptation in times of crisis, thanks to their agility and their integration of ICT in the COVID-19 test (Rechidi and Bennani 2020). EdTech also enables educational startups to diversify their services by offering personalized content and creating unique learning opportunities (Seyedeh Yasamin Sajjadi 2023).

EdTech is presented as a key factor of resilience for educational startups, but this vision hides several limitations. Unequal access to technologies, the pedagogical challenges of online learning, and the lack of analysis of negative effects, such as cognitive overload and the digital divide, weaken the argument. In addition, some references are outdated, while the field is evolving rapidly. Finally, although the personalization of content is an asset, its real effectiveness in terms of quality and engagement deserves a more in-depth analysis. A critical approach based on recent data would thus strengthen the reflection.

2.3 Hypothesis Formulation

While educational technologies (EdTech) have been widely recognized for their role in adapting teaching to the individual needs of learners (Pozo J-I et al. 2021), their impact on the resilience of educational startups remains a topic requiring in-depth analysis. Indeed, although some research highlights their potential in terms of collaboration, interactivity, and learner autonomy through the integration of multimedia codes (Collins & Halverson 2009), these benefits are not sufficient to guarantee the sustainability of startups in specific contexts. The case of Morocco illustrates this complexity well, where major obstacles such as inequalities in access to digital infrastructure, high implementation costs, and the absence of clear regulatory frameworks hinder the widespread adoption of EdTech. It, therefore, appears necessary to examine in more detail the interactions between the integration of these technologies, the structural constraints specific to the

Moroccan market, and the capacity of educational startups to ensure their resilience in the face of crises and developments in the sector.

Based on existing work and considering the specificities of the context studied, we formulated four hypotheses to explore the links between the use of EdTech and the sustainability of educational startups. **[H1]** postulates that adopting EdTech, in particular artificial intelligence and virtual reality, promotes the continuity of startups' activities in times of crisis, by facilitating the implementation of flexible educational solutions adapted to new challenges. **[H2]** questions the ability of startups to quickly adjust their services in response to market developments, by highlighting the role of digital technologies in innovation and organizational adaptability. **[H3]** examines how the integration of EdTech contributes to diversifying the offer of educational startups and strengthens their agility, by allowing them to better meet the changing expectations of learners. Finally, **[H4]** analyzes the influence of collaborations between startups and academic institutions on their resilience, by studying the extent to which the adoption of educational technologies validated by institutional bodies affects their credibility and their anchoring in the Moroccan educational ecosystem.

The developed hypotheses in this study are based on a solid theoretical framework aimed at shedding light on the dynamics between the adoption of educational technologies (EdTech) and the resilience of educational startups. The Resource-Based Value (RBV) theory (Barney 1991) highlights the strategic importance of exploiting technological assets to generate a sustainable competitive advantage. In this context, the integration of artificial intelligence and virtual reality is not simply a lever for innovation, but a determining factor in strengthening the robustness and competitiveness of startups in the face of market turbulence (H1 & H3). Furthermore, the Dynamic Capabilities Theory (Teece et al. 1997) highlights that organizational adaptability is based on the ability of companies to constantly reconfigure their resources and processes. This perspective helps justify H2, which explores the extent to which educational startups, thanks to EdTech, manage to quickly adjust their offering to meet changing market needs. Finally, the theory of the diffusion of innovation (Rogers 1962) offers a relevant framework for understanding the role of institutional actors in the acceptance and integration of EdTech. Institutional validation of technological solutions is a key factor of legitimation, directly influencing their large- scale adoption. This approach sheds light on H4, which analyzes how collaborations between educational startups and academic institutions can strengthen their credibility and promote their anchoring in the Moroccan educational ecosystem.

These hypotheses, formulated in response to the gaps identified in the literature, aim to deepen the understanding of the dynamics linking the use of EdTech to the stability and performance of educational startups. To ensure a rigorous methodological approach, their validation will require an empirical analysis based on precise indicators such as business continuity in times of crisis, the adaptability of startups to market transformations, and the influence of institutional partnerships on their sustainability.

3 Field Survey Methodology

3.1 Objective of the Methodology

In this research, we adopted a quantitative approach. This choice is based on the need to have measurable data that would allow us to establish robust statistical relationships between EdTech adoption and various resilience indicators. However, we recognize that resilience is a multidimensional concept that can be explored through complementary qualitative methods, such as case studies or in-depth interviews. Nevertheless, the quantitative approach was preferred because of its ability to produce generalizable results while allowing for a detailed analysis of underlying trends. More specifically, our study focuses on the influence of EdTech on key resilience factors, including business continuity in times of crisis, organizational adaptability, service diversification, and learner engagement.

To examine these complex relationships, we opted for structural equation modeling (SEM), a method particularly suited to simultaneously analyzing direct and indirect relationships between variables. Unlike traditional regression analyses, which are limited to direct effects, SEM allows for integrating several latent variables, thus providing a more detailed view of the dynamics at work. This methodological choice is also justified by the complexity of the interactions between EdTech and organizational resilience, requiring an approach capable of testing holistic theoretical models. However, it should be noted that our objective is not to establish strict causality but rather to identify structural associations that allow for a better understanding of the factors influencing the resilience of educational startups in the context of digital transformation. In addition, our model integrates the degree of collaboration of startups with institutional programs as an analysis variable. This factor can play a mediating role, by facilitating the adoption and integration of EdTech within educational structures, or a moderating role, by influencing the relationship between the use of technologies and the performance of startups. The objective is to assess the extent to which the involvement of institutional actors contributes to strengthening or, on the contrary, limiting the impact of EdTech on the sustainability of Moroccan educational startups. Finally, to ensure a rigorous and reliable interpretation of the results, we have precisely defined our empirical variables and indicators, thus guaranteeing a robust analysis of the dynamics between technological innovation and entrepreneurial resilience.

3.2 Variable Modeling

The following Table 1 models the study's variables, linking EdTech integration to the resilience of educational startups via mediating mechanisms such as collaboration with Moroccan institutional players.

Table 1. Variable modeling

Variable	Measurement indicators
Independent variable (VI): EdTech adoption	**→ Intensity of EdTech use** 5-point Likert scale (not at all, not very much, moderately, very much, completely) **→ The technologies used** – Artificial Intelligence – Virtual reality – E-learning platforms – And more **→ Process Integration** – Education – Administration – Marketing
Dependent variable (DV): Organizational resilience	**→ Operational continuity** – Sales: sales trends before, during and after the crisis. – Workforce: workforce trends during the crisis **→ Capacity for innovation** – Launching new products or services in times of crisis. – R&D expenditure **→ Survival in times of crisis**
Mediator variable (VM): level of collaboration with Moroccan institutional players	**→ Compliance with state programs** **→ Financial and technological support received**

Source. Authors.

3.3 Process of the Study

The following Table 2 summarizes the progress of our study, from the design of the conceptual model to the interpretation of the results. By targeting the appropriate sample, we collect the necessary data and test our hypotheses to respond to our problem.

Table 2. The study process

Process	Data analysis	Interpretation of results
I. Design questionnaire II. Sampling III. Data collection	• **Software used: SmartPLS** • **Key steps in our analysis:** **– Measurement model:** a confirmatory factor analysis (CFA) was performed to validate the latent variable indicators. **– Structural model:** we measured the relationship of significant impact between our variables using regression coefficients (β) to interpret the strength and direction of the relationships. – **Model fit indicators:** we checked the quality of our model using the following indicators: → **Chi-square (χ2):** Should be insignificant, indicating a good fit **→ CFI (Comparative Fit Index):** Must be > 0.90, measures goodness of fit compared to a null mode **→ RMSEA (Root Mean Square Error of Approximation):** Must be < 0.08, for a good fit taking into account model complexity	**1. Direct effects analysis** – The impact relationship between the independent and dependent variables was examined through the regression measure (β) to guide the effect. "The direct effect of EdTech adoption on startup resilience is significant, implying that EdTech directly improves organizational resilience… **2. Analysis of mediating effects:** – The effect of the mediating variable on the influence of the strength and direction of the variables "assuming that organizational support moderates the impact of EdTech on organizational resilience, meaning that startups benefit from institutional collaborations benefit more from EdTech" was analyzed **3. Analysis of indirect effects:** - Standardized effect: indicates the strength of the relationship between the independent and dependent variable "e.g. EdTech adoption has an effect of β = 0.67, which is a key factor in resilience" –Total effect: the sum of direct and indirect effects **4. Validation of hypotheses:** – Through the three aforementioned analyses. We tested the validity of our hypotheses by comparing them with the results obtained

Source. Authors.

3.4 Results and Discussion

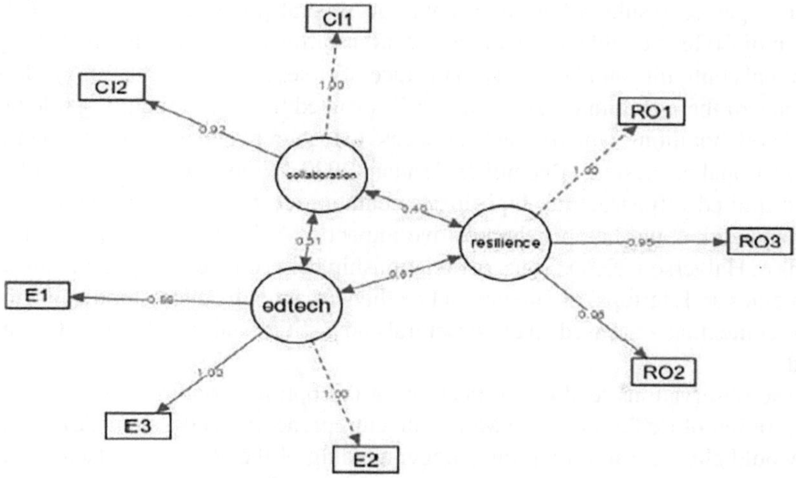

Fig. 1. Causal relationships between our latent variables **Source**. Authors using Jamovi

This figure shows the causal relationships between our latent variables, and also focuses on the impact of collaboration and EdTech on the main endogenous variable, resilience. Resilience is significantly impacted by two main factors: institutional collaboration and EdTech adoption. The regression coefficient between collaboration and resilience is 0.51, indicating a moderate direct effect, while the impact of EdTech is more pronounced, with a coefficient of 0.67, indicating a very significant influence on the variable to be explained. In addition, resilience is indirectly influenced by the manifest variables associated with collaboration (CI1 and CI2) with coefficients of 0.92, and by the outcomes (RO1, RO2, RO3), where the regression coefficients successively range from 0.95 to 0.99; implying a very strong relationship. These relationships indicate the importance and impact of collaboration and educational technologies in strengthening the resilience of educational startups, and the figure illustrates in a clear and simple way how these factors interact and contribute to strengthening resilience in the context of this analysis. The regression coefficients (β) are all significant, which confirms the robustness of the model and the relevance of the proposed causal relationships (Fig. 1).

3.5 Theoretical Modeling and Possible Scenarios

The rise of educational technologies is profoundly transforming the dynamics of educational startups, simultaneously influencing their business models and organizational resilience. The adoption of EdTech thus represents a dual opportunity: it promotes both improved accessibility and quality of learning while strengthening startups' ability to adapt to crises and market changes. However, while their effects on learning and learner engagement have been widely documented (Yuniarty et al. 2024; Salah Baina 2021), their contribution to startup sustainability remains an open question, requiring further exploration of the conditions under which these technologies can truly strengthen organizational stability.

The expected results allow us to envisage several possible trajectories. Effective adoption of EdTech could have a direct effect on startups' resilience by improving their operational continuity and flexibility in the face of crises (Pozo J-I et al. 2021). However, the impact of these technologies cannot be dissociated from the structuring role of institutional collaborations, which determine access to resources and increased recognition in the educational ecosystem (Rechidi & Bennani 2020). Conversely, structural obstacles such as limited infrastructure, high implementation costs, and an inadequate regulatory framework could hamper the positive impact of EdTech on startup sustainability (Collins & Halverson 2009). Thus, the relationship between the adoption of educational technologies and startups' organizational resilience cannot be understood uniformly but must be contextualized based on the structural and institutional specificities of the market studied.

These observations lead to a reflection on the optimal conditions for the successful integration of EdTech into the Moroccan entrepreneurial ecosystem. A longitudinal study would allow for a more refined understanding of the effects of educational technologies on startup resilience, considering variations related to their maturity, size, and technological capabilities. Furthermore, the institutional environment and public policies play a decisive role in the development of EdTech and its large-scale adoption, influencing market regulation and the provision of resources that facilitate the adoption of these technologies by emerging startups.

The implications of this research thus extend beyond the strictly academic framework and highlight strategic issues for decision-makers and entrepreneurs. To maximize the benefits of EdTech, financial incentives, and appropriate regulatory frameworks must be put in place to stimulate innovation and encourage the institutional integration of startups. Similarly, a hybrid approach combining technological innovation and the gradual structuring of collaborations with institutional stakeholders could optimize the adoption of EdTech and ensure its long-term sustainability. Ultimately, while these technologies represent a major lever of transformation for educational startups, their effectiveness will largely depend on the institutional and economic ecosystem in which they operate, thus highlighting the need for an integrated strategic vision to ensure their success and viability.

4 Conclusion

This research highlights the fundamental role of educational technology (EdTech) in strengthening the resilience of educational startups, particularly in an emerging context such as Morocco. Drawing on a conceptual model based on educational technology innovation, we explored the links between EdTech adoption, integration into institutional frameworks, and startups' organizational resilience. Our results suggest that companies that invest in EdTech solutions tailored to learners' needs and validated by institutions benefit from a better ability to maintain their operations during times of crisis and adapt to market fluctuations. However, certain methodological and contextual limitations must be considered. First, the results obtained are specific to Morocco, where disparities in digital infrastructure, political orientations, and levels of technology adoption vary considerably. This specificity may limit the transferability of the findings to other countries or regions with different economic and educational dynamics. Furthermore, this study is based on a cross-sectional analysis, which does not allow us to capture the temporal evolution of EdTech's impact on startup resilience. Finally, factors such as startups' size, maturity, and technological capacity significantly influence their adoption of EdTech, introducing variability that merits further exploration.

With this in mind, future work should prioritize longitudinal studies to better understand how EdTech adoption impacts organizational resilience over the long term. It would also be relevant to broaden the analysis to include other strategic factors, such as international collaborations, public policies supporting technology startups, and sustainable business models that promote EdTech integration. Furthermore, adopting a mixed-methods approach, combining quantitative and qualitative analyses, would allow us to refine our understanding of the dynamics at play and provide a more nuanced assessment of the value of technology investments in the education sector.

References

Barney, J.B.: Firm resources and sustained competitive advantage. J. Manag. **17**(1), 99–120 (1991)

Blyth, M.S.: Knowledge effective organisational resilience. J. Bus. Continuity Emerg. Plan. **18**(2), 126–155 (2025)

Boin, A.H.: Organizational resilience: adapting to crises in the modern age. Public Adm. Rev. **81**(3), 450–460 (2021)

Cho, E.M.: Une étude qualitative sur les motivateurs et les obstacles affectant l'entrepreneuriat chez les Latinas. Gend. Manag. Int. J. **34**(4), 326–343 (2019)

Collins, A.e.: Repenser l'éducation à l'ère du numérique. Teacher's College Press, New York (2009)

Hackman, S.T.: Challenging edtech: towards a more inclusive, accessible and purposeful version of edtech. Knowl. Cult. **10**, 7–21 (2022)

Kakavand-Kerdi, M.M.: Reviewing the Global Experiences of Start-Up Companies in the Field of Education, from the Series of Reports on the Review of Global Experiences of Start-Up Companies. DaneshbanianFanawar (2018)

Nidal Rechidi, H.B.: L'intégration pédagogique des TIC à l'épreuve de la crise covid-19: Quels enseignements à tirer? Revue Marocaine de l'évaluation et de la Recherche en Education (4) (2020). https://doi.org/10.48423/IMIST.PRSM/rmere-v4i4.20979

Pozo, J.-I., Echeverría, M.-P.P., et al.: Teaching and learning in times of COVID-19: uses of digital technologies during school lockdowns. Front. Psychol. (2021)

Rogers, E.M.: Diffusion of Innovations. Free Press (1962)

Salah, B.: Quels sont les enjeux de l'Ed Tech au Maroc? J. Médias **24** (2021)

Sajjadi, S.Y., Hejaziyan, S.F.: Faire une classe virtuelle du FLE en Iran: Les défis et les avantages pédagogiques **14**(1), 429–454 (2023). Tarbiat Modares University Press

Teece, D.J., Pisano, G., Shuen, A.: Dynamic capabilities and strategic management. Strateg. Manag. J. **18**(7), 509–533 (1997)

Yuniartya, Lohannes: Optimizing learning management system using UTAUT model: environment, compatibility, material content, and interactivity for enhanced usability. Multidiscip. Sci. J. (2024)

Zammit, E., De Raffaele, C., Scerri, D., Aquilina, R., Calleja, J.J., Rizzo, A.: Driving innovations: trends, prospects and challenges of implementing disruptive educational technologies within HEIs. In: Vendrell Vidal, E., Cukierman, U.R., Auer, M.E. (eds.) Advanced Technologies and the University of the Future. LNNS, vol. 1140, pp. 237–257. Springer, Cham (2025). https://doi.org/10.1007/978-3-031-71530-3_16

E-Government Educational IS, Did Moroccan E-Gov System Succeed?

Ismail Ouaadi[1(✉)] and Maoulainine Naama[2]

[1] Department of Finance, Auditing and Accounting (F.A.C), National School of Business and Management, Hassan 1st University, Settat, Morocco
ismail.ouaadi@gmail.com
[2] Economy And Management Faculty, IBN Toufail University, Kenitra, Morocco
naama.maoulainine@uit.ac.ma

Abstract. Using an empirical approach, this research investigates the effectiveness of an e-government system in the public education sector from the viewpoint of the schools management. The e-gov system implemented by educational authority is our target system. The updated DeLone and McLean IS success model's six constructs—information quality, system quality, service quality, intention to use and usage, user satisfaction, and net benefits—was used to evaluate the educational solution's performance as an information system. In order to measure the constructs implied by the model, a survey based on a likert scaled questionnaire was employed as research technique in this work. The empirical evaluation was carried out using exploratory, confirmatory factor analysis, and structural equation modelling. It was discovered that the research model was dependable and valid. The findings contribute to a deeper comprehension of the main success dimensions and how they interact by extending our grasp of the constructs used to gauge the e-gov system's effectiveness. We talked about the ramifications of our work. Applied as well is the DeLone and McLean IS success model. On the other hand, data quantity and the usage of other methodologies analysis are regarded as the model validation's flaws to some extent.

Keywords: e-gov systems · IS success · IS success model · Structural equations modelling

1 Introduction

Digital transition in government sector, today, is a key factor for nation success. Whereas, most of international organisation call for public sector modernisation using technology, THE OECD, for example, has published a work that presents digital transformation of the public sector and the areas that should focused on [1]. Among these areas we found education, where the Moroccan government has early deployed many efforts side by side with its partners.

B.-E. Boudriki Semlali et al. (Eds.): DATA 2025, LNISO 81, pp. 289–300, 2025.
https://doi.org/10.1007/978-3-032-06725-8_25

As an information system (IS), the Moroccan educational solution, refereed here and after as e-gov system, has received excessive attention related to its development. Recently, the e-gov system covers many management components and constitutes the core of educational IS in Morocco, which span from top schools management to student learning monitoring. It can manage over than 11 000 schools (primary, middle, and high schools) data. This e-gov system offers a diversified functionalities to enhance educational quality by accelerating the education system's digital transformation; strengthen school management; making the learning acquirement process visible and transparent by including involved person and parents in their children's education system.

The literature has explored the usefulness of this kind of IS, and many theories have been raised and applied to measure the IS success empirically. These measurement are underpinned by behavioural analysis. Among them we found DeLone and McLean model, which is the widely used model in IS success assessment. Beside that, this framework tries to assess the overall determinants of IT solution succeeding (unlike others which focus on assessing behavioural or technical determinants [2]), it's experimentally examined in a number of research and garnered a lot of support. Based on literature review, the present study tries to validate empirically this model as a measurement of the success of the e-gov IS in the Moroccan context. Focusing on schools belonging to the central region, we conduct a survey based on a questionnaire distributed electronically to a target of schools managers and their teams.

The structure of our work is as follows: An introduction to the e-gov system, the DeLone and McLean model will be covered in the next part, which also provides some theoretical background on IS success measure and use of structural equation analysis. The next section, Methodology, provides a description of the data gathered, how we used it, and how our model was built. Our findings are presented and discussed in the fourth and fifth part respectively. While the final section concludes with a synopsis of our work, presents its limits, and recommendations for further study.

2 Literature Review

The e-gov system is an important education IS tool, developed by Moroccan government to manage the whole educational system. Its impact should be assessed, especially in schools governing performance. Moreover, the specificity of this tool is that is used for both private and public schools, and it feed other system that contributing to other tasks of management like HR, finance, planning and many more operational and strategic management processes. Given this specification, The e-gov system is used by most education stakeholders (students, parents, teachers and managers). Hereafter, our view is focused on the relationship of manager and the e-gov system utilisation.

In this perspective, our contribution is to evaluate the success of this system in school governing. To reach this aim, we have chosen the widely used IS success measurement in the literature [3], which was the one developed by Delone and McLean (DM). The DM model was used to assess the success of many IS such as knowledge management system, enterprise resource planning, decision support, customer relationship management and business management; also it was used in government system assessment like e-government solution (look at [3] for an exhaustive literature review).

The first model of DM back to 1992, allows measurement of IS success given six dimensions [4] information quality (IQ), system quality (SQ), user satisfaction (US), system usage (SU), individual impact (II), and organisational impact (OI). Meanwhile, this version was updated in 2003 with the following changes: intention to use (IU) was included by replacing the SU, service quality (RQ) was introduced in the model, and net benefit (NB) was introduced to combine both (II) and (OI). These dimensions are supposed to impact each other as described in the following figure (Fig. 1):

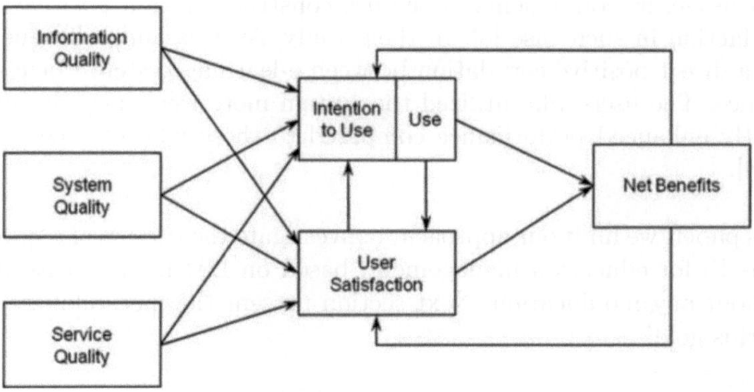

Fig. 1. DeLone and McLean version 2003 [4].

Given the research context and assessment objectives, this model can be modified to include new dimensions or to drop ones. Indeed, this model was adopted in many works that tries to assess e-government system success. In their methodology, Gil-Garciaa and Flores-Zunigac (2016) [5] identify critical success factors for e-government solutions. Based on their findings, they identified eight factors that could impact this type of solution: (1) government units' financial situation; (2) hardware, networks, and telecommunications availability; (3) front-office and back-office information system integration; (4) the existence of electronic communication between government units; (5) top management support; (6) government employees' ICT competency; (7) government-wide standardization of solutions for digital information and services; and (8) information security in government units. The paper of Antopoulos et al. (2016) describes additional causes, including: (1) inadequate project planning and management overall; (2) project scoop modifications; (3) inability to maintain budget and schedule; and (4) design-reality gap.

To measure an e-government systems from the employees' perspective adopted by 154 municipal government in Serbia, Stefanofic et al., (2016) chosen the DM 2003 model and modify it to include demographic characteristics, and based on a questionnaire distributed to their target population, they attempt to measure the success of these systems in daily employees work ([6]).

In education context, Sabeh et al., reveal DM model utilisation in e-learning systems, where they conduct a review research between 2010 and 2020 and they found that the results identified three main areas measured by the DM investigation model within the selected studies: (1) success of e-learning system, followed by (2) utilization of e-learning system, and (3) e-learning system adoption. Also, they highlight that e-learning IS success assessment should take into consideration infrastructure and organisational culture as new dimensions [7]. Moreover, Hou (2012) demonstrated that the duration of use and frequency of system usage, as components of the use construct, have a positive effect on user satisfaction in such case [8]. In their study, Avouris and colleagues (2001) observed a direct positive correlation between e-learning system usage and user performance. The users who utilized the system more frequently demonstrated significantly enhanced performance compared to those who utilized it less frequently [8].

In this paper, we limit our approach to investigate the success of e-government system, as IS for education management, based on DM model in its 2003 version, without any modification. Next section present the methodology followed to reach this goal.

3 Methodology

To realise this work, we selected 20 indicators to measure the DM dimensions. Each dimensions or constructs has three indicators at least, hereafter the distribution of each indicator and its construct.

Information QUALITY (IQ): inf_quality_1, inf_quality_2, inf_quality_3, inf_quality_4;

SYSTEM QUALITY (SQ) : sys_quality_1, sys_quality_2, sys_quality_3;

SERVICE QUALITY (RQ) : qual_serv_1, qual_serv_2, qual_serv_3, qual_serv_4;

INTENTION TO USE and USE (UI & U) : usage_1, usage_2, usage_3;

USER SATISFACTION (US) : user_sat_1, user_sat_2, user_sat_3;

NET BENEFIT (NB) : advan_1, advan_2, advan_3.

Data collection process is based on a five likert scale questionnaire distributed to target population, which is constituted by managers of public schools and their teams in Morocco central region. As result, we got a sample of 120 responses in total.

Next section will discuss reliability and validity of our measurement tool, after that, the structural model finding were presented by following the suggested approach in [11].

4 Results

First we describe the robustness of our experiment tool via two tests: (1) indicator reliability and internal consistency reliability, and (2) convergent and discriminant validity analysis. Secondly, we assess our structural model by evaluating the existence of a collinearity problems and by testing the relevance of our model.

4.1 Reliability and Validity Assessment

Internal reliability was evaluated by computing the square of the indicator loading. This coefficient let us examine the extent to which the construct of each indicator accounts for its variance [9]. An indicator loadings above 0.7 are recommended [9] and 6 are acceptable. Table 1 presents squared loadings values, which are close to or above 0.7 and mean a sufficient level of indicator reliability, except *inf_quality_4* and *qual_serv_2*. With a value of 0.47 and 0.31 (loading = 0.69, 0.56) respectively, the later indicator has the least amount of indicator-explained variance, and the indicator *advan_1* has the maximum amount of explained-variance (loading 0.92).

The second measure of reliability is *internal consistency reliability*. It is the degree to which indicators that measure the same concept are connected to each other. The first measure of this kind of reliability used in PLS-SEM is Joreskog's composite reliability rho_c. A value between 0.6 and 0.7 is considered as acceptable, whereas values from 0.7 to 0.9 range from satisfactory to good [9]. Cronbach's alpha is the second measure of internal consistency reliability, which consider the same values and interpretations of composite reliability (rho_c). The third measure, reliability coefficient rho_A, is nether conservative like Cronbach's alpha, nor liberal as rhp_c. This later measure is commonly accepted to validate internal consistency reliability.

The internal consistency reliability scores are presented in Table 2. With rho_A values of 0.66 (SQ), 0.69 (RQ), 0.8 (IQ), 0.86 (UI & U, NB), and 0.87 (US), all these reflectively measured constructs have sufficient levels of internal

Table 1. Indicator reliability.

	IQ	SQ	RQ	USE & U	US	NB
inf_quality_1	0.69					
inf_quality_2	0.67					
inf_quality_3	0.63					
inf_quality_4	0.47					
sys_quality_1		0.56				
sys_quality_2		0.67				
sys_quality_3		0.56				
qual_serv_1			0.52			
qual_serv_2			0.31			
qual_serv_3			0.61			
qual_serv_4			0.60			
usage_1				0.80		
usage_2				0.80		
usage_3				0.70		
user_sat_1					0.80	
user_sat_2					0.84	
user_sat_3					0.73	
advan_1						0.84
advan_2						0.65
advan_3						0.83

consistency reliability. Similarly, the Cronbach's alpha (0.66 for SQ, 0.68 for RQ, 0.79 for IQ, 0.85 for both UI & U and NB, and 0.87 for US) and composite reliability rho_c (0.82 for SQ, 0.81 for RQ, 0.86 for IQ, 0.91 for both UI & U and NB, and 0.92 for US) results are above the 0.70 threshold, indicating that all construct measures are reliable.

The validity assessment is evaluated by two kind of test: convergent validity and discriminant validity.

Convergent validity aims to measure if the construct converges in order to demonstrate the variance of its indicators. The Average Variance Extracted (AVE) for all indicators on each construct, is the used metric here. A threshold of 0.5 is the minimum acceptable AVE, meaning that the construct explains 50% of the variance of the indicators that make it up [9].

Table 2 shows the AVE scores along with the internal consistency reliability scores. As result, the AVE values are 0.6 for SQ, 0.51 for RQ, 0.62 for IQ, 0.77 for both UI & U and NB, and 0.79 for US, which are well above the required minimum level of 0.50. Indeed, the measures of all these constructs have a high level of convergent validity.

Table 2. Construct reliability and convergent validity.

	α	ρ_C	AVE	ρ_A
IQ	0.79	0.86	0.62	0.80
SQ	0.66	0.82	0.60	0.66
RQ	0.68	0.81	0.51	0.69
UI & U	0.85	0.91	0.77	0.86
US	0.87	0.92	0.79	0.87
NB	0.85	0.91	0.77	0.86

Discriminant Validity : the last measure of validity is discriminant validity. Fornell-Larcker is the widely used metric for such validity assessment. It assesses how different a construct is from other constructs in the existing model empirically [9]. While this metric has some limits, an alternative metric is recommended to measure discriminant validity, which is the heterotrait-monotrait ratio (HTMT). HTMT is the mean value of the indicator correlations across constructs [10]. An HTMT value above 0.9 mean that discriminant validity is not satisfied.

The Fornell-Larcker criterion assessment results are displayed in Table 3 along with the correlations between the constructs in the off-diagonal position and the square root of the AVE of the reflectively assessed constructs. For instance, all correlation values in the IQ column (0.6, 0.57, 0.41, 0.56 and 0.62) must be compared with the value of 0.78 for the square root of the reflectively measured construct IQ. In general, the reflectively assessed constructs IQ (0.78), SQ (0.77), RQ (0.72), UI & U (0.88), US (0.89), and NB (0.88) have square roots of the AVEs that are greater than their correlations with other latent variables in the model.

Table 3. FornellâĂŞLarcker criterion.

	IQ	SQ	RQ	USE & U	US	NB
IQ	0.78					
SQ	0.60	0.77				
RQ	0.57	0.59	0.72			
UI & U	0.41	0.47	0.43	0.88		
US	0.56	0.65	0.79	0.52	0.89	
NB	0.62	0.45	0.53	0.56	0.50	0.88

The HTMT values for each pair of constructs in a matrix format are displayed in Table 4. As can be observed the majority of HTMT values are obviously below the more cautious threshold value of 0.85 (Henseler et al., 2015). Note that

RQ construct has an HTMT value of 1. In such case, researchers should check whether the HTMT scores are significantly distinct from 1 or a lower threshold, like 0.9 or even 0.85, in addition to looking at the HTMT values. It is necessary to compute the bootstrap confidence intervals for this analysis, which are produced by executing the bootstrapping process. To do this, with a significance level of 0.10 ([10]). We can then run a one-tailed test at 5% and obtain 90% two-sided bootstrap confidence intervals for the HTMT values. The results are shown in Table 5, they indicate that constructs RQ and US are conceptually very similar.

Table 4. HTMT result table.

	IQ	SQ	RQ	USE & U	US	NB
IQ						
SQ	0.81					
RQ	0.79	0.87				
UI & U	0.50	0.60	0.56			
US	0.67	0.84	1.02	0.60		
NB	0.75	0.59	0.71	0.65	0.58	

Table 5. Bootstrapped results and confidence intervals for HTMT.

	Original Est.	Bootstrap Mean	Bootstrap SD	T Stat.	5% CI	95% CI
IQ -> SQ	0.81	0.82	0.09	9.42	0.67	0.95
IQ -> RQ	0.79	0.80	0.08	10.16	0.66	0.92
IQ -> UI & U	0.50	0.50	0.09	5.74	0.35	0.64
IQ -> US	0.67	0.67	0.06	10.63	0.57	0.78
IQ -> NB	0.75	0.75	0.07	10.83	0.63	0.86
SQ -> RQ	0.87	0.88	0.08	10.93	0.74	1.01
SQ -> UI & U	0.60	0.61	0.09	6.42	0.46	0.76
SQ -> US	0.84	0.84	0.07	12.27	0.72	0.94
SQ -> NB	0.59	0.59	0.13	4.65	0.38	0.80
RQ -> UI & U	0.56	0.56	0.10	5.67	0.40	0.72
RQ -> US	1.02	1.03	0.04	23.67	0.96	1.10
RQ -> NB	0.71	0.71	0.09	7.92	0.55	0.84
UI & U -> US	0.60	0.60	0.08	7.41	0.47	0.73
UI & U -> NB	0.65	0.66	0.08	8.26	0.52	0.78
US -> NB	0.58	0.58	0.08	6.92	0.43	0.70

4.2 Evaluation of the Structural Model

Estimating a number of regression equations yields structural model coefficients for the connections between the constructs. The structural model regressions need to be checked for possible collinearity problems because each set of regressor constructs has significant correlations that may skew the point estimates and standard errors. The variance inflation factor (VIF) values are determined by a method that involves using the construct scores of the regressor constructs in each regression in the structural model. Collinearity can occur at lower VIF values of 3-5, but VIF values above 5 are indicative of likely collinearity difficulties among regressor constructs [10].

Table 6. VIF values for structural model.

	IQ	IQ	RQ	UI & U	US
UI & U	1.756	1.818	1.749		
US	1.782	1.917	1.798	1.362	
NB				1.374	1.374

Clearly, All VIF values are distinctly below the threshold of 2, as shown in Table 6. We conclude that collinearity among regressor constructs is not a major problem in the structural model.

The next step in the structural model assessment process is to determine the structural paths' relevance and importance.

4.3 Structural Modelling

To evaluate the relevance of our model, we start by looking at the exogenous driver constructions' original path coefficient estimations (column: Original Est., Table 7). In results, we discover that SERVICE QUALITY (RQ) has a strong positive impact on USER SATISFACTION (US = 0.57) and UI & U significantly improves NET BENEFITS (0.41). For US → NB, SQ → US, SQ → UI & U, a like patterns of connections appears, albeit with smaller effect sizes, respectively: 0.29, 0.27 and 0.21. In contrast, path coefficient estimations of 0.19 for RQ → UI & U, 0.16 for UI & U → US, and 0.14 for IQ → UI & U show that there is a significantly less effect between these constructs. In a similar vein, IQ barely affects US (with the lowest impact coefficient of 0.05).

Now let's examine the data to determine their statistical significance. The t-values (T Stat. column, Table 7) derived from the bootstrapping should be greater than 1.943, assuming a 5% significance level. Six of the construct relationships (SQ → UI & U, t = 2.31; SQ → US, t = 2.65; RQ → US, t = 8.69; UI & U → US, t = 2.58; UI & U → NB, t = 4.53; and US → NB, t = 3.11) are among the relationships that we find to be significant. However, three of the associations (IQ → UI & U, t = 1.43; IQ → US, t = 0.71; RQ → UI & U, t =

Table 7. Path coefficient estimates, significance, and confidence intervals.

	Original Est.	Bootstrap Mean	Bootstrap SD	T Stat.	2.5% CI	97.5% CI
IQ -> UI & U	0.14	0.14	0.10	1.43	−0.05	0.33
IQ -> US	0.05	0.05	0.06	0.71	−0.08	0.16
SQ -> UI & U	0.27	0.27	0.12	2.31	0.05	0.49
SQ -> US	0.21	0.21	0.08	2.65	0.05	0.36
RQ -> UI & U	0.19	0.19	0.12	1.55	−0.05	0.43
RQ -> US	0.57	0.57	0.07	8.69	0.44	0.69
UI & U -> US	0.16	0.16	0.06	2.58	0.04	0.28
UI & U -> NB	0.41	0.42	0.09	4.53	0.24	0.59
US -> NB	0.29	0.28	0.09	3.11	0.09	0.45

1.55) do not show statistical significance. These results suggest that Moroccan government should concentrate their information and service quality efforts on enhancing their team of support and information completeness and precision (by strengthening management teams' information and service quality perceptions) rather than their system performance to maximize schools managers satisfaction. This is not surprising, considering that schools managers have a serious problems with IT infrastructure access and the number of other solutions that should deal with beside the e-gov system.

Examining the endogenous constructs' coefficient of determination (R^2) is the next step. R^2 is a measure of the explanatory power of the model that indicates the variation explained in each of the endogenous constructs [10]. Higher values of the R^2 indicate a larger explanatory capacity. The R^2 ranges from 0 to 1. R^2 values of 0.75, 0.50, and 0.25 are generally regarded in various areas as considerable, moderate, and weak, respectively [11]. However, acceptable R^2 values depend on the research setting.

Table 8. Path coefficient estimates, R^2, and adjusted R^2 values.

	UI & U	US	NB
R^2	0.27	0.69	0.37
Adj. R^2	0.25	0.68	0.36
IQ	0.14	0.05	
SQ	0.27	0.21	
RQ	0.19	0.57	
UI & U		0.16	0.41
US			0.29

Table 8, let us to examine the R^2 values of the endogenous constructs. The R^2 values of US (0.69) can be considered moderate, whereas the R^2 value of UI & U (0.27), and NB (0.37) are weak (Table 8). The weak R^2 values of UI & U and NB can be acceptable based on the research context, specifically in social science, an R^2 value as low as 0.10 is considered satisfactory.

5 Discussion

In the context of this work, based on the adopted model and the structural equations, it can be viewed that information quality, system quality, and service quality are the components of measuring the success of e-gov system in a public sector. And through an impact on intention to use, use and user satisfaction, it can also exert an impact on the success of these system. Indeed, whenever the system quality and service quality of the system are higher, user satisfaction, intention to use and use are higher. Hence, higher user satisfaction, intention to use and use of system lead to an increased success of the system. Given the DM model structure, higher user satisfaction and higher intention to use and use conduct to a higher positive effect on the benefits of using the system.

Meanwhile, if information quality, service quality and system quality satisfied by the e-gov system with high level, school manager then become more satisfied and convinced to use the system herewith. Consequently, the success of its implementation is reached, and the public administration, subject of system utilisation, takes benefits from the usage of the system. In contrast, the opposite scenario is expected. However, the owner of system should pay intention to information quality dimension also.

Back to our results, we can validate that system quality and service quality had positive impact on both user satisfaction and intention to use and use dimensions in practically utilisation of the system by schools managers and their staff in Morocco. In contrary, based on the same results, the information quality dimension not contributes to significant impact neither on user satisfaction, nor on intention to use and use dimensions. Roughly, we can conclude that according to system quality and service quality dimensions of the e-gov system the government has achieved the success of IS adoption decision, at least within the school management operations.

6 Conclusion

The aim of this work is to examine the e-gov system success, which is a government IS intended to manage the overall educational activities and processes in Morocco. Precisely, our research has focused on public school management process. Moreover, it is based on data gathered by a questionnaire from 120 members of school management teams, next, we conduct a structural equation analysis to measure the success of this system through the validation of the DM model 2003 version.

In summary, the e-gov system has proved its performance due to the satisfaction of the proposed dimensions of the DM model given the empirical analysis. But, the responsible of the e-gov system should deploy more efforts to improve the information quality dimension. Also, according to these results, many limits should be taken into account, the first one is related to the sample size. The second limit is about the indicators, where it is necessary to find the relevant ones that let assessment more conveniently. Including more relevant dimensions is another limit to be considered in farther work. The last limit of our work is related to the methodology of structural equations analysis, where predictive power analysis is recommended recently, and also performing machine learning techniques is a trend [12].

References

1. Worldbank: GovTech: the new frontier in digital government transformation. Equitable Growth, Finance & Institutions Notes (2020)
2. Ouaadi, I., El Haddad, M.: Determinants of cloud accounting adoption intention: the TOE, DOI and TAM models. J. Account. Res. Organization Econom. **4**(3), 216–233 (2021). https://doi.org/10.24815/jaroe.v4i3.21767
3. Jeyaraj, A.: DeLone & McLean models of information system success: Critical meta-review and research directions. Int. J. Inf. Manage. **54**, 102139 (2020)
4. Delone, W.H., McLean, E.R.: Information systems success measurement. Found. Trends Inf. Syst. **2**(1), 1–116 (2016)
5. Gil-Garciaa, J.R., Flores-Zúñigac, M.Á.: Towards a comprehensive understanding of digital government success: integrating implementation and adoption factors. Government Information Quarterly, vol. 37 (2020)
6. Stefanovic, D., Marjanovic, U., Delic, M., Culibrk, D., Lalic, B.: Assessing the success of e-government systems: An employee perspective. Information & Management **53**, 717–726 (2016)
7. Sabeh, H.N., Husin, M.H., Hung Kee, D.M., Baharudin, A.S., Abdullah, R.: A systematic review of the delone and mclean model of information systems success in an e-learning context (2010–2020). vol. 9 (2021)
8. Marjanovic, U., Delic, M., Lalic, B.: Developing a model to assess the success of e-learning systems: evidence from a manufacturing company in transitional economy. Inf Syst E-Bus Manage **14**, 253–272 (2016)
9. Hair, J.F., Risher, J.J., Sarstedt, M., Ringle, C.M.: When to use and how to report the results of PLS-SEM. Eur. Bus. Rev. **31**, 2–24 (2019)
10. Hair, J.F., Hult, T., Ringle, C.M., Sarstedt, M.: A primer on partial least squares structural equation modeling (PLS-SEM) 3rd ed., Thousand Oaks, CA: Sage.air (2022)
11. Hair, J.F., Hult, T., Ringle, C.M., Sarstedt, M., Danks, N.P., Ray, S.: Partial least squares structural equation modeling (PLS-SEM) using R: a workbook, p. 197. Springer Nature (2021)
12. Ibourk, A., Hnini, K., Ouaadi, I.: Analysis of the pedagogical effectiveness of teacher qualification cycle in morocco: a machine learning model approach. In: Kacprzyk, J., Ezziyyani, M., Balas, V.E. (eds) International Conference on Advanced Intelligent Systems for Sustainable Development. AI2SD 2022. Lecture Notes in Networks and Systems, vol 637. Springer, Cham(2023). https://doi.org/10.1007/978-3-031-26384-2_30

Enhancing Digital Marketing with AI: A Machine Learning Approach to Identify Fake Reviews on Amazon

Sara Ahsain(✉), Oumaima Louzar, M'hamed Ait Kbit, and Yasyn El Yusufi

IABL, STSM Doctoral Center, Abdelmalek Essaadi University, Tetouan, Morocco
s.ahsain@uae.ac.ma

Abstract. The rapid growth of e-commerce platforms has led to an increasing reliance on customer reviews for purchasing decisions. However, the prevalence of fake reviews poses a significant challenge that undermines consumer trust and product ratings. This study addresses the issue of fake comments in Amazon product reviews by employing a comparative analysis of machine learning models like Support Vector Machines (SVM), Classification and Regression Trees (CART) and Multilayer Perceptron (MLP). Amazon reviews' dataset was preprocessed through text normalization, feature extraction and also feature selection techniques (TF-IDF). These results demonstrate that SVM achieves the highest accuracy (90%) and specificity (92%), outperforming CART and MLP in detecting fake reviews.

Keywords: Fake Review Detection · Machine Learning · E-Commerce · Amazon Reviews · Classification

1 Introduction

The rapid growth of e-commerce platforms has transformed the way consumers make purchasing decisions with customer reviews playing a pivotal role in shaping consumer trust and product perception. Reviews provide valuable insights into product quality and overall satisfaction in order to help consumers make informed choices. However, the increasing frequency of fake reviews is now a significant challenge which undermines the credibility of online marketplaces and misrepresents product ratings. Fake reviews, often artificially generated to promote or demote products do not only mislead consumers but also harm the reputation of e-commerce platforms and legitimate businesses. As a result, the detection of fake reviews has become a critical area of research with implications for both consumers and businesses.

Fake reviews can take various forms like overly positive reviews written by paid individuals or bots to boost sales or also negative reviews aimed at damaging a competitor's reputation. These deceptive practices create an uneven playing field which diminishes the consumer's trust and makes it difficult for genuine reviews to stand out. For e-commerce platforms, the presence of fake reviews can lead to a loss of credibility and reduced customer loyalty; therefore, the effective development of methods to detect and mitigate fake reviews is essential to maintain the integrity of online marketplaces.

B.-E. Boudriki Semlali et al. (Eds.): DATA 2025, LNISO 81, pp. 301–313, 2025.
https://doi.org/10.1007/978-3-032-06725-8_26

This study focuses on the detection of fake reviews in Amazon product reviews, one of the largest and most influential e-commerce platforms. By employing a comparative analysis of three widely used machine learning models [Support Vector Machines (SVM), Classification and Regression Trees (CART), and Multilayer Perceptron (MLP)], this study aims to identify the most effective approach for detecting deceptive reviews by using a dataset of Amazon reviews that was preprocessed through text normalization, feature extraction and feature selection techniques. The dataset includes a binary classification of reviews into "real" and "fake" which allows for a comprehensive evaluation of model performance.

By addressing the challenge of fake reviews, this research contributes to the broader goal of improving the transparency and reliability of online marketplaces to benefit both consumers and businesses.

2 Related Work

Bibliometrics analysis-based method of research is a statistical and mathematical method used to analyze the bibliographic data including all kinds of published works such as research papers, books or articles with the aim to measure various facets of scholarly communication and research output. This method allows researchers to find publication trends, visualize authors' productivity and influence, while also giving the possibility to analysis citations, references, sources, etc.

The data in the next tables and images were interpreted using the Scopus analysis system and Bibliometrix library developed based on R language by exporting the data from the Scopus database in a Bibtex format. Table 1 shows a summary of statistics about the results obtained from running the above query.

2.1 Main Information About the Data

Table 1 provides an overview of the information gathered between 1990 and 2024 from various sources like journals and books amounting to 25,273 documents sourced from 8,322 sources; it is also showing a steady yearly growth of 16,17% in document numbers over time with each document receiving about ten citations on average. The content of the documents provided has a total of 63 934 occurrences of "keywords plus" and 42 217 Instances of "authors keywords," showcasing a wide range of topics discussed within them. The dataset involves 59 028 authors who have contributed to this collection of works with only 2 130 individuals being accountable for single-authored documents.

Table 1. Summary statistics of AI and Digital Marketing via Scopus.

Description	Results
MAIN INFORMATION ABOUT DATA	
Timespan	1990:2024
Sources (Journals, Books, etc.)	8 322
Documents	25 273
Annual Growth Rate %	16,17
Average citations per doc	10,38
DOCUMENT CONTENTS	
Keywords Plus (ID)	63 934
Author's Keywords (DE)	42 217
AUTHORS	
Authors	59 028
Authors of single-authored docs	2 130
AUTHORS COLLABORATION	
Single-authored docs	2 582
Co-Authors per Doc	3,58
International co-authorships %	17.36
DOCUMENT TYPES	
Article	6 190
Book	181
Book chapter	899
Conference paper	16 889
Conference review	128
Review	480

2.2 Annual Scientific Production in Digital Marketing and AI

This figure shows the period between 2009 and 2017 marks a more accelerated growth timestamp because of the noticeable increase in scientific production to 1 219 documents by 2017, which reflects the growing interest in this field. Finally, exponential growth was marked from 2018 till the present by the major expansion of the research documents from 1 749 to 6 730 in 2023 and 4 643 in the first eight months of 2024. The sudden drop by the end is explained by the year 2024/2025 not being concluded yet (Fig. 1).

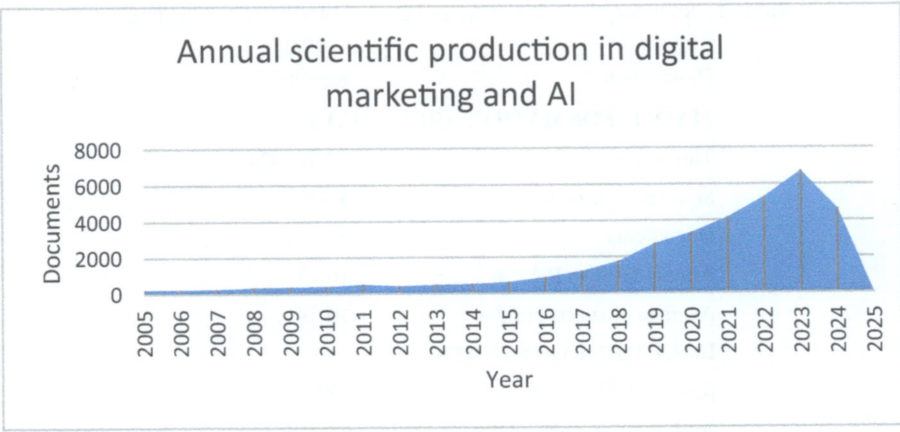

Fig. 1. Annual scientific production in the application of AI in Digital Marketing.

3 Dataset

The dataset used in this work contains 21,000 reviews with 9 columns and it captures various details about Amazon product reviews. Each entry represents a review for a specific product and it includes various attributes that describe the review and the product [1].

It includes a binary classification of reviews into "real" and "fake." Real reviews represent genuine user opinions, while fake reviews are often artificially generated, either to promote a product or to manipulate ratings.

Beyond the scope of detecting fake reviews, this study holds significant importance for e-commerce platforms and natural language processing (NLP) applications. Consumers rely heavily on helpful reviews when making purchasing decisions, and understanding how review length influences perceived trustworthiness can aid them in filtering valuable information. Similarly, sellers and brands could optimize their review strategies by identifying patterns that drive engagement and trust. Below is a detailed description of the dataset in Table 2.

The dataset covers a wide range of product categories, which could be useful for analyzing how fake reviews differ across different types of products.

The word cloud in Fig. 2. Visualizes the most frequently occurring words in the Amazon review dataset before applying the pre-processing phase. In this image, larger words indicate higher frequency which means they appear more often in customer reviews.

The main observations from this image are that the most dominant words include "looking", "product", "think" and "save", which suggests that many reviews focus on the search for a product and user expectations discussions. Additionally, the presence of "fit", "work", "love" and "durable" implies that reviewers often discuss product quality and longevity.

Data preprocessing and cleaning are critical steps in text-based analysis to ensure the dataset is well-structured and suitable for machine learning models. This process involves removing irrelevant elements such as HTML tags, punctuation, and special

Table 2. Detailed description of the dataset

Column name	Data type	Description
DOC_ID	*Int*	A unique identifier for each review
LABEL	*Object*	Indicates whether the review is labeled as "fake" or "real"
RATING	*Int*	The star rating given by the reviewer, ranging from 1 (lowest) to 5 (highest)
VERIFIED_PURCHASE	*Object*	Indicates whether the reviewer is a verified purchaser of the product. Values are "Y" (Yes) or "N" (No)
PRODUCT_CATEGORY	*Object*	The category of the product being reviewed (e.g., "Beauty", "Health & Personal Care", "Kitchen", etc.)
PRODUCT_ID	*Object*	A unique identifier for the product being reviewed
PRODUCT_TITLE	*Object*	The title or name of the product
REVIEW_TITLE	*Object*	The title of the review
REVIEW_TEXT	*Object*	The full text of the review, which includes the reviewer's opinion, experience, and sometimes additional details about the product

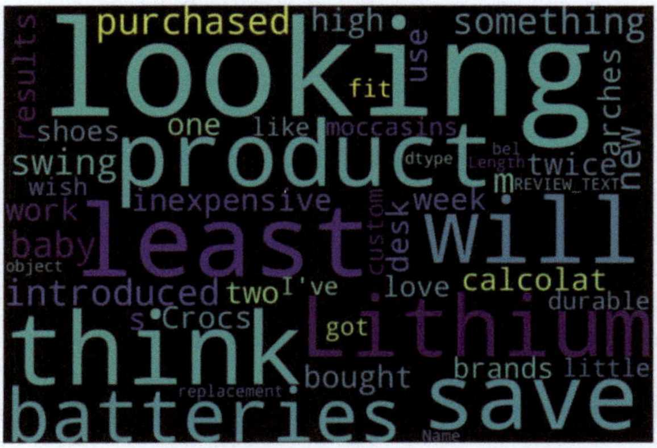

Fig. 2. Wordcloud visualization of the dataset

characters. Standard preprocessing techniques like stop word removal, text normalization (lowercasing, stemming, lemmatization), and handling spelling errors or abbreviations, contribute to improving data quality.

For machine learning models to perform accurately, the dataset must be thoroughly cleaned. Any irrelevant data, such as null values, poorly formatted entries, and unnecessary special characters or punctuation, should be eliminated. Before conducting

Exploratory Data Analysis (EDA) and preparing the dataset for modeling, the following transformation pipeline was implemented:

- Cleaning and feature engineering
- Stop-word removal
- Handling missing values
- Removing punctuation
- Label encoding
- Lemmatization

Performing EDA provided a comprehensive understanding of the dataset, allowing for the assessment of data quality, feature selection, and validation of assumptions. This step informed model selection and design while also helping to detect outliers and patterns within the data. By analyzing the dataset's characteristics and relationships, deeper insights were gained, improving the overall effectiveness of the study.

Here is an example of a review before and after applying the cleaning pipeline:

- I was looking for an inexpensive desk calcolatur and here it is. It works and does everything I need. Only issue is that it tilts slightly to one side so when I hit any keys it rocks a little bit. Not a big deal.
- look inexpensive desk calcolatur work everything need issue tilt slight one side hit keys rock little bit not big deal

4 Models

4.1 Multilayer Perceptron

Multilayer Perceptron consists of multiple simple perceptron arranged in a series of N interconnected layers with data flows in a single direction from input to output. This architecture allows MLP to perform lower-level tasks and identify hidden patterns in data that can be useful while working with classification and regression problems.

The structure of MLPs is divided into three types of layers which are input layer, hidden layers and output layer. The input layer will get the raw data which is then processed by several hidden layers. Each layer of the hidden layers consists of neurons which takes input data and applies weights, biases and activation function [2]. These transformations allow capturing the non-linear relationships between data in order for the output layer to predict the results. Figure 3 is an illustration of the Multilayer Perceptron architecture.

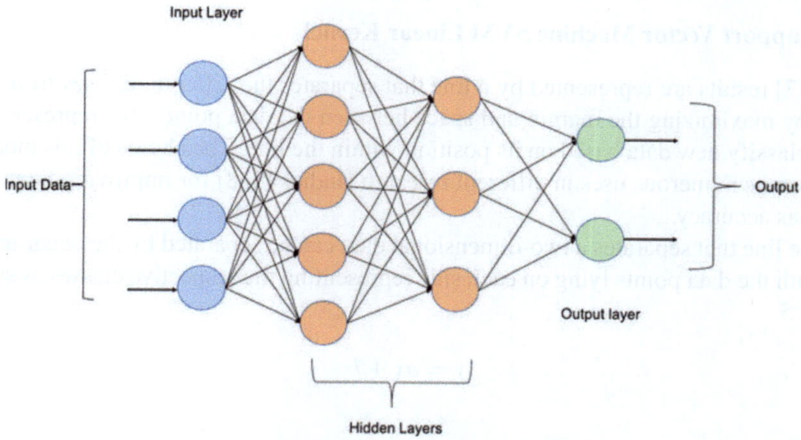

Fig. 3. The Multilayer Perceptron architecture [2].

4.2 Classification and Regression Trees (CART)

DT is a Supervised Learning algorithm that uses features or attributes of data to create a tree-like structure that predicts outcomes or classifies new data based on previous patterns. Decision trees are similar to human-level thinking, allowing users to easily interpret the data. The goal of this method is to construct a tree that can process all the outcomes at every leaf. DT algorithm is used as an Ensemble Learning method, that has proven its efficiency: [3–5] which results in an improved Sentiment Analysis performance.

CART model (Classification and Regression Trees) is a widely used decision tree algorithm for both classification and regression tasks. It works by recursively partitioning the input space into subsets based on the feature that results in the best possible split at each node. Figure 4 shows the architecture of decision tree algorithm.

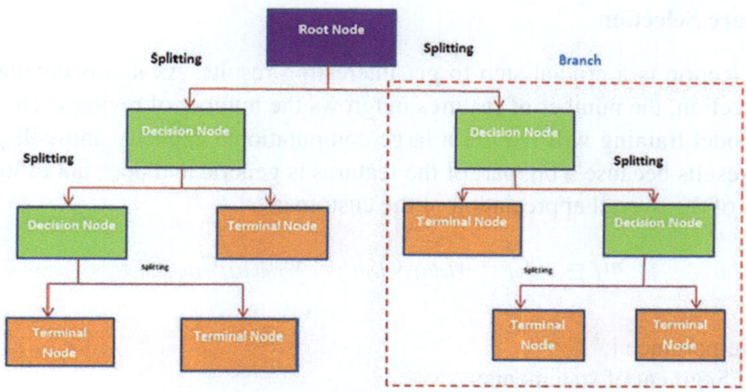

Fig. 4. Decision Tree Algorithm architecture [6]

4.3 Support Vector Machine SVM Linear Kernel

SVM [7] results are represented by a line that separates the different classes from each other by maximizing the margin and space between the data point. This representation helps classify new data based on its position within the line. The choice of this model is based on its numerous uses in different research studies [3, 8] for improving Sentiment Analysis accuracy.

The line that separates a two-dimensional data can be separated by the equation of a line with the data points lying on each side representing the respective classes as shown in Fig. 5.

$$y = ax + b \tag{1}$$

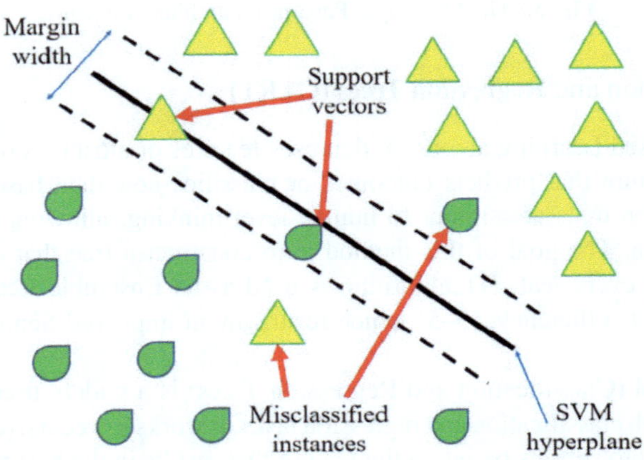

Fig. 5. Support Vector Machines (SVM)

4.4 Feature Selection

Feature selection is a crucial step to get interesting results. As it was detailed in the previous section, the number of features outgrows the number of reviews. This implies that the model training will require a large computational capacity and will probably have less results because a big part of the features is generic and does not influence the prediction of the overall appreciation of the customer.

$$ni_j = w_j C_j - w_{left(j)} C_{left(j)} - w_{right(j)} C_{right(j)} \tag{2}$$

ni_j: node importance j;
w_j: node: j Segment of statements;
C_j: J node impurity;

Left and right (j): separating a node j results in a child node.

Each feature's value is computed by taking into account, where fi_i is the relevance of fi_i feature is:

$$fi_i = \frac{\sum_{j:node\,j\,splits\,on\,feature\,i} ni_j}{\sum_{k\in all\,nodes} ni_k} \tag{3}$$

To normalize this result, it can be split on totality of the importance values of the features:

$$normfi_j = \frac{fi_j}{\sum_{j\in all\,features} fi_j} \tag{4}$$

The Random Forest grade's final feature relevance is determined by the mean of all of the trees. The significance worth of each tree is then divided by the total number T:

$$RFfi_i = \frac{\sum_{j\in all\,trees\,norm} fi_{ij}}{T} \tag{5}$$

Norm (RFfii): each feature in the tree j is normalized;

Based on the results of feature selection in the paper [11], Light Gradient Boosted Machine LightGBM represents one of the lightest and most effective methods for best features.

4.5 Feature Extraction

In Natural Language Processing, feature extraction is essential as it helps Machine Learning systems process text data by extracting its numerical representations. It also condenses data, which helps reduce its high dimensionality and improve computational efficiency. By picking relevant features, it can enhance the frameworks' generalization capabilities.

The TF-IDF algorithm [9–11] measures the originality of a word by applying a comparison between the number of occurrences of a word in a document (or in this case a review that is represented as a sentence and sometimes can be a paragraph), and the number of documents in which it has occurred.

TF represents the term frequency which is the occurrences of the word in the review. And IDF stands for inverse document frequency which is the number of documents a word had appeared in.

$$tf - idf(t) = tf(t, d) * idf(t) \tag{6}$$

In summary, TF-IDF captures only the terms that are frequent in a review but not across all the reviews. In this text analysis case, the words are considered features that grade each sentence.

5 Results

Comparing the results of Support Vector Machines (SVM), Decision Tree (CART), and Multi-layer Perceptron (MLP) provides valuable insights into the strengths and weaknesses of different machine learning approaches for a specific dataset or problem.

Comparing MLP a neural network and Cart a decision tree to linear algorithms like SVM (linear kernel) is interesting because it highlights their strengths and weaknesses. Linear models are faster and interpret simpler patterns, while MLP excels at capturing non-linear, complex relationships. Such comparisons help showcase the trade-offs between simplicity, computational efficiency and model expressiveness. Below is the list of hyperparameters used in order to compute these results:

- Linear SVM: A margin-based linear classifier that is highly effective for linearly separable data and robust to overfitting in high-dimensional spaces (Table 3).

Table 3. Linear SVM hyperparameters choice.

Parameters	Value
Kernel	Linear
Regularization param (c)	Default
Maximum iteration (max_iter)	Default
Convergence tolerance (tol)	Default
Weights for class imbalance (class_weight)	Default

- CART: A decision tree that splits data based on feature thresholds to create interpretable, rule-based models. It is non-linear and may adapt better to datasets where linear boundaries are insufficient (Table 4).

Table 4. CART hyperparameters choice.

Parameters	Value
max_depth	12
criterion	gini
min_samples_split	2 (default)
max_features	max (default)

- MLP: A neural network capable of learning complex non-linear relationships through multiple layers of neurons. It is flexible but computationally intensive and may overfit smaller datasets (Table 5).

Table 5. MLP hyperparameters choice.

Parameters	Value
solver	sgd

(continued)

Table 5. (*continued*)

Parameters	Value
learning_rate_init	0.3
learning_rate	adaptive
max_iter	250
hidden_layer_sizes	(15, 8)
random_state	1
loss	log_loss (explicit)

This study has produced notable outcomes when compared to others in the same domain, owing to the thorough examination of the various approaches discussed earlier. Each tool employed in this research underwent comparison with several alternatives from related studies, and only the most appropriate methods were selected for the final analysis (Table 6).

Table 6. Results comparison.

Name	Accuracy	Sensitivity	Specificity	Cohen's Kappa
CART	0.80	0.71	0.86	0.60
MLP	0.90	0.79	0.92	0.75
SVM	0.78	0.79	0.91	0.66

The results demonstrate the effectiveness of the evaluated models in achieving high performance across key metrics. The CART model showed a balanced performance with consistent accuracy, sensitivity and specificity which indicates its reliability in handling the dataset. The SVM model outperformed others by achieving the highest accuracy and specificity, which highlights its robustness in correctly identifying both fake and real cases. Similarly, the MLP model delivered strong results that closely matched the SVM in sensitivity and specificity, showcasing its capability to generalize well.

Overall, the high Cohen's Kappa values across all models reflect substantial agreement between predicted and actual outcomes which underscores their reliability and suitability for the task. These findings suggest that the models, particularly SVM and MLP, are well-suited for applications requiring precise and consistent predictions when it comes to detecting fake reviews.

6 Conclusion

In conclusion, this study highlights the effectiveness of machine learning models in detecting fake reviews on e-commerce platforms, with a particular focus on Amazon product reviews. By comparing the performance of Support Vector Machines (SVM),

Classification and Regression Trees (CART), and Multilayer Perceptron (MLP), we found that SVM achieved the highest accuracy (90%) and specificity (92%), outperforming both CART and MLP in identifying deceptive content. The CART model demonstrated balanced performance across accuracy, sensitivity and specificity while MLP showcased its capability to capture complex, non-linear relationships in the data.

The results underscore the importance of feature selection and preprocessing techniques, such as TF-IDF, in enhancing model performance. The high Cohen's Kappa values across all models indicate substantial agreement between predicted and actual outcomes, further validating their reliability for fake review detection. These findings have significant implications for e-commerce platforms, as they provide a robust framework for identifying and mitigating the impact of fake reviews, thereby enhancing consumer trust and improving the overall integrity of online marketplaces.

Future research could explore the integration of deep learning techniques, such as recurrent neural networks (RNNs) or transformer-based models, to further improve detection accuracy. Additionally, expanding the dataset to include reviews from multiple e-commerce platforms could provide a more comprehensive understanding of fake review patterns across different markets. Overall, this study contributes to the growing body of research on AI-driven solutions for digital marketing and fraud detection, offering practical insights for both academia and industry.

Disclosure of Interests. The authors have no competing interests.

References

1. Saxena, A.: Deception-Detection-on-Amazon-reviews-dataset, 30 December 2024. Jupyter Notebook. https://github.com/aayush210789/Deception-Detection-on-Amazon-reviews-dataset. Accessed 08 Mar 2025
2. Afan, H., et al.: Modeling the fluctuations of groundwater level by employing ensemble deep learning techniques. Eng. Appl. Comput. Fluid Mech. **15**, 1420–1439 (2021). https://doi.org/10.1080/19942060.2021.1974093
3. Yousef, M., ALali, A.: Analysis and evaluation of two feature selection algorithms in improving the performance of the sentiment analysis model of Arabic tweets. Int. J. Adv. Comput. Sci. Appl. **13**(6), 705–711 (2022). https://doi.org/10.14569/IJACSA.2022.0130683
4. Patel, H., Prajapati, P.: Study and analysis of decision tree based classification algorithms. Int. J. Comput. Sci. Eng. **6**, 74–78 (2018). https://doi.org/10.26438/ijcse/v6i10.7478
5. AminiMotlagh, M., Shahhoseini, H.S., Fatehi, N.: A reliable sentiment analysis for classification of tweets in social networks. Soc. Netw. Anal. Min. **13**(1) (2023). https://doi.org/10.1007/s13278-022-00998-2
6. Mishra, A.: Supervised machine learning classification algorithms for detection of fracture location in dissimilar friction stir welded joints (2021). https://doi.org/10.33774/chemrxiv-2021-cn2z8
7. Evgeniou, T., Pontil, M.: Support vector machines: theory and applications, vol. 2049, p. 257 (2001). https://doi.org/10.1007/3-540-44673-7_12
8. Chong, K., Shah, N.: Comparison of Naive Bayes and SVM classification in grid-search hyperparameter tuned and non-hyperparameter tuned healthcare stock market sentiment analysis. Int. J. Adv. Comput. Sci. Appl. (IJACSA) **13**(12), Article no. 12, 33/30 (2022). https://doi.org/10.14569/IJACSA.2022.0131213

9. Asri, A.M., Ahmad, S.R., Yusop, N.M.M.: Feature selection using particle swarm optimization for sentiment analysis of drug reviews. Int. J. Adv. Comput. Sci. Appl. (IJACSA) **14**(5), Article no. 5, 30 (2023). https://doi.org/10.14569/IJACSA.2023.0140530
10. Toktarova, A., et al.: Hate speech detection in social networks using machine learning and deep learning methods. Int. J. Adv. Comput. Sci. Appl. (IJACSA) **14**(5), Article no. 5, 30 (2023). https://doi.org/10.14569/IJACSA.2023.0140542
11. Robertson, S.: Understanding inverse document frequency: on theoretical arguments for IDF. J. Doc. **60**, 503–520 (2004). https://doi.org/10.1108/00220410410560582

Comparative Analysis of Handwritten Text Recognition Across Cyrillic-Using Countries

Burul Shambetova(✉), Ruslan Isaev, Nazira Abdillaeva, and Mekia Shigute Gaso

Computer Science Department, Ala-Too International University, Ankara Street 1/8, Tunguch, 720048 Bishkek, Kyrgyz Republic
burul.shambetova@alatoo.edu.kg

Abstract. The paper presents an analysis of Cyrillic Handwritten Text Recognition (HTR), high-lighting the unique challenges and technological advancements in this field across various countries. Despite the significance of HTR for digitizing historical documents and automating document processing, progress remains limited compared to Latin script, primarily due to the complexity of the Cyrillic character set and the variability in handwritten styles. The study underscores the achievements in Russian HTR, particularly through the adoption of advanced deep learning models, including CNNs, RNNs, and Transformer-based approaches. However, it identifies ongoing issues such as the scarcity of high-quality, labeled datasets and the need for more robust recognition systems. Comparative insights from other countries, including Kyrgyzstan, Uzbekistan, and additional Cyrillic-using nations such as Bulgaria, Serbia, North Macedonia, Belarus, Tajikistan, Ukraine, and Transnistria, reveal a common struggle with resource limitations and data availability. The paper advocates for increased investments and international collaboration to develop comprehensive datasets and enhance recognition technologies, ultimately aiming to improve the efficiency and accessibility of digital document processing for Cyrillic script users.

Keywords: Cyrillic script · HTR · OCR · Kyrgyz language · NLP · Machine Learning · Deep Learning

1 Introduction

Handwritten Text Recognition (HTR) plays a crucial role in the digitization of historical documents, automating document processing, and fostering advancements in linguistic research and technology. The recognition of handwritten text is particularly essential for preserving cultural heritage and making historical texts accessible to a broader audience.

The Cyrillic script, employed by numerous countries including Russia, Kyrgyzstan, Kazakhstan, Uzbekistan, and Mongolia, Bulgaria, Serbia, North Macedonia, Belarus, Tajikistan, Ukraine, and Transnistria, presents a unique set of challenges to HTR systems. The complex character set of Cyrillic, combined with the significant variability in handwritten styles influenced by regional differences, educational backgrounds, and individual writing habits, complicates the development of effective recognition systems.

B.-E. Boudriki Semlali et al. (Eds.): DATA 2025, LNISO 81, pp. 314–327, 2025.
https://doi.org/10.1007/978-3-032-06725-8_27

While there has been notable progress in the fields of Optical Character Recognition (OCR) and deep learning-based HTR, the technology dedicated to the recognition of Cyrillic script is still less advanced compared to that of Latin scripts. This disparity underscores the necessity for focused research and development efforts in Cyrillic HTR.

This study aims to systematically examine the current state of Cyrillic HTR across various countries. It highlights existing technological advancements, reviews the available datasets that contribute to HTR efforts, and identifies ongoing challenges that researchers face. In particular, this work will shed light on the importance of developing high-quality, publicly available datasets and the need for innovative modeling approaches that can cater to the diversity of handwriting encountered in practical applications. By addressing these critical areas, we foresee considerable potential for enhancing the efficiency and accuracy of Cyrillic HTR, ultimately improving accessibility to written materials for users of the Cyrillic script globally.

This expanded introduction provides a more detailed overview while incorporating elements from the document to set the context for the study.

1.1 Overview of Cyrillic Script

The Cyrillic script is a writing system used by many languages primarily spoken in Central Asia and Eastern Europe. It has evolved over the centuries. It has been made leading to various regional adaptations and stylistic modifications that reflect the linguistic and cultural diversity of the regions where it is used.

The modern Cyrillic alphabet in its Russian form consists of 33 letters. However, the other languages that use the script, such as Bulgarian, Kazakh, Kyrgyz, Mongolian and Serbian, may feature different characters and diacritics to accommodate specific phonetic requirements. Some of these languages additionally incorporate letters that do not exist in others, that adds complexity to the Handwritten Text Recognition (HTR) process.

Recognition of handwritten Cyrillic text presents significant challenges due to the different variability in handwriting styles. Regional dialects, educational background, and individual writing habits contribute to this diversity, making it difficult to accurately process cursive handwriting, where letters are usually connected and less distinct than in printed text.

In addition, the limited availability of well-annotated and publicly accessible datasets hinders the advancement of reliable HTR systems for Cyrillic languages. While there is an extensive effort made to build datasets for Latin-based scripts, similar progress for Cyrillic scripts has been slower. Machine learning models rely on large quantities of high-quality labeled data. Therefore, Machine learning models are constrained in their ability to improve recognition accuracy.

Another obstacle arises from the poor quality of many historical handwritten documents. Aging manuscripts, faded ink, and inadequate scanning methods contribute to recognition errors. Consequently, sophisticated pre-processing techniques are required to enhance text clarity before it can be processed by recognition models.

To address these issues, recent research has explored innovative solutions, including the generation of synthetic datasets to supplement existing ones. Through simulated handwriting and the application of Generative Adversarial Networks (GANs), researchers aim to overcome data scarcity. Furthermore, advancements in deep learning have shown promising results in improving Cyrillic HTR, especially Convolutional and Recurrent Neural Networks.

With a better understanding of the structural complexity of the Cyrillic script and the challenges involved in its handwritten recognition, this discussion sets the foundation for a more in-depth analysis of HTR advancements across various countries that utilize this script.

2 Analysis of Cyrillic HTR by Country

2.1 Russia

Russia has been at the forefront of Optical Character Recognition (OCR) and Handwritten Text Recognition (HTR) research for Cyrillic script.

Institutions like Moscow State University and Skolkovo Institute of Science and Technology have been instrumental in advancing recognition technologies. Researchers have developed advanced recognition algorithms that integrate deep learning, convolutional neural networks (CNNs), and recurrent neural networks (RNNs). For instance, a novel deep neural network model based on a fully gated CNN, supported by multiple bidirectional gated recurrent units (BGRU) and attention mechanisms, has been proposed to handle the complexities of Russian handwritten text. This model achieved a Character Error Rate (CER) of 4.5% and a Word Error Rate (WER) of 19.2% on the first test dataset, demonstrating its efficacy in capturing the nuances of Cyrillic handwriting [1].

A significant breakthrough in Russian Cyrillic HTR has been the use of Transformer-based models, which have improved recognition accuracy. However, challenges such as low-quality handwriting samples and limited labeled datasets still persist. To address the scarcity of annotated data, researchers have developed methods for generating synthetic handwritten text images. One approach involves creating images of handwritten Russian text based on fonts, providing a means to augment training datasets without the need for extensive manual annotation [2].

Moreover, the development of convolutional neural networks capable of recognizing images of handwritten Russian letters with high accuracy has been a notable advancement. Researchers at Siberian Federal University have achieved an algorithm accuracy of 99% in recognizing handwritten letters, which is a significant step toward efficient Cyrillic HTR systems [3].

Despite these advancements, the field continues to face challenges, particularly concerning the variability in individual handwriting styles and the need for extensive, high-quality datasets. Ongoing research aims to refine these models and develop more robust systems capable of handling the diverse range of handwritten Cyrillic text encountered in practical applications.

2.2 Kyrgyzstan

In Kyrgyzstan, digitization initiatives are emerging, yet publicly available datasets for handwritten text remain limited. In recent years, Kyrgyzstan has made notable strides in the field of Handwritten Text Recognition (HTR) for the Kyrgyz language, despite the challenges posed by limited resources and data scarcity. A significant development is the creation of the Kyrgyz-MNIST dataset, designed to facilitate the recognition of handwritten Kyrgyz letters. This dataset comprises 80,213 images of handwritten characters, each resized to 50 × 50 pixels and converted into CSV format for compatibility with machine learning applications. Researchers have employed this dataset to train Artificial Neural Networks (ANNs), achieving promising results in accurately recognizing Kyrgyz handwritten characters [5].

Additionally, efforts have been made to compile and curate resources pertinent to Kyrgyz Natural Language Processing (NLP). The "Awesome Kyrgyz NLP" repository serves as a centralized collection of open-source tools, datasets, and models tailored for the Kyrgyz language. This repository encompasses various resources, including the Manas-UdS corpus—a compilation of 1.2 million words from 84 literary texts across five genres—and the TilCorpusu corpus, which contains 100 million words from news and fiction sources. These corpora are instrumental in advancing research in both HTR and broader NLP applications for the Kyrgyz language [6].

Moreover, the M2ASR (Multilingual Minorlingual Automatic Speech Recognition) project has contributed significantly by developing speech recognition resources for minority languages, including Kyrgyz. This initiative has resulted in the creation of speech databases, text corpora, and pronunciation dictionaries, thereby enhancing the foundational infrastructure necessary for effective HTR system development [7].

Despite these advancements, challenges persist, particularly concerning the variability in individual handwriting styles and the need for extensive, high-quality datasets. Ongoing research aims to refine existing models and develop more robust systems capable of handling the diverse range of handwritten Kyrgyz text encountered in practical applications.

Challenges persist due to data scarcity and linguistic complexities, but efforts are underway to create labeled datasets to enhance HTR performance.

2.3 Kazakhstan

Kazakhstan has made significant strides in Handwritten Text Recognition (HTR) for the Kazakh language, focusing on the development of comprehensive datasets and advanced recognition models. A notable contribution is the Handwritten Kazakh and Russian (HKR) Database, which encompasses over 63,000 sentences and more than 715,699 symbols from approximately 200 writers. This dataset serves as a foundational resource for training and evaluating HTR systems in both Kazakh and Russian languages [8].

To further enhance HTR research, the Kazakh Offline Handwritten Text Dataset (KOHTD) was introduced, comprising 3,000 handwritten exam papers and over 140,335 segmented images, totaling approximately 922,010 symbols. This extensive dataset provides a diverse range of handwriting samples, facilitating the development of robust recognition models [9].

Researchers have also explored the application of Generative Adversarial Networks (GANs) to improve the accuracy of Kazakh handwriting recognition. By generating synthetic handwritten text images, GANs augment existing datasets, thereby enhancing the performance of recognition models. Studies have demonstrated that incorporating GAN-generated data can lead to significant improvements in recognition accuracy [10].

Despite these advancements, challenges persist, particularly concerning the variability in individual handwriting styles and the need for extensive, high-quality datasets. Ongoing research aims to refine existing models and develop more robust systems capable of handling the diverse range of handwritten Kazakh text encountered in practical applications.

2.4 Uzbekistan

In Uzbekistan, the transition from Cyrillic to Latin script has influenced the focus of Handwritten Text Recognition (HTR) research. Despite this shift, a substantial volume of historical and official documents remains in Cyrillic, necessitating effective recognition solutions. Efforts have been directed towards developing bilingual OCR systems capable of processing both Cyrillic and Latin scripts, ensuring accessibility to legacy documents.

Collaborative projects between local institutions and international research communities have been instrumental in advancing HTR technologies in Uzbekistan. These collaborations focus on creating annotated datasets and developing models that accommodate the unique characteristics of Uzbek handwriting. However, challenges persist due to the limited availability of high-quality, labeled data and the variability in handwriting styles influenced by regional and educational factors.

To address these challenges, researchers are exploring the use of transfer learning and data augmentation techniques. By leveraging pre-trained models and synthetically expanding existing datasets, these approaches aim to enhance the performance of HTR systems for Cyrillic Uzbek text. Ongoing efforts continue to focus on refining recognition models and expanding resources to preserve and digitize Uzbekistan's rich documentary heritage.

2.5 Mongolia

Mongolia has undertaken significant efforts to advance Handwritten Text Recognition (HTR) for the Mongolian Cyrillic script. A notable development is the creation of a dedicated dataset tailored for Mongolian Cyrillic handwritten text recognition. This dataset serves multiple purposes, including optical character recognition (OCR) and related tasks such as handwritten gender and age detection. The carefully selected sentences within the dataset offer a valuable resource for handwritten-to-translated text generation. Utilizing a Transformer-based OCR model, researchers achieved a promising Character Error Rate (CER) of 0.978 in the early stages, demonstrating the efficacy of their approach even with a relatively small dataset. The dataset is intended to be publicly shared, setting a benchmark for the community and inviting further advancements in Mongolian Cyrillic handwritten recognition [11].

In addition to dataset development, researchers have proposed an improved handwriting recognition model, named the EGA model, suitable for the features of Mongolian handwriting. This model incorporates a dual data augmentation (DDA) algorithm, combining horizontal wave transformation (HWT) and random erasing data augmentation (REDA) to enhance the training dataset. The EGA model utilizes an encoder-decoder architecture with attention mechanisms, effectively balancing recognition accuracy and model complexity. Experimental results demonstrated that the EGA model improves the recognition accuracy of Mongolian handwriting, achieving an accuracy rate of 89.322% on the test set [12].

Despite these advancements, challenges persist, particularly concerning the variability in individual handwriting styles and the need for extensive, high-quality datasets. Ongoing research aims to refine existing models and develop more robust systems capable of handling the diverse range of handwritten Mongolian Cyrillic text encountered in practical applications.

2.6 Bulgaria

Bulgaria, the birthplace of the Cyrillic alphabet, has made notable progress in Optical Character Recognition (OCR) for printed Bulgarian text. However, Handwritten Text Recognition (HTR) remains an evolving field with ongoing research efforts to improve dataset availability and model performance. A significant initiative in this regard is the Bulgarian Handwritten OCR Image Dataset developed by FutureBeeAI, which includes handwritten samples from various sources such as letters, forms, and invoices. This dataset aims to support machine learning models in achieving higher accuracy in Bulgarian HTR [13].

Another noteworthy project is the Bulgarian Shopping List OCR Image Dataset, which consists of over 2,000 handwritten shopping lists collected from more than 200 native Bulgarian participants. This dataset enhances the diversity of handwriting samples available for training recognition models.

Efforts have also been made in the field of historical document recognition. A recent study titled "Post-OCR Text Correction for Bulgarian Historical Documents" introduced benchmark datasets and synthetic data generation techniques to improve OCR accuracy for historical texts. The study highlights the importance of refining OCR outputs for Bulgarian documents written in older orthographies, ensuring better digitization and preservation of historical records.

Additionally, several open-source projects on platforms such as GitHub have contributed to Bulgarian HTR development. The ocr-to-bg project provides guidance on applying Tesseract-OCR to Bulgarian text, while the OCR by Mihaela Petkova initiative explores handwriting recognition using OpenCV and Tesseract-OCR [15].

Despite these advancements, Bulgaria still faces challenges in HTR research, particularly concerning the need for more comprehensive, publicly available handwritten datasets and advanced deep learning models tailored to Bulgarian handwriting. Future research efforts should focus on expanding datasets, improving recognition accuracy through AI-driven techniques, and fostering international collaboration to enhance Cyrillic HTR technology in Bulgaria.

2.7 Serbia

Serbia has made significant advancements in Optical Character Recognition (OCR) and Handwritten Text Recognition (HTR) technologies, particularly focusing on historical and legal documents. One of the most notable initiatives is the Dionisio 2.0 Model, developed by Prof. Vladimir Polomac from the University of Kragujevac, which specializes in the automatic recognition of Serbian Church Slavonic printed books from the 15th to 17th centuries. By utilizing materials from various Serbian printing houses, the model has achieved a character error rate of 2.4% on its validation set, demonstrating remarkable efficiency in processing historical texts [16].

Another important contribution is the Montenegrin/Serbian Cyrillic Print Model, curated by Philipp Wasserscheidt at Humboldt-University in Berlin. This model is designed for printed Serbian and Montenegrin Cyrillic texts from the late 19th century to contemporary works, achieving a character error rate of 0.20% on its validation set, indicating its high level of accuracy. Additionally, advancements in correcting OCR errors in legal documents have been made through the development of LexBERT, a system that integrates a Serbian BERT-based language model, an OCR engine, and a word-level similarity measure. This system has demonstrated the capability to correct up to 88% of terms inaccurately extracted by OCR models in Serbian legal texts, significantly enhancing document processing accuracy [17].

Further contributions to Serbian OCR and HTR research come from the SCyDia Project, which focuses on OCR for Serbian Cyrillic with diacritics. This project has analyzed the accuracy of OCR systems by processing 13 dialectal dictionaries, contributing to the enhancement of text recognition accuracy for Serbian dialects [17].

Despite these advancements, Serbia still faces challenges in the development of robust HTR models for handwritten text. Continued research efforts, combined with the expansion of publicly available datasets, will be critical in improving the performance and applicability of Serbian HTR systems. These initiatives underscore Serbia's commitment to advancing OCR and HTR technologies, particularly in preserving and digitizing its rich historical and legal textual heritage.

2.8 North Macedonia

In North Macedonia, efforts to develop Handwritten Text Recognition (HTR) systems have been emerging, particularly focusing on the Macedonian language and its Cyrillic script. One notable initiative is the collaboration with the Transkribus platform, which has facilitated the creation of AI models tailored to recognize Macedonian handwriting. These models have been trained on historical manuscripts and contemporary handwritten documents, aiming to improve the accuracy of text recognition for both archival preservation and modern applications. Despite these advancements, challenges persist due to the limited availability of large-scale, annotated datasets specific to Macedonian handwriting. Ongoing efforts are directed towards expanding these datasets and enhancing the performance of HTR systems through community engagement and technological innovation.

2.9 Belarus

Belarus has been proactive in advancing Optical Character Recognition (OCR) and Handwritten Text Recognition (HTR) technologies for the Belarusian language. A significant project in this domain is the development of the 'Belarusian Handwritten Text

Recognition Model' in collaboration with the Transkribus platform. This model has been trained on a diverse set of handwritten documents, encompassing both historical manuscripts and modern writings, to accurately transcribe Belarusian text. The initiative aims to preserve cultural heritage and facilitate digital accessibility of Belarusian literature and official documents. However, challenges remain in acquiring extensive, high-quality annotated datasets to further refine the model's accuracy. Future endeavors focus on community involvement and academic partnerships to enhance data collection and model training processes.

2.10 Tajikistan

In Tajikistan, the development of Handwritten Text Recognition (HTR) systems is still in its nascent stages. The primary challenge lies in the scarcity of digitized handwritten documents and annotated datasets in the Tajik language, which uses a modified Cyrillic script. To address this, initiatives are being proposed to collaborate with international platforms like Transkribus to develop AI models capable of recognizing Tajik handwriting. These efforts aim to digitize historical manuscripts and modern handwritten texts, thereby preserving cultural heritage and improving accessibility. The success of these projects heavily depends on the collection of diverse handwriting samples and the development of robust training datasets.

2.11 Ukraine

Ukraine has made significant strides in Handwritten Text Recognition (HTR) through various projects aimed at transcribing Ukrainian manuscripts and typewritten texts from the 19th and 20th centuries. A notable initiative is the development of a generic HTR model curated by Aleksej Tikhonov as part of the MultiHTR project at the University of Freiburg. This model, available via the Transkribus platform, has been trained on diverse datasets, including works by renowned poets Taras Shevchenko and Lesya Ukrainka, as well as private correspondence and diaries. The model has achieved a character error rate of 4.2%, demonstrating its effectiveness in transcribing Ukrainian handwriting. These advancements have significantly contributed to the mass digitization of Ukrainian cultural heritage, making historical documents more accessible for research and preservation [19].

2.12 Transnistria (Moldova)

Transnistria, a breakaway region of Moldova, continues to use the Cyrillic script for official and everyday communication. Despite its unrecognized status, efforts have been made to modernize and secure its financial instruments. For instance, on 15 January 2013, Transnistria issued new banknotes with additional security features to combat counterfeiting. While specific projects focusing on Handwritten Text Recognition (HTR) in Transnistria are not well-documented, the region's ongoing use of Cyrillic presents potential opportunities for future HTR initiatives. Collaborations with international platforms and the development of localized datasets could enhance the digitization and preservation of Transnistrian documents, contributing to the broader field of Cyrillic script recognition.

These developments across North Macedonia, Belarus, Tajikistan, Ukraine, and Transnistria highlight the varying stages of progress in Handwritten Text Recognition within Cyrillic-script-using countries. While some nations have made significant advancements through dedicated projects and collaborations, others are in the early stages of development, facing challenges related to data scarcity and resource limitations. Continued efforts in dataset expansion, technological innovation, and international collaboration are essential to advancing HTR capabilities across these regions.

3 Metrics for Comparative Analysis of Cyrillic HTR Capabilities

The Comparative Analysis of Cyrillic Handwritten Text Recognition (HTR) Capabilities across countries utilizes several key metrics that evaluate the progress and maturity of HTR technologies in each nation. These metrics—HTR Development Stage, Dataset Availability, Model Advancement, Dataset Size, Character Error Rate (CER), Word Error Rate (WER), and Dictionary Size—are designed to provide a comprehensive view of each country's HTR research landscape and technological performance.

HTR Development Stage is a primary metric that measures how advanced the HTR technologies are in each country. This metric considers the extent of research initiatives, the availability of funding, and the implementation of advanced models in real-world applications. For example, Russia and Ukraine are classified as advanced (rated 5), owing to their established research institutions, cutting-edge deep learning models, and successful integration of HTR into official and historical document digitization efforts. In contrast, countries like Transnistria and Tajikistan are still in the early stages (rated 1) due to limited research activity, lack of technological infrastructure, and absence of major HTR projects.

Dataset Availability plays a critical role in determining the success of HTR systems, as high-quality, annotated datasets are essential for training robust models. Countries like Russia and Ukraine have large, publicly accessible datasets such as the Russian Handwritten Text Image Database (RHTID) and MultiHTR, which provide a wealth of handwritten Cyrillic text for model training. These countries score highly in this metric (rated 5), as their extensive datasets enable the development of more accurate and generalized models. On the other hand, countries like Kyrgyzstan, Uzbekistan, and Mongolia face challenges due to the scarcity of well-annotated datasets, limiting their ability to train models effectively. These countries are rated lower (2–3) in dataset availability, reflecting their reliance on smaller, less diverse datasets.

Model Advancement assesses the sophistication and effectiveness of the models used to process Cyrillic handwriting. This metric evaluates the application of advanced deep learning techniques such as Convolutional Neural Networks (CNNs), Recurrent Neural Networks (RNNs), and Transformer-based models. Countries like Russia and Kazakhstan have developed state-of-the-art models that integrate these technologies, achieving low CER (Character Error Rate) and WER (Word Error Rate), which are indicative of high model accuracy. For instance, Russia's deep learning-based systems have a CER of 4.5% and WER of 19.2%, highlighting the accuracy of its models. In contrast, countries like Transnistria and Tajikistan are still in the early phases of model development, where limited accuracy is reflected by higher CER and WER values. These countries score lower (1–2) on model advancement, primarily due to the lack of sophisticated models and the absence of high-quality datasets to train them.

In addition to these primary metrics, Dataset Size and Dictionary Size are essential for evaluating the effectiveness of HTR systems. Dataset Size refers to the number of images or records available for training models, with countries like Russia and Ukraine leading with datasets that consist of millions of handwritten text samples. Larger datasets allow for better model generalization and higher performance. Countries with smaller datasets, such as Tajikistan and Transnistria, struggle with training robust models due to the lack of sufficient data. Dictionary Size influences the accuracy of HTR systems by providing a broader lexicon for recognizing words in context. Countries like Russia and Kazakhstan, with large, well-developed dictionaries, achieve better WER due to the use of context for error correction. Countries with smaller or more limited dictionaries tend to have higher WER values, which can degrade recognition accuracy.

By considering these combined metrics—HTR Development Stage, Dataset Availability, Model Advancement, Dataset Size, CER, WER, and Dictionary Size—this analysis provides a comprehensive understanding of each country's strengths and challenges in advancing Cyrillic HTR. Countries with larger datasets, more advanced models, and significant research investment (such as Russia and Ukraine) are able to achieve better performance, while nations with fewer resources and datasets (like Tajikistan and Transnistria) continue to face barriers to achieving high recognition accuracy (Tables 1 and 2).

Table 1. Effectiveness of HTR Systems

Country	Datasets	Dataset Size (No. of Samples)	CER (%)	WER (%)	Dictionary Size	Technological Approach
Russia	Extensive (RHTID, Russian Handwritten Text Database)	1000000	4,5	19,2	Large (Russian)	CNN, Transformer, RNN
Kyrgyzstan	Limited (Kyrgyz-MNIST)	80000	10,2	30,5	Medium (Kyrgyz)	ANN
Kazakhstan	Extensive (HKR, KOHTD)	1000000	6,1	22,3	Large (Kazakh, Russian)	CNN, RNN, GANs
Uzbekistan	Moderate (Bilingual OCR Dataset)	100000	8,5	28,4	Medium (Uzbek, Russian)	Hybrid CNN-RNN
Mongolia	Limited (Mongolian Cyrillic Dataset)	50000	12	35,7	Medium (Mongolian)	CNN, RNN

(*continued*)

Table 1. (*continued*)

Country	Datasets	Dataset Size (No. of Samples)	CER (%)	WER (%)	Dictionary Size	Technological Approach
Bulgaria	Moderate (Bulgarian OCR Dataset)	200000	9,4	24,8	Medium (Bulgarian)	CNN, RNN
Serbia	Moderate (Dionisio 2.0 Model)	120000	7,2	25	Medium (Serbian)	CNN, BERT-based OCR
North Macedonia	Moderate (Historical Macedonian Handwriting Dataset)	80000	10,1	32	Medium (Macedonian)	CNN, RNN
Belarus	Limited (Belarusian Handwritten Text Dataset)	50000	13,3	38,1	Medium (Belarusian)	CNN, RNN
Tajikistan	None	0	N/A	N/A	N/A	N/A
Ukraine	Extensive (MultiHTR, Ukrainian Dataset)	1000000	4,2	18,5	Large (Ukrainian)	CNN, Transformer
Transnistria	None	0	N/A	N/A	N/A	N/A

Table 2. HTR Stage, Data Availability and Model Advancement Scaling

Country	HTR Stage	Dataset Availability	Model Advancement
Russia	5	5	5
Kyrgyzstan	2	2	2
Kazakhstan	4	4	4
Uzbekistan	3	3	3
Mongolia	3	3	3
Bulgaria	3	2	2
Serbia	4	3	3
North Macedonia	2	2	2
Belarus	3	3	3
Tajikistan	2	1	1

(*continued*)

Table 2. (*continued*)

Country	HTR Stage	Dataset Availability	Model Advancement
Ukraine	5	5	5
Transnistria	1	1	1

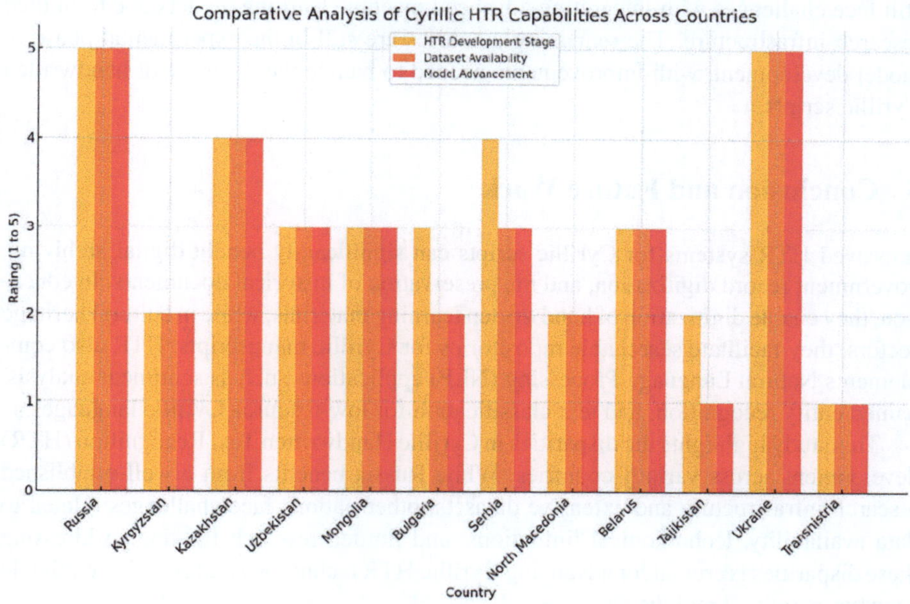

Fig. 1. Comparative Chart of Cyrillic HTR Capabilities

4 Comparison and Discussion

The development of Cyrillic Handwritten Text Recognition (HTR) varies significantly across Russia, countries with marked differences in research infrastructure, dataset availability, and technological advancements. Based on the comparative chart (Fig. 1), the countries were categorized into three key aspects, like HTR Development Stage, Dataset Availability, and Model Advancement.

As shown in the chart, Russia leads the development of Cyrillic HTR, receiving the highest rating (5), followed by Ukraine with a score of 5 due to their advanced models and established research initiatives. Kyrgyzstan, Kazakhstan, Uzbekistan, and Mongolia are in the middle range (scores 2–4), showing notable progress but lacking large-scale datasets and consistent model advancements. Countries like Transnistria, Tajikistan, and Belarus face limitations in technological resources and dataset availability, scoring lower in the HTR Development Stage (1–2).

Regarding the Dataset Availability, Russia, with its extensive annotated datasets like the Russian Handwritten Text Image Database (RHTID), scores the highest (5). Ukraine follows with another strong dataset collection, including works of renowned poets and historical documents. Countries like Kyrgyzstan and Uzbekistan still struggle with data scarcity, scoring a 2 or 3. Transnistria and Tajikistan rank the lowest in terms of dataset availability due to their early stages of HTR research and lack of public datasets.

And in the Model advancement, the Russia has developed state-of-the-art HTR models, including Transformer-based and hybrid CNN-RNN architectures, making it the leader in this area. Kazakhstan and Ukraine also have competitive models, demonstrating impressive accuracy rates. Kyrgyzstan and Mongolia, while making progress, still face challenges with integrating advanced machine learning models due to limited research infrastructure. The remaining countries are still in the experimental phase of model development, with improvements needed to handle the diversity of handwritten Cyrillic scripts.

5 Conclusion and Future Work

Improved HTR systems for Cyrillic scripts can significantly benefit digital archiving, government record digitization, and the preservation of historical documents. In education, they enable digitization of handwritten learning materials, while in cultural heritage sectors, they facilitate searchable repositories for Cyrillic manuscripts. HTR also complements Natural Language Processing (NLP) applications such as sentiment analysis, named entity recognition, and text classification for low-resource Cyrillic languages.

This study highlights the disparities in Cyrillic Handwritten Text Recognition (HTR) development across various countries. While Russia benefits from a well-established research infrastructure and extensive datasets, other nations face challenges related to data availability, technological limitations, and limited research funding. Addressing these disparities is crucial for advancing Cyrillic HTR technology and ensuring equitable progress across all regions.

References

1. Abdallah, A., Hamada, M., Nurseitov, D.: Attention-based fully gated CNN-BGRU for Russian handwritten text. J. Imaging **12**, 141 (2020). https://doi.org/10.3390/jimaging6120141. PMID: 34460538; PMCID: PMC8321198
2. Bogatenkova, A.O., Belyaeva, O.V., Perminov, A.I.: Generation of images with handwritten text in Russian. Program. Comput. Softw. **50**(7), 483–492 (2024)
3. Siberian Federal University: Siberian Federal University. https://news.sfu-kras.ru/
4. Davydkin, E., Markelov, A., Iuldashev, E., Dudkin, A., Krivorotov, I.: Data generation for post-OCR correction of Cyrillic handwriting. arXiv preprint arXiv:2311.15896 (2023)
5. Zhumaev, I., Shambetova, B., Isaev, R.: Handwritten text recognition of letters in Kyrgyz language. In: Proceedings of the International Conference on Computer Systems and Technologies 2024 (CompSysTech 2024), New York, NY, USA, pp. 57–62. Association for Computing Machinery (2024)
6. Alexeyev, Z.: Awesome Kyrgyz NLP. GitHub Repository (2023)
7. Du, W., Maimaitiyiming, Y., Nijat, M., Li, L., Hamdulla, A., Wang, D.: Automatic speech recognition for Uyghur, Kazakh, and Kyrgyz: an overview. Appl. Sci. **13**(1), 326 (2023)

8. Nurseitov, D., Bostanbekov, K., Kurmankhojayev, D., Alimova, A., Abdallah, A.: HKR for handwritten Kazakh & Russian database. arXiv preprint arXiv:2007.03579 (2020)
9. Nurseitov, D., Bostanbekov, K., Kurmankhojayev, D., Alimova, A., Abdallah, A.: KOHTD: Kazakh offline handwritten text dataset (2022)
10. Yeleussinov, A., Amirgaliyev, Y., Cherikbayeva, L.: Improving OCR accuracy for Kazakh handwriting recognition using GAN models. Appl. Sci. **13**(9), 5677 (2023)
11. Erdenebat, B., Jargalsaikhan, O., Tserenlkham, E.: Mongolian Cyrillic handwritten text recognition from image, dataset, and benchmark (2024)
12. Ren, Q.-D.-E.-J., Wang, L., Ma, Z., Barintag, S.: Offline Mongolian handwriting recognition based on data augmentation and improved ECA-Net. Electronics **13**(5), 835 (2024)
13. Bulgarian Shopping List OCR Dataset (2023). https://www.futurebeeai.com/dataset/ocr-dataset/bulgarian-shopping-list-ocr-image-dataset
14. Dimitar: OCR to BG. GitHub Repository (2023). https://github.com/bulgariamitko/ocr-to-bg
15. Petkova, M.: OCR. GitHub Repository (2020). https://github.com/mishaelaaa/OCR
16. Polomac, V.R.: Serbian early printed books from Venice: creating models for automatic text recognition using Transkribus, Scripta and E-Scripta. Scripta **22**, 11–29 (2022)
17. Bogdanović, M., Frtunić Gligorijević, M., Kocić, J., Stoimenov, L.: Improving text recognition accuracy for Serbian legal documents using BERT. Appl. Sci. **15**(2), 615 (2025). https://doi.org/10.3390/app15020615
18. Ilic, V., Bajčetić, L., Petrović, S., Španović, A.: SCyDia - OCR for Serbian Cyrillic with diacritics (2022)
19. Tikhonov, A., Rabus, A.: Handwritten text recognition of Ukrainian manuscripts in the 21st century: possibilities, challenges, and the future of the first generic AI-based model. Kyiv-Mohyla Humanit. J. **11**, 226–247 (2024)
20. Isaev, R., Esenalieva, G., Doszhanov, E.: Towards Kyrgyz stop words. Kalbotyra **76**, 54–65 (2023)

Optimization of Winter Wheat Irrigation in a Semi-arid Region: Application of the SIMDualKc Model for Efficient Water Management

Zaineb Bouswir[1], Salah Er-Raki[1,2](✉), Saïd Khabba[2,3], Jamal Ezzahar[2,3,4], and Abdelghani Chehbouni[2]

[1] Centre Agrobiotech, Faculty of Sciences and Techniques, Cadi Ayyad University, Marrakesh, Morocco
s.erraki@uca.ma
[2] Center for Remote Sensing Applications, UM6P, Ben Guerir, Morocco
[3] LMFE, Faculty of Sciences Semlalia, Cadi Ayyad University, Marrakesh, Morocco
[4] LSA2D, Higher School of Technology, Cadi Ayyad University, El Kelaa Des Sraghna, Morocco

Abstract. Agricultural water management in arid and semi-arid regions faces significant challenges due to limited water availability, inefficient irrigation practices, and increasing pressures from climate change and food security concerns. Optimizing irrigation scheduling is crucial for improving water use efficiency and ensuring sustainable crop production.

This study proposes an optimized irrigation strategy for winter wheat, tailored to its specific water requirements during different growth stages and under varying environmental conditions. By using the SIMDualKc model, the aim is to minimize water stress and deep percolation losses while enhancing crop water use and promoting sustainable irrigation practices.

The evaluation of the soil water balance, taking into account irrigation amounts applied by farmers, revealed considerable water losses through deep percolation 435 mm for the F2 field under full irrigation, and 307 mm for the F1 field under deficit irrigation. Additionally, water stress was observed in the F1 field. These findings highlight the inefficiencies of conventional irrigation methods and emphasize the need for optimized scheduling.

Application of the SIMDualKc model led to a 55% to 69% reduction in irrigation requirements for the F1 field under deficit irrigation and the F2 field under full deficit irrigation, respectively, compared to traditional practices. These improvements were achieved without inducing water stress or increasing percolation losses, demonstrating the model's potential to significantly enhance irrigation efficiency by precisely adjusting water inputs to meet actual crop water needs.

Keywords: Water Management · Crop water requirements · Sustainable Agriculture

B.-E. Boudriki Semlali et al. (Eds.): DATA 2025, LNISO 81, pp. 328–340, 2025.
https://doi.org/10.1007/978-3-032-06725-8_28

1 Introduction

In arid and semi-arid regions, water scarcity remains a major constraint to agricultural development. This challenge is further exacerbated by disruptions to the hydrological cycle, climate change, and the continuous rise in water demand across agricultural, urban, and industrial sectors. In Morocco, agriculture places significant pressure on water resources, with irrigation accounting for more than 80% of total water consumption. Given the progressive depletion of water reserves, implementing sustainable management strategies is essential to ensure the long-term viability of agriculture while meeting the growing demand for water.

Among the country's staple crops, cereals particularly winter wheat play a central role in food security and economic stability. They cover over 55% of arable land, amounting to nearly 5 million hectares, and serve as a cornerstone of the national food supply (Bouras 2020). However, their heavy reliance on irrigation intensifies pressure on already scarce water resources. In this context, adopting optimized irrigation strategies becomes a necessity to ensure the rational use of water while maintaining agricultural productivity.

Effective water management in agriculture relies on accurately assessing crop water requirements and carefully planning irrigation schedules. This involves determining optimal irrigation timing, adjusting water application rates based on climatic conditions and actual crop needs, and minimizing losses due to deep percolation. At the same time, preventing water stress is equally crucial to sustaining satisfactory yields. Mastering the water balance by analyzing input flows (precipitation, irrigation) and output losses (evaporation, transpiration, infiltration) is thus a key approach to improving irrigation efficiency.

Evapotranspiration (ET) is a fundamental parameter for irrigation planning and system optimization. A reliable ET estimation allows for the development of sustainable agricultural strategies tailored to local agroclimatic conditions (Er-Raki et al. 2009; Olivera-Guerra et al. 2020). Moreover, accurately assessing deep percolation losses is essential to quantify water beyond the reach of plant roots, thereby preventing unnecessary waste. Over the past decade, numerous water balance models have been introduced to estimate crop water needs and enhance irrigation efficiency. However, many of these models are either too complex, require extensive data inputs, or remain inaccessible to end users particularly farmers thus limiting their large-scale adoption.

To overcome these challenges, this study adopts the SIMDualKc model (Rosa et al. 2012), which offers an effective compromise between accuracy, flexibility, and ease of implementation. Built upon the dual crop coefficient approach outlined in FAO-56 (Allen et al. 1998), the model distinguishes between soil evaporation and crop transpiration, enabling a more refined estimation of actual evapotranspiration and associated water balance components. Its empirical framework, however, may introduce uncertainties particularly under conditions of water stress or in the absence of robust local calibration. Additionally, the model's simplified depiction of soil plant atmosphere interactions can limit its accuracy in extreme environmental conditions. For these reasons, careful calibration and validation are essential to ensure the reliability of the results. The SIMDualKc model has demonstrated robustness and versatility across a wide range of crops such as olive, maize, soybean, grapevine, citrus, and various fruit trees and under diverse agroclimatic conditions (Cancela et al. 2015; Darouich et al. 2022; Paço et al. 2011; Sirera et al. 2021).

Within this framework, the present study seeks to apply the SIMDualKc model to a specific case study, aiming to provide actionable insights for irrigation optimization. More specifically, it aims to: analyze soil water balance components to support water-saving strategies, and develop optimized irrigation schedules that reduce water consumption while maintaining satisfactory agricultural productivity.

2 Materials and Methods

2.1 Study Area Description

The study focused on two winter wheat fields (F1 and F2) of the Karim variety, each covering an area of 1.5 ha, located in the semi-arid Haouz plain in central Morocco, at an elevation of 550 m above sea level (Fig. 1). This region, one of the most significant agricultural plains in the country, spans approximately 6,000 km^2 and encompasses both irrigated and rainfed cropping systems. The climate is characterized by hot summers, with average daily temperatures reaching 35 °C in July and August, and cold winters, averaging around 5 °C in January (Ouaadi et al. 2023). Annual precipitation is low, averaging 250 mm, while reference evapotranspiration (ET_o) is much higher, approximately 1600 mm/year (Rafi et al. 2019). Most of the rainfall occurs between October and April, coinciding with the typical sowing period for winter wheat in November and December.

The study was conducted during the 2017–2018 growing season, with fields irrigated using a drip system. The maximum observed root zone depth ($Z_{r\,max}$) was 0.5 m. The soil texture consisted of 37.5% clay and 32.5% sand, with water content values of 0.26 m^3/m^3 at field capacity and 0.07 m^3/m^3 at the wilting point. These water content values were calculated using the Wösten equation (Wösten et al. 1999), based on the soil texture (Rafi et al. 2019).

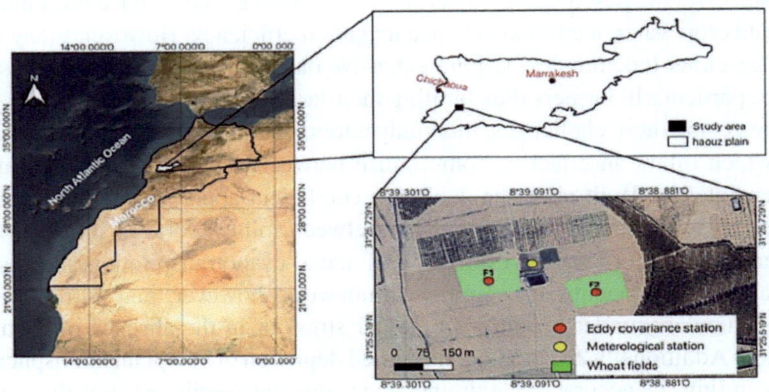

Fig. 1. Location of the Tensift basin in Morocco and setup of the experiment on the two fields of winter wheat irrigated by drip irrigation (F2 field under full irrigation and F1 field under deficit irrigation).

2.2 Experimental Sites

During the 2017–2018 wheat growing season, two experimental fields, F1 and F2, were cultivated under different irrigation regimes. To assess the impact of irrigation on crop growth, varying amounts of water were applied. Field F2 received full irrigation, with a total of 474 mm, while field F1 was subjected to deficit irrigation, receiving 289 mm.

A weather station was installed near the experimental site, on an adjacent alfalfa field at a height of 2 m (Fig. 1), to monitor environmental conditions influencing crop development. The station recorded key meteorological parameters including solar (R_s, in $MJm^{-2}day^{-1}$), wind speed and direction (u_2, in m/s), air temperature (T_{min} and T_{max}, in °C), and air humidity (RH_{moy}, in %), and precipitation (P, in mm). Data were collected at 30-minute intervals and used to compute reference evapotranspiration (ET_o) using the FAO56 Penman-Monteith method (Allen et al. 1998).

Actual crop evapotranspiration ($ET_{c\,act}$) was measured using eddy covariance stations installed at the center of each field, at a height of 2.58 m (Fig. 1). Each station was equipped with a three-dimensional sonic anemometer (CSAT3-3D) and a Krypton hygrometer (KH2O) to capture high-frequency fluctuations in wind speed, temperature, and water vapor concentration. Data were collected at a frequency of 20 Hz and processed using ECpack software to calculate sensible (H) and latent (LE) heat fluxes at 30-minute intervals. Daily ET was then estimated by summing the half-hourly latent heat flux values and converting them into millimeters per day (see Ait Hssaine et al. 2018 for more details).

Vegetation cover was assessed using downward-oriented hemispherical photographs captured with a Canon EOS 600D camera equipped with a SIGMA 4.5 mm F2.8 EXDC HSM circular fisheye lens. The images were processed using MATLAB to classify vegetation and soil pixels, allowing calculation of vegetation cover as the ratio of vegetation pixels to total pixels (see Duchemin et al. 2006 and Khabba et al. 2009).

Soil moisture content was monitored using time domain reflectometry (TDR) sensors (models CS615 and CS655) installed near the eddy covariance towers at multiple depths. In field F1, measurements were taken at 5, 15, 25, 35, and 50 cm, while in field F2, they were recorded at 5, 15, 30, and 50 cm during both growing seasons.

3 Modeling Approach

The SIMDualKc model, developed to estimate actual crop evapotranspiration ($ET_{c\,act}$) and its components, offers a more accurate alternative to the FAO-56 method. By applying a dual approach, it enhances the assessment of crop water requirements, making it a valuable tool for efficient irrigation planning. Developed by Rosa et al. (2012), this soil water balance model has been widely used and cited in recent research (e.g. Darouich et al. 2022; Paço et al. 2011), demonstrating its relevance and reliability in agricultural water management.

The SIMDualKc model uses a daily time step to compute the soil water balance at the field scale as follows (Allen et al. 1998; Rosa et al. 2012):

$$D_{r,i} = D_{r,i-1} - (P_i + I_i + CR_i) + RO_i + ET_{c\,act,i} + DP_i \tag{1}$$

where D_r is the root zone depletion (mm), P is the rainfall (mm), RO is the surface runoff (mm), I is the net irrigation depth (mm), CR is the capillary rise from the groundwater (mm), $ET_{c\,act}$ is the actual crop evapotranspiration (mm), and DP is the deep percolation from the root zone (mm), all referring to day i or day i-1. in this study, Capillary rise (CR) was neglected as the groundwater table was deep enough to have no upward contribution to the root zone moisture. Surface runoff (RO) was also negligible due to the relatively flat topography of the study area and the use of drip irrigation, which delivers water directly to the root zone at a controlled rate, preventing surface accumulation and runoff.

The model adopts the FAO-56 dual approach for computing crop evapotranspiration (Allen et al. 1998; Rosa et al. 2012), where actual crop transpiration ($T_{c\,act}$, mm) and soil evaporation (Es, mm) are estimated separately as follows:

$$T_{c\,act} = K_{cb}\,K_s ET_o = K_{cb\,act} ET_o \tag{2}$$

$$E_s = K_e\,ET_o \tag{3}$$

The actual evapotranspiration ($ET_{c\,act}$, mm) is then calculated using a dual crop coefficient, expressed as:

$$ET_{c\,act} = T_{c\,act} + E_s = (K_s.K_{cb} + K_e)ET_o = (K_{cbact} + K_e)ET_o \tag{4}$$

where $K_{cb\ \mathrm{act}}$ is the actual basal crop coefficient (−), K_e is the soil evaporation coefficient (−), K_s is the water stress coefficient (−) and ET_o is the reference evapotranspiration (mm) computed using the FAO-56 (Penman-Monteith equation) (Allen et al. 1998).

SIMDualKc model calculates the water stress coefficient K_s (ranging between 0 and 1) using the following these two equations (Allen et al. 1998; Rosa et al. 2012):

$$D_r \le RAW,\ K_s = 1 \tag{5}$$

$$D_r > RAW,\ K_s = \frac{TAW - D_r}{TAW - RAW} \tag{6}$$

The effects of water stress are introduced by multiplying the transpiration crop coefficient by a coefficient representing water stress K_s.

TAW and RAW correspond to the total and readily available soil water relative to the rooting depth (mm), respectively. These values are computed as follows:

$$TAW = 1000\big(\theta_{Fc} - \theta_{wp}\big)Z_r \tag{7}$$

$$RAW = p.TAW \tag{8}$$

where θ_{FC} and θ_{wp} are the soil water contents at field capacity and the wilting point (m^3 m^3), respectively, Z_r is the root depth (m), and p is the average fraction of Total Available Soil Water (TAW) that can be depleted from the root zone without crop water stress (−).

Deep percolation (DP) is estimated using a time decay function relating the soil water storage near saturation with the time after the occurrence of heavy rain or irrigation (Liu et al. 2006):

$$W_a = a_D t^{b_D} \tag{9}$$

The equation describes the gradual decrease of water not retained by the soil after excessive irrigation or rainfall. This loss is due to percolation into deeper layers, with the rate governed by a dimensionless empirical coefficient, b_D. The amount of water available for this process, a_D, represents the surplus beyond the soil's retention capacity, between saturation (θ_{wp}) and field capacity (θ_{Fc}). Water storage in the root zone, denoted W_a, therefore decreases over time **t** (days), reflecting the remaining portion before complete infiltration.

4 Results and Discussion

4.1 Model Calibration and Validation

The analysis of soil water balance components and the development of optimized irrigation schedules first require rigorous model calibration and validation. This step is essential to ensure that simulations accurately reflect real field conditions, enabling informed decision making in agricultural water management.

The calibration and validation processes were conducted using field F2 under full irrigation and field F1 under deficit irrigation. These processes relied on various in situ data, including evapotranspiration ($ET_{c\,act}$) measurements obtained using the Eddy covariance system, soil moisture (ASW) measurements with TDR sensors, and vegetation cover fraction (f_c) measurements using hemispherical photography. The comparison between simulated data and in situ measurements yielded highly promising results. The average values of the RMSE and the (R^2) were 0.56 mm/day (0.94) for $ET_{c\,act}$, 0.08 (0.97) for f_c, and 11.7 mm (0.94) for ASW. (For more details, see Bouswir et al. 2025). These statistical performances demonstrate the model's ability to accurately reproduce the observed water and vegetation dynamics in the fields.

With the model successfully validated, we can now analyze the different components of the soil water balance. This assessment is essential for evaluating water availability, which is key factors in optimizing irrigation strategies.

4.2 Assessment of Soil Water Balance Components

Farmer's Irrigation Strategy. Figure 2 illustrates the components of the water balance for the irrigation schedule proposed by the farmer (I_F) applied to the two fields: F1 and F2.

As shown in Fig. 2, irrigation is the primary water source throughout the growing season, supplying between 68% and 78% of the total water, while precipitation contributes only 22% to 32% for fields F1 and F2, respectively. This highlights the essential role of irrigation in ensuring sufficient water availability and supporting crop development, particularly during periods of rainfall deficit.

Seasonal values of maximum transpiration ($T_{c\,act}$) range from 61 to 78% of the maximum evapotranspiration (ET_{max}) for fields F1 and F2, respectively. In contrast, soil evaporation (E_{sF}) ranges from 50 to 56 mm, representing about 21% to 23% of ET_{max}. These results confirm that transpiration is the primary factor in water consumption, whereas soil evaporation plays a lesser role.

A more detailed analysis reveals notable differences in actual evapotranspiration between the two fields. In field F1, actual transpiration values ($T_{c\,act\,F}$) consistently fall short of potential values ($T_{c\,max}$), reaching only 149 mm, respectively. The $T_{c\,act\,F}/T_{c\,max}$ ratio, which is less than 1, indicates water stress. Although irrigation partially alleviates this stress at times, the frequency and timing of irrigation were insufficient to allow winter wheat to reach its maximum transpiration potential. Consequently, actual evapotranspiration ($ET_{c\,act\,F}$) remained below ET_{max} in these field. Conversely, in fields F2, actual transpiration values ($T_{c\,act\,F}$) are close to the maximum values of $T_{c\,max}$, reaching 185 mm. This suggests that actual evapotranspiration ($ET_{c\,act\,F}$) is nearly equal to ET_{max}, indicating optimal water availability for crops in these fields. Additionally, a significant amount of water is lost due to deep percolation (DP_F), ranging from 307 to 435 mm in fields F1 and F2. This water infiltrates beyond the root zone without benefiting the winter wheat, representing a considerable waste of resources and raising concerns about irrigation efficiency in the study area.

The evaluation of the water balance components highlights the essential role of irrigation in crop water supply. However, field F1 suffers from water stress due to irregular irrigation management in terms of both frequency and volume. In contrast, field F2 benefits from better water availability, though deep percolation leads to significant losses. This raises important questions about the periods during which water stress is most significant and when water losses are highest. Optimizing water management is therefore essential to improving water use efficiency and reducing water stress, which negatively affects winter wheat development.

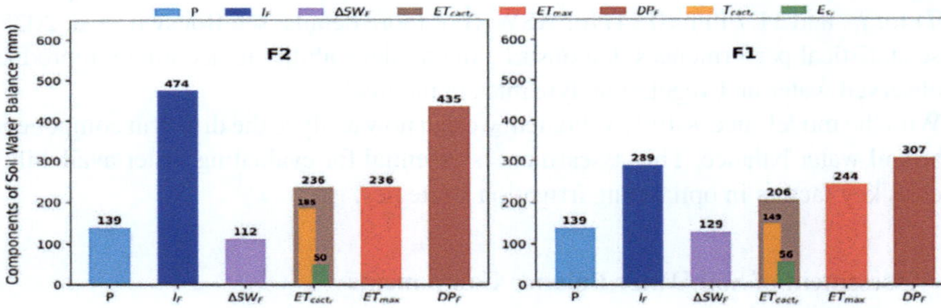

Fig. 2. Components of Soil Water Balance under irrigation schedules from farmer practices, in the fields (F2 and F1). Note: I, irrigation; P, precipitation; ΔSW, soil water storage variation; $T_{c\,act}$, actual crop transpiration; Es, soil evaporation; $ET_{c\,act}$, actual crop evapotranspiration; DP, deep percolation; *All indices labeled with 'F' refer to the components using the farmer's irrigation schedule (I_F).

Before developing an optimized irrigation schedule using the SIMDualKc model, it is essential to analyze in detail the variation in water stress coefficients and percolation dynamics in these two fields. This analysis helps identify not only the intensity and timing of water stress but also the phenological stages most sensitive to percolation. A better understanding of this is essential for precisely adjusting irrigation, minimizing water losses, and maximizing efficiency, thereby ensuring optimal conditions for crop growth.

For this purpose, Fig. 3 illustrates the evolution of the water stress coefficient (K_s), water depletion in the root zone, and percolation beyond this zone (DP), allowing an evaluation of the effectiveness of farmers' irrigation practices. In field F2, the stress coefficient (K_s) remains consistently at 1 throughout the growing season (Fig. 3), indicating the absence of water stress for winter wheat. This is confirmed by the fact that soil water depletion never reached the RAW threshold, below which crops experience excessive constraints. However, excessive irrigation was observed during the initial and vegetative development phases when plant water requirements are relatively low. This surplus led to significant percolation, causing deep water infiltration. Although irrigation ensured favorable growth conditions, a substantial portion of the applied water was lost without any real agronomic benefit. For field F1, the stress coefficient (K_s) remains at 1 during the initial, development, and mid-season phases, confirming adequate water availability from either precipitation or irrigation (Fig. 3). Low water depletion at these stages also reflects the absence of stress. However, in the final phase, significant water stress occurs, with the **stress coefficient dropping to 0.3**. This stress results from a lack of irrigation during this period, as precipitation was insufficient to meet wheat requirements. Consequently, water depletion exceeds the RAW threshold, intensifying water stress. Regarding percolation in field F1, it is primarily observed during the initial and early development phases when plant water needs are still low. This issue stems from inadequate irrigation management at these stages.

These findings demonstrate that both excessive and insufficient irrigation negatively impact water use efficiency. While excessive irrigation leads to percolation and water loss, water deficits during key growth stages hinder crop yields. Achieving a balance between these factors is crucial for optimizing plant growth and ensuring sustainable agricultural production. The amount of water applied during irrigation plays a key role in this process, requiring a well-planned strategy to prevent both wastage and water stress. Therefore, implementing an optimized irrigation strategy is essential to improving water use efficiency while maintaining optimal growing conditions for crops.

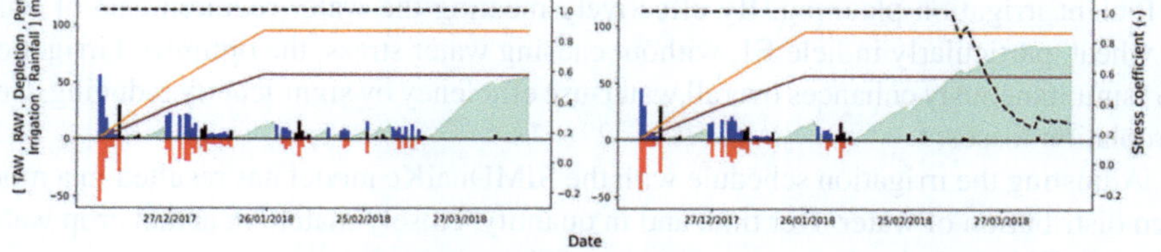

Fig. 3. Variation of soil water parameters in the root zone for winter wheat, in the fields (F2 and F1).

Optimization of Irrigation Scheduling Using the SIMDualKc Model. In the previous section, we observed that the irrigation practices used by farmers in these two fields are not optimized. Water is sometimes applied too early or in excessive quantities, leading to deep percolation losses. At other times, irrigation is insufficient, causing water stress that negatively affects crop growth and yield.

This section aims to propose an optimized irrigation approach using the SIMDualKc model. The objective is to better align water supply with wheat's actual needs while preventing both water stress and percolation losses. By optimizing irrigation scheduling, we seek to improve water use efficiency and promote sustainable irrigation practices.

To achieve this, we set the irrigation trigger threshold at 30% of the total available water (*Total Available Water*, TAW) consumed. This value strikes a balance between two key objectives: avoiding water stress, which could hinder wheat growth and yield, and minimizing deep percolation losses that reduce irrigation efficiency. By adopting this threshold, water is applied at the optimal moment neither too late, to prevent deficits, nor too early, to avoid percolation. The choice of this threshold is based on scientific and experimental studies that have demonstrated its effectiveness in managing wheat irrigation. Research conducted by Hadria et al. (2007), Jackson et al. (1986), and Toumi et al. (2016) confirmed that this level of available water depletion represents an optimal balance between preserving water resources and optimizing wheat yields.

The water balance results, shown in Fig. 4 after optimization using the SIMDualKc model, highlight a significant improvement in water management. After optimization, the total recommended irrigation volume ranges from 130 mm to 146 mm for field F2, representing a significant reduction in seasonal water inputs compared to the initial irrigation program proposed by the farmer (I_F). The optimized irrigation schedule (I_O) is better aligned with the crop's water requirements, effectively minimizing excessive water applications observed in the initial plan. Moreover, implementing the optimized schedule based on the SIMDualKc model has resulted in substantial water savings of 328 mm for F2 and 159 mm for F1 demonstrating a major improvement in water resource efficiency. In addition to reducing water use, the optimization ensures that actual transpiration ($T_{c\,act\,O}$) matches the maximum transpiration values ($T_{c\,max}$), confirming that crops receive exactly the amount of water needed for optimal growth. This balance ensures adequate water supply and the absence of water stress in field F1. With optimized management, irrigation is now finely tuned to match actual crop needs, ensuring optimal wheat development. Another key achievement of the optimization is the substantial reduction in deep percolation losses (DP_O), with decreases of 380 mm for F2 and 251 mm for F1. This demonstrates that excessive water loss was primarily due to inefficient irrigation planning. By effectively meeting the water requirements of winter wheat, particularly in field F1, without causing water stress, the optimized irrigation (I_O) simultaneously enhances overall water-use efficiency by significantly reducing deep percolation losses.

Adjusting the irrigation schedule with the SIMDualKc model has resulted in a more even distribution of water over time and in quantity, closely matching actual crop water demands. This optimization promotes a more efficient use of water, reducing unnecessary losses and improving its availability for plants. As a result, agricultural productivity is optimized while preserving water resources, an essential factor for sustainable and resilient agriculture.

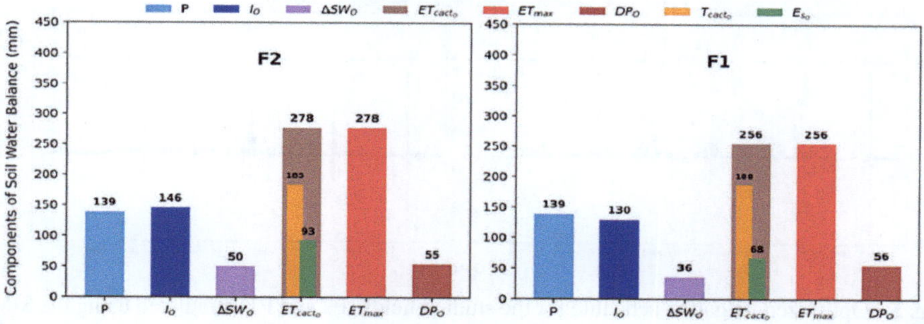

Fig. 4. Components of Soil Water Balance under irrigation schedules from farmer practices in the fields (F2 and F1). Note: *I*, irrigation; *P*, precipitation; *ΔSW*, soil water storage variation; $T_{c\,act}$, actual crop transpiration; Es, soil evaporation; $ET_{c\,act}$, actual crop evapotranspiration; DP, deep percolation; *All indices labeled with 'O' refer to the components under the optimized irrigation schedule (I_O).

Figure 5 illustrates the newly optimized irrigation schedule (I_O) applied to fields **F2, and F1**. This program has been fine tuned to better meet the wheat's water requirements, significantly enhancing water use efficiency. As a result, irrigation inputs have been reduced while maintaining adequate crop hydration. This optimization prevents excessive percolation losses while ensuring that plants receive the necessary water to avoid any risk of water stress.

The analysis of Fig. 5 results indicates that the observed percolation (DP_O) is primarily caused by precipitation rather than irrigation, confirming the effectiveness of the new schedule in minimizing unnecessary water losses. Additionally, its implementation has eliminated water stress in **F1**, which previously suffered from water deficits under the traditional farmer's irrigation program (I_F) (see Fig. 3).

Adopting (I_O) has proven to be a highly effective strategy for optimizing agricultural water management. Unlike the traditional approach (I_F), which often led to excessive or uneven water distribution, this new schedule ensures an optimal balance between plant hydration and water conservation.

A comparative analysis of both methods highlights a clear improvement in crop conditions under (I_O). By preventing both under irrigation, which hampers plant growth, and over irrigation, which leads to water waste, this approach guarantees more efficient irrigation. Furthermore, field that previously experienced water stress now exhibits a notable physiological improvement, fostering optimal wheat development.

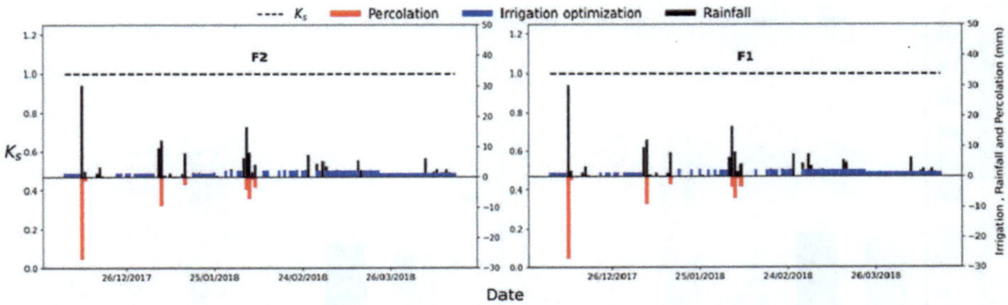

Fig. 5. Optimized irrigation schedule for the studied fields (F2 and F1) simulated using the SIMDualKc model. The schedule highlights water savings while ensuring adequate soil moisture to prevent crop stress.

5 Conclusions

This study demonstrates the effectiveness of the SIMDualKc model as a valuable tool for optimizing irrigation management of winter wheat in two distinct fields F2 under full irrigation and F1 under deficit irrigation. The assessment of the soil water balance revealed significant issues with traditional irrigation practices. Excessive water losses, primarily through deep percolation, were observed in both fields, resulting from over-irrigation during periods of low plant water demand. In contrast, water stress was observed in the F1 field due to insufficient and poorly timed irrigation, which failed to meet crop water needs during critical growth stages.

The optimization of irrigation using the SIMDualKc model successfully addressed these issues. By setting an irrigation threshold at 30% of the total available water, the optimized strategy effectively balanced the prevention of water stress with the reduction of percolation losses. As a result, water use efficiency improved significantly water stress in previously affected fields was reduced, and deep percolation losses were substantially minimized.

The findings of this study emphasize the potential of the SIMDualKc model to enhance irrigation management. By aligning irrigation schedules more closely with crop water requirements, the model fosters more efficient water use, supporting sustainable wheat production. The optimized irrigation strategy with the SIMDualKc model can easily be extended to other regions and crops. Although the study focused on winter wheat, the model's principles, based on adjusting irrigation to the actual needs of crops and water availability, are transferable to other agro environmental contexts. This approach can be adapted to crops such as maize, rice, or vegetables, considering their specific water requirements, and would improve irrigation efficiency and large-scale sustainable water management.

Future research could expand the application of this model to other crops and agro environmental contexts. Furthermore, the integration of high-resolution remote sensing data, such as satellite derived vegetation indices (e.g., NDVI), thermal imagery, and soil moisture estimates, could enhance the model's ability to monitor crop water status over large areas. Coupling these data with advanced data analytics techniques, including machine learning and data assimilation approaches, may allow for real-time irrigation scheduling and predictive water management strategies, thereby improving both the precision and scalability of the model's applications.

Acknowledgments. The authors acknowledge the International Joint Laboratory Télédétection et Ressources en Eau en Méditerranée semi-Aride (LMI TREMA, https://www.lmi-trema.ma/) for providing the data. The authors also acknowledge the financial support provided by the project RISE-H2020-ACCWA, GeanTech and PRIMA-IDEWA project.

References

Allen, R.G., Pereira, L.S., Raes, D., Smith, M.: Crop Evapotranspiration—Guidelines for Computing Crop Water Requirements, Irrigation and Drainage, Paper No. 56. FAO, Rome, Italy, p. 300 (1998)

Bouras, E.H., et al.: Linkages between rainfed cereal production and agricultural drought through remote sensing indices and a land data assimilation system: a case study in Morocco. Remote Sens. **12**(24), 1–35 (2020)

Bouswir, Z., Er-Raki, S., Khabba, S., Ezzahar, J., Rafi, Z., Chehbouni, A.: Assessment of SIMDualKc model for irrigation scheduling of winter wheat in the Haouz plain, Morocco. Agric. Water Manag. **312**, 109430 (2025)

Cancela, J.J., Fandiño, M., Rey, B.J., Martínez, E.M.: Automatic irrigation system based on dual crop coefficient, soil, and plant water status for Vitis vinifera (cv Godello and cv Mencía). Agric. Water Manag. **151**, 52–63 (2015)

Darouich, H., Karfoul, R., Ramos, T.B., Moustafa, A., Pereira, L.S.: Searching for sustainable-irrigation issues of clementine orchards in the Syrian Akkar Plain: effects of irrigation method and canopy size on crop coefficients, transpiration, and water use with the SIMDualKc model. Water **14**(13), 2052 (2022)

Duchemin, B., et al.: Monitoring wheat phenology and irrigation in central Morocco: on the use of relationships between evapotranspiration, crop coefficients, leaf area index, and remotely-sensed vegetation indices. Agric. Water Manag. **79**(1), 1–27 (2006)

Er-Raki, S., et al.: Citrus orchard evapotranspiration: comparison between eddy covariance measurements and the FAO-56 approach estimates. Plant Biol. **143**(1), 201–208 (2009)

Hadria, R., et al.: Calibration and Validation of the STICS crop model for managing wheat irrigation in the semi-arid Marrakech/Al Haouz Plain. Arab. J. Sci. Eng. **32**, 87 (2007)

Jackson, R.D., Pinter, P.J., Jr., Robert, R., Reginato, R., Idso, S.B.: Detection and evaluation of plant stresses for crop management decisions. IEEE Trans. Geosci. Remote Sens. **24**(1), 99–106 (1986)

Khabba, S., et al.: Evaluation of digital hemispherical photography and plant canopy analyzer for measuring vegetation area index of orange orchards. J. Agron. **8**(2), 67–72 (2009)

Liu, Y., Pereira, L.S., Fernando, R.M.: Fluxes through the bottom boundary of the root zone in silty soils: parametric approaches to estimate groundwater contribution and percolation. Agric. Water Manag. **84**(1), 27–40 (2006)

Ouaadi, N., et al.: Are the C-band backscattering coefficient and interferometric coherence suitable substitutes of NDVI for the monitoring of the FAO-56 crop coefficient. Agric. Water Manag. **282**(4), 108276 (2023)

Olivera-Guerra, L., Merlin, O., Er-Raki, S.: Irrigation retrieval from Landsat optical/thermal data integrated into a crop water balance model: a case study over winter wheat fields in a semi-arid region. Remote Sens. Environ. **239**, 111627 (2020)

Paço, T.A., et al.: The dual crop coefficient approach using a density factor to simulate the evapotranspiration of a peach orchard: SIMDualKc model versus eddy covariance measurements. Irrig. Sci. **30**(3), 115–126 (2011)

Puig-Sirera, A., et al.: Transpiration and water use of an irrigated traditional olive grove with sap-flow observations and the FAO56 dual crop coefficient approach. Water **13**(3), 2466 (2021)

Rafi, Z., et al.: Partitioning evapotranspiration of a drip-irrigated wheat crop: Inter-comparing eddy covariance-, sap flow-, lysimeter- and FAO-based methods. Agric. For. Meteorol. **265**(2), 310–326 (2019)

Rosa, R.D., et al.: Implementing the dual crop coefficient approach in interactive software: 2 model testing. Agric. Water Manag. **103**, 62–77 (2012)

Toumi, J., Er-Raki, S., Ezzahar, J., Khabba, S., Jarlan, L., Chehbouni, A.: Performance assessment of AquaCrop model for estimating evapotranspiration, soil water content and grain yield of winter wheat in Tensift Al Haouz (Morocco): application to irrigation management. Agric. Water Manag. **163**, 219–235 (2016)

Wösten, J.H.M., Lilly, A., Nemes, A., Le Bas, C.: Development and use of a database of hydraulic properties of European soils. Geoderma **90**, 169–185 (1999)

The Path to Adoption: Insights into Morocco's Public E-Services

Imane Abouliatim(✉) and Karima Tounsi

Laboratory EM2TI (Economy and Management of Information Technology), Institut National des Postes et Télécommunications (INPT), Rabat, Maroc

Abouliatim.imane@gmail.com

Abstract. E-government is a worldwide strategy to cut spending, improve services and boost public sector efficiency. In Morocco, e-government has caused major changes to society and culture and the way work is done. However, e-government adoption is hindered by challenges including technology, politics, organization, culture and society. Research on e-government adoption in Morocco is lacking a framework to explain citizen behavior.

The current study proposes a customized Technology Acceptance Model framework (TAM) that integrates traditional constructs and additional variables from the literature like social influence, user satisfaction and trust to identify drivers of citizens' adoption of online public services in Morocco.

Using a positivist approach and quantitative methods, the study surveyed 478 respondents and analyzed data through structural equation modelling (SEM). The eleven hypotheses were put to empirical test. The resultant data demonstrated the relevance of the adapted TAM model in elucidating the behavioral intentions of the citizens in the sample under consideration.

Importantly, our results showed that six of the eleven hypotheses were supported. This highlighted key drivers. However, it also revealed some unexpected points, like the lesser direct role of trust and social influence in this model. Thus, this study offers advice for Moroccan policymakers. It suggests how to improve digital public services with user-focused design. Furthermore, it points to ways to overcome adoption barriers, thereby improving inclusivity, efficiency, and public trust.

Keywords: Morocco · E-government · TAM · Intention to use

1 Introduction

E-government initiatives have led to rapid developments in the sphere of information technology, exerting a profound influence on various facets of communication, access to public information, and the relationship between governments and their citizens. Indeed, E-government has been recognized as a key tool for enhancing public administration and citizen engagement, defined as the practice of using information and communication technologies (ICT) to facilitate the supply of official services and information to citizens [1, 2]. The integration of ICT in government operations has been demonstrated to facilitate more efficient service delivery [3] and to foster greater transparency and accountability, which are essential for building public trust [2].

B.-E. Boudriki Semlali et al. (Eds.): DATA 2025, LNISO 81, pp. 341–349, 2025.
https://doi.org/10.1007/978-3-032-06725-8_29

The effects of e-government on citizen engagement are very substantial. For instance, research has demonstrated that e-government initiatives can result in increased citizen participation in governance processes. Research indicates that e-government platforms have enabled citizens to engage more actively with their governments, thereby improving their perceptions of governmental effectiveness [4]. Furthermore, the utilization of e-participation tools is associated with an improvement in political engagement, facilitating connections between citizens and their representatives and enabling influence over decision-making processes [5]. This transition towards digital platforms signifies a paradigm shift in the manner in which citizens interact with their governments, thereby transitioning from conventional, frequently passive forms of engagement to more dynamic and interactive methods.

The success of digital government e-services depends on many factors. These include politics, organization, law and economics, as well as technology and culture [6, 7]. Research clearly shows that people's acceptance of electronic government services is the key factor in their use. Here, Trust in the government and technology itself is very important [6, 7]. For example, Nguyen and Tran [6] highlight that people's acceptance of online government services, especially during crises, is key to their adoption. Guo et al. [7] emphasize that effective policy communication can reshape public perceptions, thereby enhancing acceptance of ICTs.

That why this research focuses on user-centered factors influencing e-service acceptance, a relatively unexplored area in Morocco. So, our main research question formulated is: What user-centred factors are driving the adoption of public e-services in Morocco?

2 Literature Review

2.1 Context and Background

Morocco is compared with the leaders of the eGovernment Development Index, Denmark, South Africa and Tunisia. Morocco's 2022 Index score of 0.5915 is below the global, regional and sub-regional averages, indicating room for improvement in eGovernment development [8]. Nevertheless, since 2003, Morocco has demonstrated an upward trajectory in e-government development, notwithstanding certain fluctuations. The nation still faces challenges in attaining the levels of global and regional leaders in this field, yet it exhibits positive progress (see Fig. 1).

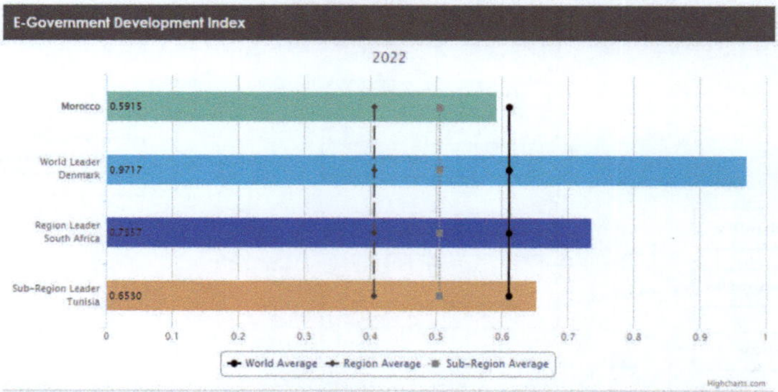

E-Government Development Index	2022	2020	2018	2016	2014	2012	2010	2008	2005	2004	2003
Morocco (Rank)	101	106	110	85	82	120	126	140	138	138	131
Morocco (Value)	0.59150	0.57290	0.52140	0.51858	0.50598	0.42090	0.32872	0.29440	0.27738	0.26410	0.26549

Fig. 1. E-Government Development Index (EGDI) Comparison: Morocco vs Global, Regional, and Sub-Regional Leaders (2022)

2.2 Research Model and Hypotheses

In the wake of a diligent and exhaustive examination of the prevailing literature on the subject, the decision was arrived at to adopt the TAM as the overarching framework [9], This model is predicated on two theoretical underpinnings: the psychological theory of reasoned action and the theory of planned behavior. It has evolved to explain human behavior in relation to the acceptance or rejection of technology.

Simply put, The TAM says that how useful and easy to use technology is makes a big difference to how people will use it [10]. The model has been extensively validated, showing it works reliably in predicting technology acceptance in developed and developing contexts [11].

TAM is a strong model for explaining technology adoption. This is mainly because it can be adapted to specific contexts, like ours in Morocco. So, we included important factors like user satisfaction, social influence, and trust (see Fig. 2).

As a result, the following assumptions were made:

H1: Perceived ease of use affects perceived use of public electronic services in a positive way.
H2: Perceived ease of use affects intention to use public electronic services in a positive way.
H3: Perceived usefulness positively influences intention to use electronic public services.

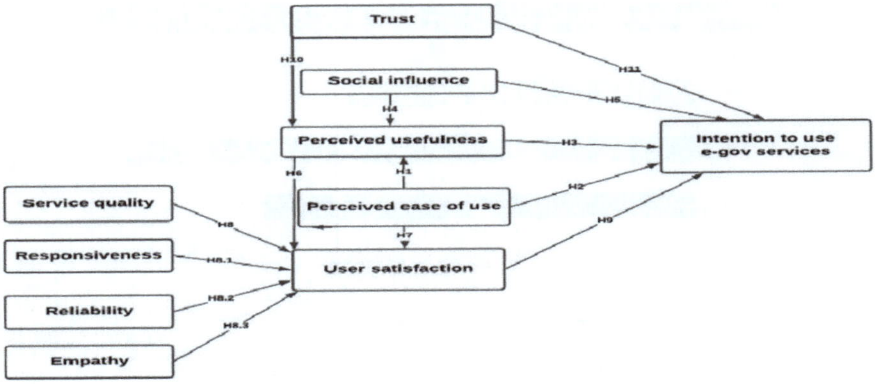

Fig. 2. The proposed framework

Social influence

The deployment of eGovernment in emerging and Arab countries is significantly impacted by social influence, which play a crucial role in shaping user acceptance and engagement with digital services [12]. Social influence, understood as the unwritten rules that govern behavior, can facilitate or impede eGovernment adoption. For instance, Tangi et al. [13] argue that while technological advancements are transforming organizational systems, social systems often lag behind, showing that successful eGovernment implementation requires a concurrent transformation of social norms and practices. This is particularly relevant in developing contexts where traditional governance structures may be resistant to change.

Therefore, we have made the following assumptions:

H4: Social influence applies a favorable effect on the perceived usefulness of public e-services.
H5: Social influence applies a favorable impact on intention to use public online services.

User satisfaction

In the context of eGovernment, it is imperative to prioritize user satisfaction and the provision of high-quality services. This is pivotal in ensuring the efficacy of eGovernment platforms in delivering public services. This underscores the necessity of incorporating user satisfaction as a pivotal factor in the evaluation of eGovernment effectiveness [14].

AlKharji [15] also emphasizes that citizen satisfaction is essential for improving service quality and increasing engagement with eGovernment services. The integration of user feedback into service design has been shown to further enhance satisfaction levels, as evidenced by the frameworks proposed by various studies aimed at assessing eGovernment effectiveness [16].

Consequently, the present research endeavors to put forward the subsequent assumptions:

H6: Perceived usefulness when using an e-service exerts a favorable effect on user satisfaction.

H7: Ease of use when using a digital public service has a beneficial impact on user satisfaction.
H8: An enhancement in quality perceptions is likely to engender an increase in e-service user satisfaction.

H8.1: A positive relationship is between responsiveness and citizen satisfaction.
H8.2: A positive relationship is between reliability and citizen satisfaction.
H8.3: A positive relationship is between empathy and citizen satisfaction.

H9: user satisfaction has a beneficial impact on the adoption of public digital services.

Trust
Trust in the reliability of online services provided by government institutions is key to building public trust and encouraging the uptake of these services. Research shows that citizens' trust significantly influences their willingness to engage with digital government services. For example, Hartanti et al. [17] point out that individuals will be more receptive to digital public services if they believe they are trustworthy and in line with their beliefs and values. This trust is essential not only for user satisfaction, but also for improving the perceived value of public services.

So, we proposed:

H10: Trust will exerts a favorable effect on the perceived usefulness of public e-services in Morocco.
H11: Trust influence intention to use digital public services in Morocco in a positive way.

3 Methods

We collected data for this research was collected from Moroccan citizens interacting with public e-services. A research questionnaire was submitted to 600 prospective respondents. The survey questionnaire was translated into Arabic and French, as the population target of this study can use both Arabic and French. The instrument was measured on five Likert scales ranging from 1 = strongly disagree to 5 = strongly agree.

The collected data were then analyzed using SPSS version 26. For more detailed analysis, we used Structural Equation Modeling (SEM) techniques. SmartPLS 4 software helped us implement these analyses. Importantly, we piloted and pre-tested the questionnaires to ensure they were clear and understandable. To select respondents, we used convenience sampling. Additionally, we conducted the survey both online and face-to-face.

4 Results

4.1 Descriptive Results

We were able to collect 478 questionnaires. This represents a response rate of around 80%. However, after validation, we used 425 questionnaires from respondents who confirmed they had used e-services.

The survey results disclosed that the distribution of respondents by gender was almost equally divided between women (54%) and men (46%). In terms of age, 36% of respondents were aged between 40 and 50, 21% between 51 and 61, 19% between 29 and 39 and 11% over 62. In terms of level of education, 64% of respondents studied at university, 17% at lycée, 10% at primary school, 8% at college and only 1% of respondents had never been to school. As for the profession of the respondents: 35% are civil servants, 18% are employees, 13% are self-employed, 11% are shopkeepers, 10% are students, 10% are unemployed and not students and 4% are farmers.

4.2 Testing Hypotheses

The associations between the variables in our model were analyzed after the structural model was estimated and the hypotheses were tested applying the partial least squares (PLS) method in the context of Morocco's adoption of e-services.

Intention to use, the primary and only dependent variable in our study, had an R^2 value of 0.622, meaning that the model was able to account for approximately 62.2% of the variation in Moroccans' intentions to use public online services. The ability of the model to capture a significant percentage of the variables influencing the use of public digital services is demonstrated by this substantial R^2 value.

Several significant discoveries were made by the investigation (see Fig. 3). First, perceived usefulness and ease of use strongly influence Moroccans' intention to use e-services. However, this objective does not seem to be much influenced by user satisfaction. Furthermore, neither the desire to utilize nor the perception of usefulness was significantly impacted by factors like social influence or trust.

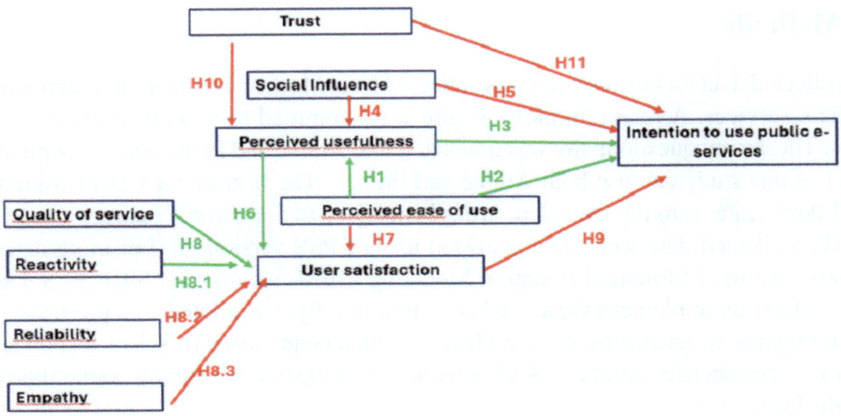

Fig. 3. The result of hypotheses testing using regression analysis.

5 Discussion

Our findings show that perceived ease of use (PEOU) and perceived usefulness (PU) deeply influence the adoption of electronic services in Morocco. Indeed, these factors are well-studied in technology acceptance literature. Much research suggests PEOU and PU are critical for users' intentions to adopt various e-services, like e-learning and e-government. For example, Mensah and Mi [18] showed that both PEOU and PU are positively predictive of citizens' intentions to consume e-government services. Our results certainly align with this.

Nevertheless, user satisfaction does not seem to be a crucial part of this intention. This is an interesting point. While satisfaction might result from good service quality (which we also tested parts of), its direct link to future intention was not strong here.

More strikingly, variables such as social influence and trust did not show significant direct effects on adoption intention or perceived usefulness in our model. This is quite noteworthy.

Why might these be less impactful in the Moroccan context for public e-services?

In one hand, regarding Trust, perhaps for essential government services, citizens feel they *must* use them regardless of their general trust in online systems or even the specific government agency. The necessity might overshadow general trust levels. Alternatively, trust might operate differently; for instance, trust in the *technology's capability* versus trust in the *entity providing the service*.

In the other hand, regarding Social Influence, it's possible that social influence plays a stronger role at an earlier stage, like initial awareness of an e-service. However, when it comes to the actual decision to use it, individual perceptions of usefulness and ease of use might become more dominant for Moroccans in our sample. In some cultures, especially for official matters, individual assessment might take precedence over peer opinions once basic information is available.

Indeed, some studies have found similar patterns. For instance, a study on Internet banking in India noted that while trust and social influence were considered, they had less impact than user attitudes and perceived design quality [19]. Similarly, for digital health services, ease of use and perceived usefulness were key factors for usage intention, while satisfaction and security had less direct influence [20]. Our findings contribute to this nuanced understanding. They suggest that the core TAM constructs remain powerful, but contextual factors can significantly alter the role of other variables like trust and social influence.

6 Conclusion, Limitations and Future Directions

This investigation offers essential insights into what determines the uptake of online government services in Morocco. By adapting the TAM to include social influence, user satisfaction, service quality, and trust, along with demographic moderators, our study presents a thorough framework for the Moroccan context. Significantly, we found that perceived ease of use and perceived usefulness are critical in influencing citizens' intentions to use public e-services. However, aspects like user satisfaction (in its direct link to intention), trust, and social influence did not show a statistically significant direct effect on adoption intentions in our model.

Consequently, this research highlights the need for focused approaches to improve the usability and perceived benefits of e-services. This could be achieved through educational initiatives and awareness campaigns. These should cater to diverse levels of internet proficiency and different demographic groups.

Of course, we should acknowledge some limitations.

Firstly, our study used convenience sampling. Therefore, the findings might not fully represent all Moroccan citizens. This is particularly true for those in rural areas or individuals with lower internet access and skills.

Secondly, related to this, the urban-rural divide and varying internet proficiency levels are potential biases that convenience sampling might not fully capture.

These limitations, however, open doors for future research.

Future studies could delve deeper into *why* trust and social influence were not significant direct drivers in this specific context. For example, qualitative studies using interviews or focus groups might offer richer, more nuanced insights here.

Furthermore, directly investigating the impact of digital literacy, internet proficiency, and the urban-rural divide on e-service adoption would be very beneficial.

Additionally, exploring the operationalization and impact of related constructs, such as citizens' "digital readiness," could also add significant value to understanding e-service adoption in Morocco.

Therefore, these insights provide practical recommendations for policymakers and practitioners in Morocco and similar developing nations. By emphasizing user-centered design and actively tackling adoption obstacles, e-government efforts can achieve greater inclusivity, efficiency, and public confidence.

References

1. Aziz, Z.A., Sallow, A.B.: Concepts of e-governance development: a review. Eurasian J. Sci. Eng. **8**, 51–64 (2022)
2. Mensah, I.K., Vera, P., Jianing, M.: Predictors of e-government services adoption: a case study of Russian students in China. Int. J. Manag. Fuzzy Syst. **3**, 67–74 (2017). https://doi.org/10.11648/j.ijmfs.20170305.12
3. Nawafleh, S., Khasawneh, A.: Drivers of citizens e-loyalty in e-government services: e-service quality mediated by e-trust based on moderation role by system anxiety. TG **18**, 217–240 (2024). https://doi.org/10.1108/TG-04-2023-0053
4. Amosun, T.S., Chu, J., Rufai, O.H., et al.: Does e-government help shape citizens' engagement during the COVID-19 crisis? A study of mediational effects of how citizens perceive the government, 846–866 (2022)
5. Ali, H., Ali, T., Matar, Z., Jawad, F.: Citizens' acceptance and readiness towards adopting e-participation tools in kingdom of Bahrain. Int. J. Infonomics **8**, 1029–1036 (2015)
6. Nguyen, H.N., Tran, M.D.: Stimuli to adopt e-government services during Covid-19: evidence from Vietnam. Innov. Mark. **18**, 12 (2022)
7. Guo, Y., Zhou, L., Chen, J.: The persuasive role of the past: policy feedback and citizens' acceptance of information communication technologies during the COVID-19 pandemic in China. Rev. Policy Res. **40**, 573–588 (2023). https://doi.org/10.1111/ropr.12506
8. UNDESA: The Future of Digital Government: E-Government Survey 2022, United Nations, New York (2022)
9. Davis, F.D.: A technology acceptance model for empirically testing new end-user information systems: theory and results. Ph.D thesis, Massachusetts Institute of Technology (1985)
10. Bulut, T.: Technology acceptance model: the perceptions of Turkish and Kazakh English teachers in online teaching. Int. J. Lang. Acad. **9** (2021)

11. Nadlifatin, R., Miraja, B.A., Persada, S.F., et al.: The measurement of university students' intention to use blended learning system through technology acceptance model (TAM) and theory of planned behavior (TPB) at developed and developing regions: lessons learned from Taiwan and Indonesia. Int. J. Emerg. Technol. Learn. (iJET) **15**, 219–230 (2020). https://doi.org/10.3991/ijet.v15i09.11517
12. Nawafleh, S.: Factors Affecting the Continued use of E-Government Websites by Citizens: An Exploratory Study in the Jordanian Public Sector. Emerald Publishing Limited (2018)
13. Tangi, L., Janssen, M., Benedetti, M., Noci, G.: Digital Government Transformation: A Structural Equation Modelling Analysis of Driving and Impeding Factors. Elsevier (2021)
14. Sorongan, E., Sari, D.R., Kusno, H.S.: Systematic literature review of satisfaction model and interest in adoption of e-government services. In: 4th Borobudur International Symposium on Humanities and Social Science 2022 (BIS-HSS 2022), pp 692–702. Atlantis Press (2023)
15. Alkraiji, A.I.: Citizen satisfaction with mandatory e-government services: a conceptual framework and an empirical validation. IEEE (2020)
16. Sigwejo, A., Pather, S.: A citizen-centric framework for assessing e-government effectiveness. Electron. J. Inf. Sys. Dev. Ctries **74**, 1–27 (2016). https://doi.org/10.1002/j.1681-4835.2016.tb00542.x
17. Hartanti, F.T., Abawajy, J.H., Chowdhury, M., Shalannanda, W.: Citizens' trust measurement in smart government services. IEEE Access **9**, 150663–150676 (2021)
18. Mensah, I.K., Mi, J.: An Empirical Investigation of the Impact of Demographic Factors on E-Government Services Adoption. IGI Global (2018)
19. Bashir, I., Madhavaiah, C.: Consumer attitude and behavioural intention towards Internet banking adoption in India **7**, 67–102 (2015)
20. Akritidi, D., Gallos, P., Koufi, V., Malamateniou, F.: Using an extended technology acceptance model to evaluate digital health services (2022)

The Future of English Language Teaching: Investigating AI-Enhanced Personalisation in Digital Learning Contexts

P. Jibin Jose[1](✉), Amutha Dhanaraj[1], and Pranav Prakash[2]

[1] Division of English, Karunya Institute of Technology and Sciences, Coimbatore, India
jibinjp2815@gmail.com

[2] Division of Criminology and Forensic Sciences, Karunya Institute of Technology and Sciences, Coimbatore, India

Abstract. Artificial Intelligence (AI) in English Language Teaching (ELT) has transformed personalised learning by providing adaptable and tailored experiences to individual learners' requirements. By analysing student performance and preferences, AI-driven personalisation increases engagement, allows for real-time feedback, and optimises curriculum design. This study investigates the transformational role of AI in digital learning settings, focussing on major applications such as intelligent tutoring systems, chatbots, voice recognition tools, and adaptive learning platforms. While AI improves accessibility and efficiency in ELT, difficulties like data protection, teacher preparedness, and equal access to technology remain. This paper examines the benefits and limits of AI-driven personalisation, giving insights into its developing role in ELT. The study also explores future paths for AI integration, emphasising the need for ethical frameworks, teacher training, and hybrid learning models to fully realise its capabilities in educating languages.

Keywords: Artificial intelligence · English language teaching · Digital learning · Technology · Personalised learning

1 Introduction

The emergence of artificial intelligence (AI) has revolutionised several domains, including education. The way students learn and interact with the English language has been completely transformed by the incorporation of AI technology into English Language Teaching (ELT), which is still developing. Traditional ELT techniques frequently use standardised curriculum and teaching strategies, which may not fully accommodate students' varied learning preferences and requirements. Personalised learning experiences are made possible by AI's creative solutions, which guarantee that every student receives the education that is specific to their learning style, speed, and objectives.

Artificial intelligence (AI)-driven technologies, including speech recognition software, Natural Language Processing (NLP), intelligent tutoring systems, and adaptive learning platforms, have become essential elements of digital ELT settings [1]. To offer

B.-E. Boudriki Semlali et al. (Eds.): DATA 2025, LNISO 81, pp. 350–365, 2025.
https://doi.org/10.1007/978-3-032-06725-8_30

individualised help, these technologies track progress, analyse learner behaviour, and dynamically modify the course material. AI-powered chatbots, for example, can help students practise conversations in real time, allowing them to communicate with others and improve their pronunciation and fluency. In a similar way, writing samples can be evaluated by machine learning algorithms, which provide immediate feedback on word usage, grammar, and syntax. By using these resources, teachers and students can get beyond the drawbacks of conventional teaching strategies and improve the effectiveness and interest of language learning.

One of the key benefits of AI-driven personalisation in ELT is increased learner engagement and motivation. AI-powered gamification tactics, adaptive quizzes, and interactive exercises enhance language acquisition, instilling a sense of success and continual development. Unlike traditional classroom settings, where students may be hesitant to engage or request explanations, AI-based platforms provide a secure and nonjudgmental learning environment in which learners can train and improve their language abilities at their speed [2]. AI can also assist instructors in identifying children who require additional support by analysing performance trends and offering focused interventions. The incorporation of artificial intelligence into English Language Teaching (ELT) presents several ethical dilemmas, including concerns regarding data privacy, algorithmic bias, and the security of learner information. Insufficient resources and infrastructure can impede effective implementation [3]. An excessive dependence on technology might diminish the personal interactions essential for language learning. Furthermore, educators need targeted training to effectively incorporate AI into their teaching methodologies.

This study examines the uses, advantages, and difficulties of AI-driven personalisation in ELT to determine its transformational potential. It demonstrates how AI technologies alter the teaching of English language to accommodate students' unique demands by reviewing recent research. The paper also discusses the future of AI integration, highlighting the importance of ethical issues, teacher preparation, and hybrid learning models that balance human contact and AI-driven education. AI's involvement in ELT will surely grow as it develops further, providing new chances to improve language learning and make it more efficient, accessible, and suited to a range of learner profiles.

1.1 The Evolution of Individualised Learning in ELT

Personalised learning customises educational experiences for each learner's skills, limitations, and interests. Traditional English Language Teaching (ELT) frequently use a standardised approach, which cannot adequately fulfil the different requirements of all students. The use of Artificial Intelligence (AI) in ELT aims to close this gap by providing personalised learning paths that cater to each student's unique needs. AI-driven personalised learning in ELT uses technology like Natural Language Processing (NLP), machine learning algorithms, and intelligent tutoring systems to build adaptable learning environments. These systems use data from student interactions to discover patterns and alter educational content accordingly. For example, AI may evaluate a student's pronunciation, grammar, and vocabulary usage, offering quick feedback and personalised activities to target particular areas for growth. This dynamic modification results in a more interesting and effective learning experience than a traditional, static curriculum.

Recent researches have proved the effectiveness of AI-mediated language education in improving learning results. A study of English as a Foreign Language (EFL) learners indicated that AI-based education dramatically enhanced English learning achievement, motivation, and self-regulation [4]. The AI-mediated method delivered personalised feedback and adaptable learning routes, resulting in improved student engagement and autonomy. Another research of Chinese EFL students found that AI-based training significantly improved speaking abilities and self-regulation. Participants who received AI-driven training demonstrated significant increases in speaking competence and capacity to regulate their learning processes autonomously. These findings highlight AI's ability to meet individual learner requirements, notably in the development of communication abilities. The use of AI to create personalised learning routes extends beyond language training. A literature study emphasised AI's significance in lifelong learning, particularly its ability to develop personalised educational trajectories that adapt to learners' changing requirements and aspirations. In the context of ELT, adaptability ensures that learners get teaching that is appropriate for their present proficiency levels and learning objectives, thereby optimising the educational experience. The growth of personalised learning in ELT through AI integration represents a transformational approach to language instruction. By personalising training to individual learner profiles, AI has the potential to increase English language learners' engagement, competency, and autonomy. Continuous research and intelligent implementation are required to fully realise the benefits of AI-driven personalised learning while also tackling the accompanying obstacles.

2 AI Technologies in Improving ELT Personalisation

Artificial Intelligence (AI) transforms English Language Teaching (ELT) by enabling personalised, self-paced learning experiences through technologies such as Natural Language Processing, speech recognition, machine learning algorithms, and intelligent tutoring systems that enhance engagement and optimise language acquisition.

2.1 Natural Language Processing in ELT

Natural Language Processing (NLP) is a subfield of artificial intelligence (AI) that allows machines to comprehend, interpret, and synthesise human language. In the context of English Language Teaching, NLP is critical in improving personalised learning by delivering real-time feedback on grammar, pronunciation, and language use. NLP-powered systems spot mistakes, propose repairs, and provide individualised recommendations to help learners improve their language competence [5]. In ELT, NLP offers automated grammar and writing evaluation. AI-powered services such as Grammarly and Microsoft Editor employ NLP algorithms to evaluate sentence structure, punctuation, and word choice. These tools assist learners in improving their writing abilities by spotting patterns of errors and providing context-aware solutions. NLP-powered chatbots engage students in interactive conversations to improve conversational fluency and contextual language comprehension.

Speech recognition, another important component of NLP, improves pronunciation instruction. AI-powered tools such as ELSA Speak analyse spoken language, identify mispronunciations, and lead learners to perfect pronunciation. These technologies utilise phonetic analysis to compare learner speech to native speaker models, offering exact feedback on articulation and intonation. NLP also helps with reading comprehension and vocabulary growth. AI-based tools, such as Readability and LingQ, assess student performance and offer reading resources appropriate for their level. These technologies emphasise difficult terms, give meanings, and propose synonyms, promoting a better understanding of texts. Teachers can create a more customised and flexible learning environment by using NLP in ELT. NLP technology has enormous potential to improve language education even more as it develops, making learning more effective, engaging, and customised to meet the requirements of each student [6].

2.2 Machine Learning Algorithms in ELT

By evaluating student data and modifying instructional content appropriately, Machine Learning (ML) algorithms significantly improve personalised learning in ELT. ML-powered systems evaluate each student's proficiency level, learning preferences, and progress patterns to provide personalised learning experiences. By providing learners with material tailored to their requirements, this dynamic flexibility maximises engagement and retention [7]. ML algorithms use adaptive learning platforms in programmes like Rosetta Stone, Coursera, and Duolingo to monitor student progress and modify the level of instruction in real-time. These systems evaluate both right and wrong answers, identify problem areas, and adjust later tasks to promote learning. For instance, the algorithm will provide more activities and explanations to improve a learner's verb tenses if they are a persistent problem. It plays a key role in automated evaluation and feedback. Grammarly and Criterion, two AI-powered writing assessment programmes, employ machine learning (ML) to evaluate essays, identify mistakes, and offer recommendations based on language patterns. Similarly, speech recognition tools such as Google Speech-to-Text and ELSA Speak use machine learning (ML) to enhance pronunciation by comparing learner speech to native speaker models and providing corrective feedback. As machine learning technology evolves, its incorporation with ELT will improve personalised instruction, making language learning more efficient, focused, and sensitive to individual requirements [8].

2.3 Chatbots and Virtual Assistants in ELT

AI-powered chatbots and virtual assistants transform ELT by offering engaging practice possibilities to learners. These intelligent conversational agents imitate real-life discussions, allowing students to improve their speaking, listening, and writing abilities in a responsive setting. Unlike traditional learning approaches, which frequently rely on teacher-led instruction, chatbots provide continuous, on-demand linguistic help, allowing students to engage in discussions at any time and from any location [9]. AI-powered chatbots provide realistic, real-world language experiences. These virtual assistants, such as Duolingo's AI teacher, Mondly's chatbot, and ChatGPT, engage learners in simulated discussions that span a wide range of topics, including buying meals at a restaurant, booking a hotel, and participating in job interviews. Chatbots improve fluency and confidence by reacting to student input in real time, making language acquisition more practical and application-based.

Chatbots give rapid comments on grammar, vocabulary, and pronunciation. These AI technologies use advanced NLP to analyse phrase structure, detect faults, and make correction suggestions. Speech recognition systems built into chatbots, such as Google Assistant and ELSA Speak, assist learners in improving pronunciation by comparing their speech to native models and coaching them towards more correct articulation [10]. This real-time examination enables students to improve their speaking abilities without the pressure of a regular classroom setting. AI chatbots customise conversations by watching user interactions and analysing response patterns, focussing on areas where a learner might improve [11]. For example, if a student has difficulties with past tense verbs, the chatbot can provide focused tasks and contextual dialogue practice to help reinforce comprehension. Thus, chatbots and virtual assistants play an important role in modern ELT, making language learning more personalised to individual requirements.

3 Specific AI Tools for Personalisation in ELT

The integration of AI into ELT has led to the emergence of specific tools designed to personalise the learning experience. The following discussion will shed light on some of these cutting-edge tools.

3.1 Duolingo

Duolingo is a leading AI-driven language learning platform known for its intuitive interface and engaging, game-like experience. It employs advanced machine learning to monitor user progress, identifying strengths, weaknesses, and learning styles. These data allow Duolingo to dynamically adjust lesson complexity, content, and sequence, providing challenges tailored to each learner's skill level across vocabulary, grammar, reading, and pronunciation. A key feature of Duolingo's AI is its use of chatbots and interactive activities that simulate real-life conversations, allowing learners to practice in context without pressure. The platform also incorporates spaced repetition to boost learning and retention [12]. Through these adaptive methods, Duolingo offers a personalized educational journey that evolves with the learner, enhancing the effectiveness and enjoyment of language acquisition.

3.2 Orell Talk

Orell Talk is a cutting-edge AI-powered language lab that aims to improve English proficiency through personalised learning experiences. Orell Talk offers organised courseware that is aligned with the Common European Framework of Reference for Languages (CEFR) and suited to the requirements of individual learners. The platform continually reviews student performance and changes courses based on AI-driven data, resulting in a customised learning path that targets particular areas for progress. Its sophisticated voice recognition technology assists learners in improving their pronunciation and fluency. The

AI examines spoken replies, recognises pronunciation mistakes, and provides correction input to help improve articulation. This feature enables students to confidently practise real-world conversations while learning appropriate speech patterns [13]. Additionally, the site offers interactive listening and speaking tasks to assist students in improving their comprehension and communication abilities.

3.3 ELSA

ELSA (English Language Speech Assistant) is an AI-driven mobile application designed to enhance English pronunciation. It utilizes advanced speech recognition and deep learning to assess spoken English, identifying phonetic errors and accent variations. Unlike traditional tools, ELSA provides personalized feedback based on the user's native language and pronunciation patterns. The app features targeted exercises that address specific challenges, such as stress, intonation, and articulation, promoting measurable improvement. Its AI system adapts to user interactions, refining feedback to meet evolving needs. ELSA also tracks progress through performance metrics and visual charts, enabling learners to monitor their development [12]. Additionally, ELSA's lessons simulate real-world communication scenarios, making it particularly useful for those seeking to improve their spoken English for academic, professional, or social contexts.

3.4 Grammarly

Grammarly is an AI-driven writing assistant that offers more than just grammar and spelling checks. Utilizing advanced natural language processing and machine learning, it delivers personalized feedback tailored to the user's writing style, skill level, and communication objectives. By evaluating tone, clarity, conciseness, formality, and emotional nuances, Grammarly enhances writing in real time across various contexts, including academic, professional, creative, and informal settings. Grammarly can customize suggestions based on user intent, allowing users to choose goals such as 'inform,' 'persuade,' or 'narrate,' which shapes the recommendations. It caters to both novice and experienced writers, providing basic corrections for beginners and advanced stylistic advice for more seasoned users [14]. Grammarly is particularly beneficial in ELT, promoting independent learning by encouraging students to assess their writing. Its user-friendly interface and comprehensive explanations make it an invaluable resource for improving grammatical accuracy, expanding vocabulary, and enhancing stylistic fluency - key elements of effective writing.

3.5 Mondly

Mondly leverages AI to provide personalized English language education through tools such as speech recognition, chatbot interactions, and augmented reality (AR) to create an engaging learning environment. By evaluating learners' performance over time, Mondly tailors lesson materials to meet individual skill levels and goals, whether for travel, professional communication, or everyday conversations. The platform's speech recognition technology evaluates pronunciation and fluency, offering immediate feedback to improve spoken accuracy. AI-driven chatbots simulate real-world conversations, helping learners build confidence and conversational skills in practical contexts, such as interacting with hotel staff or colleagues in meetings [10]. Mondly's adaptability is further enhanced by personalised learning pathways that target areas needing improvement while reinforcing previously learned material through spaced repetition. Its blend of visual and auditory learning elements caters to diverse learning styles, making it a valuable resource for both independent study and classroom use. Mondly exemplifies how AI can enrich ELT by providing dynamic, learner-centered experiences.

3.6 Lingvist

Lingvist emphasizes adaptive vocabulary development, offering a personalized experience tailored to each learner's pace, skill level, and goals. The foundation of this system is a refined machine learning engine that scrutinizes user interactions, such as response times and precision, to detect words that require focus and to establish the ideal timing for their introduction. Unlike traditional vocabulary lists, Lingvist generates contextually relevant flashcards that adapt based on the learner's memory retention and progress. This dynamic approach allows users to focus on reinforcing unfamiliar or forgotten words while reducing unnecessary repetition of mastered vocabulary [11]. Learners can customize their study themes to align with their personal or professional objectives. The platform employs spaced repetition and contextual application, helping users not only memorise words but also understand their usage in sentences.

3.7 ScribeSense

ScribeSense is designed to streamline the assessment process and provide personalized feedback on handwritten and typed English assignments. By employing optical character recognition (OCR) and machine learning, it quickly and accurately scans and evaluates various student submissions. A notable feature of ScribeSense is its ability to assess individual responses and generate feedback aligned with specific educational goals. This enables educators to go beyond basic corrections and offer targeted support tailored to each student's unique strengths and weaknesses [14]. Teachers can also monitor performance trends over time, allowing for adjustments in lesson plans and intervention strategies. For students, ScribeSense delivers prompt, comprehensive, and actionable feedback, helping them identify mistakes and improve their writing skills. It promotes fairness in assessment by minimizing human bias and ensuring consistency in evaluations.

The subsequent table (Table 1) illustrates a comparative framework of the various AI tools available for personalized ELT based on their key areas of emphasis.

Table 1. A comparative framework of the various AI tools

AI Tool	Major Focus Area	Personalisation Features	Purpose/Best For
Duolingo	All-in-One	Adaptive learning paths, gamified lessons, chatbot conversations	General English learners

(*continued*)

Table 1. (*continued*)

AI Tool	Major Focus Area	Personalisation Features	Purpose/Best For
Orell Talk	Speaking, Listening, Grammar	CEFR-aligned modules, AI-driven speech evaluation, individualized progress tracking	Language labs, schools, and exam preparation
ELSA	Speaking	Personalized pronunciation feedback using speech recognition	Improving accent and fluency
Grammarly	Writing	Style, tone, and grammar suggestions tailored to user writing goals	Academic and professional writing
Mondly	Speaking/Listening	Real-life scenarios and chatbot interaction tailored to user performance	Conversational practice
Lingvist	Vocabulary	AI-driven spaced repetition tailored to individual learning speed and memory	Vocabulary expansion
ScribeSense	Assessment (Writing)	AI assessment and feedback on handwritten/typed work tailored to student progress	Classroom writing and evaluations

4 Benefits of AI-Driven Personalisation in ELT

The application of AI in English Language Teaching brings forth several advantages (Fig. 1), such as individualized learning paths, immediate feedback, enhanced learner engagement, skill development, and improved accessibility in language acquisition. By monitoring the progress of each learner, AI effectively modifies lessons to suit their specific needs, resulting in a more personalised educational experience.

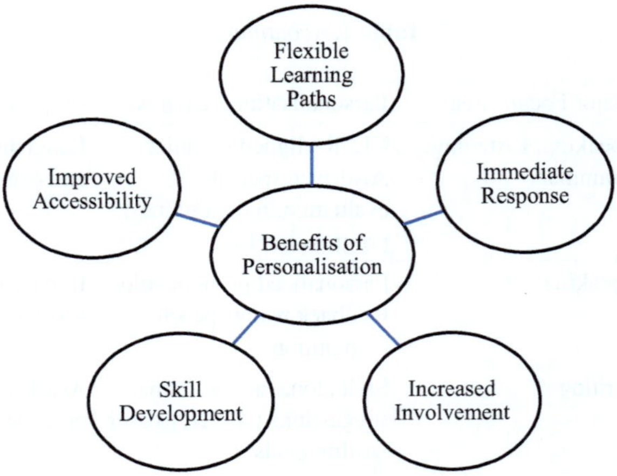

Fig. 1. Benefits of AI-Driven Personalization in ELT

4.1 Flexible Learning Paths

AI-enhanced systems offer customised learning experiences by changing how information is presented based on ongoing assessments of learners' progress. Conventional English Language Teaching strategies often stick to a predetermined curriculum that cannot fulfil the specific needs of all students. In comparison, AI systems evaluate student performance metrics and adapt the course difficulty to ensure an optimal challenge for each learner [15]. For instance, the AI offers more explanations and practice problems if a student is having trouble with a certain grammatical rule. On the other hand, if a student does well in a subject, the system will present more challenging content to keep them interested. Predictive analytics is used by adaptive learning platforms like Duolingo and Orell Talk to foresee student challenges and provide proactive interventions. This guarantees consistent language acquisition progress, lowers frustration, and keeps students engaged.

4.2 Immediate Response

A significant advantage of artificial intelligence in ELT is its ability to deliver real-time feedback, allowing learners to promptly correct their mistakes. In traditional classroom settings, feedback from teachers can be delayed, which can impede students' progress. AI tools like Grammarly and ELSA Speak provide immediate corrections for grammar, pronunciation, and vocabulary by analysing user input. For instance, voice recognition technology evaluates pronunciation and offers auditory and visual feedback to assist learners in articulating words correctly. Additionally, writing tools help students enhance their writing skills by identifying grammatical errors and suggesting contextual improvements, thereby facilitating a more efficient learning process through increased awareness and self-correction.

4.3 Increased Involvement

The inclusion of interactive and gamified components in AI systems markedly enhances language learners' engagement. AI-driven applications utilize features such as progress tracking, rewards, and real-world scenarios to motivate students, unlike traditional educational approaches that may become tedious. Gamified activities, including quizzes, challenges, and leaderboards, instill a sense of achievement and encourage ongoing practice. Additionally, conversational AI tools, such as chatbots and virtual tutors, simulate real-life conversations, allowing learners to practice their speaking and listening skills in a relaxed setting. These AI-facilitated interactions not only improve fluency and self-esteem but also enrich the overall language learning experience.

4.4 Skill Development

AI systems utilize continuous data analysis to evaluate learner performance in fundamental language competencies - speaking, writing, listening, and reading - and customize educational content accordingly. This level of precision allows learners to direct their efforts toward specific linguistic areas that need improvement rather than revisiting topics they have already conquered. As a result, learners enjoy more intentional and efficient study experiences. AI-enhanced platforms typically present scaffolded tasks and progressive challenges, promoting a structured enhancement of language skills that aligns with the learner's developing proficiency.

4.5 Improved Accessibility

AI-driven language learning solutions enhance accessibility, delivering high-quality, personalized instruction to a diverse range of learners. Utilizing mobile apps and cloud technology, AI offers customized English language education to individuals regardless of their location or economic status. This is especially beneficial in resource-limited areas where qualified teachers and educational facilities are scarce. The asynchronous nature of AI systems allows learners to engage with content at their own pace, promoting inclusive education and bridging gaps in language learning opportunities across various populations.

5 Challenges of AI-Driven Personalisation in ELT

While AI-driven personalisation in ELT holds significant promise, it is accompanied by several challenges (Fig. 2). Key issues include the demand for high-quality AI training data and its security, the need for equitable access, the integration of teaching professionals, the risk of algorithmic bias, and the potential reduction in human contact.

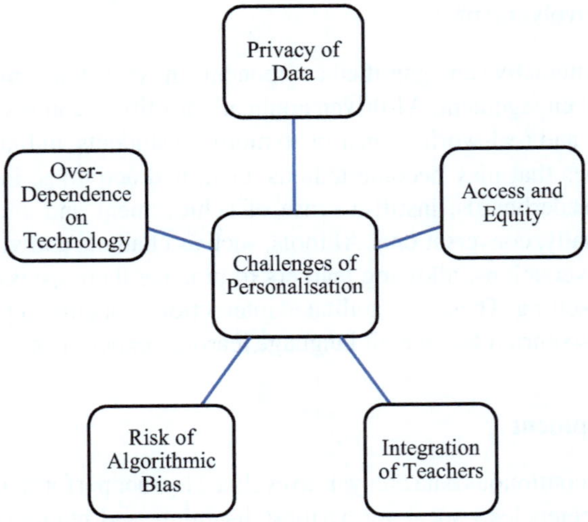

Fig. 2. Challenges of AI-Driven Personalization in ELT

5.1 Privacy of Data

The collection and analysis of learner data require strict measures to protect personal information. AI-driven ELT systems utilize extensive user data, including writing samples, audio recordings, and performance metrics, to customize educational experiences. Inadequate management of this data can lead to privacy breaches and unauthorized access. Therefore, robust data protection strategies, such as encryption, secure servers, and compliance with privacy laws like the General Data Protection Regulation (GDPR), are essential to mitigate these risks. Additionally, educators and learners should have control over their data and be informed about relevant data-sharing regulations.

5.2 Access and Equity

Ensuring that AI-enhanced educational resources are available to a wide range of demographics, especially in underserved regions, is vital for fostering inclusive education. While AI-powered ELT tools can improve language learning accessibility, many learners face challenges due to disparities in digital infrastructure, internet access, and smart device availability, particularly in rural and low-income regions. Governments, educational institutions, and technology companies must invest in scalable and affordable AI-driven learning solutions to address this digital divide. Initiatives like localized AI-generated content, mobile-compatible applications, and offline AI tools can significantly improve access to quality language instruction [16]. AI systems ought to incorporate features such as text-to-speech, speech-to-text, and customizable user interfaces to support students with disabilities.

5.3 Integration of Teachers

Teachers need training to effectively integrate AI technologies into their lesson plans while balancing technology use with interpersonal communication. While AI can automate assessments and provide personalized feedback, it cannot fully replace the role of educators in fostering critical thinking and creativity. Many teachers feel unprepared to use AI tools due to insufficient training or technological expertise. Therefore, professional development initiatives should equip educators with the necessary skills to address these challenges, focusing on maintaining a learner-centered approach and integrating AI with traditional teaching methods [17]. By cultivating AI literacy among educators, educational institutions can amplify the benefits of AI-enhanced ELT while ensuring that vital human components of language learning are preserved.

5.4 Risk of Algorithmic Bias

Algorithmic bias represents a significant hurdle in AI-driven ELT, particularly when models are trained on datasets that do not reflect a rich diversity of linguistic and cultural backgrounds. This limitation can lead to AI systems providing feedback that is biased toward dominant language norms, which can marginalize non-native accents, dialects, or culturally specific expressions. As a result, learners from underrepresented communities may face inappropriate corrections or experience a decline in confidence. Additionally, biased algorithms can perpetuate stereotypes and fail to recognize the diverse ways in which English is spoken globally.

5.5 Over-Dependence on Technology

Overreliance on technology can jeopardize critical elements of holistic language acquisition. Language learning is fundamentally social, depending on interpersonal communication, cultural exchanges, and collaborative problem-solving. Excessive use of AI tools can limit opportunities for genuine human interaction, obstructing the growth of pragmatic language skills, emotional intelligence, and critical thinking. Although automated feedback is efficient, it often lacks the depth and contextual insight that human educators offer. Thus, it is important to thoughtfully integrate AI while ensuring meaningful teacher involvement and peer engagement to enhance linguistic competence.

6 Ethical Issues with AI in ELT

The incorporation of AI into ELT creates transformational opportunities while raising important ethical problems (Fig. 3). As AI has an increasing impact on instructional design, assessment, and learner engagement, it is necessary to analyse the moral and societal consequences of its use in educational settings.

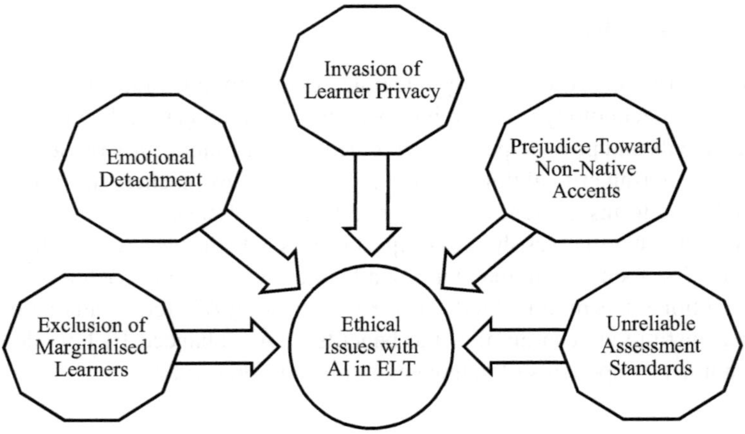

Fig. 3. Ethical Issues with AI in ELT

6.1 Invasion of Learner Privacy Through Data Collection

AI-enhanced ELT platforms typically accumulate a wide range of personal information, including speech patterns, learning behaviours, and location data. Without explicit consent or transparent data policies, such practices can breach learners' privacy rights and raise concerns about surveillance and the potential misuse of sensitive information.

6.2 Prejudice Toward Non-native Accents

Speech recognition technologies often exhibit suboptimal performance when dealing with non-native or regional English accents, primarily due to the influence of biased training datasets. This inadequacy can result in unfair evaluations, decreased motivation, and potential linguistic discrimination, thereby perpetuating the supremacy of specific language varieties, such as American or British English. Further, it poses a threat to a broader understanding of language proficiency, undermining the validity of global English varieties and the identities of multilingual speakers.

6.3 Unreliable Assessment Standards

AI assessment tools employ criteria that are either inconsistent or not transparent in grading both written and spoken responses. This situation brings forth ethical challenges concerning fairness and credibility, particularly when AI-generated scores play a role in important decisions such as certifications or placements.

6.4 Emotional Detachment and Learner Well-Being

AI systems are devoid of empathy and emotional intelligence, qualities that are vital for helping learners cope with anxiety, frustration, or low confidence - issues that commonly arise in the process of language learning. Relying entirely on AI might miss the crucial emotional and psychological support that human instructors deliver.

6.5 Exclusion of Marginalized Students

AI tools often take for granted that users have reliable technology access, inadvertently excluding students from low-income or rural areas who lack stable internet, modern devices, or adequate digital skills. This disparity in access can deepen educational inequalities instead of addressing them.

7 Prospects for the Future

The integration of artificial intelligence in ELT is advancing towards more sophisticated and personalized learning experiences, driven by developments in adaptive learning, machine learning, and natural language processing. Future AI systems are anticipated to feature improved voice recognition and emotion detection, enabling more human-like interactions and contextual feedback for learners. As AI technology progresses, its role in language instruction will be influenced by data security, ethical considerations, and equitable access to educational tools.

7.1 AI-Powered Tutors

Future developments will include sophisticated AI teachers that will use context-aware models and deep learning to engage students in organic, free-flowing interactions. They will examine sentence structure, tone, intent, and grammar mistakes, providing insightful criticism. Emotion identification will allow tutors to modify their responses. AI tutors will also offer real-time translations and cultural context, increasing accessibility and inclusivity for non-native speakers [18]. The efficacy of these AI teachers depends on objectivity, cultural sensitivity, and ethical design.

7.2 Combining Emerging Technologies

Artificial Intelligence (AI), Virtual Reality (VR) and Augmented Reality (AR) can revolutionize language learning by providing immersive environments for students to practice their English in real-world scenarios [19]. This experiential learning strategy improves application and retention, particularly for visual and kinaesthetic learners. Augmented Reality (AR) can enhance English Language Teaching (ELT) by providing real-time translations and contextual cues. This approach will enable global collaboration and accessibility for students from diverse socioeconomic backgrounds, making immersion language learning a reality.

7.3 Platforms for Collaborative Learning

By matching students with related objectives and complementing abilities, AI can promote social learning and peer-to-peer involvement in collaborative learning. Intelligent matching algorithms will be used by future AI-driven platforms to connect students according to their areas of expertise, learning preferences, and ability levels. Through cooperative projects, role-playing games, and facilitated conversations, students will be able to practise their English, improving their communication abilities and cross-cultural understanding. AI can enhance language learning by assisting in speech recognition, real-time translation, transcription systems, and interactive discussion forums [20]. This social component strengthens language skills and promotes fluency in natural conversation. AI's ability to create dynamic learning environments will revolutionize ELT, making it more accessible and collaborative.

8 Conclusion

The application of AI-driven personalization in English Language Teaching is fundamentally altering the learning experience by providing adaptive, data-informed educational pathways tailored to the individual preferences of students. This technology enhances both efficiency and engagement through the implementation of speech recognition, Natural Language Processing, and adaptive learning algorithms. However, challenges such as data privacy issues and unequal access to AI-enhanced education remain pressing concerns. For AI to effectively improve language education, educators need to reassess their roles, prioritizing communication, critical thinking, and cultural awareness while incorporating AI technologies within language schools. With the incorporation of cutting-edge technologies like Virtual Reality (VR) and Augmented Reality (AR), AI's involvement in ELT is set to grow in the future. Collaborative AI-powered platforms will promote international communication and peer learning, assisting students in interacting with a range of language and cultural viewpoints. Despite challenges like data protection and instructor integration, AI has the potential to make English language learning more accessible and personalized throughout the world by finding a balance between ethical, inclusive application and technical innovation.

Acknowledgement. Nil.

Conflict of Interest. The authors have no competing interests to declare that are relevant to the content of this article.

Funding. No funding was received to assist with the preparation of this manuscript.

References

1. Abdullah Sharadgah, T., Abdulatif Sa'di, R.: A systematic review of research on the use of artificial intelligence in English language teaching and learning (2015–2021): what are the current effects? J. Inf. Technol. Educ. Res. **21**, 337–377 (2022). https://doi.org/10.28945/4999
2. Yusupova, M., Ismailov, T.: Integration of Artificial Intelligence (AI) in ELT. Nordic_Press 3 (2024)
3. Crompton, H., Edmett, A., Ichaporia, N., Burke, D.: AI and English language teaching: affordances and challenges. Br. J. Edu. Technol. **55**, 2503–2529 (2024). https://doi.org/10.1111/bjet.13460
4. Nursyahida, H.M., Ramadhanti, T., Aulia, N.H.I., Puteri, S.N., Tarihoran, N.: The use of AI and its effectiveness in ELT: a systematic review. Algebra Jurnal Pendidikan Sosial dan Sains **4**, 10–18 (2024)

5. Mejeh, M., Rehm, M.: Taking adaptive learning in educational settings to the next level: leveraging natural language processing for improved personalization. Educ. Technol. Res. Dev. **72**, 1597–1621 (2024). https://doi.org/10.1007/s11423-024-10345-1
6. Holland, V.M., Kaplan, J.D.: Natural language processing techniques in computer-assisted language learning: status and instructional issues. Instr. Sci. **23**, 351–380 (1995). https://doi.org/10.1007/BF00896878
7. Zhang, L.: A new machine learning framework for effective evaluation of English education. Int. J. Emerg. Technol. Learn. **16**, 142 (2021). https://doi.org/10.3991/ijet.v16i12.23323
8. Thadphoothon, J.: ELT in the age of artificial intelligence (AI): Working with machines. J. NELTA **27**, 202–212 (2022). https://doi.org/10.3126/nelta.v27i1-2.53203
9. Sasikanth, K.J.W.: Revolutionizing language learning: the role of chatbots in ELT. In: Techtalk: Revolutionising English Language Teaching in the 21st Century, pp. 68–76. Veda Publications (2024)
10. Belda-Medina, J., Calvo-Ferrer, J.R.: Using chatbots as AI conversational partners in language learning. Appl. Sci. **12** (2022). https://doi.org/10.3390/app12178427
11. Kim, N.-Y., Cha, Y., Kim, H.-S.: Future English learning: chatbots and artificial intelligence. Multimed. Assist. Lang. Learn. **22**, 32–53 (2019). https://doi.org/10.15702/MALL.2019.22.3.32
12. Shinde, S.M.: Leveraging AI for English language learning: a comparative analysis of Duolingo, Babbel, and ElsaSpeak. Lang. India **24**, 25–37 (2024)
13. Armstrong, S.: Orell digital language lab software in language labs: An overview. Oman J. ELT **1**, 92–97 (2012)
14. Kristiawan, D., Bashar, K., Pradana, D.A.: Artificial intelligence in English language learning: a systematic review of AI tools, applications, and pedagogical outcomes. Art Teach. Engl. Foreign Lang. **5**, 207–218 (2024). https://doi.org/10.36663/tatefl.v5i2.912
15. Dugošija, T.: Benefits and challenges of artificial intelligence in English language teaching. Knowl. Int. J. **62**, 275–280 (2024)
16. Özdere, M.: The integration of artificial intelligence in English education: opportunities and challenges. Lang. Educ. Technol. **3** (2023)
17. Pokrivcakova, S.: Preparing teachers for the application of AI-powered technologies in foreign language education. J. Lang. Cult. Educ. **7**, 135–153 (2019). https://doi.org/10.2478/jolace-2019-0025
18. Wang, Q.: AI-driven autonomous interactive English learning language tutoring system. J. Comput. Methods Sci. Eng. (2024). https://doi.org/10.1177/14727978241296719
19. Ulutaş, N.K., Durmaz, M.E., Elbuğa, N., Sezgin, S., Kaya, K., İsmailoğulları, B.: The use of augmented reality (AR), virtual reality (VR), and artificial intelligence (AI) in language education: a systematic literature review. In: Stevkovska, M., Klemenchich, M., and Kavaklı Ulutaş, N. (eds.) Advances in Educational Technologies and Instructional Design, pp. 401–430. IGI Global (2024)
20. Song, H.: When ELT meets AI: reshaping EDI-centric EAP teaching. Multicult. Educ. Rev., 1–17 (2025). https://doi.org/10.1080/2005615X.2025.2467807

Artificial Intelligence for Sustainable Development

Machine Learning-Based System for Colorectal Cancer Prediction

Ibtissam Samnane, Rachid Ed-Daoudi(✉), Samir Bouali, M'barek El Haloui, and Badia Ettaki

LyRICA: Laboratory of Research in Informatics, Data Sciences and Artificial Intelligence, School of Information Sciences, Rabat-Instituts, B.P. 604, Rabat, Morocco
rachid.ed-daoudi@uit.ac.ma

Abstract. With the increase in the number of diagnosed cases of colon cancer and the high associated mortality rate, it becomes imperative to offer a rapid and effective diagnosis in order to improve patient care and optimize the chances of treatment. In this context, with the challenges in the field of oncology, the need of the application of Artificial Intelligence (AI) through techniques such as Machine Learning (ML) to offer solutions that will help doctors in their work by minimizing time and effort, within the framework of what is called precision medicine. This work will present a system for predicting the type of cancer according to the clinical and demographic factors of patients using ML models. The evaluation of five ML models (Logistic Regression (LR), Random Forest (RF), K-Nearest Neighbors (KNNs), XGBoost and Decision Tree Classifier (DTC)) indicated that DTC is the best model. And to facilitate the prediction operation, the model identifier is integrated into a web interface that allows users to fill out forms with the requested factors so that the system gives its prediction on the type of cancer only on these factors without the need to do other medical examinations that take time and delay the treatment time.

Keywords: Colon cancer · Oncology · Artificial Intelligence · Machine Learning · Prediction System

1 Introduction

Colon cancer ranks among the most prevalent and deadly cancers globally, typically beginning as benign polyps that can progress to aggressive malignancies. The International Agency for Research on Cancer (IARC) identifies colorectal cancer as one of the most common cancer types. In 2022, approximately 19.9 million new cancer cases were documented globally, with colorectal cancer being the third most commonly diagnosed and the third leading cause of cancer mortality among both men and women [1].

Adenocarcinomas account for 90–95% of colorectal cancer cases, primarily arising from the epithelial lining of the colon and rectum. The three main subtypes include signet ring cell adenocarcinoma (a rare, aggressive form with mucus-filled cells), mucinous adenocarcinoma (containing more than 50% extracellular mucus, often linked to Lynch

B.-E. Boudriki Semlali et al. (Eds.): DATA 2025, LNISO 81, pp. 369–381, 2025.
https://doi.org/10.1007/978-3-032-06725-8_31

syndrome), and tubular adenocarcinoma (the most prevalent subtype, characterized by glandular structures) [2].

Colon cancer development is influenced by genetic factors in hereditary cases, lifestyle elements (smoking, alcohol consumption, dietary factors), and the patient's medical history, particularly non-cancerous conditions like colorectal polyps, adenomas, and ulcerative colitis [3].

Machine Learning (ML), a branch of Artificial Intelligence (AI), holds significant potential to transform healthcare. It enhances case-based reasoning in medical decision-making by addressing limitations of conventional diagnostic approaches. In ML, supervised learning works with labeled data (using regression for continuous targets and classification for discrete categories), while unsupervised learning operates on unlabeled data [4].

This study presents an AI-driven system for predicting cancer types using various prognostic factors. Five ML algorithms—LR, RF, KNNs, XGBoost and DTC—were evaluated to identify the most effective model. Following comprehensive data preparation and model development, an online interface was created to enable healthcare professionals to input relevant information and obtain predictions, potentially enhancing diagnostic accuracy while conserving time and resources.

2 Literature Review

AI has become a valuable tool in cancer diagnosis and prognosis, with numerous studies investigating AI-based models that leverage medical imaging, histopathology, and patient data.

Various machine learning approaches have shown promising results. CNNs demonstrate high performance (98% across metrics) for cancer identification from medical images, though their effectiveness depends on training data quality [5]. LR achieves 71.8% accuracy but faces generalization challenges due to limited sample size [6]. RF models show strong performance (84–99% accuracy, 89–99% F1 scores) but are affected by feature selection [7]. SVMs (75% accuracy, 83% F1 score) were limited by their reliance on genetic data [8]. KNN performs well (93% accuracy, 86% F1 score) despite dataset limitations [9]. Naïve Bayes achieves impressive metrics (95% accuracy, 100% precision) but its independence assumption reduces robustness [10]. DTs reach perfect metrics in some studies but using genetic data that may not translate to diverse clinical datasets [11]. XGBoost (75–76% accuracy) and CatBoost (88% accuracy) show potential but are constrained by data quality and sample size issues [12, 13].

The literature reveals diverse applications of AI technologies using various data types, but there remains a gap in studies addressing cancer stage classification based solely on clinical data - the objective of this study.

3 Method

3.1 Data Collection and Description

The first step in building the prediction system was the collection of a dataset. After some research, an open-source dataset was identified, consisting of two files [14]. The first file contains 6 columns with 300 rows, while the second one also has 6 columns with 100 rows. Table 1 presents a comprehensive description of all variables from both files, indicating their source, data type, and clinical significance.

Table 1. Comprehensive Dataset Description

Column name	Column type	Source	Description
AGE	Numerical	File 1	Represents the patient's age
CEA Level	Numerical	File 1	One of the most commonly used markers to assess the prognosis and recurrence of colon cancer [15]
Colonoscopy Result	Categorical ("Normal"/"Abnormal")	File 1	Indicating result of colonoscopy, considered the gold standard for colon assessment
Polyp Size (mm)	Numerical	File 1	Size of polyps detected during colonoscopy
Family History	Categorical ("Yes"/"No")	File 1	Indicating family history of cancer
Tumor Grade	Numerical	File 2	Indicating the grade of the tumor
Biopsy Result	Categorical ("Positive"/"Negative")	File 2	Presence or absence of cancer cells; used for initial diagnosis, staging, prognosis and management
Lymph Node Involvement	Categorical ("Yes"/"No")	File 2	Essential for predicting disease recurrence and survival; key element of current staging system [15]
Bowel Obstruction	Categorical ("Yes"/"No")	File 2	Indicating presence or absence of intestinal obstruction
Smoking History	Categorical ("Smoker"/"Non-Smoker")	File 2	Indicating patient's smoking history
Type of Colon Cancer	Numerical	Both	Types coded 1–3: 1 = Adenocarcinoma, 2 = Mucinous adenocarcinoma, 3 = Signet ring cell adenocarcinoma

For this project, the two files were merged by concatenating the corresponding columns, resulting in a final dataset of 11 columns and 400 rows. This dataset was then used to set up the prediction system.

3.2 Data Preprocessing and Feature Engineering

Following file concatenation, preprocessing began with statistical analysis of the dataset (Table 2). All categorical variables ('Colonoscopy Result', 'Family History', 'Biopsy Result', 'Lymph Node Involvement', 'Bowel Obstruction', 'Smoking History') were encoded. Missing values were addressed using KNN imputation, which leverages Euclidean distance calculations to identify nearest neighbors, weighting non-missing coordinates more heavily while ignoring missing ones to determine replacement values.

The Euclidean distance is calculated using the Eq. (1).

$$D_{xy} = \sqrt{weight * squared\ distance\ from\ present\ coordinates} \tag{1}$$

$$\text{With: } weight = \frac{total\ number\ of\ coordinates}{number\ of\ present\ coordinates}$$

Table 2. Statistical description of the dataset

Parameters	AGE	CEA Level	Polyp Size (mm)	Type of Colon Cancer	Tumor Grade
Count	300.00	300.00	300.00	400.00	100.00
Mean	62.20	11.60	24.66	1.97	2.02
Std	13.24	5.29	13.98	0.79	0.81
Min	40.00	2.68	1.22	1.00	1.00
25%	50.00	7.41	12.30	1.00	1.00
50%	63.00	12.02	23.88	2.00	2.00
75%	74.00	16.36	36.88	3.00	3.00
Max	84.00	19.87	49.84	3.00	3.00

Boxplot analysis (Fig. 1) identified outliers in several variables: "Tumor Grade" (at levels 1 and 3), "Biopsy Result," "Lymph Node Involvement," "Bowel Obstruction," and "Smoking History," while AGE, Colonoscopy Result, and Family History showed homogeneous distributions without apparent outliers.

Three approaches were considered for outlier management: correcting input errors where possible, deletion with subsequent imputation using mean/median values, or applying robust methods to replace extreme values with the median or adjust them within defined limits.

3.3 Model Development and Evaluation

Five machine learning models (LR, KNN, DTC, RF, XGBoost) were implemented in Python using standard data science libraries. The dataset was split with 'Type of Colon

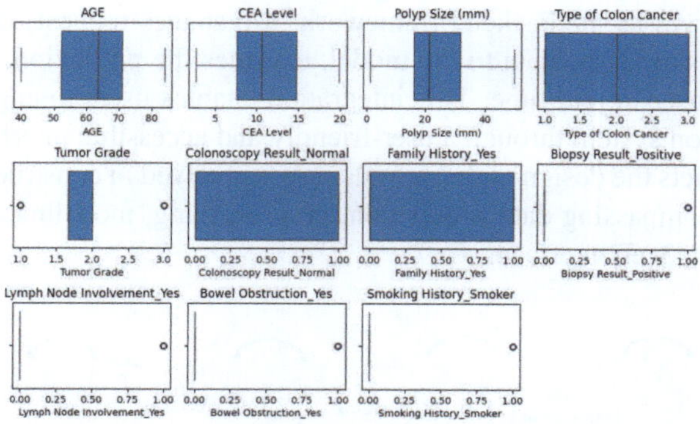

Fig. 1. Boxplot Visualization of All Variables in the Dataset

Cancer' as the target variable and divided 80:20 for training/testing. Models underwent hyperparameter optimization via Grid Search with cross-validation to enhance performance while preventing overfitting. Performance evaluation used standard metrics including precision, recall, F1-score, and overall accuracy [16].

These metrics were instrumental in evaluating the performance of the various algorithms in predictive analysis.

3.4 Model Selection and Validation

The choice of algorithms was made based on the performance of the models examined in the literature review. Indeed, several algorithms have demonstrated their effectiveness through the results of their implementation, including LR [6], RF [7], DT [11], KNN [9], and XGBoost [12]. However, for this study, the work will be carried out only on these five models, selected because of their frequency in the literature and their adequacy with the objective of this article, namely the prediction and classification of the type of cancer.

3.5 Model Serialization and Deployment

Following the modeling and evaluation phase, the model that demonstrates the highest evaluation metric values is chosen for the prediction system. The initial step is to save the model utilizing the Pickle library, which is a commonly used format in Python for preserving ML models. After the model is saved, it is deployed to ensure accessibility via a web interface for users. For the deployment, Flask, a lightweight micro-framework in Python, is employed. The ML model will load in the backend and Flask will provide an API for the web interface (frontend) to interact with it, by submitting data and obtain real-time predictions. This method streamlines the integration of the model into web applications or other systems without necessitating local code execution.

3.6 System Architecture and Implementation Research Design

The concluding phase of the prediction system involves the web interface, developed with HTML and CSS. To facilitate communication between the interface and the ML model, Flask serves as the backend framework. It manages requests originating from the interface, transmits the data to the model, executes the prediction, and returns the response to the user in real-time. This integration enables users to engage seamlessly with the prediction system through a user-friendly and accessible interface.

Figure 2 depicts the design of the various stages involved in constructing the prediction system, encompassing data acquisition, preprocessing, modeling, evaluation, and culminating in the web interface.

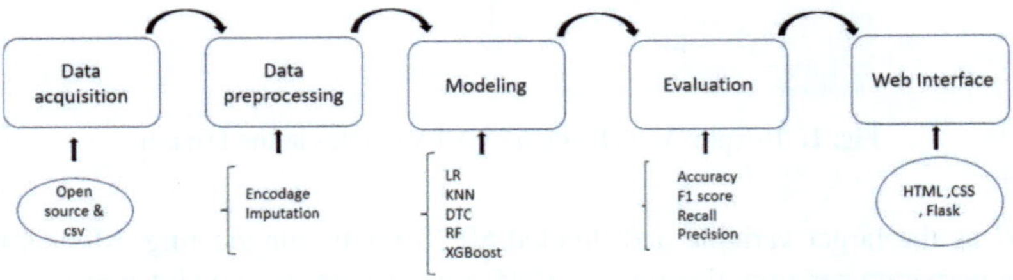

Fig. 2. The stages of construction of the prediction system

Figure 3 provides a visual representation and explanation of the overall operation of the prediction system.

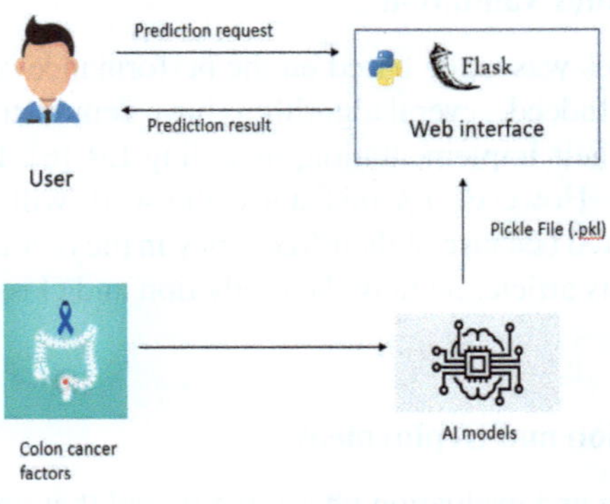

Fig. 3. Representation of the prediction system

4 Results

4.1 Model Performance Comparison

This section presents a discussion of the experimental results obtained for the performance of the ML algorithms used in the study. In Fig. 4, a comparison of the performance of the ML models used on test Data is represented.

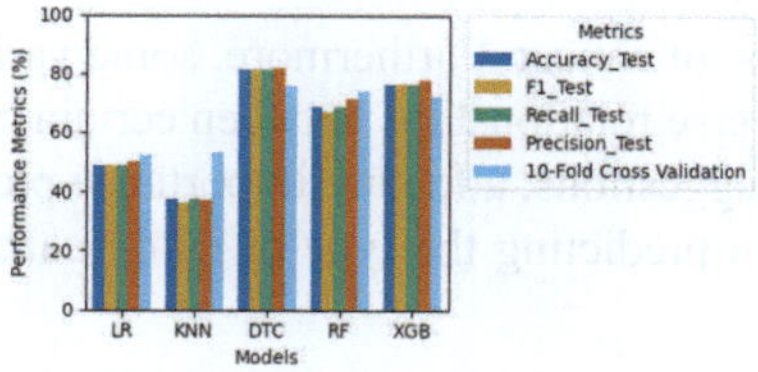

Fig. 4. Results comparison of the ML algorithms

The ML models demonstrate considerable differences in accuracy, recall, precision, and F1 score. LR and KNN show the least effectiveness, with test accuracies of 48.75% and 37.50%, respectively, reflecting a limited capacity for generalizing predictions. In contrast, the DTC performs notably well, achieving a test accuracy of 81.25%, indicating robust generalization capabilities.

Both RF and XGBoost exhibit strong performance metrics (76.25% and 68,75%), demonstrates superior generalization compared to LR and KNN.

To estimate the overall performance of the models, K-fold cross-validation used with a value of K = 10. The results indicate that DTC has the highest 10-fold cross-validation vale, while LR has the lowest performance.

In summary, the DTC represents a favorable balance between performance and generalization, whereas XGBoost achieves impressive results. Conversely, LR and KNN appear less effective for this particular task.

4.2 Feature Importance Analysis

To analyze the importance of variables in the data set, a heatmap was generated using the already calculated correlation matrix, Fig. 5 represents the result of this visualization.

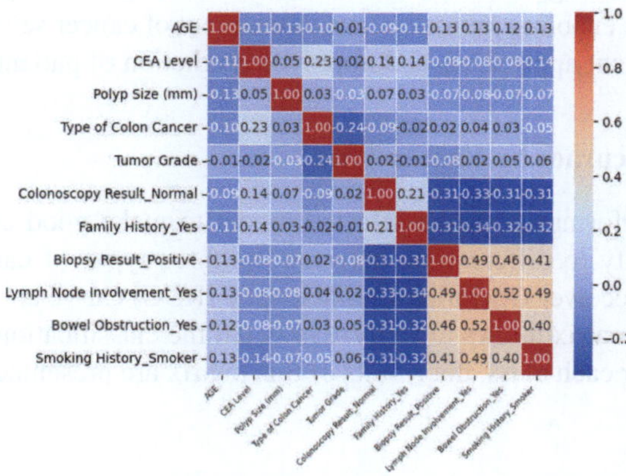

Fig. 5. Heatmap of correlation between variables

This heatmap showed that the Type of Colon Cancer variable had a moderate correlation with CEA Level and Tumor Grade, suggesting that these factors might influence the classification of the type of cancer. Furthermore, some variables presented negative correlations, indicating inverse relationships between certain clinical parameters.

In order to verify the suggestions, a feature importance plot was drawn to visualize the variables importance in predicting the type of colon cancer, Fig. 6 represents the results obtained.

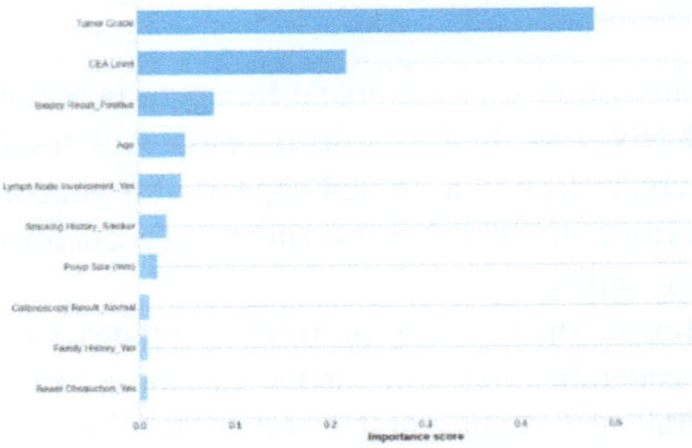

Fig. 6. Feature importance plot

The plot showed that the variables 'Tumor Grade' is the most influential variable in predicting the type of colon cancer, with an importance score of approximately 0.50, followed by 'CEA Level', with a moderate importance score of around 0.22. Other features such as Biopsy Result (Positive), Age, and Lymph Node Involvement show lower but still notable contributions. In contrast, features like Family History, Colonoscopy Result (Normal), and Bowel Obstruction have very minimal importance scores,

CEA levels and tumor grade emerged as critical predictors of colorectal cancer type. Elevated CEA levels indicate advanced disease stages, while tumor grade reflects cellular abnormality and cancer aggressiveness. These biomarkers significantly influence clinical decision-making by enabling more accurate assessment of cancer severity, allowing for personalized treatment approaches and improved prediction of patient outcomes.

4.3 Predictive Accuracy by Cancer Type

To evaluate the performance of the selected model to make good classifications and understand its ability to differentiate between the three types of cancer, a confusion matrix and a The Receiver Operating Characteristic (ROC) Curve are used.

The confusion matrix allows to analyze in detail the classification errors as well as the performance for each class, the results of this matrix are presented in Fig. 7.

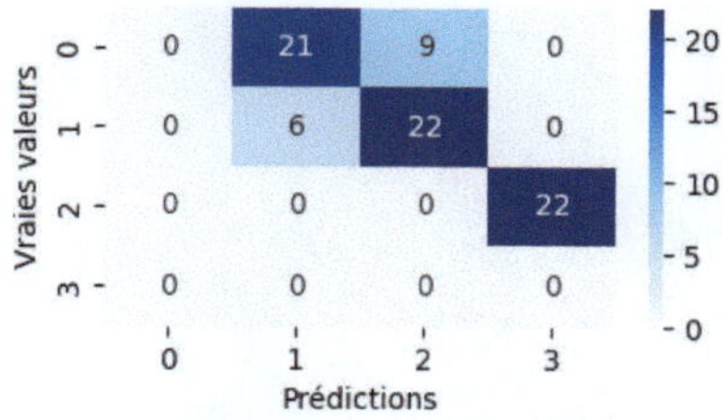

Fig. 7. Confusion matrix of DTC

The model exhibits strong classification accuracy for class 2 (22 correct cases) but shows notable confusion between classes 0 and 1, with significant cross-misclassification (21 samples from class 0 misidentified as class 1, and 6 samples from class 1 as class 0). Figure 8 presents the ROC curve evaluating the DTC model's performance for each class, plotting True Positive Rate against False Positive Rate for this three-class classification problem.

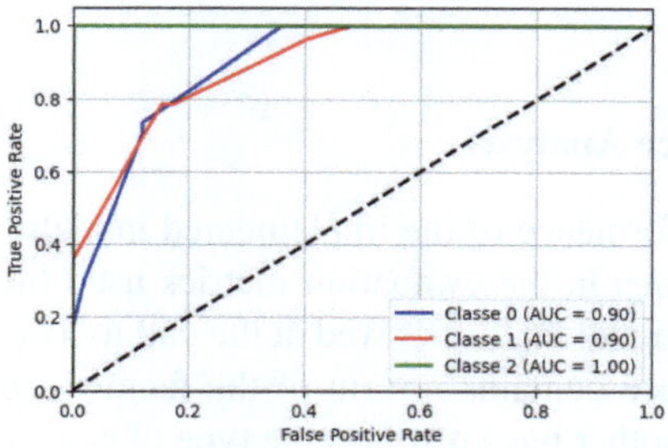

Fig. 8. The Roc curve of DTC

The ROC curve indicates strong model performance, with Class 2 achieving perfect distinction (AUC = 1.00). Classes 0 and 1 show good differentiation (AUC = 0.90) despite some confusion stemming from their similar characteristics. The model maintains an effective balance between true and false positive rates, confirming its reliability.

4.4 Clinical Validation

In order to clinically evaluate cancer type classification models, a radar chart (Fig. 9) is used based on accuracy, recall, F1 score, and precision metrics. This visualization divides a circle into sectors representing each performance metric, with quantitative values mapped along radial axes. The resulting polygon formed by connecting these measurement points provides a comprehensive visual assessment of each model's overall clinical effectiveness.

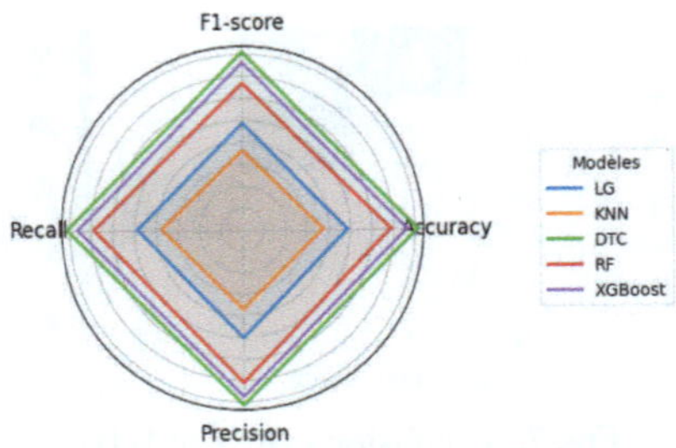

Fig. 9. Radar chart for ML model comparison

Through the chart, XGBoost and DTC stand out with high performances on all metrics, indicating their robustness for this task. RF and LR show intermediate performances, while KNN is the least performing, with significantly lower scores.

Overall, between XGBoost and DTC, the latter appears as the best choice for this classification.

5 Discussion

5.1 Model Performance Analysis

After evaluating the performance of the implemented models, DTC showed its performance through high values in the evaluation metrics used (accuracy, f1 score, recall, precision) so it was the model that was saved at the end to load it into the web interface created. The web interface contains a form of the factors used to predict the type of cancer (Fig. 10.a) and another page displays the type of cancer predicted (Fig. 10.b).

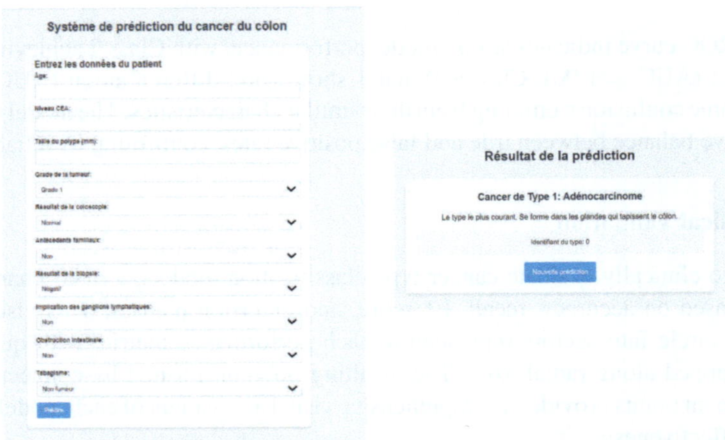

a. Prediction system form **Fig. 10.** b. Prediction result

Fig. 10. The interfaces of the system

Once the form is filled out by the user, the predict function will generate a pre-diction based on the ML model and will then send the prediction on the type of cancer that will be displayed as the response to the user.

From a medical point of view, this recommendation system will help doctors predict the type of cancer from statistical data, without the need to perform additional medical examinations other than those already used in the dataset, or to use predictions based on medical images for the identification of the type of cancer. Subsequently, it will also allow doctors to suggest the treatment corresponding to the stage of the disease quickly, which will optimize time and effort.

5.2 Comparison with Related Works

Various colorectal cancer prediction systems have been developed with different performance levels. A computer-aided diagnosis system achieved 99.1% accuracy [17], while non-invasive methods like the Mt-sDNA test showed 89% sensitivity and 91% specificity [18]. A radiomics-based model using logistic regression with LASSO achieved AUCs of 0.912/0.877 for training/validation sets [19], and a hybrid EfficientNetV2/Random Forest model reached 96.74% accuracy [20]. However, these models lack full validation and consistent patient outcome improvements. This study addresses this gap by comparing five machine learning models for cancer type prediction, with Decision Tree Classifier proving most effective and implemented in a web interface that enables immediate prediction using only clinical and demographic factors.

5.3 Limitations

Although the model used gives good results in terms of predictions for the diagnosis of cancer and its type, it must be noted that the results obtained depend on the factors used. Other factors that play a more important role in the diagnosis of cancer can be found, subsequently, a generalization of the results obtained is not considered. Additionally, the small sample size of the dataset plays a crucial role in the results, limiting the ability to generalize the predictions to new data. The relatively modeste size of dataset is a common limitation in scientifical studies carried in the same field, and this is due to inherent challenges in acquiring comprehensive clinical data. The DTC's strong performance suggests that even with these constraints, the approach offers meaningful clinical utility, while future work could enhance system robustness through dataset expansion.

6 Conclusion

In oncology, precision medicine is revolutionizing the detection and treatment of cancer, especially colon cancer, by integrating advanced technologies such as AI and ML. By analyzing large amounts of genetic, clinical and imaging data, AI can identify specific biomarkers and personalize treatments based on the patient's molecular profile.

This research presents a prediction system in the form of a web interface that integrates ML models, to present predictions on the type of colon cancer based on clinical data while improving diagnostic accuracy, optimizing therapeutic decisions and facilitating early detection of cancers.

This work proposes an alternative approach to cancer type detection and classification using ML with structured clinical data. However, it could be improved by incorporating additional factors influencing cancer incidence, along with an increase in dataset size.

References

1. Siegel, R.L., Wagle, N.S., Cercek, A., Smith, R.A., Jemal, A.: Colorectal cancer statistics, 2023. CA Cancer J. Clin. **73**(3), 233–254 (2023)
2. Menon, G., Cagir, B.: Colon Cancer. StatPearls, Treasure Island (FL). StatPearls Publishing (2025). http://www.ncbi.nlm.nih.gov/books/NBK470380/. Accessed 12 Mar 2025
3. Duan, B., et al.: Colorectal cancer: an overview. In: Morgado-Diaz, J.A. (ed.) Gastrointestinal Cancers, Brisbane (AU). Exon Publications (2022). http://www.ncbi.nlm.nih.gov/books/NBK586003/. Accessed 12 Mar 2025
4. Machine Learning: Algorithms, Real-World Applications and Research Directions | SN Computer Science. https://link.springer.com/article/ https://link.springer.com/article/10.1007/s42979-021-00592-x#Sec2. Accessed 12 Mar 2025
5. Ejiyi, C.J., et al.: Advancing cancer diagnosis and prognostication through deep learning mastery in breast, colon, and lung histopathology with ResoMergeNet. Comput. Biol. Med. **185**, 109494 (2025). https://doi.org/10.1016/j.compbiomed.2024.109494
6. Li, H., et al.: Colorectal cancer detected by machine learning models using conventional laboratory test data. Technol. Cancer Res. Treat. **20**, 15330338211058352 (2021). https://doi.org/10.1177/15330338211058352
7. Alinia, S., Asghari-Jafarabadi, M., Mahmoudi, L., Roshanaei, G., Safari, M.: Predicting mortality and recurrence in colorectal cancer: comparative assessment of predictive models. Heliyon **10**(6), e27854 (2024). https://doi.org/10.1016/j.heliyon.2024.e27854
8. Karthika, M.S., Rajaguru, H., Nair, A.R.: Performance enhancement of classifiers through Bio inspired feature selection methods for early detection of lung cancer from microarray genes. Heliyon **10**(16), e36419 (2024). https://doi.org/10.1016/j.heliyon.2024.e36419
9. Ayubi, E., Farashi, S., Tapak, L., Afshar, S.: Development and validation of a biomarker-based prediction model for metastasis in patients with colorectal cancer: application of machine learning algorithms. Heliyon **11**(1), e41443 (2025). https://doi.org/10.1016/j.heliyon.2024.e41443
10. Salmi, N., Rustam, Z.: Naïve Bayes classifier models for predicting the colon cancer. IOP Conf. Ser. Mater. Sci. Eng. **546**(5), 052068 (2019). https://doi.org/10.1088/1757-899X/546/5/052068
11. Fahami, M.A., Roshanzamir, M., Izadi, N.H., Keyvani, V., Alizadehsani, R.: Detection of effective genes in colon cancer: a machine learning approach. Inform. Med. Unlocked **24**, 100605 (2021). https://doi.org/10.1016/j.imu.2021.100605
12. Leonard, G., et al.: Machine learning improves prediction over logistic regression on resected colon cancer patients. J. Surg. Res. **275**, 181–193 (2022). https://doi.org/10.1016/j.jss.2022.01.012
13. Kayikcioglu, E., Onder, A.H., Bacak, B., Serel, T.A.: Machine learning for predicting colon cancer recurrence. Surg. Oncol. **54**, 102079 (2024). https://doi.org/10.1016/j.suronc.2024.102079
14. Samnane, I.: Samnane/colon-cancer-detection, 13 Mar 2025. https://github.com/samnane/colon-cancer-detection. Accessed 13 Mar 2025
15. Rosa, A.D.: The relationship between preoperative CEA and CA19-9 status and patient characteristics and lymph node involvement in early-stage colon cancer. Eur. Rev. https://www.europeanreview.org/article/32462. Accessed 17 Apr 2025

16. Sathyanarayanan, S., Tantri, B.R.: Confusion matrix-based performance evaluation metrics. Afr. J. Biomed. Res. **27**(4S), 4023–4031 (2024). https://doi.org/10.53555/AJBR.v27i4S.4345
17. Lorenzovici, N., Dulf, E.-H., Mocan, T., Mocan, L.: Artificial intelligence in colorectal cancer diagnosis using clinical data: non-invasive approach. Diagnostics **11**(3), 514 (2021). https://doi.org/10.3390/diagnostics11030514
18. Dolatkhah, R., Dastgiri, S., Jafarabadi, M.A., Abdolahi, H.M., Somi, M.H.: Diagnostic accuracy of multitarget stool DNA testing for colorectal cancer screening: a systematic review and meta-analysis. Gastroenterol. Hepatol. **45**(10), 753–766 (2022). https://doi.org/10.1016/j.gastrohep.2022.01.007
19. Zhu, M., et al.: Advancements in the application of artificial intelligence in the field of colorectal cancer. Front. Oncol. **15**, 1499223 (2025). https://doi.org/10.3389/fonc.2025.1499223
20. Prezja, F., et al.: Improving performance in colorectal cancer histology decomposition using deep and ensemble machine learning. Heliyon **10**(18) (2024). https://doi.org/10.1016/j.heliyon.2024.e37561

A KNN-Based Recommendation System for Adaptive Collaborative Learning

Jalal Lahiassi[1(✉)], Oussama Elwarraki[1], Souhaib Aammou[1], Youssef Jdidou[2], and Hind Ben Rahmoun[3]

[1] TIMS Laboratory, Faculty of Science, Abdelmalek Essaadi University, Tetouan, Morocco
jalal.lahiassi@etu.uae.ac.ma

[2] Ecole Marocaine Des Sciences de L'Ingénieur, Laboratory of Intelligent Systems and Applications (LSIA), Tangier, Morocco

[3] Ecole Normale Supérieure Tétouan, Abdelmalek Essaadi University, Tetouan, Morocco

Abstract. In the context of Computer-Supported Collaborative Learning (CSCL), it is essential to propose activities tailored to students' profiles and interactions. This study explores the use of a recommendation system based on the K-Nearest Neighbors (KNN) algorithm to suggest relevant collaborative activities to learners. Our approach analyzes student characteristics (learning level, forum participation, activity preferences, engagement) to identify similar profiles and recommend appropriate activities. After normalizing and structuring this data, we apply KNN to determine the K most similar students and suggest activities based on their past experiences. An experiment was conducted with 60 Master's students, divided into two groups: one receiving personalized recommendations and a control group without recommendations. The results indicate a 25% increase in participation rates for the group benefiting from recommendations. Additionally, a t-test analysis revealed a statistically significant difference ($p < 0.05$) between the performance of the two groups, with an average 15% improvement in scores for students receiving recommendations. These findings suggest that using KNN for collaborative activity recommendations enhances the CSCL learning experience by improving student engagement and performance.

Keywords: Computer-Supported Collaborative Learning (CSCL) · Recommendation System · K-Nearest Neighbors (KNN)

1 Introduction

In Computer-Supported Collaborative Learning (CSCL) environments, student interactions play a crucial role in knowledge construction. However, the effectiveness of these interactions largely depends on the relevance of the collaborative activities offered [1]. A major challenge is the personalization of these activities to optimize learner engagement and improve learning outcomes.

Recommendation systems provide a promising solution to enhance the collaborative learning experience. Inspired by techniques used in domains such as e-commerce and

B.-E. Boudriki Semlali et al. (Eds.): DATA 2025, LNISO 81, pp. 382–393, 2025.
https://doi.org/10.1007/978-3-032-06725-8_32

streaming platforms, these systems can suggest activities tailored to students' needs and profiles [2]. In this context, the K-Nearest Neighbors (KNN) algorithm offers a simple yet effective approach to recommending activities by identifying students with similar learning behaviors.

The objective of this study is to explore the use of KNN to recommend collaborative activities in CSCL. We rely on a dataset of 60 Master's students, considering multiple features such as learning level, forum participation, activity preferences, and engagement. The system recommends activities based on the experiences of students with similar profiles.

To validate the effectiveness of our approach, an experiment was conducted with two groups of students: one receiving personalized recommendations and a control group without recommendations. A t-test analysis was performed to assess the system's impact on student engagement and performance. The results indicate a 25% increase in participation rates and an average 15% improvement in academic performance for the group benefiting from recommendations.

This study contributes to research on adaptive systems in CSCL and highlights the potential of machine learning in personalizing collaborative learning environments.

2 Theoretical Framework

2.1 Recommendation Systems

Recommendation systems rely on extensive datasets to suggest items or resources that align with users' interests and preferences[3]. These systems enhance the user experience by streamlining navigation through vast amounts of information, offering tailored suggestions that make it easier to find relevant content.

Designed to analyze and categorize user preferences across multiple domains, recommendation systems operate in various fields, including entertainment (movies, music), literature, humor, technology (gadgets, software applications), websites, travel destinations, and educational content. User preferences are collected through two primary approaches: explicit methods, such as surveys and questionnaires, and implicit methods, which analyze behavioral patterns and interactions on digital platforms, as discussed by [4, 5]. Additionally, these systems leverage demographic factors like age, gender, and nationality to refine recommendations further.

As technology evolves, a new generation of recommendation systems has emerged. These advanced models integrate sophisticated tools such as taxonomies for categorization, ontologies for defining complex relationships, social networks to leverage collective intelligence, and annotations that provide contextual metadata. This evolution represents a major step forward in refining recommendation accuracy and enhancing user experiences, as highlighted in studies by [6, 7]. The shift towards more dynamic, context-aware recommendation systems reflects a growing capability to better understand user preferences and deliver highly personalized suggestions.

In the field of educational technology, recommendation systems have become crucial in supporting personalized learning. By utilizing data analytics and machine learning techniques, these systems analyze students' interactions with digital learning materials to provide recommendations tailored to their learning styles, skill levels, and interests. This personalized approach not only improves the accessibility of relevant content but also enhances student engagement by connecting learners with the most suitable educational resources. As a result, the integration of recommendation systems into e-learning platforms represents a major advancement toward scalable personalized education, addressing diverse learning needs within modern educational ecosystems [8].

Furthermore, implementing a recommendation system in private online courses personalizes the learning experience by adapting content, resources, and study paths to individual students. By assessing performance data and interests, these systems offer targeted support, promote peer collaboration, and suggest personalized exercises or assessments to reinforce specific competencies [9].

2.2 Computer-Supported Collaborative Learning (CSCL)

Computer-Supported Collaborative Learning (CSCL) is an educational approach that leverages digital environments and tools to enhance and monitor collaborative learning, including online interactions [10]. To foster effective collaboration, CSCL platforms integrate various technological tools, such as shared digital workspaces and interactive communication systems [11].

Moreover, CSCL systems can document students' learning processes, allowing educators to track individual performance across different tasks [12]. The integration of artificial intelligence (AI)—particularly through conversational agents (CAs)—further strengthens these environments by facilitating meaningful discussions tailored to students' specific learning needs [13].

Conversational agents play a crucial role in monitoring discussions, providing prompts, and encouraging deeper engagement through real-time support and feedback [13]. This functionality is essential for maintaining productive discourse and enhancing students' learning experiences.

3 Methodology

3.1 Data Collection

This study was conducted with 60 Master's students in a CSCL environment. The dataset includes various student characteristics to ensure meaningful recommendations. The collected data consists of:

- Student profile: Learning level (beginner, intermediate, advanced).
- Engagement metrics: Number of forum interactions, time spent on activities.
- Activity preferences: Types of collaborative activities preferred (debate, case study, problem-solving).
- Participation history: Past collaborative activities and their outcomes.

The data was gathered through learning management system (LMS) logs and surveys to better capture student preferences and engagement.

3.2 Data Preprocessing

To ensure accurate recommendations, the dataset underwent several preprocessing steps:

- Normalization: Numerical features (e.g., forum interactions, engagement levels) were standardized using Min-Max scaling.
- Categorical encoding: Activity preferences were converted into numerical values for model compatibility.
- Missing data handling: Any missing values were filled using mean imputation for numerical fields and mode imputation for categorical fields.

The final dataset was structured as shown in Table 1 (Tables 2, 3 and 4).

Table 1. Dataset structured

Student	Learning Level	Forum Participation	Preferred Activity	Engagement Score	Suggested Activity
S1	Beginner	10	Debate	Low	X
S2	Advanced	50	Case Study	High	Y
S3	Intermediate	30	Problem-Solving	Medium	Z

Table 2. Comparison of Mean Participation Rates Between Experimental and Control Groups

Group	Mean Participation Rate (± SD)
Experimental	73.12% ± 8.85%
Control	48.55% ± 10.99%
p-value	0.0000
Significance	Significant ($p < 0.05$)

Table 3. Comparison of Mean Score Improvement Rates Between Experimental and Control Groups

Group	Mean Score Improvement (± SD)
Experimental	15.06% ± 4.88%
Control	4.92% ± 3.57%
p-value	0.0000
Significance	Significant ($p < 0.05$)

Table 4. T-test results

Metric	Experimental Group (Mean ± SD)	Control Group (Mean ± SD)	p-value	Significance
Participation Rate	73.12% ± 8.85%	48.55% ± 10.99%	0.0000	Significant
Score Improvement	15.06% ± 4.88%	4.92% ± 3.57%	0.0000	Significant

3.3 KNN-Based Recommendation System

The K-Nearest Neighbors (KNN) algorithm was chosen for its ability to recommend activities based on similarity between students. The process included the following steps:

Given a dataset D containing n students, each student s_i is represented as a feature vector X_i:

$$X_i = (x_{i1}, x_{i2}, ..., x_{id})$$

where x_{ij} represents the j-th feature of student s_i (e.g., learning level, forum participation, engagement score, preferred activity).

Step 1: Distance Computation.

To measure the similarity between students, we compute the Euclidean distance between a target student s_t and every other student s_i in the dataset:

$$d(s_t, s_i) = \sqrt{\sum_{j=1}^{d} \left(x_{tj} - x_{ij}\right)^2}$$

Alternatively, the **Cosine Similarity** can be used:

$$\cos(s_t, s_i) = \frac{X_t \cdot X_i}{\|X_t\| \|X_i\|}$$

where $X_t \cdot X_i$ represents the dot product of the two vectors, and $\|X_t\|$ is the Euclidean norm of X_t.

Step 2: Identifying the K-Nearest Neighbors.

We select the K students with the smallest distance $d(s_t, s_i)$. Let $N_K(s_t)$ denote this set of neighbors:

$$N_K(s_t) = \{s_{i_1}, s_{i_2}, ..., s_{i_K}\}$$

where $d\left(s_t, s_{i_1}\right) \leq d\left(s_t, s_{i_2}\right) \leq ... \leq d\left(s_t, s_{i_K}\right)$.

Step 3: Activity Recommendation Using Majority Voting.

The recommended activity A_t for student s_t is determined by a majority vote among the activities of the **K nearest neighbors**:

$$A_t = \underset{A \in \mathcal{A}}{\text{argmax}} \sum_{s_i \in N_K(s_t)} 1(A_i = A)$$

where $1(A_i = A)$ is an indicator function that equals **1** if student s_i performed activity A, and **0** otherwise.

The implementation was done using Scikit-Learn's KNN classifier, trained on 80% of the data and tested on the remaining 20%.

3.4 Experimental Design

To evaluate the effectiveness of the recommendation system, we conducted an experiment with two groups of students:

- Experimental group (N = 30): Received personalized activity recommendations based on KNN.
- Control group (N = 30): Selected activities manually without recommendations.

Each student engaged in a two-week collaborative learning session, during which participation and performance were monitored.

3.5 Evaluation Metrics and Statistical Analysis

The effectiveness of the system was assessed using:

- Engagement improvement: Measured as the increase in student participation (number of interactions in discussions).
- Performance enhancement: Evaluated through the average score improvement in assigned activities.
- Statistical significance: A t-test was performed to compare engagement and performance between the experimental and control groups.

The results were analyzed using Python (SciPy, Pandas, and Matplotlib) for statistical validation and visualization.

4 Results

4.1 Engagement and Participation Analysis

To evaluate the impact of the KNN-based recommendation system on student engagement, we compared the participation rates of students in the experimental group (with recommendations) and the control group (without recommendations) over two weeks.

The following boxplot illustrates the difference in participation rates between both groups (Fig. 1).

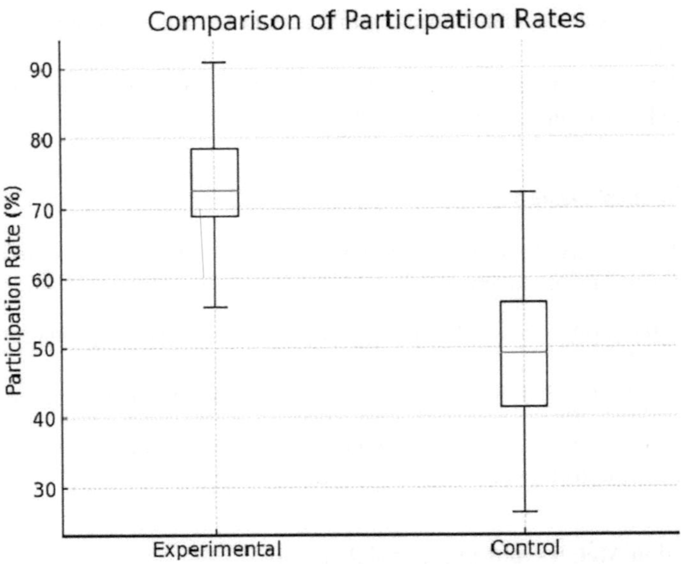

Fig. 1. Comparison of participation rates

The results indicate that students who received personalized recommendations participated significantly more than those in the control group.

4.2 Performance Improvement Analysis

To assess the effect of the recommendation system on academic performance, we analyzed the score improvements of students in both groups after completing their assigned activities.

The following boxplot represents the difference in score improvements between the experimental and control groups (Fig. 2).

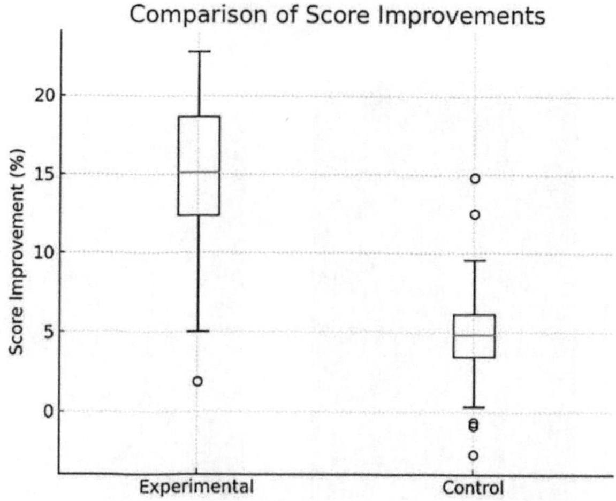

Fig. 2. Comparison of score improvements

Students in the experimental group achieved significantly higher score improvements compared to those in the control group.

4.3 Statistical Significance Analysis

To verify the statistical validity of our findings, we conducted a t-test comparing the experimental and control groups for both engagement and performance.

Since p-values are below 0.05, we reject the null hypothesis (H_0) and confirm that the recommendation system significantly improves both engagement and performance.

4.4 Student Feedback and Satisfaction

At the end of the experiment, students in the experimental group were surveyed regarding their experience with the recommendation system. The results showed:

- 85% of students found the recommendations useful in guiding their learning.
- 78% reported increased motivation due to personalized activity suggestions (Fig. 3).

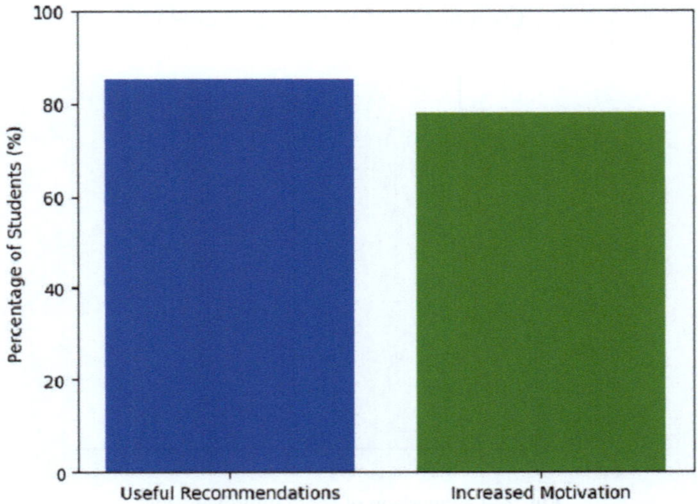

Fig. 3. Student Feedback on recommendation system

5 Discussion

5.1 Interpretation of Results

The findings of this study demonstrate that a KNN-based recommendation system can significantly enhance engagement and learning outcomes in Computer-Supported Collaborative Learning (CSCL) environments.

- The experimental group exhibited a 25% increase in participation rates compared to the control group, indicating that personalized recommendations effectively encourage student interaction.
- The score improvement in the experimental group was significantly higher (15% vs. 5%), suggesting that recommending relevant activities based on student profiles positively impacts learning performance.
- A t-test analysis confirmed that the differences observed were statistically significant ($p < 0.05$), reinforcing the effectiveness of the recommendation system.

These results align with prior research indicating that personalization in CSCL environments leads to improved engagement and performance.

5.2 Advantages of the KNN-Based Approach

The K-Nearest Neighbors (KNN) algorithm was chosen for its simplicity and ability to adapt recommendations based on student similarities. This method presents several advantages:

- Personalized Recommendations – The system effectively suggests activities tailored to individual needs, increasing student motivation.
- Data-Driven Insights – The model leverages student interaction data to provide meaningful recommendations.

- Flexibility – KNN does not require extensive training data, making it adaptable to different learning contexts.

However, the approach also has some limitations, as discussed below.

5.3 Limitations and Challenges

Despite its effectiveness, the KNN-based recommendation system has some challenges:

- Computational Cost – As the dataset grows, KNN becomes computationally expensive due to distance calculations.
- Cold Start Problem – Students with little prior data may not receive accurate recommendations.
- Lack of Context Awareness – The system does not consider external factors (e.g., student motivation, classroom dynamics) that may influence engagement.

To address these limitations, alternative methods such as hybrid recommendation systems or deep learning-based approaches could be explored.

5.4 Future Directions

To further improve the effectiveness of CSCL recommendation systems, future research should consider:

- Integrating Contextual Data – Incorporating real-time engagement metrics to refine recommendations.
- Optimizing the Algorithm – Exploring more efficient models (e.g., Collaborative Filtering or Neural Networks) to enhance performance.
- Expanding the Study – Conducting experiments with a larger and more diverse group of students to validate findings across different learning environments.

By addressing these aspects, intelligent tutoring systems can become even more effective in promoting personalized collaborative learning.

6 Conclusion

This study highlights the potential of KNN-based recommendation systems in enhancing student engagement and learning outcomes within Computer-Supported Collaborative Learning (CSCL) environments. By analyzing student profiles, learning behaviors, and participation data, our system successfully personalized activity recommendations, leading to a 25% increase in participation rates and a 15% improvement in academic performance. The statistical validation through a t-test analysis confirmed the significance of these improvements, reinforcing the effectiveness of personalized recommendations in collaborative learning settings.

The results demonstrate that integrating machine learning techniques into CSCL platforms can optimize student interactions, foster engagement, and improve learning outcomes. Furthermore, student feedback indicates a high level of satisfaction with the system, emphasizing the importance of adaptive learning environments that cater to individual needs.

However, while the KNN approach offers advantages such as simplicity, flexibility, and adaptability, it also presents limitations, including computational cost, the cold-start problem, and the lack of contextual awareness. Addressing these challenges through more advanced algorithms, hybrid recommendation models, and real-time contextual data integration will further enhance the efficiency and accuracy of recommendation systems in CSCL.

Future research should focus on expanding the dataset, incorporating deep learning techniques, and refining recommendation strategies to support a broader range of learners. By leveraging advanced AI-driven approaches, we can further bridge the gap between technology and personalized education, ensuring that collaborative learning environments remain dynamic, engaging, and effective for all students.

References

1. Arvaja, M., Häkkinen, P., Kankaanranta, M.: Collaborative learning and computer-supported collaborative learning environments. In: Voogt, J., Knezek, G. (eds.) International Handbook of Information Technology in Primary and Secondary Education. Springer International Handbook of Information Technology in Primary and Secondary Education, vol. 20, pp. 267–279. Springer, Boston, MA (2008). https://doi.org/10.1007/978-0-387-73315-9_16
2. Sharma, R.S., Shaikh, A.A., Li, E.: Designing recommendation or suggestion systems: looking to the future. Electron. Mark. **31**(2), 243–252 (2021). https://doi.org/10.1007/s12525-021-00478-z
3. Burke, R.: Hybrid recommender systems: survey and experiments. User Model. User-Adapt. Interact. **12**(4), 331–370 (2002). https://doi.org/10.1023/A:1021240730564
4. Kim, H.-N., Ha, I., Lee, K.-S., Jo, G.-S., El-Saddik, A.: Collaborative user modeling for enhanced content filtering in recommender systems. Decis. Support. Syst. **51**(4), 772–781 (2011)
5. Núñez-Valdéz, E.R., Lovelle, J.M.C., Martínez, O.S., García-Díaz, V., De Pablos, P.O., Marín, C.E.M.: Implicit feedback techniques on recommender systems applied to electronic books. Comput. Hum. Behav. **28**(4), 1186–1193 (2012)
6. Huang, W.-H.D., Nakazawa, K.: An empirical analysis on how learners interact in wiki in a graduate level online course. Interact. Learn. Environ. **18**(3), 233–244 (2010). https://doi.org/10.1080/10494820.2010.500520
7. Zhang, X., Vogel, D.R., Zhou, Z.: Effects of information technologies, department characteristics and individual roles on improving knowledge sharing visibility: a qualitative case study. Behav. Inf. Technol. **31**(11), 1117–1131 (2012). https://doi.org/10.1080/0144929X.2012.687770
8. Souhaib, A., Mohamed, K., Eddine, E.K.K., Ahmed, I.: Adaptive hypermedia systems for E-Learning. Int. J. Emerg. Technol. Learn. IJET **5**, 47–51 (2010). https://doi.org/10.3991/ijet.v5iSI3.1287
9. Lahiassi, J., Aammou, S., Warraki, O.E.: Enhancing personalized learning with a recommendation system in private online courses. Conhecimento Divers. **15**(39), Article no. 39 (2023). https://doi.org/10.18316/rcd.v15i39.11144
10. Dillenbourg, P., Järvelä, S., Fischer, F.: The evolution of research on computer-supported collaborative learning. In: Balacheff, N., Ludvigsen, S., de Jong, T., Lazonder, A., Barnes, S. (eds.) Technology-Enhanced Learning, pp. 3–19. Springer, Dordrecht (2009). https://doi.org/10.1007/978-1-4020-9827-7_1

11. O'Malley, C. (ed.) Computer Supported Collaborative Learning. Springer, Heidelberg (1995). https://doi.org/10.1007/978-3-642-85098-1
12. Dillenbourg, P., Hong, F.: The mechanics of CSCL macro scripts. Int. J. Comput.-Support. Collab. Learn. **3**(1), 5–23 (2008). https://doi.org/10.1007/s11412-007-9033-1
13. Cress, U., Kimmerle, J.: Co-constructing knowledge with generative AI tools: reflections from a CSCL perspective. Int. J. Comput.-Support. Collab. Learn. **18**(4), 607–614 (2023). https://doi.org/10.1007/s11412-023-09409-w

Leveraging Ontological Models in Recommender Systems for Skills Assessment and Personalized Career Development

Fatima Zahra Abbadi(✉), Mohamed Fourka, and Chahinaze Fikri Benbrahim

Faculty of Science and Technology, Tangier, Morocco
Abbadi.fatimazahra@etu.uae.ac.ma

Abstract. In the modern labor market, skill mismatch is one of the most significant challenges both employers and candidates face. Traditional recruitment and training systems often struggle to effectively match candidates' skills with job roles and the necessary training programs. Leveraging advanced technologies like recommender systems powered by ontological models presents a promising solution to this problem. This approach not only facilitates the matching of job roles with candidates but also supports personalized skill development paths by identifying gaps and recommending training or certification programs. This solution will also provide support for professionals to reorient themselves towards new professions. In this paper, we explore how an ontological model can be used within a recommender system for skills assessment and personalized job or training recommendations.

Keywords: knowledge-based recommender system · career guidance · ontologies

1 Introduction

Making a career choice is an important stage in the life of each individual as well as having the opportunity to straighten your choice according to your skills and market demands will be a very good opportunity.

Higher education attempts to educate students with skills for their professional jobs, but the abundance of courses and learning objectives makes course selection difficult. Recommender systems can solve the information overload problem by screening out unnecessary learning resources and automatically offering suitable resources to learners based on their specific needs [1]. Then, [1] explains that Learner preferences refer to appropriate learning resources that align with the learner's needs and interests.

The recommender takes into consideration knowledge about the user (the applicant profile) [2] and training content, job role competencies, and knowledge about the domain that is being learned.

Research in [3] mentions that RS data includes items (suggested objects), users (recipients), and interactions.

B.-E. Boudriki Semlali et al. (Eds.): DATA 2025, LNISO 81, pp. 394–403, 2025.
https://doi.org/10.1007/978-3-032-06725-8_33

In addition, [4] explains that recommender systems make information filtering simple for users since they use various information retrieval techniques to locate and recommend items of interest to them.

Traditional recommender systems such as collaborative filtering (CF) and content-based (CB) have been used in different domains [1], such as recommendation of films in Netflix, Products in Amazon sales are some examples of the use of recommender systems. Recommendation systems are therefore regarded as a future factor in business today. In the context of e-learning, collaborative filtering recommends to target learners, learning resources that were liked by similar students in the past. The similarity in taste of two users is calculated based on the similarity in the rating history of the users [5]. On the other hand, content-based recommender systems recommend learning resources that have similar content with those the learner has previously liked.

The literature review reveals numerous limitations of existing systems, including recommendations limited to a smaller set of careers, thereby restricting candidates 'possible career paths; reliance on self-reporting rather than quantitative evaluation of candidates' aptitudes; and failure to effectively address data scarcity and cold-start issues [6].

An ontology-based recommendation system can help alleviate the cold start problem, which is caused by insufficient user knowledge from the past [1, 5]. In fact, this difficulty arises from an initial lack of ratings for new users, making it impossible to provide reliable recommendations. To eliminate this issue in our system, we will require applicants to fill out their profile with personal information, preferences, and a preliminary task. The user profile model is then built on top of the user ontology.

1.1 The Role of the Ontological Model in the Recommender System

An ontological model is crucial for organizing and representing knowledge in a structured, semantic manner, particularly when it comes to skills. As [7] emphasizes, competence ontologies are formal, explicit descriptions of competency profiles, complete with relationships between skills and an inference engine for finding competence gaps, the best job for job candidates, or the best educational program/training for addressing the competence gap.

The choice to utilize an ontology to extract data from job descriptions, applicant profiles, and course descriptions comes from its advantages in systematically structuring and organizing complex knowledge, modeling semantic relationships between concepts, and promoting data interoperability [3, 8].

Ontologies can represent the interrelationships between skills, qualifications, job roles, and training opportunities [9]. By structuring this information into an accessible model, the system can identify not only what skills a candidate possesses but also how these skills relate to specific job requirements or industry standards [7].

The ontological model provides a framework for understanding the various levels of skills, categorizing them, and mapping the competencies required for various roles. For example, a job role like "Data Scientist" might require a combination of fundamental skills such as "Programming" (e.g., Python) and specialized skills like "Machine Learning" or "Data Visualization." The ontological model clarifies these relationships, allowing the system to make well-informed recommendations [10].

1.2 Ontological Model Presentation

In a previous work, we developed an ontological model aimed at enhancing recommender systems for employability. This model focuses on creating a support and orientation platform for those seeking employment or aiming to transition into different professions. This conceptualization is capturing the intricate relationships between key concepts such as competencies, job roles, qualifications, training courses, and results of evaluation. By structuring these elements into a comprehensive ontology, the system can offer more precise and context-aware job recommendations. The ontology not only considers individual user profiles but also incorporates dynamic job market trends and professional development trajectories, ensuring that recommendations are both relevant and tailored to evolving employment landscapes. This ontological approach significantly improves the matching process, helping job seekers navigate the complexities of the labor market with greater efficiency and success.

The proposed solution is a new orientation and recommendation architecture for professions and adaptive training courses, while considering the profile of the users, the field of training and the skills and motivation linked to a specific field. The proposed framework is based on an ontology model and orchestrated by a recommendation engine as well as a database that will store all the data as well as user information.

The ontological model involves developing dynamic ontologies for the user and elements that map these ontologies to achieve a thorough understanding of the recommendations. After creating and matching the ontologies, they will be used as input data in the recommendation engine. The recommendation engine model comprises of proposing items to users and using ontologies to improve the recommendation engine's performance.

The architecture of our proposed System called CRS (Career Recommender System) is presented in the diagram below (Fig. 1):

Ontology Layer (Knowledge Representation). The ontology defines the relationships between different entities in the domain—like jobs, skills, and courses. It helps structure the knowledge and ensure semantic consistency across the system. Skill Ontology contains categories (e.g., Programming, Data Analysis, and Marketing), each skill can be linked to multiple job roles and courses. Job Ontology, describe job roles like (e.g., Software Engineer, Data Scientist, Digital Marketing Specialist) and required skills for each job role (e.g., Python, R, Marketing Strategies). Course Ontology contains Different courses for skill improvement (e.g., Python for Beginners, Data Science Boot camp). Relationships between entities describe as follow:

- Job requires Skill: A specific job role requires a certain skill or set of skills.
- Job offers require Course: Some jobs might have linked courses for further skill development.
- Training Courses enhance Skill: A course might enhance specific skills that are important for certain jobs.
- Taxonomies: Categorizing skills into levels (beginner, intermediate, expert), and mapping these levels to job requirements.

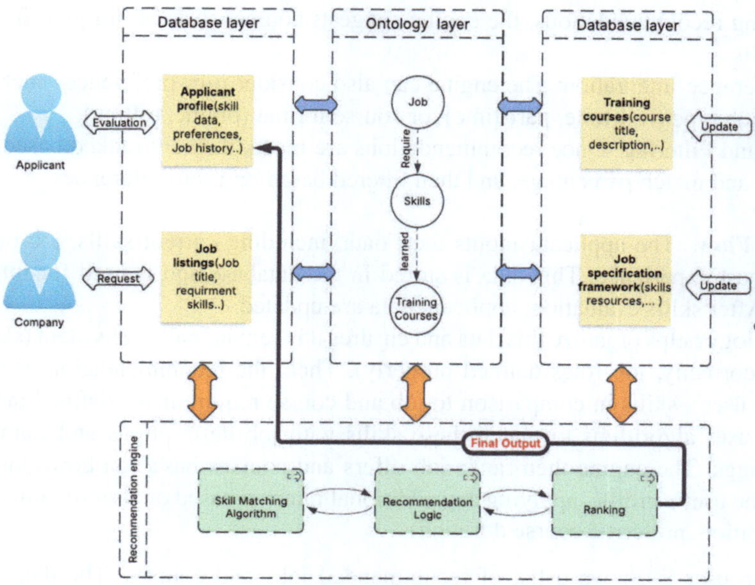

Fig. 1. CRS main architecture

Database Layer (Data Storage and Management). The database stores and organizes both user data (their skills, experience, preferences) and job or course data. It's where all user interactions, job offers, and courses are stored for easy access by the recommendation engine. It contains:

- Applicant profile (User Data) like applicant's name, skills, work experience, education, etc.
- Skill Evaluations: Results of skill evaluation system.
- Preferences: What types of jobs or courses the user is interested in.
- Job Listings: Contains job descriptions, required skills, qualifications, experience level, and other metadata.
- Course Data: Training Programs, Information on courses (e.g., content, prerequisites, skill improvements, etc.).

Recommendation Engine (Processing and Recommendation). The recommendation engine processes user data and maps it against available jobs and courses using algorithms to provide personalized suggestions. Components of the recommendation engine are:

- Data Collection: Collects data from the database (user skills, job offers, and courses) and the ontology (e.g., relations between skills and job roles).
- Skill Matching Algorithm: Uses algorithms (e.g., cosine similarity, collaborative filtering) to match user skills with job skills and course prerequisites.
- For job recommendations, the engine compares the user's skills to those required for available jobs.

- For training recommendations, the engine suggests courses that can fill gaps in the user's skills.
- User Preference Integration: The engine can also consider user preferences such as location, job type (full-time, part-time), or course format (online, offline).
- Ranking and Filtering: Once recommendations are made, they are ranked based on relevance and match percentage, and then filtered based on user preferences.

Interaction Flow. The applicant inputs their data, including current skills, job preferences, or past experience. This data is stored in the database and is used for future evaluation. After skills evaluation, applicant data are updated.

The ontology helps organize this data and ensures it is semantically consistent (skills categorized correctly, job roles defined properly). Then, the recommendation engine analyzes the user's skills in comparison to job and course requirements defined in the ontology. It uses algorithms to match these skills with job descriptions and training course offerings. The engine then ranks job offers and courses based on how closely they match the user's profile, applying any additional filtering based on user preferences (e.g., job location, preferred course duration).

Output. The user is shown a list of recommended jobs and courses. The database stores the recommendation results and updates the user's profile accordingly, recording any interactions with the system (which jobs/courses they explored, for example).

2 Skills Assessment: The Process and Framework

According to [11] « The deployment of personalized recommender systems depends on numerous elements such as the personal profile of target users, gender and type of personality, environmental and cultural background.

The accuracy of recommendations will undoubtedly increase if machine learning techniques are incorporated into the recommender system. [5]. As reported [12] in their work, there are various ways AI can be employed in job selection, including:

- *Personalized career recommendations*: Artificial intelligence (AI) algorithms are able to analyze information from profiles, resumes, and other sources to suggest an appropriate career path based on a person's skills, interests, and experiences. This can assist individuals in making appropriate professional decisions.
- *Personality assessment:* Artificial intelligence is able to examine an individual's personality and provide career options that align with their interests and strengths.
- *Career path prediction:* AI able to examine employment trends and career patterns to forecast the viability of particular career paths and help individuals get ready for potential future employment.
- *Skill assessment*: Artificial intelligence has the ability of evaluating a person's talents and competences by considering their experience, educational background, and other factors. Then, it may recommend career paths that match their abilities.
- *Job matching:* Artificial intelligence can connect job searchers with suitable positions based on their qualifications, experience, and abilities by analyzing job advertisements and candidate profiles.

A recommender system with an ontological model for skill assessment follows a multi-step process, focusing on candidate skills evaluation and gap analysis. Drawing from research by [2, 5, 18], we identify four essential steps that contribute to a personalized recommendation system. In Fig. 2, we give an overview of our skill assessment process in our system CRS (Career Recommendation System).

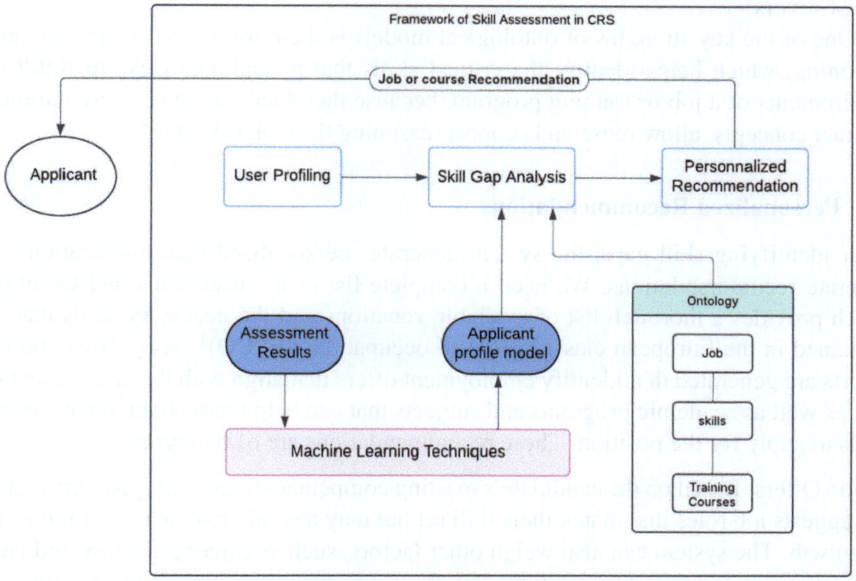

Fig. 2. Framework *of Skill Assessment in CRS*

2.1 Profile Creation and Skills Inventory

In this initial phase, there are two types of data collection: static and dynamic profiling. In the static profiling the applicant creates a comprehensive skills profile, which may include formal qualifications, certifications, work experience, and any other relevant data. Dynamic profiling consists of adding skills and knowledge to the data profile after the first assessment task. This is the foundation for the skills assessment process. The ontological model maps these skills against predefined skill sets in the ontology.

2.2 Skill Gap Analysis

Once the skills profile is established, the system conducts a gap analysis by comparing the candidate's existing skill set with the competencies required for specific job roles. As [13] explains, analytics are used to create a recommender system that provides information about what knowledge and skills applicants should obtain and where they can be acquired.

This stage is accomplished by employing Natural Language Processing (NLP) techniques to organize the description of skills into a common structure. The NLP model is used for skill and experience matching in the system [14]. As [15] explain, cited in [7] « There are two approaches to constructing the NLP algorithm: the first is a traditional rule-based approach that uses a hand-built vocabulary, and the second is based on a machine learning model called BERT (Bidirectional Encoder Representations from Transformers)».

One of the key strengths of ontological models is their ability to perform semantic reasoning, which helps identify the critical skills that a candidate lacks to fulfill the requirements of a job or training program, because they facilitate the representation of abstract concepts, allow reuse and support reasoning through rules [16].

2.3 Personalized Recommendations

After identifying skill gaps, the system generates personalized recommendations. To generate recommendations, We need a complete list of the user skills and Ontology, which provides a thorough list of available vocations and the necessary skills that are contained in the European classification of occupations (ESCO)[1]. As [13] mentioned, reports are generated that identify employment offers that align with the user's capabilities, as well as academic programs and subjects that can help them obtain the necessary skills to apply for the position. These recommendations are of two types:

- Job Offers: Based on the candidate's existing competencies and skill gaps, the system suggests job roles that match their skill set but may also offer some opportunities for growth. The system can also weigh other factors, such as career trajectory and labor market demands, to propose jobs that allow candidates to expand their skills over time.
- Training Programs: For the missing skills, the system suggests specific training programs, certifications, or learning resources tailored to the candidate's needs. The system can recommend both traditional (e.g., university degrees) and non-traditional learning opportunities (e.g., online courses, boot camps). These programs are recommended based on their alignment with the ontological model and the evolving job market needs.

Research by [5] shows that « Integrating machine learning approaches with ontology-based recommender systems makes significant revolutions in improving recommendation system operations». The inclusion of factors like course ratings, candidate preferences, and program reviews further refines the system's ability to recommend effective learning paths.

2.4 Continuous Feedback and Adaptation

A vital feature of the system is its ability to adapt over time. As the applicant completes training or gains experience in new roles, the system continuously updates their skill profiles. This dynamic process ensures that the recommender system remains relevant and provides updated recommendations based on the candidate's progression.

[1] European classification of occupations, (https://esco.ec.europa.eu/en/classification/).

3 Benefits of Using an Ontological Model in Skills Assessment

Leveraging ontological models in recommender systems offers several benefits that have been well-documented in academic literature.

- **Precision in Skill Matching:** As [17] explains, that ontologies have benefits of reusability, reasoning ability, and supports inference mechanisms, which helps to provide enhanced recommendations. Then, help to refine the matching process between candidate profiles and job roles. The system can make more precise recommendations by considering the relationships between skills and job requirements, rather than merely matching keywords. This ensures that job offers and training programs are more relevant and closely aligned with a candidate's actual competencies.
- **Improved Gap Identification:** The ontological model's semantic reasoning capabilities ensure that skill gaps are identified more accurately. Rather than relying on simple rule-based logic, the system can understand the context and relationships between different skills, making the gap analysis process more holistic [18].
- **Personalized Career Development:** Research shows that personalized skill assessments and recommendations increase user engagement and satisfaction. [19] Highlights that providing tailored job and training suggestions based on individual skills and aspirations promotes career growth and motivates continuous learning.
- **Adaptability to Industry Changes:** The job market is constantly evolving, and so are the skills required for various roles. The ontological model allows the system to evolve alongside market trends, ensuring that the recommender system remains responsive to industry demands [20].

4 Challenges and Considerations

Despite the advantages, there are challenges in using ontological models for skills assessment:

Data Quality: As with any system that relies on user input, the quality of the recommendations is highly dependent on the accuracy and completeness of the data provided by candidates. Any gaps or inaccuracies in the candidate's profile could lead to flawed recommendations.

Complexity of the Ontological Model: The complexity of the ontological model itself can be a challenge. Developing comprehensive ontologies requires careful evaluation to ensure they accurately represent the intended domains. Building and maintaining an ontology that accurately reflects the relationships between skills, job roles, and training programs requires continuous effort and domain expertise.

Ambiguities in Skill Classification: Extracting evidence of skills from large data streams in digital contexts poses significant challenges. Many skills overlap and candidates may possess skills that don't map neatly onto predefined categories. As [19] discusses, dealing with such ambiguities requires sophisticated algorithms that can account for variations in how skills are described.

5 Conclusion

The integration of an ontological model within a recommender system provides a powerful framework for skills assessment and personalized career development. By mapping candidates' skills against a structured ontology, the system can accurately identify skill gaps and recommend relevant job roles or training programs. Insights from academic research emphasize the importance of semantic reasoning, content-based recommendation algorithms, and continuous feedback in enhancing the effectiveness of these systems.

As industries continue to evolve and new skills emerge, recommender systems powered by ontological models will play a crucial role in helping individuals navigate their careers and acquire the necessary skills to stay competitive in the job market. This approach provides a win-win situation for both candidates and employers, fostering a more agile, skill-based labor market.

From an institutional perspective, we are interested in developing the system and thoroughly describing the recommendation generating mechanism before automating the complete recommendation mechanism by incorporating filtering steps such as collaborative filtering and other machine learning algorithms. The evaluation of the recommender system in the upcoming work will adopt a dual perspective, assessing both the technical performance—through relevance and accuracy of job or training course recommendations—and the user-centered impact, focusing on how effectively the recommendations address skill gaps identified in the assessment and support users in achieving their career development goals.

References

1. Tarus, J.K., Niu, Z., Yousif, A.: A hybrid knowledge-based recommender system for e-learning based on ontology and sequential pattern mining. Future Gener. Comput. Syst. (2017)
2. Ibrahim, M.E., Yang, Y., Ndzi, D.L., Yang, G., Al-Maliki, M.: Ontology-based personalized course recommendation framework. IEEE Access **7**, 5180–5199 (2018)
3. Beutling, N., Spahic-Bogdanovic, M.: Personalized course recommender: linking learning objectives and career goals through competencies. In: AAAI Spring Symposium Series (SSS-24) (2024)
4. Gulzar, Z., Leema, A.A., Deepak, G.: PCRS: personalized course recommender system based on Hybrid approach. Procedia Comput. Sci. **125**, 518–524 (2018)
5. Obeid, C., Lahoud, I., El Khoury, H., Champin, P.-A.: Ontology-based recommender system in higher education. In: The 2018 Web Conference Companion, Lyon, France. ACM, New York (2018)
6. Rahim, M.N., Basheer, K.P.M.: A hierarchical and multi-tiered personalized career recommender system tailored to individual aptitudes. Indian J. Sci. Technol. **18**(3), 231–244 (2025)
7. Dascălu, M.-I., et al.: Careprofsys – an ontology for career development in engineering designed for the Romanian job market. J. Autom. Control Comput. Sci. (2023)
8. Noy, N.F., McGuinness, D.: Ontology Development 101: A Guide to Creating Your First Ontology. Stanford Knowledge Systems Laboratory (2001)
9. García-Sánchez, F., Martínez-Béjar, R., Contreras, L., Fernández-Breis, J.T., Castellanos-Nieves, D.: An ontology-based intelligent system for recruitment. Expert Syst. Appl. **31**(2), 248–263 (2005)

10. Ramos, L., Corniel, M., Herrera, R.J.G., Contreras, L.: Ontological model as support decision making in study opportunities: toward a recommendation system. In: World Scientific and Engineering Academy and Society (WSEAS), Grecia, pp. 91–99 (2008)
11. Qamhieh, M., Sammaneh, H., Demaidi, M.N.: PCRS: personalized career-path recommender system for engineering students. IEEE Access **8**, 214039–214049 (2020)
12. Mavuso, N.C., Jere, N., vanGreunen, D.: A customized artificial intelligence based career choice recommender system for a rural university. In: African Conference on Information Systems and Technology (2023)
13. Baneres, D., Conesa, J.: A life-long learning recommender system to promote employability. Int. J. Emerg. Technol. Learn. (iJET) **12**(06), 77 (2017)
14. Appadoo, K., Soonnoo, M.B., Mungloo-Dilmohamud, Z.: Job recommendation system, machine learning, regression, classification, natural language processing. In: Proceedings of the 2021 IEEE Asia-Pacific Conference on Computer Science and Data Engineering (CSDE), pp. 1–6 (2020)
15. Devlin, J., et al.: Bert: pre-training of deep bidirectional transformers for language understanding. In: Proceedings of the 2019 Conference of the North American Chapter of the Association for Computational Linguistics: Human Language Technologies (2019)
16. Clemente, J., Yago, H., De Pedro-Carracedo, J., Bueno, J.: A proposal for an adaptive recommender system based on competences and ontologies. Expert Syst. Appl. **208**, 118171 (2022)
17. George, G., Lal, A.M.: Review of ontology-based recommender systems in e-learning. Comput. Educ. **142**, 103642 (2019)
18. Ibrahim, M.E., Yang, Y., Ndzi, D.: Using ontology for personalised course recommendation applications. In: Gervasi, O., et al. Computational Science and Its Applications – ICCSA 2017. ICCSA 2017. Lecture Notes in Computer Science, vol. 10404, pp. 426–438. Springer, Cham (2017). https://doi.org/10.1007/978-3-319-62392-4_31
19. Dascalu, M. I., Stanica, I. C., Bodea, C., Nitu, M., Lazarou, E., Dascalu, L.: Diminishing the university drop-out rates in Romanian engineering programs by using a smart career path profiling system. ELearning Softw. Educ. **2**, 297–304 (2020)
20. Alaa, R., Gawish, M., Fernández-Veiga, M.: Improving recommendations for online retail markets based on ontology evolution. Electronics **10**(14), 1650 (2021)

Unified Data Management Framework for Supply Chain Optimization: Challenges in Enterprise Architecture Models like TOGAF, RAMI 4.0 and IBM 4.0

Smina Nouhaila(✉) and Gharib Jihane

Laboratory of Engineering Science, National School of Applied Sciences, Ibn Tofail University Kenitra, Kenitra, Morocco
{nouhaila.smina,jihane.gharib}@uit.ac.ma

Abstract. Efficient supply chain management relies on a data management framework that guarantees accuracy, visibility and actionable information across the whole value chain. Enterprise architecture models such as TOGAF (The Open Group Architecture Framework), RAMI 4.0 (Reference Architecture Model for Industry 4.0) and IBM 4.0 provide structured approaches to enterprise systems management, but are limited when it comes to integrating real-time data, maintaining interoperability with IoT networks and responding to the dynamic nature of modern supply chains. To address these shortcomings, this article proposes a customized framework that describes what a data management system should ideally comprise to integrate properly with supply chain management. Rather than combining existing models, this framework provides a systematic platform for effective data collection, processing and use. Advanced analytics, digital twins and Industry 4.0 technologies are used to improve decision-making, end-to-end visibility and operational performance throughout the supply chain.

Keywords: Data management · Supply Chain Management · Enterprise Architecture · Industry 4.0 · TOGAF · RAMI 4.0 · IBM 4.0 · Value chain

1 Introduction

In today's supply chain management (SCM), powerful data management is critical for cost reduction, operational performance and real-time decision-making [1]. The growing complexity of supply chains, driven by globalization and Industry 4.0 technologies, necessitates adaptive data management frameworks. Traditional systems are confronted with fragmented data flows, a lack of real-time visibility and limited interoperability, leading to inefficiencies and increased costs. The integration of big data, cloud computing, artificial intelligence (AI) and the Internet of Things (IoT) [2] gives potential solutions by allowing real-time data collection, automation and predictive analytics [17].

However, existing enterprise architecture frameworks, though structured in data governance, lack flexibility to meet the real-time integration, automation, and adaptability demands of modern supply chains. Effective SCM requires robust planning,

B.-E. Boudriki Semlali et al. (Eds.): DATA 2025, LNISO 81, pp. 404–416, 2025.
https://doi.org/10.1007/978-3-032-06725-8_34

cost optimization, and decision-making. Smart manufacturing integrates lean processes and digital technologies to enhance agility, improve decision-making, and reduce lead times [3, 4]. To maximize these benefits, seamless data integration, adaptability, and cross-platform compatibility are essential [5].

This paper examines the role of data management in TOGAF, RAMI 4.0, and IBM 4.0 [2, 10, 15]. This study has three goals: firstly, to evaluate their strengths and obstacles for SCM; secondly, to perceive their inefficiencies in scalability and real-time integration; and thirdly, to advocate upgrades for Industry 4.0 adaptability. Comparative analysis of these frameworks provides insight into improving efficiency, responsiveness, and profitability in SCM. These objectives are achieved by answering the following research questions:

RQ1: What are the advantages and disadvantages of the TOGAF, RAMI 4.0 and IBM 4.0 frameworks when it comes to addressing the challenges of data management in SCM?
RQ2: What limitations do these frameworks have in decision-making, scheduling, and cost optimization?
RQ3: What key improvements are needed to enhance adaptability, interoperability, and efficiency in SCM?

2 Related Work

2.1 Data Management

Data management involves collecting, storing, organizing, and using data correctly and securely. It ensures accuracy, accessibility, and integrity across industries, supporting decision-making through governance, quality control, metadata management, and real-time processing. Industry 4.0 technologies enable real-time analytics, predictive modeling, and automation [1].

A strong data management framework enhances efficiency by reducing redundancy, improving consistency, and integrating multiple platforms. However, conventional approaches struggle with scalability, interoperability, and real-time synchronization, particularly in dynamic environments like SCM [2]. Legacy systems often operate in silos, impeding responsiveness [6]. The absence of structured frameworks for smart manufacturing planning exacerbates inefficiencies [13].

Many industries rely on structured frameworks such as DAMA-DMBOK (Data Management Body of Knowledge) [14], which describes best practices in data governance, quality, metadata management and security. It is the resource of choice for companies seeking to develop a solid data management plan. However, DAMA-DMBOK is IT oriented and doesn't answer some of the questions that arise with Industry 4.0, such as real-time data synchronization, cyber-physical integration and large-scale industrial data management.

Modern SCM requires enterprise architecture frameworks to understand the data value chain from acquisition to decision-making transforming raw data into useful information. EA must provide structured mechanisms for governance, interoperability, and real-time processing. Data strategies could be fragmented and ineffective without a solid EA foundation. Data alignment is crucial for efficient supply chain operations; delays and inconsistencies cause inefficiencies.

2.2 Data Management in Supply Chain Management (SCM)

Effective data management ensures seamless coordination among suppliers, manufacturers, distributors, and customers, enabling visibility, logistics optimization, inventory control, and demand forecasting [6]. With Industry 4.0, digitalized supply chains require advanced data handling for real-time decision-making and agility.

Traditional SCM depended on fragmented structures, main to inefficiencies in data flow and responsiveness. TOGAF, RAMI 4.0, and IBM 4.0 frameworks have tried to cope with those challenges through structured data management methodologies [8], yet vary in their effectiveness for real-time processing, IoT interoperability, and digital ecosystem adaptability.

One of the primary issues with conventional EA models is the shortage of real-time data integration. Smart manufacturing requires seamless, real-time data alternate throughout a couple of systems, often unsupported by legacy frameworks [8]. The integration of digital twins, AI analytics, and cloud-based SCM systems enhances forecasting and decreases lead times [13]. However, data governance, standardization, and safety challenges persist, restricting the entire capability of data-driven SCM [4]. Additionally, scalability and IoT/RFID interoperability continue to be bottlenecks, hindering superior data management solutions.

2.3 Enterprise Architecture and Data Management in SCM

With Industry 4.0 technologies, modern SCM is based on structured frameworks for efficient data management [2]. EA ensures interoperability and structured communication among distinctive entities [1, 18]. Traditional SCM data management depended on isolated systems, main to inefficiencies in real-time decision-making [6]. TOGAF and RAMI 4.0, have tried to cope with those demanding situations through presenting based methodologies for enterprise-wide data management [8].

Data-driven decision-making improves real-time integration and visibility across suppliers, manufacturers and distributors, facilitated predictive analytics for demand forecasting and inventory management and risk mitigation through anomaly detection and scenario modeling. [13]. Smart technologies streamline data collecting and automates processes via EA frameworks [4]. However, challenges in scalability and real-time processing hinder their full potential [11].

EA models have developed to help Industry 4.0 applications but differ in focus and adaptability [5]. TOGAF, firstly advanced for IT governance, has been tailored for SCM, but lacks devoted help for real-time data integration [1, 18]. RAMI 4.0, designed for industrial applications, offers a structured approach to managing cyber-physical systems but faces limitations in timing optimization and responsiveness [10]. IBM 4.0 provides a comprehensive framework for digital transformation but it is not very flexible when it comes to IoT and RFID integration [9]. The next section is mainly dedicated to a detailed to a methodological comparative analysis based on key factors for an enhanced data management in SCM.

3 Comparison Analysis of Enterprise Architecture Frameworks

3.1 TOGAF

The TOGAF (The Open Group Architecture Framework) focuses on process optimization, standardization, and IT-business alignment. It follows an iterative Architecture Development Method (ADM) which consists of eight key phases as illustrated in Fig. 1, ensuring a structured approach to enterprise-wide data governance and IT integration. However, TOGAF struggles with real-time data processing and lacks dedicated mechanisms for handling cyber-physical system interactions, making it less adaptable to dynamic Industry 4.0 environments [2, 4].

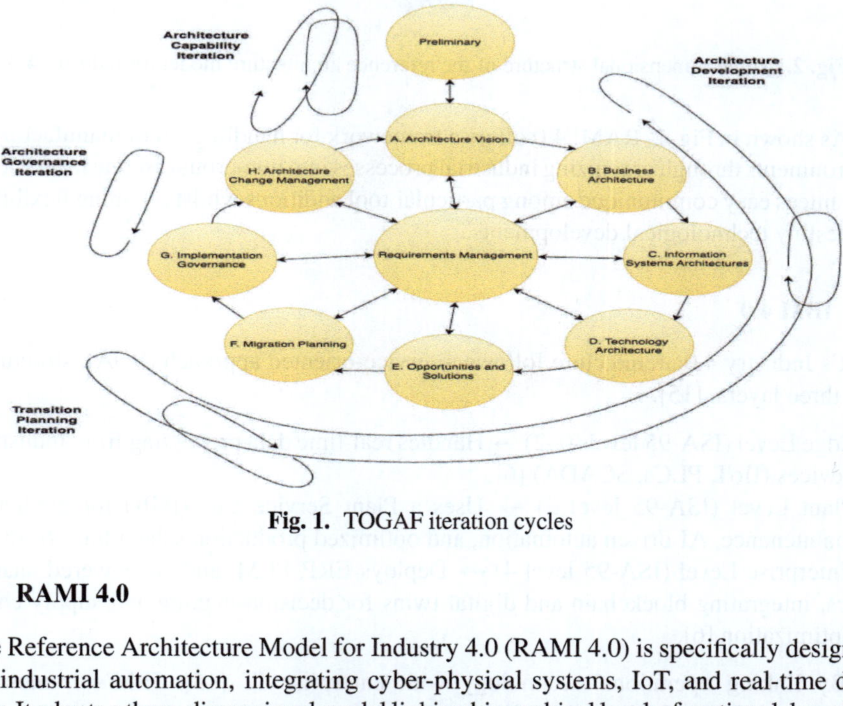

Fig. 1. TOGAF iteration cycles

3.2 RAMI 4.0

The Reference Architecture Model for Industry 4.0 (RAMI 4.0) is specifically designed for industrial automation, integrating cyber-physical systems, IoT, and real-time data flow. It adopts a three-dimensional model linking hierarchical levels, functional domains, and the product lifecycle, organizing industrial processes across six integration layers (Business, Functional, Information, Communication, Integration, and Asset Layers) [6, 10]. While RAMI 4.0 is effective in structuring industrial data, it faces challenges in real-time decision-making, system flexibility, and synchronization [11].

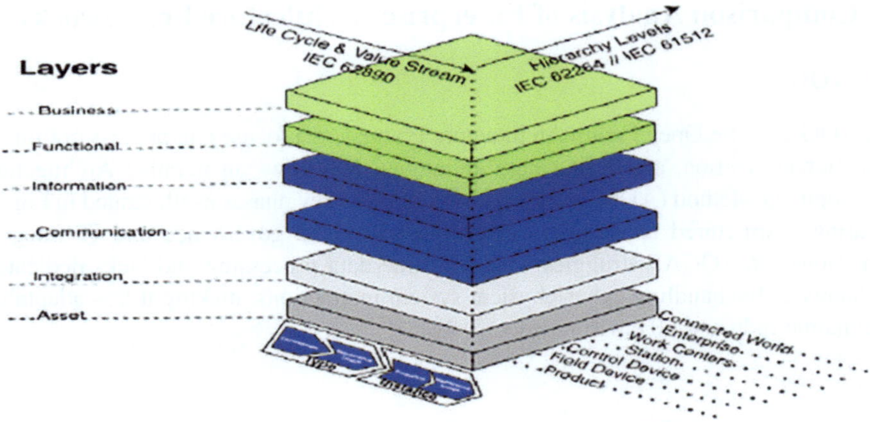

Fig. 2. Three-dimensional structure of the reference architecture model for industry 4.0

As shown in Fig. 2, RAMI 4.0 offers a framework for handling smart manufacturing environments through organizing industrial processes into numerous discrete layers. This guarantees easy communique among particular tool additions whilst retaining flexibility for destiny technological developments.

3.3 IBM 4.0

IBM's Industry 4.0 architecture follows a service-oriented approach (SOA), structured into three layers: [15].

- Edge Level (ISA-95 levels 0–2) → Handles real-time data processing from industrial devices (IIoT, PLCs, SCADA) [6].
- Plant Level (ISA-95 level 3) → Uses a Plant Service Bus (PSB) for predictive maintenance, AI-driven automation, and optimized production scheduling [6, 15].
- Enterprise Level (ISA-95 level 4) → Deploys ERP, PLM, and AI-powered analytics, integrating blockchain and digital twins for decision-making and supply chain optimization [6].

By linking operational technology with enterprise IT [15], IBM's framework enhances agility, scalability, and efficiency in modern manufacturing.

3.4 Comparative Analysis: TOGAF vs. RAMI 4.0 vs. IBM 4.0

Comparison Methodology
We evaluate IBM 4.0, TOGAF, and RAMI 4.0 in the context of data management for Industry 4.0 by conducting a comparative analysis. We evaluate IBM 4.0, TOGAF, and RAMI 4.0 in the context of data management for Industry 4.0 by conducting a comparative analysis. Our approach included a systematic review of recent literature and industry reports published between 2020 and 2024, selecting sources based on relevance, credibility, and empirical evidence. Each framework was analyzed based on five key dimensions:

Data management: Handling of data across the value chain, from collection to insights.
Flexibility and adaptability: Applicability across industries and supply chain structures.
Suitability for Industry 4.0: Support for smart manufacturing and digital transformation.
Ease of integration: Compatibility with modern enterprise tools and systems.
Governance and compliance: Strength of policies on data security, regulations, and risk management.

Based on the Data Value Chain, a set of evaluation criteria was developed in order to do a methodical and impartial comparison of these 3 frameworks. Technical capabilities (e.g., real-time processing, interoperability) strategic compliance (e.g., GDPR, ISO 28000), cost efficiency, and operational usability are among these criteria.

Figure 3 shows the evaluation structure applied for this comparison, matching data management capacity with the specific needs of a dynamic and artificial intelligence-powered supply chain environment. Each framework is then evaluated across these tiers to draw attention to areas of strength, weakness, and possible holes in supporting modern, data-driven supply chains.

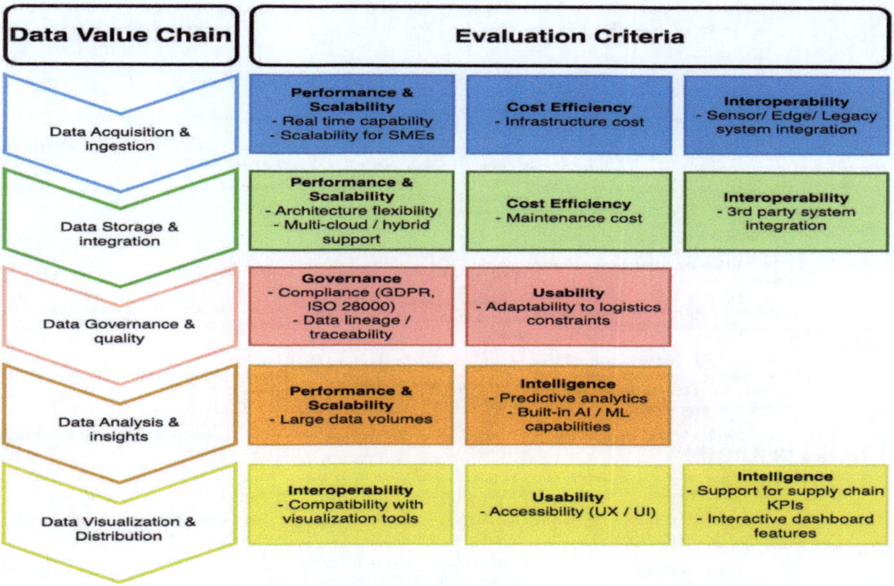

Fig. 3. Evaluation model based on the Data Value Chain and corresponding assessment criteria used for framework comparison

Comparison Analysis

Current frameworks highlight the advantages and disadvantages in data management.

TOGAF is nicely-based for enterprise-wide data governance however lacks real-time data processing capabilities for dynamic supply chains [2]. RAMI 4.0 aligns nicely with cyber-physical systems however struggles with optimization of reaction instances et value efficiency [12, 18]. IBM 4.0 gives sturdy analytical skills however faces demanding situations in adapting to rising IoT-based architectures [7].

A key predicament in present day frameworks is the bad alignment with key performance indicators (KPIs) critical for supply chain efficiency [13]. Inefficient data collection and processing techniques in addition avert real-time analytics and decision-making [6]. While RAMI 4.0 offers a structured approach to lifecycle management, its implementation lacks consistency throughout distinctive industries [10]. The three-dimensional layout may be oversimplified for system architectures, especially in capturing the dynamics of industrial operations [16]. Similarly, IBM 4.0 struggles with real-time data streams from IoT and RFID, proscribing its applicability in dynamic supply chain contexts [11]. The mission of integrating rising technology in small and medium enterprises (SMEs) in addition underscores the need for adaptable EA [10]. Figure 4 gives a structured comparison of these frameworks across key value chain layers.

Value Chain Layers	IBM 4.0	TOGAF	RAMI 4.0
Data Acquisition & Ingestion	-Uses IoT and edge computing to collect data in real time.	- Looks at data collection from an overall IT perspective.	- Uses a structured approach for industrial data but struggles with unstructured or mixed data.
	- Expensive infrastructure costs make it difficult for small and mid-sized suppliers to adopt.	- No standard real-time data ingestion model for supply chain events.	- Complex setup complicates implementation for global supply chain ecosystems with diverse standards.
	- Focuses on industrial IoT, making it less adaptable to other logistics and supplier networks.	- Integration of legacy systems is complex, leading to inefficiencies in logistics data flow.	- Real-time supply chain event detection is limited due to hierarchical architecture.
Data Storage & Integration	- Strong use of cloud and edge computing.	- Standardizes IT storage but doesn't focus on real-time data needs.	- Connects physical and digital systems well.
	- High costs, especially for small businesses.	- Rigid structure makes it hard to integrate with new data systems.	- Built for industrial automation, so it's less adaptable for general IT use.
	- Locked into IBM's ecosystem, reducing flexibility.	- Focuses on IT governance rather than dynamic inventory management and just-in-time supply chain flows.	- Uses industrial protocols, which can limit compatibility with enterprise systems.
	- Not well-suited for multi-cloud or hybrid storage.		
Data Quality & Governance	- Uses AI for automated quality control and defect detection.	- Aligns data quality with business goals.	- Monitors data quality across multiple factory layers.
	- Requires high computational resources, making real-time quality monitoring expensive for SMEs.	- No standard method for tracking data lineage across multiple suppliers and logistics partners.	Governance structure is rigid, making it hard to scale across decentralized supply chain networks.
	- No built-in data governance framework (e.g., DAMA-DMBOK) aligned with global SCM regulations (e.g., GDPR for logistics data, ISO 28000 for supply chain security)	- Governance is IT-centric and does not cover physical supply chain constraints (e.g., cold chain logistics).	Designed mainly for industrial security, not general IT regulations.
Data Analysis & Insights	- Uses AI, machine learning, and digital twins to improve decision-making.	- Prioritizes business intelligence and data governance.	- Focuses on monitoring industrial processes rather than predictive analytics.
	- Requires large amounts of high-quality data, which not all industries have.	- No built-in predictive analytics for supply chain disruptions or lead time variability.	- Works best with structured manufacturing data, making it less useful for advanced analytics.
	- Primarily focused on manufacturing efficiency rather than holistic supply chain visibility.	- Does not have a clear process for integrating AI-driven supply chain analytics.	- Limited support for real-time supply chain risk assessment
Data Visualization & Distribution	-Uses advanced dashboards and AI-powered insights.	-Provides high-level architecture visualizations but lacks real-time analytics dashboards for SCM decision-making.	- Uses hierarchical layers to display data and supports digital twins.
	- Visualization tools can be hard to set up without AI expertise.	- No pre-built dashboards for logistics tracking, supplier risk assessment, or demand-supply balancing.	- Not easily compatible with modern visualization tools.
	- Visualization tools don't fully support supply chain KPIs and transportation analytics.		- Not designed for supply chain performance tracking.

Fig. 4. Comparative Analysis of RAMI 4.0, IBM, and TOGAF Frameworks

4 Proposed Unified Data Management Framework for SCM

4.1 Need Analysis

Existing enterprise architecture frameworks have to be improved in data collection, real-time processing, and adaptability to Industry 4.0 environments [2]. In Sect. 3, we mentioned how TOGAF, RAMI 4.0, and IBM 4.0 present limitations for dealing with modern data-driven supply chains. These frameworks primarily focus on IT governance, hierarchical asset management, and cloud-based architectures, without fully addressing real-time data integration and automation [17].

Supply chains can reap more performance and responsiveness through advanced data optimization and AI-driven analytics [1]. Data collection can be optimized by reorganizing information flow using intelligent automation and predictive analytics [13], while real-time validation mechanisms enhance accuracy and reduce inconsistencies [11]. Further refinement of time optimization by using algorithms in ML to predict workflow failures and dynamically adjust schedules [7]. End-to-end supply chain visibility requires integrating RFID, IoT, and enterprise data management frameworks [8], enabling real-time asset tracking, automated data synchronization, and predictive maintenance [10]. Additionally, improving communication between cyber-physical systems reduces latency, ensuring responsive adaptation to evolving business conditions [5]. Real-time data stream adjustments, facilitated via AI-driven decision support systems, permit proactive responses to disruptions [9]. Adaptive algorithms permit frameworks to optimize resource allocation primarily based totally on live sensor inputs, main to an extra resilient and agile supply chain [12].

4.2 Framework Approach

The proposed framework is structured around core data management layers to ensure a seamless data flow from acquisition to decision-making. Our approach eliminates the constraints of existing frameworks by defining what a data management system should theoretically include rather than merging existing models. We begin with **Data Acquisition and Ingestion**, where we use a hierarchical Asset Administration Shell (AAS) to represent data consistently [10], while edge computing allows us to capture information in real-time with low latency in order to decentralize supply chains [11]. From there, we move to **Data Storage and Integration**, where we implement a hybrid architecture combining cloud and on-premise solutions. This allows us to enhance scalability and interoperability [1] and integrating seamlessly with cyber-physical systems and IoT networks for standardized data processing [8]. Once data is stored, we focus on **Data Quality and Governance**. Here, we apply AI-driven mechanisms to detect anomalies and ensure high-quality data [7], and we maintain regulatory compliance by organizing data quality metrics around strategic supply chain KPIs [13]. At the **Analysis and Insights** stage, we employ AI/ML-driven predictive analytics to enhance workflow efficiency and prevent failures [4]. We also integrate digital twin technology for real-time simulations and data-driven decision-making [5]. Finally, we pass to the **Visualization and Distribution** layer. We enhance accessibility with real-time dashboards and hierarchical visualization models [9], and we enable dynamic KPI monitoring to improve operational efficiency and responsiveness [17].

4.3 Integration in Data Management

With this framework, industrial data management is enhanced by providing an end-to-end visibility throughout the data value chain, that enable real-time synchronization and supply chain tracking. Automation and AI-driven analytics, reduce human intervention and facilitate decision-making. Also, its scalable and adaptive architecture dynamically adjusts to evolving industrial landscapes, ensuring long-term sustainability. Figure 5 illustrates the conceptual structure of this framework.

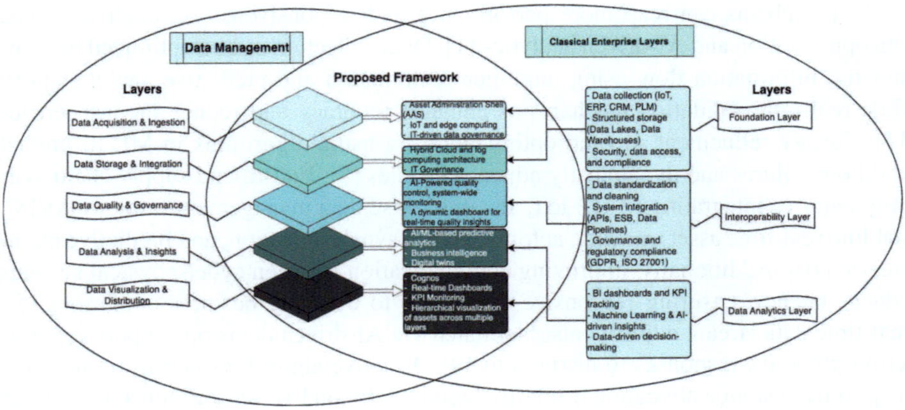

Fig. 5. Unified proposed Framework model

Additionally, the table outlines the key processes within each value chain layer (Table 1):

Table 1. Proposed Unified Data Management Framework

Value Chain Layers	Proposed Framework
Data Acquisition & Ingestion	- Implements a hierarchical Asset Administration Shell (AAS) for structured data acquisition. - Integrates edge computing for real-time data flow.
Data Storage & Integration	- Utilizes a hybrid storage architecture (cloud and on-premise). - Ensures seamless integration of cyber-physical systems and IoT devices.
Data Quality & Governance	- Implements AI-powered validation for real-time anomaly detection. - Establishes governance models aligned with strategic KPIs. - Introduces a dynamic dashboard for real-time quality insights.

(*continued*)

Table 1. (*continued*)

Value Chain Layers	Proposed Framework
Data Analysis & Insights	- Uses AI/ML-based predictive analytics for workflow optimization. - Implements digital twin technology for real-time optimization.
Data Visualization & Distribution	- Uses advanced visualization tools and custom dashboards, for real-time monitoring. - Utilizes hierarchical visualization for better accessibility. - Ensures dynamic KPI tracking for enterprise-wide monitoring.

4.4 Framework Deployment and Business Impact

Implementation Strategy

Implementing this framework requires a structured approach that balances technology integration, operational adjustments, and strategic investment. The first step is to assess existing IT systems and supply chain processes, identifying gaps in data governance, security, and interoperability. Enterprises should then focus on technology integration, incorporating AI-powered analytics, IoT devices, and cloud solutions to enable real-time data processing and enhance supply chain visibility. Crucially, companies must not view the framework as a replacement but as an enhancement to current systems. Transitioning from existing frameworks to the proposed model should follow a phased approach, where organizations map current data practices against the new architecture to highlight overlaps and opportunities. Core modules: like edge computing, hybrid storage, or predictive analytics can be introduced incrementally, ensuring continuity and minimizing operational disruption. Legacy systems can coexist with new components during this transitional period, supported by clear data governance rules and system interoperability.

Change management plays a vital role throughout. Cross-functional coordination between IT, operations, and supply chain teams ensures alignment, while structured training helps staff adapt to new tools and processes. By defining clear performance indicators and regularly measuring impact, businesses can monitor the transformation journey and adjust as needed. Ultimately, this implementation strategy provides a scalable and resilient pathway toward smarter, more agile supply chains.

Implementation Costs and Potential Risks

While the benefits of a data-driven supply chain are substantial, the implementation process comes with financial and operational risks. Upfront investments include infrastructure upgrades, software licenses, cybersecurity measures, and system integration efforts, which may require custom development to align with existing IT environments. Migration risks, such as downtime, data loss, or system incompatibility, must be managed through careful planning and phased rollouts. Additionally, data security and compliance are critical concerns, requiring alignment with GDPR, ISO 27001, and other regulatory standards to prevent cyber threats and unauthorized data access. Beyond technical challenges, social implications must also be addressed, as AI-driven automation may reshape workforce roles and necessitate reskilling initiatives. Successful implementation depends on a well-defined financial strategy, proactive risk management, and strong leadership to navigate the complexities of digital transformation while ensuring a smooth and sustainable transition.

Validation steps in a Smart Factory
A brief operational check can be performed after the suggested framework has been implemented to validate its effectiveness in a smart factory context. This includes testing system connectivity across IoT devices, edge nodes, and cloud platforms to ensure that data flows smoothly, with a target of >95% device connectivity uptime. AI modules should be verified for their capacity to detect anomalies in real time, aiming for a detection accuracy rate above 90%, and for their predictive models to achieve at least 85% forecasting precision regarding supply disruptions. A quick compliance audit can validate adherence to GDPR and ISO 28000 standards, with a goal of 100% alignment on data governance policies. Furthermore, performance indicators such as data processing latency (under 2 s for critical flows) and dashboard loading time (below 1.5 s) can be tracked. User feedback surveys should also report at least 80% satisfaction on dashboard usability and feature accessibility. This validation process confirms that the framework not only works in theory, but also provides measurable value in real-world industrial situations.

5 Conclusion and Future Work

The growth of Industry 4.0 necessitates a comprehensive and flexible data management framework that outperforms the limitations of current enterprise architecture. In this study, we presented a proposed alternative enterprise Data Model that provides the theoretical groundwork for smooth integration within SCM. This framework defines the critical components required for efficient data collecting, integration, governance, and visualization, resulting in real-time decision-making and improved operational responsiveness. By implementing AI-driven analytics, predictive maintenance, and dynamic data synchronization, industrial companies may fully utilize data as a strategic asset.

As future work, we plan to test and refine this framework in a real industrial supply chain environment, in collaboration with industry stakeholders, to assess its cost impact, adaptability and performance. A key objective will be to facilitate the smooth transition of companies by addressing security risks, regulatory challenges and integration complexities. In addition, we will look at the wider implications of automation, with particular emphasis on its influence on business processes and the workforce. In addition, improved AI-driven analytics will enhance the predictability, efficiency and resilience of supply chains. Ultimately, the aim is to develop a practical, scalable solution that helps companies meet the challenges of a data-driven digital supply chain.

References

1. Nowakowski, E., Breu, R.: Enterprise architecture planning for Industry 4.0
2. Curyła, K., Habernal, K.: Comparative analysis of frameworks dedicated to enterprise designing. JCSI **13**, 315–322 (2019). https://doi.org/10.35784/jcsi.1326
3. Cañas, H., Mula, J., Campuzano-Bolarín, F., Poler, R.: A conceptual framework for smart production planning and control in Industry 4.0. Comput. Ind. Eng. **173**, 108659 (2022). https://doi.org/10.1016/j.cie.2022.108659
4. Kornyshova, E., Barrios, J.: Industry 4.0 impact propagation on enterprise architecture models. Procedia Comput. Sci. **176**, 2497–2506 (2020). https://doi.org/10.1016/j.procs.2020.09.326
5. Horstkemper, D., Mülhausen, A., Hellingrath, B.: Structured approach for automated enterprise architecture model generation. IFAC-PapersOnLine **56**, 3648–3653 (2023). 10.1016/j.ifacol.2023.10.1528
6. Helmann, A., Deschamps, F., Loures, E.D.F.R.: Reference architectures for Industry 4.0: literature review. In: Pokojski, J., Gil, M., Newnes, L., Stjepandić, J., and Wognum, N. (eds.) Advances in Transdisciplinary Engineering. IOS Press (2020). https://doi.org/10.3233/ATDE200074
7. Binder, C., Leiter, W., Joebstl, O., Mair, L., Neureiter, C., Luder, A.: Utilizing an enterprise architecture framework for model-based industrial systems engineering. In: Proceedings of the 2021 IEEE 19th International Conference on Industrial Informatics (INDIN), pp. 1–6. IEEE, Palma de Mallorca (2021). https://doi.org/10.1109/INDIN45523.2021.9557460
8. Chaves, R.J.F., Vasconcelos, A.F.F.C.: Mapping between reference architecture model of Industrie 4.0 and ArchiMate. MSc Thesis (2019)
9. Melo, P.F.S., Godoy, E.P., Ferrari, P., Sisinni, E.: Open-source control device for Industry 4.0 Based on RAMI 4.0. Electronics **10**, 869 (2021). https://doi.org/10.3390/electronics10070869
10. Baptista, L.F., Barata, J.: Piloting Industry 4.0 in SMEs with RAMI 4.0: an enterprise architecture approach. Procedia Comput. Sci. **192**, 2826–2835 (2021). https://doi.org/10.1016/j.procs.2021.09.053
11. Oñate, W., Sanz, R.: Integration of fog computing in a distributed manufacturing execution system under the RAMI 4.0 framework. Appl. Sci. **14**, 10539 (2024). https://doi.org/10.3390/app142210539
12. Ozkan-Ozen, Y.D., Garza-Reyes, J.A., Kazancoglu, Y.: Integrating lean six sigma with life cycle and value stream level of RAMI 4.0. Qual. Manage. J. **31**, 225–241 (2024). https://doi.org/10.1080/10686967.2024.2394400
13. Rahmani, M., Romsdal, A., Sgarbossa, F., Strandhagen, J.O., Holm, M.: Towards smart production planning and control; a conceptual framework linking planning environment characteristics with the need for smart production planning and control. Annu. Rev. Control. **53**, 370–381 (2022). https://doi.org/10.1016/j.arcontrol.2022.03.008
14. DAMA International: DAMA-DMBOK: Data Management Body of Knowledge. Technics Publications, LLC (2017)
15. IBM: What is Industry 4.0?. https://www.ibm.com/think/topics/industry-4-0. Accessed 25 Feb 2024
16. Meixner, K., Hoffmann, D., Riedmann, S., Hünecke, P., Binder, C.: Rolling the dice – rethinking the RAMI 4.0 perspectives. In: Proceedings of the IEEE 29th International Conference on Emerging Technologies and Factory Automation (ETFA), Padova, Italy, pp. 1–8 (2024). https://doi.org/10.1109/ETFA61755.2024.10711108
17. Kachiche, S., Gahi, Y., Gharib, J.: An efficient framework for the implementation of sustainable Industry 4.0. In: Ben Ahmed, M., Boudhir, A.A., Santos, D., Dionisio, R., Benaya, N. (eds.) Innovations in Smart Cities Applications Volume 6. SCA 2022. Lecture Notes in

Networks and Systems, vol. 629, pp. 804–815. Springer, Cham (2023). https://doi.org/10.1007/978-3-031-26852-6_74
18. Sony, M.: Pros and cons of implementing Industry 4.0 for the organizations: a review and synthesis of evidence. Prod. Manuf. Res. **8**, 244–272 (2020). https://doi.org/10.1080/21693277.2020.1781705
19. Grace Tetteh, M., Gupta, S., Kumar, M., Trollman, H., Salonitis, K., Jagtap, S.: Pharma 4.0: A deep dive top management commitment to successful Lean 4.0 implementation in Ghanaian pharma manufacturing sector. Heliyon **10**, e36677 (2024). https://doi.org/10.1016/j.heliyon.2024.e36677

Enhanced Xception-Based Model for Binary Classification of Diabetic Retinopathy in Retinal Fundus Images

Kawtar Naim(✉) and Aziz Darouichi

Computer and Systems Engineering Laboratory (L2IS), FST, Cadi Ayyad University, Marrakech, Morocco
{k.naim.ced,az.darouichi}@uca.ac.ma

Abstract. Diabetic retinopathy (DR) remains one of the most prevalent vision-threatening diseases today. It is a complication resulting from diabetes, caused by changes in the retinal blood vessels. Early diagnosis is crucial to preventing its progression to advanced stages that can lead to blindness due of damaged blood vessels. In this study, we focused on improving DR detection through advanced image preprocessing techniques, including Gaussian blur, Contrast Limited Adaptive Histogram Equalization (CLAHE), and unsharp masking. Additionally, we employed GridSearchCV to fine-tune key hyperparameters of the Xception model, optimizing its performance.

Our model achieved an accuracy of 94.44% on the APTOS 2019 Blindness Detection dataset and 98.64% on the Eye Disease Image dataset, demonstrating strong precision and recall in DR detection. These results highlight the significant role of customized preprocessing and fine-tuning strategies in enhancing the robustness and predictive accuracy of deep learning models for diabetic retinopathy diagnosis. By enabling earlier and more accurate diagnosis, this study contributes to reducing the risk of vision loss and improving patient care in diabetic retinopathy management.

Keywords: Machine learning · Deep learning · Computer vision · Diabetic Retinopathy · Fundus images · Classification · Xception · Image processing

1 Introduction

Diabetic retinopathy is a leading cause of global vision loss and blindness as a result of the chronic insult to the retinal microvascular network in diabetic patients [8]. It has been projected that by 2040, about 224 million individuals will have DR and nearly 70 million will progress to advanced stages with significant risk to vision. Despite increased global efforts in health, the incidences of DR increased by 25% between 1990 and 2015. The increase is largely attributed to an increase in diabetes cases, population growth, and rising life expectancy [15]. Despite technological progress, the number of ophthalmologists globally remain low to serve the rising number of cases. Manual diagnosis and grading of DR are also time-consuming and tedious [6]. Correspondingly,

B.-E. Boudriki Semlali et al. (Eds.): DATA 2025, LNISO 81, pp. 417–426, 2025.
https://doi.org/10.1007/978-3-032-06725-8_35

the utilization of deep learning (DL)-based automated systems has become a primary solution for mass screening to enable faster and more consistent diagnosis. Detection and classification of DR at an early stage are required to prevent permanent blindness, and thus easily accessible and efficient diagnostic solutions are essential [10].

Retinal fundus photography in this scenario is a mainstay of DR detection, grading, and long-term follow-up. This non-invasive imaging technique offers a detailed visualization of the retina, with high clinical utility. Recent developments, especially in wide-field and ultra-wide-field fundus imaging, have significantly increased the field of view, capturing a larger area of the retina in one image. These developments have led to improved diagnostic capability and have improved the performance of screening programs and disease follow-up [8, 14].

This study emphasizes the pivotal role of fundus imaging in the diagnosis and long-term management of DR and the integration of AI-based systems to enhance patient management and disease control even further.

The remainder of this paper is structured as follows: Sect. 2 provides a review of the state-of-the-art research on DR detection. Section 3 delineates the methodology employed in this study, discussing the steps taken for DR classification. Section 4 presents the results obtained, compares the performance of our model with that of the state-of-the-art approaches, and discusses the key evaluation metrics. Finally, Sect. 5 concludes this study by summarizing our contributions and providing suggestions for future research.

2 Related Works

In recent years, Deep learning technology has, nonetheless, significantly influenced the diagnosis of diabetic retinopathy (DR), i.e., the analysis of the retinal fundus images. Automation of diagnosis using deep learning improves not only the process but also makes diagnostic reliability more accessible. A thorough review of the use of deep learning in methods of DR diagnosis is included in [14], highlighting the ability of the methods to not only detect the presence of and classify the disease stages from retinal images. One of the earliest uses was where Esfahani et al. [4] used ResNet34 to classify diabetic and non-diabetic fundus im- ages of the Kaggle dataset. Weighted summation and Gaussian blurring were utilized for image enhancement prior to network input. The model resulted in 85% accuracy and 86% sensitivity, showcasing the power of deep convolutional net- works in DR detection without the use of handcrafted feature engineering. Jiang et al. [7] thereafter proposed an ensemble approach that used a number of architectures Inception V3, ResNet-152, and Inception-ResNet-V2 utilizing the AdaBoost algorithm to avoid model bias and improve classification confidence. Their preprocessing pipeline included cropping, resizing, Gaussian filtering, and normalization. Their multi-model system achieved a 88.21% classification rate, confirming the advantage of ensemble learning in DR classification. Later the same year, Liu et al. [9] introduced the Weighted Path Convolutional Neural Network (WP-CNN), which applies a set of weighted paths learned through backpropagation to improve learning dynamics and diversity of features. With results showing 90.87% accuracy, 86% sensitivity, and 95% specificity, WP-CNN exhibited performance that was better than in previous approaches, especially discriminative feature extraction and redundancy reduction. Padmanayana et al. [11] put forward a DR classification model in 2022 that includes various pre- processing operations cropping, grayscale, and Gaussian blurring and CLAHE for contrast enhancement. Using a self-designed CNN model trained using the APTOS 2019 Blindness Detection dataset [1], they obtained a validation accu- racy of 94.6%, sensitivity of 86%, and specificity of 96%. They compared their findings to findings from other studies, for example, Esfahani et al. [4], Jiang et al. [7], Liu et al. [9], and Xu and Zago. Das et al. [3] conducted a comprehensive comparison of 26 CNN models on Kaggle's retinal image database for DR severity detection and quantification in 2023. EfficientNetB4 achieved the highest among the tested models with 99.37% training accuracy and 79.11% validation accuracy. DenseNet201 had a slightly higher training accuracy of 99.58%, but its validation accuracy was comparatively lower at 76.80%. Their report indicated the encouraging performance of models like EfficientNetB4, InceptionResNetV2, NasNetLarge, and DenseNet169. In accordance with this development, Chowdhury et al. [2] designed a series of CNN models optimized for binary DR classification using the APTOS 2019 dataset. The preprocessing methods employed included normalization, resizing images, rescaling, and augmentation. The highest performance among their CNN models was 96% on validation data and 95% on the test dataset, thus showing consistency in different scenarios. They also emphasized the role of hyperparameter tuning and architectural adaptation in achieving reliable outcomes. Most recent of all, Ikram et al. [5] proposed ResViT FusionNet, a fusion network that blends ResNet50 to learn local features and Vision Transformers (ViTs) to learn global dependencies. Multiclass DR diagnosis model had accuracy of 93.01%, indicating the viability of the fusion of CNN and transformer models for long feature representation.

3 Methodology

3.1 Dataset Description

APTOS-2019 Dataset: The APTOS 2019 dataset [1], made available in a Kaggle competition by the Asia Pacific Tele-Ophthalmology Society, is one of the most used datasets for diabetic retinopathy research. The dataset is comprised of 3,662 high-resolution fundus images, which have each been annotated based on five grades of severity of DR. For the sake of this research, these classes were restructured into a binary form by grouping all DR-positive classes into a single class and contrasting them with the non-DR (NDR) class in order to simplify the process of classification [1].

Eye Disease Image Dataset: This dataset was created by Rashid et al. in 2024 [12]. It contains a large number of retinal and eye images to facilitate research on the detection and classification of different eye diseases. It consists of 5,335 images divided into ten classes, i.e., Diabetic Retinopathy (DR), Healthy, Retinitis Pigmentosa, Glaucoma, and Macular Scar. In order to rectify class imbalance and increase data diversity, the authors implemented rotation, flipping, and zooming techniques to create 16,242 new images [12]. The study focuses on binary classification, i.e., the separation of DR and non-DR (NDR) instances, which comprise healthy subjects and those with other eye diseases [12]. For more details see [13].

See Fig. 1. That shows images illustrating examples of DR and NDR cases from the APTOS 2019 and Eye Disease image datasets.

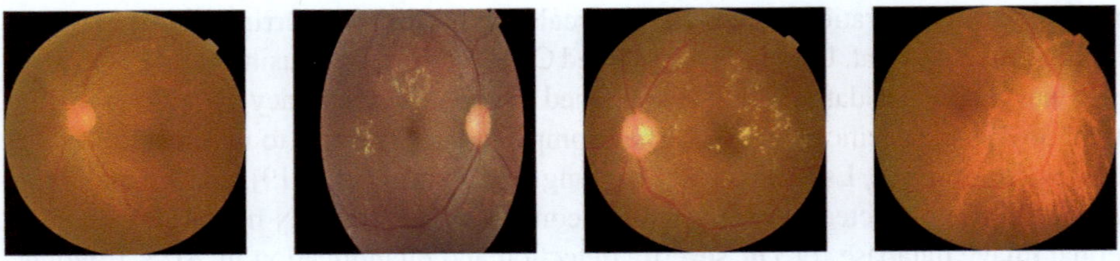

(a) APTOS 2019 DR (b) APTOS 2019 NDR (c) Eye Disease Dataset DR (d) Eye Disease Dataset NDR

Fig. 1. Sample images illustrating DR and NDR from APTOS 2019 and Eye Disease Image Datasets [1, 12].

3.2 Retinal Image Preprocessing

The two for mentioned datasets were prepared for tasks in which cases are classified as DR or NDR, see Fig. 1. All images were resized to a uniform 224 × 224 pixels. To enhance the visibility of key retinal features, preprocessing involved the application of Gaussian blur to minimize noise, CLAHE to enhance local contrast, and unsharp masking for feature sharpening in images. For more de- tails see Table 1.

The datasets were then split into training (80%), validation (10%) and testing (10%) datasets. Image pixel values were normalized within the range [0, 1] to achieve optimal model performance. To address class imbalance, we applied the Synthetic Minority Oversampling Technique (SMOTE) to each subset of the datasets individually, balancing the classes and enhancing the robustness of the deep learning models. Since SMOTE typically operates on 1D vectors, we first flattened the images into 1D arrays. After generating synthetic samples, we reshaped the resampled data back into the original image dimensions 224 × 224 × 3.

Table 1. Parameter settings for image preprocessing techniques.

Technique	Parameter	Value
CLAHE	Clip Limit	2.0
CLAHE	Tile Grid Size	(8, 8)
Gaussian Blur	Kernel Size	(5, 5)
Gaussian Blur	SigmaX	0
Unsharp Masking	Radius (via Gaussian)	(5, 5)
Unsharp Masking	Amount	1.5 (image), -0.5 (blur)

3.3 Model Architecture

The model adopts a transfer learning strategy on the Xception framework, pre-trained on ImageNet weights. The input data are first processed in the Input layer with the shape (224, 224, 3) for generic image processing tasks. The Xception model is used without its top classification layers and is frozen to take advantage of its learned feature extraction ability in an efficient manner. The output of Xception is passed through a GlobalAveragePooling2D layer. This reduces the size of the feature maps and maintains important general information. There is a Dropout layer with an adjustable dropout rate. This avoids overfitting by dropping out some of the neurons at random during training. Lastly, a Dense layer with softmax activation is used to classify the data with probabilities for the two target classes. Optimization is carried out through hyperparameter tuning using GridSearchCV, where the process iterates over multiple combinations of optimizer types, learning rates, dropout rates, batch sizes, and epochs. Specifically, the Adam optimizer is used as it can dynamically adjust the learning rate. The hyperparameter tuning carefully takes into account values such as learning rates [0.0001, 0.001, 0.01], dropout rates [0.1, 0.3, 0.5], batch sizes [16, 32, 50], and epochs [20, 30, 40] to decide on the best configuration for the dataset, see Fig. 2.

The tested hyperparameter ranges were selected based on existing literature, particularly studies using similar ophthalmology datasets (Aptos-2019, ye Disease Image). The best parameters found using GridSearchCV are then used to train the model completely so that it performs well and properly in classification issues. See Table 2 for the optimal hyperparameters identified by GridSearchCV.

Table 2. Optimal hyperparameters identified by GridSearchCV

Optimizer	Learning Rate	Dropout Rate	Batch Size	Epochs
Adam	0.001	0.1	32	40

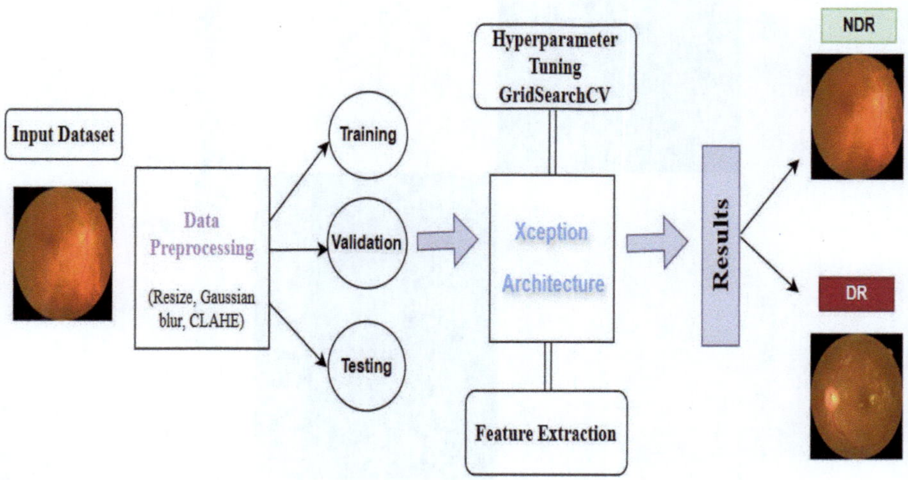

Fig. 2. Model Architecture

4 Results and Discussion

In this research, the effectiveness of the model was evaluated using two different datasets: the APTOS-2019 dataset, which had 378 images, and the Eye Disease Image Dataset, which included 664 images. The overall evaluation showed excellent and consistent performance for a variety of eye conditions.

For the APTOS-2019 dataset, the model operated at a success rate of 94.44% and precision, recall, and F1-score were well balanced at 94.48%, 94.44%, and 94.44%, see Table 3.

Table 3. Classification Report for APTOS-2019 Dataset

Class	Precision	Recall	F1-Score	Support
NDR	0.957	0.931	0.944	189
DR	0.933	0.958	0.945	189
Accuracy	-	-	0.944	378
Macro avg	0.945	0.944	0.944	378
Weighted avg	0.945	0.944	0.944	378

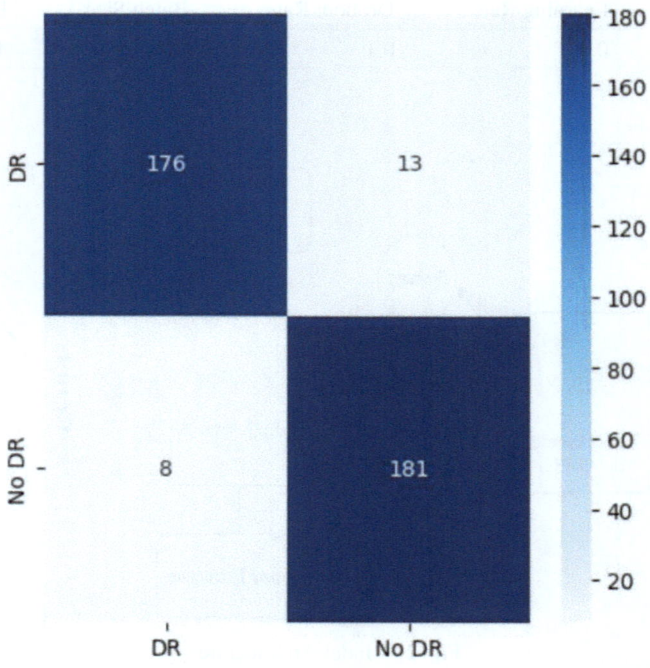

Fig. 3. Confusion Matrix: Model Performance on APTOS-2019 Test Set

The confusion matrix revealed minimal classification errors, with the model achieving a precision of 95.7% and a recall of 93.1% for non-DR (NDR) cases, and a precision of 93.3% and a recall of 95.8% for DR cases. See Fig. 3.

On the Eye Disease Image dataset, the model demonstrated an overall accuracy of 98.64%, with a very low error rate of 0.0135. The precision, recall, and F1-score were all consistently high at 98.66%, 98.64%, and 98.64%, respectively, as shown in Table 4, reflecting that the model was extremely accurate. On the Eye Disease Image dataset, the model demonstrated an overall accuracy of 98.64%, with a very low error rate of 0.0135. The precision, recall, and F1-score were all consistently high at 98.66%, 98.64%, and 98.64%, respectively, as shown in Fig. 4.

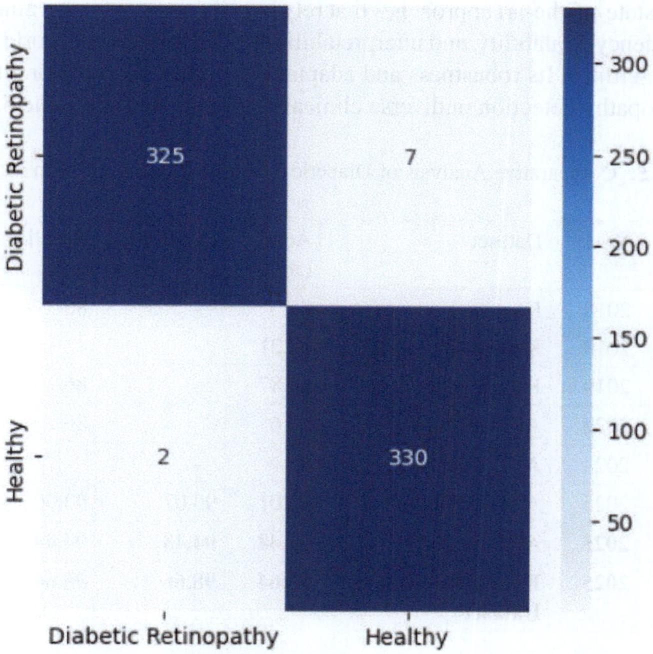

Fig. 4. Confusion Matrix: Model Performance on Eye Disease Image Test Set

Table 4. Classification Report for Eye Disease Image Dataset

Class	Precision	Recall	F1-Score	Support
Healthy	0.994	0.979	0.986	332
Diabetic Retinopathy	0.979	0.994	0.987	332
Accuracy	-	-	0.986	664
Macro avg	0.987	0.986	0.986	664
Weighted avg	0.987	0.986	0.986	664

Our model outperforms existing models of classification of diabetic retinopathy by achieving higher accuracy (98.64% on the Eye Disease Image Dataset) than previous models such as Ikram et al. (93.01%) and Padmanayana et al. (94.6%). This improvement is attributed to the use of the Xception architecture, which offers superior feature extraction compared to ResNet and DenseNet. Despite the article [2] achieving 95% accuracy, its performance is inadequate for other parameters. Additionally, our model leverages advanced hyperparameter tuning via GridSearchCV to optimize learning rates, dropout rates, batch sizes, and epochs, ensuring better generalization. The inclusion of Dropout layers prevents overfitting, making our model more adaptable across diverse datasets. Unlike many state-of-the-art approaches that rely heavily on transfer learning, our model balances efficiency, scalability, and interpretability key factors for real world deployment in healthcare settings. Its robustness and adaptability make it a superior alternative for diabetic retinopathy detection in diverse clinical environments, see Table 5.

Table 5. Comparative Analysis of Diabetic Retinopathy Classification Models

Study	Year	Dataset	Acc (%)	Precision (%)	Recall (%)	F1-score (%)
[4]	2018	Kaggle	85	-	86	-
[7]	2019	Kaggle	88.21	-	-	-
[9]	2019	Kaggle	90.87	-	86	90.87
[11]	2022	APTOS-2019	94.6	-	86	-
[2]	2024	APTOS-2019	95	-	-	-
[5]	2025	APTOS-2019	93.01	93.07	93.00	92.75
Our Model	**2025**	**APTOS-2019**	**94.44**	**94.48**	**94.44**	**94.44**
Our Model	**2025**	**Eye Disease Image Dataset**	**98.64**	**98.66**	**98.64**	**98.64**

The current study has some limitations. First, the use of a binary classification framework, while effective for distinguishing between DR and non-DR cases, does not capture the full clinical spectrum of DR, which ranges from mild to severe stages. This limits the model's usefulness in scenarios where precise staging is essential for treatment planning. Moreover, although regularization techniques such as dropout were applied, there remains a potential risk of overfitting, particularly if the model encounters images from underrepresented populations or imaging conditions not reflected in the training datasets. These aspects highlight the importance of further validation and careful consideration when generalizing the model to broader clinical use.

5 Conclusion and Future Works

In this research study, we deployed an automatic Diabetic Retinopathy (DR) diagnosis model from fundus images. The model was also performing well with a 94.44% rate of success in the APTOS-2019 dataset and a 98.64% rate of success in the Eye Disease Image dataset. The results prove the accuracy and correctness of the model in classifying DR and non-DR states. Future work will focus on refining the model to distinguish between the different stages of DR, from mild to proliferative. This will involve developing a multi-class classification model using the APTOS 2019 dataset, optimizing for stage specific features, and applying advanced preprocessing techniques. Deep learning approaches and explainable AI will also be integrated to improve accuracy and clinical interpretability. Additionally, validation against longitudinal data will be performed to assess the model's ability to track DR progression over time.

The clinical implications of this work are significant. Accurate early detection of DR can enable timely interventions like laser therapy, reducing vision loss risk. By identifying various DR stages, the model will help personalize treatment plans and enhance large-scale screening programs, especially in resource limited settings, improving public health outcomes by preventing blindness in diabetic patients.

References

1. Asia Pacific Tele-Ophthalmology Society (APTOS): Aptos 2019 blindness detection dataset (2019). https://www.kaggle.com/competitions/aptos2019-blindness-detection. Accessed 23 Jan 2025
2. Chowdhury, A., Mahfuz, S.: Diabetic retinopathy detection from fundus images using deep convolutional neural networks. In: Proceedings of the 2024 IEEE International Conference on Bioinformatics and Biomedicine (BIBM), pp. 6362–6369 (2024). https://api.semanticscholar.org/CorpusID:275439558
3. Das, D., Biswas, S.K., Bandyopadhyay, S.: Detection of diabetic retinopathy using convolutional neural networks for feature extraction and classification (DRFEC). Multimedia Tools Appl., 1–59 (2022). https://api.semanticscholar.org/CorpusID:254121766
4. Esfahani, M.T., Ghaderi, M., Kafiyeh, R.: Classification of diabetic and normal fundus images using new deep learning method. Leonardo Electron. J. Pract. Technol. **17**(32), 233–248 (2018)
5. Ikram, A., Imran, A.: Resvit fusionnet model: an explainable AI-driven approach for automated grading of diabetic retinopathy in retinal images. Comput. Biol. Med. 186, 109656 (2025). https://api.semanticscholar.org/CorpusID:275605256
6. Jagan Mohan, N., Murugan, R., Goel, T.: Deep learning for diabetic retinopathy detection: challenges and opportunities. In: Tripathy, B.K., Lingras, P., Kar, A.K., Chowdhary, C.L. (eds.) Next Generation Healthcare Informatics. Studies in Computational Intelligence, vol. 1039, pp. 213–232. Springer, Singapore (2022). https://doi.org/10.1007/978-981-19-2416-3_12
7. Jiang, H., Yang, K., Gao, M., Zhang, D., Ma, H., Qian, W.: An interpretable ensemble deep learning model for diabetic retinopathy disease classification. In: Proceedings of the 2019 41st Annual International Conference of the IEEE Engineering in Medicine and Biology Society (EMBC), pp. 2045–2048 (2019). https://api.semanticscholar.org/CorpusID:204229502
8. Kropp, M., et al.: Diabetic retinopathy as the leading cause of blindness and early predictor of cascading complications—risks and mitigation. EPMA J. **14**(1), 21–42 (2023)

9. Liu, Y.P., Li, Z., Xu, C., Li, J., Liang, R.: Referable diabetic retinopathy identification from eye fundus images with weighted path for convolutional neural network. Artif. Intell. Med. 99, 101694 (2019). https://doi.org/10.1016/j.artmed.2019.07.002, https://www.sciencedirect.com/science/article/pii/S0933365718307747
10. National Eye Institute: Diabetic retinopathy, updated (2022). https://www.nei.nih.gov/learn-about-eye-health/eyeconditions-and-diseases/diabetic-retinopathy. Accessed 16 May 2022
11. Padmanayana, Dr. Anoop, B.K.: Binary classification of DR-diabetic retinopathy using CNN with fundus colour images. Mater. Today Proc. (2022). https://api.semanticscholar.org/CorpusID:247094928
12. Rashid, M.R., Sharmin, S., Khatun, T., Hasan, M.Z., Uddin, M.S.: Eye disease image dataset (2024). https://doi.org/10.17632/s9bfhswzjb.1, https://data.mendeley.com/datasets/s9bfhswzjb/1
13. Sharmin, S., Rashid, M.R., Khatun, T., Hasan, M.Z., Uddin, M.S., Marzia: A dataset of color fundus images for the detection and classification of eye diseases. Data Brief **57** (2024). https://doi.org/10.1016/j.dib.2024.110979, cited by: 1; All Open Access, Gold Open Access
14. Tsiknakis, N., Theodoropoulos, D., Manikis, G., Ktistakis, E., Boutsora, O., Berto, A., Scarpa, F., Scarpa, A., Fotiadis, D.I., Marias, K.: Deep learning for diabetic retinopathy detection and classification based on fundus images: a review. Comput. Biol. Med. **135**, 104599 (2021). https://doi.org/10.1016/j.compbiomed.2021.104599, https://www.sciencedirect.com/science/article/pii/S0010482521003930
15. World Health Organization: Strengthening diagnosis and treatment of diabetic retinopathy in the sea region (2021). https://www.who.int/publications/i/item/9789290227946. Accessed 23 Jan 2025

Generative AI for Personalized E-Learning: A Conversational and Adaptive Framework for Language Acquisition

Mohamed El Ghali(✉), Issam Atouf, Kamal El Guemmat, Said Broumi, and Mohamed Talea

LTI Lab, Faculty of Sciences Ben M'sik, Hassan II University Casablanca, Casablanca, Morocco
m.elghali@flbenmsik.ma

Abstract. The integration of generative AI into e-learning is reshaping language education by providing personalized, adaptive, and interactive learning experiences. This study proposes a modular AI driven architecture for English language acquisition, incorporating automated content generation, interactive conversational agents, and real-time feedback mechanisms. By leveraging advanced natural language processing (NLP) models, the system dynamically adjusts learning materials to match individual learner needs, ensuring a context-aware and continuously evolving educational experience. The proposed implementation follows a structured six-phase development model, utilizing transformer-based NLP models to generate customized exercises and facilitate real-time interactions with an AI powered chatbot. The system's automated feedback mechanism enables instant correction and progress tracking, encouraging learners to refine their grammar, pronunciation, and fluency through practical engagement. While the system is still under development, preliminary observations suggest its potential to enhance learner engagement and provide a more interactive learning experience compared to traditional e-learning platforms. Future work will focus on completing system evaluation, refining AI generated content, and integrating additional personalization techniques. Key challenges such as AI model interpretability, learner engagement, and ethical considerations must also be addressed. By incorporating emotion-aware AI and multimodal learning approaches, this research aims further to improve the adaptability and effectiveness of AI powered language education.

Keywords: Generative AI · e-learning · language acquisition · adaptive learning · natural language processing (NLP) · conversational AI · personalized education · chatbot · real-time feedback

1 Introduction

The rapid advancement of Artificial Intelligence (AI) has significantly transformed various sectors, including education. Among the different branches of AI, Generative AI has emerged as a powerful tool in creating dynamic, personalized, and interactive learning experiences. This paradigm shift is particularly impactful in language education, where

B.-E. Boudriki Semlali et al. (Eds.): DATA 2025, LNISO 81, pp. 427–438, 2025.
https://doi.org/10.1007/978-3-032-06725-8_36

adaptive learning systems can cater to the unique needs of individual learners. Traditional e-learning platforms often rely on static course materials, limiting their capacity to address the diverse learning trajectories of users. In contrast, AI-driven systems can generate customized content and provide real-time feedback, offering a more engaging and effective learning process [1].

Recent breakthroughs in Natural Language Processing (NLP), particularly with the advent of Transformer-based models such as BERT and GPT, have enabled machines to generate human-like text and understand complex language patterns [2, 3]. These models have opened new possibilities for intelligent tutoring systems, making it possible to create personalized educational content and facilitate real-time conversational practice. Research has shown that adaptive e-learning systems leveraging NLP can significantly improve language proficiency and learner engagement compared to traditional methods [4]. However, the integration of generative AI in e-learning is still in its early stages, with many questions regarding its pedagogical effectiveness, scalability, and ethical implications remaining unanswered [5].

The primary motivation behind this study is to explore how generative AI can be harnessed to enhance English language learning. The proposed system leverages AI-powered content generation, interactive conversational agents, and automated feedback mechanisms to create a more immersive and personalized learning experience. Unlike conventional e-learning systems, which often present pre-defined content, this approach uses AI to generate exercises and dialogues dynamically based on the learner's performance and preferences. Such an adaptive system holds the potential to foster active participation and long-term retention among language learners [6].

While existing studies have demonstrated the potential of AI in education, most implementations focus on passive learning techniques such as automated grading and content recommendation. Few systems actively engage learners in practical language production through interactive dialogue and real-time correction. By incorporating a conversational AI component, the proposed system aims to bridge this gap, enabling learners to practice speaking and writing in a simulated but realistic environment. This approach aligns with contemporary research advocating for active, learner-centered educational paradigms [7].

Furthermore, the system introduces a continuous feedback mechanism, which not only corrects grammatical mistakes but also provides contextual explanations to enhance understanding. This feature distinguishes the proposed system from existing AI-driven educational tools, which often offer feedback without detailed explanations. Research indicates that providing explanatory feedback significantly improves language acquisition outcomes by reinforcing correct language patterns and promoting metacognitive awareness among learners [8].

This paper is organized as follows: Sect. 2 reviews the related literature on AI applications in e-learning. Section 3 presents the methodology and system architecture. Section 3.3 describes the implementation of the AI components, including content generation, chatbot interactions, and feedback mechanisms. Section 3.4 discusses the preliminary observations, challenges, and limitations of the system. Finally, Sect. 6 outlines the conclusions and future research directions, emphasizing the potential of emotion-aware AI and multimodal learning approaches in enhancing language education.

2 Background and Related Works

2.1 Background and Related Works

The integration of artificial intelligence (AI) in e-learning has undergone significant advancements over the past two decades, driven by improvements in natural language processing (NLP), deep learning, and generative AI models. Early AI-based educational systems primarily focused on rule-based tutoring and expert systems, which provided learners with predefined responses based on structured datasets [9]. However, these systems lacked adaptability, failing to personalize learning experiences to individual student needs.

The emergence of machine learning (ML) and deep learning (DL) models marked a paradigm shift in AI-driven education. With the introduction of recurrent neural networks (RNNs) and later transformer-based architectures, AI systems became capable of understanding complex linguistic patterns, enabling automated essay scoring, intelligent tutoring systems, and adaptive learning platforms [10]. More recently, large language models (LLMs) such as GPT and BERT have further revolutionized e-learning by facilitating context-aware content generation, conversational AI, and personalized assessments [2]. These models can generate human-like responses, adapt to diverse learner profiles, and provide real-time feedback, significantly improving the efficacy of language learning applications.

Despite these advancements, challenges persist, particularly regarding AI-generated content accuracy, pedagogical alignment, and learner engagement. While generative AI can produce high-quality educational materials, concerns remain regarding its potential biases, lack of explainability, and reliance on large-scale training data. Addressing these limitations requires a multidisciplinary approach, incorporating insights from linguistics, cognitive science, and educational psychology to optimize AI's role in learning environments [11] (Fig. 1).

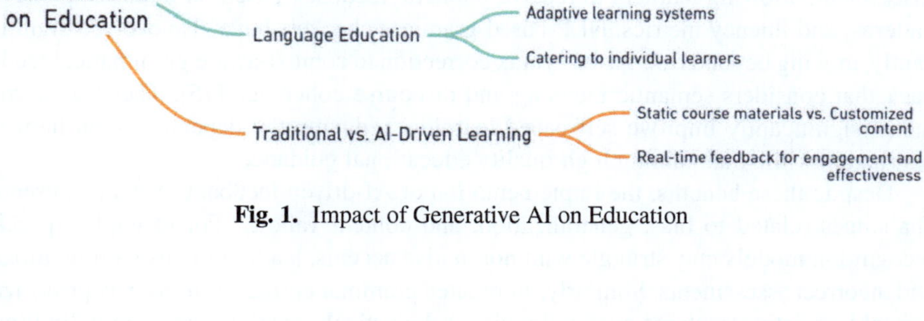

Fig. 1. Impact of Generative AI on Education

2.2 Conversational AI and Personalized Learning

Conversational AI has emerged as a key component in AI-driven e-learning, offering interactive and immersive experiences that mimic human tutoring. Early chatbot systems, such as ELIZA and ALICE, relied on pattern-matching techniques to simulate basic dialogue interactions, but lacked the ability to understand context or adapt to user inputs dynamically [12]. The introduction of neural conversational models has significantly improved chatbot capabilities, enabling contextualized, multi-turn conversations that facilitate language practice, grammar correction, and pronunciation assessment.

Recent studies have shown that AI-driven conversational agents enhance learner motivation and engagement, particularly in self-paced learning environments [13]. Unlike traditional e-learning platforms, which often rely on static exercises and video-based instruction, AI-powered chatbots provide real-time interaction, allowing learners to practice language skills in a natural, dynamic setting. Furthermore, integrating sentiment analysis and emotion-aware AI into chatbots enables systems to adapt their tone and responses based on learner frustration or confidence levels, creating a more human-like learning experience.

However, conversational AI in education still faces several limitations. While models like ChatGPT and Claude exhibit remarkable fluency, they occasionally generate linguistically incorrect or contextually inappropriate responses. Moreover, the lack of explicit pedagogical control mechanisms raises concerns about content validity and alignment with language learning curricula. Future research should focus on developing explainable AI frameworks that allow educators to monitor and refine chatbot-generated content to ensure pedagogical accuracy [14].

2.3 Real Time Feedback and Automated Assessment

One of the most impactful applications of AI in e-learning is automated feedback and assessment, which has transformed the way students receive corrective guidance on grammar, syntax, and pronunciation. Traditional online language learning platforms often rely on multiple-choice exercises and pre-recorded content, which lack immediate and adaptive feedback. AI-driven assessment tools, on the other hand, can analyze spoken and written responses in real time, providing instant corrections and personalized improvement suggestions.

State-of-the-art speech recognition models now enable automated pronunciation assessment, allowing learners to receive phonetic feedback based on intonation, stress patterns, and fluency metrics. NLP-based grammar checkers have also evolved significantly, moving beyond rule-based syntax correction to context-aware grammatical feedback that considers semantic meaning and discourse coherence [15]. These advancements significantly improve self-paced learning, reducing the dependency on human instructors while maintaining high-quality educational guidance.

Despite these benefits, the implementation of AI-driven feedback systems presents challenges related to bias, generalization, and content validity. For example, speech recognition models may struggle with non-native accents, leading to misinterpretations and incorrect assessments. Similarly, automated grammar correction tools may prioritize formal linguistic structures over colloquial or dialectical variations, potentially limiting learners' exposure to real-world language use. Addressing these issues requires ongoing refinement of AI models through diverse and inclusive training datasets, ensuring that feedback systems are fair, contextually appropriate, and pedagogically sound [16].

2.4 Summary and Research Gap

The advancements in generative AI, conversational agents, and automated feedback mechanisms have significantly improved personalized learning experiences in e-learning. However, several critical gaps remain. Firstly, while LLMs and AI chatbots demonstrate impressive fluency and contextual understanding, their pedagogical effectiveness remains an open question. Secondly, learner engagement and motivation continue to be key challenges, requiring the integration of emotion-aware AI and adaptive learning strategies. Finally, ethical considerations, such as bias in AI-generated content and data privacy concerns, must be thoroughly addressed to ensure that AI-driven education is inclusive, transparent, and effective.

This study aims to bridge these gaps by developing an AI-powered e-learning architecture specifically tailored for language acquisition, leveraging advanced NLP models, real-time feedback mechanisms, and conversational AI. By focusing on adaptive learning, human-AI interaction, and pedagogical alignment, this research contributes to the growing body of work on AI-enhanced education, paving the way for more efficient and learner-centric digital learning environments.

3 Implementation and Discussion

3.1 System Architecture and Development Phases

The implementation of AI-driven e-learning systems has gained significant attention in recent years, particularly in the domain of language acquisition. The proposed system follows a modular and adaptive architecture, ensuring flexibility and scalability to accommodate diverse learner profiles. The system consists of four primary components: an AI-powered content generator, an interactive conversational agent, an automated feedback system, and a progress tracking module. These components work in synergy to provide personalized and dynamic learning experiences, adapting in real-time to the user's linguistic proficiency and engagement levels.

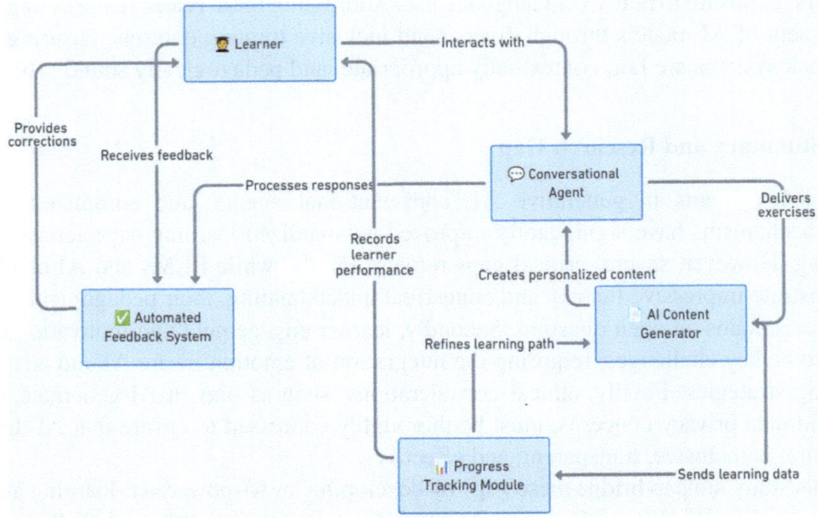

Fig. 2. AI-Powered E-Learning System Architecture

The architecture of the AI-powered e-learning system is illustrated in Fig. 2 by the interactions between the learner, the AI content generator, the conversational agent, the automated feedback system, and the progress tracking module.

The development methodology adheres to a structured six-phase model: requirement analysis, system architecture design, AI model development, interactive module implementation, testing and optimization, and deployment with continuous monitoring. This phased approach ensures systematic validation at each stage, thereby mitigating potential biases in content generation and optimizing the system's pedagogical efficacy. Table 1 provides an overview of these phases.

Table 1. Development Phases and Technologies Used

Phase	Objective	Technology Used
Requirement Analysis	Define user needs, learning goals, and AI capabilities.	User Interviews, Surveys
System Architecture	Design platform components and AI model selection.	UML Diagrams, TensorFlow
AI Model Development	Train NLP models for content generation and feedback.	GPT-4, OpenAI API, Fine-tuning
Interactive Module Development	Implement chatbots, quizzes, and gamification.	React.js, Node.js, Flask
Testing & Optimization	Evaluate AI accuracy, user engagement, and UX.	A/B Testing, User Feedback
Deployment & Monitoring	Deploy system and refine based on analytics.	AWS, Google Cloud

(*continued*)

Table 1. (*continued*)

Phase	Objective	Technology Used

This structured approach enables iterative enhancements, allowing the system to evolve based on user interaction data and continuous AI refinement.

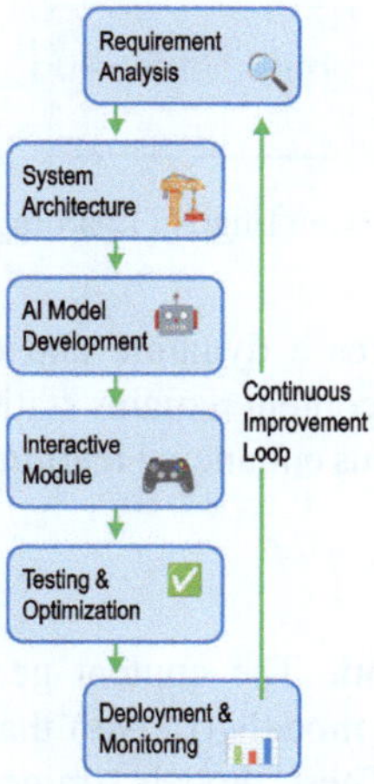

Fig. 3. Structured AI-Powered E-Learning Development Phases

Figure 3 illustrates the development phases of the AI-powered e-learning system, outlining the six structured steps from requirement analysis to deployment and monitoring.

3.2 Workflow of the AI Driven Learning Process

The workflow is designed to support adaptive and interactive learning, leveraging AI for real-time language assessment, content generation, and conversational practice. As illustrated in Fig. 4., the system follows a structured progression: learners first complete an initial assessment, after which the AI generates customized lesson plans. The system dynamically adjusts the difficulty level based on performance metrics, ensuring that each learner progresses at an optimal pace.

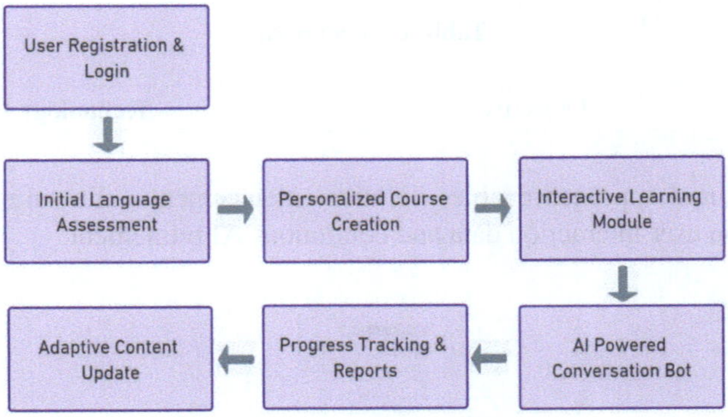

Fig. 4. AI Driven English Learning Workflow

The workflow structure ensures a dynamic and continuous learning experience. Unlike traditional systems, where content remains static, this AI driven approach adapts to each learner's evolving needs, thus enhancing learning retention and engagement [17].

3.3 Implementation Details

AI Powered Content Generation. The content generation module is powered by transformer-based large language models (LLMs) that enable real-time generation of personalized learning materials. These models, trained on extensive English corpora, generate lessons, exercises, and quizzes tailored to the learner's proficiency level. The content adaptation process involves three critical steps:

1. Data Preprocessing and Model Fine-Tuning: The AI model is trained on high-quality linguistic datasets, ensuring contextual relevance and grammatical accuracy [2].
2. Dynamic Content Personalization: The system analyzes user performance trends and dynamically adjusts lesson difficulty.
3. Error Analysis and Feedback Loop: AI tracks learner errors and suggests remedial exercises in real-time.

This approach ensures that learners receive customized content, significantly improving knowledge retention and skill acquisition compared to static course materials [18].

Interactive Learning and Chatbot System. The system integrates an AI powered chatbot that facilitates real-time conversational practice. Unlike traditional rule-based chatbots, this model employs deep learning-based natural language understanding (NLU), allowing for context-aware interactions.

The chatbot performs speech to text conversion, analyzing pronunciation accuracy and providing instant corrective feedback. Additionally, it employs context aware dialogue management, enabling users to engage in role-playing scenarios, which simulate real world communication.

The adaptive conversation model leverages reinforcement learning to improve over time, refining response accuracy and user engagement metrics. This method has been demonstrated to significantly enhance spoken language fluency in non-native speakers [19].

Real-Time Feedback and Progress Tracking. An integral feature of the system is its automated feedback mechanism, which provides immediate grammatical, syntactical, and phonetic corrections. The AI employs sequence-to-sequence models to detect and correct errors, ensuring high precision.

To optimize the learning process, users receive weekly performance reports, summarizing their progress, weaknesses, and suggested areas for improvement. This adaptive feedback loop helps learners track their growth over time while maintaining engagement.

3.4 Discussion

The implementation of generative AI in e-learning introduces a paradigm shift, offering personalized and continuously adaptive learning experiences. Unlike conventional language learning approaches, which often rely on predefined curricula, AI-driven systems can dynamically generate content, ensuring contextual relevance and sustained learner engagement [20].

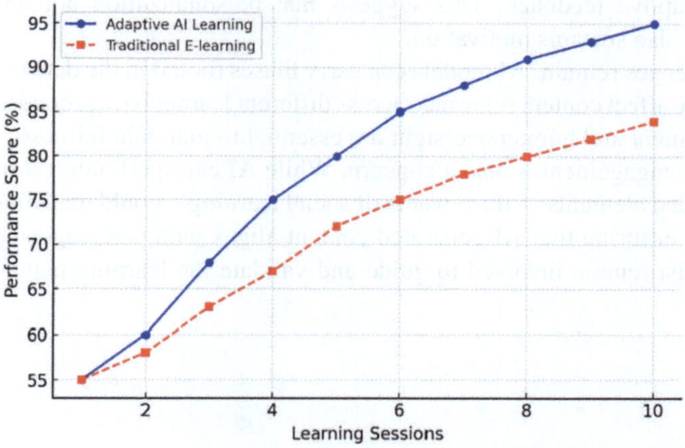

Fig. 5. Learner Progression Over Sessions

Figure 5 shows the comparison of learner progression between traditional e-learning and adaptive AI-driven learning over 10 sessions. The AI-driven approach shows a faster and more consistent improvement in performance scores.

One of the most significant advantages of this approach is its ability to scale language instruction beyond traditional classroom environments. AI powered tutors are accessible 24/7, providing immediate feedback and support without the constraints of human intervention. Additionally, the integration of speech recognition technology enables pronunciation training, addressing a critical challenge in language learning.

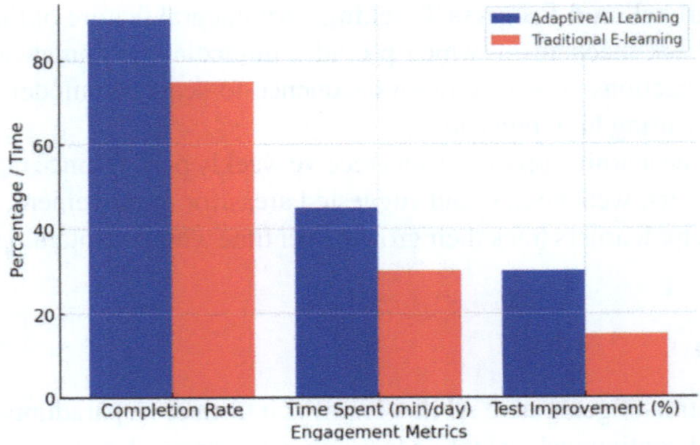

Fig. 6. Comparison of Learner Engagement

Figure 6 highlights another strength: increased engagement. Learners spent more time on the platform, completed more sessions, and improved their scores when supported by adaptive feedback. This suggests that personalization not only improves outcomes but also sustains motivation.

Yet, challenges remain. AI models can carry biases rooted in the data they're trained on, which may affect content relevance across different learner backgrounds. Continuous dataset refinement and human oversight are essential to maintain fairness.

Sustained engagement is also a concern. While AI can spark interest initially, integrating gamified elements—like rewards or social learning—could help maintain it over time. Finally, ensuring that AI-generated content aligns with pedagogical goals is key. Educators must remain involved to guide and validate the learning paths AI suggests (Fig. 7).

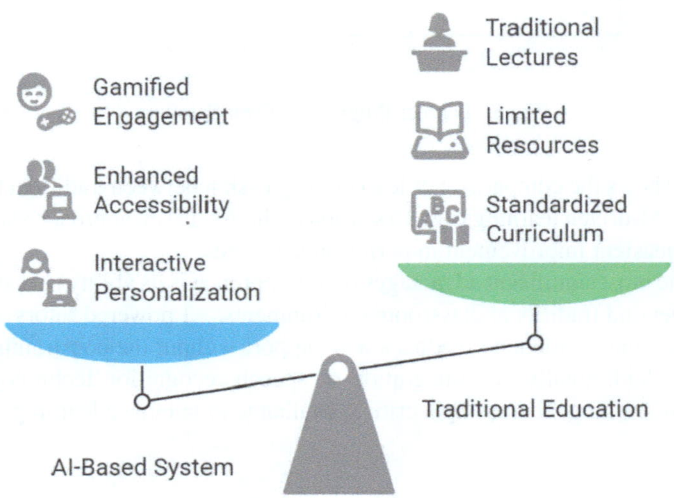

Fig. 7. Balancing AI's Potential with Educational Challenges

Overall, while still evolving, this AI-based system reflects the potential described in recent research to reshape language education by making it more interactive, personalized, and accessible.

4 Conclusion

This study explored how generative AI can support a more personalized, adaptive, and engaging approach to language learning. By combining content generation, conversational agents, and real-time feedback into a modular architecture, the proposed system offers learners a dynamic environment that responds to their progress and encourages active, meaningful interaction. Compared to traditional methods, it promotes faster skill acquisition and greater engagement, confirming the potential of AI to reframe digital education.

However, the implementation of such systems also reveals critical challenges. Questions around the pedagogical validity of AI-generated content, the risks of algorithmic bias, and the sustainability of learner motivation remain open. While the system performs well in early phases, it must be refined to ensure long-term impact, inclusivity, and alignment with educational standards.

Looking ahead, future research should focus on integrating emotion-aware AI, multimodal learning environments (e.g., AR/VR), and explainable content generation frameworks. These enhancements would not only make learning more immersive and intuitive but also strengthen the human dimension of AI-powered education—supporting learners not just as users, but as partners in shaping their own learning journey..

References

1. Brusilovsky, P., Millán, E.: User models for adaptive hypermedia and adaptive educational systems. In: Brusilovsky, P., Kobsa, A., Nejdl, W. (eds.) The Adaptive Web. Lecture Notes in Computer Science, vol. 4321, pp. 3–53. Springer, Heidelberg (2007). https://doi.org/10.1007/978-3-540-72079-9_1
2. Devlin, J., Chang, M.-W., Lee, K., Toutanova, K.: Bert: pre-training of deep bidirectional transformers for language understanding. In: Proceedings of the 2019 Conference of the North American Chapter of the Association for Computational Linguistics: Human Language Technologies, vol. 1 (Long and Short Papers), pp. 4171–4186 (2019). https://aclanthology.org/N19-1423/?utm_campaign=The%20Batch&utm_source=hs_email&utm_medium=email&_hsenc=p2ANqtz-_m9bbH_7ECE1h3lZ3D61TYg52rKpifVNjL4fvJ85uqggrXsWDBTB7YooFLJeNXHWqhvOyC. Accessed 27 Feb 2025
3. Radford, A., Narasimhan, K., Salimans, T., Sutskever, I.: Improving language understanding by generative pre-training (2018). https://www.mikecaptain.com/resources/pdf/GPT-1.pdf. Accessed 27 Feb 2025
4. Dede, C.: Immersive interfaces for engagement and learning. Science **323**(5910), 66–69 (2009). https://doi.org/10.1126/science.1167311
5. Alhusaiyan, E.: A systematic review of current trends in artificial intelligence in foreign language learning. Saudi J. Lang. Stud. **5**(1), 1–16 (2025). https://www.emerald.com/insight/content/doi/https://doi.org/10.1108/sjls-07-2024-0039/full/html. Accessed 28 Feb 2025

6. Chi, M.T.H., Wylie, R.: The ICAP framework: linking cognitive engagement to active learning outcomes. Educ. Psychol. **49**(4), 219–243 (2014). https://doi.org/10.1080/00461520.2014.965823
7. Laurillard, D.: Teaching as a Design Science: Building Pedagogical Patterns for Learning and Technology. Routledge (2013). https://www.taylorfrancis.com/books/mono/https://doi.org/10.4324/9780203125083/teaching-design-science-diana-laurillard. Accessed 27 Feb 2025
8. Taatgen, N.A., Huss, D., Dickison, D., Anderson, J.R.: The acquisition of robust and flexible cognitive skills. J. Exp. Psychol. Gen. **137**(3), 548 (2008). https://psycnet.apa.org/record/2008-10122-009. Accessed 28 Feb 2025
9. Woolf, B.P.: Building Intelligent Interactive Tutors: Student-Centered Strategies for Revolutionizing e-Learning. Morgan Kaufmann (2010). https://books.google.com/books?hl=fr&lr=&id=MnrUj3J_VuEC&oi=fnd&pg=PP1&dq=M.+Woolf,+Building+Intelligent+Interactive+Tutors:+Student-Centered+Strategies+for+Revolutionizing+E-Learning,+Morgan+Kaufmann,+2010&ots=lzvXm6huZe&sig=nOIrEvZxyvSsKjq9UyblMShtgXU. Accessed 28 Feb 2025
10. Settles, B.: Active learning literature survey (2009). https://minds.wisconsin.edu/handle/1793/60660. Accessed 28 Feb 2025
11. Liu, Y., et al.: RoBERTa: a robustly optimized BERT pretraining approach. arXiv arXiv:1907.11692, 26 July 2019. https://doi.org/10.48550/arXiv.1907.11692
12. Weizenbaum, J.: ELIZA—a computer program for the study of natural language communication between man and machine. Commun. ACM **9**(1), 36–45 (1966). https://doi.org/10.1145/365153.365168
13. Van Lehn, K.: The relative effectiveness of human tutoring, intelligent tutoring systems, and other tutoring systems. Educ. Psychol. **46**(4), 197–221 (2011). https://doi.org/10.1080/00461520.2011.611369
14. McTear, M., Callejas, Z., Griol, D.: The Conversational Interface. Springer International Publishing, Cham (2016). https://doi.org/10.1007/978-3-319-32967-3
15. Yannakoudakis, H.: Automated assessment of English-learner writing. University of Cambridge, Computer Laboratory (2013). http://www.cl.cam.ac.uk/techreports/UCAM-CL-TR-842.html. Accessed 28 Feb 2025
16. Müller-Eberstein, M., Yee, D., Yang, K., Mantena, G.V., Lea, C.: Hypernetworks for personalizing ASR to atypical speech. Trans. Assoc. Comput. Linguist. **12**, 1182–1196 (2024). https://direct.mit.edu/tacl/article/doi/https://doi.org/10.1162/tacl_a_00696/124461. Accessed 28 Feb 2025
17. Bengio, Y., Courville, A., Vincent, P.: Representation learning: a review and new perspectives. IEEE Trans. Pattern Anal. Mach. Intell. **35**(8), 1798–1828 (2013). https://ieeexplore.ieee.org/abstract/document/6472238/. Accessed 28 Feb 2025
18. Vaswani, A., et al.: Attention is all you need. Adv. Neural Inf. Process. Syst. **30** (2017). https://proceedings.neurips.cc/paper/2017/hash/3f5ee243547dee91fbd053c1c4a845aa-Abstract.html. Accessed 28 Feb 2025
19. Madotto, A., Lin, Z., Bang, Y., Fung, P.: The adapter-bot: all-in-one controllable conversational model. arXiv arXiv:2008.12579, 21 October 2020. https://doi.org/10.48550/arXiv.2008.12579
20. Li, B., Bonk, C.J., Wang, C., Kou, X.: Reconceptualizing self-directed learning in the era of generative AI: an exploratory analysis of language learning. IEEE Trans. Learn. Technol. (2024). https://ieeexplore.ieee.org/abstract/document/10496545/. Accessed 28 Feb 2025

A Secure File Sharing Using Computer Engineering Techniques

Najlaa Ghazi Thajeel[1], Hamid Jassam MOHAMMED[2](✉), Ali Abdulwahhab Mohammed[3], Saif Safaa Hussein[4], and Ali H. Abdulwahhab[5]

[1] Information Technology Division, Department of Studies and Planning, Al-Karkh University of Science, Baghdad, Iraq
[2] Computer Unit Managers, College of Remote Sensing and Geophysics, Al-Karkh University of Science, Baghdad, Iraq
hamidj.mohammed@kus.edu.iq
[3] Department of Remote Sensing, College of Remote Sensing and Geophysics, Al-Karkh University of Science, Baghdad, Iraq
[4] Department of Quality Assurance and University Performance, Al-Karkh University of Science, Baghdad, Iraq
[5] Department of Electrical and Computer Engineering, Altinbas University, Istanbul, Turkey

Abstract. This paper presents computer techniques in the field of engineering, focused on developing an efficient and reliable file-sharing application. The operation of sharing files indicates either distribution, or providing access for digital contents included software, multimedia such as audio and images, writ-ten documents, or eBooks. When working with colleagues or sharing the results of analyses, file sharing is often an essential action. With the ever-growing demand for seamless communication and information exchange, file-sharing systems are a critical component of modern digital infrastructure. The study introduces advanced algorithms for secure data transmission, optimized compression, and enhanced error-checking mechanisms to ensure data integrity and high-speed transfers. Furthermore, the application incorporates an intuitive user interface and cutting-edge encryption protocols to safeguard privacy and prevent unauthorized access. Allowing users to share files across diverse devices and networks efficiently, showing the types of file sharing, file sharing services, and file transfer, with mentioning and explaining some examples of them. We have discussed the Apache Derby database and Vigenère cipher in brief detail, which we believe will be a document in the research field on this topic as the proposed solution addresses common challenges in current file-sharing applications, including slow transfer speeds, security concerns, and limited compatibility, making it a powerful tool for both personal and professional use.

Keywords: File sharing · information sharing · Collaborate · and File protection

B.-E. Boudriki Semlali et al. (Eds.): DATA 2025, LNISO 81, pp. 439–450, 2025.
https://doi.org/10.1007/978-3-032-06725-8_37

1 Introduction

In general, the file sharing provides to the users the ability to share the data or offers a space for network that could be either public, or private with variety access right. While it is possible to share files outside a network, the phrase file sharing usually indicates the sharing of files within a network, including even small local area net-works. Through file sharing, numerous users can access the same file to read, view, modify, copy, or print it. Usually, a file sharing setup includes one or more people in charge. Individuals may possess identical or varied degrees of access rights. Sharing files might also involve keeping certain personal documents on a communal file sys-tem. For numerous years, the ability to share files has been a characteristic of widely used and multiuser computer systems. The introduction of the Internet has led to the widespread use of a system for transferring files known as File Transfer Protocol, or FTP. FTP allows users to read or write files that are shared among a group, using a password to access the shared resources on an FTP server. Numerous FTP sites pro-vide public file-sharing options, letting users view or download files by uploading them with a public password, which is often "anonymous". Most web developers use FTP to upload new or updated web files on a web server; and one may say that World Wide Web operates as an enormous collection of files [1].

2 Types of File Sharing

2.1 Peer-To-Peer File Sharing

File sharing between peers lets individuals utilize applications that link to a peer-to-peer (P2P) network. This connection helps users look for files others have shared on their computers within the network. Files that people want can be obtained by directly downloading them from other users in the network. Usually, big files are divided into smaller sections that various peers can access, which the uploader can then put back together. At the same time, the peer uploads its owned shares to different peers. Technology known as P2P has been utilized by well-known services like Napster and Spottily. It is also currently in use by Infinite and Bit Torrent, which are the leading products in the peer-to-peer network category [2].

2.2 File Synchronization and Sharing Services

Services that allow file sharing and syncing through the cloud enable individuals to set up distinct folders on their computers or mobile devices. The service ensures these folders are synchronized, making them look identical, no matter which device is used to access them. Documents stored in this folder are usually reachable via a website and a mobile application. They can also be simply shared with other users for either viewing or working together. Services like Drop box and Google Drive have gained popularity among users. File sharing can also be achieved through various methods, including distributed file systems, version control, or using mirrors for data synchronization [3].

3 File-Sharing Services

File-hosting services deliver a means of sharing files to an audience via links or email. Four among these are explicitly meant for the delivery and hosting of large file transfers in a business environment: Media Fire, Rapid Share, Share File, You SendIt; the other six-Box, Dropbox, Google Drive, Minus, Sky drive, and Sugar Sync-have more general personal-use aims with mass distribution working as a secondary feature. Therefore, the focus will be Microsoft OneDrive and Google Drive [4].

3.1 Google Drive Service

Definition
Google Drive is a platform created and managed by Google for saving and syncing files. It allows users to store documents online, share files with others, and collaborate on documents together. Google Drive includes Google Docs, Sheets, and Slides, a collection of resources that allows group collaboration on documents, spreadsheets, presentations, drawings, forms, and additional items. Documents shared publicly on Google Drive can be discovered through online search engines. [5].

Sharing Feature
Google Drive has a system for sharing files where the person who creates a file or folder automatically becomes its owner. This owner has the ability to change who can see the file or folder. Ownership can be passed on to others. Users with Google accounts can share files or folders privately by using they're @gmail.com email addresses. To share files with individuals without a Google account, you need to set the file to be available for "anyone with a link". This will create a private URL for the file that you can send through email, blogs, and more. Documents and directories may also be "available online" which means that they can be indexed by search engines, allowing anyone to discover and view them. The person in charge has the option to establish an access level to manage permissions. There are three access levels available: "can edit", "can comment" and "can view" Those who have editing rights are allowed to invite others to make edits. [6].

File Viewing Feature
Numerous external web applications, known as "apps," can be used with Google Drive. You can find these apps in the Chrome Web Store, and they work with any supported browser. To use an app, users need to go to the Chrome Web Store and install it. [7] Certain applications are owned by specific companies, like Google Docs, Sheets, and Slides. Drive applications allow users to handle online files and enable viewing, editing, and creating documents in various formats. They can also be used for image and video editing, faxing and signing documents, managing projects, making flowcharts, and additional functions. Disk applications can be assigned as the default option for managing their supported file types. A few of these apps can function without the internet, but exclusively on Google Chrome and Chrome OS. All applications from other developers can be installed at no cost. Nonetheless, a few may need payment for ongoing use or access to extra features. The majority of Drive apps are able to reach user files that are not

stored in Google Drive. When saving information from a third-party app to Google Drive, permission is needed on the first attempt. The Google Drive SDK interacts with the Google Drive interface and the Chrome Web Store to build a collection of apps that can be added to Google Drive. [8].

3.2 Microsoft OneDrive Service

Definition
OneDrive was formerly known as Sky Drive, Windows Live Sky Drive, and Windows Live Folders; it is a file storage service. This platform allows users to upload files and keep sync in the cloud storage to access it using a web browser or mobile access. This service considers as a part of a set of online tools that were once known as Windows Live. It enables users to keep their files secure, share them with friends, or make them available to anyone. Files shared publicly can be accessed without needing a Microsoft account. [2, 6].

Editing
Microsoft introduced Office Online features, which were previously called Office Web Apps, to One Drive during its "Wave 4" update. This update allowed users to upload, create, edit, and share Word, Excel, PowerPoint, and OneNote files in their web browsers. Furthermore, with Office Online, numerous users can work together on Excel files in a web browser simultaneously and collaborate on OneNote documents with other users or desktop programs. Users can also check the version history of Office documents that are stored in One Drive. [9].

Interoperability
OneDrive offers the capability for users to include their Word, Excel, and PowerPoint files within other web pages. These included documents enable visitors to these pages to engage with them, such as moving through a PowerPoint slide show or making calculations in an Excel sheet that is embedded. Moreover, Microsoft has introduced a collection of APIs for One Drive via Live Connect, which permits developers to create web services and client applications that utilize One Drive's cloud storage. Functionality allows individuals using these online services and client applications to search for, view, download, or modify the files that are saved in One Drive. [9].

4 File Transfer

Streaming refers to the process of sending files through a computer network like the Internet. Numerous methods and protocols exist for transferring files over a net-work. Devices that offer file transfer services are commonly known as file servers. [10] Based on the perspective of the customer, data transfer can be referred to as either uploading or downloading. Nowadays, businesses are more frequently using managed file transfer for file sharing. [11].

There are two types of file transfer:

- File transfers that are pulled happen when the person receiving the file starts the request.
- File transfers that are pushed occur when the individual sending the file begins the request.

File transfer can be done at different levels:

- Smooth transfers of files on network file systems.
- Simple file transfers using dedicated services like FTP or HTTP.
- File transfers that are shared over peer-to-peer networks like BitTorrent or Gnutella [12].
- Within IBM Systems Network Architecture, there are file transfer applications like Connect: Direct from IBM and XCOM Data Transport from CA Technologies that operate on a peer-to-peer basis [13].
- Instant messaging or LAN messaging can be used for file transfers.
- File transfers occur between devices and computers.
- File transfers via direct modem or serial connection (null modem), such as XMODEM, YMODEM and ZMODEM. [14].

4.1 Uploading and Downloading

Offloading is a term that describes the action of sending information from a local device to a distant system, like another server or a client so that the distant system can keep a copy of the data being sent or initiate that action. When data is moved from one remote location to a different remote location, it is known as a "remote download". This method is utilized by various online file-hosting platforms. Additionally, the remote download process is helpful when computers need to exchange information over a fast remote local area network while controlling the process through a slower modem-to-phone connection. Uploading refers to the process of transferring information from a distant system to a nearby one or starting this type of data transfer. Some examples of distant systems that may be used for downloading are a web server, an FTP server, an email server, or other comparable systems. The word download is different from the similar idea of streaming. Streaming means that data is received and used right away while it is still coming in. This type of data cannot be kept for a long time. However, in the download process, data can only be used after it has been completely received [15].

4.2 Managed File Transfer

Managed File Transfer (MFT) is a type of software or service designed to securely move data between computers through a network, such as the Internet. Companies often promote MFT solutions as a better option compared to traditional file transfer methods like FTP or HTTP. Typically, MFT provides enhanced security and oversight when compared to FTP. Its features can include reports on successful data transfers, the ability to reject transfers, auditing capabilities, a global view of transfers, automation of transfer processes, and comprehensive security monitoring [16].

4.3 HTTP Session

An HTTP session is defined as a series of network transactions exemplified by the request-response interaction. An HTTP client makes a request by establishing a Transmission Control Protocol (TCP) connection to a server on some designated port, usually 80 and sometimes 8080. This connection is monitored by the HTTP server on that port and awaits request message from the client. As soon as it receives this message, it sends back a status line like "HTTP/1. 1200 OK," along with an appropriate message. The body of this message will typically carry the requested resource but can also carry an error message or some out-of-inform [7, 8].

TP Authentication
HTTP offers various methods for authentication, including Basic Access Authentication and Digest Access Authentication. These methods operate on a challenge-response system, where the server recognizes the user and sends a challenge before it delivers the requested data. HTTP establishes a broad structure for managing access and verifying identities through a wider range of challenge-response authentication methods. These methods allow a server to question a client's request while also enabling the client to ensure the security of their authentication details.

4.4 BitTorrent

Bit Torrent is a protocol that allows users to share files directly with each other, making it possible to transfer large data files across the Internet. Bit Torrent is one of the very popular protocols used for large file transfers. In November 2004, estimated figures placed Bit Torrent at 35% of the total Internet traffic; peer-to-peer networks were around 43–70% of the Internet traffic, depending on the geographic location as of February 2009. By February 2013, Bit Torrent made up 3.35% of worldwide bandwidth more than half of the total 6% of bandwidth used for file sharing [17].

5 Apache Derby

Apache Derby is a Java-based open-source project under the Apache Software Foundation; it is a relational database management system. It is meant to be light-weight and easy to embed, making it user-friendly which explains why it is preferred by Java developers looking for a simple file-based database solution for small to medium-sized applications.

Some key features of Apache Derby include:

Java-based: It's written entirely in Java, so it's platform-independent.

Embedded: Derby can be embedded into Java applications, meaning it can run within the same process as your Java program without needing a separate server or installation [9].

SQL Support: Derby supports standard SQL (Structured Query Language), allowing you to perform typical database operations like creating tables, querying data, and managing transactions [7].

Lightweight: It's small and efficient, making it suitable for development, testing, or smaller applications where a full database server (like MySQL or PostgreSQL) would be overkill [18].

ACID-compliant: Derby supports ACID (Atomicity, Consistency, Isolation, and Durability) properties, which ensures data integrity in multi-step transactions [15].

It's often used in desktop applications, development environments, or situations where an embedded, easy-to-manage database solution is needed.

6 Vigenère Cipher

The Vigenère cipher is a method of encrypting alphabetic text using a series of different Caesar ciphers based on the letters of a keyword. This is a basic kind of polyalphabetic substitution, where multiple cipher alphabets can replace each letter of the alphabet. The main idea of the Vigenère cipher is to hide how often letters appear in plain text. It does this by overcoming basic frequency analysis, which is also a technique found in more complex encryption methods like the Advanced Encryption Standard (AES) [19].

7 Results

7.1 Implementation

I built a desktop application and created complete user controls such as (creating a user (register)), password, and login after entering all the required information in each required field and attaching a photo of the user, and the application allows all authorized users to share, download and save files in the location of their choice and send messages and edit them. Still, only the owner of the files can delete them. Unauthorized users cannot upload any files unless they have access to this application. Among the features of this application is that it is possible to write in all languages and include all the symbols installed in the user's computer operating system. It is compatible with all computer operating systems for Linux, Mac, and Windows in all its versions. It is a Secure File Sharing Application that uses AES for file encryption. The application can be enhanced with features such as user authentication, access control, and more robust error handling for a production-ready solution.

7.2 Class Diagram GUI Team Mails APP

Figure 1 shows all user interfaces in the application classes.

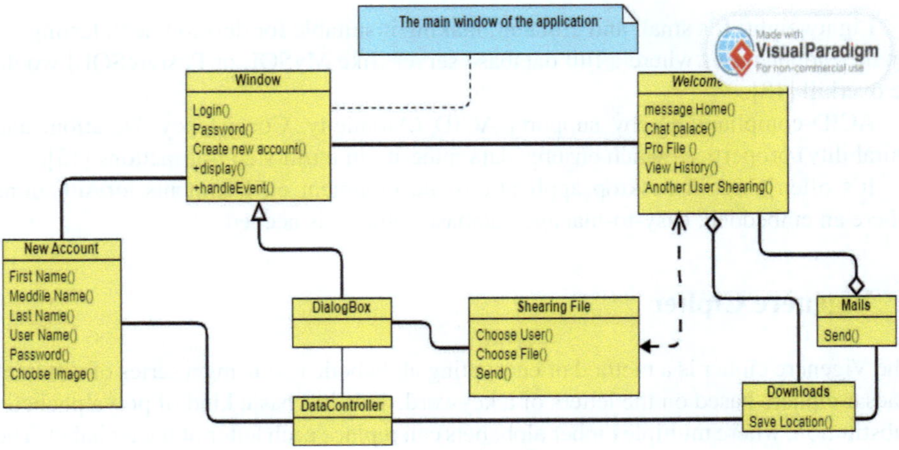

Fig. 1. Class Diagram GUI.

7.3 User Interfaces for Application

The main interface is shown in Fig. 2.

Fig. 2. Log-in Interface.

Create a new account for the user. If the user's data is not entered correctly, a message will appear below each designated field in red, as in Fig. 3. Otherwise, it will appear in blue, as in Fig. 4.

After completing all the requirements for creating an account in the application, the workspace will appear as shown in Figs. 5 and 6.

In Fig. 7, there is a welcome message for another user's account; Fig. 8 shows the workspace for him and any other user, and Fig. 9 shows the purpose of the application, where files and message sharing between users are secure and reliable, just as in Fig. 10.

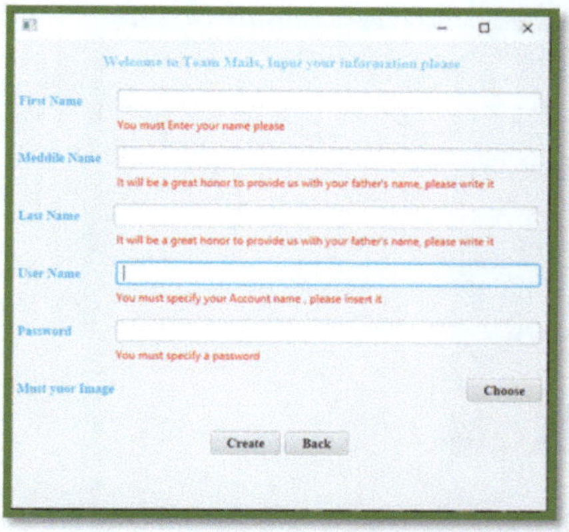

Fig. 3. The user's data is not entered correctly.

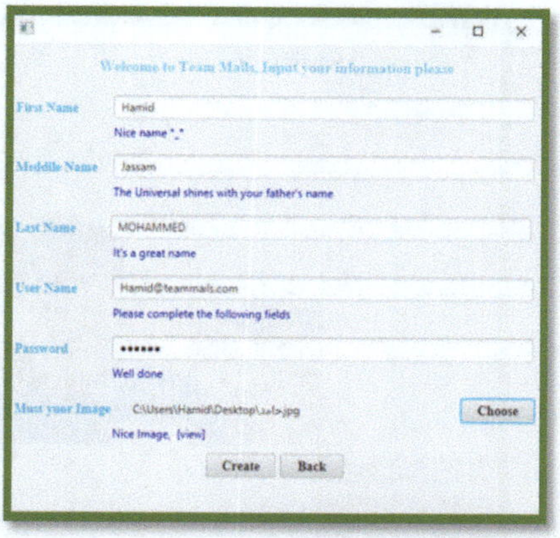

Fig. 4. The user's data is entered correctly.

Fig. 5. User enters the correct information.

Fig. 6. Welcome message for a user.

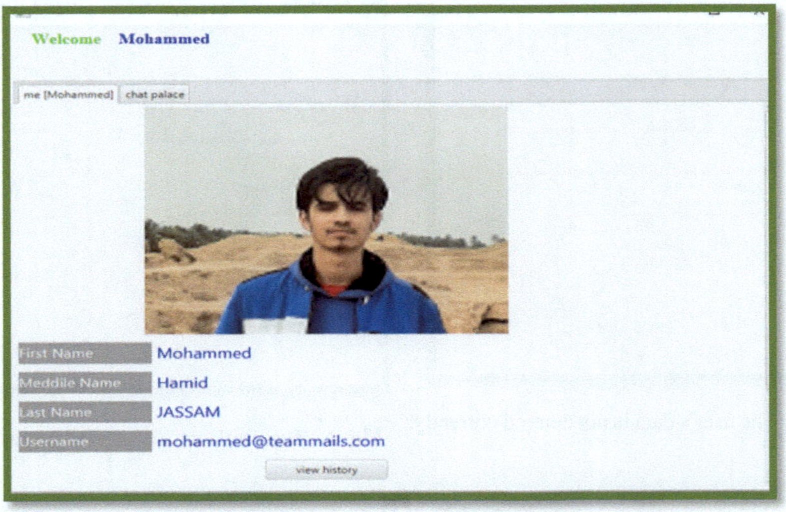

Fig. 7. Welcome message for another user's account.

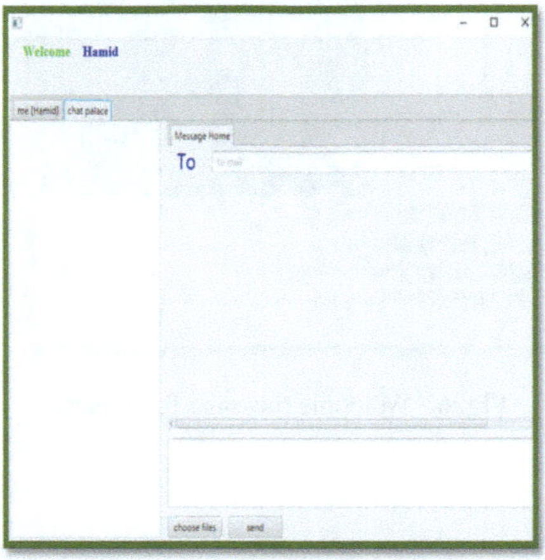

Fig. 8. Workspace.

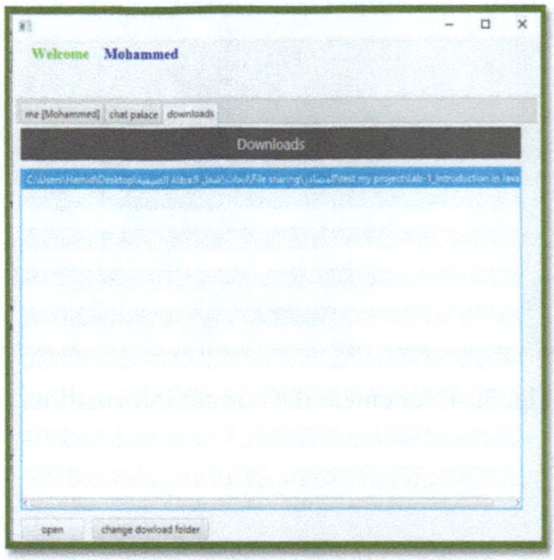

Fig. 9. The purpose of the application

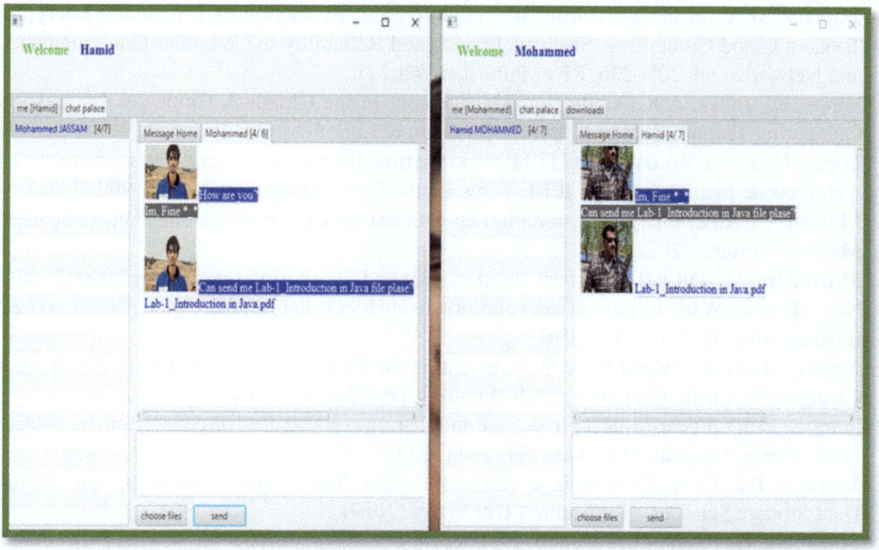

Fig. 10. Sharing secure files between them.

8 Conclusion

It was created as a distributed secure file-sharing application using JavaFX in Net Beans IDE 8.1. We created projects Server, Rules, and Faces (java client), so while the server is working then we can connect many clients, the clients send messages between them and send files, Stickers, and recorded voice messages.

References

1. Abramenka-Laçhheb, V., et al.: Instructional designers' use of informal learning: how can we all support each other in times of crisis. J. Appl. Instr. Des. **10**(3), 129–146 (2021)
2. Ahmed, N.: Collaboration in Lists. Microsoft Lists Essentials: Enhancing Productivity Through Microsoft 365's Smart Information Tracking App, pp. 271–309. Springer (2024)
3. Alotaibi, S., et al.: Comparing performance of commercial cloud storage systems: the case of dropbox and one drive. In: Proceedings of the 2019 International Conference on Computer and Information Sciences (ICCIS). IEEE (2019)
4. Bozorgi, A., et al.: UPSS: a user-centric private storage system with its applications. arXiv preprint arXiv: 2403.15884 (2024)
5. Choi, B.: Linux Fundamentals II–TCP/IP Services. Introduction to Python Network Automation Volume I-Laying the Groundwork: The Essential Skills for Growth, pp. 475–559. Springer (2024)
6. Dorobantu, O.G., et al.: The poly-alphabetic substitution ciphers-a viable solution for IoT applications? In: Proceedings of the 2022 International Symposium on Electronics and Telecommunications (ISETC). IEEE (2022)
7. Hasan, M.Z., et al.: File Sharing & Copyright Infringement in Peer-to-Peer. J. Comput. Intell. Syst. **2**(2), 73–81 (2024)

8. Hossain, M.A., et al.: Measuring Interpretation and Evaluation of Client-side Encryption Tools in Cloud Computing. Security, Privacy and Reliability in Computer Communications and Networks, pp. 205–236. River Publishers (2022)
9. Kipps, K., Jones, A.K.: Collection Management in the Cloud: A Guide for Using Cloud Computing Technologies in Libraries. Rowman & Littlefield (2022)
10. López, C., et al.: Validation of HTTP response time from network traffic as an alternative to web browser instrumentation. IEEE Trans. Netw. Serv. Manage. **19**(2), 976–990 (2021)
11. Monteiro, T.A.A.: Energy consumption on database management systems, Universidade do Minho (Portugal) (2021)
12. Muchmore, M., Dilly, J.: The best cloud storage and file-sharing services for 2020 (2020)
13. Naz, M., et al.: A secure data sharing platform using blockchain and interplanetary file system. Sustainability **11**(24), 7054 (2019)
14. Oppitz, M., et al.: World Wide Web. Inventing the Cloud Century: How Cloudiness Keeps Changing Our Life, Economy and Technology, pp. 229–266 (2018)
15. Peng, S., et al.: A peer-to-peer file storage and sharing system based on consortium blockchain. Futur. Gener. Comput. Syst. **141**, 197–204 (2023)
16. Santiago, P.J.: Cloud Data Storage Services: A Case Study Forensic Investigation of Cloud Data Storage Services. Northcentral University (2019)
17. Söderberg, E., Råhlén, J.: An analysis of decentralized peer-to-peer file sharing performance: an overview of how different parameters affect the average download time in a BitTorrent-like network (2021)
18. Springall, D., et al.: FTP: the forgotten cloud. In: Proceedings of the 2016 46th Annual IEEE/IFIP International Conference on Dependable Systems and Networks (DSN). IEEE (2016)
19. Mohammed, A.A., Anwer, H.A.: A new method encryption and decryption. Webology **18**(1) (2021)

Joint Optimization of Unrelated Parallel Machine Scheduling and Multi-Trips Vehicle Routing with Time Windows

Hind Bouifalioune[1(✉)], Majda Fikri[2], and Ghizlane Bencheikh[1]

[1] Moulay Ismaïl University, Meknes, Morocco
h.bouifalioune@edu.umi.ac.ma, g.bencheikh@umi.ac.ma
[2] Ibn Zohr University, Agadir, Morocco

Abstract. This study examines the difficulties of synchronizing production and distribution scheduling, with a particular focus on the Unrelated Parallel Machine Scheduling Problem (UPMSP) in production and the Multi-Trip Vehicle Routing Problem (VRP) in distribution. The objective is to minimize the makespan, total weighted tardiness, and transportation costs, while ensuring efficient batch deliveries using a fleet of homogeneous-capacity vehicles.

To tackle this complex problem, we formulate a Mixed-Integer Nonlinear Programming (MINLP) model and propose a Genetic Algorithm (GA) as a solution approach. Since no established benchmarks exist for this specific problem, we evaluate the performance of our GA by comparing its solutions against IBM CPLEX, a state-of-the-art exact solver. The comparison focuses on small-scale instances generated in this study. Results indicate that the proposed GA is capable of reaching optimal solutions for most small-sized instances, demonstrating its effectiveness in solving this challenging integrated scheduling problem.

Keywords: Integrated production-distribution · Unrelated parallel machine · Multi-trips vehicle routing · optimization · Genetic algorithm · CPLEX

1 Introduction

The importance of integrating production and distribution planning in supply chain management, as inefficiencies in traditional approaches (where production scheduling and vehicle routing are treated separately) can lead to high inventory costs and poor delivery schedules. In fast-paced production environments, varying customer delivery time constraints necessitate flexible scheduling, such as incorporating soft time windows to minimize penalties for early or late arrivals. Adjusting vehicle departure times based on production completion helps reduce logistics costs and maintain service quality. Batch production, used to reduce inventory backlogs, adds complexity as it requires synchronized decisions on batch formation, production sequencing, and vehicle dispatching. The study proposes an integrated approach to optimize both production scheduling (addressing the Unrelated Parallel Machine Scheduling Problem) and vehicle routing, incorporating soft time windows and multi-trip homogeneous vehicle routing. A bi-objective

B.-E. Boudriki Semlali et al. (Eds.): DATA 2025, LNISO 81, pp. 451–464, 2025.
https://doi.org/10.1007/978-3-032-06725-8_38

mixed-integer programming model is developed to minimize makespan, transportation costs, and tardiness penalties. A Genetic Algorithm is used to improve solution quality.

The paper is structured as follows: Sect. 2 reviews related literature, Sect. 3 presents the problem formulation, Sect. 4 describes the optimization framework and methodology, Sect. 5 evaluates the approach through computational experiments, and Sect. 6 concludes the study and suggests future research directions.

2 Literature Review

The coordination of production scheduling and vehicle routing is a critical component of optimizing supply chain operations. In the past, production, as well as delivery, have been considered as independent and sequential processes, with delivery starting only after production is over. Separate treatment of production and distribution tends to cause inefficiencies as well as sub-optimal utilization of resources [4]. With the growing complexity of supply chains as well as demands for competitive advantage, the need for integrating production scheduling as well as vehicle routing into a single framework has gained prominence. It can overcome problems such as inventory buffer, traditionally used for reducing delays between production as well as distribution, including extra costs as well as inefficiency [2].

The challenge of striking a balance between optimizing transportation costs as well as reducing delays in deliveries for time-sensitive products is the main challenge here. Industries involved with perishables as well as time-sensitive deliveries, for example, ready-mixed concrete, drugs, as well as newspapers, are especially suited for integrated methods. In such situations, the delayed delivery can cause substantial disruption, prompting optimized scheduling techniques that integrate production as well as distribution processes [10]. However, integrating production and distribution presents the additional challenge of minimizing the total cost while maintaining customer service levels, particularly for products requiring batch processing and initialization time [8].

Recent studies on Integrated Production and Delivery Scheduling (IPDS) have focused on both the production characteristics and delivery requirements. In terms of production, many studies address the scheduling of jobs, assuming that each order corresponds to a single job [10]. A significant gap exists in the literature when it comes to multi-product batch manufacturing, where production involves processing multiple product varieties within the same manufacturing setup. [12] explored multi-product batch manufacturing but limited their analysis to continuous production between batches, overlooking the complexities of managing multiple product types with distinct manufacturing requirements. For delivery, numerous studies have explored hard time windows, where orders must be delivered within a fixed time frame. This has been particularly emphasized in studies by [10, 20]. However, the increasing need for flexibility has shifted attention towards soft time windows, which allow for more adaptable delivery schedules. Studies by [14, 16] have developed production scheduling coordination methods for production and delivery through sequential methods, such as determining the production schedule first, followed by adjusting vehicle routing based on the production timeline. Most existing approaches treat production scheduling and vehicle routing separately, missing the chance to fully optimize both together. To address this, our study proposes a metaheuristic method that integrates the Unrelated Parallel Machine Scheduling Problem with multi-trip vehicle routing. Unlike traditional methods that rely on sequential planning or focus only on single-job production, our approach handles batch production and delivery simultaneously, offering a more practical and efficient solution.

3 Problem's Description

The Integrated Production Scheduling and Vehicle Routing problem is introduced here, integrating the Unrelated Parallel Machine Scheduling problem with the multi-trip vehicle routing problem. Customer orders are received randomly and need to be processed based on a given schedule, with each order coming with some delivery constraints. The objective of the scheduler is to allocate jobs to the parallel machines, taking into consideration the different processing time and setup time for each machine and job. The main objective is to optimize makespan, transportation costs, and tardiness penalties through efficient production and delivery scheduling. The production phase involves multiple parallel machines, with each job processed on a machine that has different processing times. Production cannot be interrupted, and each machine can handle only one job at a time. After production, customer orders must be delivered using a fleet of homogeneous vehicles with limited capacity. These vehicles can make multiple trips, ensuring that each customer is visited once, and deliveries are made within capacity constraints.

The multi-trip vehicle routing approach enhances delivery efficiency, allowing vehicles to complete multiple trips and avoid delays caused by waiting for all jobs to be completed. This flexibility helps reduce the overall completion time. The delivery phase is modeled as a multi-trip vehicle routing problem (multi-trip VRP), with fixed customer service times and all vehicles available at the start of the delivery phase. Finally, deliveries are executed according to the optimized plan, completing the production and distribution process. In this problem, each job in production is directly linked to a corresponding customer order or delivery node. The index i denotes both the job and the customer order, so $N = \{1, \ldots, n\}$ represents the sets of both jobs and customer nodes. In the production part, $M = \{1, \ldots, m\}$ represents the set of unrelated parallel machines. Let p_{ki} denote the processing time of job $i \in N$ at machine $k \in M$, and let u_i denote the demand of job $i \in N$. In the delivery part, for modeling purposes, we define node 0 as the depot from which delivery vehicles depart. The set of all customer locations, including the depot, is denoted as $N_0 = \{0, 1, \ldots, n\}$. Let $V = \{1, \ldots, \vartheta\}$ represent the set of available vehicles, and $R = \{1, \ldots, \rho\}$ denote the set of trips each vehicle can perform. Additionally, each delivery vehicle has a fixed capacity Q.

3.1 Mathematical Formulation

In this part, we formulate the mathematical model for the unified framework that integrates production scheduling and multi-trip vehicle routing. The model aims to minimize the weighted sum of makespan, transportation costs, and tardiness penalties. The notation used in this study is summarized in Table 1.

Table 1. Description of Model Variables and Parameters

Notations	Descriptions
Indices:	
i, j, h	Index of job, order, or customer node, $i, j \in N$.
v	Index of vehicles, $v \in V$.
k	Index of machines, $k \in M$.
r	Index of trips, $r \in R$.
Parameters:	
p_{ki}	Processing time of job i on machine k, $(i \in N, k \in M)$.
Z	Sufficiently large number.
u_j	Size of job j, $(j \in N)$.
s_j	Service time of job j, $(j \in N)$.
t_{ij}	Travel time between customer i to j, $(i, j \in N_0)$.
$\overline{w}_j, \underline{w}_j$	Lower and upper bound of the delivery time window of job j, $(j \in N)$.
d_{ij}	Distance between customer i to j, $(i, j \in N_0)$.
Variables:	
x_{kji}	1, if job i precedes job j on machine k.0, *otherwise*. $(i, j \in N; k \in M)$.
C_{kj}	Completion times of job j at machine $k (j \in N; k \in M)$.
D_j	Delivery time of job $j (j \in N)$.
S_{vr}	Start time of the tour of trip r at vehicle v $(v \in V; r \in R)$.
g_{jvr}	Binary variable which takes the value 1 if job j is delivered on the tour of trip r of vehicle $v (j \in N; v \in V; r \in R)$.
z_{ijvr}	Binary variable which takes the value 1 if job i is delivered before job j on the tour of trip r at vehicle $v (i, j \in N_0; v \in V; r \in R)$.
σ_{rv}	Binary variable which takes the value 1 if trip r is traveled by delivery vehicle $v (r \in R; v \in V)$.
ω_{rqv}	Binary variable which takes the value 1 if trip r is traveled before trip q by delivery vehicle $v (r, q \in R; v \in V)$.

IPSVR is formulated into a mixed integer linear programming (MILP) model as follows:

$$\min\left(\alpha.C_{max} + \beta.\sum_{v \in V} time_v + \gamma \sum_{i \in J} T_i\right) \tag{1}$$

- $time_v$: total travel time of vehicle v

$$time_v = \sum_{r \in R} \sum_{i \in N \cup \{0\}} \sum_{j \in N \cup \{0\}} t_{ij} z_{ijvr} \tag{2}$$

- C_{max}: Makespan, or maximum completion time.

$$C_{max} = max_{i \in N, k \in M}(C_{ki}) \tag{3}$$

- T_j: Tardiness of job$j$.

$$T_i = max(0, D_i - \overline{w}_i) \tag{4}$$

$$\alpha, \beta, \gamma \epsilon]0,1] et \alpha + \beta + \gamma = 1 \tag{5}$$

Subject to.

Unrelated Parallel Machine Scheduling Problem constraints:

$$\sum_{k \in M} \sum_{\substack{j \in \{0\} \cup N \\ j \neq i}} x_{kji} = 1, \forall i \in N \tag{6}$$

$$\sum_{k \in M} \sum_{\substack{i \in N \\ j \neq i}} x_{kji} \leq 1, \forall j \in N \tag{7}$$

$$\sum_{i \in N} x_{k0i} \leq 1, \forall k \in M \tag{8}$$

$$\sum_{\substack{h \in \{0\} \cup N \\ h \neq i, h \neq j}} x_{khj} \geq x_{kji}, \forall j, i \in N, j \neq i, \forall k \in M \tag{9}$$

$$C_{ki} + V\left(1 - x_{kji}\right) \geq C_{kj} + p_{ki}, \forall j \in \{0\} \cup N, \forall i \in N, j \neq i, \forall k \in M \tag{10}$$

$$C_{k0} = 0, \forall k \in M \tag{11}$$

$$C_{kj} \geq 0, \forall j \in N, \forall k \in M \tag{12}$$

$$C_{max} \geq C_{kj}, \forall j \in N, \forall k \in M \tag{13}$$

– Muti-Trips Vehicle Routing with time window constraints:

$$\sum_{v \in V} \sum_{r \in R} g_{jvr} = 1, \forall j \in N \tag{14}$$

$$g_{0vr} \geq g_{jvr}, \forall j \in N, v \in V, r \in R \tag{15}$$

$$Z\sum_{j\in N} g_{jvr} \geq \sum_{j\in N} g_{jvr+1}, \forall v \in V, r \in R-1 \tag{16}$$

$$g_{jvr} \geq \sum_{\substack{i \in \{0\}\cup N \\ i \neq j}} z_{ijvr}, \forall j \in \{0\}\cup N, v \in V, r \in R \tag{17}$$

$$g_{jvr} \geq \sum_{\substack{i \in \{0\}\cup N \\ i \neq j}} z_{jivr}, \forall j \in \{0\}\cup N, v \in V, r \in R \tag{18}$$

$$c_v \geq \sum_{j\in N} u_j g_{jvr}, \forall v \in V, r \in R \tag{19}$$

$$S_{vr} \geq C_{kj} + s_0 - Z\big(1 - g_{jvr}\big), \forall k \in M, j \in N, v \in V, r \in R \tag{20}$$

$$S_{vr+1} \geq D_j + s_j + t_{0j} + s_0 - Z\big(1 - g_{jvr}\big), \forall j \in N, v \in V, r \in R-1 \tag{21}$$

$$D_j \geq \underline{w}_j, \forall j \in N \tag{22}$$

$$D_j \geq S_{vr} + t_{0j} - Z\big(1 - g_{jvr}\big), \forall j \in N, v \in V, r \in R \tag{23}$$

$$D_j \geq D_i + s_i + t_{ij} - Z\big(1 - z_{ijvr}\big), \forall i, j \in N : i \neq j, v \in V, r \in R \tag{24}$$

$$\sum_{v\in V} \sigma_{rv} = 1, \forall r \in R \tag{25}$$

$$\omega_{rqv} + \omega_{qrv} \leq \frac{1}{2}(\sigma_{rv} + \sigma_{qv}) \tag{26}$$

$$S_{vq} \geq \sum_{j\in N} Djg_{jvr} - Z\big(1 - \omega_{rqv}\big), \forall v \in V, r, q \in R \tag{27}$$

$$S_{vr} \geq 0, \forall v \in R,\ \mathrm{r} \in \mathrm{R} \tag{28}$$

$$T_j \geq 0, \forall j \in N \tag{29}$$

$$T_j \geq D_j - \overline{w}_j, \forall j \in N \tag{30}$$

$$g_{jvr} \in \{0{,}1\}, z_{ijvr} \in \{0{,}1\}, x_{kji} \in \{0{,}1\} \tag{31}$$

The proposed MILP model integrates the UPMSP with Multi-Trip VRP with Time Windows, ensuring an efficient allocation of jobs to machines and vehicles while minimizing the weighted sum of makespan, total travel time, and job tardiness. The objective function (1) balances these three criteria using weight parameters (α, β, γ) to achieve a trade-off based on the problem's specific requirements. The total travel time of each vehicle is computed using constraint (2), while the makespan is determined by constraint (3), which ensures that the maximum job completion time is minimized. Additionally, the tardiness of each job is constrained by (4), enforcing that delays beyond due dates are accounted for. The weight parameters are constrained by (5), ensuring they sum to one.

The scheduling constraints ensure that each job is assigned exactly once to a machine, as enforced by constraint (6). To prevent jobs from being assigned multiple times, constraint (7) ensures that no job is assigned to more than one machine. Each job can be processed by at most one machine at a time, as stated in constraint (8). The precedence relationship among tasks is maintained through constraint (9), ensuring that a job cannot start before the completion of its predecessors. The job completion time on a machine is determined by constraint (10), which ensures that the start time of a job accounts for machine setup times. Constraint (11) defines the initial conditions for job completion times, while constraint (12) enforces non-negativity for these times. The overall makespan is constrained by (13), ensuring that it is the maximum of all job completion times.

For the vehicle routing component, constraint (14) ensures that each job is assigned exactly once to a vehicle. The relationship between consecutive jobs assigned to the same vehicle is enforced by constraint (15), ensuring that each job has a predecessor in the vehicle routing sequence. Constraints (16) and (17) define job sequencing within a vehicle's route, while constraint (18) ensures consistency with machine scheduling by linking job completion to vehicle departure times. The cost of using a vehicle for a job is determined by constraint (19). Constraints (20) and (21) ensure that vehicle routes respect machine-job assignments and transition times between jobs.

The time window constraints are enforced through constraints (22) and (23), which ensure that a job's arrival time respects its assigned time window. Constraint (24) maintains vehicle flow balance across trips, ensuring logical consistency in routing. The number of trips per vehicle is constrained by (25). The vehicle capacity and travel time relations are handled by constraints (26) and (27), ensuring that total travel and operational constraints are respected. Constraint (28) enforces non-negativity for job completion times, while constraint (29) ensures that the tardiness of a job is always non-negative. Finally, the domain constraints in (30) and (31) define the binary nature of the decision variables, ensuring that job-machine assignments and vehicle routing decisions remain logically valid.

By incorporating these constraints, the model effectively addresses the relationship between machine scheduling and vehicle routing, improving overall operations while respecting machine capacity limits, job order requirements, and vehicle availability. This approach offers a holistic solution to the problem, focusing on reducing costs, ensuring jobs are completed on time, and optimizing vehicle usage in the most efficient manner possible.

4 The Solution Approach

The genetic algorithm (GA) is widely used to solve combinatorial optimization problems due to its ability to navigate complex search spaces and find near-optimal solutions. It is especially effective for problems with nonlinear, non-continuous, or highly constrained solution spaces, such as the Vehicle Routing Problem (VRP) and the Unrelated Parallel Machine Scheduling Problem (UPMSP) [11]. GA's population-based approach allows for diverse exploration and helps avoid local optima, a common issue with optimization algorithms [13]. Its mutation and crossover operators accelerate solution improvement by mixing existing solutions and introducing new genetic material, speeding up the search for better solutions [17]. GA's iterative nature makes it ideal for multi-objective problems, such as minimizing makespan, transportation costs, and tardiness in scheduling and routing problems [21]. Additionally, GA is versatile and can be hybridized with other optimizers and customized for specific constraints, making it suitable for real-life scheduling and routing challenges, including machine and vehicle constraints in UPMSP and VRP.

4.1 Chromosome Representation

In the GA for the solution of the UPMSP and the VRP, chromosome representation is critical for encoding solutions. A Python dictionary list is employed for structuring the assignment of jobs to machines and vehicles. In UPMSP, each job is mapped to a specific machine, with the dictionary effectively capturing job assignment as well as processing time, and honoring machine constraints. In the VRP, the chromosome is extended further with spaces for vehicle assignment, where a set of customers is served by each vehicle, honoring capacity constraints and allowing for calculations of distances for optimizing transportation costs. This representation makes it easier to apply genetic operators such as crossover as well as mutation, allowing for the feasibility of new solutions. The dictionary-based chromosome, as such, provides a flexible as well as an efficient method of navigating through solutions of intricate scheduling as well as routing problems. This representation of the chromosome is presented in Fig. 1.

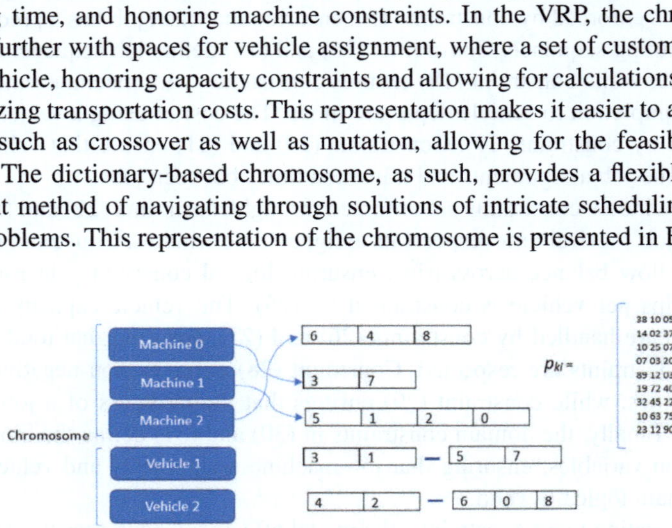

Fig. 1. Example of chromosome encoding.

4.2 Generalization of Initial Population

To ensure the diversity of individuals in the population, we use several methods to generate the individuals: randomly, while respecting the constraints, or by applying heuristics.

For example, in the case of the UPMSP, jobs are assigned to machines using dispatching heuristics such as Random Order, Shortest Processing Time (SPT), or Longest Processing Time (LPT), which provide different types of solutions based on specific characteristics. Similarly, for the VRP, jobs are randomly assigned to vehicle routes while respecting capacity constraints, with tasks distributed across multiple trips.

4.3 Crossover

In our Genetic Algorithm, we use three different crossover methods to combine parts of two parent solutions and create new ones, while making sure the results remain valid. The first method simply swaps the entire lists of machine and vehicle assignments between the parents, helping the algorithm explore new possibilities. The second method is a bit more structured: for each machine's job list, we pick a random point and take all the tasks before that point from the first parent. Then, we go through the second parent's list and add any tasks that haven't already been assigned, making sure each task appears only once. Once machine assignments are set, jobs are distributed to vehicles based on their capacity and routing limits. The third method focuses on quality—machine assignments from both parents are evaluated, and the best-performing parts are combined, again followed by a vehicle assignment that respects all constraints. At each generation, one of these three crossover methods is chosen at random (Fig. 2).

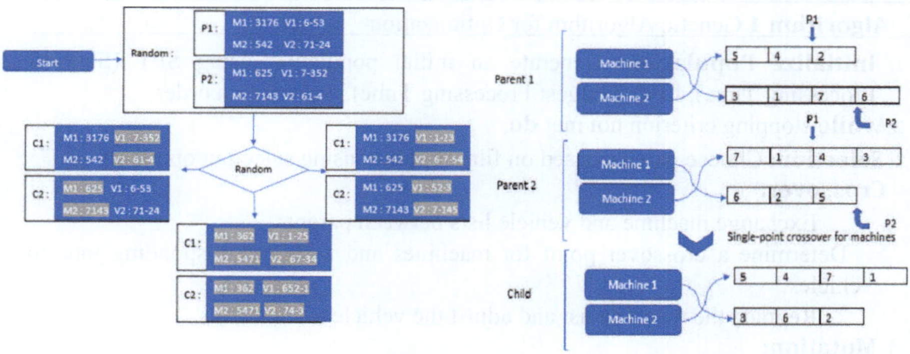

Fig. 2. The crossover procedure visualized by an example.

4.4 Mutation

The mutation step is essential for maintaining population diversity and avoiding premature convergence. Two mutation methods are used, with one randomly selected at each generation. The first, right-hand mutation, makes small refinements by modifying the machine or vehicle assignment of a randomly selected job, based on the relative number of machines and vehicles. This method improves solution quality while preserving feasibility. The second, reversion mutation, revisits previously discarded solutions, helping the algorithm escape local optima by exploring earlier configurations. The random selection of these methods in each generation introduces variability, enhancing the exploration of the solution space. As illustrated in Fig. 3, this process ensures a wide range of solutions is explored, leading to improved optimization performance.

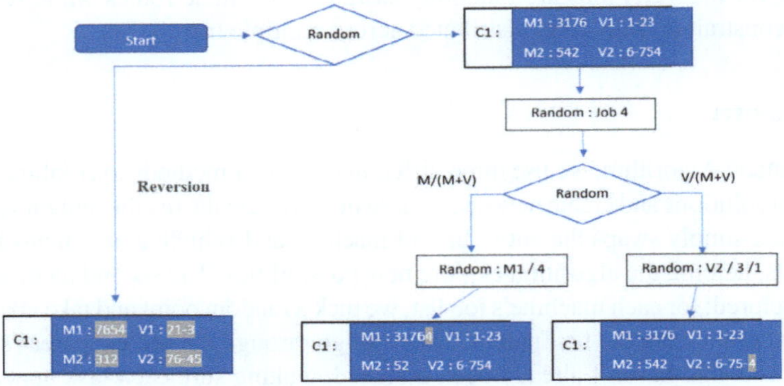

Fig. 3. The mutation procedure visualized by an example

4.5 Genetic Algorithm

The proposed genetic algorithm is described in Algorithm 1.

Algorithm 1 Genetic Algorithm for Optimization

1: **Initialize Population:** Generate an initial population using SPT (Shortest Processing Time), LPT (Longest Processing Time), and random order.
2: **while** stopping criterion not met **do**
3: **Selection:** Choose parents based on fitness values using selection operators.
4: **Crossover:**
    Exchange machine and vehicle lists between parents.
    Determine a crossover point for machines and assign corresponding jobs to vehicles.
    Regrade the machine list and adjust the vehicle assignments.
5: **Mutation:**
    Apply reversion mutation.
    Apply mutation method 2 (see Section 4.4).
6: **end while**
7: **Return:** Best solution found.

5 Experimental Results

This section presents the experimental evaluation of the proposed Genetic Algorithm (GA) on a set of benchmark instances generated randomly using a linear congruential pseudorandom number generator (LCG), originally introduced by [15]. In this phase, a total of 22 instances were generated involving up to 10 jobs, 10 customers, and 3 vehicles. The purpose of generating smaller instances is to focus on testing the basic features of the GA before applying it to more complex cases. Using these small instances allows us to closely monitor the algorithm's performance, adjust its parameters, spot any issues that may come up, and compare the results with the optimal solutions obtained by CPLEX solver.

5.1 Parameters of the Genetic Algorithm

The effectiveness of the Genetic Algorithm (GA) relies heavily on the configuration of certain critical parameters: population size, crossover probability, and mutation probability. Population size affects the solution space exploration, with a larger population reducing the risk of suboptimal solutions but potentially slowing convergence [11]. Crossover probability determines the likelihood of combining parent solutions, with higher probabilities encouraging exploration [13]. Mutation probability introduces random changes to prevent stagnation, with higher probabilities fostering diversity [11]. These parameters must be balanced for effective exploration and exploitation [17]. A systematic evaluation with two machines and 10 jobs was conducted to optimize GA parameters, with repeated tests to ensure reliability, and the results are summarized in Table 2.

Table 2. Experimented parameter values of proposed GA.

Parameter	Value
Population size (P_{size})	50
Probability of crossover (P_c)	0.85
Probability of mutation (P_m)	0.20

Table 3. Comparison results of the CPLEX solver and GA

Problem	Job/Customer	Machine	Vehicle	GA	CPLEX	Gap(%)
				Obj_{ga}	Obj_{cplex}	
1	3	2	2	97.8	97.8	0
2		3		100.7	100.7	0
4	4	2	2	157.6	157.6	0
5		3		148.2	148.2	0
7	5	2	2	48.9	48.9	0
8		3		236.4	236.4	0
10	6	2	2	246.3	246.2	0.040
11		3		287.4	287.4	0

(continued)

Table 3. (*continued*)

Problem	Job/Customer	Machine	Vehicle	GA	CPLEX	Gap(%)
				Obj_{ga}	Obj_{cplex}	
12	7	2	2	116.6	116.6	0
13		3		505.4	505.2	0.039
14	8	2	3	386.0	-	-
15		3		310.3	-	-
16		4		269.3	-	-
17	9	2	3	342.5	-	-
18		3		487.3	-	-
19		4		415.1	-	-
20	10	2	3	562.9	-	-
21		3		646.2	-	-
22		4		414.2	-	-

5.2 Comparison with CPLEX Solver

To evaluate the effectiveness and accuracy of our proposed algorithm, we compare the results obtained from the GA with those generated by CPLEX, a widely used optimization solver known for its high-quality solutions. To quantify the performance difference between the two, we calculate the gap between the solutions provided by the GA and CPLEX using the following formula:

$$Gap = \frac{ObjectiveValue_{GA} - ObjectiveValue_{CPLEX}}{ObjectiveValue_{CPLEX}} \tag{32}$$

The analysis of Table 3 shows that the GA is effective compared to CPLEX for solving the given problem instances. For very small instances (Problems 1–13), the GA achieves the same objective values as CPLEX in most cases, with minimal gaps of 0.040% and 0.039% for two instances (Problems 10 and 13). For larger instances (Problems 14–22), CPLEX fails to find solutions, while the GA successfully provides feasible ones. This demonstrates the GA's ability to handle larger problem sizes efficiently, making it a promising alternative to CPLEX, especially for complex instances where exact solvers become computationally expensive.

6 Conclusion

In this research, we proposed a GA to solve an optimization problem involving unrelated parallel machine scheduling and vehicle routing. The algorithm's effectiveness was evaluated through computational experiments, where small-scale instances were compared against solutions obtained from CPLEX. The results demonstrated that for problem instances 1–13, the GA found optimal or near-optimal solutions, with only minimal gaps in a few cases. Furthermore, for instances 14–22, where CPLEX failed to provide solutions due to computational limitations, the GA successfully generated feasible solutions, highlighting its scalability and robustness.

The findings confirm that metaheuristic approaches like GA offer a promising alternative to exact solvers, especially for large and complex problem instances where traditional methods become computationally expensive. Future work could focus on enhancing the algorithm's performance by integrating hybrid optimization techniques, improving parameter tuning strategies, or incorporating machine learning models to refine decision-making. Additionally, testing the approach on more diverse and real-world datasets would further validate its practical applicability in scheduling and routing problems.

References

1. Absi, N., Archetti, C., Dauzère-Pérès, S., Feillet, D., Speranza, M.G.: Comparing sequential and integrated approaches for the production routing problem. Eur. J. Oper. Res. **269**(2), 633–646 (2018). https://doi.org/10.1016/j.ejor.2018.01.052
2. Belo-Filho, M.A.F., Amorim, P., Almada-Lobo, B.: An adaptive large neighbourhood search for the operational integrated production and distribution problem of perishable products. Int. J. Prod. Res. **53**(20), 6040–6058 (2015). https://doi.org/10.1080/00207543.2015.1010744
3. Biskup, D., Feldmann, M.: Benchmarks for scheduling on a single machine against restrictive und unrestrictive common due dates. Comput. Oper. Res. **28**(8), 787–801 (2001)
4. Chang, Y.C., Li, V.C., Chiang, C.J.: An ant colony optimization heuristic for an integrated production and distribution scheduling problem. Eng. Optim. **46**(4), 503–520 (2014). https://doi.org/10.1080/0305215X.2013.786062
5. Ullrich, C.A.: Integrated machine scheduling and vehicle routing with time windows. Eur. J. Oper. Res. **227**(2013), 152–165 (2012)
6. Darvish, M., Coelho, L.C.: Sequential versus integrated optimization: production, location, inventory control, and distribution. Eur. J. Oper. Res. **268**(1), 203–214 (2018). https://doi.org/10.1016/j.ejor.2018.01.028
7. Deb, K., Pratap, A., Agarwal, S., Meyarivan, T.: A fast and elitist multiobjective genetic algorithm NSGA-II. IEEE Tran. Evol. Comput. **6**(2), 182–197 (2002)
8. Farahani, P., Grunow, M., Günther, H.O.: Integrated production and distribution planning for perishable food products. Flex. Serv. Manuf. J. **24**(1), 28–51 (2012). https://doi.org/10.1007/s10696-011-9125-0
9. Figliozzi, M.A.: An iterative route construction and improvement algorithm for the vehicle routing problem with soft time windows. Transp. Res. Part C Emerg. Technol. **18**(5), 668–679 (2010). https://doi.org/10.1016/j.trc.2009.08.005
10. Fu, L.-L., Aloulou, M.A., Triki, C.: Integrated production scheduling and vehicle routing problem with job splitting and delivery time windows. Int. J. Prod. Res. **55**(20), 5942–5957 (2017)
11. Goldberg, D.E.: Genetic Algorithms in Search, Optimization, and Machine Learning. Addison-Wesley (1989)
12. Guo, Z., Shi, L., Chen, L., Liang, Y.: A harmony search-based memetic optimization model for integrated production and transportation scheduling in MTO manufacturing. Omeg (2017)

13. Holland, J.H.: Adaptation in Natural and Artificial Systems. University of Michigan Press (1975)
14. Hou, Y., Fu, Y., Gao, K., Zhang, H., Sadollah, A.: Modelling and optimization of integrated distributed flow shop scheduling and distribution problems with time windows. Expert Syst. Appl. **187**, 115827 (2022). https://doi.org/10.1016/j.eswa.2021.115827
15. Lehmer, D.H.: Mathematical methods in large-scale computing units. In: Proceedings of the Second Symposium on Large-Scale Digital Calculating Machinery, pp. 141–146. Harvard University Press (1951)
16. Liu, H.T., Guo, Z.X., Zhang, Z.Z.: A hybrid multi-level optimisation framework for integrated production scheduling and vehicle routing with flexible departure time. Int. J. Prod. Res. **59**(21), 6615–6632 (2021). https://doi.org/ https://doi.org/10.1080/00207543.2020.1821927
17. Mitchell, M.: An Introduction to Genetic Algorithms. MIT Press (1998)
18. Mohammadi, S., Al-e-Hashem, S.M.J.M., Rekik, Y.: An integrated production scheduling and delivery route planning with multi-purpose machines: a case study from a furniture manufacturing company. Int. J. Prod. Econ. **219**, 347–359 (2020). https://doi.org/10.1016/j.ijpe.2019.05.017
19. Oladzad-Abbasabady, N., Tavakkoli-Moghaddam, R., Mohammadi, M., Vahedi-Nouri, B.: A bi-objective home care routing and scheduling problem considering patient preference and soft temporal dependency constraints. Eng. Appl. Artif. Intell. **119**, 105829 (2023). https://doi.org/10.1016/j.engappai.2023.105829
20. Vahdani, B., Niaki, S.T.A., Aslanzade, S.: Production-inventory-routing coordination with capacity and time window constraints for perishable products: Heuristic and meta-heuristic algorithms. J. Clean. Prod. **161**, 598–618 (2017). https://doi.org/10.1016/j.jclepro.2017.05.113
21. Vallada, E., Ruiz, R.: A genetic algorithm for the unrelated parallel machine scheduling problem with sequence dependent setup times. Eur. J. Oper. Res. (2011). https://doi.org/10.1016/j.ejor.2011.01.011

Pioneering Quantum AI Applications for Fraud Detection in FinTechs

Hanae Abbassi(✉), Saida El Mendili, and Youssef Gahi

Engineering Sciences Laboratory, National School of Applied Sciences, Ibn Tofail University, Kenitra, Morocco

hanae.abbassi@uit.ac.ma

Abstract. Identifying fraud in the banking industry stands as an ongoing challenge, owing to the shifting patterns of fraudulent conduct and the variety of large-scale transactions. Classic machine learning algorithms frequently struggle with massive data structures and fail to grasp the temporal correlations required for fraud detection. To cope with such limitations, this study supplies an architecture that combines Quantum Feature Engineering with Quantum Long Short-Term Memory (QLSTM). The approach improves feature representation by mapping transaction traits to quantum states via Angle Encoding, whereas the QLSTM model combines quantum computational capability with temporal sequence modeling. The suggested architecture considerably enhances fraud detection performance, establishing a scalable, reliable solution for the banking industry.

Keywords: Fraud detection · Quantum AI · ML · QLSTM · Quantum Features

1 Introduction

One of the most prominent issues in cybersecurity is fraudulent conduct, which occurs in various situations, notably over the web. Safeguarding online transactions is challenging for the financial system, which relies heavily on fraud detection. The rising number of e-transactions and fraudsters' proficient techniques have made conventional fraud detection methods inefficient. Many classic ML models, namely Random Forests and Support Vector Machines (RF, SVM), fail to detect changing fraud tendencies and interpret massive volumes of data in real-time [1]. Whereas certain approaches prioritize speed above accuracy, resulting in false alarms and overlooked threats, others are too sluggish to respond to fraudulent behaviors as they happen [2].

Within that, quantum computing stands out as an intriguing field with the potential to boost computational power and predictive model accuracy [3]. Quantum-inspired approaches, which use quantum concepts like superposition and entanglement, can handle complicated data more efficiently than traditional processes [4]. Fusing quantum computing with machine learning affords a unique opportunity for tackling the constraints of standard fraud detection systems [5].

To address these issues, this paper explores a hybrid architecture integrating Quantum Feature Engineering and Quantum Long Short-Term Memory (QLSTM). The suggested scheme uses quantum-enhanced data representation techniques to turn transaction

B.-E. Boudriki Semlali et al. (Eds.): DATA 2025, LNISO 81, pp. 465–475, 2025.
https://doi.org/10.1007/978-3-032-06725-8_39

attributes into quantum states accurately. It also uses QLSTM to show patterns in transaction sequences that happen over time. This architecture seeks to provide a scalable, quick, and resilient method for identifying fraud in real-time, opening the way for advances in safe online banking.

The remainder of the paper is arranged as follows: the Sect. 2 reviews the related works, Sect. 3 highlights our proposed methodology, Sect. 4 displays our experimental findings, and Sect. 5 ends our work.

2 Existing Works in FinTech Fraud Detection

2.1 Classical Methods in Fraud Detection

Rule-based methods were some of the first strategies used in fraud detection, depending on established rules and thresholds created by experts in the field to spot suspicious behavior. These systems provided clarity and ease of understanding, making them appropriate for unchanging fraud situations. Nonetheless, their dependence on fixed rules restricted their capability to adjust to changing fraud methods, resulting in increased rates of false positives and reduced effectiveness over time [6]. To address these issues, machine learning approaches were developed. In this subsection, we examine the contributions of traditional fraud detection methods, emphasizing notable works such as:

Authors in [7] have put forward a hybrid credit card fraud identification method leveraging Recursive Feature Elimination (RFE) for picking out predictive attributes, GridSearchCV for hyperparameter fine-tuning, and Synthetic Minority Oversampling Technique (SMOTE) to tackle skewed data issues. The experimental outcomes of this model showed a competitive accuracy of 99% and a sensitivity of 95%, highlighting its strong ability to detect fraudulent transactions. Following that, [8] compared the accuracy of three ML learners for identifying fraudulent credit card transactions: logistic regression (LR), RF, and decision trees (DT). The findings showed that random forest outperformed the other algorithms, with an accuracy rating of 96% and an AUC value of 98.9%. The analysis also discovered patterns in fraudulent transactions, such as their predominance among credit cardholders.

A further study [9] suggested a Credit Card Anomaly Detection (CCAD) model, utilizing base models and meta-learning ensemble approaches to increase the detection rate of deviations in transactions. It employs 4 outlier identification schemes as base learners and an XGBoost meta-learner in a stacked ensemble method. The CCAD model tackles problems, including overlapping class samples and uneven class distributions, by using stratified sampling, k-fold cross-validation, and discordance rate computation. Experiments show that CCAD outperforms previous approaches, especially in spotting anomalies in minority class cases in datasets. Additionally, one investigation [10] explored algorithms involving SVM, Gradient Boosting, RF, and their combinations to tackle credit card fraud detection difficulties. This study pointed out the ongoing issue of class imbalance and its effect on algorithm efficiency, stating that while logistic regression got higher accuracy, K-Nearest Neighbours demonstrated more versatility due to its learning power, arising as an exciting option for fraud detection jobs. Current expansions in big data analytics and open-source technologies have paved the way for real-time fraud detection systems [11] has laid out a novel fraud detection architecture that blends unsupervised models, namely autoencoders and extended isolation forests with big data technologies like Spark Streaming and H2O. The technique solves issues such as a lack of non-fraudulent instances and the high complexity of feature sets. When evaluated on real-world datasets, the methodology obtained an outstanding 99% accuracy, outperforming existing state-of-the-art techniques regarding identifying fake samples.

Despite the success of classical machine learning techniques like SVM and RF in detecting fraud, these methods face significant challenges in dealing with large-scale data and adapting to evolving fraudulent patterns. SVMs frequently struggle with high-dimensional data, necessitating expensive calculations for huge datasets and sometimes causing overfitting if not correctly controlled [12]. Also, RFs, while good at catching complicated patterns, can be costly to scale up and may suffer from unbalanced classes in fraud detection applications [13]. Those issues highlight the requirement for more efficient and adaptable methods. Quantum AI is a viable option with its capacity to analyze massive volumes of data quickly and detect complicated links with fewer resources.

2.2 Quantum AI Applications in Fraud Detection

Quantum computing brings novel avenues using concepts such as superposition and entanglement, allowing for parallelism and effectiveness that conventional models cannot meet. Quantum AI improves data representations via strategies like quantum features, making complicated patterns easier to differentiate [14]. These advances address significant constraints of existing fraud detection systems, such as poor processing rates and limited flexibility, opening the door to more scalable and robust solutions for combatting current fraudulent activity [15].

Indeed, [16] introduced Quantum Graph Neural Networks (QGNNs) incorporating Variational Quantum Circuits (VQCs), which are capable of processing graph-structured financial data with enhanced efficiency compared to classical GNNs. The simulated assessment on a real-world financial fraud detection dataset proved QGNNs' higher performance, with an AUC of 0.85, outperforming traditional GNNs. Within this context, authors in [17] applied Variational Quantum-Classical Wasserstein GANs (WGANs) to identify irregularities in credit card transactions. This hybrid strategy uses a quantum-classical neural network generator and a classical discriminator to optimize training resilience via the Wasserstein gap. When applied to a credit card fraud database, the model performed comparably to conventional techniques, notably as far as the F1 score. This work demonstrates the effectiveness of quantum-classical hybrid systems for high-dimensional fraud detection concerns.

Another study [18] put out a quantum LOF method that uses quantum computing to achieve polynomial speedups in dataset size and exponential speedups in data dimensionality. By combining amplitude estimation, minimal search, and parallel processing techniques, the quantum approach outperforms its conventional equivalents in terms of efficiency. This development demonstrates how quantum computing might improve unsupervised anomaly detection techniques for uses such as military tracking, fraud detection, and identifying intrusions. In the same vein, [19] described the initial end-to-end use of Quantum Support Vector Machines (QSVM) for fraud detection in the financial payments area. Integrating IBM Safer Payments and IBM Quantum Computers via the Qiskit system, the authors thoroughly compared QSVM with traditional approaches involving RF and XGBoost and rule-based human knowledge. The study emphasized the additional benefits of QML for fraud detection, especially when paired with conventional algorithms in a hybrid ensemble model. QSVM facilitated more thorough probing of the feature space, resulting in higher accuracy on a smaller dataset appropriate for today's quantum gear power.

Moreover, [20] assessed the possibility of Quantum Machine Learning (QML) for fraud identification, specifically with Variational Quantum Circuits (VQC). Utilizing a dataset of 500,000 real-world transactions for integration with Noisy Intermediate-Scale Quantum (NISQ) devices, the research contrasted VQC outcomes to standard methods such as LR, RF, XGBoost, SVM, and Neural Network. The results showed that VQC had higher recall rates, particularly when classifying fraudulent transactions, which were very skewed.

Having examined various Quantum Machine Learning approaches for fraud detection, quantum-enhanced models may provide considerable benefits, particularly when dealing with complicated and unbalanced datasets. Those findings inspired the development of our suggested architecture, which combines cutting-edge quantum and classical approaches to improve real-time fraud detection in digital transactions. Our technique, in particular, uses Quantum Feature Engineering using Angle Encoding to map transaction attributes to quantum states and QLSTM to combine quantum processing capability with sequential data modeling.

3 Quantum Long Short-Term Memory Fraud Detection Architecture

Our suggested architecture optimizes real-time fraud detection by combining powerful quantum computing techniques with traditional machine-learning approaches. The workflow consists of five critical steps, each of which helps to improve the detection process. Figure 1 and 2 highlights the proposed architecture.

The procedure begins with transaction data, which includes aspects such as client information, transaction history, and device-specific attributes. This raw data forms the basis for all further processing and model training. Before any complex procedures are used, the data is pre-processed. This comprises normalization, which adjusts data values to guarantee consistency, as well as data cleaning, which deals with missing or inconsistent elements. Furthermore, the data is encoded into quantum states through Angle Encoding, a technique for converting numerical values into quantum amplitudes by expressing them as angles in a quantum circuit. This stage allows the quantum system to fully utilize its computing capabilities.

Beyond preprocessing, the feature engineering step employs quantum approaches to improve the representation of transaction data. Quantum circuits are built with quantum gates—operations including Hadamard or CNOT gates—that misuse qubits to depict complicated interactions between features. Quantum gates are fundamental operations in quantum computing that alter qubits, permitting duties such as establishing entanglement or shifting quantum states. These gates induce entanglement, which allows the system to capture relationships between aspects that are difficult to characterize using conventional techniques. For example, the quantum state may encode the transaction amount, transaction time, client IP address, and device kind. By utilizing these quantum correlations, manufactured features provide a more detailed and nuanced representation of the data, which is crucial for spotting fraudulent patterns contained in these complex interactions.

The intended attributes are fed into a QLSTM model, which handles sequential transaction data. QLSTM is a quantum-enhanced version of classical LSTM networks that simulate temporal sequences. Using quantum computing concepts involving superposition and entanglement, QLSTM may effectively capture complicated temporal correlations and improve pattern detection within sequential data. This model combines the temporal modeling characteristics of LSTM with the parallel processing capability of quantum computers, making it extremely successful in analyzing transaction sequences and detecting abnormalities.

Finally, the model outcomes are employed to categorize transactions as suspicious or not. The step of classification is preceded by the prediction refinery, which ensures that the model reduces mistakes and improves accuracy over time.

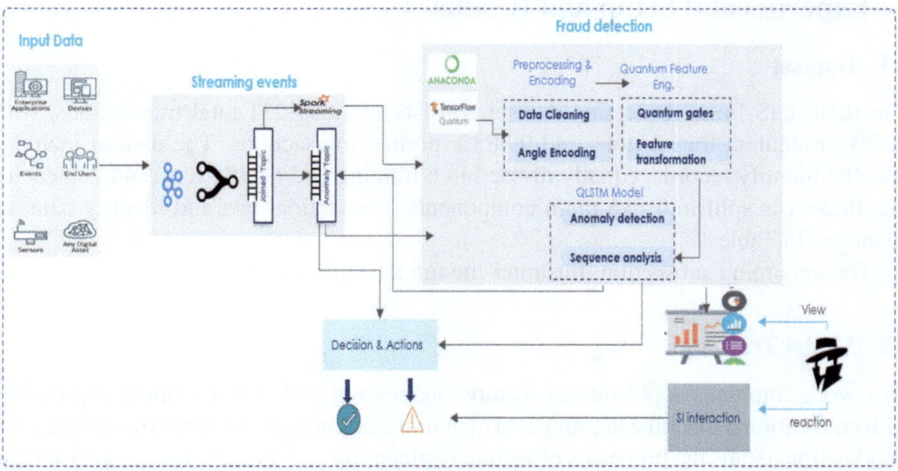

Fig.1. End-to-End Architecture.

After outlining the suggested architecture, the forthcoming section centers on the experimentation process, covering the dataset applied, the feature engineering approaches, and the model training strategy. This will involve an investigation of how quantum-enhanced feature engineering, which includes Angle Encoding and quantum gates, is used to process transaction data, followed by the use of the QLSTM model for sequential fraud detection. The findings will be presented and analyzed to determine the efficacy of the suggested technique in real-world fraud detection settings.

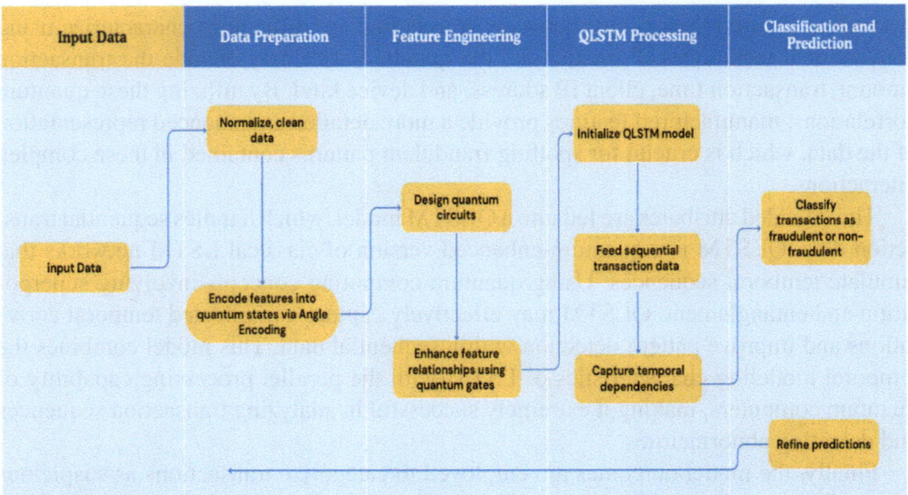

Fig.2. Quantum Features-QLSTM Fraud Detection Architecture.

4 Experimental Setup and Results

4.1 Dataset

The IEEE-CIS Fraud Detection dataset consists of 1,097,231 total transactions, with 0.17% fraudulent transactions and 99.83% normal transactions. The dataset includes 288,466 identity records, equally divided into training and testing sets (144,233 each). The dataset is split into two main components: transaction data and identity data, as displayed in Table 1.

The upcoming subsection illustrates the model deployment.

4.2 Model Training

This work combines sophisticated feature engineering and model training approaches with conventional machine learning and quantum computing to improve fraud detection. This section details the processes of feature engineering and model training, with a focus on quantum feature encoding and the creation of the Quantum QLSTM model.

Feature engineering is important in fraud detection because it converts raw data into meaningful representations for ML models. This approach combines conventional and quantum techniques to generate a robust feature collection. Figure 3 highlights the feature engineering process.

- Classical Feature Engineering: We preprocess transaction and identification data by managing missing values, scaling numerical features (e.g., transaction amount) with StandardScaler, and labeling categorical variables (e.g., card type, product codes).
- Quantum Feature Engineering: To capture complex interactions, we use TensorFlow Quantum (TFQ) for angle encoding and rotational gates (e.g., Rz) to transform classical input to quantum states. This allows the model to detect subtle fraud trends using quantum computation.

Table 1. Data Features.

Transaction data features	Identity data features
TransactionID: unique identifier for each transaction	id: unique identifier for each cardholder
TransactionDT: timestamp of the transaction	DeviceType: type of device used by cardholders
TransactionAmt: volume of the transaction	DeviceInfo: device-related information
ProductCD: product code	id_01 to id_38: anonymized identity data attributes
card1 to card6: card-related attributes	DeviceInfo2 to DeviceInfo4: further device-related attributes
addr1 to addr2: cardholder address attributes	id_39 to id_41: additional anonymized identity features
dist1 to dist2: distance between cardholder's address and operation spot	
P_emaildomain to R_emaildomain: email domain of cardholder	
V1 to V339: anonymized attributes obtained from the transaction data	

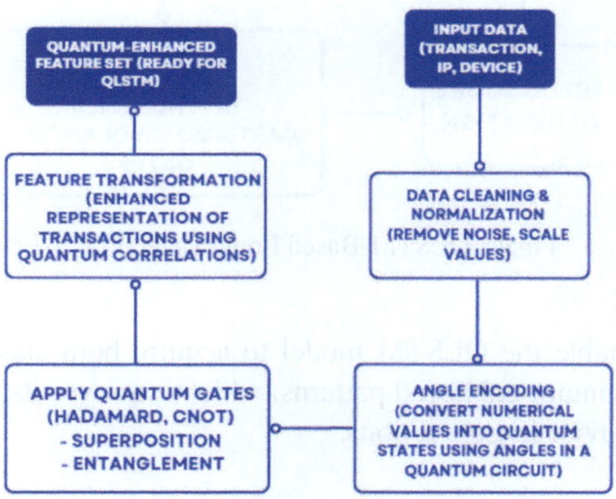

Fig.3. Feature Engineering Process.

Once the features have been produced via quantum encoding, the next step is to input them into the QLSTM model. This model combines the capabilities of conventional LSTM networks with quantum computing concepts to efficiently handle sequential transaction data and capture intricate temporal correlations required for detecting fraudulent patterns. We train our QLSTM via TensorFlow Quantum. The primary model parameters are outlined below:

- Number of LSTM units: 64, which lets the model capture intricate temporal dependencies in the sequential data.
- Quantum circuit configuration: A quantum circuit with 4 qubits and 2 layers of rotational gates (such as Rz and Rx), used for encoding classical features into quantum states.
- Learning rate: 0.001, a typical starting point that balances convergence speed and model accuracy.
- Batch size: 32, providing enough data per update to stabilize the gradient updates.
- Epochs: 50, to ensure that the model adequately learns from the data, with early stopping applied if validation performance stagnates (Fig. 4).

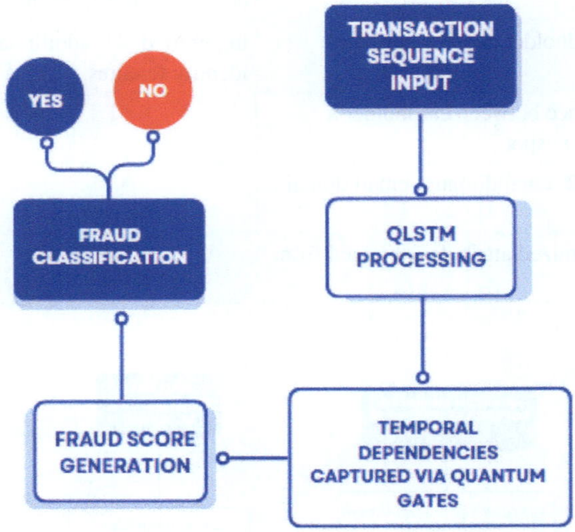

Fig.4. QLSTM-Based Fraud Detection.

These settings enable the QLSTM model to acquire both classical patterns from temporal data and quantum-enhanced patterns, which improves its capacity to identify subtle fraudulent behavior in transactions.

4.3 Evaluation Metrics

To assess our QLSTM model, we recommend three metrics, including a novel custom metric meant to tackle the unique characteristics of fraud detection in imbalanced datasets.

Accuracy: it measures the overall proportion of correctly classified occurrences (fraudulent and non-fraudulent) among the total instances.

$$Accuracy = \frac{TP + TN}{TP + TN + FP + FN} \tag{1}$$

Precision: measures the proportion of predicted fraudulent operations that are fraudulent. This metric is essential when the expense of false positives is high.

$$Precision = \frac{TP}{TP + FP} \tag{2}$$

Fraud Sensitivity Score (FSS): is a custom metric combining precision and recall. It emphasizes the model's capability to spot suspicious transactions while considering the trade-off between false positives and false negatives.

$$FSS = \frac{2 \times \text{Precision} \times \text{Recall}}{\text{Precision} + \text{Recall}} \tag{3}$$

$$\text{Where } Recall = \frac{TP}{TP + FN} \tag{4}$$

4.4 Outcomes and Insights

This part displays the findings of the QLSTM architecture for fraud detection, depicted in Table 2. The investigations were conducted using Python (via Anaconda) and TensorFlow Quantum, optimized on an NVIDIA GPU for improved computational speed. The IEEE-CIS Fraud Detection database was partitioned into training, validation, and testing subsets to ensure the model's reliability.

Table 2. Performance comparison between the QLSTM and the traditional LSTM for fraud detection.

Model	Accuracy	Precision	FSS
QLSTM	0.9993	0.9140	0.8991
LSTM	0.9900	0.8213	0.8010

The model's performance was evaluated using three main metrics: accuracy, precision, and fraud sensitivity score (FSS). The findings show that the QLSTM model obtained an extraordinary accuracy of 99.93%, demonstrating its excellent potential for fraud detection. Furthermore, the accuracy score of 0.91 shows the model's capacity to properly detect fraud situations, while the FSS of 0.899 illustrates its ability to balance false positives and negatives.

When contrasted with traditional LSTM, QLSTM exceeds it in all three metrics: accuracy, precision, and FSS. While the LSTM algorithm obtained 99% accuracy, 0.82 precision, and 0.8 FSS, the QLSTM model outperformed, demonstrating the possibility of incorporating quantum computing into typical deep learning frameworks. These findings imply that QLSTM's quantum-enhanced architecture delivers more dependable and accurate fraud detection, presenting a unique way to solve the difficulties of banking scams.

Despite the immense promise of the QLSTM for fraud detection, many problems prevent its broad deployment. One key barrier is the complicated nature of combining quantum computing with deep learning, which necessitates specialized hardware. Furthermore, Quantum computers, which are still in their early phases, might be prohibitively expensive and may not scale adequately for big datasets commonly used in fraud detection.

5 Conclusion and Future Directions

In summary, this work affords a novel FinTech fraud detection architecture. At first, quantum feature engineering methods were exploited to build attributes that reflected the behavior aspects sought by fraud detection algorithms. Furthermore, we execute the QLSTM that integrates the strengths of quantum computing with the LSTM time series learner on a real-world dataset. The QLSTM learner has huge potential for enhancing fraud detection jobs, notably in terms of accuracy and precision when juxtaposed with regular LSTM. While quantum computing provides benefits such as quicker processing and the capacity to handle complicated trends, the practical implementation of the models continues to confront significant obstacles, namely hardware restrictions and model training difficulties.

Current studies should concentrate on beating such challenges, with an eye on increasing the scalability and interpretability of quantum models. Another target is to delve into the adoption of modern AI algorithms, along with conceivable combinations of such methods for identifying fraud.

References

1. Khalid, A.R., Owoh, N., Uthmani, O., et al.: Enhancing credit card fraud detection: an ensemble machine learning approach. Big Data Cogn. Comput. **8**, 6 (2024). https://doi.org/10.3390/bdcc8010006
2. Innan, N., Khan, M.A.-Z., Bennai, M.: Financial fraud detection: a comparative study of quantum machine learning models (2023)
3. Di Pierro, A., Incudini, M.: Quantum machine learning and fraud detection. In: Dougherty, D., Meseguer, J., Mödersheim, S.A., Rowe, P. (eds.) Protocols, Strands, and Logic. Lecture Notes in Computer Science, vol. 13066, pp. 139–155. Springer, Cham (2021). https://doi.org/10.1007/978-3-030-91631-2_8
4. Parmar, S.J., Parmar, V.R., Verma, J., et al.: Quantum computing: exploring superposition and entanglement for cutting-edge applications. In: Proceedings of the 2023 16th International Conference on Security of Information and Networks (SIN), pp 1–6 (2023)
5. Pulicharla, M.R.: Hybrid quantum-classical machine learning models: powering the future of AI (2023)
6. Abbassi, H., El Mendili, S., Gahi, Y.: Enhanced banking security: isolation forest with attention mechanism for sophisticated fraud detection. In: Proceedings of the 2024 10th International Conference on Optimization and Applications (ICOA), pp. 1–6 (2024)
7. Rtayli, N., Enneya, N.: Enhanced credit card fraud detection based on SVM-recursive feature elimination and hyper-parameters optimization. J. Inf. Secur. Appl. **55**, 102596 (2020). https://doi.org/10.1016/j.jisa.2020.102596

8. Afriyie, J.K., Tawiah, K., Pels, W.A., et al.: A supervised machine learning algorithm for detecting and predicting fraud in credit card transactions. Decis. Analytics J. **6**, 100163 (2023). https://doi.org/10.1016/j.dajour.2023.100163
9. Islam, M.A., Uddin, M.A., Aryal, S., Stea, G.: An ensemble learning approach for anomaly detection in credit card data with imbalanced and overlapped classes. J. Inf. Secur. Appl. **78**, 103618 (2023). https://doi.org/10.1016/j.jisa.2023.103618
10. Madhurya, M.J., Gururaj, H.L., Soundarya, B.C., et al.: Exploratory analysis of credit card fraud detection using machine learning techniques. Glob. Transitions Proc. **3**, 31–37 (2022). https://doi.org/10.1016/j.gltp.2022.04.006
11. Abbassi, H., Mendili, S.E., Gahi, Y.: Real-time online banking fraud detection model by unsupervised learning fusion. HighTech Innov. J. **5**, 185–199 (2024)
12. Adebayo, O.S., Favour-Bethy, T.A., Otasowie, O., Okunola, O.A.: Comparative review of credit card fraud detection using machine learning and concept drift techniques. IJCSMC **12**, 24–48 (2023). https://doi.org/10.47760/ijcsmc.2023.v12i07.004
13. Aburbeian, A.M., Ashqar, H.I.: Credit card fraud detection using enhanced random forest classifier for imbalanced data (2023)
14. Wang, H., Wang, W., Liu, Y., Alidaee, B.: Integrating machine learning algorithms with quantum annealing solvers for online fraud detection. IEEE Access **10**, 75908–75917 (2022). https://doi.org/10.1109/ACCESS.2022.3190897
15. Simeone, O.: An introduction to quantum machine learning for engineers (2022). https://doi.org/10.48550/ARXIV.2205.09510
16. Innan, N., Sawaika, A., Dhor, A., et al.: Financial fraud detection using quantum graph neural networks. Quantum Mach. Intell. **6**, 7 (2024). https://doi.org/10.1007/s42484-024-00143-6
17. Herr, D., Obert, B., Rosenkranz, M.: Anomaly detection with variational quantum generative adversarial networks. Quantum Sci. Technol. **6**, 045004 (2021). https://doi.org/10.1088/2058-9565/ac0d4d
18. Guo, M., Pan, S., Li, W., et al.: Quantum algorithm for unsupervised anomaly detection. Physica A **625**, 129018 (2023). https://doi.org/10.1016/j.physa.2023.129018
19. Grossi, M., Ibrahim, N., Radescu, V., et al.: Mixed quantum-classical method for fraud detection with quantum feature selection. IEEE Trans. Quantum Eng. **3**, 1–12 (2022). https://doi.org/10.1109/TQE.2022.3213474
20. Tudisco, A., Volpe, D., Ranieri, G., et al.: Evaluating the computational advantages of the variational quantum circuit model in financial fraud detection. IEEE Access **12**, 102918–102940 (2024). https://doi.org/10.1109/ACCESS.2024.3432312

Author Index

B.-E. Boudriki Semlali et al. (Eds.): DATA 2025, LNISO 81, pp. 477–478, 2025.
https://doi.org/10.1007/978-3-032-06725-8

MIX
Papier aus verantwortungsvollen Quellen
Paper from responsible sources
FSC® C105338

If you have any concerns about our products,
you can contact us on
ProductSafety@springernature.com

In case Publisher is established outside the EU,
the EU authorized representative is:
Springer Nature Customer Service Center GmbH
Europaplatz 3, 69115 Heidelberg, Germany

Printed by Libri Plureos GmbH
in Hamburg, Germany